T0215738

Lecture Notes in Computer Science   12138

More information about this series at http://www.springer.com/series/7407

Valeria V. Krzhizhanovskaya ·
Gábor Závodszky · Michael H. Lees ·
Jack J. Dongarra · Peter M. A. Sloot ·
Sérgio Brissos · João Teixeira (Eds.)

Computational Science – ICCS 2020

20th International Conference
Amsterdam, The Netherlands, June 3–5, 2020
Proceedings, Part II

 Springer

Editors
Valeria V. Krzhizhanovskaya ⓘ
University of Amsterdam
Amsterdam, The Netherlands

Michael H. Lees
University of Amsterdam
Amsterdam, The Netherlands

Peter M. A. Sloot ⓘ
University of Amsterdam
Amsterdam, The Netherlands

ITMO University
Saint Petersburg, Russia

Nanyang Technological University
Singapore, Singapore

João Teixeira
Intellegibilis
Setúbal, Portugal

Gábor Závodszky ⓘ
University of Amsterdam
Amsterdam, The Netherlands

Jack J. Dongarra ⓘ
University of Tennessee
Knoxville, TN, USA

Sérgio Brissos
Intellegibilis
Setúbal, Portugal

ISSN 0302-9743   ISSN 1611-3349   (electronic)
Lecture Notes in Computer Science
ISBN 978-3-030-50416-8   ISBN 978-3-030-50417-5   (eBook)
https://doi.org/10.1007/978-3-030-50417-5

LNCS Sublibrary: SL1 – Theoretical Computer Science and General Issues

This Springer imprint is published by the registered company Springer Nature Switzerland AG
The registered company address is: Gewerbestrasse 11, 6330 Cham, Switzerland

Preface

Twenty Years of Computational Science

Welcome to the 20th Annual International Conference on Computational Science (ICCS – https://www.iccs-meeting.org/iccs2020/).

During the preparation for this 20th edition of ICCS we were considering all kinds of nice ways to celebrate two decennia of computational science. Afterall when we started this international conference series, we never expected it to be so successful and running for so long at so many different locations across the globe! So we worked on a mind-blowing line up of renowned keynotes, music by scientists, awards, a play written by and performed by computational scientists, press attendance, a lovely venue... you name it, we had it all in place. Then corona hit us.

After many long debates and considerations, we decided to cancel the physical event but still support our scientists and allow for publication of their accepted peer-reviewed work. We are proud to present the proceedings you are reading as a result of that.

ICCS 2020 is jointly organized by the University of Amsterdam, NTU Singapore, and the University of Tennessee.

The International Conference on Computational Science is an annual conference that brings together researchers and scientists from mathematics and computer science as basic computing disciplines, as well as researchers from various application areas who are pioneering computational methods in sciences such as physics, chemistry, life sciences, engineering, arts and humanitarian fields, to discuss problems and solutions in the area, to identify new issues, and to shape future directions for research.

Since its inception in 2001, ICCS has attracted increasingly higher quality and numbers of attendees and papers, and 2020 was no exception, with over 350 papers accepted for publication. The proceedings series have become a major intellectual resource for computational science researchers, defining and advancing the state of the art in this field.

The theme for ICCS 2020, "Twenty Years of Computational Science", highlights the role of Computational Science over the last 20 years, its numerous achievements, and its future challenges. This conference was a unique event focusing on recent developments in: scalable scientific algorithms, advanced software tools, computational grids, advanced numerical methods, and novel application areas. These innovative novel models, algorithms, and tools drive new science through efficient application in areas such as physical systems, computational and systems biology, environmental systems, finance, and others.

This year we had 719 submissions (230 submissions to the main track and 489 to the thematic tracks). In the main track, 101 full papers were accepted (44%). In the thematic tracks, 249 full papers were accepted (51%). A high acceptance rate in the thematic tracks is explained by the nature of these, where many experts in a particular field are personally invited by track organizers to participate in their sessions.

ICCS relies strongly on the vital contributions of our thematic track organizers to attract high-quality papers in many subject areas. We would like to thank all committee members from the main and thematic tracks for their contribution to ensure a high standard for the accepted papers. We would also like to thank Springer, Elsevier, the Informatics Institute of the University of Amsterdam, the Institute for Advanced Study of the University of Amsterdam, the SURFsara Supercomputing Centre, the Netherlands eScience Center, the VECMA Project, and Intellegibilis for their support. Finally, we very much appreciate all the Local Organizing Committee members for their hard work to prepare this conference.

We are proud to note that ICCS is an A-rank conference in the CORE classification.

We wish you good health in these troubled times and hope to see you next year for ICCS 2021.

June 2020

Valeria V. Krzhizhanovskaya
Gábor Závodszky
Michael Lees
Jack Dongarra
Peter M. A. Sloot
Sérgio Brissos
João Teixeira

Organization

Thematic Tracks and Organizers

Advances in High-Performance Computational Earth Sciences: Applications and Frameworks – IHPCES

Takashi Shimokawabe
Kohei Fujita
Dominik Bartuschat

Agent-Based Simulations, Adaptive Algorithms and Solvers – ABS-AAS

Maciej Paszynski
David Pardo
Victor Calo
Robert Schaefer
Quanling Deng

Applications of Computational Methods in Artificial Intelligence and Machine Learning – ACMAIML

Kourosh Modarresi
Raja Velu
Paul Hofmann

Biomedical and Bioinformatics Challenges for Computer Science – BBC

Mario Cannataro
Giuseppe Agapito
Mauro Castelli
Riccardo Dondi
Rodrigo Weber dos Santos
Italo Zoppis

Classifier Learning from Difficult Data – CLD^2

Michał Woźniak
Bartosz Krawczyk
Paweł Ksieniewicz

Complex Social Systems through the Lens of Computational Science – CSOC

Debraj Roy
Michael Lees
Tatiana Filatova

Computational Health – CompHealth

Sergey Kovalchuk
Stefan Thurner
Georgiy Bobashev

Computational Methods for Emerging Problems in (dis-)Information Analysis – DisA

Michal Choras
Konstantinos Demestichas

Computational Optimization, Modelling and Simulation – COMS

Xin-She Yang
Slawomir Koziel
Leifur Leifsson

Computational Science in IoT and Smart Systems – IoTSS

Vaidy Sunderam
Dariusz Mrozek

Computer Graphics, Image Processing and Artificial Intelligence – CGIPAI

Andres Iglesias
Lihua You
Alexander Malyshev
Hassan Ugail

Data-Driven Computational Sciences – DDCS

Craig C. Douglas
Ana Cortes
Hiroshi Fujiwara
Robert Lodder
Abani Patra
Han Yu

Machine Learning and Data Assimilation for Dynamical Systems – MLDADS

Rossella Arcucci
Yi-Ke Guo

Meshfree Methods in Computational Sciences – MESHFREE

Vaclav Skala
Samsul Ariffin Abdul Karim
Marco Evangelos Biancolini
Robert Schaback

Rongjiang Pan
Edward J. Kansa

Multiscale Modelling and Simulation – MMS

Derek Groen
Stefano Casarin
Alfons Hoekstra
Bartosz Bosak
Diana Suleimenova

Quantum Computing Workshop – QCW

Katarzyna Rycerz
Marian Bubak

Simulations of Flow and Transport: Modeling, Algorithms and Computation – SOFTMAC

Shuyu Sun
Jingfa Li
James Liu

Smart Systems: Bringing Together Computer Vision, Sensor Networks and Machine Learning – SmartSys

Pedro J. S. Cardoso
João M. F. Rodrigues
Roberto Lam
Janio Monteiro

Software Engineering for Computational Science – SE4Science

Jeffrey Carver
Neil Chue Hong
Carlos Martinez-Ortiz

Solving Problems with Uncertainties – SPU

Vassil Alexandrov
Aneta Karaivanova

Teaching Computational Science – WTCS

Angela Shiflet
Alfredo Tirado-Ramos
Evguenia Alexandrova

Uncertainty Quantification for Computational Models – UNEQUIvOCAL

Wouter Edeling
Anna Nikishova
Peter Coveney

Program Committee and Reviewers

Ahmad Abdelfattah
Samsul Ariffin
   Abdul Karim
Evgenia Adamopoulou
Jaime Afonso Martins
Giuseppe Agapito
Ram Akella
Elisabete Alberdi Celaya
Luis Alexandre
Vassil Alexandrov
Evguenia Alexandrova
Hesham H. Ali
Julen Alvarez-Aramberri
Domingos Alves
Julio Amador Diaz Lopez
Stanislaw
   Ambroszkiewicz
Tomasz Andrysiak
Michael Antolovich
Hartwig Anzt
Hideo Aochi
Hamid Arabnejad
Rossella Arcucci
Khurshid Asghar
Marina Balakhontceva
Bartosz Balis
Krzysztof Banas
João Barroso
Dominik Bartuschat
Nuno Basurto
Pouria Behnoudfar
Joern Behrens
Adrian Bekasiewicz
Gebrai Bekdas
Stefano Beretta
Benjamin Berkels
Martino Bernard

Daniel Berrar
Sanjukta Bhowmick
Marco Evangelos
   Biancolini
Georgiy Bobashev
Bartosz Bosak
Marian Bubak
Jérémy Buisson
Robert Burduk
Michael Burkhart
Allah Bux
Aleksander Byrski
Cristiano Cabrita
Xing Cai
Barbara Calabrese
Jose Camata
Mario Cannataro
Alberto Cano
Pedro Jorge Sequeira
   Cardoso
Jeffrey Carver
Stefano Casarin
Manuel Castañón-Puga
Mauro Castelli
Eduardo Cesar
Nicholas Chancellor
Patrikakis Charalampos
Ehtzaz Chaudhry
Chuanfa Chen
Siew Ann Cheong
Andrey Chernykh
Lock-Yue Chew
Su Fong Chien
Marta Chinnici
Sung-Bae Cho
Michal Choras
Loo Chu Kiong

Neil Chue Hong
Svetlana Chuprina
Paola Cinnella
Noélia Correia
Adriano Cortes
Ana Cortes
Enrique
   Costa-Montenegro
David Coster
Helene Coullon
Peter Coveney
Attila Csikasz-Nagy
Loïc Cudennec
Javier Cuenca
Yifeng Cui
António Cunha
Ben Czaja
Pawel Czarnul
Flávio Martins
Bhaskar Dasgupta
Konstantinos Demestichas
Quanling Deng
Nilanjan Dey
Khaldoon Dhou
Jamie Diner
Jacek Dlugopolski
Simona Domesová
Riccardo Dondi
Craig C. Douglas
Linda Douw
Rafal Drezewski
Hans du Buf
Vitor Duarte
Richard Dwight
Wouter Edeling
Waleed Ejaz
Dina El-Reedy

Amgad Elsayed
Nahid Emad
Chriatian Engelmann
Gökhan Ertaylan
Alex Fedoseyev
Luis Manuel Fernández
Antonino Fiannaca
Christos
    Filelis-Papadopoulos
Rupert Ford
Piotr Frackiewicz
Martin Frank
Ruy Freitas Reis
Karl Frinkle
Haibin Fu
Kohei Fujita
Hiroshi Fujiwara
Takeshi Fukaya
Wlodzimierz Funika
Takashi Furumura
Ernst Fusch
Mohamed Gaber
David Gal
Marco Gallieri
Teresa Galvao
Akemi Galvez
Salvador García
Bartlomiej Gardas
Delia Garijo
Frédéric Gava
Piotr Gawron
Bernhard Geiger
Alex Gerbessiotis
Ivo Goncalves
Antonio Gonzalez Pardo
Jorge
    González-Domínguez
Yuriy Gorbachev
Pawel Gorecki
Michael Gowanlock
Manuel Grana
George Gravvanis
Derek Groen
Lutz Gross
Sophia
    Grundner-Culemann

Pedro Guerreiro
Tobias Guggemos
Xiaohu Guo
Piotr Gurgul
Filip Guzy
Pietro Hiram Guzzi
Zulfiqar Habib
Panagiotis Hadjidoukas
Masatoshi Hanai
John Hanley
Erik Hanson
Habibollah Haron
Carina Haupt
Claire Heaney
Alexander Heinecke
Jurjen Rienk Helmus
Álvaro Herrero
Bogumila Hnatkowska
Maximilian Höb
Erlend Hodneland
Olivier Hoenen
Paul Hofmann
Che-Lun Hung
Andres Iglesias
Takeshi Iwashita
Alireza Jahani
Momin Jamil
Vytautas Jancauskas
João Janeiro
Peter Janku
Fredrik Jansson
Jirí Jaroš
Caroline Jay
Shalu Jhanwar
Zhigang Jia
Chao Jin
Zhong Jin
David Johnson
Guido Juckeland
Maria Juliano
Edward J. Kansa
Aneta Karaivanova
Takahiro Katagiri
Timo Kehrer
Wayne Kelly
Christoph Kessler

Jakub Klikowski
Harald Koestler
Ivana Kolingerova
Georgy Kopanitsa
Gregor Kosec
Sotiris Kotsiantis
Ilias Kotsireas
Sergey Kovalchuk
Michal Koziarski
Slawomir Koziel
Rafal Kozik
Bartosz Krawczyk
Elisabeth Krueger
Valeria Krzhizhanovskaya
Pawel Ksieniewicz
Marek Kubalcík
Sebastian Kuckuk
Eileen Kuehn
Michael Kuhn
Michal Kulczewski
Krzysztof Kurowski
Massimo La Rosa
Yu-Kun Lai
Jalal Lakhlili
Roberto Lam
Anna-Lena Lamprecht
Rubin Landau
Johannes Langguth
Elisabeth Larsson
Michael Lees
Leifur Leifsson
Kenneth Leiter
Roy Lettieri
Andrew Lewis
Jingfa Li
Khang-Jie Liew
Hong Liu
Hui Liu
Yen-Chen Liu
Zhao Liu
Pengcheng Liu
James Liu
Marcelo Lobosco
Robert Lodder
Marcin Los
Stephane Louise

Frederic Loulergue
Paul Lu
Stefan Luding
Onnie Luk
Scott MacLachlan
Luca Magri
Imran Mahmood
Zuzana Majdisova
Alexander Malyshev
Muazzam Maqsood
Livia Marcellino
Tomas Margalef
Tiziana Margaria
Svetozar Margenov
Urszula
   Markowska-Kaczmar
Osni Marques
Carmen Marquez
Carlos Martinez-Ortiz
Paula Martins
Flávio Martins
Luke Mason
Pawel Matuszyk
Valerie Maxville
Wagner Meira Jr.
Roderick Melnik
Valentin Melnikov
Ivan Merelli
Choras Michal
Leandro Minku
Jaroslaw Miszczak
Janio Monteiro
Kourosh Modarresi
Fernando Monteiro
James Montgomery
Andrew Moore
Dariusz Mrozek
Peter Mueller
Khan Muhammad
Judit Muñoz
Philip Nadler
Hiromichi Nagao
Jethro Nagawkar
Kengo Nakajima
Ionel Michael Navon
Philipp Neumann

Mai Nguyen
Hoang Nguyen
Nancy Nichols
Anna Nikishova
Hitoshi Nishizawa
Brayton Noll
Algirdas Noreika
Enrique Onieva
Kenji Ono
Eneko Osaba
Aziz Ouaarab
Serban Ovidiu
Raymond Padmos
Wojciech Palacz
Ivan Palomares
Rongjiang Pan
Joao Papa
Nikela Papadopoulou
Marcin Paprzycki
David Pardo
Anna Paszynska
Maciej Paszynski
Abani Patra
Dana Petcu
Serge Petiton
Bernhard Pfahringer
Frank Phillipson
Juan C. Pichel
Anna
   Pietrenko-Dabrowska
Laércio L. Pilla
Armando Pinho
Tomasz Piontek
Yuri Pirola
Igor Podolak
Cristina Portales
Simon Portegies Zwart
Roland Potthast
Ela Pustulka-Hunt
Vladimir Puzyrev
Alexander Pyat
Rick Quax
Cesar Quilodran Casas
Barbara Quintela
Ajaykumar Rajasekharan
Celia Ramos

Lukasz Rauch
Vishal Raul
Robin Richardson
Heike Riel
Sophie Robert
Luis M. Rocha
Joao Rodrigues
Daniel Rodriguez
Albert Romkes
Debraj Roy
Katarzyna Rycerz
Alberto Sanchez
Gabriele Santin
Alex Savio
Robert Schaback
Robert Schaefer
Rafal Scherer
Ulf D. Schiller
Bertil Schmidt
Martin Schreiber
Alexander Schug
Gabriela Schütz
Marinella Sciortino
Diego Sevilla
Angela Shiflet
Takashi Shimokawabe
Marcin Sieniek
Nazareen Sikkandar
   Basha
Anna Sikora
Janaína De Andrade Silva
Diana Sima
Robert Sinkovits
Haozhen Situ
Leszek Siwik
Vaclav Skala
Peter Sloot
Renata Slota
Grazyna Slusarczyk
Sucha Smanchat
Marek Smieja
Maciej Smolka
Bartlomiej Sniezynski
Isabel Sofia Brito
Katarzyna Stapor
Bogdan Staszewski

Jerzy Stefanowski
Dennis Stevenson
Tomasz Stopa
Achim Streit
Barbara Strug
Pawel Strumillo
Dante Suarez
Vishwas H. V. Subba Rao
Bongwon Suh
Diana Suleimenova
Ray Sun
Shuyu Sun
Vaidy Sunderam
Martin Swain
Alessandro Taberna
Ryszard Tadeusiewicz
Daisuke Takahashi
Zaid Tashman
Osamu Tatebe
Carlos Tavares Calafate
Kasim Tersic
Yonatan Afework
  Tesfahunegn
Jannis Teunissen
Stefan Thurner

Nestor Tiglao
Alfredo Tirado-Ramos
Arkadiusz Tomczyk
Mariusz Topolski
Paolo Trunfio
Ka-Wai Tsang
Hassan Ugail
Eirik Valseth
Pavel Varacha
Pierangelo Veltri
Raja Velu
Colin Venters
Gytis Vilutis
Peng Wang
Jianwu Wang
Shuangbu Wang
Rodrigo Weber
  dos Santos
Katarzyna
  Wegrzyn-Wolska
Mei Wen
Lars Wienbrandt
Mark Wijzenbroek
Peter Woehrmann
Szymon Wojciechowski

Maciej Woloszyn
Michal Wozniak
Maciej Wozniak
Yu Xia
Dunhui Xiao
Huilin Xing
Miguel Xochicale
Feng Xu
Wei Xue
Yoshifumi Yamamoto
Dongjia Yan
Xin-She Yang
Dongwei Ye
Wee Ping Yeo
Lihua You
Han Yu
Gábor Závodszky
Yao Zhang
H. Zhang
Jinghui Zhong
Sotirios Ziavras
Italo Zoppis
Chiara Zucco
Pawel Zyblewski
Karol Zyczkowski

Contents – Part II

Modified Binary Tree in the Fast PIES
for 2D Problems with Complex Shapes

Andrzej Kużelewski$^{(\boxtimes)}$, Eugeniusz Zieniuk , Agnieszka Bołtuć ,
and Krzysztof Szerszeń

Institute of Informatics, University of Bialystok,
Ciolkowskiego 1M, 15-245 Bialystok, Poland
{akuzel,ezieniuk,aboltuc,kszerszen}@ii.uwb.edu.pl

Abstract. The paper presents a modified binary tree in the fast multipole method (FMM) included into the modified parametric integral
equations system (PIES), called the fast PIES, in solving potential 2D
boundary value problems with complex shapes. The modified binary tree
proposed in this paper is built based on a one-dimensional reference system contrary to a quad-tree (based on a two-dimensional reference system) which is applied in the fast multipole boundary element method
(FM-BEM). Application of the proposed tree allows reducing the number of numerical computations performed during its construction and
fast multipole calculations in the fast PIES. The proposed modification
of the tree in the fast PIES allows obtaining accurate solutions in engineering problems with complex shapes on a standard personal computer
in a short time.

Keywords: Parametric integral equations system · Fast multipole
method · Boundary value problems

1 Introduction

The fast multipole method (FMM) was initially proposed by Rokhlin [1] to accelerate the solution of 2D potential problems using boundary integral equations
(BIE). Its main advantage is the reduction of the computational complexity of
the matrix-vector multiplication from $O(N^2)$ to $O(N)$ (where N is the size of
the matrix) combined with an iterative solver. In papers [2,3], Greengard refined
the algorithm by adding decomposition of the domain using the hierarchical
tree structure. It significantly reduces the utilization of random access memory
(RAM) in computer. The use of the FMM for solving 2D and 3D boundary value
problems (BVPs) is well-documented [4–6], also in application to the boundary
element method (BEM) [7].

Mentioned above BEM together with the finite element method (FEM) [8] are
well-established (might be called classic or conventional) methods of modelling
and solving boundary problems. However, new approaches are also being developed to eliminate the disadvantages of classical methods (the time-consuming

© Springer Nature Switzerland AG 2020
V. V. Krzhizhanovskaya et al. (Eds.): ICCS 2020, LNCS 12138, pp. 1–14, 2020.
https://doi.org/10.1007/978-3-030-50417-5_1

discretization of a boundary or a domain resulting in a large number of finite or boundary elements). That group includes, among others, meshless methods [9] and still being developed the parametric integral equations system (PIES) [10].

The authors of this paper still working on development and application of the PIES in modelling and solving BVPs. The PIES has been used to solve potential [11], elasticity [12] or acoustics problems [13]. The PIES includes in its mathematical formalism the shape of the boundary of considered problem [10], therefore it does not require discretization of the domain or the boundary, contrary to element methods. The shape of the boundary is directly included into the PIES kernels using well-known functions from computer graphics - curves for 2D and surface patches for 3D problems. The accuracy of solutions can be improved by changing the number of collocation points only, without any interference in the shape of the modelled boundary. The efficiency of modelling and high accuracy of solutions obtained using the PIES has been confirmed in previous studies (e.g. [11–13]). The authors of this paper also proposed extensions of the PIES method for uncertainly defined [14,15] or transient [16] problems.

Conventional PIES, similarly to the BEM, produces non-symmetrical dense matrices using $O(N^2)$ operations and to solve the problem using direct solver, it needs another $O(N^3)$ operations (where N is the number of system equations). Therefore, solving engineering problems with complex shapes requires a lot of RAM and time-consuming computations. Application of OpenMP [17] or acceleration of solving the PIES using CUDA [18,19] allows for a significant reduction of time of computations. Unfortunately, the problem of limited resources of RAM in a personal computer (PC) still exists. Therefore, mentioned above parallelization techniques do not allow for efficient solving of problems on a PC.

The authors of this paper presented a way of including the FMM, based on a binary tree, into modified PIES in [20]. Application of the FMM increased the difficulty of implementation of so-called the fast PIES. However, the proposed approach gives accurate solutions in a short time for examples with a quite simple shape of the boundary. Some assumptions of the FMM may result in obtaining incorrect solutions, especially for complex shapes of a boundary. Our research has shown the need to modify the binary tree and to change the FMM algorithm to consider assumptions mentioned above for complex shapes of a boundary.

The main goal of this paper is to present the modified binary tree in the FMM used to accelerate numerical calculations in the PIES and to reduce utilization of RAM for BVPs with complex shapes of a boundary. The efficiency and accuracy of the proposed fast PIES are tested on 2D potential BVPs.

2 Formulation of the Fast PIES Method

Conventional PIES for 2D potential problems is presented by the following formula [10]:

$$\frac{1}{2}u_l(\overline{s}) = \sum_{j=1}^{n} \int_{s_{j-1}}^{s_j} \overline{U}_{lj}^*(\overline{s}, s)p_j(s)J_j(s)ds - \sum_{j=1}^{n} \int_{s_{j-1}}^{s_j} \overline{P}_{lj}^*(\overline{s}, s)u_j(s)J_j(s)ds, \quad (1)$$

where: $l = 1, 2, ..., n$, $s_{l-1} \leq \overline{s} \leq s_l$, $s_{j-1} \leq s \leq s_j$, s_{l-1} and s_{j-1} correspond to the beginning of l-th and j-th segment, while s_l and s_j to their ends, $J_j(s)$ is the Jacobian, n is the number of parametric segments that creates boundary of domain in parametric reference system s and $\overline{s}$ (presented in Fig. 1).

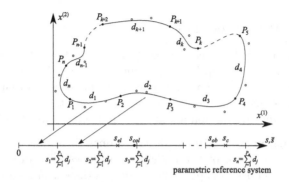

Fig. 1. Defining the shape of the boundary in the PIES on a straight line in parametric reference system $s, \overline{s}$

Integrands $\overline{U}_{lj}^*(\overline{s}, s)$ and $\overline{P}_{lj}^*(\overline{s}, s)$ in (1) are presented in the following form:

$$\overline{U}_{lj}^*(\overline{s}, s) = \frac{1}{2\pi} \ln \frac{1}{\sqrt{\left(S^{(1)}\right)^2 + \left(S^{(2)}\right)^2}},$$

$$\overline{P}_{lj}^*(\overline{s}, s) = \frac{1}{2\pi} \frac{S^{(1)} n_j^{(1)}(s) + S^{(2)} \cdot n_j^{(2)}(s)}{\left(S^{(1)}\right)^2 + \left(S^{(2)}\right)^2},$$

$$(2)$$

where: $S^{(1)} = S_l^{(1)}(\overline{s}) - S_j^{(1)}(s)$ and $S^{(2)} = S_l^{(2)}(\overline{s}) - S_j^{(2)}(s)$, $n_j^{(k)}(s)$ $(k = \{1, 2\})$ are the components of normal vector to segment j. Expressions $S_k^{(i)}(s_n)$ $\{i = 1, 2\}, \{k = j, l\}, s_n = \{\overline{s}, s\}$ are parametric functions (curves or lines), which describe particular segments of a boundary (j or l), i is the number of coordinate in Cartesian reference system.

Boundary functions $u_j(s)$ and $p_j(s)$ in (1) are approximated by the following series:

$$u_j(s) = \sum_{k=0}^{N} u_j^{(k)} L_j^{(k)}(s), \quad p_j(s) = \sum_{k=0}^{N} p_j^{(k)} L_j^{(k)}(s), \quad (3)$$

where $u_j^{(k)}$ and $p_j^{(k)}$ are unknown or given values of boundary functions in defined points of the segment j, N - is the number of terms in series, $L_j^{(k)}(s)$ - are the base functions (Lagrange polynomials).

The pseudospectral method was applied to solve the PIES (1). Therefore, the PIES is transformed into the system of algebraic equations $\mathbf{Ax} = \mathbf{b}$. The oldest implementation of the PIES uses Gaussian elimination with pivot to solve the system. Newer version, as well as the fast PIES, uses very fast and popular iterative solver based on the generalized minimal residual method (GMRES) [21]. Application of the FMM allows reducing computational time of matrix-vector multiplication (frequently used by GMRES solver) from order $O(N^2)$ to $O(N)$. There is also no need to store in the RAM entire matrix [6].

2.1   The Fast PIES Procedures

The FMM in the fast PIES is composed of two main steps: the upward and the downward pass (a full description is presented in [20]). The upward pass uses procedures of kernels expansion and moment-to-moment translation, while in the downward pass moment-to-local and local-to-local translations are applied. The most important information about these procedures are described below.

Kernels Expansion and Multipole Moments. The first procedure of the fast PIES is an expansion of kernels in Taylor series. The direct inclusion of the FMM [1] into the PIES is problematic due to the calculation of subsequent derivatives of kernels (2) for the Taylor series approximation. In the paper [20], the authors presented a modification of the PIES kernels by complex analysis. The PIES with modified kernels have the following form [20]:

$$
\frac{1}{2} u_l(\bar{s}) = \sum_{j=1}^{n} \Re \left\{ \int_{s_{j-1}}^{s_j} \overline{U}_{lj}^{*(c)}(\bar{\tau}, \tau) \, p_j(s) \, J_j(s) ds \right\}
$$
$$
- \sum_{j=1}^{n} \Re \left\{ \int_{s_{j-1}}^{s_j} \overline{P}_{lj}^{*(c)}(\bar{\tau}, \tau) \, u_j(s) \, J_j(s) ds \right\}, \tag{4}
$$

where: $\Re$ - is the real part of complex number, the parametric functions, which describe the shape of the boundary in the PIES, can be defined in complex notation as $\bar{\tau} = S_l^{(c)}(\bar{s}) = S_l^{(1)}(\bar{s}) + i S_l^{(2)}(\bar{s})$, $\tau = S_j^{(c)}(s) = S_j^{(1)}(s) + i S_j^{(2)}(s)$, (c) - means complex variable and an indeterminate (the imaginary unit) $i = \sqrt{-1}$. The complex form of kernels is as follows [20]:

$$
\overline{U}_{lj}^{*(c)}(\bar{\tau}, \tau) = -\frac{1}{2\pi} \ln(\bar{\tau} - \tau), \quad \overline{P}_{lj}^{*(c)}(\bar{\tau}, \tau) = \frac{1}{2\pi} \frac{n^{(c)}}{\bar{\tau} - \tau}, \tag{5}
$$

where $n^{(c)} = n^{(1)} + i n^{(2)}$ - the complex notation of normal vector to the curve created segment j.

We assume that the point s_c (corresponding to the complex point τ_c) is close to the observation point s_{ob} and the point s_{el} (corresponding to the complex point τ_{el}) is close to the collocation point s_{col} (presented in Fig. 2). If $|s_{ob} - s_c| \ll |s_{col} - s_c|$ and $|\tau_{ob} - \tau_c| \ll |\tau_{col} - \tau_c|$, then kernels (5) can be expanded about the point τ_c using the Taylor series expansion. Therefore, we obtained the multipole moments $M_k(\tau_c)$ and $N_k(\tau_c)$ [20]:

$$M_k(\tau_c) = \int_{s_{j-1}}^{s_j} \frac{(\tau - \tau_c)^k}{k!} p_j(s) J_j(s) ds,$$

$$N_k(\tau_c) = \int_{s_{j-1}}^{s_j} \frac{(\tau - \tau_c)^{k-1}}{(k-1)!} n^{(c)} u_j(s) J_j(s) ds \tag{6}$$

and the following form of approximated Eq. (4) [20]:

$$\frac{1}{2} u_l(\bar{s}) = \sum_{j=1}^{n} \Re \left\{ \frac{1}{2\pi} \sum_{k=0}^{\infty} U_k(\bar{\tau}, \tau_c) M_k(\tau_c) \right\}$$

$$- \sum_{j=1}^{n} \Re \left\{ \frac{1}{2\pi} \sum_{k=1}^{\infty} P_k(\bar{\tau}, \tau_c) N_k(\tau_c) \right\}, \tag{7}$$

where:

$$U_k(\bar{\tau}, \tau_c) = \begin{cases} -\ln(\bar{\tau} - \tau_c) & \text{for } k = 0 \\ \frac{(k-1)!}{(\bar{\tau} - \tau_c)^k} & \text{for } k \geq 1 \end{cases}, \quad P_k(\bar{\tau}, \tau_c) = \frac{(k-1)!}{(\bar{\tau} - \tau_c)^k} \quad \text{for } k \geq 1.$$

Moments are calculated once only and they are independent of $\bar{\tau}$ (that is also from $\bar{s}$).

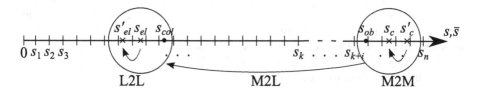

Fig. 2. Location of specific FMM points in the parametric reference system

Moment-to-Moment Translation (M2M). Moments (6) can be very efficiently recalculated for new point s_c' (corresponding to the complex point τ_c') close to the point s_c (presented in Fig. 2) without reuse of integration. For this purpose, moment-to-moment translation was used [20]:

$$M_k(\tau_c') = \sum_{m=0}^{k} \frac{(\tau_c - \tau_c')^{(k-m)}}{(k-m)!} M_m(\tau_c),$$

$$N_k(\tau_c') = \sum_{m=0}^{k} \frac{(\tau_c - \tau_c')^{(k-m)}}{(k-m)!} N_m(\tau_c). \tag{8}$$

Moment-to-Local Translation (M2L) and Local Expansion. The next step of the fast PIES is local expansion around the point s_{el} (corresponding to the complex point τ_{el}) (presented in Fig. 2). If $|s_{col} - s_{el}| \ll |s_c - s_{el}|$ and $|\tau_{col} - \tau_{el}| \ll |\tau_c - \tau_{el}|$, then U_k and P_k in (7) can be expanded about the point τ_{el} using the Taylor series expansion, hence [20]:

$$
\begin{aligned}
\frac{1}{2}u_l(\bar{s}) = &\sum_{j=1}^{n} \Re\left\{ \frac{1}{2\pi} \sum_{l=0}^{\infty} L_l^U(\tau_{el}, \tau_c) \frac{(\tau_{col} - \tau_{el})^l}{l!} \right\} \\
&- \sum_{j=1}^{n} \Re\left\{ \frac{1}{2\pi} \sum_{l=0}^{\infty} L_l^P(\tau_{el}, \tau_c) \frac{(\tau_{col} - \tau_{el})^l}{l!} \right\},
\end{aligned}
\tag{9}
$$

where:

$$
L_l^U(\tau_{el}, \tau_c) = \begin{cases} -\ln(\tau_{el} - \tau_c)M_0(\tau_c) + \sum_{k=1}^{\infty} \frac{(k-1)! \cdot M_k(\tau_c)}{(\tau_{el}-\tau_c)^k} & \text{for } l = 0 \\ (-1)^l \sum_{k=0}^{\infty} \frac{(k+l-1)! \cdot M_k(\tau_c)}{(\tau_{el}-\tau_c)^{k+l}} & \text{for } l \geq 1, \end{cases}
$$

$$
L_l^P(\tau_{el}, \tau_c) = (-1)^l \sum_{k=1}^{\infty} \frac{(k+l-1)! \cdot N_k(\tau_c)}{(\tau_{el} - \tau_c)^{k+l}} \quad \text{for } l \geq 1.
$$

This procedure allows the transformation of moments collected at a point s_c (τ_c) to a local expansion point s_{el} (τ_{el}) and is called moment-to-local translation (M2L) [20].

During modelling complex shapes of the boundary, fulfilment of the condition $|\tau_{col} - \tau_{el}| \ll |\tau_c - \tau_{el}|$ might be not possible to meet for some cells. The results obtained in this case are subject to large errors. The authors of this paper proposed the way of elimination such cases - in the FMM algorithm they are considered as neighbouring cells (a more detailed description of the neighbourhood of the cell is described in Sect. 2.3).

Local-to-Local Translation (L2L). Similarly to the M2M, moments at a point s_{el} (τ_{el}) can be efficiently recalculated for a nearby point s'_{el} (corresponding to the complex point τ'_{el}) (presented in Fig. 2). For this purpose, a transformation called local-to-local (L2L) translation (similar to M2M) is used [20]:

$$
\begin{aligned}
L_l^U(\tau'_{el}, \tau_c) &= (-1)^l \left\{ \sum_{k=0}^{\infty} \sum_{m=l}^{\infty} \frac{(k+m-1)! \cdot M_k(\tau_c)}{(\tau_{el} - \tau_c)^{k+m}} \cdot \frac{(\tau'_{el} - \tau_{el})^{m-l}}{(m-l)!} \right\}, \\
L_l^P(\tau'_{el}, \tau_c) &= (-1)^l \cdot \left\{ \sum_{k=1}^{\infty} \sum_{m=l}^{\infty} \frac{(k+m-1)! \cdot N_k(\tau_c)}{(\tau_{el} - \tau_c)^{k+m}} \cdot \frac{(\tau'_{el} - \tau_{el})^{m-l}}{(m-l)!} \right\}.
\end{aligned}
\tag{10}
$$

2.2   Tree Structure in the Fast PIES

In classic FMM, the binary tree for 1D, quad-tree for 2D and octa-tree for 3D problems are applied. In this paper, 2D problems are discussed, hence in classic FMM implementation quad-tree presented in Fig. 3 [6] is used.

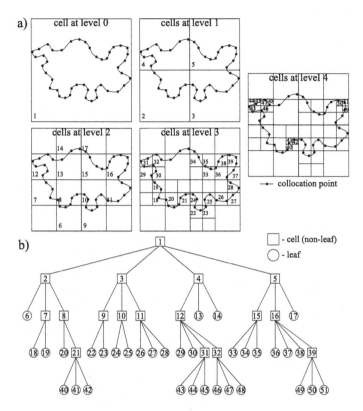

Fig. 3. a) Constructing the FMM tree in the FM-BEM, b) structure of obtained quad-tree for 2D problem

A square surrounding the entire domain of the problem (Fig. 3a) is called level 0 cell. This cell is the parent of four cells of level 1 (so-called children cells) obtained as a result of dividing the parent cell into four identical squares. Only the cell crossing the boundary of the problem is considered to be a child cell. The parent cell of level l ($l \geq 0$) is divided into children cells of a level $l + 1$. The division is performed until a predetermined number of elements are inside a cell (in the example in Fig. 3a - max. 2 elements whose centre is marked as a collocation point) or the assumed maximum level l has been reached. A cell without a child is called a leaf. The quad-tree presented in Fig. 3b is obtained in the described way. The presented method of the FMM tree construction is applied, among others, in the FM-BEM [6].

In the PIES applied for the 2D problems, it is possible to implement a suitably modified and improved binary tree. The improvement is related to the way of defining problems in the PIES - the boundary of the problem is defined in the 1D parametric reference system (presented in Fig. 1). It is a different way of modelling the shape of a boundary than in the BEM. Also, the number of segments describing the boundary in the PIES is smaller than the number of elements in the BEM (see Fig. 4). It is connected with the way of defining the shape of the boundary in the PIES [10].

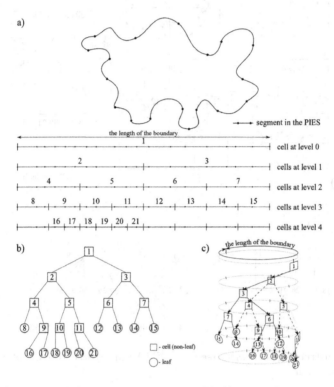

Fig. 4. a) Constructing the FMM tree in the PIES, b) classic binary tree for the 1D problem, c) modified binary tree in the PIES for the 2D problem

The proposed binary tree structure assumes that the boundary is described in the 1D reference system. A level 0 cell covers the entire boundary of the problem (presented in Fig. 4a). This cell is the parent of two children cells of level 1 obtained as a result of dividing the parent cell into two identical segments. The parent cell of level l ($l \geq 0$) is divided into children cells of a level $l + 1$. The division is performed until a predetermined number of segments are inside a cell (in the example in Fig. 4a - max. 2 segments) or the assumed maximum level l has been reached. In the classic FMM, the binary tree for 1D problems has the form presented in Fig. 4b. However, in the PIES 2D problem is reduced to the 1D parametric reference system, hence the first and the last boundary segment are very close to each other. Therefore, we propose to close the binary tree in the form presented in Fig. 4c. For all levels, the first and the last cells are treated as adjacent.

2.3 Algorithm of Solving the Fast PIES

The algorithm of the PIES with modified kernels proceeds in several steps (flow chart presented in Fig. 5). The first step after initialization of the algorithm is to determine the structure of the modified binary tree. Then right-hand sided vector **b** is computed and GMRES is called to find solution of $\mathbf{Ax} = \mathbf{b}$. These procedures uses the modified fast multipole algorithm presented in Algorithm 1. At the end of algorithm results are presented on the screen and written to the output file.

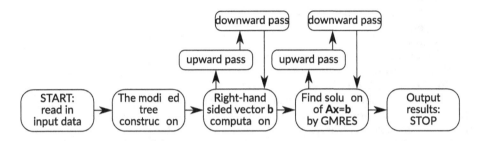

Fig. 5. Flow chart for the fast PIES

The first run of the fast multipole procedure is used for calculating the right-hand sided vector **b**. The fast multipole procedure is composed of two steps: the upward pass and the downward pass. At the upward pass, all moments in leaves are calculated (line 4 in Algorithm 1) for the level lev. Then, tracing the tree structure upward, moments in all parent cells are calculated up to level 2 using M2M (line 6).

In the downward pass, we trace the tree structure downward, and previously calculated moments are used in calculations. First of all, we should remind cells neighbourhood to clarify this step [20]. Two cells are $adjacent$ at level i, if they have a common end at level i. Two cells are $well-separated$ at level i, if they are not adjacent at level i, but their parent cells are adjacent at level $i-1$. The $interaction\ list$ of cell $K-th$ is the list of cells $well-separated$ from cell $K-th$ at level i. At last, two cells are far cells, if their parent cells are not adjacent.

In modelling complex shapes of the boundary, we should modify the FMM algorithm described in [20] due to the possibility of too small distance between cells with τ_c and τ_{el} points, hence the assumption of $|\tau_{col} - \tau_{el}| \ll |\tau_c - \tau_{el}|$ used for the M2L translation is not fulfilled. The authors of this work propose to modify the algorithm by marking such cells as $adjacent$ and not to enter them on $interaction\ list$ (calculations are performed as in the case of $adjacent$ cells). Including this modification to the fast PIES, solutions with the same accuracy as in conventional PIES are obtained.

Starting from level 2 and tracing the tree structure downward to all leaves coefficients of local expansion (the line 24 in the Algorithm 1) are computed. Coefficients at cell $K-th$ at level i are computed as the sum of two elements:

Algorithm 1. Modified fast multipole procedure

Require:
  lev - the number of tree levels
  $\min c_i$ - the lowest number of a cell on the level i
  $\max c_i$ - the highest number of a cell on the level i
  //upward pass
1: **for** $i \leftarrow lev$ to 2 **do**
2:   **for** $icell \leftarrow \min c_i$ to $\max c_i$ **do**
3:     **if** $i == lev$ **then**
4:       $multipole\_moments(icell)$
5:     **else**
6:       $moment\_expansion(icell)$
7:     **end if**
8:   **end for**
9: **end for**
  //downward pass
10: **for** $i \leftarrow lev$ to 2 **do**
11:   **for** $icell \leftarrow \min c_i$ to $\max c_i$ **do**
12:     **if** $i \neq 2$ **then**
13:       $local\_to\_local(icell)$
14:     **end if**
15:     **for** $jcell \leftarrow \min c_i$ to $\max c_i$ **do**
16:       **if** $parent(icell)$ & $parent(jcell)$ are neighbours **then**
17:         **if** $(cell(icell)$ & $cell(jcell)$ are neighbours$)$ $\|$ $!(|\tau_{col} - \tau_{jcell}| \ll |\tau_{icell} - \tau_{jcell}|)$ **then**
18:           $direct(icell, jcell)$
19:         **else**
20:           $moment\_to\_local(icell, jcell)$
21:         **end if**
22:       **end if**
23:     **end for**
24:     $local\_expansion(icell)$
25:   **end for**
26: **end for**

---

contributions from all far cells (computed using L2L - the line 13) and from the cells in the interaction list of cell $K - th$ (computed using M2L - the line 20). There are no far cells to a cell K at level 2, therefore only M2L is used to compute coefficients. Contributions from adjacent cells of leaf K at the lowest level are computed directly (line 18), as in conventional PIES. Finally, the FMM procedure produces a right-hand vector **b**.

To solve the system of algebraic equations $\mathbf{Ax} = \mathbf{b}$, iterative GMRES solver is used. The method requires the application of multiplication of the matrix **A** by the vector of unknowns **x**, therefore the solver can be directly integrated with the fast PIES. The FMM in GMRES is performed in the same way as for vector **b**.

3 Tests of the Fast PIES with Modified Binary Tree

In the paper [20], the authors presented preliminary results of the fast PIES solutions. However, solving the problem presented in this paper by the fast PIES without modification of the binary tree gives unsatisfactory results. Obtained solutions were far from expected.

The problem of temperature distribution (modelled by Laplace's equation) in heat-sink is considered. The shape of the boundary and boundary conditions are shown in Fig. 6. The boundary is composed of 716 linear segments. The same number of collocation points (from 4 to 8) is defined on each segment, and finally, we should solve the system of 2864 to 5728 algebraic equations. The problem is solved by conventional and fast PIES. PC based on Intel Core i5-4590S with 8 GB RAM and g++ 7.4.0 compiler with -O2 optimization on 64-bit Linux operation system (Ubuntu, kernel 5.0.0) and LAPACK 3.10.3 library [22] is used during tests.

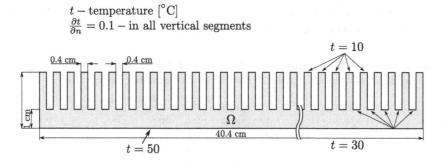

Fig. 6. The shape of considered heat-sink

First of all, a comparison of CPU time and RAM utilization between a different number of the modified tree levels in the fast PIES is performed to find the optimal value of tree levels. Taylor series composed of 25 terms approximated the fast PIES kernels. The value of tolerance (convergence criterion) of the GMRES was equal to 10^{-8}.

As can be seen from Fig. 7, both CPU time and memory utilization decrease with the growing number of tree levels reaching the minimum value for a tree with 5–6 levels regardless of the number of collocation points. In our studies, the tree with 6 levels is adopted.

The research involved a comparison of the speed and RAM utilization between conventional, the fast PIES and the fast multipole BEM (application from [6]). In the fast multipole BEM two meshes are used: 2864 and 5728 elements. The value of tolerance (convergence criterion) of the GMRES and the number of terms in Taylor series are the same as previous, i.e. 10^{-8} and 25 respectively. Solutions are presented for two versions of conventional PIES

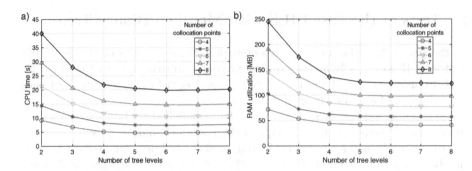

Fig. 7. Influence of the number of tree levels on a) computational time, b) RAM utilization of the fast PIES

(PIES$_d$ with direct solver (Gaussian elimination method) and PIES$_i$ with iterative solver GMRES), the fast PIES (fPIES) and the fast multipole BEM (the fmBEM). The accuracy of the solutions is calculated as the mean square error (MSE) between the results of conventional (PIES$_d$) and the fast PIES (P-P) and the fast PIES and the fast multipole BEM (P-B).

Table 1. Comparison of computational time and RAM utilization between the fast multipole BEM, conventional and the fast PIES with modified binary tree.

Number of		CPU time [s]				RAM utilization [MB]				MSE	
col. pts	eqs	fPIES	PIES$_d$	PIES$_i$	fmBEM	fPIES	PIES$_d$	PIES$_i$	fmBEM	P-P	P-B
4	2864	4.63	30.10	27.82	9.56	40.96	254	175	105.9	2.31e−10	0.051
5	3580	7.28	53.58	45.07	–	58.25	395	207	–	1.54e−10	–
6	4296	10.43	94.17	67.20	–	78.08	567	297	–	2.04e−10	–
7	5012	14.44	140.94	98.49	–	98	770	398	–	2.29e−10	–
8	5728	19.69	214.57	128.66	46.18	124	1005	517	107.5	2.18e−10	0.063

As can be seen from Table 1, the fast PIES is significantly faster than both versions of conventional PIES and about 2 times faster than the fast multipole BEM. The fast PIES also needs up to 4 times less RAM than the PIES$_i$ and about 8 times less than the PIES$_d$. Obtained solutions are practically the same as in conventional versions of the PIES - MSE between the PIES$_d$ and the fast PIES does not exceed $2.31 \cdot 10^{-10}$. The MSE between the fast PIES and the fast multipole BEM has higher value. However, our previous studies proved that the PIES is more accurate than the BEM.

Graphical comparison of CPU time and RAM utilization between all the PIES methods is presented in Fig. 8.

4   Conclusions

The paper presents the fast PIES based on the FMM with the modified binary tree in solving potential BVPs with complex shapes. The proposed modified

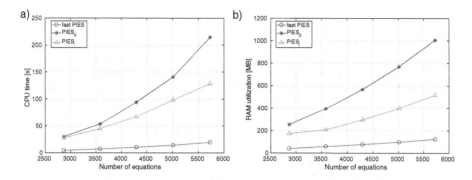

Fig. 8. Comparison of a) CPU time, b) RAM utilization between conventional and the fast PIES

binary tree is built based on the 1D parametric reference system. Such modification of the tree in the fast PIES allows obtaining very accurate solutions for engineering problems with complex shapes on standard PC in a short computational time. It also allows solving problems in which the condition of the appropriate distance between the centres of the leaves is not fulfilled.

The numerical test shows a reduction of the computation time and RAM utilization of the fast PIES compared to conventional ones as well as the fast multipole BEM. The speed-up of computations between the fast and conventional PIES increases with the size of solving problem, while the accuracy of solutions is almost the same.

Acknowledgments. The scientific work is founded by resources of Ministry of Science and Higher Education for research, granted to the Institute of Informatics, University of Bialystok.

References

1. Rokhlin, V.: Rapid solution of integral equations of classical potential theory. J. Comput. Phys. **60**(2), 187–207 (1985)
2. Greengard, L.F., Rokhlin, V.: A fast algorithm for particle simulations. J. Comput. Phys. **73**(2), 325–348 (1987)
3. Greengard, L.F.: The Rapid Evaluation of Potential Fields in Particle Systems. The MIT Press, Cambridge (1988)
4. Blankrot, B., Leviatan, Y.: FMM-accelerated source-model technique for many-scatterer problems. IEEE Trans. Antennas Propag. **65**(8), 4379–4384 (2017)
5. Nishimura, N.: Fast multipole accelerated boundary integral equation methods. Appl. Mech. Rev. **55**(4), 299–324 (2002)
6. Liu, Y.J., Nishimura, N.: The fast multipole boundary element method for potential problems: a tutorial. Eng. Anal. Boundary Elem. **30**(5), 371–381 (2006)
7. Brebbia, C.A., Telles, J.C.F., Wrobel, L.C.: Boundary Element Techniques, Theory and Applications in Engineering. Springer, New York (1984). https://doi.org/10.1007/978-3-642-48860-3

8. Zienkiewicz, O.C.: The Finite Element Method. McGraw-Hill, London (1977)
9. Perazzo, F., Lohner, R., Perez-Pozo, L.: Adaptive methodology for meshless finite point method. Adv. Eng. Softw. **39**(3), 156–166 (2008)
10. Zieniuk, E.: Hermite curves in the modification of integral equations for potential boundary-value problems. Eng. Comput. **20**(1–2), 112–128 (2003)
11. Zieniuk, E., Kapturczak, M.: Modeling the shape of boundary using NURBS curves directly in modified boundary integral equations for Laplace's equation. Comput. Appl. Math. **37**(4), 4835–4855 (2018)
12. Zieniuk, E., Bołtuć, A., Kuzelewski, A.: Algorithms of identification of multi-connected boundary geometry and material parameters in problems described by Navier-Lamé equation using the PIES. In: Pejaś, J., Saeed, K. (eds.) Advances in Information Processing and Protection, pp. 409–418. Springer, Boston (2007). https://doi.org/10.1007/978-0-387-73137-7_37
13. Zieniuk, E., Szerszeń, K.: Triangular Bézier patches in modelling smooth boundary surface in exterior Helmholtz problems solved by PIES. Arch. Acoust. **34**(1), 51–61 (2009)
14. Zieniuk, E., Kużelewski, A., Kapturczak, M.: The influence of interval arithmetic on the shape of uncertainly defined domains modelled by closed curves. Comput. Appl. Math. **37**(2), 1027–1046 (2016). https://doi.org/10.1007/s40314-016-0382-0
15. Zieniuk, E., Kapturczak, M., Kużelewski, A.: Modification of interval arithmetic for modelling and solving uncertainly defined problems by interval parametric integral equations system. In: Shi, Y., et al. (eds.) ICCS 2018. LNCS, vol. 10862, pp. 231–240. Springer, Cham (2018). https://doi.org/10.1007/978-3-319-93713-7_19
16. Zieniuk, E., Sawicki, D., Bołtuć, A.: Parametric integral equations systems in 2D transient heat conduction analysis. Int. J. Heat Mass Transf. **78**, 571–587 (2014)
17. Kużelewski, A., Zieniuk, E.: OpenMP for 3D potential boundary value problems solved by PIES. In: 13th International Conference of Numerical Analysis and Applied Mathematics ICNAAM 2015, AIP Conference Proceedings, vol. 1738, Article number 480098. AIP Publishing LLC (2016)
18. Kuzelewski, A., Zieniuk, E., Boltuc, A.: Application of CUDA for acceleration of calculations in boundary value problems solving using PIES. In: Wyrzykowski, R., Dongarra, J., Karczewski, K., Waśniewski, J. (eds.) PPAM 2013. LNCS, vol. 8385, pp. 322–331. Springer, Heidelberg (2014). https://doi.org/10.1007/978-3-642-55195-6_30
19. Kużelewski, A., Zieniuk, E., Kapturczak, M.: Acceleration of integration in parametric integral equations system using CUDA. Comput. Struct. **152**, 113–124 (2015)
20. Kużelewski, A., Zieniuk, E.: The fast parametric integral equations system in an acceleration of solving polygonal potential boundary value problems. Adv. Eng. Softw. **141**, 102770 (2020)
21. Saad, Y., Schultz, M.H.: GMRES: a generalized minimal residual algorithm for solving nonsymmetric linear systems. SIAM J. Sci. Stat. Comput. **7**, 856–869 (1986)
22. Anderson, E., et al.: LAPACK Users' Guide. Society for Industrial and Applied Mathematics, Philadelphia (1999)

Generating Random Floating-Point Numbers by Dividing Integers: A Case Study

Frédéric Goualard[⊠] [iD]

University of Nantes and LS2N UMR CNRS 6004, Nantes, France
Frederic.Goualard@univ-nantes.fr
http://frederic.goualard.net/

Abstract. A method widely used to obtain IEEE 754 binary floating-point numbers with a standard uniform distribution involves drawing an integer uniformly at random and dividing it by another larger integer. We survey the various instances of the algorithm that are used in actual software and point out their properties and drawbacks, particularly from the standpoint of numerical software testing and data anonymization.

Keywords: Floating-point number · Random number · Error analysis

1 Introduction

In his oft-quoted 1951 paper [18], John von Neumann asserts that *"[i]f one wants to get random real numbers on $(0, 1)$ satisfying a uniform distribution, it is clearly sufficient to juxtapose enough random binary digits."* That dismissive opinion seems to have been so largely shared to this day that the implementation of methods to compute random floating-point numbers (or *floats*, for short) almost always feels like an afterthought, even in respected numerical software and programming languages. Take a look at your favorite software package documentation: chances are that it will describe the algorithm —or algorithms, since many provide more than one method— used to draw integers at random; on the other hand, it will often fail to give any precise information regarding the way it obtains random floats. Such information will have to be gathered directly from the source code when it is available. Besides, software offering the most methods to compute random integers will almost always provide only one means to obtain random floats, namely through the division of some random integer by another integer. It is the first method proposed by Knuth in *The Art of Computer Programming volume 2* [8, sec. 3.2, p. 10]; also the first method in *Numerical Recipes in C* [21, chap. 7.1, pp. 275–276] and in many other resources (e.g., [1] and [5, pp. 200–201]).

Despite its pervasiveness, that method has flaws, some of them well known, some less so or overlooked. In particular, some implementations may return values outside of the intended domain, or they may only compute values whose

© Springer Nature Switzerland AG 2020
V. V. Krzhizhanovskaya et al. (Eds.): ICCS 2020, LNCS 12138, pp. 15–28, 2020.
https://doi.org/10.1007/978-3-030-50417-5_2

binary representations all share some undesirable properties. That last flaw is a baleful one for applications that rely on random number generators (RNGs) to obtain *differential privacy* for numerical data [16], or to ensure a proper coverage in numerical software testing.

The lack of awareness to these issues may be underpinned by the fact that the libraries meant to test random number generators do not implement means to check for them properly: the procedure ugfsr_CreateMT19937_98() based on the Mersenne Twister [15] MT19937 to generate floats in TestU01 [10] passes the *Small Crush* battery of tests without failure, even though all double precision numbers produced have the 32nd bit of their fractional part always set to "1," and their 31st bit has a 75% chance of being "0"; worse, the Kiss99 [13] generator umarsa_CreateKISS99() used in TestU01 passes all *Small Crush*, *Crush* and *Big Crush* batteries while the probability to be "0" of each bit of the fractional part of the numbers produced increases steadily from 0.5 for the leftmost one to the 32nd, and is then equal to 1 for the remaining rightmost bits.

We briefly present in Sect. 2 the details of the IEEE 754 standard for binary floating-point numbers [7] that are relevant to our study. In Sect. 3, we will consider various implementations of the RNGs that compute random floating-point numbers through some division; that section is subdivided into two parts, depending on the kind of divisor used. The actual implementation in widely available software is considered in Sect. 4. Lastly, we summarize our findings in Sect. 5, and we assess their significance depending on the applications targeted; alternative implementations that do not use a division are also considered.

2   Floating-Point Numbers

The IEEE 754 standard [7] is the ubiquitous reference to implement binary floating-point numbers and their associated operators on processors. IEEE 754 binary floating-point numbers are of varying formats, depending on the number of bits used to represent them in memory. A format is completely defined by a pair $(p, emax)$ of two natural integers. Let $\mathbb{F}_p^{emax}$ be the set of floats with format $(p, emax)$. We will also note y_k the kth bit of the binary representation of the number y (with y_0 being the rightmost least significant bit).

A floating-point number $x \in \mathbb{F}_p^{emax}$ can be viewed as a binary fractional number (the *significand*), a sign and a scale factor. There are five classes of floats: ± 0, *normal* floats with p significant bits, *subnormal* floats with less than p significant bits, *infinities* and *Not A Numbers*. Only the first three classes are of any concern to us here. Floats from these classes are represented with three fields (s, E, f) with the interpretation:

$$x = \begin{cases} (-1)^s \times 1.f_{p-2}f_{p-3} \cdots f_0 \times 2^E & \text{, if } x \text{ is normal;} \\ (-1)^s \times 0.f_{p-2}f_{p-3} \cdots f_0 \times 2^{1-emax} & \text{, if } x \text{ is subnormal or zero.} \end{cases} \tag{1}$$

with s the sign bit, $E \in [1 - emax, emax]$ the *exponent*, and f the *fractional part*.

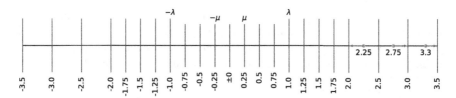

Fig. 1. The real line with $\mathbb{F}_3^1$ finite floating-point numbers. Subnormals are displayed with shorter purple segments. Reals materialized by light blue dots are rounded to nearest-even floats according to red arrows. (Color figure online)

The IEEE 754 standard defines three binary formats of which we will consider only the two most popular: *single precision* (aka binary32) $\mathbb{F}_{24}^{127}$ and *double precision* (aka binary64) $\mathbb{F}_{53}^{1023}$.

From Eq. (1), one can anticipate the *wobbling effect* that is illustrated in Fig. 1: starting with λ —the smallest positive normal float— the gap from a representable number to the next doubles every 2^{p-1} floating-point numbers, and the same goes for the negative side.

Reals in-between floating-point numbers have to be *rounded* to be represented. Given some floating-point format, let $\mathrm{fl}(x)$ be the floating-point number that is nearest to the real x. When x is at the same distance of two floats, $\mathrm{fl}(x)$ is the float whose rightmost bit b_0 of the fractional part is equal to 0. This is the *rounding to nearest-even* policy, which is usually the default and which will be assumed throughout this paper. We will also write $\mathrm{fl}\langle expr \rangle$ to denote the rounding of an expression (e.g.: $\mathrm{fl}\langle a + b \times c \rangle = \mathrm{fl}(\mathrm{fl}(a) + \mathrm{fl}(\mathrm{fl}(b) \times \mathrm{fl}(c)))$) for $(a, b, c) \in \mathbb{R}^3$).

3   Dividing Random Integers

"If you want a random float value between 0.0 and 1.0 you get it by an expression like $x = rand()/(RAND\_MAX+1.0)$." This method, advocated here in *Numerical Recipes in C* [21, pp. 275–276], is used in many libraries for various programming languages to compute random floating-point numbers with a standard uniform distribution, the variation from one library to the next being the algorithm used to compute the random integer in the numerator and the value of the fixed integer as denominator.

For a floating-point set $\mathbb{F}_p^{emax}$, there are $emax \times 2^{p-1}$ floats in the domain $[0, 1)$, to compare with the $(2emax + 1)2^p$ finite floats overall; that is, almost one fourth of all finite floats are in $[0, 1)$. Dividing a random nonnegative integer a by an integer b strictly greater than a can only return at most b distinct floats, less than that if two fractions round to the same float. Let $\mathbb{D}_b$ be the set $\{\mathrm{fl}\langle x/b \rangle \mid x = 0, 1, \ldots, b-1\}$ with b a strictly positive integer. Two necessary and sufficient conditions for $\mathrm{fl}\langle x/b \rangle$ to be equal to x/b for any x in $\{0, 1, \ldots, b-1\}$ are that b be:

1. a power of 2, otherwise some x/b are bound to not be dyadic rationals;
2. smaller or equal to 2^p, since not all integers greater than 2^p are representable in $\mathbb{F}_p^{emax}$. Besides, the largest gap between two consecutive floats from $[0,1)$ being 2^{-p}, x/b might not be representable for the large values of x otherwise.

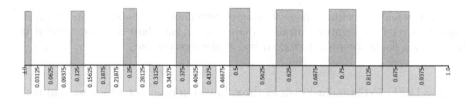

Fig. 2. Drawing 10 000 floats at random from $\mathbb{F}_4^3$ by drawing random integers from $[0,7]$ and dividing by 8 (top purple bars), or by drawing random integers from $[0,15]$ and dividing by 16 (bottom orange bars). (Color figure online)

All members of $\mathbb{D}_b = \{k/b \mid k = 0, \ldots b - 1\}$, for any b in $\{2^i \mid i = 0, \ldots, p\}$, are then representable in $\mathbb{F}_p^{emax}$ and uniformly distributed on $[0,1)$. Figure 2 shows such situations on $\mathbb{F}_4^3$: the top purple bars correspond to drawing $10,000$ random integers from $[0,7]$ and dividing them by 2^3, while the bottom orange bars correspond to drawing the same number of integers from $[0,15]$ and dividing them by $2^p = 16$. Rectangle heights indicate the number of times the corresponding float was obtained; the width of each rectangle indicates the set of reals that would all round to that same floating-point value. We see in Fig. 2 that the floats obtainable are computed with the same frequency, and that they split neatly the real line on $[0,1)$ into eight (resp. sixteen) same-sized segments.

To ensure a uniform spreading on $[0,1)$, we have to meet the two abovementioned conditions, the second one of which means that we cannot expect to be able to draw uniformly more than a fraction $2^p/(emax \times 2^{p-1}) = 2/emax$ of all the floats in $[0,1)$. To put that in perspective, it means that, for the double precision format $\mathbb{F}_{53}^{1023}$, we can only draw less than 0.2% $(2/1023)$ of all floating-point numbers in $[0,1)$.

If we violate one of the two conditions, the set $\mathbb{D}_b$ contains floats that are no longer spread evenly on $[0,1)$ (see the examples in Fig. 3). When $b \leqslant 2^p$, each float in $\mathbb{D}_b$ has the same probability of being chosen (since then, no two values a_1/b and a_2/b can round to the same float when $a_1 \neq a_2$). On the other hand, when $b > 2^p$, several fractions a/b may round to the same float, leading to it being over-represented. This is the case as soon as the distance $2^E 2^{1-p}$ between two consecutive floats is greater than b^{-1} (See Fig. 3 for $b = 24$).

As previously pointed out, the gap between adjacent floats increases from 0 to 1, with the largest gap being 2^{-p} between 1 and prev (1), its predecessor. We have seen that a division-based method to compute random floats that are evenly spread on $[0,1)$ cannot offer more than a very small proportion of all the floats in that domain. In particular, if $emax$ is not smaller than p, it is not

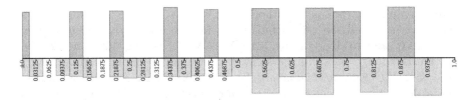

Fig. 3. Drawing 10 000 floats at random from $\mathbb{F}_4^3$ by drawing random integers in $[0, 8]$ and dividing by 9 (top purple), and by drawing from $[0, 23]$ and dividing by 24 (bottom orange). (Color figure online)

possible to draw subnormal floats by division while preserving uniform spreading of the floats we can draw. Provided we use a procedure that draws integers in $[0, b - 1]$ uniformly at random —an assumption we will always make in this paper—, all floats from $\mathbb{D}_b$ are also drawn uniformly at random for $b \leqslant 2^p$. This is different from claiming that the division-based procedure draws the floats from $[0, 1)$ uniformly at random since only one in $2/emax$ can ever be drawn. We must also keep in mind that it is also markedly different from claiming that the floats computed are the rounded results of drawing real values in $[0, 1)$ uniformly at random: since the space between floats doubles regularly from 0 to 1, the floats in $\mathbb{D}_b$ should be drawn following a geometric distribution for their exponent to accurately represent that. Several works [3,4,24] and [17, Chap. 9] have investigated means to do it, even though some of their authors are not yet convinced that there are real applications waiting for their method.

After analyzing a set of prominent libraries and programming languages, we have identified essentially two classes of division-based methods to compute floating-point numbers with a standard uniform distribution:

1. Division by a power of 2;
2. Division by a Mersenne number.

3.1   Dividing by a Power of 2

Since Lehmer's work on *Linear Congruential Generators* [11], many methods have been devised to compute random integers in $[0, m - 1]$ using modular arithmetic with some modulus m. Initially, it was very convenient to choose m as a power of 2, preferably matching the size of the integers manipulated —since the modulo was then essentially free— even though that choice may not offer the largest period for the RNG [6]. People quickly proceeded to divide by m the integers obtained to get random numbers in $[0, 1)$ [9], which is also very convenient when m is a power of 2 as the division then amounts to a simple manipulation of the float's exponent.

To this day, many libraries still draw floats by dividing random integers by a power of 2 to obtain random floats in $[0, 1)$ or $(0, 1)$ in order to get the benefit of avoiding a costly "real" division.

For example, the canonical method to get random floats in $[0, 1)$ for the ISO C language is to call the rand() function, which returns a random integer in $[0, \texttt{RAND\_MAX}]$, and to divide it by $\texttt{RAND\_MAX}+1$ (see, e.g., [21] and [23, Q. 13.21]), even though one of its flaws is well known: according to Section 7.22.2.1 of the *ISO/IEC 9899:2011 standard for C*, RAND_MAX is only constrained to be at least $2^{15} - 1$, and the GNU C library sets it to $2^{31} - 1$ only [12]. As shown above, the theoretical maximum value of the denominator to get uniformly spaced double precision floats in $[0, 1)$ is 2^{53}. It is then wasteful to only divide by 2^{15} or even 2^{31}, leading to very few possible floating-point numbers.

A property expected from RNGs that compute integers is that each bit of their binary representation be equally likely a "0" or a "1." The left graph in Fig. 4 shows the probability computed by drawing $1\,000\,000$ integers in $[0, 2^{31}-1]$ using a Mersenne Twister-based RNG. We see that all bits from the zeroth in the rightmost position to the thirtieth in the leftmost position have a probability approximately equal to 0.5 to be a "1," which is expected.

However, when we divide these integers from $[0, 2^{31} - 1]$ by 2^{31} to get floats in the domain $[0, 1)$, we lose that property. The picture on the right in Fig. 4 shows the probability to be "1" of each bit of the fractional part of double precision floats obtained by such a process: we see that the 22 rightmost bits of the fractional part have probability equal to 0 to be a "1." The probability then increases steadily until it peeks at around 0.5 for the leftmost bits.

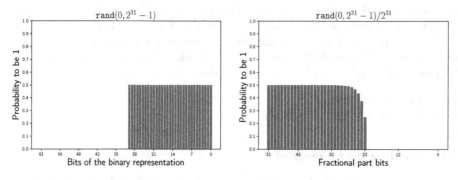

Fig. 4. Probability for each bit to be 1 (left: when drawing an integer in $[0, 2^{31} - 1]$; right: when drawing a double precision float by dividing an integer in $[0, 2^{31} - 1]$ by 2^{31}). Simulation with $1\,000\,000$ draws.

Indeed, if we note $P[e]$ the probability of the event e, we have:

Proposition 1. *Given $\mathbb{F}_p^{emax}$ a set of floats and k an integer in $[1, p]$, let $x = x_{k-1}x_{k-2}\ldots x_1 x_0$ be a strictly positive random integer in $[1, 2^k - 1]$ with $P[x_i = 1] = \frac{1}{2}$ for $i = 0, 1, \ldots, k - 1$. Given $y = y_{p-1}.y_{p-2}\cdots y_0$ the significand of the normal float $\frac{x}{2^k}$, with $y \in \mathbb{F}_p^{emax}$, we have:*

$$\begin{cases} P[y_i = 1] = 0 & \forall i \in [0, p - k - 1] \\ P[y_i = 1] = \sum_{j=0}^{i-p+k} \dfrac{1}{2^{j+2}} = \dfrac{1}{2} - 2^{p-k-i-2} & \forall i \in [p - k, p - 1] \end{cases} \qquad (2)$$

Proof. The division by a power of 2 corresponds to a simple shift of the binary point. Considering a set of floats $\mathbb{F}_p^{emax}$, if we draw a non-null integer $x = x_{k-1}x_{k-2}\ldots x_1x_0$ in the domain $[1, 2^k - 1]$ (with $k \leqslant p$) and divide it by 2^k, we get:

$$\frac{x}{2^k} = 0.x_{k-1}x_{k-2}\ldots x_1x_0$$

That number needs to be normalized to be stored as an IEEE 754 float, and we get its significand y:

$$y = 1.x_{k-2}\ldots x_0 \underbrace{0\cdots 0}_{p-k} \qquad \text{if } x_{k-1} = 1$$

$$= 1.x_{k-3}\ldots x_0 \underbrace{0\cdots 0}_{p-k+1} \qquad \text{if } x_{k-1} = 0 \wedge x_{k-2} = 1$$

$$= \quad \ddots$$

$$= 1.x_{k-(j+1)}\ldots x_0 \underbrace{0\cdots 0}_{p-k+j-1} \quad \text{if } (\forall l \in [1, j+1]: x_{k-l} = 0) \wedge x_{k-j} = 1$$

Obviously, we get $P[y_i = 1] = 0$ *for all* $i \in [0, p - k - 1]$. *In addition:*

$$\begin{aligned} P[y_{p-k} = 1] \ &= \ P[x_{k-1} = 1] \times P[x_0 = 1] \\ P[y_{p-k+1} = 1] &= \ P[x_{k-1} = 1] \times P[x_1 = 1] + \\ & \qquad P[x_{k-1} = 0] \times P[x_{k-2} = 1] \times P[x_0 = 1] \\ P[y_{p-k+2} = 1] &= \ P[x_{k-1} = 1] \times P[x_2 = 1] + \\ & \qquad P[x_{k-1} = 0] \times P[x_{k-2} = 1] \times P[x_1 = 1] + \\ & \qquad P[x_{k-1} = 0] \times P[x_{k-2} = 0] \times P[x_{k-3} = 1] \times P[x_0 = 1] \end{aligned}$$

$$\vdots$$

$$P[y_{p-k+j} = 1] = \sum_{l=0}^{j} \prod_{i=1}^{l} \left(P[x_{k-i} = 0] \right) \times P[x_{k-(l+1)} = 1] \times P[x_{j-l} = 1]$$

Considering that $P[x_i = 0] = P[x_i = 1] = \frac{1}{2}$ *for all* $x \in [1, 2^k - 1]$, *the result ensues.* □

    If the value for k is greater than p, some values in $\mathbb{D}_b$ will need rounding, and the *round-to-nearest-even* rule will slightly favor results with a "0" as rightmost bit of the fractional part, as can be readily seen in Fig. 5.

    Another problem when k is greater than p is that $x/2^k$ may round to 1, even though we claim to compute floats in $[0, 1)$. This is what happens in software that divides a 32-bit random integer x by 2^{32} to get a single precision float (see Sect. 4): as soon as $x/2^{32}$ is greater or equal to $1 - 2^{-25}$, the fraction rounds to 1 when rounding to nearest-even.

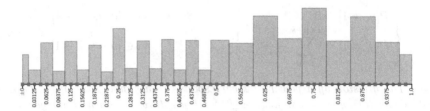

Fig. 5. Drawing 10 000 integers uniformly at random in $[0, 2^6 - 1]$ and dividing them by 2^6 in $\mathbb{F}_4^3$. Values $i/2^6$ (for $i = 0, 1, \ldots, 2^6 - 1$) are represented before rounding as dots or squares on the real line, where the color and form correspond to the float they round to.

3.2   Dividing by a Mersenne Number

The division of a random integer in $[0, m - 1]$ by m gives a float in $[0, 1)$ (with 1 excluded, provided m is not too large, as seen above). To obtain values in $[0, 1]$ instead, one can divide by $m - 1$. Since m is often a power of 2 for the reasons explained in Sect. 3.1, the divisor is now a Mersenne number. That algorithm is proposed in many sources and in particular by Nishimura and Matsumoto —the authors of the Mersenne Twister— with a contribution by Isaku Wada [19, 20]. Due to the popularity of the Mersenne Twister, their work on drawing floats has been integrated in many numerical software *as is*, along with Mersenne Twister-based RNGs for integers.

Nishimura *et al.* propose two versions of their algorithm to compute double precision floats in $[0, 1]$: one that uses the 32-bit version of MT19937 [19] to compute a 32-bit integer that is divided by $2^{32} - 1$, and the other that uses the 64-bit version of MT19937 [20] to compute a 53-bit integer that is divided by $2^{53} - 1$. In order to avoid some costly division, both versions replace the division with a multiplication by the inverse (which is precomputed at compile time).

The problem with both methods is that $\mathsf{fl}\langle 1/(2^k - 1)\rangle$ has a special structure that may induce some correlation between bits of the fractional part of the floats produced and a non-uniform probability of each bit to be "1." Indeed, we have:

Proposition 2. *Given a set of floats $\mathbb{F}_p^{emax}$ and an integer $k \in [2, p]$:*

$$
\mathsf{fl}\left\langle \frac{1}{2^k - 1} \right\rangle = \begin{cases} \left(1 + 2^{1-p} + \sum_{i=1}^{\lfloor \frac{p}{k} \rfloor - 1} 2^{-ik}\right) \times 2^{-k}, & \text{if } p \equiv 0 \pmod{k}, \\ \left(1 + \sum_{i=1}^{\lfloor \frac{p}{k} \rfloor} 2^{-ik}\right) \times 2^{-k}, & \text{otherwise.} \end{cases} \tag{3}
$$

Proof. We have:

$$
2^k - 1 = \underbrace{11\cdots 1}_{k}
$$

Then:

$$
\frac{1}{2^k-1} = 0.\underbrace{0\cdots 0}_{k-1}1\underbrace{0\cdots 0}_{k-1}1\cdots
$$
$$
= \left(1 + \sum_{i=1}^{\infty} 2^{-ik}\right) \times 2^{-k}
$$

If p is a multiple of k, the first bit outside the stream of bits we can represent in the fractional part is a "1," which means we have to round upward to represent $\mathrm{fl}\langle 1/(2^k - 1)\rangle$:

$$\frac{1}{2^k - 1} = \overbrace{1.\underbrace{0\cdots0}_{k-1}1\underbrace{0\cdots0}_{k-1}1\underbrace{0\cdots0}_{k-1}1\cdots}^{p} \times 2^{-k}$$

Otherwise, we must round downward:

$$\frac{1}{2^k - 1} = \overbrace{1.\underbrace{0\cdots0}_{k-1}1\underbrace{0\cdots0}_{k-l}0\underbrace{0\cdots0}_{l}1\underbrace{0\cdots0}_{k-1}1\cdots}^{p} \times 2^{-k}$$

$\square$

For $k = 32$ —the first method by Nishimura *et al.* [19]—, we get:

$$\mathrm{fl}\left\langle \frac{1}{2^{32} - 1} \right\rangle = 1.\underbrace{0\cdots0}_{31}1 \times 2^{-32}$$

When multiplying $x = x_{31}\cdots x_0$ by $\mathrm{fl}\left\langle \frac{1}{2^{32}-1} \right\rangle$, we get the real z:

$$z = x \times \mathrm{fl}\left\langle \frac{1}{x^{32} - 1} \right\rangle = 0.x_{31}\cdots x_0 x_{31} \cdots x_0$$

Normalizing the result, we get:

$$\text{if } x_{31} = 1: z = 1.\overbrace{x_{30}\cdots x_0 1 x_{30} \cdots x_{11}}^{52} x_{10} \cdots x_0 \times 2^{-1}$$
$$\text{if } x_{31} = 0 \wedge x_{30} = 1: z = 1.\underbrace{x_{29}\cdots x_0 01 x_{29} \cdots x_{10}}_{52} x_9 \cdots x_0 \times 2^{-2}$$

$$\cdots$$

Notice that:

- Bit 20 of the fractional part of the restriction to double precision of z is always equal to 1;
- The probability of being "1" of Bits 21 through 51 follows the same law as in Proposition 1;
- Bits of the fractional part are highly correlated since some bits of x occur twice (e.g., when $x_{31} = 1$, x_{30} occur as z_{51} and z_{19}, and so on).

The first two properties can readily be seen in Fig. 6.

The second method, which divides integers by $2^{53} - 1$ exhibits a different behavior since then $k = p$ and, according to Proposition 2, the rounded value of $1/(2^{53} - 1)$ has a different structure, viz.:

$$\mathrm{fl}\left\langle \frac{1}{2^{53} - 1} \right\rangle = 1.\underbrace{0\cdots0}_{51}1 \times 2^{-53}$$

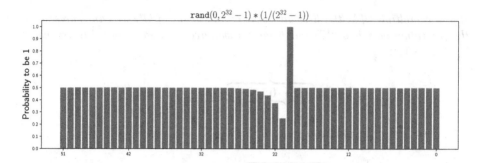

Fig. 6. Computing $1\,000\,000$ random double precision floats in $[0,1]$ according to the MT19937-32-based procedure by Nishimura *et al.* [19].

Note how there are now only $k-2$ "0s" in the fractional part instead of $k-1$ for $\mathsf{fl}\big\langle \frac{1}{2^{32}-1} \big\rangle$. As a consequence, there is some overlap of the bits during the multiplication by $x = x_{52}\cdots x_0$, and we get:

$$z = x \times \mathsf{fl}\big\langle \frac{1}{2^{53}-1} \big\rangle = 0.x_{52}\cdots x_1(x_0 + x_{52})x_{51}\cdots x_0$$

That structure of z seems sufficient to remove the most glaring flaws compared to the first method (See Fig. 7 on the right). Unfortunately, we have not yet been able to explain the slight dip in the probability of bits 2 to 5 to be "1."

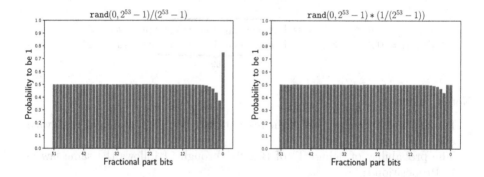

Fig. 7. Division by a Mersenne number vs. multiplication by its inverse

Since $1/(2^{53}-1)$ needs rounding to double precision, there is a difference in dividing x by $2^{53}-1$ or multiplying it by $\mathsf{fl}\big\langle 1/(2^{53}-1) \big\rangle$ as Nishimura *et al.* do. That difference also shows in the probability to be "1" of each bit of the fractional part of the resulting floats (compare the left and right graphics in Fig. 7). Since we never encountered implementations that used directly a division by a Mersenne number, we have not investigated the difference further.

Lastly, note that the division by a Mersenne number, as well as the multiplication by its inverse, involve some rounding of the result. As a consequence, both methods may introduce some non-uniformity when drawing floats, even when b is smaller or equal to p.

4   Implementations in Software

We have studied the implementation of floating-point RNGs in some major programming languages and libraries. Except when explicitly stated otherwise, we focus here on the RNGs that compute double precision floats in $[0, 1)$. Due to the popularity of the Mersenne Twister [15], many software implement one of the algorithms proposed by Nishimura, Matsumoto and Wada [19,20], of which we have shown some of the flaws in the preceding sections. Some few software do not resort to a division by a power of 2 or by a Mersenne number; they are discussed only in the last section.

We have seen in Sect. 3.1 that the ISO C language does not offer any standard function to compute random floats in $[0, 1)$, the accepted algorithm being to divide the result of rand() by RAND_MAX+1, where RAND_MAX is only $2^{31} - 1$ in the *GNU Compiler Collection* (GCC) library. This is also the approach used in the Lua language, whose library is based on C.

The GNU Scientific Library (GSL) [5] strives to be the go-to numerical library for C in the absence of a true standardized one. Unfortunately, the default implementation to compute a random float is not much better than using rand() as it uses a Mersenne Twister to compute a 32-bit random integer and multiply it by 2^{-32}. The same method is also used by Scilab 6.0.2.

The C++ language has been offering random number generation in its standard library since the C++11 standard. The generate_canonical() function is used to draw uniformly at random a floating-point number in $[0, 1)$. Unfortunately, from the C++11 standard up to the latest draft for the C++20 standard included [22, p. 1156], the algorithm mandated to compute a random float requires to divide a random integer by a value that may be much larger than 2^p, with p the size of the significand of the type of floats produced. As a consequence, the value 1.0 may be returned, in violation of the definition of the function itself. This is a known error that has been addressed formally by the *C++ Library Working Group* in 2017 only by proposing to recompute a new random float whenever the one computed is equal to 1.0. The C++ library for GCC had already implemented that workaround in its 6.1.0 release in 2015. Oddly enough, it was changed in 2017 for the 7.1.0 release in favor of returning prev (1) in that case, which breaks the uniformity requirement, even if only slightly.

The Go language, at least in its most current version as of 2019, generates a double precision number in $[0, 1)$ by computing a 63-bit random integer and dividing it by 2^{63}. As a consequence, the implementation of its Float64() function is plagued by the same problems as C++ generate_canonical(), and the current method to avoid returning 1.0 is to loop generating a new number until getting a value different from 1.0. Since 63 is larger than $p = 53$, there are also uniformity problems in drawing floats, as seen in the preceding sections.

In Java, the Random.nextDouble() method to obtain a random double precision float in $[0, 1)$ is implemented, both in OpenJDK and in the Oracle JDK, by computing a 53-bit random integer and multiplying it by 2^{-53}. An older method used to computed a 54-bit integer and multiply it by 2^{-54}; ironically, it was discarded in favor of the new method because it was discovered that, due to rounding, it introduced a bias that skewed the probability of the rightmost bit of the significand to be "0" or "1," even though the new method presents the same bias for different reasons, as seen previously. The Rust language does it in exactly the same way. It is also the method used in both the standard library for Python, as well as in the Numpy package.

GNU Fortran 9.2.0 introduces a twist in the computation of a random double precision number in that it uses the *xoshiro256*** algorithm by Blackman and Vigna [2] to compute a 64-bit random integer in $[0, 2^{64} - 1]$, forces the 11 lowest bits to 0 to get an integer that is a multiple of 2^{11}, then divides the result by 2^{64} to get multiples of 2^{-53}. As a result, all values are representable and uniformly spread in $[0, 1)$. The same effect would be obtained by computing a 53-bit random integer and dividing it by 2^{53}. Consequently, the algorithm suffers from the flaws shown in Sect. 3.1.

The default algorithm in MATLAB [14] computes a double precision float in $(0, 1)$, not $[0, 1)$, by creating a 52-bit random integer, adding 0.5 to it, and dividing the sum by 2^{52} to obtain a number in $[2^{-53}, 1 - 2^{-53}]$. Evidently, that algorithm shares the same flaws as the algorithm that directly divides an integer by 2^{53}. GNU Octave, a MATLAB-like software, computes both single precision and double precision random numbers in $(0, 1)$. Single precision floats are generated by computing a 32-bit integer, adding 0.5 and dividing the sum by 2^{32}. Since the divisor is greater than 2^{23}, the distribution is no longer uniform, due to rounding. Additionally, the RNG may return exactly 1.0 in violation of its specification. The implementation of the double precision generator seems also flawed, at least in the latest version as of 2019 (5.1.0), as it consists in generating a 53-bit integer, adding 0.4 [sic] (which has to be rounded) and dividing the result by 2^{53}.

5   Conclusion

The motivation for this work stems from our attempt to study empirically the behavior of some complex arithmetic expression in the Julia language: while evaluating the expression numerous times with random floats, we discovered that the number of results needing rounding was much smaller than anticipated. Only when studying the structure of the floats produced by the rand() Julia function did we notice that the least significant bit of their fractional part was consistently fixed to 0 when requiring floats in some domains, but not in others. Investigating the RNGs producing floating-point numbers in other programming languages and libraries, we found they were almost all lacking in some respect when used to test numerical code.

Many studies are devoted to the analysis of RNGs producing integers; they are much fewer to consider RNGs producing floats, and we are not aware of other works considering the structure of the floats produced at the bit level. In truth, such a dearth of work on that subject might lead one to wonder whether the systemic irregularities in the structure of the fractional part produced when using a division really matters. We have identified two domains of relevance:

- When investigating numerical software empirically (see our war story directly above). In addition, the inability of the division-based methods to generate subnormals may be redhibitory for some applications;
- When anonymizing numerical data.

That last use case should take on more and more importance in the *Big Data* era, and problems with floating-point random number generators have already been reported [16], particularly when the probabilities to be "1" or "0" of the bits of the fractional part are skewed.

Some programming languages and libraries do not use a division to compute random floats. Amusingly enough, it is the case with Julia, the very language that prompted our investigation, even though its algorithm exhibits other flaws. After some experimentation, we believe it is also the case of Mathematica, at least in its version 12 (anecdotally, its default implementation exhibits the same flaws as Julia's). We have already started a larger study on the other classes of algorithms to generate random floating point numbers in the hope of isolating the better algorithms that offer both good performances, uniformity, and a perfect regularity of the structure of the fractional part of the floats produced.

Acknowledgments. We would like to thank Dr. Alexandre Goldsztejn for providing us with the means to test the random engine of Mathematica v. 12.

References

1. Beebe, N.H.F.: The Mathematical-Function Computation Handbook: Programming Using the MathCW Portable Software Library. Springer, Heidelberg (2017). https://doi.org/10.1007/978-3-319-64110-2
2. Blackman, D., Vigna, S.: Scrambled linear pseudorandom number generators, August 2019. https://arxiv.org/abs/1805.01407, revision 2
3. Campbell, T.R.: Uniform random floats, April 2014. https://mumble.net/~campbell/2014/04/28/uniform-random-float, unpublished note
4. Downey, A.B.: Generating pseudo-random floating-point values, July 2007. http://allendowney.com/research/rand, unpublished note
5. Galassi, M., et al.: The GNU Scientific Library documentation, August 2019. https://www.gnu.org/software/gsl/doc/html/
6. Hull, T., Dobell, A.: Random number generators. SIAM Rev. 4(3), 230–254 (1962)
7. IEEE Standards Association: IEEE standard for floating-point arithmetic. IEEE Standard IEEE Std 754–2008. IEEE Computer Society (2008)
8. Knuth, D.E.: The Art of Computer Programming: Seminumerical Algorithms, vol. 2, 3rd edn. Addison-Wesley Professional, Boston (1997)

9. L'Ecuyer, P.: History of uniform random number generation. In: 2017 Winter Simulation Conference (WSC), pp. 202–230, December 2017
10. L'Ecuyer, P., Simard, R.: TestU01: a C library for empirical testing of random number generators. ACM Trans. Math. Softw. **33**(4), 22:1–22:40 (2007)
11. Lehmer, D.H.: Mathematical methods in large-scale computing units. In: Proceedings of the Second Symposium on Large Scale Digital Computing Machinery, pp. 141–146. Harvard University Press, Cambridge (1951)
12. Loosemore, S., Stallman, R., McGrath, R., Oram, A., Drepper, U.: The GNU C library reference manual for version 2.30 (2019)
13. Marsaglia, G.: Random numbers for C: the END? Post at sci.stat.math and sci.math, January 1999. http://www.ciphersbyritter.com/NEWS4/RANDC.HTM#36A5FC62.17C9CC33@stat.fsu.edu
14. MathWorks: MATLAB documentation: Creating and controlling a random number stream (2019). https://www.mathworks.com/help/matlab/math/creating-and-controlling-a-random-number-stream.html. Accessed 28 Dec 2019
15. Matsumoto, M., Nishimura, T.: Mersenne twister: a 623-dimensionally equidistributed uniform pseudo-random number generator. ACM Trans. Model. Comput. Simul. **8**(1), 3–30 (1998)
16. Mironov, I.: On significance of the least significant bits for differential privacy. In: Proceedings of the 2012 ACM Conference on Computer and Communications Security, CCS 2012, pp. 650–661. ACM, New York (2012)
17. Moler, C.: Numerical Computing with MATLAB. SIAM (2004)
18. von Neumann, J.: Various techniques used in connection with random digits. Natl. Bureau Stand. Appl. Math. Ser. **12**, 36–38 (1951)
19. Nishimura, T., Matsumoto, M.: A C-program for MT19937, with initialization improved 2002/1/26, January 2002. http://www.math.sci.hiroshima-u.ac.jp/~m-mat/MT/MT2002/CODES/mt19937ar.c (with a contribution by Isaku Wada). Accessed 30 Dec 2019
20. Nishimura, T., Matsumoto, M., Wada, I.: A C-program for MT19937-64 (2004/9/29 version), September 2004. http://www.math.sci.hiroshima-u.ac.jp/~m-mat/MT/VERSIONS/C-LANG/mt19937-64.c. Accessed 30 Dec 2019
21. Press, W.H., Teukolsky, S.A., Vetterling, W.T., Flannery, B.P.: Numerical Recipes in C, 2nd edn. Cambridge University Press, Cambridge (1992)
22. Smith, R., Köppe, T., Maurer, J., Perchik, D.: Working draft, standard for programming language C++. Technical report. N4842, ISO/IEC, November 2019, committee draft for C++20
23. Summit, S.: C programming FAQs: Frequently Asked Questions. http://c-faq.com/. Accessed 21 Dec 2019
24. Walker, A.J.: Fast generation of uniformly distributed pseudorandom numbers with floating-point representation. Electron. Lett. **10**(25), 533–534 (1974)

An Effective Stable Numerical Method for Integrating Highly Oscillating Functions with a Linear Phase

Leonid A. Sevastianov[ID], Konstantin P. Lovetskiy$^{(\boxtimes)}$[ID], and Dmitry S. Kulyabov[ID]

Peoples' Friendship University of Russia (RUDN University), 6 Miklukho-Maklaya Street, Moscow 117198, Russian Federation
{sevastianov_la, lovetskiy_kp, kulyabov_ds}@rudn.ru

Abstract. A practical and simple stable method for calculating Fourier integrals is proposed, effective both at low and at high frequencies. An approach based on the fruitful idea of Levin, which allows the use of the collocation method to approximate the slowly oscillating part of the antiderivative of the desired integral, allows reducing the calculation of the integral of a rapidly oscillating function (with a linear phase) to solving a system of linear algebraic equations with a triangular or Hermitian matrix.

The choice of Gauss-Lobatto grid nodes as collocation points let to increasing the efficiency of the numerical algorithm for solving the problem. To avoid possible numerical instability of the algorithm, we proceed to the solution of a normal system of linear algebraic equations.

Keywords: Oscillatory integral · Chebyshev interpolation · Numerical stability

1 Introduction

The initial formulation of the method of numerical integration of highly oscillating functions by Levin and his followers suggests a possible ambiguity in finding the antiderivative: any solution to the differential equation without boundary (initial) conditions can be used to calculate the desired value of the integral.

Levin's approach [1] to the integration of highly oscillating functions consists in the transition to the calculation of the antiderivative function from the integrand using the collocation procedure in physical space. In this case, the elements of the degenerate [2] differentiation matrix of the collocation method [3] are a function of the coordinates of the grid points, the matrix elements are calculated using very simple formulas. In books [3, 4] various options for the implementation of this method are considered, many applied problems are solved.

The method proposed by Levin both in the one-dimensional and in the multidimensional case was published by him in articles [1, 9], and then he was thoroughly studied in [10]. The method is presented in great detail in the famous monograph [4], which describes the evolution of numerical methods for integrating rapidly oscillating functions over the past fifteen years.

© Springer Nature Switzerland AG 2020
V. V. Krzhizhanovskaya et al. (Eds.): ICCS 2020, LNCS 12138, pp. 29–43, 2020.
https://doi.org/10.1007/978-3-030-50417-5_3

There are a large number of works using various approaches in order to propose fast and stable methods for solving systems of linear algebraic equations (SLAE) that arise when implementing the collocation method. However, many of them [5–7] encounter difficulties in solving the corresponding systems of linear equations.

In particular, the use in specific implementations of the Levin collocation method in the physical space of degenerate Chebyshev differentiation matrices, which also have eigenvalues differing by orders of magnitude, makes it impossible to construct a stable numerical algorithm for solving the resulting SLAEs. The approach to solving the differential equation of the Levin method, described in [5, 6, 8], is based on the approximation of the solution, as well as the integrand phase and amplitude functions in the form of expansion into finite series in Chebyshev polynomials. Moreover, to improve the properties of the algorithms, and hence the matrices of the corresponding SLAEs, three-term recurrence relations are used that connect the values of Chebyshev polynomials of close orders. However, these improvements are not enough to ensure stable calculation of integrals with large matrix dimensions.

In our work, we consider a method of constructing a primitive, based on the spectral representation of the desired function.

We propose increasing the efficiency of the algorithm by reducing the corresponding system of linear equations to a form that is always successfully solved using the LU-decomposition method with partial selection of the leading element.

Consider the integral that often occurs in Fourier analysis - in applications related to signal processing, digital images, cryptography and many other areas of science and technology.

$$I_\omega[f] = \int_a^b f(x)e^{i\omega g(x)}dx \tag{1}$$

In accordance with the Levin method, the calculation of this integral reduces to solving an ordinary differential equation

$$p'(x) + i\omega g'(x)p(x) = f(x), x \in [a, b] \tag{2}$$

As argued in [1], the system (2) has a particular solution which is not rapidly oscillatory, and we shall look for an approximation to this particular solution by collocation with 'nice' functions, e.g. polynomials. If the unknown function p(x) is a solution of Eq. (2), then the result of integration can be obtained according to the formula

$$I_\omega(f, g) = \int_a^b (p'(x) + i\omega g'(x)p(x))e^{i\omega g(x)}dx = p(b)e^{i\omega g(b)} - p(a)e^{i\omega g(a)}. \tag{3}$$

Below we will consider the special case of integration of a highly oscillating function with a linear phase, reduced to the standard form

$$I_\omega[f] = \int_{-1}^{1} f(x)e^{i\omega x}dx = p(1)e^{i\omega} - p(-1)e^{-i\omega}. \tag{4}$$

This can be justified, in particular, by the fact that in many well-known publications [7, 11, 12] stable transformations are discussed in detail, which make it possible to proceed from a general integral with a nonlinear phase to an integral in standard form (on the interval $[-1, 1]$) with a linear phase.

In the paper by Levin [1], to automatically exclude the rapidly oscillating component $ce^{-i\omega x}$ of the general solution $p(x) = p_0(x) + ce^{-i\omega g(x)}$, it is proposed to search for a numerical solution (2) based on the collocation method, using its expansion in a basis of slowly oscillating functions, rather than using difference schemes (or methods of the Runge-Kutta type).

In this case, the following statement is true [2]:

Statement. The solution of Eq. (2) obtained using the Levin collocation method is a slowly oscillating function $\mathcal{O}(\omega^{-1})$ for $\omega \gg 1$.

2 Approximation of the Antiderivative. Calculation Method

Let us consider in more detail the problem of finding the antiderivative integrand, or rather, the approximating polynomial $p(x)$, satisfying condition (2) in a given number of points on the interval $[-1, 1]$. Consider the spectral method of finding an approximating function in the form of expansion in a finite series

$$p(x) = \sum_{k=0}^{n} c_k T_k, x \in [-1, 1] \tag{5}$$

in the basis of Chebyshev polynomials of the first kind $\{T_k(x)\}_{k=0}^{\infty}$, defined in the Hilbert space of functions on the interval $[-1, 1]$.

The application of the collocation method to solve the problem $p'(x) + i\omega p(x) = f(x)$ leads to the need to fulfill the following equalities for the desired coefficients $c_k, k = 1, \ldots, n$

$$\sum_{k=0}^{n} c_k T_k'(x_j) + i\omega \sum_{k=0}^{n} c_k T_k(x_j) = f(x_j), j = 0, \ldots, n \tag{6}$$

at the collocation points $\{x_0, x_1, \ldots, x_n\}$.

The last statement is equivalent to the fact that the coefficients $c_k, k = 0, \ldots, n$ should be a solution to the system of linear algebraic equations of the collocation method:

$$\begin{cases} p'(x_0) + i\omega p(x_0) = f(x_0), \\ p'(x_1) + i\omega p(x_1) = f(x_1). \\ \qquad \cdots \\ p'(x_n) + i\omega p(x_n) = f(x_n). \end{cases} \tag{7}$$

We represent the values of the derivative of the desired function (polynomial) at the collocation points in the form of the product $Dp = p'$ of the matrix D by the vector of values of p. Recall that the Chebyshev differentiation matrix D has the standard representation in the physical space [3]

$$D_{kj} = \begin{cases} \frac{r_k}{r_j}(-1)^{k+j}/(x_k - x_j) & k,j = 0,\ldots n, k \neq j \\ -\sum_{l=0,l\neq k}^{n} D_{kl} & k = j, \end{cases} \tag{8}$$

where $r_j = \begin{cases} 2 & j = 0,n \\ 1 & 1,\ldots,n-1. \end{cases}$

Substituting $p' = Dp$ into Eq. (7) we reduce it to a system of linear algebraic equations

$$(\mathbf{D} + i\omega\mathbf{E})\mathbf{p} = \mathbf{f}. \tag{9}$$

Here $\mathbf{E}$ is an identity matrix, $\mathbf{f}$ is a vector of values of the amplitude function on the grid. Denote by $\mathbf{B}$ the differentiation matrix in the frequency (spectral) space [13], whose coefficients are explicitly expressed as

$$\mathbf{B}_{ij} = \begin{cases} (1/r_j)2j & \text{if } j > i, i+j \text{ odd} \\ 0 & \text{otherwise} \end{cases} \tag{10}$$

where $0 \leq i,j \leq n$ and $r_i = \begin{cases} 2 & i = 0 \\ 1 & i > 0. \end{cases}$

Denote by T the Chebyshev matrix of mapping a point (vector) from the space of coefficients to the space of values of the function [14]. Given that $\mathbf{p} = \mathbf{Tc}$ is the vector of values of the desired function (also in physical space), the components of the derivative vector can be written as $\mathbf{Dp} = \mathbf{TBc}$ [14]. As a result, we obtain the system of linear algebraic equations equivalent to system (9),

$$(\mathbf{TBc} + i\omega\mathbf{Tc}) = \mathbf{f} \tag{11}$$

which is valid for an arbitrary grid on the interval $[-1, 1]$. We write Eq. (11) in detail

$$\begin{bmatrix} T_{00} & T_{10} & T_{20} & \vdots & T_{n0} \\ T_{01} & T_{11} & T_{21} & \vdots & T_{n1} \\ T_{02} & T_{12} & T_{22} & \vdots & T_{n2} \\ \cdots & \cdots & \cdots & \ddots & \cdots \\ T_{0n} & T_{1n} & T_{2n} & \vdots & T_{nn} \end{bmatrix} \left(\begin{bmatrix} 0 & 1 & 0 & 3 & \vdots \\ & 0 & 4 & 0 & \vdots \\ & & 0 & 6 & \vdots \\ & & & \ddots & \vdots \\ & & & & 0 \end{bmatrix} + i\omega\mathbf{E} \right) \begin{bmatrix} c_0 \\ c_1 \\ c_2 \\ \cdots \\ c_n \end{bmatrix} = \begin{bmatrix} f_0 \\ f_1 \\ f_2 \\ \cdots \\ f_n \end{bmatrix}$$

$$\tag{12}$$

where to reduce the formulas we used the notation $T_{kj} = T_k(x_j), k,j = 0,\ldots,n$.

The product of a non-degenerate matrix T by a non-degenerate triangular matrix $B + i\omega E$ is a non-degenerate matrix. Therefore, the system of linear algebraic Eqs. (12) has a unique solution.

Statement 1. The solution of this system of linear algebraic equations with respect to the coefficients $c = (c_0, c_1, \ldots, c_n)$ allows us to approximate the antiderivative function in the form of a series (5) and calculate the approximate value of the integral by formula (4).

3 Modification of the Calculation Method

System (12) is valid for an arbitrary grid on the interval $[-1, 1]$. However, consideration of the collocation problem on a Gauss-Lobatto grid allows significant simplification of this system of linear algebraic equations. First, we multiply the first and last equations from (12) by $1/\sqrt{2}$ to obtain an equivalent "modified" system with a new matrix $\tilde{T}$ (instead of T), which is good because it has the property of discrete "orthogonality" and, therefore, is non-degenerate. Therefore, multiplying it on the left by its transposed one gives the diagonal matrix:

$$
\tilde{T}^T \tilde{T} =
\begin{bmatrix}
n & 0 & 0 & : & 0 \\
0 & n/2 & 0 & : & 0 \\
0 & 0 & n/2 & : & 0 \\
\cdots & \cdots & \cdots & \ddots & \cdots \\
0 & 0 & 0 & : & n
\end{bmatrix}
$$

We use this property and multiply the reduced (modified) system (12) on the left by the transposed matrix $\tilde{T}^T$, thereby reducing it to the upper triangular form. Indeed, the matrix of the resulting system is calculated as the product of the diagonal matrix by the triangular matrix, which, in turn, is the sum of the Chebyshev differentiation matrix in the spectral space and the diagonal matrix.

Since the matrix $\tilde{T}^T$ is non-degenerate, the new system of linear algebraic equations is equivalent to system (12) and has a unique solution.

Taking into account the specific values of the Chebyshev polynomials *on the Gauss-Lobatto grid* [15], simplifies the system, bringing it to the form

$$\mathbf{Ac} = \begin{bmatrix} i\omega & 1 & 0 & 3 & \vdots & n=1 \\ 0 & i\omega & 2 & 0 & \vdots & 0 \\ 0 & 0 & i\omega & 3 & \vdots & n-1 \\ 0 & 0 & 0 & i\omega & \vdots & 0 \\ \cdots & \cdots & \cdots & \cdots & \ddots & n-1 \\ 0 & 0 & 0 & 0 & \vdots & i\omega \end{bmatrix} \begin{bmatrix} c_0 \\ c_1 \\ c_2 \\ c_3 \\ \cdots \\ c_n \end{bmatrix} = \begin{bmatrix} \tilde{f}_0/2 \\ \tilde{f}_1 \\ \tilde{f}_2 \\ \tilde{f}_3 \\ \cdots \\ \tilde{f}_n/2 \end{bmatrix} \tag{13}$$

where $\tilde{f}_j = \frac{1}{n}\sum''_{k=0,n} T_j(x_k)f(x_k), j = 0,\ldots,n$ and symbol $\sum''$ denotes a sum in which the first and last terms are additionally multiplied by 1/2.

By the Kronecker-Capelli theorem, the system of linear algebraic Eqs. (13) with a square matrix and a non-zero determinant is not only solvable for any vector of the right-hand side, but also has a unique solution.

Statement 2. For $|\omega| > 2n$ the SLAE (13) has a stable solution.

Statement 3. To solve system (13), no more than $(\sim n^2/4)$ operations of addition/subtraction and multiplication/division with a floating point are required.

4 Efficient Method for Solving the Problem

Algorithms for solving systems of linear equations such as the Gauss method or the LU-decomposition work well when the matrix of the system has the property of diagonal dominance. Otherwise, standard solution methods lead to the accumulation of rounding errors.

A stable solution to the system may be provided by the LU-decomposition method with a partial choice of a leading element. For the triangular matrix (13), there is no forward pass (in which the leading element is selected) of the LU-decomposition, and the back pass of the method is always implemented without the selection of leading element.

A solution to system (13) can still be unstable in the case when $|i\omega| \leq 2n$. Passing to the solution of the normal system, that is, to the problem of minimizing the residual $\|\mathbf{Ac} - \tilde{f}\|^2$, multiplying the system of Eqs. (13) on the left by the Hermitian conjugate matrix

$$\mathbf{A}^*\mathbf{Ac} = \mathbf{A}^*\tilde{f} \tag{14}$$

we transform the matrix of the system (13) to the Hermitian form.

Although the system of linear equations became more filled, since instead of upper triangle three-diagonal matrix a system of linear equations with all matrix elements filled appeared, its computational properties are cardinally improved. The resulting matrix of a system of linear algebraic equations is Hermitian, its eigenvalues are real, and the eigenvectors form an orthonormal system. The method of LU-decomposition with a partial choice of the leading element, due to the properties of the resulting matrix, provides [17] the stability of the numerical algorithm for finding the only solution to the system.

5 Description of the Algorithm

Let us describe the sequence of operations of the presented algorithm for calculating the integral of a rapidly oscillating function of the form (1) with a linear phase.

Input data preprocessing.

1. If the integral is given on the interval $[a, b]$, we pass to the standard domain of integration $[-1, 1]$ by changing the variables $x = \frac{b-a}{2}t + \frac{b+a}{2}, t \in [-1, 1]$.
2. Fill by columns the Chebyshev transformation matrix T from (12) using only one pass of the recursive method for calculating the values of Chebyshev polynomials of the first kind of the n-th order.

Antiderivative algorithm

3. Calculate the vector of the right-hand side of system (13).
4. Fill in the elements of the sparse matrix (13), which depend only on the dimension n and the phase value ω.
5. If $|\omega| > 2n$, then go to step 6. Otherwise go to step 7.
6. The matrix of system (13) is a matrix with a diagonal dominance and can be stably solved. The solution values at the boundary points are used to determine the desired antiderivative values. Go to step 8.
7. Multiply relation (13) on the left by the conjugate matrix to obtain a Hermitian matrix (14) with diagonal dominance. In this case, to determine the values of the antiderivative at the boundary points the normal solution is stably determined using the LU-decomposition with a partial choice of the leading element.
8. We calculate the values of the antiderivative at the ends of the interval using the formulas $p(1) = \sum_{j=0}^{n} c_j$, and $p(-1) = \sum_{j=0, j-even}^{n} c_j - \sum_{j=0, j-odd}^{n} c_j$. The desired value of the integral is obtained using the formula $I(f, \omega) = p(1)e^{i\omega} - p(-1)e^{-i\omega}$.

6 Numerical Examples

Example 1. We give an example of calculating the integral when, for a good polynomial approximation of a slowly oscillating factor of the integrand, it is necessary to use polynomials of high degrees.

$$I_\omega\left[\frac{1}{x+2}\right] = \int_{-1}^{1} \frac{1}{x+2} e^{i\omega x} dx \tag{15}$$

This integral is given by Olver ([16], p. 6) as an example of the fact that the GMRES method allows one to calculate the integral much more accurately than the Levin collocation method. However, in his article, solving the resulting system of linear algebraic equations requires $\mathcal{O}(n^3)$ operations, as in the Levin collocation method using the Gaussian elimination algorithm (Table 1).

Table 1. The following table shows the values of the integral calculated by us for various values of the parameter ω with an accuracy of 17 significant digits.

ω	Real part	Image
$\omega = 1$	0.9113301035062809891	−0.1775799622517861791
$\omega = 10$	−0.07854759997855625023	−0.04871911238563061052
$\omega = 50$	−0.00665013790168713	0.0129677770647216
$\omega = 100$	−0.00667389328931381	0.00580336592710437

A comparison of our results at 40 interpolation points with the results of [16] shows a significant gain in accuracy: the deviation from the exact solution is of the order of 10^{-17} compared with the deviation of the order of 10^{-7} in Olver's article. The proposed algorithm to achieve an accuracy of 10^{-13} in the calculation of the integral uses no more than 30 points ($n \leq 30$) for $\omega = 1,\ldots, 100$.

Example 2. As a second example, we consider the integral

$$\int_{-1}^{1} \frac{1}{x^2+1} e^{i\omega\sin(x+1/4)}dx \tag{16}$$

from [16], where the results of calculating the integrals depending on the number of approximation points are illustrated on Fig. 1 [16].

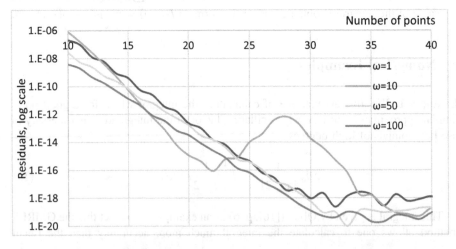

Fig. 1. The error in approximating integral (15) for n = 10–40 and for different choices of ω

To reduce this integral to the (standard form of the Fourier integral) form of integral with the linear phase, we change the variables $y = \sin\left(x + \frac{1}{4}\right)$. Then $dx = \frac{1}{\sqrt{1-y^2}}dy$, $x = \arcsin(y) - 1/4$, the integration limits are changed to $-\sin\left(\frac{3}{4}\right), \sin\left(\frac{5}{4}\right)$ and the integral can be written as:

$$\int_{-\sin\left(\frac{3}{4}\right)}^{\sin\left(\frac{5}{4}\right)} \frac{1}{\sqrt{1-y^2}\left(\left(\arcsin(y) - \frac{1}{4}\right)^2 + 1\right)} e^{i\omega y}dy \tag{17}$$

Let us consider the calculation of this integral for various values of the parameter ω using an algorithm that takes into account the linearity of the phase function (Table 2).

Table 2. The table shows the values of the integral for various values of the parameter ω.

ω	Re Int	Im Int
$\omega = 0.1$	1.5687504317409	0.0337582105322438
$\omega = 1$	1.3745907842843	0.305184104407599
$\omega = 3$	0.311077689499021	0.339612459676631
$\omega = 10$	0.00266714972608754	0.180595659138141
$\omega = 30$	0.00706973992290492	0.0455774930833239
$\omega = 50$	−0.00620005944852318	0.0155933115982172
$\omega = 100$	0.00460104072965418	−0.00790563176002816

The proposed algorithm to achieve an accuracy of 10^{-16} when calculating the integral uses no more than 90 points ($n \leq 90$) with $\omega = 0.1, \ldots, 100$. A significant gain in the number of addition/subtraction and multiplication/division operations is achieved when the frequency value is greater than the number n, which ensures the diagonal dominance of the system of linear algebraic equations in the matrix (13) (Fig. 2).

It is useful to compare the algorithm we developed for finding the integrals of rapidly oscillating functions with the results of [5], which presents various and carefully selected numerical examples for various classes of amplitude functions.

Example 3. Consider the calculation of the integral with an exponential function as the amplitude

$$I(\alpha, \omega) = \int_{-1}^{1} e^{\alpha(x-1)}e^{i\omega x}dx, \alpha = 16, 64; \omega = 20, 1000. \tag{18}$$

The exact value of the integral can be calculated by the formula $I(\alpha, \omega) = \frac{2*e^{-\alpha}\sinh(\alpha+i\omega)}{(\alpha+i\omega)}$ [5]. The plot of the deviation of the integral calculated by us from the exact one depending on the number of collocation points (absolute error) is shown in Fig. 3.

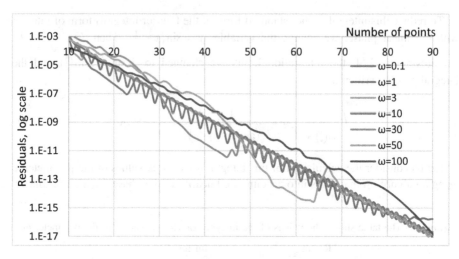

Fig. 2. The figure shows the absolute error in approximating integral (17) for n = 10–90 and for different choices of ω: 0.1, 1, 3, 10, 30, 50, 100.

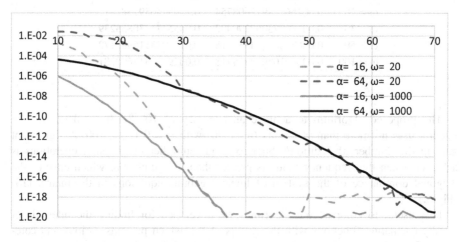

Fig. 3. Plot of the absolute error of the approximation of the integral (18) with $\alpha = 16.64$; at $\omega = 20$ and $\omega = 1000$ depending on the number of nodes of the collocation method. Logarithmic scale.

Comparison with the results of [5] shows that the accuracy of calculating the integrals practically coincides with that of [5]. Our advantage is the much simpler form of the matrix of a system of linear equations. In the best case, when $|\omega| > n$ matrix of the system is a triangular matrix with a dominant main diagonal. If $n > |\omega|$, then the transition to the search for a normal solution to a system with a positively defined Hermitian matrix allows us to create a numerically stable solution scheme.

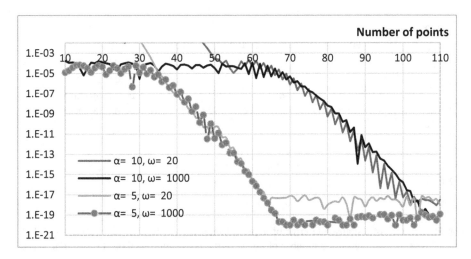

Fig. 4. The graph of the absolute error of the approximation of the integral (19) with $\alpha = 5, 10; \omega = 20, 1000$ depending on the number of nodes of the collocation method. Logarithmic scale.

Example 4. In this example [5], the rapidly oscillating function $e^{i2\pi\alpha x}$ is considered as the amplitude one. It is clear that in this case, to achieve the same accuracy in calculating the integral as in the previous example, a larger number of collocation points will be required (Fig. 4).

$$I(\alpha, \omega) = \int_{-1}^{1} e^{i2\pi\alpha x} e^{i\omega x} dx, \alpha = 5, 10; \omega = 20, 1000. \tag{19}$$

Example 5. The example demonstrates the calculation of the integral in the case when the amplitude function is the generating function of the Chebyshev polynomials of the first kind.

$$I(\alpha, \omega) = \int_{-1}^{1} \frac{1 - \alpha^2}{1 - 2\alpha x + \alpha^2} e^{i\omega x} dx, \alpha = 0.8, 0.9; \omega = 20, 1000. \tag{20}$$

The behaviour of the amplitude function should lead to an almost linear dependence of the approximation accuracy on the number of points for various values of the parameters α and ω. Numerical experiments carried out confirm this assertion (Fig. 5).

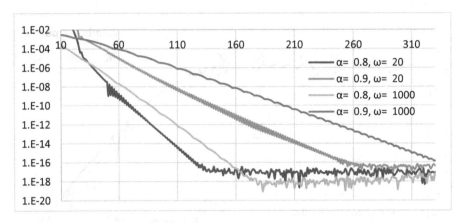

Fig. 5. The plot of the absolute error of approximation of the integral (22) with α = 0.8, 0.9; ω = 20,1000 depending on the number of nodes of the collocation method. Logarithmic scale.

Moreover, the accuracy of calculating the integrals is not inferior to the accuracy of the methods of [5].

Example 6. Amplitude is a bell-shaped function

$$I(\alpha, \omega) = \int_{-1}^{1} \frac{1}{x^2 + \alpha^2} e^{i\omega x} dx, \alpha = 1/4, 1/8; \omega = 20, 1000. \tag{21}$$

The example is rather complicated for interpolation by Chebyshev polynomials. To achieve acceptable accuracy (10^{-18}), the deviation of the calculated value of the integral from the exact one requires about 300 approximation points both for small values of $\omega = 20$ and for large $\omega = 1000$ (Fig. 6).

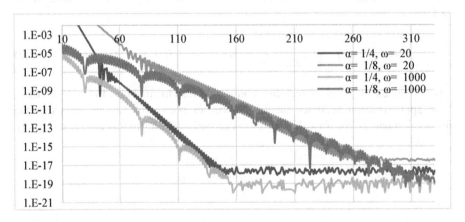

Fig. 6. Plot of the absolute error of approximation of the integral (21) with α = 1/4, 1/8; ω = 20,1000 depending on the number of nodes of the collocation method. Logarithmic scale.

Example 7. We give an example of integration when the amplitude function has second-order singularities at both ends of the integration interval

$$I(\omega) = \int_{-1}^{1} \left(1 - t^2\right)^{3/2} e^{i\omega x} dx, \omega = 20, 1000. \tag{22}$$

The value of this integral can be calculated in an analytical form: $f(\omega) = 3\pi J_2(\omega)/\omega^2$. We present the numerical values of the integral for various values of the frequency: $I(20) = -0.00377795409950960, I(1000) = -2.33519886790130 \times 10^{-7}$ (Fig. 7).

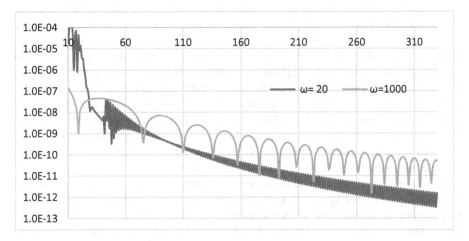

Fig. 7. The plot of the absolute error of the approximation of the integral (22) with $\omega = 20, 1000$ depending on the number of nodes of the collocation method. Logarithmic scale.

Similar to the previous example, to achieve good accuracy in calculating the integral, it is necessary to consider a large number of collocation points. However, for this type of amplitude functions, the method presented in the article works reliably both in the case of low and high frequencies.

The given examples demonstrate that the dependence of the solution on the number of approximation points is similar to the dependence demonstrated in Olver [17] and Hasegawa [5]. The advantage of our approach is the simplicity of the algorithm and the high speed of solving the resulting very simple system of linear algebraic equations. If it is necessary to repeatedly integrate various amplitude functions at a constant frequency, multiple gains are possible due to the use of the same LU-decomposition backtracking procedure.

7 Conclusion

A simple, effective, and stable method for calculating the integrals of highly oscillating functions with a linear phase is proposed. It is based on Levin's brilliant idea, which allows the use of the collocation method to approximate the antiderivative of the desired integral. Using the expansion in slowly oscillating polynomials provides a slowly changing solution of the differential equation.

The transition from a solution in physical space to a solution in spectral space makes it possible to effectively use the discrete orthogonality property of the Chebyshev mapping matrix on a Gauss-Lobatto grid. With this transformation, the uniqueness of the solution of the studied system is preserved, and its structure from a computational point of view becomes easier.

There are a large number of works using various approaches aimed to offer fast and effective methods for solving SLAEs that arise when implementing the collocation method. However, many methods [5, 6] encounter instability when solving the corresponding systems of linear equations. When using Chebyshev differentiation matrices in physical space, instability is explained primarily by the degeneracy of these matrices and the huge spread of eigenvalues of the matrix of the collocation method system. The approach to solving the differential equation based on the representation of the solution, as well as the phase and amplitude functions, in the form of expansion in finite series by Chebyshev polynomials and the use of three-term recurrence relations [5, 6, 8] also does not provide a stable calculation for $n > |\omega|$. To overcome instability, various methods of regularizing the systems under study are proposed.

In our work, we propose a new method for improving computational properties by preconditioning of the system in the spectral representation and by searching for its pseudo-normal solution. The proposed method has been reduced to solving a SLAE with a Hermitian matrix. A number of numerical examples demonstrates the advantages of the proposed effective stable numerical method for integrating rapidly oscillating functions with a linear phase.

Acknowledgement. The publication has been prepared with the support of the "RUDN University Program 5-100".

References

1. Levin, D.: Procedures for computing one- and two-dimensional integrals of functions with rapid irregular oscillations. Math. Comput. **38**(158), 531–538 (1982)
2. Deaño, A., Huybrechs, D., Iserles, A.: Filon and Levin methods In: Computing Highly Oscillatory Integrals, pp. 29–57 (2018)
3. Mason, J.C., Handscomb, D.C.: Chebyshev Polynomials. Chapman and Hall/CRC Press, Boca Raton (2002)
4. Deaño, A., Huybrechs, D., Iserles, A.: Computing Highly Oscillatory Integrals. Society for Industrial and Applied Mathematics, Philadelphia (2017)
5. Hasegawa, T., Sugiura, H.: A user-friendly method for computing indefinite integrals of oscillatory functions. J. Comput. Appl. Math. **315**, 126–141 (2017)

6. Domínguez, V., Graham, I.G., Smyshlyaev, V.P.: Stability and error estimates for Filon-Clenshaw-Curtis rules for highly oscillatory integrals. IMA J. Numer. Anal. **31**(4), 1253–1280 (2011)
7. Li, J., Wang, X., Wang, T., Xiao, S.: An improved Levin quadrature method for highly oscillatory integrals. Appl. Numer. Math. **60**(8), 833–842 (2010)
8. Ma, J., Liu, H.: A well-conditioned Levin method for calculation of highly oscillatory integrals and its application. J. Comput. Appl. Math. **342**, 451–462 (2018)
9. Levin, D.: Fast integration of rapidly oscillatory functions. J. Comput. Appl. Math. **67**(1), 95–101 (1996)
10. Olver, S.: Moment-free numerical integration of highly oscillatory functions. IMA J. Numer. Anal. **26**(2), 213–227 (2006)
11. Evans, G.A., Webster, J.R.: A comparison of some methods for the evaluation of highly oscillatory integrals. J. Comput. Appl. Math. **112**(1–2), 55–69 (1999)
12. Evans, G.A.: An alternative method for irregular oscillatory integrals over a finite range. Int. J. Comput. Math. **52**(3–4), 185–193 (1994)
13. Fornberg, B.: A Practical Guide to Pseudospectral Methods. Cambridge University Press, Cambridge (1996)
14. Lovetskiy, K., Sevastianov, L., Nikolaev, N.: Regularized computation of oscillatory integrals with stationary points. Procedia Comput. Sci. **108**, 998–1007 (2017)
15. Lovetskiy, K.P., Sevastyanov, L.A., Sevastyanov, A.L., Mekeko, N.M.: Integration of highly oscillatory functions. Math. Model. Geom. **3**(3), 11–24 (2014)
16. Olver, S.: Fast, numerically stable computation of oscillatory integrals with stationary points. BIT Numer. Math. **50**, 149–171 (2010). https://doi.org/10.1007/s10543-010-0251-y
17. Olver, S.: GMRES shifted for oscillatory integrals. Numer. Math. **114**, 607–628 (2010). https://doi.org/10.1007/s00211-009-0264-0

Fitting Penalized Logistic Regression Models Using QR Factorization

Jacek Klimaszewski$^{(\boxtimes)}$ [iD] and Marcin Korzeń [iD]

Faculty of Computer Science and Information Technology,
West Pomeranian University of Technology in Szczecin, Szczecin, Poland
{jklimaszewski,mkorzen}@wi.zut.edu.pl

Abstract. The paper presents improvement of a commonly used learning algorithm for logistic regression. In the direct approach Newton method needs inversion of Hessian, what is cubic with respect to the number of attributes. We study a special case when the number of samples m is smaller than the number of attributes n, and we prove that using previously computed QR factorization of the data matrix, Hessian inversion in each step can be performed significantly faster, that is $\mathcal{O}\left(m^3\right)$ or $\mathcal{O}\left(m^2 n\right)$ instead of $\mathcal{O}\left(n^3\right)$ in the ordinary Newton optimization case. We show formally that it can be adopted very effectively to ℓ^2 penalized logistic regression and also, not so effectively but still competitively, for certain types of sparse penalty terms. This approach can be especially interesting for a large number of attributes and relatively small number of samples, what takes place in the so-called extreme learning. We present a comparison of our approach with commonly used learning tools.

Keywords: Newton method · Logistic regression · Regularization · QR factorization

1 Introduction

We consider a task of binary classification problem with n inputs and with one output. Let $\mathbf{X} \in \mathbb{R}^{m \times n}$ be a dense data matrix including m data samples and n attributes, and $\boldsymbol{y}_{m \times 1}$, $y_i \in \{-1, +1\}$ are corresponding targets. We consider the case $m < n$. In the following part bold capital letters $\mathbf{X}, \mathbf{Y}, \ldots$ denote matrices, bold lower case letters $\boldsymbol{x}, \boldsymbol{w}$ stand for vectors, and normal lower case x_{ij}, y_i, λ for scalars. The paper concerns classification, but it is clear that the presented approach can be easily adopted to the linear regression model.

We consider a common logistic regression model in the following form:

$$Pr(y = +1|\boldsymbol{x}, \boldsymbol{w}) \equiv \sigma(\boldsymbol{x}, \boldsymbol{w}) = \frac{1}{1 + e^{-\sum_{j=1}^{n} x_j w_j}}. \tag{1}$$

This work was financed by the National Science Centre, Poland. Research project no.: 2016/21/B/ST6/01495.

V. V. Krzhizhanovskaya et al. (Eds.): ICCS 2020, LNCS 12138, pp. 44–57, 2020.
https://doi.org/10.1007/978-3-030-50417-5_4

Learning of this model is typically reduced to the optimization of negative log-likelihood function (with added regularization in order to improve generalization and numerical stability):

$$L(\boldsymbol{w}) = \lambda P(\boldsymbol{w}) + \sum_{i=1}^{m} \log(1 + e^{-y_i \cdot \sum_{j=1}^{n} x_{ij} w_j}), \tag{2}$$

where $\lambda > 0$ is a regularization parameter. Here we consider two separate cases:

1. rotationally invariant case, i.e. $P(\boldsymbol{w}) = \frac{1}{2}\|\boldsymbol{w}\|_2^2$,
2. other (possibly non-convex) cases, including $P(\boldsymbol{w}) = \frac{1}{q}\|\boldsymbol{w}\|_q^q$.

Most common approaches include IRLS algorithm [7,15] and direct Newton iterations [14]. Both approaches are very similar — here we consider Newton iterations:

$$\boldsymbol{w}^{(k+1)} = \boldsymbol{w}^{(k)} - \alpha \mathbf{H}^{-1} \boldsymbol{g}, \tag{3}$$

where step size α is chosen via backtracking line search [1]. Gradient $\boldsymbol{g}$ and Hessian $\mathbf{H}$ of $L(\boldsymbol{w})$ have a form:

$$\boldsymbol{g} = \lambda \frac{\partial P}{\partial \boldsymbol{w}} + \sum_{i=1}^{m} y_i \cdot (\sigma(\boldsymbol{x}_i, \boldsymbol{w}) - 1) \cdot \boldsymbol{x}_i \equiv \lambda \frac{\partial P}{\partial \boldsymbol{w}} + \mathbf{X}^T \boldsymbol{v}, \tag{4}$$

$$\mathbf{H} = \lambda \frac{\partial^2 P}{\partial \boldsymbol{w} \partial \boldsymbol{w}^T} + \mathbf{X}^T \mathbf{D} \mathbf{X} \equiv \mathbf{E} + \mathbf{X}^T \mathbf{D} \mathbf{X}, \tag{5}$$

where $\mathbf{D}$ is a diagonal matrix, whose i-th entry equals $\sigma(\boldsymbol{x}_i, \boldsymbol{w}) \cdot (1 - \sigma(\boldsymbol{x}_i, \boldsymbol{w}))$, and $v_i = y_i \cdot (\sigma(\boldsymbol{x}_i, \boldsymbol{w}) - 1)$.

Hessian is a sum of the matrix $\mathbf{E}$ (second derivative of the penalty function multiplied by λ) and the matrix $\mathbf{X}^T \mathbf{D} \mathbf{X}$. Depending on the penalty function P, the matrix $\mathbf{E}$ may be: 1) scalar diagonal ($\lambda \mathbf{I}$), 2) non-scalar diagonal, 3) other type than diagonal. In this paper we investigate only cases 1) and 2).

Related Works. There are many approaches to learning logistic regression model, among them there are direct second order procedures like IRLS, Newton (with Hessian inversion using linear conjugate gradient) and first order procedures with nonlinear conjugate gradient as the most representative example. A short review can be found in [14]. The other group of methods includes second order procedures with Hessian approximation like L-BFGS [21] or fixed Hessian, or truncated Newton [2,13]. Some of those techniques are implemented in `scikit-learn` [17], which is the main environment for our experiments. QR factorization is a common technique of fitting the linear regression model [9,15].

2   Procedure of Optimization with QR Decomposition

Here we consider two cases. The number of samples and attributes leads to different kinds of factorization:

- LQ factorization for $m < n$,
- QR factorization for $m \geqslant n$.

Since we assume $m < n$, we consider LQ factorization of matrix $\mathbf{X}$:

$$\mathbf{X} = \mathbf{LQ} = [\hat{\mathbf{L}} \; \mathbf{0}] \cdot \begin{bmatrix} \hat{\mathbf{Q}} \\ \tilde{\mathbf{Q}} \end{bmatrix} = \hat{\mathbf{L}}\hat{\mathbf{Q}}, \tag{6}$$

where $\hat{\mathbf{L}}$ is $m \times m$ lower triangular matrix, $\mathbf{Q}$ is $n \times n$ orthogonal matrix and $\hat{\mathbf{Q}}$ is $m \times n$ semi-orthogonal matrix ($\hat{\mathbf{Q}}\hat{\mathbf{Q}}^T = \mathbf{I}$, $\tilde{\mathbf{Q}}\hat{\mathbf{Q}}^T = \mathbf{0}$). The result is essentially the same as if QR factorization of the matrix $\mathbf{X}^T$ was performed.

Finding the Newton direction from the Eq. (3):

$$\boldsymbol{d} = \mathbf{H}^{-1}\boldsymbol{g} \tag{7}$$

involves matrix inversion, which has complexity $\mathcal{O}(n^3)$. A direct inversion of Hessian can be replaced (and improved) with a solution of the system of linear equations:

$$\mathbf{H}\boldsymbol{d} = \boldsymbol{g}, \tag{8}$$

with the use of the conjugate gradient method. This Newton method with Hessian inversion using linear conjugate gradient is an initial point of our research. We show further how this approach can be improved using QR decomposition.

2.1   The ℓ^2 Penalty Case and Rotational Invariance

In the ℓ^2-regularized case solution has a form:

$$\boldsymbol{d} = \left(\mathbf{X}^T\mathbf{DX} + \lambda\mathbf{I}\right)^{-1}\left(\mathbf{X}^T\boldsymbol{v} + \lambda\boldsymbol{w}\right). \tag{9}$$

Substituting $\mathbf{LQ}$ for $\mathbf{X}$ and $\hat{\mathbf{Q}}^T\hat{\mathbf{Q}}\boldsymbol{w}$ for $\boldsymbol{w}$:

$$\frac{\partial}{\partial\boldsymbol{w}}\left(\frac{1}{2} \cdot \|\hat{\mathbf{Q}}\boldsymbol{w}\|_2^2\right) = \hat{\mathbf{Q}}^T\hat{\mathbf{Q}}\boldsymbol{w} \tag{10}$$

in the Eq. (9) leads to:

$$\begin{aligned} \boldsymbol{d} &= \left(\mathbf{Q}^T\mathbf{L}^T\mathbf{DLQ} + \lambda\mathbf{I}\right)^{-1}\left(\mathbf{Q}^T\mathbf{L}^T\boldsymbol{v} + \lambda\hat{\mathbf{Q}}^T\hat{\mathbf{Q}}\boldsymbol{w}\right) \\ &= \left[\mathbf{Q}^T\left(\mathbf{L}^T\mathbf{DL} + \lambda\mathbf{I}\right)\mathbf{Q}\right]^{-1}\left(\mathbf{Q}^T\mathbf{L}^T\boldsymbol{v} + \lambda\hat{\mathbf{Q}}^T\hat{\mathbf{Q}}\boldsymbol{w}\right) \\ &= \mathbf{Q}^T\left(\mathbf{L}^T\mathbf{DL} + \lambda\mathbf{I}\right)^{-1}\mathbf{Q}\left(\mathbf{Q}^T\mathbf{L}^T\boldsymbol{v} + \lambda\hat{\mathbf{Q}}^T\hat{\mathbf{Q}}\boldsymbol{w}\right) \\ &= [\hat{\mathbf{Q}}^T \; \tilde{\mathbf{Q}}^T] \cdot \begin{bmatrix} \hat{\mathbf{L}}^T\mathbf{D}\hat{\mathbf{L}} + \lambda\mathbf{I} & \mathbf{0} \\ \mathbf{0} & \lambda\mathbf{I} \end{bmatrix}^{-1} \cdot \left(\begin{bmatrix} \hat{\mathbf{L}}^T \\ \mathbf{0} \end{bmatrix} \cdot \boldsymbol{v} + \begin{bmatrix} \lambda\hat{\mathbf{Q}}\boldsymbol{w} \\ \mathbf{0} \end{bmatrix}\right) \\ &= \hat{\mathbf{Q}}^T\left(\hat{\mathbf{L}}^T\mathbf{D}\hat{\mathbf{L}} + \lambda\mathbf{I}\right)^{-1}\left(\hat{\mathbf{L}}^T\boldsymbol{v} + \lambda\hat{\mathbf{Q}}\boldsymbol{w}\right). \end{aligned} \tag{11}$$

Algorithm 1. Newton method for ℓ^2 penalized Logistic Regression with QR factorization using transformation into a smaller space (L2-QR).

Input: $\mathbf{X} = \mathbb{R}^{m \times n}$, $y_{m \times 1}$, $m < n$
Initialization: $[\hat{\mathbf{L}}, \hat{\mathbf{Q}}] = \mathrm{lq}(\mathbf{X})$, $\hat{w}_{m \times 1} = \mathbf{0}$
repeat
    Compute $\hat{g}$ and $\mathbf{D}$ for $\hat{w}^{(k)}$.
    Solve $\left(\hat{\mathbf{L}}^T \mathbf{D} \hat{\mathbf{L}} + \lambda \mathbf{I} \right) \cdot \hat{d} = \hat{g}$.
    $\hat{w}^{(k+1)} = \hat{w}^{(k)} - \hat{d} \cdot \arg\min_\alpha L \left(\hat{w}^{(k)} - \alpha \hat{d} \right)$.
until $\|\hat{g}\|_2^2 < \epsilon$
Output: $w = \hat{\mathbf{Q}}^T \hat{w}$

First, multiplication by $\hat{\mathbf{Q}}$ transforms w to the smaller space, then inversion is done in that space and finally, multiplication by $\hat{\mathbf{Q}}^T$ brings solution back to the original space. However, all computation may be done in the smaller space (using $\hat{\mathbf{L}}$ instead of $\mathbf{X}$ in the Eq. (9)) and only final solution is brought back to the original space — this approach is summarized in the Algorithm 1. In the experimental part this approach is called L2-QR.

This approach is not new [8,16], however the use of this trick does not seem to be common in machine learning tools.

2.2 Rotational Variance

In the case of penalty functions whose Hessian $\mathbf{E}$ is a non-scalar diagonal matrix, it is still possible to construct algorithm, which solves smaller problem via QR factorization.

Consider again (5), (6) and (7):

$$
\begin{aligned}
d &= \left(\mathbf{Q}^T \mathbf{L}^T \mathbf{D} \mathbf{L} \mathbf{Q} + \mathbf{E} \right)^{-1} g \\
&= \left[\mathbf{Q}^T \left(\mathbf{L}^T \mathbf{D} \mathbf{L} + \mathbf{Q} \mathbf{E} \mathbf{Q}^T \right) \mathbf{Q} \right]^{-1} g \\
&= \mathbf{Q}^T \left(\mathbf{L}^T \mathbf{D} \mathbf{L} + \mathbf{Q} \mathbf{E} \mathbf{Q}^T \right)^{-1} \mathbf{Q} g.
\end{aligned}
\tag{12}
$$

Let $\mathbf{A} = \mathbf{Q} \mathbf{E} \mathbf{Q}^T$, $\mathbf{B} = \mathbf{L}^T \mathbf{D} \mathbf{L}$, so $\mathbf{A}^{-1} = \mathbf{Q} \mathbf{E}^{-1} \mathbf{Q}^T$. Using Shermann-Morrison-Woodbury formula [5] we may write:

$$
(\mathbf{A} + \mathbf{B})^{-1} = \mathbf{A}^{-1} - \mathbf{A}^{-1} \left(\mathbf{I} + \mathbf{B} \mathbf{A}^{-1} \right)^{-1} \mathbf{B} \mathbf{A}^{-1}.
\tag{13}
$$

Let $\mathbf{C} = \mathbf{I} + \mathbf{B} \mathbf{A}^{-1}$. Exploiting the structure of the matrices $\mathbf{L}$ and $\mathbf{Q}$ (6) yields:

$$
\begin{aligned}
\mathbf{C}^{-1} &= \left(\mathbf{I} + \begin{bmatrix} \hat{\mathbf{L}}^T \mathbf{D} \hat{\mathbf{L}} & 0 \\ 0 & 0 \end{bmatrix} \cdot \begin{bmatrix} \hat{\mathbf{Q}} \mathbf{E}^{-1} \hat{\mathbf{Q}}^T & \hat{\mathbf{Q}} \mathbf{E}^{-1} \tilde{\mathbf{Q}}^T \\ \tilde{\mathbf{Q}} \mathbf{E}^{-1} \hat{\mathbf{Q}}^T & \tilde{\mathbf{Q}} \mathbf{E}^{-1} \tilde{\mathbf{Q}}^T \end{bmatrix} \right)^{-1} \\
&= \begin{bmatrix} \hat{\mathbf{L}}^T \mathbf{D} \mathbf{X} \mathbf{E}^{-1} \hat{\mathbf{Q}}^T + \mathbf{I} & \hat{\mathbf{L}}^T \mathbf{D} \mathbf{X} \mathbf{E}^{-1} \tilde{\mathbf{Q}}^T \\ 0 & \mathbf{I} \end{bmatrix}^{-1} \\
&= \begin{bmatrix} \mathbf{C}_1 & \mathbf{C}_2 \\ 0 & \mathbf{I} \end{bmatrix}^{-1} = \begin{bmatrix} \mathbf{C}_1^{-1} & -\mathbf{C}_1^{-1} \mathbf{C}_2 \\ 0 & \mathbf{I} \end{bmatrix}.
\end{aligned}
\tag{14}
$$

Algorithm 2. Newton method for Logistic Regression with QR factorization and general regularizer.

Input: $\mathbf{X} = \mathbb{R}^{m \times n}$, $y_{m \times 1}$, $m < n$
Initialization: $[\hat{\mathbf{L}}, \hat{\mathbf{Q}}] = \text{lq}(\mathbf{X})$
repeat
    Compute g, $\mathbf{E}$ and $\mathbf{D}$ for $w^{(k)}$.
    Compute $\mathbf{C}_1^{-1}$.
    Compute d according to eq. (15) or eq. (20)
    $w^{(k+1)} = w^{(k)} - d \cdot \arg\min_\alpha L\left(w^{(k)} - \alpha d\right)$.
until $\|g\|_2^2 < \epsilon$
Output: w

Hence only matrix $\mathbf{C}_1 = \hat{\mathbf{L}}^T \mathbf{D} \mathbf{X} \mathbf{E}^{-1} \hat{\mathbf{Q}}^T + \mathbf{I}$ of the size $m \times m$ needs to be inverted — inversion of the diagonal matrix $\mathbf{E}$ is trivial. Putting (14) and (13) into (12) and simplifying obtained expression results in:

$$d = \left(\mathbf{E}^{-1} - \mathbf{E}^{-1}\hat{\mathbf{Q}}^T\mathbf{C}_1^{-1}\hat{\mathbf{L}}^T\mathbf{D}\mathbf{X}\mathbf{E}^{-1}\right) g. \tag{15}$$

This approach is summarized in the Algorithm 2.

Application to the Smooth ℓ^1 Approximation. Every convex twice continuously differentiable regularizer can be put in place of ridge penalty and above procedure may be used to optimize such a problem. In this article we focused on the smoothly approximated ℓ^1-norm [12] via integral of hyperbolic tangent function:

$$\|x\|_{1\text{soft}} = \sum_{j=1}^n \frac{1}{a} \log\left(\cosh\left(ax_j\right)\right), \ a \geqslant 1, \tag{16}$$

and we call this model `L1-QR-soft`. In this case

$$\mathbf{E} = \text{diag}\{\lambda a \left(1 - \tanh^2\left(aw_1\right)\right), \ldots, \lambda a \left(1 - \tanh^2\left(aw_n\right)\right)\}.$$

Application to the Strict ℓ^1 Penalty. Fan and Li proposed a unified algorithm for the minimization problem (2) via local quadratic approximations [3]. Here we use the idea presented by Krishnapuram [11], in which the following inequality is used:

$$\|w\|_1 \leq \frac{1}{2}\sum_{j=1}^n \left(\frac{w_j^2}{|w_j'|} + |w_j'|\right), \tag{17}$$

what is true for any w' and equality holds if and only if $w' = w$.

Cost function has a form:

$$L(w) = \sum_{i=1}^m \log(1 + e^{-y_i \cdot \sum_{j=1}^n x_{ij} w_j}) + \frac{\lambda}{2}\sum_{j=1}^n \left(\frac{w_j^2}{|w_j'|} + |w_j'|\right). \tag{18}$$

If we differentiate penalty term, we get:

$$\frac{\lambda}{2}\frac{\partial P}{\partial \boldsymbol{w}} = \frac{\partial}{\partial \boldsymbol{w}}\left(\frac{\lambda}{2}\sum_{j=1}^{n}\frac{w_j^2}{|w_j'|} + |w_j'|\right) = \mathbf{E}\boldsymbol{w}, \tag{19}$$

where

$$\mathbf{E} = \operatorname{diag}\left\{\frac{\lambda}{|w_1'|}, \ldots, \frac{\lambda}{|w_n'|}\right\} = \lambda\frac{\partial^2 P}{\partial \boldsymbol{w}\partial \boldsymbol{w}^T}.$$

Initial $\boldsymbol{w}$ must be non zero (we set it to $\mathbf{1}$), otherwise there is no progress. If $|w_j|$ falls below machine precision, we set it to zero.

Applying the idea of the QR factorization leads to the following result:

$$\boldsymbol{d} = \left(\mathbf{E}^{-1} - \mathbf{E}^{-1}\hat{\mathbf{Q}}^T\mathbf{C}_1^{-1}\hat{\mathbf{L}}^T\mathbf{D}\mathbf{X}\mathbf{E}^{-1}\right)\left(\mathbf{X}^T\boldsymbol{v} + \mathbf{E}\boldsymbol{w}\right)$$
$$= \left(\mathbf{I} - \mathbf{E}^{-1}\hat{\mathbf{Q}}^T\mathbf{C}_1^{-1}\hat{\mathbf{L}}^T\mathbf{D}\mathbf{X}\right)\cdot\left(\mathbf{E}^{-1}\mathbf{X}^T\boldsymbol{v} + \boldsymbol{w}\right). \tag{20}$$

One can note that when $\boldsymbol{w}$ is sparse, corresponding diagonal elements are 0. To avoid unneccessary multiplications by zero, we rewrite product $\mathbf{X}\mathbf{E}^{-1}\hat{\mathbf{Q}}^T$ as a sum of outer products:

$$\mathbf{X}\mathbf{E}^{-1}\hat{\mathbf{Q}}^T = \sum_{j=1}^{n}e_{jj}^{-1}\hat{\boldsymbol{x}}_j \otimes \hat{\boldsymbol{q}}_j, \tag{21}$$

where $\hat{\boldsymbol{x}}_j$ and $\hat{\boldsymbol{q}}_j$ are j-th columns of matrices $\mathbf{X}$ and $\hat{\mathbf{Q}}$ respectively. Similar concept is used when multiplying matrix $\mathbf{E}^{-1}\hat{\mathbf{Q}}^T$ by a vector e.g. $\boldsymbol{z}$: j-th element of the result equals $e_{jj}^{-1}\hat{\boldsymbol{q}}_j \cdot \boldsymbol{z}$. We refer to this model as L1-QR.

After obtaining direction $\boldsymbol{d}$ we use backtracking line search[1] with sufficient decrease condition given by Tseng and Yun [19] with one exception: if a unit step is already decent, we seek for a bigger step to ensure faster convergence.

Application to the $\ell^{q<1}$ Penalty. The idea described above can be directly applied to the $\ell^{q<1}$ "norms" [10] and we call it Lq-QR. Cost function has a form:

$$L(\boldsymbol{w}) = \sum_{i=1}^{m}\log(1 + e^{-y_i\cdot\sum_{j=1}^{n}x_{ij}w_j}) + \frac{\lambda}{2}\sum_{j=1}^{n}\left(\frac{qw_j^2}{|w_j'|^{2-q}} + (2-q)\,|w_j'|^q\right), \tag{22}$$

where

$$\mathbf{E}^{-1} = \operatorname{diag}\left\{\frac{|w_1|^{2-q}}{\lambda q}, \ldots, \frac{|w_n|^{2-q}}{\lambda q}\right\}.$$

3 Complexity of Proposed Methods

Cost of each iteration in the ordinary Newton method for logistic regression equals $k\cdot\left(2n^2 + n\right)$, where k is the number of conjugate gradient iterations. In general $k \leq n$, so in the worst case its complexity is $\mathcal{O}\left(n^3\right)$.

---

[1] In the line search procedure we minimize (2) with $P(\boldsymbol{w}) = \|\boldsymbol{w}\|_1$.

Rotationally Invariant Case. QR factorization is done once and its complexity is $\mathcal{O}\left(2m^2 \cdot \left(n - \frac{m}{3}\right)\right) = \mathcal{O}\left(m^2 n\right)$. Using data transformed to the smaller space, each step of the Newton procedure is much cheaper and it requires about km^2 operations (cost of solving system of linear equations using conjugate gradient, $k \leq m$), what is $\mathcal{O}\left(m^3\right)$ in general.

As it is shown in the experimental part, this approach dominates other optimization methods (especially exact second order procedures). Looking at the above estimations, it is clear that the presented approach is especially attractive when $m \ll n$.

Rotationally Variant Case. In the second case the most dominating operation comes from computation of the matrix $\mathbf{C}_1$ in the Eq. (15). Due to dimensionality of matrices: $\hat{\mathbf{L}}_{m \times m}, \mathbf{X}_{m \times n}$ and $\hat{\mathbf{Q}}_{m \times n}$, the complexity of computation $\mathbf{C}_1$ is $\mathcal{O}(m^2 n)$ — cost of inversion of the matrix $\mathbf{C}_1$ is less important i.e. $\mathcal{O}(m^3)$. In the case of ℓ^1 penalty taking sparsity of w into account reduces this complexity to $\mathcal{O}(m^2 \cdot \#\text{nnz})$, where $\#\text{nnz}$ is the number of non-zero coefficients.

Therefore theoretical upper bound on iteration for logistic regression with rotationally variant penalty function is $\mathcal{O}\left(m^2 n\right)$, what is better than direct Newton approach. However, looking at (15), we see that the number of multiplications is large, thus a constant factor in this estimation is large.

4   Experimental Results

In the experimental part we present two cases: 1) learning ordinary logistic regression model, and 2) learning a 2-layer neural network via extreme learning paradigm. We use following datasets:

1. Artificial dataset with 100 informative attributes and 1000 redundant attributes, informative part was produced by function make_classification from package scikit-learn and whole set was transformed introducing correlations.
2. Two micro-array datasets: leukemia [6], prostate cancer [18].
3. Artificial non-linearly separable datasets: chessboard 3×3 and 4×4, and two spirals — used for learning neural network.

As a reference we use solvers that are available in the package scikit-learn for LogisticRegression model i.e. for ℓ^2 penalty we use: LibLinear [4] in two variants (primal and dual), L-BFGS, L2-NEWTON-CG; For sparse penalty functions we compare our solutions with two solvers available in the scikit-learn: LibLinear and SAGA.

For the case ℓ^2 penalty we provide algorithm L2-QR presented in the Sect. 2.1. In the "sparse" case we compare three algorithms presented in the Sect. 2.2: L1-QR-soft, L1-QR and Lq-QR. Our approach L2-QR (Algorithm 1) is computationally equivalent to the L2-NEWTON-CG meaning that we solve an identical optimization problem (though in the smaller space). In the case of ℓ^2 penalty all models should converge theoretically to the same solution, so differences in

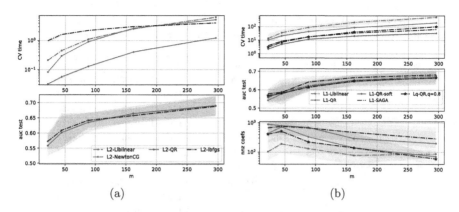

Fig. 1. Comparison of algorithms for learning ℓ^2 (a) and sparse (b) penalized logistic regressions on the artificial (300×1100) dataset. Plots present time of cross-validation procedure (CV time), AUC on test set (auc test), and number of non-zero coefficients for sparse models (nnz coefs).

the final value of the objective function are caused by numerical issues (like numerical errors, approximations or exceeding the number of iterations without convergence). These differences affect the predictions on a test set.

The case of ℓ^1 penalty is more complicated to compare. The L1-QR Algorithm is equivalent to the L1-Liblinear i.e. it minimizes the same cost function. Algorithm L1-QR-soft uses approximated ℓ^1-norm, and algorithm Lq-QR uses a bit different non-convex cost function which gives similar results to ℓ^1 penalized regression for $q \approx 1$. We also should emphasize that SAGA algorithm does not optimize directly penalized log-likelihood function on the training set, but it is stochastic optimizer and it gives sometimes qualitatively different models. In the case L1-QR-soft final solution is sparse only approximately (and depends on a (16)), whereas other models produce strictly sparse models. The measure of sparsity is the number of non-zero coefficients. For L1-QR-soft we check the sparsity with a tolerance of order 10^{-5}.

All algorithms were started with the same parameters: maximum number of iterations (1000) and tolerance ($\epsilon = 10^{-6}$), and used the same learning and testing datasets. All algorithms depend on the regularization parameter C (or $1/\lambda$). This parameter is selected in the cross-validation procedure from the same range. During experiments with artificial data we change the size of training subset. Experiments were performed on Intel Xeoen E5-2699v4 machine, in the one threaded envirovement (with parameters n_jobs=1 and MKL_NUM_THREADS=1).

Learning Ordinary Logistic Regression Model. In the first experiment, presented in the Fig. 1, we use an artificial highly correlated dataset (1). We used training/testing procedure for each size of learning data, and for each classifier we select optimal value of parameter $C = 1/\lambda$ using cross-validation. The number of samples varies from 20 to 300. As we can see, in the case ℓ^2 penalty our solution using QR decomposition L2-QR gives better times of fitting than

Table 1. Experimental results for micro-array datasets and ℓ^2 penalized logistic regressions. All solvers converge to the same solution, there are only differences in times.

Dataset	Classifier	$TIME_{FIT}$[s]	COST FCN.	AUC_{TEST}	ACC_{TEST}
Golub (38 × 7129)	L2-NEWTON-CG	0.0520	1.17e+11	0.8571	0.8824
	L2-QR	0.0065	1.17e+11	0.8571	0.8824
	SAG	1.2560	1.17e+11	0.8571	0.8824
	LIBLINEAR L2	0.0280	1.17e+11	0.8571	0.8824
	LIBLINEAR L2 dual	0.0737	1.17e+11	0.8571	0.8824
	L-BFGS	0.0341	1.17e+11	0.8571	0.8824
Singh (102 × 12600)	L2-NEWTON-CG	0.6038	5.14e+11	0.9735	0.9706
	L2-QR	0.0418	5.14e+11	0.9735	0.9706
	SAG	5.2822	5.13e+11	0.9735	0.9706
	LIBLINEAR L2	0.1991	5.14e+11	0.9735	0.9706
	LIBLINEAR L2 dual	0.6083	5.14e+11	0.9735	0.9706
	L-BFGS	0.1192	5.14e+11	0.9735	0.9706

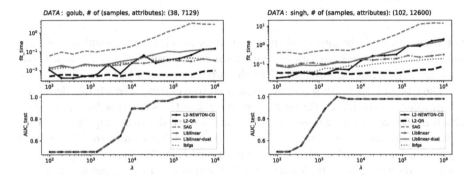

Fig. 2. Comparison of algorithms learning ℓ^2 penalized logistic regression on micro-array datasets for a sequence of λs; mean values are presented in the Table 1.

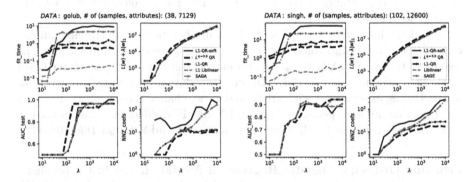

Fig. 3. Detailed comparison of algorithms learning ℓ^1 penalized logistic regression on micro-array datasets for a sequence of λs. Mean values for this case are presented in the Table 2.

Table 2. Experimental results for micro-array datasets and ℓ^1 penalized logistic regressions. `L1-QR` solver converges to the same solution as `L1-Liblinear`, there are only difference in times. `SAGA` and `L1-QR-soft` gives different solution.

Dataset	Classifier	$TIME_{FIT}$ [s]	Cost Fcn.	AUC_{TEST}	ACC_{TEST}	NNZ coefs.
Golub	L1-QR-soft	8.121	2.74e+07	0.8929	0.9118	90.1
	$L^{q=0.9}$-QR	0.544	2.80e+07	0.9393	0.95	9.1
	L1-QR	1.062	2.28e+07	0.8679	0.8912	10.2
	Liblinear	0.042	2.28e+07	0.8679	0.8912	10.4
	SAGA	4.532	2.78e+07	0.8857	0.9059	46.7
Singh	L1-QR-soft	51.042	6.74e+07	0.8753	0.8794	91.2
	$L^{q=0.9}$-QR	3.941	8.65e+07	0.8893	0.9	13.4
	L1-QR	6.716	6.52e+07	0.8976	0.8912	20.1
	Liblinear	0.225	6.52e+07	0.8976	0.8912	20.2
	SAGA	21.251	7.11e+07	0.8869	0.8912	65.9

ordinary solvers available in the `scikit-learn` and all algorithms work nearly the same, only `L2-lbfgs` gives slightly different results. In the case of sparse penalty our algorithm `L1-QR` works faster than `L1-Liblinear` and obtains comparable but not identical results. For sparse case `L1-SAGA` gives best predictions (about 1–2% better than other sparse algorithms), but it produces the most dense solutions similarly like `L1-QR-soft`.

In the second experiment we used micro-array data with an original train and test sets. For those datasets quotients (samples/attributes) are fixed (about 0.005–0.01). The results are shown in Table 1 (ℓ^2 case) and in Table 2 (ℓ^1 case). Tables present mean values of times and cost functions, averaged over λs. Whole traces over λs are presented in the Fig. 2 and Fig. 3. For the case of ℓ^2 penalty we notice that all tested algorithms give identical results looking at the quality of prediction and the cost function. However, time of fitting differs and the best algorithm is that, which uses QR factorization.

For the case of sparse penalty functions only algorithms `L1-Liblinear` and `L1-QR` give quantitatively the same results, however `L1-Liblinear` works about ten times faster. Other models give qualitatively different results. Algorithm `Lq-QR` obtained the best sparsity and the best accuracy in prediction and was also slightly faster than `L1-QR`. Looking at the cost function with ℓ^1 penalty we see that `L1-Liblinear` and `L1-QR` are the same, `SAGA` obtains worse cost function than even `L1-QR-soft`. We want to stress that `scikit-learn` provides only solvers for ℓ^2 and ℓ^1 penalty, not for general case ℓ^q.

Application to Extreme Learning and RVFL Networks. Random Vector Functional-link (RVFL) network is a method of learning two (or more) layer neural networks in two separate steps. In the first step coefficients for hidden neurons are chosen randomly and are fixed, and then in the second step learning algorithm is used only for the output layer. The second step is equivalent to learning the logistic regression model (a linear model with the sigmoid output

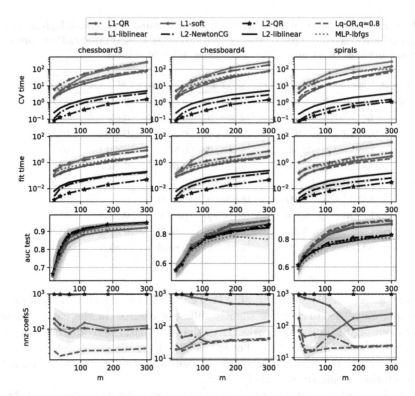

Fig. 4. Experimental results for the extreme learning. Comparison on artificial datasets. CV time is the time of cross-validation procedure, fit time is the time of fitting for the best λ, auc test is the area under ROC on test dataset, and nnz coefs5 is the number of non-zero coefficients.

function). Recently, this approach is also known as "extreme learning" (see: [20] for more references).

The output of neural network with a single hidden layer is given by:

$$y(\boldsymbol{x}; \mathbf{W}^1, \boldsymbol{b}^1, \boldsymbol{w}^2, b^2) = \varphi\left(\sum_{j=1}^{Z} w_i^{(2)} \varphi(\boldsymbol{x}; \mathbf{W}^1(:, j), \boldsymbol{b}^1(j)) + b^2\right), \qquad (23)$$

where: Z is the number of hidden neurons, $\varphi(\boldsymbol{x}; \boldsymbol{w}, b) = \tanh\left(\sum_{k=1}^{n} w_k x_k + b\right)$ is the activation function.

In this experiment we choose randomly hidden layer coefficients $\mathbf{W}^1$ and $\boldsymbol{b}^1$, with number of hidden neurons $Z = 1000$ and next we learn the coefficients of the output layer: $\boldsymbol{w}^2$ and b^2 using the new transformed data matrix:

$$\Phi_{m \times Z} = \varphi\big(\mathbf{X}(i,:); \mathbf{W}^1(j,:), \boldsymbol{b}^1(j)\big).$$

For experiments we prepared the class `ExtremeClassier` (in `scikit-learn` paradigm) which depends on the number of hidden neurons Z, the kind of linear output classifier and its parameters. In the fitting part we ensure the same

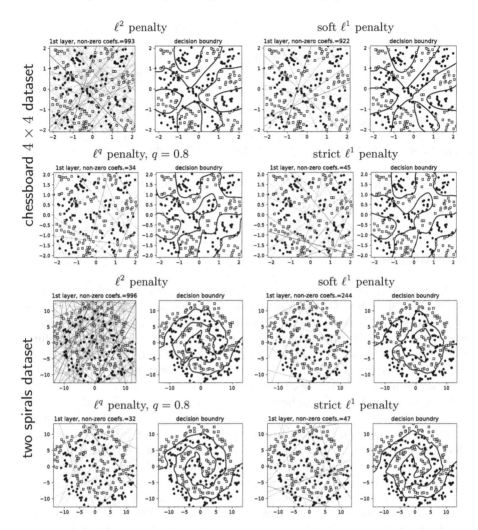

Fig. 5. Exemplary decision boundaries for different penalty functions (ℓ^2, ℓ^1 with a smooth approximation of the absolute value function, $\ell^{q=0.8}$, ℓ^1) on used datasets. In the figure there are coefficients of the first layer of the neural network represented as lines — intensity and color represents magnitude and sign of the particular coefficient. (Color figure online)

random part of classifier. In this experiment we also added a new model — multi-layer perceptron with two layers and with Z hidden neurons fitted in the standard way using L-BFGS algorithm (MLP-lbfgs).

Results of the experiment are presented in the Fig. 4. For each size of learning data and for each classifier we select optimal value of parameter $C = 1/\lambda$ using cross-validation. The number of samples varies from 20 to 300. As we can see, in both cases (ℓ^2 and sparse penalties) our solution using QR decomposition gives always better times of fitting than ordinary solvers available in the `scikit-learn`. Time of fitting of `L1-QR` is 2–5 times shorter than `L1-Liblinear`, especially for the case chessboard 4×4 and two spirals. Looking at quality we see that sparse models are similar, but slightly different. For two spirals the best one is `Lq-QR` and it is also the sparsest model. Generally sparse models are better for two spirals and chessboard 4×4. The MLP model has the worst quality and comparable time of fitting to sparse regressions.

The experiment shows that use of QR factorization can effectively implement learning of RVFL network with different regularization terms. Moreover, we confirm that such learning works more stable than ordinary neural network learning algorithms, especially for the large number of hidden neurons. Exemplary decision boundaries, sparsity and found hidden neurons are shown in the Fig. 5.

5   Conclusion

In this paper we presented application of the QR matrix factorization to improve the Newton procedure for learning logistic regression models with different kind of penalties. We presented two approaches: rotationally invariant case with ℓ^2 penalty, and general convex rotationally variant case with sparse penalty functions. Generally speaking, there is a strong evidence that use of QR factorization in the rotational invariant case can improve classical Newton-CG algorithm when $m < n$. The most expensive operation in this approach is QR factorization itself, which is performed once at the beginning. Our experiments showed also that this approach, for $m \ll n$ surpasses also other algorithms approximating Hessian like L-BFGS and truncated Newton method (used in Liblinear). In this case we have shown that theoretical upper bound on cost of Newton iteration is $\mathcal{O}\left(m^3\right)$.

We showed also that using QR decomposition and Shermann-Morrison-Woodbury formula we can solve a problem of learning the regression model with different sparse penalty functions. Actually, improvement in this case is not as strong as in the case of ℓ^2 penalty, however we proved that using QR factorization we obtain theoretical upper bound significantly better than for general Newton-CG procedure. In fact, the Newton iterations in this case have the same cost as the initial cost of the QR decomposition i.e. $\mathcal{O}\left(m^2 n\right)$. Numerical experiments revealed that for more difficult and correlated data (e.g. for extreme learning) such approach may work faster than L1-Liblinear. However, we should admit that in a typical and simpler cases L1-Liblinear may be faster.

References

1. Boyd, S., Vandenberghe, L.: Convex Optimization. Cambridge University Press, New York (2004)
2. Dai, Y.H.: On the nonmonotone line search. J. Optim. Theory Appl. **112**(2), 315–330 (2002)

3. Fan, J., Li, R.: Variable selection via nonconcave penalized likelihood and its oracle properties. J. Am. Stat. Assoc. **96**(456), 1348–1360 (2001)
4. Fan, R.E., Chang, K.W., Hsieh, C.J., Wang, X.R., Lin, C.J.: LIBLINEAR: a library for large linear classification. J. Mach. Learn. Res. **9**, 1871–1874 (2008)
5. Golub, G., Van Loan, C.: Matrix Computations. Johns Hopkins Studies in the Mathematical Sciences, Johns Hopkins University Press (2013)
6. Golub, T.R., et al.: Molecular classification of cancer: class discovery and class prediction by gene expression monitoring. Science **286**(5439), 531–537 (1999)
7. Green, P.J.: Iteratively reweighted least squares for maximum likelihood estimation, and some robust and resistant alternatives (with discussion). J. R. Stat. Soc. Ser. B Methodol. **46**, 149–192 (1984)
8. Hastie, T., Tibshirani, R.: Expression arrays and the $p \gg n$ problem (2003)
9. Hastie, T., Tibshirani, R., Friedman, J.: The Elements of Statistical Learning. Springer Series in Statistics. Springer, New York (2001). https://doi.org/10.1007/978-0-387-21606-5
10. Kabán, A., Durrant, R.J.: Learning with $L_{q<1}$ vs L_1-norm regularisation with exponentially many irrelevant features. In: Daelemans, W., Goethals, B., Morik, K. (eds.) ECML PKDD 2008. LNCS, vol. 5211, pp. 580–596. Springer, Heidelberg (2008). https://doi.org/10.1007/978-3-540-87479-9_56
11. Krishnapuram, B., Carin, L., Figueiredo, M.A.T., Hartemink, A.: Sparse multinomial logistic regression: fast algorithms and generalization bounds. IEEE Trans. Pattern Anal. Mach. Intell. **27**(6), 957–968 (2005)
12. Lee, Y.J., Mangasarian, O.: SSVM: a smooth support vector machine for classification. Comput. Optim. Appl. **20**, 5–22 (2001)
13. Lin, C.J., Weng, R.C., Keerthi, S.S.: Trust region Newton method for logistic regression. J. Mach. Learn. Res. **9**, 627–650 (2008)
14. Minka, T.P.: A comparison of numerical optimizers for logistic regression (2003). https://tminka.github.io/papers/logreg/minka-logreg.pdf
15. Murphy, K.P.: Machine Learning: A Probabilistic Perspective. MIT Press, Cambridge (2013)
16. Ng, A.Y.: Feature selection, L_1 vs. L_2 regularization, and rotational invariance. In: Proceedings of the Twenty-First International Conference on Machine Learning, ICML 2004, pp. 78–85. ACM, New York (2004)
17. Pedregosa, F., et al.: Scikit-learn: machine learning in python. J. Mach. Learn. Res. **12**, 2825–2830 (2011)
18. Singh, S., Skanda, S., Scott, S., Arie, B., Sujata, P., Gurmit, S.: Overexpression of vimentin: role in the invasive phenotype in an androgen-independent model of prostate cancer. Cancer Res. **63**(9), 2306–2311 (2003)
19. Tseng, P., Yun, S.: A coordinate gradient descent method for nonsmooth separable minimization. Math. Program. **117**, 387–423 (2009)
20. Wang, L.P., Wan, C.R.: Comments on "the extreme learning machine". IEEE Trans. Neural Netw. **19**(8), 1494–1495 (2008)
21. Zhu, C., Byrd, R.H., Lu, P., Nocedal, J.: Algorithm 778: L-BFGS-B: Fortran subroutines for large-scale bound-constrained optimization. ACM Trans. Math. Softw. **23**(4), 550–560 (1997)

Uncertainty Quantification in Fractional Stochastic Integro-Differential Equations Using Legendre Wavelet Collocation Method

Abhishek Kumar Singh[ID] and Mani Mehra[✉][ID]

Indian Institute of Technology Delhi, 110016 New Delhi, India
assinghabhi@gmail.com, mmehra@maths.iitd.ac.in

Abstract. The paper aims to present an efficient numerical scheme to quantify the uncertainty in the solution of stochastic fractional integro-differential equations. The numerical scheme presented here is based on Legendre wavelets combined with block pulse functions using their deterministic and stochastic operational matrix of integration. The operational matrices are utilized to convert the stochastic fractional integro-differential equation to a linear system of algebraic equation. Finally, the accuracy and efficiency of the proposed scheme are investigated through numerical experiments.

Keywords: Legendre polynomial · Legendre wavelets · Stochastic operational matrix · Itô integral · Integro-differential equations

1 Introduction

A stochastic fractional integro-differential equation (SFIDE), where order of derivative is non integer, is a generalization of the fractional Folkker-Plank equation which describes the random walk of a particle [2]. This model has the following form

$$D^\alpha u(t) = f(t) + \int_0^t u(s)k_1(s,t)ds + \int_0^t u(s)k_2(s,t)dW(s), \ t \in [0,T], \tag{1}$$
$$u(0) = u_0,$$

where D^α, $0 < \alpha < 1$, denotes the Caputo fractional derivative, $W(s), s \in [0,T]$ is the standard Wiener process and the integral with respect to it is the Itô integral. Presence of the Itô integral in Equation (1) causes randomness in the solution and hence it becomes non deterministic. In this paper, we develop a novel approach to quantify this uncertain behavior in the numerical solution.

In recent decade, the need to obtain the numerical solution of SFIDE has increased significantly. However, in literature, only a handful of papers are

Supported by University Grants Commission, New Delhi-110002, India.

V. V. Krzhizhanovskaya et al. (Eds.): ICCS 2020, LNCS 12138, pp. 58–71, 2020.
https://doi.org/10.1007/978-3-030-50417-5_5

available that actually discuss about the numerical solution of SFIDE. In [8], Maleknejad *et al.* provided an operational matrix method based on block pulse functions to solve stochastic Volterra integral equations. In [12], Taheri *et al.* formulated the spectral collocation method based on shifted Legendre polynomials to solve SFIDE. In [10], Mirzaee and Samadyar constructed an efficient scheme to solve SFIDE based on Bernstein polynomials. In [11], Mirzaee and Samadyar provide a meshless discrete collocation method based on radial basis functions to solve SFIDE.

In this paper, a new scheme is derived based on Legendre wavelet collocation method and block pulse function involving the operational matrix for solving SFIDE (1). In Sect. 2, we give basic definition of fractional calculus and construction of Legendre wavelet based on Multi-resolution analysis. Then in Sect. 3, operational matrix of fractional order integration and integration operational matrix are derived. The proposed scheme for the SFIDE is discussed in Sect. 4, while Sect. 5 provides numerical experiments performed to showcase the effectiveness of the approach. In Sect. 6, we present various applications of SFIDE. Finally, Sect. 7 gives the brief conclusion.

2 Preliminaries

In this section, we discuss the mathematical preliminaries of fractional calculus and construction of wavelet which are required for subsequent development.

Definition 1. *[5] The left Riemann-Liouville fractional integral of order $\alpha \geq 0$ of a function $f(t)$, $t \in (a, b)$ is defined as follows*

$$\begin{aligned} {}_aI_t^\alpha f(t) &= \frac{1}{\Gamma(\alpha)} \int_a^t (t-s)^{\alpha-1} f(s)ds, \\ {}_aI_t^0 f(t) &= f(t). \end{aligned} \tag{2}$$

Similar to integer order integration, the left Riemann-Liouville fractional integral operator is a linear operator

$$ {}_aI_t^\alpha(\lambda f(t) + \mu g(t)) = \lambda\, {}_aI_t^\alpha f(t) + \mu\, {}_aI_t^\alpha g(t), $$

where λ and μ are constants.

Definition 2. *[5] The left Caputo derivative with order $\alpha > 0$ of the given function $f(t)$, $t \in (a, b)$ is defined as*

$$ D_{a,t}^\alpha f(t) = \frac{1}{\Gamma(m-\alpha)} \int_a^t (t-s)^{m-\alpha-1} f^{(m)}(s)ds, \tag{3}$$

where m is a positive integer satisfying $m - 1 < \alpha \leq m$.

2.1   Multi-resolution Analysis (MRA)

An MRA is an increasing family of closed subspace $V^j \subset L^2(\mathbb{R})$ which satisfies the following axioms [9] :

1. $V^j \subset V^{j+1}$
2. $\overline{\bigcup_{j \in \mathbb{Z}} V^j} = L^2(\mathbb{R})$
3. $\{\phi(x - k) : k \in \mathbb{Z}\}$ is an orthonormal basis of V^0
4. $f(\cdot) \in V^j$ if and only if $f(2(\cdot)) \in V^{j+1}$ for all $j \in \mathbb{Z}$.

For given nested sequence subspace V^j, define the space W^j as the orthogonal complement of V^j in V^{j+1}, i.e., $V^j \perp W^j$ and

$$V^{j+1} = V^j \oplus W^j, \tag{4}$$

applying recursively, we get

$$V^j = V^{j_0} \oplus \bigoplus_{k=j_0}^{j-1} W^k, \quad j > j_0. \tag{5}$$

Now, based on the above analysis, to construct a wavelet define a space V_M^J of piecewise polynomial functions as follows :

$$V_M^J := \{\phi \ : \ \text{the restriction of } \phi \text{ to the interval } [2^{-J+1}(k-1), 2^{-J+1}k)$$
$$\text{is a polynomial of degree less than M for } k = 1, 2, \cdots, 2^{(J-1)}, \tag{6}$$
$$\text{and } \phi \text{ vanishes elsewhere}\}.$$

The space V_M^J has dimension $2^{(J-1)}M$ and

$$V_M^1 \subset V_M^2 \subset \cdots V_M^J \subset \cdots \subset L^2([0,1)).$$

Next, consider the $2^{(J-1)}M$-dimensional space W_M^J which is an orthogonal complement of V_M^J in V_M^{J+1}, i.e.,

$$V_M^{J+1} = V_M^J \oplus W_M^{J+1}.$$

Inductively, one can obtain

$$V_M^J = V_M^1 \oplus \bigoplus_{j=1}^{J-1} W_M^j. \tag{7}$$

Unlike Haar, the element of the space W_M^J do not have a general form. To construct the elements of W_M^J one can refer [1].
Further, if

$$V_M^J = \text{span}\{\phi_{k,m}^J, \ m = 0, 1, \cdots, M-1, \ k = 1, 2, \cdots, 2^{(J-1)}\},$$

then we define the projection operator $P_{V_M^J} : L^2[0,1] \to V_M^J$ as

$$P_{V_M^J}(f(x)) := \sum_{k=1}^{2^{J-1}} \sum_{m=0}^{M-1} c_{k,m}^J \phi_{k,m}^J(x), \tag{8}$$

where $c_{k,m}^J = \int_{\frac{(k-1)}{2^{(J-1)}}}^{\frac{k}{2^{(J-1)}}} f(x)\phi_{k,m}^J(x)dx$. Set

$$W_M^J = \text{span}\{\psi_{k,m}^J(x), \ m = 0,1,\cdots, M-1, \ k = 1,2,\cdots, 2^{(J-1)}\},$$

where $\psi_{k,m}^J(x) = 2^{\frac{(J-1)}{2}}\psi_m(2^{(J-1)}x - k + 1)$. The support of $\psi_{k,m}^J$ is $[\frac{k-1}{2^{J-1}}, \frac{k}{2^{J-1}})$ and $\psi_m(x)$ satisfies the following property (vanishing moment property)

$$\int_0^1 x^i \psi_m(x)dx = 0, \ i = 0,1,\cdots, M-1. \tag{9}$$

Now, using V_M^J, we introduced the subspace $V_M^{J,2}$ of $L^2([0,) \times [0,1))$ defined by

$$V_M^{J,2} := \{\phi \mid \phi = \phi_1\phi_2 \text{ where } \phi_1, \phi_2 \in V_M^J\}.$$

Moreover,

$$V_M^{J,2} = \text{span}\{\phi_{k,m,k',m'}^J = \phi_{k,m}^J\phi_{k',m'}^J \ : \ k,k' = 1,\cdots, 2^{J-1} \text{ and } m, m' = 0,\cdots, M-1\}.$$

Then, define the projection operator $P_{V_M^{J,2}} : L^2([0,1] \times [0,1]) \to V_M^{J,2}$ as

$$P_{V_M^{J,2}} f(s,t) = \sum_{k=1}^{2^{J-1}} \sum_{m=0}^{M-1} \sum_{k'=1}^{2^{J-1}} \sum_{m'=0}^{M-1} c_{k,m,k',m'}^J \phi_{k,m,k',m'}^J(s,t), \tag{10}$$

where $c_{k,m,k',m'}^J = \int_{\frac{(k-1)}{2^{(J-1)}}}^{\frac{k}{2^{J-1}}} \int_{\frac{(k'-1)}{2^{(J-1)}}}^{\frac{k'}{2^{J-1}}} f(s,t)\phi_{k,m,k',m'}^J(s,t)dsdt$.

Next, introduce the space $W_M^{J,2}$ which is defined by

$$W_M^{J,2} = \{\psi \mid \psi = \psi_1\psi_2 \text{ where } \psi_1, \psi_2 \in W_M^J\},$$

and

$$W_M^{J,2} = \text{span}\{\psi_{k,m,k',m'}^J = \psi_{k,m}^J\psi_{k',m'}^J \ : \ k,k' = 1,\cdots, 2^{J-1} \text{ and } m, m' = 0,\cdots, M-1\},$$

where

$$\psi_{k,m,k',m'}^J(s,t) = 2^{J-1}\psi_{m,m'}(2^{J-1}s - k + 1, 2^{J-1}t - k' + 1)$$
$$:= 2^{J-1}\psi_m(2^{J-1}s - k + 1)\psi_{m'}(2^{J-1}t - k' + 1)$$

and $\psi_{m,m'}(s,t)$ satisfies the following property

$$\int_0^1 \int_0^1 s^i t^j \psi_{m,m'}(s,t)dsdt = 0, \ i,j = 0,\cdots, M-1. \tag{11}$$

The subspace $W_M^{J,2}$ is orthogonal complement of $V_M^{J,2}$ in $V_M^{J+1,2}$. Therefore, one can write

$$V_M^{J+1,2} = V_M^{J,2} \oplus W_M^{J,2},$$

and hence

$$V_M^{J,2} = V_M^{1,2} \oplus \bigoplus_{j=1}^{J-1} W_M^{j-1,2}. \tag{12}$$

Now, if we choose $\phi_{k,m}^J$ as in [13], i.e., for $m = 0, 1, 2, \cdots, M-1$, and $\hat{k} = 2k-1$ with $k = 1, 2, \cdots, 2^{(J-1)}$

$$\phi_{k,m}^J(t) = \begin{cases} \sqrt{m + \frac{1}{2}} 2^{J/2} P_m(2^J t - \hat{k}) & \text{for } \frac{\hat{k}-1}{2^J} \leq t < \frac{\hat{k}+1}{2^J} \\ 0 & \text{otherwise}, \end{cases} \tag{13}$$

where $P_m(t)$ is a Legendre polynomials of order m are defined in the interval $[-1, 1]$ and given by the following recurrence formulas

$$P_0(t) = 1, \ P_1(t) = t,$$

$$P_{m+1}(t) = \left(\frac{2m+1}{m+1}\right) t P_m(t) - \left(\frac{m}{m+1}\right) P_{m-1}(t), \ m = 1, 2, 3, \cdots.$$

The wavelet constructed above using Legendre polynomials are called as Legendre wavelet [7].

2.2 Function Approximation

A function $f(t)$ defined over $L^2[0, 1)$ can be expanded with Legendre scaling functions $\phi_{k,m}^J(t)$ as

$$f(t) = \sum_{k=1}^{2^{J-1}} \sum_{m=0}^{\infty} c_{k,m}^J \phi_{k,m}^J(t), \tag{14}$$

where $c_{k,m}^J = \int_0^1 f(t) \phi_{k,m}^J(t) dt$ and $J \to \infty$. If the infinite series in (14) is truncated, then (14) can be written as

$$f(t) \approx P_{V_M^J}(f(t)) = \sum_{k=1}^{2^{J-1}} \sum_{m=0}^{M-1} c_{k,m}^J \phi_{k,m}^J(t) = C^T \Phi(t), \tag{15}$$

where C and $\Phi(t)$ are $2^{J-1}M \times 1$ matrices given by

$$C = [c_{1,0}^J, c_{1,1}^J, \cdots, c_{1,M-1}^J, c_{2,0}^J, \cdots, c_{2,M-1}^J, \cdots, c_{2^{J-1},0}^J, \cdots, c_{2^{J-1},M-1}^J]^T,$$

$$\Phi_{2^{J-1}M}(t) = [\phi_{1,0}^J(t), \phi_{1,1}^J(t), \cdots, \phi_{1,M-1}^J(t), \cdots, \phi_{2^{J-1},M-1}^J(t)]^T$$

$$= [\phi_1^J(t), \ldots, \phi_{2^{J-1}M}^J(t)]^T.$$

In similar way, a bivariate function $f(s,t) \in L^2[[0,1) \times [0,1)]$ can be expanded with Legendre wavelets as

$$f(s,t) \approx \sum_{i=1}^{2^{J-1}M} \sum_{j=1}^{2^{J-1}M} \phi_i^J(s) f_{ij} \phi_j^J(t) = \Phi^T(s) F \Phi(t), \tag{16}$$

where

$$f_{ij} = \int_0^1 \int_0^1 f(s,t) \phi_i^J(s) \phi_j^J(t) ds dt.$$

3 Legendre Wavelet Matrix and Block Pulse Operational Matrix

Let the collocation points be

$$t_i = \frac{2i-1}{2^J M}, \quad i = 1, 2, \ldots, 2^{J-1}M.$$

We denote the Legendre wavelet matrix as $\phi_{2^{J-1}M \times 2^{J-1}M}$ and define it as the combination of $\phi_{k,m}^J(t_i)$ at the collocation points (t_i) as

$$\phi_{2^{J-1}M \times 2^{J-1}M} = \begin{pmatrix} \phi_{1,0}^J(t_1) & \phi_{1,1}^J(t_2) & \cdots & \phi_{2^{J-1},M-1}^J(t_{2^{J-1}M}) \\ \phi_{1,0}^J(t_1) & \phi_{1,1}^J(t_2) & \cdots & \phi_{2^{J-1},M-1}^J(t_{2^{J-1}M}) \\ \vdots & \vdots & \ddots & \vdots \\ \phi_{1,0}^J(t_1) & \phi_{1,1}^J(t_2) & \cdots & \phi_{2^{J-1},M-1}^J(t_{2^{J-1}M}) \end{pmatrix}.$$

3.1 Legendre Wavelet Operational Matrix of Fractional Order Integration

If $f(t)$ is expanded as in Eq. (14), then the Riemann-Liouville fractional order integration is given by

$$_0I_t^\alpha f(t) \approx \frac{1}{\Gamma(\alpha)} \int_0^t (t-\tau)^{\alpha-1} C^T \Phi(\tau) d\tau = C^T (_0I_t^\alpha \Phi_{2^{J-1}M}(t)).$$

The $2^{J-1}M-$set of block pulse functions (BPFs) are also defined as

$$b_i(t) = \begin{cases} 1 & \frac{(i-1)}{2^{J-1}M} \leq t < \frac{i}{2^{J-1}M} \\ 0 & \text{otherwise}, \end{cases} \tag{17}$$

where $i = 1, 2, \ldots, 2^{J-1}M$. The function $b_i(t)$ has the following disjoint and orthogonal properties

- $b_i(t)b_j(t) = \delta_{ij}b_i(t)$,
- $\int_0^1 b_i(t)b_j(t) = \frac{\delta_{ij}}{2^{J-1}M}$,

where δ_{ij} is the Kronecker delta. The Legendre wavelet can be expanded into $2^{J-1}M$ - term block pulse function as

$$\Phi_{2^{J-1}M}(t) = \phi_{2^{J-1}M \times 2^{J-1}M} \mathbf{b}_{2^{J-1}M}(t), \qquad (18)$$

where $\mathbf{b}_{2^{J-1}M}(t) = [b_1(t), \dots, b_{2^{J-1}M}(t)]^T$. The block pulse operational matrix of fractional-order integration G^α is given in [6] as follows

$$_0I_t^\alpha \mathbf{b}_{2^{J-1}M}(t) \approx G^\alpha \mathbf{b}_{2^{J-1}M}(t), \qquad (19)$$

where

$$G^\alpha = \frac{1}{(2^{J-1}M)^\alpha} \frac{1}{\Gamma(\alpha+2)} \begin{pmatrix} 1 & \xi_1 & \xi_2 & \xi_3 & \cdots & \xi_{2^{J-1}M-1} \\ 0 & 1 & \xi_1 & \xi_2 & \cdots & \xi_{2^{J-1}M-2} \\ 0 & 0 & 1 & \xi_1 & \cdots & \xi_{2^{J-1}M-3} \\ \vdots & \vdots & \vdots & \vdots & \ddots & \vdots \\ 0 & 0 & 0 & 0 & \cdots & 1 \end{pmatrix},$$

with $\xi_i = (i+1)^{\alpha+1} - 2i^{\alpha+1} + (i-1)^{\alpha+1}$.
Let

$$_0I_t^\alpha \Phi_{2^{J-1}M}(t) = P_{2^{J-1}M \times 2^{J-1}M}^\alpha \Phi_{2^{J-1}M}(t), \qquad (20)$$

where the matrix $P_{2^{J-1}M \times 2^{J-1}M}^\alpha$ is called the Legendre wavelet operational matrix of fractional order integration. Using Eqs. (18) and (19) in (20), we get

$$P_{2^{J-1}M \times 2^{J-1}M}^\alpha \approx (\phi_{2^{J-1}M \times 2^{J-1}M}) G^\alpha (\phi_{2^{J-1}M \times 2^{J-1}M})^{-1}.$$

3.2   Deterministic Integration Operational Matrix

Let $m = 2^{J-1}M$ and compute $\int_0^t b_i(s)ds$ as follows

$$\int_0^t b_i(s)ds = \begin{cases} 0 & 0 \le t < \frac{i-1}{m} \\ t - \frac{i-1}{m} & \frac{i-1}{m} \le t < \frac{i}{m} \\ \frac{1}{m} & \frac{i}{m} \le t < 1. \end{cases} \qquad (21)$$

We approximate $t - \frac{i-1}{m}$, for $\frac{i-1}{m} \le t < \frac{i}{m}$, by $\frac{1}{2m}$ and express $\int_0^t b_i(s)ds$ in terms of BPFs as follows

$$\int_0^t b_i(s)ds \approx \left(0, \dots, 0, \frac{1}{2m}, \frac{1}{m}, \dots, \frac{1}{m}\right) \mathbf{b}_m(t), \qquad (22)$$

where $\frac{1}{2m}$ is the ith component of vector. Therefore

$$\int_0^t \mathbf{b}_m(s)ds \approx P\mathbf{b}_m(t), \qquad (23)$$

where the operational matrix of integration is given by

$$P = \frac{1}{2m} \begin{pmatrix} 1 & 2 & 2 & \cdots & 2 \\ 0 & 1 & 2 & \cdots & 2 \\ 0 & 0 & 1 & \cdots & 2 \\ \vdots & \vdots & \vdots & \ddots & \vdots \\ 0 & 0 & 0 & \cdots & 1 \end{pmatrix}_{m \times m}$$

3.3   Stochastic Integration Operational Matrix

The Itô integral of each single BPFs $b_i(t)$ can be computed as follows

$$\int_0^t b_i(s)dW(s) = \begin{cases} 0 & 0 \le t < \frac{i-1}{m} \\ W(t) - W(\frac{i-1}{m}) & \frac{i-1}{m} \le t < \frac{i}{m} \\ W(\frac{i}{m}) - W(\frac{i-1}{m}) & \frac{i}{m} \le t < 1. \end{cases} \qquad (24)$$

We can approximate $W(t) - W(\frac{i-1}{m})$, for $\frac{i-1}{m} \le t < \frac{i}{m}$, by $W(\frac{i-0.5}{m}) - W(\frac{i-1}{m})$ and express $\int_0^t b_i(s)dW(s)$, in terms of BPFs as follows

$$\int_0^t b_i(s)dW(s) \approx \left(0, \ldots, 0, W(\frac{i-0.5}{m}) - W(\frac{i-1}{m}), W(\frac{i}{m}) - W(\frac{i-1}{m}), \right.$$
$$\left. \ldots, W(\frac{i}{m}) - W(\frac{i-1}{m})\right)b_m(t), \qquad (25)$$

where $W(\frac{i-0.5}{m}) - W(\frac{i-1}{m})$ is the ith component of vector. Therefore, we obtain the following expression (for details, see [8])

$$\int_0^t \mathbf{b}_m(s)dW(s) \approx P_s\mathbf{b}_m(t), \qquad (26)$$

where stochastic operational matrix of integration is given by

$$P_s = \begin{pmatrix} W(\frac{1}{2m}) & W(\frac{1}{m}) & W(\frac{1}{m}) & \cdots & W(\frac{1}{m}) \\ 0 & W(\frac{3}{2m}) - W(\frac{1}{m}) & W(\frac{2}{m}) - W(\frac{1}{m}) & \cdots & W(\frac{2}{m}) - W(\frac{1}{m}) \\ 0 & 0 & W(\frac{5}{2m}) - W(\frac{2}{m}) & \cdots & W(\frac{3}{m}) - W(\frac{2}{m}) \\ \vdots & \vdots & \vdots & \ddots & \vdots \\ 0 & 0 & 0 & \cdots & W(\frac{2m-1}{2m}) - W(\frac{m-1}{m}) \end{pmatrix}_{m \times m}$$

4   Description of Numerical Method

Here we present the wavelet collocation method based on the Legendre wavelets for solving SFIDE (1). We use the relation between the fractional derivative and integral to obtain the solution $u(t)$ derived as follows

– Let $D^\alpha u(t) \approx C^T \Phi(t)$, this implies that

$$u(t) \approx C^T {}_0 I^\alpha \Phi(t) + u_0. \qquad (27)$$

– Let $k_1(s,t) \in L^2([0,1) \times [0,1))$. It can be expanded with respect to Legendre wavelet as

$$k_1(s,t) \approx \Phi^T(s)K_1\Phi(t) = \Phi^T(t)K_1^T\Phi(s), \qquad (28)$$

where $K_1 = (k_1)_{ij}$, $i = 1, 2, \ldots, m$, $j = 1, 2, \ldots, m$ is the $m \times m$ Legendre wavelets coefficient matrix with

$$(k_1)_{ij} = \int_0^1 \int_0^1 k_1(s,t)\phi_i(s)\phi_j(t)\,ds\,dt.$$

Similarly

$$k_2(s,t) \approx \Phi^T(s)K_2\Phi(t) = \Phi^T(t)K_2^T\Phi(s), \tag{29}$$

where $K_2 = (k_2)_{ij}$, $i = 1, 2, \ldots, m$, $j = 1, 2, \ldots, m$ is the $m \times m$ Legendre wavelets coefficient matrix with

$$(k_2)_{ij} = \int_0^1 \int_0^1 k_2(s,t)\phi_i(s)\phi_j(t)\,ds\,dt.$$

Substituting the above approximation in (1), we get

$$
\begin{aligned}
C^T\Phi(t) &= f(t) + \int_0^t C^T({}_0I^\alpha\Phi(s))\Phi^T(s)K_1\Phi(t)ds + u_0\int_0^t \Phi^T(s)K_1\Phi(t)ds \\
&\quad + \int_0^t C^T({}_0I^\alpha\Phi(s))\Phi^T(s)K_2\Phi(t)dW(s) + u_0\int_0^t \Phi^T(s)K_2\Phi(t)dW(s) \\
&= f(t) + C^T P^\alpha\Big(\int_0^t \Phi(s)\Phi^T(s)ds\Big)K_1\Phi(t) + u_0\Big(\int_0^t \Phi^T(s)ds\Big)K_1\Phi(t) \\
&\quad + C^T P^\alpha\Big(\int_0^t \Phi(s)\Phi^T(s)dW(s)\Big)K_2\Phi(t) + u_0\Big(\int_0^t \Phi^T(s)dW(s)\Big)K_2\Phi(t) \\
&= f(t) + C^T P^\alpha\phi\Big(\int_0^t \mathbf{b}(s)\mathbf{b}^T(s)ds\Big)\phi^T K_1\Phi(t) + u_0\Big(\int_0^t \mathbf{b}^T(s)ds\Big)\phi^T K_1\Phi(t) \\
&\quad + C^T P^\alpha\phi\Big(\int_0^t \mathbf{b}(s)\mathbf{b}^T(s)dW(s)\Big)\phi^T K_2\Phi(t) + u_0\Big(\int_0^t \mathbf{b}^T(s)dW(s)\Big)\phi^T K_2\Phi(t).
\end{aligned}
$$

Let $P^\alpha\phi = Q_1$, $\phi^T K_1 = Q_2$, $\phi^T K_2 = Q_3$, Q_2^i be the ith row of constant matrix Q_2, Q_3^i be the ith row of constant matrix Q_3, R^i be the ith row of the integration operational matrix P and R_s^i be the ith row of the stochastic operational matrix P_s. We have

$$\Big(\int_0^t \mathbf{b}(s)\mathbf{b}^T(s)ds\Big)Q_2\Phi(t) = \begin{pmatrix} R^1\mathbf{b}(t)Q_2^1 \\ \vdots \\ R^m\mathbf{b}(t)Q_2^m \end{pmatrix}\Phi(t) := B_1\Phi(t).$$

Also

$$\Big(\int_0^t \mathbf{b}(s)\mathbf{b}^T(s)dW(s)\Big)Q_3\Phi(t) = \begin{pmatrix} R_s^1\mathbf{b}(t)Q_3^1 \\ \vdots \\ R_s^m\mathbf{b}(t)Q_3^m \end{pmatrix}\Phi(t) := B_2\Phi(t).$$

Then

$$C^T(I - Q_1 B_1 - Q_1 B_2)\Phi(t) \approx f(t) + u_0(\mathbf{b}^T(t)P^T Q_2 + \mathbf{b}^T(t)P_s^T Q_3)\Phi(t) \tag{30}$$

So, by setting

$$A = \begin{pmatrix} \Phi^T(t_1)(I - Q_1 B_1 - Q_1 B_2)^T \\ \vdots \\ \Phi^T(t_m)(I - Q_1 B_1 - Q_1 B_2)^T \end{pmatrix}$$

and

$$F = \begin{pmatrix} f(t_1) + u_0(\mathbf{b}^T(t_1)P^T Q_2 + \mathbf{b}^T(t_1)P_s^T Q_3)\Phi(t_1) \\ \vdots \\ f(t_m) + u_0(\mathbf{b}^T(t_m)P^T Q_2 + \mathbf{b}^T(t_m)P_s^T Q_3)\Phi(t_m) \end{pmatrix},$$

where t_i are the collocation points. We have

$$AC = F, \tag{31}$$

which is a linear system of equations that gives the Legendre wavelets coefficient.

5 Numerical Experiments

To illustrate the proposed method discussed in Sect. 4, we consider the following examples.

Example 51. *Consider the SFIDE (1) with* $f(t) = \frac{t^2}{2} + \frac{\Gamma(2)}{\Gamma(2-\alpha)}t^{1-\alpha}$, $k_1(s,t) = 1$, $k_2(s,t) = 0$, *and* $u_0 = 0$. *For* $\alpha = 0$, $u(t) = -(2+t) + 2e^t$ *is exact solution of (1).*

Example 52. *Consider the SFIDE (1) with* $f(t) = -\frac{t^5 e^t}{5} + \frac{6t^{2.25}}{\Gamma(3.25)}$, $k_1(s,t) = e^t s$, $k_2(s,t) = 0$, *and* $u_0 = 0$. *For* $\alpha = 0.75$, $u(t) = t^3$ *is exact solution of (1).*

Example 53. *Consider the SFIDE (1) with* $f(t) = 0$, $k_1(s,t) = s^2$, $k_2(s,t) = s^3$, *and* $u_0 = 1$. *For* $\alpha = 0$, $u(t) = e^{\frac{t^3}{3} + \int_0^t s^3 dW(s)}$ *is exact solution of (1).*

Let $u_{num}(t_i, l)$ denotes the approximate solution of lth simulation at t_i and $u_{exact}(t_i, l)$ denotes the exact solution of lth simulation at t_i. The efficiency of the proposed method, for Examples 51, 52 and 53 are highlighted in Tables 1, 2 and 3, respectively, which showcase the values of maximum absolute error and root mean square (RMS) error that are defined as

$$\mathbb{E}\|e\|_\infty = \frac{1}{N} \sum_{l=1}^N \max_{1 \le i \le m} |u_{exact}(t_i, l) - u_{num}(t_i, l)|, \tag{32}$$

$$\mathbb{E}\|e\|_{2,m} = \frac{1}{N} \sum_{l=1}^N \sqrt{\frac{1}{m} \sum_{i=1}^m |u_{exact}(t_i, l) - u_{num}(t_i, l)|^2} \tag{33}$$

respectively, where N is total number of simulation and $\mathbb{E}$ is mathematical expectation. For deterministic function $\mathbb{E}\|.\|_\infty = \|.\|_\infty$ and $\mathbb{E}\|.\|_{2,m} = \|.\|_{2,m}$. Table 1 shows the maximum absolute errors obtained for Example 51 via the proposed method discussed in Sect. 4 for $\alpha = 0$ and different values of J, M and m. Table 2 shows the comparison of our method with Gaussian radial basis function (GA RBF) and thin plate splines radial basis function (TBS RBF) in terms of the absolute maximum error and RMS-error obtained for Example 52 with $\alpha = 0.75$ and different values of m. These methods (GA RBF and TBS RBF [11]) need smaller value of shape parameter for higher accuracy which increases the condition number of coefficient matrix and as a result the methods become unstable. However, the method proposed in Sect. 4 has no such behavior. Finally, Table 3 shows the calculation of the mean and standard deviation which are denoted by $\mathbb{E}\|e\|_\infty$ and S_e, respectively, of the maximum absolute error for Example 53 with $\alpha = 0$ and different number of simulation trajectories (N). For different values of N, the upper and lower limit of 95% confidence interval (C.I.) are also listed in Table 3. In Fig. 1, we plot the mean approximate solution and mean exact solution of Example 53 along with 95% confidence interval region for $\alpha = 0$ and $m = 32$ with different values of N.

Table 1. Maximum absolute error in $u(t)$ corresponding to $\alpha = 0$ with different value of J and M.

J	2	2	2	3	3	3	4	4	4
M	2	3	4	2	3	4	2	3	4
$m = 2^{J-1}M$	4	6	8	8	12	16	16	24	32
$\|e\|_\infty$	0.0412	0.0194	0.0113	0.0113	0.0052	0.0030	0.0030	0.0013	0.0008

Table 2. Results of Example 52 for different values of m and $\alpha = 0.75$

m	GA RBF [11]		TBS RBF [11]		Present method			
	$\|e\|_\infty$	$\|e\|_{2,m}$	$\|e\|_\infty$	$\|e\|_{2,m}$	J	M	$\|e\|_\infty$	$\|e\|_{2,m}$
4	–	–	–	–	2	2	$4.39e-2$	$2.53e-2$
10	$8.60e-2$	$7.62e-2$	$3.70e-2$	$4.63e-2$	2	5	$8.30e-3$	$4.10e-3$
20	$4.11e-3$	$7.30e-3$	$5.60e-3$	$4.89e-3$	3	5	$2.20e-3$	$1.0e-3$
32	–	–	–	–	5	2	$9.0052e-4$	$4.0463e-4$
40	$3.5988e-4$	$8.6523e-3$	$4.3341e-4$	$2.3569e-4$	4	5	$3.7848e-4$	$1.5841e-4$

6 Applications

In this section, we present some special cases of the proposed model and their applications in real life examples. SFIDE (1) have many practical applications in scientific field such as physics, finance and biology etc. When $\alpha = 0$, $f(t) = 0$, $k_1(s, t) = \mu$ and $k_2(s, t) = \sigma$, the proposed model reduces in the following form

$$u(t) = u_0 + \int_0^t \mu u(s)ds + \int_0^t \sigma u(s)dW(s), \qquad (34)$$

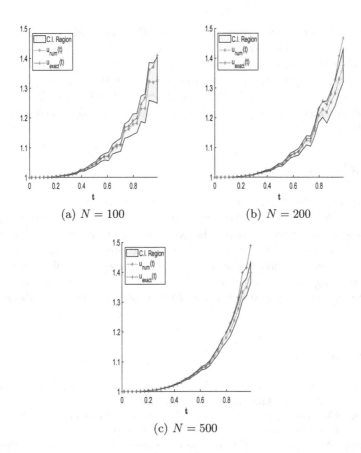

Fig. 1. The trajectory of the approximate solution and exact solution of Example 53 along with 95% confidence interval (C.I) for $M = 2$, $J = 5$, $m = 32$ and $\alpha = 0$.

Table 3. Mean, Standard deviation and mean confidence interval for maximum absolute error in Example (53) with $m = 32$, $J = 5$, $M = 2$ and $\alpha = 0$.

N	$\mathbb{E}\|e\|_\infty$	S_e	95% confidence interval for χ_e	
			Lower	Upper
30	0.2905	0.2246	0.2101	0.3709
100	0.2836	0.2074	0.2429	0.3243
200	0.2712	0.1898	0.2449	0.2975
500	0.2658	0.1687	0.2510	0.2806

or,

$$du(t) = \mu u(t) + \sigma u(t)dW(t). \tag{35}$$

The above stochastic differential equation is called geometric Brownian motion model which is used for modeling the stock prices in the finance [3]. Also in physics, (35) is the Langevin equation with multiplicative noise [4]. The Langevin equation is very important tool in physics for describing many physical process. Suppose $u(t)$ is any physical process described by the (35) then we always get probability distribution $p(u, t|u_0)$ corresponding to (35) which satisfies the Fokker-Plank equation.

7   Conclusion

In this paper, we develop a wavelet collocation method based on the Legendre wavelets to solve SFIDE (1). For this purpose, we compute the deterministic and stochastic operational matrices based on block pulse function. The SFIDE (1) is then converted to a system of linear equations by employing the collocation method and making use of operational matrices. The solution of SFIDE is obtained using our proposed method. In Sect. 5, we solve several examples to indicate the accuracy and efficiency of the proposed method as discussed in Sect. 4.

References

1. Alpert, B.K.: A class of bases in L^2 for the sparse representation of integral operators. SIAM J. Math. Anal. **24**(1), 246–262 (1993)
2. Denisov, S.I., Hänggi, P., Kantz, H.: Parameters of the fractional Fokker-planck equation. EPL (Europhys. Lett.) **85**(4), 40007 (2009)
3. Etheridge, A., Baxter, M.: A Course in Financial Calculus. Cambridge University Press, Cambridge (2002)
4. Kwok, S.F.: Langevin equation with multiplicative white noise: transformation of diffusion processes into the wiener process in different prescriptions. Ann. Phys. **327**(8), 1989–1997 (2012)
5. Li, C., Zeng, F.: Numerical Methods for Fractional Calculus. Chapman and Hall/CRC, Abingdon (2015)
6. Li, Y., Sun, N.: Numerical solution of fractional differential equations using the generalized block pulse operational matrix. Comput. Math. Appl. **62**(3), 1046–1054 (2011)
7. Maleknejad, K., Khademi, A., Lotfi, T.: Convergence and condition number of multi-projection operators by legendre wavelets. Comput. Math. Appl. **62**(9), 3538–3550 (2011)
8. Maleknejad, K., Khodabin, M., Rostami, M.: Numerical solution of stochastic-volterra integral equations by a stochastic operational matrix based on blockpulse functions. Math. Comput. Model. **55**(3–4), 791–800 (2012)
9. Mehra, M.: Wavelets Theory and Its Applications. Springer, Heidelberg (2018). https://doi.org/10.1007/978-981-13-2595-3

10. Mirzaee, F., Samadyar, N.: Application of orthonormal Bernstein polynomials toconstruct a efficient scheme for solving fractional stochasticintegro-differential equation. Optik **132**, 262–273 (2017)
11. Mirzaee, F., Samadyar, N.: On the numerical solution of fractional stochastic integro-differential equations via meshless discrete collocation method based on radial basis functions. Eng. Anal. Bound. Elements **100**, 246–255 (2019)
12. Taheri, Z., Javadi, S., Babolian, E.: Numerical solution of stochastic fractional integro-differential equation by the spectral collocation method. J. Comput. Appl. Math. **321**, 336–347 (2017)
13. Venkatesh, S., Ayyaswamy, S., Balachandar, S.R.: The legendre wavelet method for solving initial value problems of Bratu-type. Comput. Math. Appl. **63**(8), 1287–1295 (2012)

A Direct High-Order Curvilinear Triangular Mesh Generation Method Using an Advancing Front Technique

Fariba Mohammadi[1(✉)], Shusil Dangi[2], Suzanne M. Shontz[1], and Cristian A. Linte[2]

[1] The University of Kansas, Lawrence, KS 66045, USA
{fariba_m,shontz}@ku.edu
[2] Rochester Institute of Technology, Rochester, NY 14623, USA
{sxd7257,calbme}@rit.edu

Abstract. In this paper, we propose a novel method of generating high-order curvilinear triangular meshes using an advancing front approach. Our method relies on a direct approach to generate meshes on geometries with curved boundaries. Our advancing front method yields high-quality triangular elements in each iteration which omits the need for post-processing steps. We present several numerical examples of second-order curvilinear triangular meshes of patient-specific anatomical models generated using our technique on boundary meshes obtained from biomedical images.

Keywords: High-order mesh generation · Advancing front · Curvilinear triangular mesh

1 Introduction

The use of high-order methods has attracted the interest of the scientific computing community, thanks to their ability to deliver highly-accurate solutions of partial differential equations (PDEs) at a low computational cost. However, while working with curved boundaries, the mesh used with high-order PDE solvers needs to be a high-order mesh that accurately captures the curvature of the geometries [19,22]. A high-order mesh is composed of both straight-sided and curved elements, depending on the curvature of the geometric domain. One major challenge lies in generating high-order meshes that perfectly capture the curved boundaries; thus, to date, there are not many methods that can generate robust high-order meshes [22].

The work of the first author was funded by NSF OAC grant 1808553. The work of the second and fourth authors was supported by NIH grant NIGMS R35GM128877 and NSF grant OAC 1808530. The work of the third author was funded in part by NSF grants OAC 1808553 and CCF 1717894.

V. V. Krzhizhanovskaya et al. (Eds.): ICCS 2020, LNCS 12138, pp. 72–85, 2020.
https://doi.org/10.1007/978-3-030-50417-5_6

There are two categories of methods for generating a high-order mesh. The first category consists of direct methods, where a high-order mesh is generated directly from the curved geometry. To the best of our knowledge, no direct methods are currently available.

The second category includes *a posteriori* methods, which are the most commonly used approaches for generating high-order meshes. Here additional nodes are first added to the low-order mesh, then the newly-added boundary nodes are moved to conform to the curved boundary. Finally, the interior nodes are moved to their new positions [10,19–21]. These methods deform the linear mesh either using optimization [9,11,20,21] or based on the solution of PDEs [8,16,19,24], e.g., a linear elasticity approach [24], a nonlinear elasticity approach [19], or other strategies [8,16]. The main challenge associated with this approach is to obtain a valid high-order mesh [22], since the boundary curving step can create tangled elements in the mesh. In this approach, the geometry of the desired high-order mesh is required to represent the curved boundary. This is often obtained from computer-aided design (CAD) files, but in the case of patient-specific anatomical models, such CAD files are not available.

Our proposed method uses a direct approach to generate high-order curvilinear triangular meshes. Our aim is to be able to generate meshes not only from CAD files, but also from other types of boundary representations, such as patient-specific 1D boundary meshes obtained from medical images. Several algorithms for generating unstructured triangular meshes have been developed over the years; among them, the Delaunay triangulation-based methods and the advancing front-based methods are most popular [17]. Here we use an advancing front approach [12–15] to generate high-order curvilinear triangular meshes.

The novelty of our work lies in our method's ability to generate high-order meshes directly from curved boundaries. This is the first direct approach for high-order mesh generation. Our method does not require a post-processing step, such as mesh untangling, as it generates each element as a valid, high-quality element. Since our method uses a direct approach instead of an *a posteriori* approach, it can generate high-quality meshes on patient-specific models obtained from medical images where no CAD representation is available, serving as a basis for generating meshes for more complex geometries. Hence, our patient-specific meshes can accurately represent complex anatomies, and using these meshes, one will be able to deliver highly-accurate solutions when solving PDEs.

In Sect. 2, we describe our mesh generation method. Section 3 shows the numerical results of our method on several examples. Finally, in Sect. 4, we summarize our results and discuss limitations and future directions for this work.

2   High-Order Curvilinear Triangular Mesh Generation

In this section, we describe a high-order curvilinear triangular mesh generation algorithm using an advancing front approach. Our method is currently designed to yield second-order curvilinear triangular meshes. In contrast to the traditional high-order mesh generation methods, where post-processing is often a required step, our direct high-order mesh generation method aims to generate high-quality elements in each iteration, so that post-processing is not required.

In the proposed algorithm, we start with a high-order 1D boundary mesh and use it to generate a curvilinear high-order triangular mesh. First, we assign the initial boundary mesh as the initial active front. As the method progresses and new elements are generated, we update the active front by deleting the edges that are already used to generate the triangles and adding the new edges that are created after generating the triangles. Next, we calculate the lengths of the boundary mesh edges. Since the edges of the boundary mesh are curved, we numerically approximate the lengths of the curved edges by dividing them into smaller sections. Then we use shape functions for a one-dimensional second-order Lagrange element to calculate the length of each section. Since the edges to be approximated are quite short, dividing them into smaller sections essentially yields linear segments, therefore rendering this approximation method more accurate and more efficient than calculating the edge's arc length.

To ensure better quality elements from the start, our goal is to generate triangles as close to equilateral triangles as possible. To this end, we first average the lengths of all the curved boundary edges and denote this average length by L_{avg}. We set an upper bound L_{max} on the average length and calculate the upper bound as $L_{max} = b L_{avg}$, where b is a constant. After calculating L_{max}, we use that as the side of an ideal equilateral triangle and calculate the height h of that triangle. This height can be changed by varying the b value in L_{max}. The higher this value, the longer the height of the triangle will be. For our meshes, we use b values ranging from 0.78–0.8.

Once we have calculated h, we start generating the triangular elements from the boundary edges. To this end, we select the first edge from the active front and insert a vertex a at a distance h from the midpoint of the selected edge. To ensure that the vertex is inserted on the correct side of the boundary, we calculate direction normals for the edges and insert the vertex in the direction of the inward normal vector of the boundary mesh. Next, we search for other suitable candidate vertices within a specific radius, $r = \alpha h$, of a. Here, α is a constant that can be varied according to the size of the geometry and mesh. Since a fixed h is used to generate all triangular elements, element size uniformity is ensured.

Figure 1(a) shows the vertex a, height h, and search radius r. The area shown by the gray circle represents the search area for more candidate vertices. Figure 1(b) shows the first curvilinear triangular element generated in the mesh. Figure 1(c) shows a candidate triangle that intersects with an existing triangular element. Note that only the low-order vertices can serve as candidates for the third vertex of the triangle. For each candidate triangle, we perform several validity checks and calculations to ensure we generate the best possible triangle from the candidate vertices available. Moreover, we also perform an intersection test to ensure that we do not have triangular elements that intersect with an existing edge or triangle. We then calculate the scaled Jacobian and equiangular skewness to measure distortion of the curved and straight-sided triangular

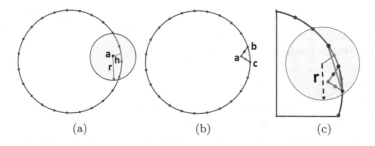

Fig. 1. Boundary mesh of a circle: (a) vertex a is inserted at a distance h, and a search for more candidate vertices is conducted within radius r; (b) first triangular element bac, and (c) an intersecting candidate triangle shown in orange. (Color figure online)

elements. Considering the boundary edge as the base of our triangular element, we measure the lengths of the two sides of the triangle to make sure one side of the selected triangle is not very long compared to the other side. These tests are described in Sect. 2.1 in detail.

If a candidate triangle passes the validity and quality checks, that vertex is inserted (as a new vertex) or selected (as a pre-existing vertex) as the third vertex of the triangle. For every triangle generated, if it has an equiangular skewness value greater than a specific value β, we then perform an edge swap on that triangle to ensure that there are no skinny triangles. Once the low-order vertex is finalized, we generate the high-order vertices for the newly-created triangle edges. For a second-order mesh, the high-order vertices will be the midpoints of each edge. Next, we update the active front by deleting the initially selected edge and adding the newly-created triangle edges. We repeat these steps until our active front is empty. Algorithm 1 gives the pseudocode of our second-order curvilinear triangular mesh generation method.

2.1 Triangle Validity and Quality Checks

Our method searches for suitable candidate vertices and uses several validity and quality checks to select the vertex that will generate the best quality triangle. We conduct the checks in the specified order, so that we can remove the unsuitable candidates one-by-one and preserve the best possible candidates. The unsuitable candidates are those which would generate triangles that intersect with an existing edge or triangle, or have a negative scaled Jacobian. Once we delete all the invalid candidate vertices, we use our triangle selection algorithm described in Algorithm 2 to select the best quality triangle from the remaining candidates. In this section, we summarize these checks.

Intersection Check: Once we identify the candidate vertices from our search, we conduct an intersection check on each triangle generated with those vertices. Our aim is to make sure we do not have candidate triangles that intersect with an existing edge or triangle. We use the polyshape overlap function of Matlab R2018a to perform this check. If we find a candidate vertex that, if selected as the

---

Algorithm 1: High-order curvilinear triangular mesh generation

---

Input: Boundary edges as active front
Output: Second-order curvilinear triangular mesh
Calculate L, L_{avg}, L_{max}, and h
if active front is not empty **then**
    for each edge **do**
        if the edge exists in the active front **then**
            1. Calculate the direction normals
            2. Insert a point a at distance h inside the geometry (Fig. 1)
            3. Search for more candidate vertices within radius r of a (Fig. 1)
            4. Run intersection test
            5. Calculate scaled Jacobian
            6. Calculate equiangular skewness
            7. Considering the selected edge as the base of the triangle,
               calculate the two side lengths of each candidate triangle
            if triangle selection criteria are met as shown in Algorithm 2 **then**
               8. Generate triangle
               if triangle skewness $> \beta$ and edge swap keeps adjacent triangle
               skewness < 0.85, **then**
                | 9. Perform edge swap
               end
               10. Insert high-order vertices
               11. Update active front
            end
        end
    end
end

---

third vertex of the triangle, would generate a triangular element that intersects with an existing edge or triangle in the mesh, we discard that candidate.

Scaled Jacobian Calculation: High-order meshes consist of both straight-sided and curved elements depending on the geometry. To measure the distortion or quality of our curvilinear triangular elements, we use the scaled Jacobian quality metric [19]. The scaled Jacobian is defined as:

$$\frac{\min J(\boldsymbol{\xi})}{\max J(\boldsymbol{\xi})}, \tag{1}$$

where $J(\boldsymbol{\xi}) = \det(\partial \mathbf{x}/\partial \boldsymbol{\xi})$. This is the Jacobian of the mapping from the reference coordinate $\boldsymbol{\xi}$ to the physical coordinate $\mathbf{x}$. Figure 2 shows a second-order triangular element in both physical coordinates and reference coordinates. Scaled Jacobian values can range from $-\infty$ to 1. For a straight-sided element, the scaled Jacobian value is 1. A scaled Jacobian value of 1 does not necessarily indicate a good quality element, since a skinny, straight-sided triangle can also have a scaled Jacobian of 1. A negative scaled Jacobian indicates an inverted element. While the scaled Jacobian is constant for straight-sided elements, for curved high-order elements, a positive near-zero scaled Jacobian value would indicate

---

Algorithm 2: Selection of the best triangular element

Input: Candidates that pass validity checks
Output: The most suitable candidate vertex
if scaled Jacobian $\in$ (0,1) **then**
    1. Select the candidate that generates the shortest triangle side length
    if selected candidate's skewness $> \beta$ **then**
        2. Select the candidate that generates the triangle with highest scaled
        Jacobian
    end
else
    if scaled Jacobian $= 1$ **then**
        1. Select the candidate that generates the shortest triangle side length
        if selected candidate's skewness $> \beta$ **then**
            2. Select the candidate that generates the triangle with lowest
            skewness
        end
    end
end
3. A candidate vertex is selected
if there is another suitable candidate vertex within a distance, l, of the
selected vertex **then**
    4. Check Delaunay empty circumcircle property
    if the selected vertex is inside the circumcircle of the triangle made with
    the newly found vertex **then**
        5. Discard the selected vertex and select the other candidate
    end
end

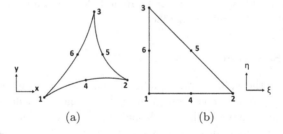

(a)   (b)

Fig. 2. Second-order triangular element: (a) in physical coordinates; (b) in reference coordinates.

significant distortion. We calculate the scaled Jacobian using the shape functions for a second-order Lagrange triangle and the high-degree Gaussian quadrature rules developed by Dunavant [6] for triangles. We use a polynomial of degree 8 with 16 Gaussian points and weights. We perform the scaled Jacobian calculation on the updated candidates that we obtain after performing the intersection check. If we obtain a candidate with a negative scaled Jacobian, that candidate is no longer considered.

Equiangular Skewness Calculation: To detect skinny triangles and to measure the distortion of straight-sided elements, we use equiangular skewness. This angular measure of element quality assesses how close a triangular element is to an equilateral triangle [2]. Since we have curvilinear triangles, we first measure the angles between the tangent lines of the curves using an analysis similar to that described in [22], then use those angles to measure the equiangular skewness, which is given by:

$$\max \left[\frac{\theta_{max} - \theta_e}{180 - \theta_e}, \frac{\theta_e - \theta_{min}}{\theta_e} \right], \tag{2}$$

where

θ_{min} = smallest angle of the element,

θ_{max} = largest angle of the element, and

θ_e = angle for equiangular element, i.e., 60° for equilateral triangles.

For triangular elements, the equiangular skewness should not exceed 0.85.

Triangle Side Length Calculation: We consider the boundary edge selected from the active front as the base of the triangular element and measure the two side lengths of the candidate triangles. To ensure that the triangular elements maintain a uniform size throughout the mesh, we use the relationship $l_1 \leq \gamma \, l_2$, where l_1 and l_2 are the lengths of the two non-base sides of the triangle, and γ is a constant. The value of γ can be changed according to the geometry and mesh element size. If one side of a candidate triangle is longer than γ times the other side, that triangle is no longer considered.

2.2   Triangle Selection

Once we complete the validity and quality checks, we use the results to select the best triangular element based on the scaled Jacobian, skewness, and triangle side lengths. Since we have both straight-sided and curved elements, we cannot use only the scaled Jacobian to measure the quality of the elements. To avoid skinny triangles, we consider candidate triangles that have a skewness value less than β. The value of β can be changed according to the geometry. If performing an edge swap makes the skewness value of an adjacent triangle greater than 0.85, then we do not perform one. Our method selects the best quality triangle based on the following two cases.

Curvilinear Triangles: First, we select the triangle with the shortest side length that meets the skewness requirement. If that triangle does not meet the requirement, we select the triangle with the maximum scaled Jacobian. If there is another candidate vertex very close to the selected triangle that meets the skewness requirement and also has a scaled Jacobian higher than the previously selected one, we select this other candidate. To find such vertices, we search within a distance, l, of the currently selected vertex. This search distance can be varied according to the size and shape of the elements. We do this to avoid creating skinny triangles in the future.

Straight-sided Triangles: Here, we also first select the triangle that has the shortest side length, provided it meets the skewness requirement. If that triangle does not meet the requirement, we select the triangle with minimum skewness. Here, we cannot rely only on the scaled Jacobian, as for all straight-sided elements the scaled Jacobian will be 1. Again, if there is another candidate vertex very close to the selected triangle that meets the skewness requirement, we select this other candidate vertex instead.

For both cases, we check the edge lengths of the two non-base sides of the triangles to make sure one is not too long or short compared to the other side. We prioritize selecting vertices that already exist in the mesh over inserting new vertices if multiple vertices pass the validity and quality checks and if the vertices are very close to each other. If our selected vertex is a new vertex, then we search within a distance l of that vertex to determine whether there is another suitable candidate vertex that already exists in the mesh. If yes, then we construct the circumcircle of the triangle made with the pre-existing vertex and check to see if the Delaunay circumcircle for that triangle is empty. If the new vertex lies inside the circumcircle, we select the triangle made with the pre-existing vertex. Algorithm 2 gives the pseudocode for the triangle selection process.

3   Numerical Results

In this section, we demonstrate the results from applying our mesh generation algorithm to generate several second-order curvilinear triangular meshes on various patient-specific models. We show how the initial front advances to create the final mesh, as well as how performing edge swaps avoid the generation of skinny triangles in the mesh. We also report the wall-clock time required to generate the meshes. The method was run using Matlab R2018a, and the execution times were measured on a machine with 16 GB of RAM and an Intel(R) Core(TM) i7-6700HQ CPU. All mesh visualizations were conducted using Gmsh [10].

For our examples, we use two different types of patient-specific geometries obtained from medical images. Our first set of examples consists of patient-specific cardiac geometries made available through several medical image segmentation challenges - the Left Ventricle Segmentation Challenge (LVSC) [7,23] available through the Statistical Atlases and Computational Modeling of the Heart and the Automatic Cardiac Diagnostic Challenge (ACDC) [3]. The cardiac image dataset consisted of a stack of 2D image slices and their associated endocardial and epicardial contours at two cardiac phases - diastole and systole - extracted using the distance map regularized convolutional neural network formulation by Dangi et al. in [5]. We use 1D surface meshes of the patient myocardium obtained from magnetic resonance imaging (MRI). We show meshes of the myocardium both at the maximum contraction phase (systole) and maximum expansion phase (diastole) of the heart. Also, for both cases, we show results for a few different MRI slices.

For our second set of examples, we use boundary meshes of the brain ventricles of a patient with hydrocephalus [18] obtained from computed tomography (CT) scan images. The final meshes show one pre-treatment and two post-treatment brain ventricles for a hydrocephalus patient who was treated by shunt insertion.

Finally, we use our method to generate triangular meshes of a pair of normal human lungs. The 1D boundary mesh for the lungs are generated from a chest CT scan image [1] using Seg3D [4].

Since our method takes a high-order curved surface mesh as input and the initial patient heart and brain meshes were straight-sided, low-order meshes, we use Gmsh [10] to generate the second-order 1D meshes from the low-order 1D meshes and to refine the meshes if necessary. Next, we use cubic spline interpolation to obtain a curved boundary mesh. We then determine the new positions of the high-order vertices on the curved boundary. For a second-order mesh, these are the midpoints of the newly-curved edges. We use this updated high-order curvilinear boundary mesh as the input for our method. Figure 3(a) shows a straight-sided, low-order mesh; Fig. 3(b) shows a second-order curvilinear coarse boundary mesh, and Fig. 3(c) shows a second-order curvilinear fine boundary mesh of the myocardium.

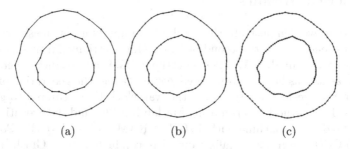

(a)   (b)   (c)

Fig. 3. 1D boundary mesh of patient myocardium: (a) low-order straight-sided coarse mesh; (b) second-order curvilinear coarse mesh; (c) second-order curvilinear fine mesh.

Figure 4 shows an example of how the active front is advancing in a counter-clockwise (CCW) direction to generate the triangular elements. The high-order 1D boundary mesh is used as the active front. As a new triangular element is generated, the active front advances, and this continues until all the vertices are connected in the mesh and the active front is empty. The red arrow in Fig. 4(a) represents the CCW direction in which the initial front is advancing. Figure 4(b) shows the first ring of triangular elements generated in the mesh. Figure 4(c) shows the two opposing directions of the fronts before they are merging.

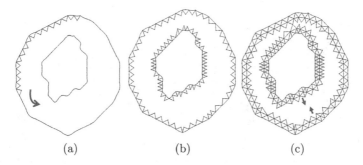

(a)   (b)   (c)

Fig. 4. Advancing front high-order mesh generation: (a) the active front is advancing in a CCW direction to create elements; (b) first ring of triangular elements; (c) the red arrows represent the directions of the two merging fronts, and the method ensures that the merging fronts do not cause element intersections. (Color figure online)

Depending on the geometry of the input mesh, the active fronts can progress from different directions. When the fronts start to merge, the different-sized edges on the various fronts make it challenging to maintain good quality elements. This can create skinny triangles, like needles and caps. We use edge swaps to avoid creating such elements in our mesh, the success of which is illustrated in Fig. 5. Figure 5(a, c) show two regions of the mesh before edge swapping is performed. The potential skinny triangles can be observed. Figure 5(b, d) show the same regions after edge swapping is performed to avoid skinny triangle generation.

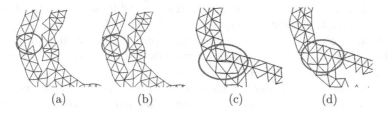

(a)   (b)   (c)   (d)

Fig. 5. Edge swapping to avoid skinny triangles: (a) before edge swap; (b) after edge swap; (c) before edge swap; (d) after edge swap.

Figure 6 shows the results of our mesh generation algorithm for various myocardia, at expansion (diastole) and contraction (systole) of the heart from four different MRI images. For diastole, we use a search radius of $1.5h$, $\gamma = 1.5$, and $\beta = 0.5$. For systole, we use a search radius of $1.4h$, $\gamma = 1.2$, and $\beta = 0.5$. Edge swapping is performed if a triangular element has a skewness greater than 0.5. Figure 6(a, b) show the meshes at diastole, and Fig. 6(c, d) show the meshes at systole. The meshes represent the patient-specific geometries accurately, and there are no inverted elements or skinny triangles in the meshes. The runtimes and element quality information are shown in Fig. 6(e).

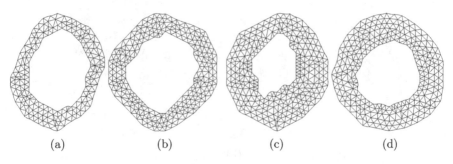

(a)   (b)   (c)   (d)

Example	# elements	Runtime(s)	Scaled Jacobian		Skewness	
			Min	Max	Min	Max
Patient 1 at diastole (slice 16)	320	303	0.416	1.000	0.000	0.705
Patient 2 at diastole (slice 5)	486	383	0.392	1.000	0.000	0.623
Patient 1 at systole (slice 16)	542	478	0.424	1.000	0.000	0.688
Patient 3 at systole (slice 3)	504	422	0.696	1.000	0.000	0.662

(e)

Fig. 6. Second-order triangular meshes of three patients' myocardia at various times in the heartbeat cycle: (a) patient 1 at diastole; (b) patient 2 at diastole; (c) patient 1 at systole; (d) patient 3 at systole; (e) mesh quality metrics and algorithm runtime statistics.

Figure 7 shows our triangular meshes for the brain ventricles of a hydrocephalus patient before and after the shunt insertion treatment. Before the treatment was performed, the enlarged brain ventricles due to the build-up of cerebrospinal fluid (CSF) inside the ventricles (i.e., the white area) can be observed in Fig. 7(a). Post treatment, the condition of the ventricles was observed at two different time points: six months and one year post treatment. It is observed in Fig. 7(b, c) that the ventricle sizes are gradually reducing post treatment. For these examples, we use a search radius of $1.5h$, $\gamma = 1.5$, and $\beta = 0.5$ to generate the meshes. There are no inverted elements in these meshes. For the two post-treatment meshes, there are a total of four triangular elements that are close to being skinny triangles, but none have a skewness value greater than 0.85. The runtimes and element quality information are shown in Fig. 7(d). Comparing our pre-treatment mesh with the mesh generated in [18], we observe that the low-order mesh had 8166 elements, whereas our high-order mesh has 1194 elements. This indicates that solving PDEs will require less computational time when employing our meshes. However, since each of our meshes has a different number of elements and different vertex connectivity, solving a PDE for the dynamic problem would require re-interpolation of the solution to go from one mesh to another. Hence, while our meshes will reduce the computational cost and deliver accurate results while solving PDEs on static meshes, their use is not designed for dynamic problems.

Figure 8 shows the triangular meshes of the right and left lungs. For these meshes, we use a search radius of $1.5h$, $\gamma = 1.2$, and $\beta = 0.5$. The lung meshes

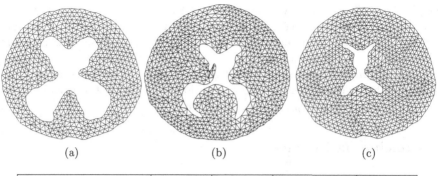

(a)   (b)   (c)

			Scaled Jacobian		Skewness	
Example	# elements	Runtime(s)	Min	Max	Min	Max
Pre-treatment	1194	1002	0.544	1.000	0.000	0.684
Post-treatment (period 1)	1695	1632	0.097	1.000	0.000	0.842
Post-treatment (period 2)	1703	1623	0.266	1.000	0.000	0.738

(d)

Fig. 7. Second-order triangular mesh of the brain ventricles of a hydrocephalus patient: (a) pre-treatment; (b) post-treatment period 1; (c) post-treatment period 2; (d) mesh quality metrics and algorithm runtime statistics.

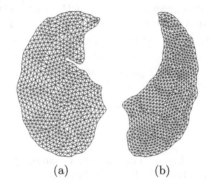

(a)   (b)

			Scaled Jacobian		Skewness	
Example	# elements	Runtime(s)	Min	Max	Min	Max
Right lung	1271	1212	0.353	1.000	0.000	0.750
Left lung	1450	1427	0.291	1.000	0.000	0.667

(c)

Fig. 8. Second-order triangular mesh of right and left lungs: (a) right lung; (b) left lung; (c) mesh quality metrics and algorithm runtime statistics.

shown in Fig. 8(a, b) have no inverted elements or skinny triangles, and they accurately capture the patient-specific geometry of the lungs.

The quality of our meshes can be controlled by changing the search radius r, skewness threshold β, and the γ value, which controls the lengths of the triangle sides. This versatility is an important feature, since for patient-specific geometries obtained from medical images, the boundary mesh can have different-sized elements. We have observed that for finer meshes, a smaller search radius and a smaller γ value produce good results, whereas for coarser meshes, we need to search within a larger radius to obtain reasonable candidate vertices (and accordingly triangles). There is also the option of changing the height of the triangles by altering the L_{max} value, if necessary.

4   Concluding Remarks

In this paper, we present a new method of generating high-order curvilinear triangular meshes directly from curved geometries. Our method does not require a post-processing step, such as mesh untangling, since we generate each element as a high-quality valid element. Our proposed method can be used to generate curvilinear triangular meshes from patient-specific epicardial and endocardial contours, brain ventricular contours, and lung contours, among others, extracted via segmentation from medical images, such as MRI or CT images. Our mesh generation method can be used as a basis to generate more complex meshes on challenging geometries. To this end, we plan to extend this method to 3D to solve more real-world problems.

We note that, since we implemented our method in Matlab, our implementation has a larger than necessary runtime. Our future work will focus on implementing the method in C++ to reduce the runtime and to be able to work with larger meshes. Also, we only used edge swapping to remove possible skinny triangles from the mesh; an edge collapse operation can also be included to further avoid such triangles.

References

1. Radiologic Images of the Lungs, The Internet Pathology Laboratory for Medical Education. http://ar.utmb.edu/webpath/radiol/pulmrad/pulm004.htm. Accessed 14 Apr 2020
2. Skewness Calculation for 2D Elements. https://www.engmorph.com/skewness-finite-elemnt. Accessed 14 Apr 2020
3. Bernard, O., Lalande, A., Zotti, C., et al.: Deep learning techniques for automatic MRI cardiac multi-structures segmentation and diagnosis: Is the problem solved? IEEE Trans. Med. Imaging **37**(11), 2514–2525 (2018)
4. CIBC: Seg3D: Volumetric Image Segmentation and Visualization. Scientific Computing and Imaging Institute (SCI) (2016). http://www.seg3d.org
5. Dangi, S., Yaniv, Z., Linte, C.: A distance map regularized CNN for cardiac cine MR image segmentation. Med. Phys. **46**(12), 5637–5651 (2019)
6. Dunavant, D.: High degree efficient symmetrical Gaussian quadrature rules for the triangle. Int. J. Numer. Methods Eng. **21**(6), 1129–1148 (1985)
7. Fonseca, C.G., Backhaus, M., Bluemke, D.A., et al.: The cardiac atlas project-an imaging database for computational modeling and statistical atlases of the heart. Bioinformatics **27**(16), 2288–2295 (2011)

8. Fortunato, M., Persson, P.O.: High-order unstructured curved mesh generation using the Winslow equations. J. Comput. Phys. **307**, 1–14 (2016)
9. Gargallo-Peiró, A., Roca, X., Peraire, J., Sarrate, J.: Optimization of a regularized distortion measure to generate curved high-order unstructured tetrahedral meshes. Int. J. Numer. Methods Eng. **103**(5), 342–363 (2015)
10. Geuzaine, C., Remacle, J.F.: Gmsh: a 3-D finite element mesh generator with built-in pre-and post-processing facilities. Int. J. Numer. Methods Eng. **79**(11), 1309–1331 (2009)
11. Karman, S.L., Erwin, J.T., Glasby, R.S., Stefanski, D.: High-order mesh curving using WCN mesh optimization. In: Proceedings of the 46th AIAA Fluid Dynamics Conference, p. 3178 (2016)
12. Lo, S.H.: Volume discretization into tetrahedra-ii. 3D triangulation by advancing front approach. Comput. Struct. **39**(5), 501–511 (1991)
13. Löhner, R., Parikh, P.: Generation of three-dimensional unstructured grids by the advancing-front method. Int. J. Numer. Methods Fluids **8**(10), 1135–1149 (1988)
14. Mavriplis, D.J.: An advancing front Delaunay triangulation algorithm designed for robustness. J. Computat. Phys. **117**(1), 90–101 (1995)
15. Merriam, M.: An efficient advancing front algorithm for Delaunay triangulation. In: Proceedings of the 29th Aerospace Sciences Meeting, p. 792 (1991)
16. Moxey, D., Ekelschot, D., Keskin, Ü., Sherwin, S.J., Peiró, J.: High-order curvilinear meshing using a thermo-elastic analogy. Comput.-Aided Des. **72**, 130–139 (2016)
17. Owen, S.J.: A survey of unstructured mesh generation technology. In: Proceedings of the 7th International Meshing Roundtable, pp. 239–267 (1998)
18. Park, J., Shontz, S.M., Drapaca, C.S.: A combined level set/mesh warping algorithm for tracking brain and cerebrospinal fluid evolution in hydrocephalic patients. In: Image-Based Geometric Modeling and Mesh Generation, pp. 107–141. Springer (2013). https://doi.org/10.1007/978-94-007-4255-0_7
19. Persson, P.O., Peraire, J.: Curved mesh generation and mesh refinement using Lagrangian solid mechanics. In: Proceedings of the 47th AIAA Aerospace Sciences Meeting including the New Horizons Forum and Aerospace Exposition, p. 949 (2009)
20. Roca, X., Gargallo-Peiró, A., Sarrate, J.: Defining quality measures for high-order planar triangles and curved mesh generation. In: Proceedings of the 20th International Meshing Roundtable, pp. 365–383. Springer (2011). https://doi.org/10.1007/978-3-642-24734-7_20
21. Ruiz-Gironés, E., Sarrate, J., Roca, X.: Generation of curved high-order meshes with optimal quality and geometric accuracy. In: Proceedings of the 25th International Meshing Roundtable, Procedia Engineering, vol. 163, pp. 315–327 (2016)
22. Stees, M., Dotzel, M., Shontz, S.M.: Untangling high-order meshes based on signed angles. In: Proceedings of the 28th International Meshing Roundtable, pp. 267–282. Zenodo (2020)
23. Suinesiaputra, A., Cowan, B.R., Al-Agamy, A.O., et al.: A collaborative resource to build consensus for automated left ventricular segmentation of cardiac MR images. Med. Image Anal. **18**(1), 50–62 (2014)
24. Xie, Z.Q., Sevilla, R., Hassan, O., Morgan, K.: The generation of arbitrary order curved meshes for 3D finite element analysis. Comput. Mech. **51**(3), 361–374 (2013)

Data-Driven Partial Differential Equations Discovery Approach for the Noised Multi-dimensional Data

Mikhail Maslyaev, Alexander Hvatov[✉], and Anna Kalyuzhnaya

ITMO University, Kronsersky pr. 49, 197101 St. Petersburg, Russia
alex_hvatov@corp.ifmo.ru

Abstract. Data-driven methods provide model creation tools for systems, where the application of conventional analytical methods is restrained. The proposed method involves the data-driven derivation of a partial differential equation (PDE) for process dynamics, which can be helpful both for process simulation and studying. The paper describes the progress made within the PDE discovery framework. The framework involves a combination of evolutionary algorithms and sparse regression. Such an approach gives more versatility in comparison with other commonly used methods of data-driven partial differential derivation by making fewer restrictions on the resulting equation. This paper highlights the algorithm features which allow the processing of data with noise, which is more similar to the real-world applications of the algorithm.

Keywords: Data-driven modelling · PDE discovery · Evolutionary algorithms · Sparse regression · Spatial fields · Physical measurement data

1 Introduction

The increasing quality and quantity of measurement techniques and the emerging reliable sets of high-resolution data of physical processes in the current years give a hand to the advances of data-driven modelling (DDM). The ability to simulate complex processes, neglecting a lack of knowledge about the underlying structure of the system, can be vital for the development of models in such spheres of science as biology, medicine, materials technology, and metocean studies. In general, data-driven modelling involves the development of complete models from various fields of measurements, using means of statistics and machine learning algorithms. However, on some occasions, DDM can enhance the existing models with the addition of supplementary expressions or by a refinement of weight values [1]. In the fluid dynamics science and hydrometeorology, development of surrogate models is the most common application of data-driven algorithms.

The majority of modern data-driven algorithms of equation derivation are based on the artificial neural networks [4,8,9]. However, this approach in the majority of cases leads to the non-interpretable models. It means that the model

© Springer Nature Switzerland AG 2020
V. V. Krzhizhanovskaya et al. (Eds.): ICCS 2020, LNCS 12138, pp. 86–100, 2020.
https://doi.org/10.1007/978-3-030-50417-5_7

that is represented by the neural network acts as the black box, while the results of the prediction or the modeling overall may be used to solve the particular problem.

On the contrary, the regression models may be interpretable, while the problem is solved with moderate quality. In this paper, the approach that involves the derivation of governing differential equations for the dynamic system is described. This approach may be considered as the compromise between the quality of the neural networks and the simplicity of the pure regression models. The predictions for the future state of the system can be acquired by solving the resulting equation. This statement of the problem corresponds with the specifics of the metocean study, in which the systems can usually be defined by single or series of partial differential equations.

In contrast to the other similar algorithms, which also use sparse regression [2,10–12], the proposed one does not require the construction of token libraries, from which the equation terms are selected. The developed method is focused on the construction of terms from simple tokens that contain the original function and its derivatives of orders selected in a specified range. Additionally, unlike other groups of methods, primarily presented by artificial neural networks, the proposed method does not assume the presence of the first-order time derivative, giving it both versatility in selecting the type of time dynamics and the opportunity to select the steady-state of a system, if it fits data the best. In the current work, the algorithm for a single equation case is developed.

The previous works [5] have shown the ability of the described method to derive the correct structure of the equation for the case of the evolution equations with one spatial coordinate, even when the noise is added, and for the synthetic noiseless case of the two-dimensional wave equation. In this paper, the algorithm is extended to the cases with higher dimensionality and for the noised input fields. This problem statement mimics one of the possible application of the framework, that is connected with the derivation of governing equations for data-driven modelling of the fluids dynamics, that are connected with hydrometeorological studies.

In the paper, the particular problem of the noise in the higher-dimensional data is considered. Usually, the fact that the differentiation error increases exponentially when the dimensionality of the problem grows is ignored in the references, and only the one-dimensional cases are considered. Paper is organised as follows: in Sect. 2 the problem of the data-driven PDE discovery is briefly described. Sect. 3 presents the additions to the method described in the previous article [5] that allow dealing with the higher-dimension data-driven PDE discovery. In Sect. 4, numerical examples on the synthetic data, as well as on the real data, are shown. Sect. 5 concludes the paper.

2   Problem Statement

The class of problems, which can be solved by the described method, can be summarized as follows: the process, which involves scalar field u, occurring in the area Ω, is governed by the partial differential Eq. 1. However, there is no a priori information about the dynamics of the process except the fact, that it can be described by some form of PDE (for simplicity, we consider temporally varying 2D field case, however, the problem could be formulated for an arbitrary field).

$$\begin{cases} F(u, \frac{\partial u}{\partial x_1}, \frac{\partial u}{\partial x_2}, ..., \frac{\partial u}{\partial t}, \frac{\partial^2 u}{\partial x_1^2}, \frac{\partial^2 u}{\partial x_2^2}, ..., \frac{\partial^2 u}{\partial t^2}, ...) = 0; \\ G(u) = 0, \ u \in \Gamma(\Omega) \times [0, T]; \end{cases} \tag{1}$$

From the area $\Omega \times [0, T]$ a set of samples $U = \{u_1, u_2, ..., u_n\}$, where $u_i = u(x_1^{(i)}, x_2^{(i)}, t_i)$ is the function value at the arbitrary point $(x_1^{(i)}, x_2^{(i)}, t_i) \in \Omega \times [0, T]$, is collected. There are no strict limitations for the distribution of the sample collection points in the area, but the further requirement of the derivative calculation makes the case of stationary points, located on the grid, more preferable than any other situation. The main task of the algorithm is the derivation of the Eq. 1, using measurements from the set of discrete measurements U with some externally defined limitations, including a range of the derivative orders, a number of terms in the equations and number of factors in the term.

The noise in this paper is assumed to be directional, where the noise in the direction x_j can be described as

$$E_{x_j}(u(x_1, ..., x_n; t)) = u(\bar{x}_1, ..., x_j, ...\bar{x}_n; t) + \epsilon(x_j) \tag{2}$$

With $\bar{x}_i$ "fixed" variables are denoted and $\epsilon(x_j)$ is the noise, which in the paper assumed to be distributed normally $N(0; \sigma)$ with the expected value of 0 and the variance of σ. It should be emphasized that poly-directional noise forms as the superposition of the unidirectional noise operators, i.e. $E_{x_j, x_k} = E_{x_j} \circ E_{x_k}$. In what follows $\bar{\sigma} = \sigma \max(u)$ is chosen as the multiplier of maximal magnitude of the measured value in this direction. The noise level is defined in the same way. In the text below bar over the variance, σ is omitted.

3   Method Description

In this section, the details of the evolutionary method of partial differential equation derivation are described. The proposed method involves a combination of evolutionary algorithm and sparse regression for the detection of the equation structure. The former is aimed at the construction of equation terms set, while the latter is focused on the selection of significant terms from the created set and calculation of weights that will be present in the resulting equation.

3.1   Data Preprocessing

To initialize algorithm, time, and spatial derivatives, which will later form the desired equation, must be calculated. In specific situations, the derivatives by themselves can be measured and, therefore, this step can be skipped, but often, only the raw value of the studied function is available for the research. For the more straightforward further computations, it can be assumed without loss of the generality, that the measurements are held on the rectangular (but not necessarily uniform) grid. Unlike the instances of single space dimension, in some experiments, even on data with moderate noise levels, the finite-difference method of derivative calculation can lead to satisfactory results, the multi-dimensional case requires advanced methods of obtaining derivatives. It is important to note that the quality of the taken derivatives is crucial for acquiring the correct structure of the equation. In some cases, when the data has high noise values, the error of the equation $\|F\|$ can be lower with an utterly incorrect set of terms.

In general, the calculation of derivatives is the operation that is vulnerable to the noise in the data. On the other hand, the result of the algorithm, in general, depends on the quality of the input derivatives: if they are computed with high errors, the alterations of the resulting structure of the equation can vary from incorrect coefficients to the entirely wrong structure. For these reasons, several noise-resistant methods of partial derivative calculations have been introduced. Notably, they include such commonly used methods as kernel smoothing [3], derivation of polynomials, fitted over sets of points, and more uncommon ones like Kalman filtering [7].

Therefore, in the framework, data clearance and noise-resistant derivative calculations have been combined to achieve decent smoothness. First of all, Gaussian smoothing kernels are applied for the data field in each time frame. This approach can reduce the significant outliers in the data and corresponds to the nature of the studied metocean processes, where the fields tend to be smooth. In the case of the time-dependent two-dimensional field, the smoothing is applied for each of the time frames. Two-dimensional Gaussian smoothing with selected bandwidth σ has the structure Eq. 3 and kernel Eq. 4, where s is the point, for which the smoothing is done, and s' - point, that value is utilized in smoothing.

$$\tilde{u}(s,t) = \int K_\sigma(s - s')u(s')ds'; \tag{3}$$

$$K_\sigma(s - s') = \frac{1}{2\pi\sigma^2} \exp\left(\frac{1}{2\sigma^2} \sum_{i=1}^{2}(s - s')_i\right); \tag{4}$$

In addition to smoothing, a noise-stable numerical differentiation scheme is applied. The derivative is taken by differentiation of polynomials, constructed over the set of points in some window. The coefficients of the polynomials, utilized in this step, are defined by linear regression. Despite all these measures, as it is presented on Table 1, derivatives of higher orders tend to have significant errors even after smoothing and polynomial derivation.

A particular example of the noised measure fields is shown in Fig. 1. For the clarity matters, only the spatial field center slice is provided. However, it should be emphasized that the entire spatial field is smoothed out to obtain the spatial derivative field.

Table 1. Noise levels (%) for the raw noised data and for the smoothed data

	u	$\frac{\partial u}{\partial t}$	$\frac{\partial^2 u}{\partial t^2}$
Noised function	15.5	260.1	12973.8
Smoothed function	12.3	10.78	458.2

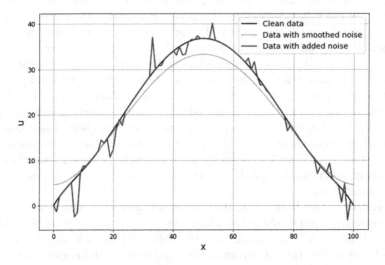

Fig. 1. Graph of a section over one spatial dimension for synthetic input function (solution of wave equation with 2 spatial dimensions) in original state, with added Gaussian noise, and after the noise was smoothed by Gaussian kernel

Differentiation of the three different fields shown in Fig. 1 gives the derivative fields that have values of the different orders. Thus, they are shown in the different graphs in Fig. 2.

In the majority of the real-world processes, derivative orders are limited to the first or second order. Derivative field of the slices shown in Fig. 1 represented in Fig. 2 show that the proposed algorithm of noise reduction in derivatives not only achieves values, close to the values of the derivatives on clear data, but also the structure of the fields are similar, which is vital for the evolutionary algorithm.

3.2 Evolutionary Algorithm

After the differentiation process commits, the evolutionary algorithm is initiated. Individual derivatives are taken as tokens (Eq. 5), combinations of which form the terms of the searched equation. An example of such a combination is presented in the Eq. 6. The vector, that will after further modifications are used as the feature for the regression, is composed as the elementwise product (that is denoted with ⊙ symbol) of vectors containing original function and its derivative along the x-axis.

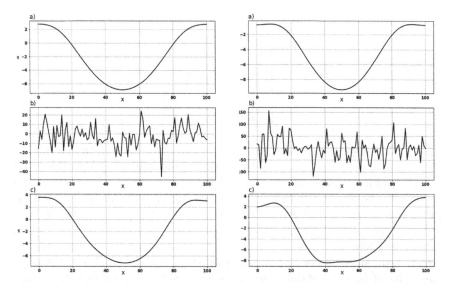

Fig. 2. Graph of first (left column) and second (right column) order time derivative, calculated on input function (a), noisy function (b), and function with noise, smoothed by Gaussian kernel (c)

$$
\mathbf{f_1} =
\begin{bmatrix} 1 \\ \vdots \\ 1 \\ \vdots \\ 1 \end{bmatrix}
\; ; \mathbf{f_2} =
\begin{bmatrix} u\,(t_0,\; x_0) \\ \vdots \\ u\,(t_i,\; x_j) \\ \vdots \\ u\,(t_m,\; x_n) \end{bmatrix}
\; ; \mathbf{f_3} =
\begin{bmatrix} u_x\,(t_0,\; x_0) \\ \vdots \\ u_x\,(t_i,\; x_j) \\ \vdots \\ u_x\,(t_m,\; x_n) \end{bmatrix}
\; ; ... \tag{5}
$$

$$
\mathbf{F'_k} =
\begin{bmatrix} u(t_0, x_0) * u_x(t_0, x_0) \\ \vdots \\ u(t_i, x_j) * u_x(t_i, x_j) \\ \vdots \\ u(t_m, x_n) * u_x(t_m, x_n) \end{bmatrix}
= \mathbf{f_2} \odot \mathbf{f_3}; \tag{6}
$$

The normalization of terms values is held for each of the time frames, passed into the algorithm, for the correct operation of regularized regression during the further weight calculation phases. In this step, various norms can be used, but most commonly, L_2 norm or L_∞ norms are applied, as it is represented in the Eq. 7.

$$\mathbf{F}_k = \begin{bmatrix} \dfrac{u(t_0,x_0) \cdot u_x(t_0,x_0)}{\|f_2(t_0) \odot f_3(t_0)\|} \\ \vdots \\ \dfrac{u(t_i,x_j) \cdot u_x(t_i,x_j)}{\|f_2(t_i) \odot f_3(t_i)\|} \\ \vdots \\ \dfrac{u(t_m,x_n) \cdot u_x(t_m,x_n)}{\|f_2(t_m) \odot f_3(t_m)\|} \end{bmatrix} ; \tag{7}$$

The evolutionary part of the algorithm is aimed at the selection of a shortlist of the terms. In the shortlist, one of the terms is randomly selected as the target. The target term is described with the weighted combination of the other terms in the list. In the beginning, a randomized collection of possible solutions, which is called population, is declared. Each of these individuals represents one set of terms with the selected target, that can be interpreted as the right part of the equation, and features, a linear combination of which composes the left part. In order to perform selection, the fitness function is introduced in Eq. 8 as the inverse value of L_2-norm of differences between target $\mathbf{F}_{target}$ and selected combination of features $\mathbf{F}$ with weighs α, obtained by the sparse regression. Therefore, the task of the evolutionary algorithm can be reduced to the obtainment of the equation structure with the highest value of fitness-function.

$$f_{fitness} = \frac{1}{\|\mathbf{F} \cdot \alpha - \mathbf{F}_{target}\|_2} \longrightarrow max \tag{8}$$

The genotype of the individual is represented by the composition of the encoded structures for each term. These encodings contain powers of each token in the term. The evolution of individuals is performed both by mutation and by a crossover in every iteration step. The mutation for an individual is introduced as the random change addition, deletion of alteration of factors in its terms. For example, this can result in shift of equation term u_t to $u_{xx} * u_t$ or $u_x * u_{tt}$ to $u_x * u_t$. The elitism that helps to preserve the best-detected candidate consists of the preservation of some population individuals with the highest fitness function value during the mutation step.

Crossover is the part of the evolutionary mechanism of gene exchange between two individuals in order to produce offsprings that can have higher fitness values. In the task of data-driven equation derivation, it can be introduced by the exchange of terms between equations. In order to produce units with higher fitness values, the crossover must be held between selected individuals. Various tests have proved that the fastest convergence to the desired solution can be achieved with the tournament selection. In this policy, several tournaments, where the unit with the highest fitness value is selected for a further crossover, is held between individuals of the population. After that, parents for the offsprings are randomly chosen between the tournament winners. In contrast to the simple selection of several individuals with the highest fitness function values for reproduction, this approach can let the offsprings take good qualities for less-valuable individuals of the population.

The next essential element of the proposed data-driven algorithm is sparse regression. Its main application is the detection of the equation structure among the set of possible terms. With no original information about the equation structure and the correct number of terms in it, that is more secure to allow the equation to have a higher number of possible term candidates. Therefore some form of filtration has to take place. The main instrument in this phase is the Least Absolute Shrinkage and Selection Operator (LASSO). In contrast to other types of regression, LASSO can reduce the number of non-zero elements of the weights vector, giving zero coefficients to the features, that are not influential to the target.

The minimized functional of the LASSO regression (Eq. 9) takes the form of the sum of two terms. First is the squared error between vectors of target, denoted as $\mathbf{F_{target}}$, and vector of predictions, obtained as the dot product of matrix of features $\mathbf{F}$ and vector of weights α, while second in the L_1-norm of the weights vector, taken with sparsity constant λ:

$$\|\mathbf{F}\alpha - \mathbf{F}_{target}\|_2^2 + \lambda\|\alpha\|_1 \to \min_{\alpha} \qquad (9)$$

The main imperfection of the LASSO regression is its disability to acquire actual values of the coefficients. To obtain the actual coefficients of the resulting PDE, final linear regression over discovered influential terms is performed. In the final step, non-zero weights from the LASSO are rescaled with original unnormalized data as features and the target.

The pseudo-code for the resulting algorithm is provided in Appendix A

4   Numerical Experiments

4.1   Synthetic Data

The analysis of the algorithm performance, just as in the case of one space dimension or on the clean data, first of all, shall be held on the synthetic data. This simplification can show the response of the result to various types and magnitude of noise, which is generally unknown on the measurement data. As in the previous studies, the solution of the wave equation with two spatial variables Eq. 10, where t - time, x, y - spatial coordinates, u - studied function (for example, vertical displacement of membrane), and $\alpha_1 = \alpha_1 = 1$ was taken as the synthetic data. The algorithm has proved to be able to detect the correct structure of the equation with the clean data, while on the noisy data, additional terms or completely wrong structures have been detected.

$$\frac{\partial^2 u}{\partial t^2} = \alpha_1 \frac{\partial^2 u}{\partial x^2} + \alpha_2 \frac{\partial^2 u}{\partial y^2} \qquad (10)$$

Several noise addition experiments were held on the synthetic data: first of all, in a fraction of points (40% of total number) the noise of various magnitudes have been added: ($\mu = 0; \sigma = n * ||u(t)||$, $n = \overline{0.1, 0.8}$). After that, the algorithm has been applied to this data. The results of the experiment are as follows: the method is successfully able to detect the structure of the equation for the interval of noise levels up to 14.9%, which corresponds to the standard deviation of Gaussian noise in the interval $[0, 0.35]$, multiplied by a norm of the field in the time frame. The weights errors in this interval are minor, as it is shown in the Table 2. With higher noise levels (in the interval between 14.9% and 15.67%), the algorithm detects additional terms that are not present in the original equation, which may result both in the distortion of equation structure and incorrect weights calculation. Finally, with high noise levels, the proposed algorithm can lose grasp of the correct structure of the equation.

Table 2. Discovered structures of Eq. 10 for the specific noise levels

Noise level of input data (%)	Equation
0	$\frac{\partial^2 u}{\partial t^2} = 1.00\frac{\partial^2 u}{\partial x^2} + 1.00\frac{\partial^2 u}{\partial y^2}$
8.3	$\frac{\partial^2 u}{\partial t^2} = 1.02\frac{\partial^2 u}{\partial x^2} + 1.01\frac{\partial^2 u}{\partial y^2}$
10.9	$\frac{\partial^2 u}{\partial t^2} = 1.04\frac{\partial^2 u}{\partial x^2} + 0.99\frac{\partial^2 u}{\partial y^2}$
13.1	$\frac{\partial^2 u}{\partial t^2} = 0.96\frac{\partial^2 u}{\partial x^2} + 0.99\frac{\partial^2 u}{\partial y^2}$
14.9	$\frac{\partial^2 u}{\partial t^2} = 0.95\frac{\partial^2 u}{\partial x^2} + 1.2\frac{\partial^2 u}{\partial y^2}$
15.67	$\frac{\partial^2 u}{\partial t^2} = 0.84\frac{\partial^2 u}{\partial x^2} + 0.63\frac{\partial^2 u}{\partial y^2} + 0.12\frac{\partial u}{\partial y}$
16.45	$\frac{\partial^2 u}{\partial x^2}\frac{\partial^2 u}{\partial y^2} = 0$
17.88	$\frac{\partial^2 u}{\partial x^2}\frac{\partial^2 u}{\partial y^2} = 0$

In Fig. 3, the influence of the noise level added to the measured field on the derivative fields is shown.

In the other experiment with the same data set, the noise of relatively high magnitudes was added to a minor fraction of points (5% of the total number). In this case, the framework has shown similar results to the previous experiment: until the noise level of approximately 15%, the discovered structure was correct. On the data with higher noise magnitudes, the errors in the structure of the equation occurred. This experiment has shown that in the studied cases, the main limiting factor for the performance of the algorithm with implemented preprocessing for noise reduction is the noise level and not the distribution of noise across the studied field.

4.2   Physical Measurements Data

For the validation of the model, the dynamics of the two-dimensional field of sea surface height (SSH) data from the NEMO ocean model for the Arctic region for a modelling month with the resolution of an hour has been used. The part of the area in the center of the Barents Sea was selected for the numerical experiments. The area is known to have strong tides, which can lead to the discovery of the time-dependent equation. It is necessary to emphasise that despite existing Tidal equations, there is no single analytical equation for the specific case of the dynamics of the SSH in this region due to the overlapping of processes of different natures.

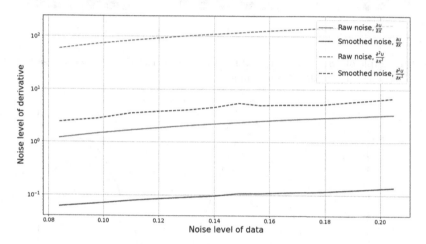

Fig. 3. Noise levels of calculated first and second (dashed line) time derivatives, related to noise levels of input data

After the application of the framework to the data, the structure of the equation in form Eq. 11 has been acquired.

$$\frac{\partial u}{\partial x} = -0.0506\frac{\partial u}{\partial t} - 0.0053\frac{\partial^2 u}{\partial t^2} \tag{11}$$

To validate the result of the algorithm, Eq. 11 was solved, and the calculated field was compared with the initial one, as shown in Fig. 4. Since there is second-order time derivative and first spatial derivative, the initial conditions (first two time frames to represent the field and its first time derivative for the beginning of studied period) and the boundary condition on one edge of the studied area are set. The graphs of daily sea surface height dynamics from reanalysis and equation solution is presented in Fig. 5. The metrics of quality show that the discovered equation can describe the equation well: $RMSE = 0.0434, MAE = 0.0446$ for the field with values in interval between approximately 0.5 and 0.9.

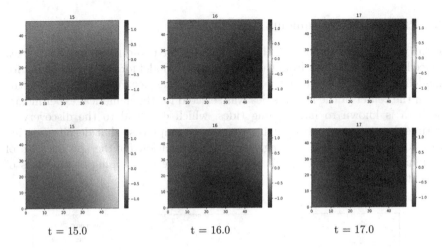

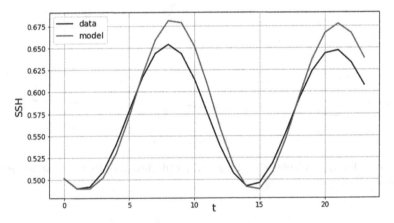

Fig. 4. Example of SSH field, obtained from reanalysis (upper row) and the same field from Eq. 11, (lower row) for 3 time frames

Fig. 5. Dynamics of sea surface height for September 18, 2013: reanalysis (denoted as data) and solution of the equation, obtained from framework, denoted as model, for the center of the studied area

4.3   Comparison with Other Methods

The experiments with conditions, similar to the one in [4], have been performed in order to compare the proposed algorithm with existing state-of-the-art methods. Due to limitations of the framework, namely the ability to derive only a single equation, instead of the system Eq. 12, that are utilized in the referenced article, a test was performed on similar Eq. 13. Additional difficulties to the comparison were contributed by unknown initial and boundary conditions in the referenced experiment.

$$\begin{cases} \frac{\partial U}{\partial t} = -U\nabla U + \nu\Delta U, U = (u,v)^T \\ U|_{t=0} = U_0(x,y) \end{cases} \quad (12)$$

$$\frac{\partial u}{\partial t} = (\frac{\partial^2 u}{\partial x^2} + \frac{\partial^2 u}{\partial y^2}) + u(\frac{\partial u}{\partial x} + \frac{\partial u}{\partial y}) \quad (13)$$

Equation 13 was solved using finite differences, and the noise from normal distribution was added to simulate the aforementioned experiment. The preprocessing phase, described in detail in previous sections and involving smoothing and derivatives calculation, was performed to reduce the influence of noise on the resulting equation (Table 3).

The added noise was created from $k \times \max_{x,y,t} u(x,y,t) \times N(0,1)$. The experiments have been conducted with the value of k, equals to 0.001, as in the compared study [4], and the discovered equations had correct structures. As the framework output, we will consider the closest to the correct structure of the equation, obtained on the grid of sparsity constant values.

Table 3. Discovered structures of Eq. 13 for the specific noise levels

k	Noise level of input data (%)	Equation
0.0005	0.25%	$\frac{\partial u}{\partial t} = (0.999\frac{\partial^2 u}{\partial x^2} + 1.000\frac{\partial^2 u}{\partial y^2})+$ $+u(1.001\frac{\partial u}{\partial x} + 1.000\frac{\partial u}{\partial y})$
0.00075	0.49%	$\frac{\partial u}{\partial t} = (1.001\frac{\partial^2 u}{\partial x^2} + 1.001\frac{\partial^2 u}{\partial y^2})+$ $+u(0.999\frac{\partial u}{\partial x} + 0.999\frac{\partial u}{\partial y})$
0.001	0.97%	$\frac{\partial u}{\partial t} = (1.000\frac{\partial^2 u}{\partial x^2} + 0.999\frac{\partial^2 u}{\partial y^2})+$ $+u(1.000\frac{\partial u}{\partial x} + 1.000\frac{\partial u}{\partial y})$
0.00125	4.2%	$\frac{\partial u}{\partial t} = (1.001\frac{\partial^2 u}{\partial x^2} + 1.002\frac{\partial^2 u}{\partial y^2})+$ $+u(0.998\frac{\partial u}{\partial x} + 0.999\frac{\partial u}{\partial y})$
0.0015	6.3%	$\frac{\partial u}{\partial t} = (0.996\frac{\partial^2 u}{\partial x^2} + 0.998\frac{\partial^2 u}{\partial y^2})+$ $+u(1.000\frac{\partial u}{\partial x} + 1.003\frac{\partial u}{\partial y}) - 0.0034\frac{\partial^2 u}{\partial y^2}\frac{\partial u}{\partial x}$

In these tests, shown noise resistance corresponds with other framework applications and is somewhat better than in the experiment in the compared experiment. Despite the insignificant difference in the coefficient k (0.001 versus 0.00015), the noise level difference is significant (1% versus 6.3%). For the noise levels approximately below 5% the correct equations were detected. After that, the structure of the equation deteriorates, which manifests in wrong weights and the presence of additional terms/lack of mandatory ones.

5   Conclusion

The proposed method has proven to be suitable for the data-driven derivation of equations, that can be used for modelling of various physical processes. The robustness of the algorithm to the noise in the input data that is provided by improved preprocessing of data, which included the application of kernel smoothing to the input data matrices, and derivative calculation by differentiating polynomials, fit over points inside a selected window, allows the framework applicable to the real-world problems. Even in the cases of substantial noise in the input data, the resulting equations had the correct structures and, therefore, can correctly describe the studied system. Other notable points about the algorithm operation can be stated:

- To achieve a good quality of the resulting processes, the areas, localizing different processes, should be separated and studied on their own. The example of such approach to the problem was presented in the case of real-world data processing when the area that is already known to have strong time dependencies of sea surface height was separated, and the equation for it was derived;
- The meta-parameters of the algorithm have a strong influence on the final result. For example, low values of sparsity constant can lead to the presence of additional terms in the equation, while its higher than optimal values can completely distort the equation structure. Therefore, mechanisms of meta-parameter selection should be implemented in the further development of the method.

Areas of the further development of the framework can include the derivation of a more generalized class of equations, using similar techniques, not limiting the results in a class of partial differential equations. Additionally, the equations for vector variables or even systems of equations can be the next targets for the work.

Source code is publicity available at GitHub [6].

Acknowledgements. This research is financially supported by The Russian Scientific Foundation, Agreement #19-71-00150.

A Pseudo-code of the Algorithm

Input: set of elementary tokens T, symbolically representing constant, initial function, and its various derivatives; set of function measurements from the studied field

Parameters : M - number of token combinations in a single individual; k - number of elementary tokens in a combination; n_pop - number of candidate solutions in the population; evolutionary algorithm parameters: number of epochs n_{epochs}, mutation $r_{mutation}$ & crossover rates $r_{crossover}$, part of the population, allowed for procreation a_{proc}, number of individuals, refrained from mutation (elitism) a_{elite}; sparse regression parameter - sparsity constant λ

Result: The structure of the partial differential equation with the corresponding weights, best fitting the input field

Apply smoothing to the measurements & calculate the derivatives; Generate population **P** of individuals, representing equation, of size n_pop, with M - random permutations of k tokens to form sets C^j;

for $epoch = 1$ to n_{epochs} **do**

   for *individual in population* **do**

      Apply sparse regression to individual to calculate weights;

      Calculate fitness function to individual;

   end

   Hold tournament selection and crossover;

   for *individual in population except* $n\_pop \times a_{elite}$ *"elite" ones* **do**

      Mutate individual;

   end

end

Select the individual with highest fitness function value as the final structure of the solution to the problem;

Calculate true weights of the equation, using linear regression.

      Algorithm 1: The pseudo-code of the algorithm operation

References

1. Berg, J., Nyström, K.: Neural network augmented inverse problems for PDEs (2018). https://arxiv.org/abs/1712.09685
2. Chang, H., Zhang, D.: Machine learning subsurface flow equations from data. Comput. Geosci. **23**(5), 895–910 (2019). https://doi.org/10.1007/s10596-019-09847-2
3. Knowles, I., Le, T., Yan, A.: On the recovery of multiple flow parameters from transient head data. J. Comput. Appl. Math. **169**(1), 1–15 (2004)
4. Long, Z., Lu, Y., Ma, X., Dong, B.: PDE-net: Learning PDEs from data (2017). https://arxiv.org/abs/1710.09668
5. Maslyaev, M., Hvatov, A., Kalyuzhnaya, A.: Data-driven partial derivative equations discovery with evolutionary approach. In: Rodrigues, J.M.F., et al. (eds.) ICCS 2019. LNCS, vol. 11540, pp. 635–641. Springer, Cham (2019). https://doi.org/10.1007/978-3-030-22750-0_61

6. NSS Team: Fedot E* algotirhms (2020). https://github.com/ITMO-NSS-team/FEDOT.Algs
7. Piche, R.: Automatic numerical differentiation by maximum likelihood estimation of state-space model (2016). https://arxiv.org/abs/1610.04397v1
8. Qin, T., Wu, K., Xiu, D.: Data driven governing equations approximation using deep neural networks (2018). https://arxiv.org/abs/1811.05537
9. Raissim, M.: Deep hidden physics models: Deep learning of nonlinear partial differential equations (2018). https://arxiv.org/abs/1801.06637
10. Rudy, S.H., Brunton, S.L., Proctor, J.L., Kutz, J.N.: Data-driven discovery of partial differential equations. Sci. Adv. **3**(4), e1602614 (2017)
11. Schaeffer, H., Caflisch, R., Hauck, C.D., Osher, S.: Learning partial differential equations via data discovery and sparse optimization. Proc. Roy. Soc. A: Math. Phys. Eng. Sci. **473**(2197), 20160446 (2017)
12. Schaeffer, H.: Learning partial differential equations via data discovery and sparse optimization. Proc. Roy. Soc. A **473**(2197), 20160446 (2017)

Preconditioning Jacobian Systems by Superimposing Diagonal Blocks

M. Ali Rostami[1] and H. Martin Bücker[1,2(✉)]

[1] Institute for Computer Science, Friedrich Schiller University Jena,
07737 Jena, Germany
`martin.buecker@uni-jena.de`

[2] Michael Stifel Center Jena for Data-driven and Simulation Science,
07737 Jena, Germany

Abstract. Preconditioning constitutes an important building block for the solution of large sparse systems of linear equations. If the coefficient matrix is the Jacobian of some mathematical function given in the form of a computer program, automatic differentiation enables the efficient and accurate evaluation of Jacobian-vector products and transposed Jacobian-vector products in a matrix-free fashion. Standard preconditioning techniques, however, typically require access to individual nonzero elements of the coefficient matrix. These operations are computationally expensive in a matrix-free approach where the coefficient matrix is not explicitly assembled. We propose a novel preconditioning technique that is designed to be used in combination with automatic differentiation. A key element of this technique is the formulation and solution of a graph coloring problem that encodes the rules of partial Jacobian computation that determines only a proper subset of the nonzero elements of the Jacobian matrix. The feasibility of this semi-matrix-free approach is demonstrated on a set of numerical experiments using the automatic differentiation tool ADiMat.

Keywords: Combinatorial scientific computing · Partial Jacobian computation · Partial graph coloring · Sparsity exploitation · ADiMat

1 Introduction

Large sparse systems of linear equations are critical to computational methods in science, technology, and society. A key characteristic of iterative methods for the solution of such systems is that they can be implemented in a matrix-free fashion [12]. That is, given an N-dimensional right-hand side vector $\mathbf{b}$ and an $N \times N$ nonsingular coefficient matrix J, these methods aim to solve systems of the form

$$J\mathbf{y} = \mathbf{b} \tag{1}$$

by making use of J solely in the form of matrix-vector products, $J\mathbf{z}$, or transposed matrix-vector products, $J^T\mathbf{z}$, where the symbol $\mathbf{z}$ denotes some given

© Springer Nature Switzerland AG 2020
V. V. Krzhizhanovskaya et al. (Eds.): ICCS 2020, LNCS 12138, pp. 101–115, 2020.
https://doi.org/10.1007/978-3-030-50417-5_8

N-dimensional vector. Therefore, there is no need to assemble the coefficient matrix in some sparse data storage format. We consider a rather typical situation in computational science where the coefficient matrix J is the Jacobian of some mathematical function given in the form of a computer program. Jacobian-vector products as well as transposed Jacobian-vector products can be efficiently and accurately computed by automatic differentiation (AD) [6,11] without explicitly setting up the Jacobian matrix. Thus, the major computational kernels of iterative methods match to the functionality that is provided by AD.

In practice, iterative methods involve preconditioning techniques [1,12] that transform (1) into an equivalent system of the form

$$M^{-1}J\mathbf{y} = M^{-1}\mathbf{b}, \tag{2}$$

whose solution $\mathbf{y}$ is the same as the solution of (1). Here, the $N \times N$ nonsingular matrix M is the preconditioner that is to be constructed such that M is somehow close to J, i.e.,

$$M \approx J.$$

Preconditioning techniques typically need access to individual nonzero elements of the coefficient matrix [1,12]. However, in a matrix-free approach, such accesses to individual Jacobian entries are computationally expensive, not only in automatic differentiation but also in numerical differentiation.

To bridge the gap between preconditioned iterative methods and AD, we propose a novel approach that is based on superimposing two diagonal block schemes. The first scheme consists of nonoverlapping diagonal blocks of size r that represent a sparsification operation. These blocks are used to define the required nonzero elements [3] of a partial Jacobian computation [9]. The required nonzeros are then determined by AD employing the solution of a suitably defined graph coloring problem [5] that colors a subset of the vertices encoding the rules of partial Jacobian computation.

The second scheme consists of nonoverlapping diagonal blocks of size d that define a simple preconditioner. A standard preconditioning approach is taken that applies ILU decomposition separately on each diagonal block. Here, we deliberately choose $d \geq r$ enabling to incorporate a maximal number of nonrequired nonzero elements outside of the $r \times r$ diagonal blocks of the Jacobian that are produced as by-products of the partial Jacobian computation.

The structure of this article is as follows. In Sect. 2, the overall approach is sketched that consists of a problem arising from scientific computing. It involves the computation of a subset of the nonzero elements of the Jacobian matrix by AD. This partial Jacobian computation problem is then modeled by a suitably defined graph coloring problem in Sect. 3. In Sect. 4 implementation details of the approach are given. Numerical experiments are reported in Sect. 5 and concluding remarks are presented in Sect. 6.

2   Preconditioning via Two Block Schemes

The novel preconditioning approach is inspired by the semi-matrix free preconditioning technique introduced in [3]. The approach in [3] for the solution of (2)
is summarized as follows:

- Carry out Jacobian-vector products $J\mathbf{z}$ or transposed matrix-vector products
  $J^T\mathbf{z}$ by applying AD with a seed matrix that is identical to the vector $\mathbf{z}$.
- Choose a block size r and get the sparsified matrix of the Jacobian J denoted
  by $\rho_r(J)$. Here, the sparsification $\rho_r(J)$ consists of the nonzero elements of
  the $r \times r$ diagonal blocks of J. Assemble $\rho_r(J)$ via AD and store it explicitly.
- Construct a preconditioner M from $\rho_r(J)$ by performing an ILU(0) decomposition [12] on each block of $\rho_r(J)$. That is, no fill-in elements are allowed
  during the decomposition.

The nonzero elements of J that are selected by the sparsification $\rho_r(J)$ are
called *required* nonzero elements. The symbols used for the block size r and the
sparsification $\rho_r(J)$ indicate that these quantities define the *r*equired elements.
We also denote the nonzero pattern of the required elements by the set $\mathbf{R}$. As
usual for sparsity patterns, we use the binary matrix and the set of the positions of the nonzero elements interchangeably. That is, symbols like $\mathbf{R}$ denoting
sparsity patterns are either matrices or sets, depending on the context.

The novel approach borrows the first and the second item of the previous
list and replaces the third item by a different preconditioning scheme. The new
idea is that AD does not only compute the required elements $\mathbf{R}$, but also certain
additional information at no extra computational cost. However, only parts of
this additional information is immediately useful for preconditioning. This useful information is called *by-product* and is denoted by the set $\mathbf{B}$. The overall
approach is detailed in the remaining part of this section.

Like the previous approach in [3], the new approach is based on computing
only a proper subset of the nonzero elements of the Jacobian J, which is referred
to as *partial Jacobian computation* [5,7–10]. We summarize partial Jacobian
computation by considering Fig. 1 taken from [3]. Suppose that we are interested
in computing the nonzeros of J on all 2×2 diagonal blocks, but are not interested
in the remaining nonzeros. In this example, all nonzeros on the diagonal blocks
of size $r = 2$ are the required nonzeros, which are denoted by black disks in
the sparsity pattern of the Jacobian depicted in this figure left. All remaining
nonzeros of J are called *nonrequired* elements, represented by black circles.

The relative computational cost associated with the forward mode of AD
computing the matrix-matrix product $J \cdot S$ is given by the number of columns of
the seed matrix S, see [6,11]. We stress that AD does not assemble the matrix J,
but computes the product $J \cdot S$ for a given S directly. The symbol $\mathrm{cp}(J) := J \cdot S$
represents this so-called *compressed Jacobian matrix*.

The exploitation of sparsity has a long tradition in AD; see the survey [5]. The
main idea behind sparsity exploitation is to form groups of columns of J. This
grouping is denoted by colors in the middle of the Fig. 1. If J is an $N \times N$ matrix,
all (zero and nonzero) elements of J are computed by setting the seed matrix

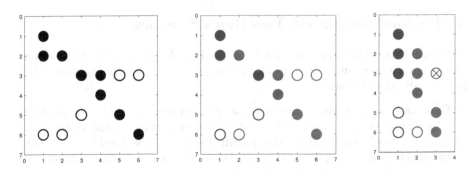

Fig. 1. Sparsity patterns of a 6×6 Jacobian J shown left and its compressed version cp(J) shown right. Grouping of columns of J is denoted by colors in the middle. (Figure taken from [3].) (Color figure online)

to the identity of order N. The relative computational cost of this approach is then the number of columns of the identity given by N. However, exploiting the grouping of columns it is possible to find a seed matrix with fewer than N columns. In the middle of Fig. 1, there are three colors representing three groups of columns. Each group of columns in J corresponds to a single column in the compressed Jacobian matrix depicted in the right. More precisely, a column of cp(J) with a certain color c is the linear combination of those columns of J that belong to the group of columns with the color c. Equivalently, there is a binary seed matrix S whose number of columns corresponds to the number of colors such that all required nonzero elements $\mathbf{R}$ of J also appear in cp(J).

In the semi-matrix-free approach [3], given the sparsity pattern of J and the set of required elements $\mathbf{R}$, the problem of assembling the required nonzero elements with a minimal relative computational cost is as follows.

Problem 1 (Block Seed). Let J be a sparse $N \times N$ Jacobian matrix with known sparsity pattern and let $\rho_r(J)$ denote its sparsification using $r \times r$ blocks on the diagonal of J. Find a binary $N \times p$ seed matrix S with a minimal number of columns, p, such that all nonzero entries of $\rho_r(J)$ also appear in the compressed matrix cp(J) := $J \cdot S$.

The compressed Jacobian cp(J) contains by definition all required elements of J. However, by inspecting the example in Fig. 1, it also contains additional nonzero elements. These additional nonzero elements decompose into two different classes. There are nonzero elements of cp(J) that are nonrequired elements of J. In the example, the three nonzeros at the positions $(5, 1)$, $(6, 1)$ and $(6, 2)$ belong to this class. The other class of nonzero elements of cp(J) consists of those nonzeros that are linear combinations of nonzero entries of J. For instance, the nonzero at the position $(3, 3)$ in cp(J) is the sum of $J(3, 5)$ and $J(3, 6)$.

The overall idea of the novel approach is to incorporate into the preconditioning not only the required elements of J, but also a certain subset of the nonzero elements of cp(J) that are nonrequired elements of J. To this end, another sparsification operator $\rho_d(\cdot)$ is introduced that extracts from cp(J) the nonzero elements of the $d \times d$ diagonal blocks of J that are not required. The set of by-products $\mathbf{B}$ is then defined as those nonzero elements of the compressed Jacobian $\mathrm{cp}(J)$ that are nonzeros within these $d \times d$ blocks of J and that are not contained in the set of required elements $\mathbf{R}$. In other words, the by-products $\mathbf{B}$ are obtained from the compressed Jacobian $\mathrm{cp}(J)$ by removing all entries that are linear combinations of nonzeros of J and by additionally removing all (required and nonrequired) nonzeros of J that are outside the $d \times d$ diagonal blocks. The preconditioner M that approximates J is then constructed by assembling the nonzeros $\mathbf{R} \cup \mathbf{B}$ in a matrix denoted as $\mathrm{rc}(J)$ and using an ILU decomposition on the $d \times d$ diagonal blocks. The symbols used for the block size d and the sparsification operator $\rho_d(\cdot)$ indicate that these quantities are used to carry out a *decomposition* on each block.

We remark that the sparsification operators, $\rho_r(\cdot)$ and $\rho_d(\cdot)$, that extract the diagonal blocks reduce the size of the bottom right block accordingly if the order of the matrix is not a multiple of the block size. For instance, returning to the example in Fig. 1 with $r = 2$ and assuming that $d = 5$, then the operator $\rho_d(\cdot)$ leads to a top left 5×5 block and a bottom right 1×1 block. The set of by-products $\mathbf{B}$ then consists of the single nonzero entry $J(5,3)$ which is stored in $\mathrm{cp}(J)$ at position (5,1).

In summary, a high-level description of the new preconditioning approach that uses two diagonal block schemes of size r and of size d is given as follows:

- Carry out Jacobian-vector products $J\mathbf{z}$ or transposed matrix-vector products $J^T\mathbf{z}$ using AD.
- Choose a block size r, solve Problem 1, and compute $\mathrm{cp}(J)$ using AD.
- Choose a block size d and assemble the required elements $\mathbf{R}$ as well as the by-products $\mathbf{B}$ from $\mathrm{cp}(J)$ using the sparsification operator $\rho_d(\cdot)$. Store $\mathbf{R} \cup \mathbf{B}$ explicitly in a matrix $\mathrm{rc}(J)$.
- Construct a preconditioner M from $\mathbf{R} \cup \mathbf{B}$ by performing an ILU decomposition on each diagonal $d \times d$ block of $\mathrm{rc}(J)$.

The only other work that is related to our approach is the preconditioning technique introduced in [4], which is also based on partial matrix computation, but differs in formulating balancing problems.

The purpose of the following section is to reformulate the combinatorial problem from scientific computing given by Problem 1 in terms of an equivalent graph coloring problem.

3   Modeling via Partial Graph Coloring

Recall from the previous section that the exploitation of sparsity is a well-studied topic in derivative computations [5]. Interpreting these scientific computing problems in the language of graph theory does not only give us a better insight to the abstract problem structure but also offers an intimate connection to the rich history of research in graph theory that can lead to efficient algorithms for the solution of the resulting problems. In this section, we consider the graph problem corresponding to the scientific computing problem that was introduced in the previous section.

In the spirit of [3], we define a combinatorial model that handles the decomposition of the nonzero elements of J into two sets called required and nonrequired elements. The following new definition introduces the concept of structurally ρ_r-orthogonal columns.

Definition 1 (Structurally ρ_r-Orthogonal). *A column $J(:,i)$ is structurally ρ_r-orthogonal to column $J(:,j)$ if and only if there is no row position ℓ in which $J(\ell,i)$ and $J(\ell,j)$ are nonzero elements and at least one of them belongs to the set of required element $\rho_r(J)$.*

Next, we define the ρ_r-column intersection graph which will be used to reformulate Problem 1 arising from scientific computing.

Definition 2 (ρ_r-Column Intersection Graph). *The ρ_r-column intersection graph $G_{\rho_r} = (V, E_{\rho_r})$ associated with a pair of $N \times N$ Jacobians J and $\rho_r(J)$ consists of a set of vertices $V = \{v_1, v_2, \ldots, v_N\}$ whose vertex v_i represents the ith column $J(:,i)$. Furthermore, there is an edge (v_i, v_j) in the set of edges E_{ρ_r} if and only if the columns $J(:,i)$ and $J(:,j)$ represented by v_i and v_j are not structurally ρ_r-orthogonal.*

That is, the edge set E_{ρ_r} is constructed in such a way that columns represented by two vertices v_i and v_j need to be assigned to different column groups if and only if $(v_i, v_j) \in E_{\rho_r}$.

Using this graph model, Problem 1 from scientific computing is transformed into the following equivalent graph theoretical problem.

Problem 2 (Minimum Block Coloring). Find a coloring of the ρ_r-column intersection graph G_{ρ_r} with a minimal number of colors.

The solution of this graph coloring problem corresponds to a seed matrix S which is then used to compute the compressed Jacobian $\mathrm{cp}(J) = J \cdot S$ using AD. Recall from the previous section that the required elements of J are contained in $\mathrm{cp}(J)$. However, we already pointed out that some additional useful information $\mathbf{B}$ is also contained in $\mathrm{cp}(J)$. In the following section, we discuss how to recover these by-products $\mathbf{B}$ from $\mathrm{cp}(J)$ and how to use it for preconditioning.

4   Implementation Details

Given the sparsity pattern $\mathbf{P}$ of the Jacobian matrix J, the following pseudocode summarizes the new preconditioning approach:

1: $\mathbf{R} = \rho_r(\mathbf{P})$
2: $S = \mathrm{partial\_coloring}(\mathbf{P}, \mathbf{R})$
3: Compute $\mathrm{cp}(J) = J \cdot S$ by AD
4: $\mathrm{rc}(J) = \rho_d(\mathrm{partial\_recover}(\mathbf{P}, S, \mathrm{cp}(J), \mathbf{R}))$
5: Construct M as the ILU decomposition of $\mathrm{rc}(J)$
6: Solve the preconditioned linear system (2)

In this pseudocode, we first compute the required elements $\mathbf{R}$ using the sparsification operator $\rho_r(\cdot)$. The required elements $\mathbf{R}$ are taken as an input to solve Problem 2 using a partial graph coloring algorithm [3]. The solution of this graph coloring problem corresponds to a seed matrix S that is used by the AD tool ADiMat [2,14] to compute the compressed Jacobian $\mathrm{cp}(J)$. Then, we need a function `partial_recover()` to recover the nonzero elements $\mathbf{R} \cup \mathbf{B}$ of J from the compressed Jacobian $\mathrm{cp}(J)$. The preconditioner M is constructed by a blockwise ILU decomposition of $\mathrm{rc}(J)$ and the preconditioned system is solved by Jacobian-vector products using ADiMat.

To introduce the function `partial_recover()`, it is convenient to consider the standard approach of recovering the nonzeros of a Jacobian from its compressed version [6]. The standard approach assumes that all nonzeros of a sparse Jacobian are to be determined, whereas in a partial Jacobian approach we are interested in a subset of the nonzeros. In the standard approach, the nonzeros are recovered using the following MATLAB-like pseudocode:

```
1: procedure RECOVER(P, S, cp(J))
2:     J = zeros(size(P))
3:     for i = 1 : size(cp(J), 1) do
4:         I = P(i, :) ~= 0
5:         S_I = S(I, :)
6:         [row, col] = find(S_I)
7:         [rs, perm] = sort(row)
8:         J(i, I) = cp(J)(i, col(perm))
```

Given the pattern $\mathbf{P}$ of a sparse Jacobian J, the seed matrix S, and the compressed Jacobian $\mathrm{cp}(J) = J \cdot S$, this procedure recovers the Jacobian matrix J. It reconstructs every row i of J step by step. In each step, it first computes the indices I of the nonzeros of the row i of J. Then, it considers a reduced seed matrix $S_I = S(I, :)$. Here, S_I is a matrix containing those rows of S that correspond to the nonzeros of J in the row i. Suppose that there is a nonzero element in J in position (i, k). We then need the column index of the entry 1 in the row k of the reduced seed matrix. With this column index, the corresponding nonzero is extracted from $\mathrm{cp}(J)$. Because of MATLAB's implementation of `find()`, the row indices in row have to be sorted in increasing order.

For partial Jacobian computation where only a subset of nonzeros is determined, we need to extend the previous procedure to recover the Jacobian matrix using the seed matrix which is computed by the partial coloring. The following pseudocode introduces the new procedure `partial_recover()` which computes the Jacobian J from its compressed version $\mathrm{cp}(J)$ in partial Jacobian computation. Compared to the previous procedure `recover()`, this procedure needs the pattern of the required elements $\mathbf{R}$ as an additional input.

```
1: procedure PARTIAL_RECOVER(P, S, cp(J), R)
2:      NR = P − R
3:      J = zeros(size(P))
4:      for i = 1 : size(cp(J), 1) do
5:          I = P(i, :) ∼= 0
6:          S_I = S(I, :)
7:          [row, col] = find(S_I)
8:          [rs, perm] = sort(row)
9:          J(i, I) = cp(J)(i, col(perm))
10:         colS = ones(1, size(S(I, :), 1)) · S(I, :)
11:         positions = find(colS > 1)
12:         if ∼ isempty(positions) then
13:             for p = 1 : length(positions) do
14:                 r_i = find(S(:, positions(p)))
15:                 if sum(NR(i, r_i)) ∼= 1 then
16:                     J(i, r_i) = 0
```

The first steps up to the step 9 of this procedure are similar to the previous procedure `recover()`. The new procedure, however, needs to take into account the nonrequired elements. So, it looks for the columns of S_I which have more than one nonzero (in steps 10 and 11) since there are the columns in which the addition of two nonrequired elements can happen. Then, it goes through all of those columns, if any, and checks if any nonrequired elements is added. If such an addition happens in a column r_i, we put a zero in the corresponding entry $J(i, r_i)$ in the recovered Jacobian. The variable NR in this algorithm contains the positions of the nonrequired elements in P.

After recovering the Jacobian matrix J via the procedure `partial_recover()`, we need to make sure that only those elements will remain that are inside the diagonal blocks of size d. That is, we need to compute the by-products B using the sparsification operator $\rho_d(\cdot)$ which, in the current implementation, is carried out outside of the procedure `partial_recover()`; see also the sparsification operator $\rho_d(\cdot)$ in the algorithm sketched at the beginning of this section.

5   Numerical Experiments

Here, we employ the semi-matrix-free approach for the solution of a system of linear equations of the form (2). This system arises in the solution of an optimal boundary control problem for radiative transfer. Throughout the following experiments, the resulting coefficient matrix has the order $N = 1,944$ and contains $49,856$ nonzero elements. Its nonzero pattern is depicted in the left of Fig. 2. In the middle of this figure, the pattern of the sparsification $\rho_r(J)$ is depicted for a block size of $r = 100$. The $\lceil N/r \rceil = 20$ blocks are visible and are highlighted using a gray background. Notice that the last block is considerably smaller than the remaining blocks. To illustrate the preconditioning approach, the pattern of $\text{rc}(J)$ is also plotted for a block size of $d = 500$ in the right of this figure.

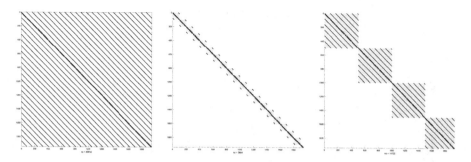

Fig. 2. Left: The nonzero pattern of the Jacobian J corresponding to the linear system (1) with the problem size $N = 1,944$. Middle: The pattern of the sparsified Jacobian $\rho_r(J)$ illustrating the required elements for a block size of $r = 100$. Right: The pattern of the required and by-product elements $rc(J)$ for a block size of $d = 500$.

The present experiments are carried out in MATLAB, R2019b. All derivative computations are computed by ADiMat. The right-hand side **b** of the linear system is chosen as the sum of all columns of J such that the exact solution **y** to (2) is given by the vector containing ones in all positions.

The linear system is solved using the Generalized Minimal RESidual method (GMRES) [13] with restart parameter of 20. We always take $\mathbf{y}_0 = \mathbf{0}$ as the initial guess. For the unpreconditioned system, the iteration is stopped in the nth step if

$$||\mathbf{b} - J\mathbf{y}_n||_2/||\mathbf{b}||_2 \le \varepsilon. \tag{3}$$

For the preconditioned system, convergence is obtained if

$$||M^{-1}(\mathbf{b} - J\mathbf{y}_n)||_2/||M^{-1}\mathbf{b}||_2 \le \varepsilon. \tag{4}$$

The tolerance $\varepsilon = 10^{-13}$ is chosen for both cases. All tests are carried out on an Intel Core i7-8550U CPU with a clock rate of 1.80 GHz and 16 GB RAM.

In Fig. 3, the convergence behavior using GMRES is plotted versus the number of matrix-vector products. The convergence is monitored by the residual vector of the nth iteration defined by $\mathbf{r}_n = \mathbf{b} - J\mathbf{y}_n$. More precisely, we show the norm of the residual scaled by the initial residual norm $||\mathbf{r}_0||_2$. We do not report the convergence versus the number of iterations for two reasons. Firstly, the number of matrix-vector products is known to be a better indication of the computing time than the number of iterations [12]; secondly, the number of matrix-vector products directly corresponds to the number of colors and thus makes it easy to relate the convergence to the cost of computing $cp(J)$ that is once needed to set up the preconditioner. This aspect is crucial in applications such as Newton-like methods for nonlinear systems where a sequence of linear systems with the same Jacobian sparsity pattern arises and the cost of solving a single coloring problem is amortized over solving multiple linear systems.

On the other hand, the number of matrix-vector products is only an approximation of the computing time, in particular for GMRES without restarts, where

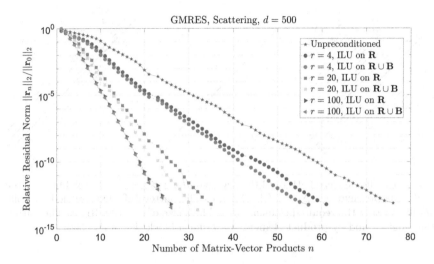

Fig. 3. Convergence behavior of GMRES. (Color figure online)

the number of operations carried out in an iteration linearly increases with the iteration number. In the first set of experiments, where the block size for the sparsification operator $\rho_d(\cdot)$ is fixed to $d = 500$, the computing time needed to converge the preconditioned iteration is always smaller than for the unpreconditioned method, if the time for partial coloring and computing $cp(J)$ is neglected. Taking this time into account so that the complete process of setting up the preconditioner is included, the preconditioned method is faster than the unpreconditioned method for all experiments where $r > 10$.

The unpreconditioned method exhibits the slowest convergence using the largest number of matrix-vector products needed to converge to the desired tolerance. This figure also contains six additional graphs by varying the block size $r = 4$, $r = 20$ and $r = 100$ and by employing two different preconditioning approaches. The approach advocated in this article is based on the blockwise ILU(0) decomposition of $rc(J)$, the matrix that contains the nonrequired elements as well as the by-products. This approach is denoted by $\mathbf{R} \cup \mathbf{B}$. For the sake of comparison, we also investigate another approach that is identical to the previously mentioned approach with a single exception. Rather than using the information $\mathbf{R} \cup \mathbf{B}$ to construct the blockwise ILU(0) preconditioner, the diagonal blocks involve only the information $\mathbf{R}$. That is, the by-products $\mathbf{B}$ are discarded from the preconditioning process. This latter approach is similar to our previous work reported in [3]. However, in [3], we do not use two different block schemes with different block sizes.

For the three block size $r = 4$, $r = 20$ and $r = 100$, the two preconditioning techniques based on $\mathbf{R} \cup \mathbf{B}$ and $\mathbf{R}$ both converge faster than the unpreconditioned method. This statement is true for GMRES as well as for other Krylov solvers that we tested but whose results are omitted due to the lack of space. It is also interesting that the convergence is improved by increasing the block sizes from

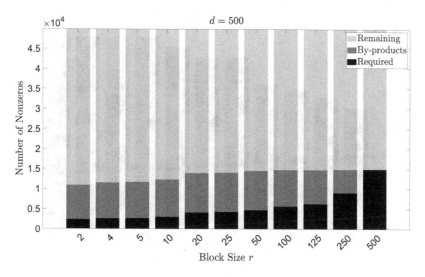

Fig. 4. Number of nonzeros varying the block size r for a fixed block size $d = 500$.

$r = 4$ via $r = 20$ up to $r = 100$. Furthermore, keeping the block size r fixed, the convergence of the approach $\mathbf{R} \cup \mathbf{B}$ tends to be faster than the approach using only $\mathbf{R}$. This observation is valid for the two block sizes $r = 4$ and $r = 20$. For large block sizes, however, it is unlikely that there will be a large set of by-products $\mathbf{B}$. So, the differences in the convergence behavior between an approach using $\mathbf{R} \cup \mathbf{B}$ and an approach using $\mathbf{R}$ tend to be small.

To better understand the preconditioning approach, we now focus on the number of nonzero elements when increasing the block size r. Figure 4 illustrates the number of required elements, $|\mathbf{R}|$, using black bars as well as the number of by-products, $|\mathbf{B}|$, using dark gray bars. The vertical axis (ordinate) is scaled to the number of nonzeros in J given by 49, 856. That is, the light gray bars denote the number of nonzero elements of J that are not taken into account when the preconditioner is constructed. For a block size of $d = 500$, this diagram shows that the number of required elements increase only mildly when increasing the block size r up to moderate values. However, when increasing r significantly, there is also a corresponding increase in the number of required elements; see the block sizes at the right of this figure.

The sum $|\mathbf{R}| + |\mathbf{B}|$ is rather constant when increasing the block size r. So, the approach tends to be relevant in particular for small block sizes r where the number of by-products is comparatively large. In this situation, the information available in $\mathbf{B}$ is particularly attractive since it comes from the partial Jacobian computation without any extra computational cost.

Next, we consider the number of colors needed for the solution of the partial graph coloring problem that is formally specified by Problem 2. This number of colors is depicted in Fig. 5. Here, the block size r is varied in the same range as in Fig. 4. Since the number of colors is an estimate for the relative computational

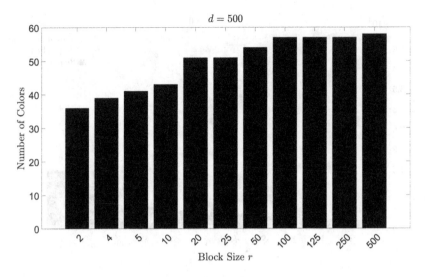

Fig. 5. Number of colors varying the block size r.

cost to compute the compressed Jacobian $\mathrm{cp}(J) = J \cdot S$ using AD, a slight increase in the number of colors can be harmful. This figure illustrates that the number of colors increases with the block size. Once more, this is an indication that the preconditioning approach is particularly relevant for small block sizes. Also, for small block sizes the storage requirement tends to be lower than for larger block sizes which corresponds to the overall setting in which a sparse data structure for the Jacobian is assumed to exceed the available storage capacity.

Finally, we analyze the number of nonzeros and the number of colors not only for a varying block size r, but also when varying the block size d. In Fig. 6, the results are depicted for the three block sizes $d = 300$, $d = 400$ and $d = 500$. For each value of d, this set of experiments involves those block sizes r that are divisors of d. The legend contains the union of all divisors of the three block sizes d. In the layout of this figure, a number of nonzeros is indicated by a bar to which the left ordinate is associated. A number of colors is denoted by a disk whose value is specified on the right ordinate. As in Fig. 4, black bars here denote the number of required nonzeros. In contrast, the by-products are now given by bars whose colors correspond to the values of r. The results show a similar behavior for all three values of d. The number of required nonzeros increase with increasing r. Compared to the number of required nonzeros, the number of by-products is in a similar magnitude, except when r approaches d. (By construction, there cannot be any by-product for $r = d$.) The number of by-products tends to by reasonably large, even for small values of r. At the same time, the number of colors is small for small block sizes r. In other words, small block sizes r are not only attractive because (i) they deliver additional information (represented by nonzero elements) that is useful for preconditioning without any extra computational cost and, at the same time, (ii) they lead to a low relative computational cost associated with AD (represented by colors).

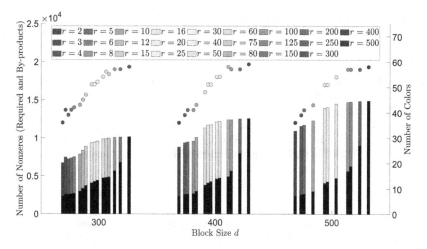

Fig. 6. Number of required nonzeros (black bars) and by-products (colored bars) on the left axis and the number of colors (colored disks) on the right axis varying the two block sizes r (color) and d (groups of bars and disks). (Color figure online)

6 Concluding Remarks

While matrix-free iterative methods and (transposed) Jacobian-vector products computed by automatic differentiation match well to each other, today, there is still a gap between preconditioning and automatic differentiation. The reason is that, in a matrix-free approach, accesses to individual nonzero entries of the Jacobian coefficient matrix which are needed by standard preconditioning techniques are computationally expensive. This statement holds not only for automatic differentiation but also for numerical differentiation.

The major new contribution of this article is a semi-matrix-free preconditioning approach that uses two separate diagonal block schemes partitioning the coefficient matrix into smaller submatrices. In both schemes, the diagonal blocks do not overlap. The first scheme employs blocks that define the required nonzero elements of a partial Jacobian computation. This scheme is relevant for minimizing the relative computational cost of the partial Jacobian computation. The resulting minimization problem is equivalent to a partial graph coloring problem. The second scheme is based on blocks whose sizes are larger than those of the first scheme. The blocks of this second scheme define the positions from which by-products of the partial Jacobian computation are extracted. Together with the required nonzero elements these by-products are used to construct a preconditioner that applies ILU decompositions to each of these blocks. Numerical experiments using the automatic differentiation tool ADiMat are reported demonstrating the feasibility of the new preconditioning technique.

There is room for further investigations that aim at bridging the gap between preconditioning and automatic differentiation. For instance, it is interesting to study more advanced preconditioning techniques and analyze to what extent they are capable of exploiting the information available in the by-products of the partial Jacobian computation.

Acknowledgements. Several computational experiments were performed on resources of Friedrich Schiller University Jena supported in part by Deutsche Forschungsgemeinschaft (DFG, German Research Foundation) – INST 275/334-1 FUGG; INST 275/363-1 FUGG. These resources were additionally supported by Freistaat Thüringen grant 2017 FGI 0031 co-funded by the European Union in the framework of Europäische Fonds für regionale Entwicklung (EFRE).

References

1. Benzi, M.: Preconditioning techniques for large linear systems: a survey. J. Comput. Phys. **182**(2), 418–477 (2002). https://doi.org/10.1006/jcph.2002.7176
2. Bischof, C.H., Bücker, H.M., Lang, B., Rasch, A., Vehreschild, A.: Combining source transformation and operator overloading techniques to compute derivatives for MATLAB programs. In: Proceedings of 2nd IEEE International Workshop on Source Code Analysis and Manipulation (SCAM 2002), pp. 65–72. IEEE Computer Society, Los Alamitos (2002). https://doi.org/10.1109/SCAM.2002.1134106
3. Bücker, H.M., Lülfesmann, M., Rostami, M.A.: Enabling implicit time integration for compressible flows by partial coloring: a case study of a semi-matrix-free preconditioning technique. In: 2016 Proceedings of 7th SIAM Workshop on Combinatorial Scientific Computing, pp. 23–32. SIAM, Philadelphia (2016). DOI: https://doi.org/10.1137/1.9781611974690.ch3
4. Cullum, K.J., Tůma, M.: Matrix-free preconditioning using partial matrix estimation. BIT Numer. Math. **46**(4), 711–729 (2006). https://doi.org/10.1007/s10543-006-0094-8
5. Gebremedhin, A.H., Manne, F., Pothen, A.: What color is your Jacobian? Graph coloring for computing derivatives. SIAM Rev. **47**(4), 629–705 (2005). https://doi.org/10.1137/S0036144504444711
6. Griewank, A., Walther, A.: Evaluating Derivatives: Principles and Techniques of Algorithmic Differentiation. No. 105 in Other Titles in App. Math. 2nd edn. SIAM, Philadelphia (2008). https://doi.org/10.1137/1.9780898717761
7. Lülfesmann, M.: Partielle Berechnung von Jacobi-Matrizen mittels Graphfärbung. In: Informatiktage 2007, Fachwissenschaftlicher Informatik-Kongress. Lecture Notes in Informatics - Seminars, vol. S-5, pp. 21–24. Gesellschaft für Informatik e.V. (2007). http://dl.gi.de/handle/20.500.12116/4920
8. Lülfesmann, M.: Graphfärbung zur Berechnung benötigter Matrixelemente. Informatik-Spektrum **31**(1), 50–54 (2008). https://doi.org/10.1007/s00287-007-0199-8
9. Lülfesmann, M.: Full and partial Jacobian computation via graph coloring: Algorithms and applications. Dissertation, Dept. Computer Science, RWTH Aachen University (2012). https://d-nb.info/1023979144/34
10. Petera, M., Lülfesmann, M., Bücker, H.M.: Partial Jacobian computation in the domain-specific program transformation system ADiCape. In: Proceedings of Internat. Multiconference on Computer Science and Information Technology, vol. 4, pp. 595–599. IEEE Computer Society, Los Alamitos (2009). https://doi.org/10.1109/IMCSIT.2009.5352778

11. Rall, L.B. (ed.): Automatic Differentiation: Techniques and Applications. LNCS, vol. 120. Springer, Heidelberg (1981). https://doi.org/10.1007/3-540-10861-0
12. Saad, Y.: Iterative Methods for Sparse Linear Systems, 2nd edn. SIAM, Philadelphia (2003). https://doi.org/10.1137/1.9780898718003
13. Saad, Y., Schultz, M.H.: GMRES: a generalized minimal residual algorithm for solving nonsymmetric linear systems. SIAM J. Sci. Stat. Comput. **7**(3), 856–869 (1986). https://doi.org/10.1137/0907058
14. Willkomm, J., Bischof, C.H., Bücker, H.M.: A new user interface for ADiMat: toward accurate and efficient derivatives of Matlab programs with ease of use. Int. J. Comput. Sci. Eng. **9**(5/6), 408–415 (2014). https://doi.org/10.1504/IJCSE.2014.064526

NURBS Curves in Parametric Integral Equations System for Modeling and Solving Boundary Value Problems in Elasticity

Marta Kapturczak[✉][iD], Eugeniusz Zieniuk[iD], and Andrzej Kużelewski[iD]

Institute of Informatics, University of Białystok, Białystok, Poland
{mkapturczak,ezieniuk}@ii.uwb.edu.pl

Abstract. This paper presents a way to improve the boundary shape modeling process in solving boundary value problems in elasticity. The inclusion of NURBS curves into the mathematical formalism of the parametric integral equations system method (PIES) is proposed. The advantages of such an application are widely discussed. Recently, the Bezier curves, mainly the cubic curves (of third-degree), were used. The segments of the boundary shape were modeled by such curves (with ensuring continuity at the connection points). Using NURBS curves, the boundary shape can be modeled with only one curve. So, continuity is automatically ensured. Additionally, the second degree NURBS curve is enough to obtain the shape with high accuracy (better than cubic Bezier curves). The NURBS curve is defined by points, their weights and knots vector. Such parameters significantly improve the shape modification process, which can directly improve e.g. the shape identification process. To examine the impact of modeling accuracy on the final PIES solutions, examples described by the Navier-Lamé equations were used. To improve calculations, the PIES method using NURBS curves was implemented as a computer program. Then, it was decided to verify the accuracy of the obtained solutions. For comparison, the solutions were also obtained using analytical solutions, boundary element method, and PIES method (with the Bezier curves). An improvement in the boundary shape modeling was noticed. It significantly affects the accuracy of solutions. As a result, the consumption of computer resources was reduced, while the process of boundary shape modeling and the accuracy of the obtained results were improved.

Keywords: Boundary value problems · NURBS · BEM · PIES

1   Introduction

In practice, the boundary value problems can be defined with a very large variety of boundary shapes. It is very difficult to use analytical methods to solve such problems. Therefore, scientists started to use and develop numerical methods.

© Springer Nature Switzerland AG 2020
V. V. Krzhizhanovskaya et al. (Eds.): ICCS 2020, LNCS 12138, pp. 116–123, 2020.
https://doi.org/10.1007/978-3-030-50417-5_9

Well known from the literature and widely used methods: finite element method (FEM) [3,7] and boundary element method (BEM) [2,8], use respectively the finite elements for modeling the domain or the boundary elements to define the boundary. Such methods tend to be less effective because, to improve the accuracy of modeling (what improves the solutions accuracy), the number of elements should be increased (what means greater consumption of computer resources).

In this paper, as an alternative to the classical discretization of the domain or boundary (occurring in the mentioned methods), a method of parametric integral equations system (PIES) was proposed [4,9,10]. This method uses the curves (known from computer graphics) directly included in the mathematical formalism of PIES. It makes modeling and modification of the shape much easier. Till now, in the PIES method, the Bézier curves were mainly used. These curves are defined using polynomial segments with an appropriate class of analytical and geometric continuity at their connection points.

With the development of computer graphics, the NURBS curves have appeared [5,6]. These curves give much more possibilities and significantly improve the accuracy of shape modeling. Therefore, they will be used for modeling the boundary shape of the boundary value problem solved by the PIES method. Now, the shape of the boundary can be modeled by one closed curve automatically (without additional care of the connection of segments). The definition of such a curve needs a set of points, weights, and knots. The point's weight determines its influence on the curve and makes it easier to accurately model curves such as the circle and ellipse. The knots allow to obtain corners and change the curve's degree. More information can be found in the paper about the NURBS curves in Laplace's equation [11]. Application of NURBS curves unifies the shape definition in PIES and also influences the modeling accuracy, which finally improves the accuracy of the solutions.

To highlight the benefits of using NURBS curves in PIES, the definition of the same shape for FEM, BEM, and PIES (using Bezier and NURBS curves) was presented on Fig. 1. Using NURBS curves here, it is enough to use two curves of the second degree (for inner and outer shape). Additionally, the 8 boundary points (4 for each shape) are enough to define properly such shape. Although the number of input data is reduced, the accuracy of the modeling is improved. So the solutions are also obtained with high accuracy.

2 Inclusion of the NURBS Curves into PIES

The general form of the PIES (for any shape of the boundary) for Navier-Lamé equation without mass forces is presented by the following formula [1]:

$$0.5\boldsymbol{u}_l(s_1) = \sum_{j=1}^{n} \int_{\widehat{s}_{j-1}}^{\widehat{s}_j} \left\{ \boldsymbol{U}_{lj}^*(s_1,s)\boldsymbol{p}_j(s) - \boldsymbol{P}_{lj}^*(s_1,s)\boldsymbol{u}_j(s) \right\} J_j(s)ds, \qquad (1)$$

where $\widehat{s}_{l-1} \leq s_1 \leq \widehat{s}_l$, $\widehat{s}_{j-1} \leq s \leq \widehat{s}_j$.

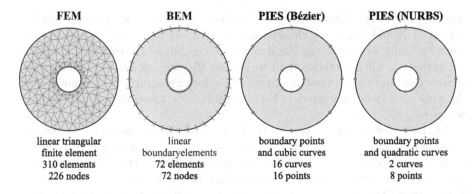

FEM	BEM	PIES (Bézier)	PIES (NURBS)
linear triangular finite element 310 elements 226 nodes	linear boundaryelements 72 elements 72 nodes	boundary points and cubic curves 16 curves 16 points	boundary points and quadratic curves 2 curves 8 points

Fig. 1. Comparison of modeling the shape of boundary for different methods.

Functions $\boldsymbol{p}_j(s) = [p_j^{(1)}(s), p_j^{(2)}(s)]$ and $\boldsymbol{u}_j(s) = [u_j^{(1)}(s), u_j^{(2)}(s)]$ are the boundary functions, that include the boundary condition. Function $J_j(s)$ is the Jacobian for segment of the curve $\boldsymbol{S}_j(s)$ and can be calculated as follows:

$$J_j(s) = \left[\left(\frac{\partial S_j^{(1)}(s)}{\partial s} \right)^2 + \left(\frac{\partial S_j^{(2)}(s)}{\partial s} \right)^2 \right]^{0.5}, \tag{2}$$

where $S_j^{(1)}(s), S_j^{(2)}(s)$ are the vector components of a curve segment $\boldsymbol{S}_j = [S_j^{(1)}(s), S_j^{(2)}(s)]^T$, and depend on parameter s.

First integrand $\boldsymbol{U}_{lj}^*(s_1, s)$, for the plane state of strain, is defined as follows:

$$U_{lj}^*(s_1, s) = -\frac{1}{8\pi(1-\nu)\mu} \begin{bmatrix} (3-4\nu)\ln(\eta) - \frac{\eta_1^2}{\eta^2} & -\frac{\eta_1\eta_2}{\eta^2} \\ -\frac{\eta_1\eta_2}{\eta^2} & (3-4\nu)\ln(\eta) - \frac{\eta_2^2}{\eta^2} \end{bmatrix}, \tag{3}$$

where $l, j = 1, 2, ..., n$, ν is the Poisson's ratio, μ is the Lamé constant (depends on the material constants). The shape of boundary is included by functions η, η_1 and η_2 defined as follows:

$$\eta = [\eta_1^2 + \eta_2^2]^{0.5}, \tag{4}$$

$$\eta_1 = S_l^{(1)}(s_1) - S_j^{(1)}(s), \quad \eta_2 = S_l^{(2)}(s_1) - S_j^{(2)}(s). \tag{5}$$

Second integrand $\boldsymbol{P}_{lj}^*(s_1, s)$ is described by formula:

$$P_{lj}^*(s_1, s) = -\frac{1}{4\pi(1-\nu)\eta} \begin{bmatrix} P_{11} & P_{12} \\ P_{21} & P_{22} \end{bmatrix}, \tag{6}$$

where $l, j = 1, 2, ..., n$, and $\boldsymbol{P}_{ik}$ (where i = 1, 2 and k = 1, 2) are defined as follows:

$$P_{ii} = \left\{ (1-2\nu) + 2\frac{\eta_i^2}{\eta^2} \right\} \frac{\partial \eta}{\partial \boldsymbol{n}}, \tag{7}$$

$$P_{ik} = \left\{ 2\frac{\eta_i \eta_k}{\eta^2} \frac{\partial \eta}{\partial n} - (1-2\nu) \left[\frac{\eta_i}{\eta} n_k(s) + \frac{\eta_k}{\eta} n_i(s) \right] \right\}, \tag{8}$$

$$\frac{\partial \eta}{\partial n} = \frac{\eta_1}{\eta} n_1(s) + \frac{\eta_2}{\eta} n_2(s). \tag{9}$$

where $n_1(s), n_2(s)$ are the components of the normal vector n_j to boundary segment S_j.

Till now the curvilinear segments were defined by cubic Bézier curves. The equation describing corresponding segments of such a curve is presented as follows:

$$S_m(s) = V_0(1-s)^3 + V_1 3(1-s)^2 + V_2(1-s)s^2 + V_3 s^3, \tag{10}$$

where $m = l, j$. The shape of such curve is defined by four control points divided into: approximation points (V_1, V_2) and interpolation points (V_0, V_3).

The inclusion of the NURBS curves in PIES (1) requires in functions (5) the substitution of the following formula describing segments [6]:

$$S_m(s) = \frac{\sum\limits_{i=0}^{n} w_i P_i N_i^k(s)}{\sum\limits_{i=0}^{n} w_i N_i^k(s)} \quad \text{dla} \quad t_k \leq s \leq t_{n+1}, \tag{11}$$

where $P_i (i = 0, 1, \ldots, n)$ are the control points, $w_i (i = 0, 1, \ldots, n)$ are weights corresponding to points, and $N_i^k(s)$ is the function of k-degree, defined by the recursive formula:

$$N_i^0(s) = \begin{cases} 1 & s \in \langle t_i, t_{i+1} \rangle \\ 0 & \text{otherwise} \end{cases}, \tag{12}$$

$$N_i^k(s) = \frac{s - t_i}{t_{i+k} - t_i} N_i^{k-1}(s) + \frac{t_{i+k+1} - s}{t_{i+k+1} - t_{i+1}} N_{i+1}^{k-1}(s), \tag{13}$$

where $0 \leq i \leq n - k - 1, 1 \leq k \leq n - 1, \frac{0}{0} := 0$. The $t_i (i = 0, 1, \ldots, m + n + 1)$ are the elements of the knot vector.

Additionally, to obtain values of Jacobian (2) and normal vector, the derivative with respect to parameter s of the NURBS curve function have to be calculated [6]:

$$\frac{\partial S_m(s)}{\partial s} = \frac{\sum\limits_{i=0}^{n} w_i P_i (N_i^k(s))' \sum\limits_{i=0}^{n} w_i N_i^k(s) - \sum\limits_{i=0}^{n} w_i P_i N_i^k(s) \sum\limits_{i=0}^{n} w_i (N_i^k(s))'}{\left(\sum\limits_{i=0}^{n} w_i N_i^k(s) \right)^2} \tag{14}$$

where $(N_i^k(s))'$ is defined as follow:

$$(N_i^k(s))' = \frac{k}{t_{i+k} - t_i} N_i^{k-1}(s) - \frac{k}{t_{i+k+1} - t_{i+1}} N_{i+1}^{k-1}(s), \tag{15}$$

The calculations processed using the above formulas cause significant computational time delay. Therefore, in the program, only NURBS curves of second degree were implemented. For calculation of the shape function and derivative, the analytical formulas were obtained (assuming the curve of the second degree).

3   Verification and Advantages of NURBS Curves in PIES

Example 1 - Circular Hole

First example is the circular hole with radius $R = 10$. Inside the cylinder, defined in the plane state of strain, the force $p = 15$ is applied. The material parameters are: $E = 21$ and $\nu = 0.1$. The shape modeling with Bézier curves, recently used in PIES, is presented in Fig. 2a. Only eight boundary points P_i are shown, for better clarity (two additional approximation points should be defined between the boundary points in the cubic curve). For the same purpose, the one NURBS curve of the second degree (4 boundary points) is enough (Fig. 2b). Moreover, despite the smaller amount of input data, the NURBS curve modeled the shape more accurately than previously used cubic Bézier curves (the average radius over the entire circumference is respectively $R_N = 9.9999$ and $R_B = 9.9944$).

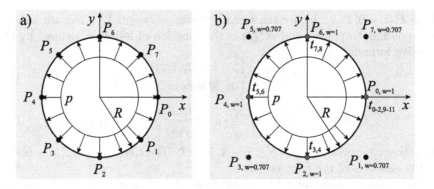

Fig. 2. Modeling the shape of boundary using a) Béziera, and b) NURBS, curves.

Solutions obtained using PIES (with NURBS curves) are compared with analytical [8] and numerical (BEM using 24 linear boundary elements [2] and PIES using Bézier curves) solutions in Table 1. Comparing the PIES (with NURBS) solutions with analytical ones, the maximum average relative error 0.06% is obtained for σ_x. It is a significant improvement in comparison with PIES (using Bézier curves) or BEM, where the errors for σ_x are respectively 0.37% and 2.04%.

Table 1. Solutions in the domain ($y = 0$) of boundary value problem from Fig. 2.

x	Analytical		PIES (NURBS)		BEM		PIES (Bézier)	
	σ_x	σ_y	σ_x	σ_y	σ_x	σ_y	σ_x	σ_y
12	−10.417	10.417	−10.426	10.415	−10.222	10.222	−10.359	10.421
15	−6.667	6.667	−6.670	6.667	−6.525	6.525	−6.642	6.654
20	−3.75	3.75	−3.751	3.750	−3.67	3.67	−3.742	3.743
Average relative error [%]			0.06	0.01	2.04	2.04	0.37	0.14

3.1   Example 2 - Multiply Connected Domain

The next example is the Lamé problem. The Fig. 3a presents only 16 boundary points of the shape modeled using Bézier curves. The additional 32 approximation points should be defined. Figure 3b presents modeling using NURBS curves.

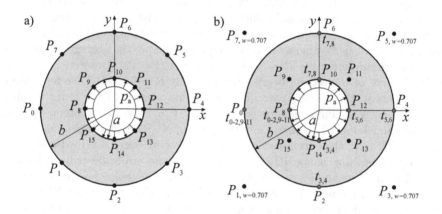

Fig. 3. Modeling the shape of boundary using a) Béziera, and b) NURBS, curves.

The cylindrical tube under inner hydrostatic pressure $p = 100MPa$ (in the plane state of strain) is considered. The inner radius is $a = 10$ cm, and the outer is $b = 25$ cm. The material parameters are: $E = 2*10^5$ MPa i $\nu = 0.25$. Analytical solutions are defined as [8]:

$$\sigma_x = \frac{p_a a^2 - p_b b^2}{b^2 - a^2} - \frac{(p_a - p_b)a^2 b^2}{r^2(b^2 - a^2)}, \qquad \sigma_y = \frac{p_a a^2 - p_b b^2}{b^2 - a^2} + \frac{(p_a - p_b)a^2 b^2}{r^2(b^2 - a^2)}, \quad (16)$$

where r is the one of polar coordinates and $r^2 = x^2 + y^2, a < r < b$. Analytical and PIES (with Bézier and NURBS curves) solutions are presented in Table 2. The example confirms the accuracy of the solutions using NURBS curves.

Table 2. Comparison PIES solutions with analytical ones - example from Fig. 3.

x	Analytical		PIES (Bézier)		Error [%]		PIES (NURBS)		Error [%]	
	σ_x	σ_y	σ_x	σ_y	σ_x	σ_y	σ_x	σ_y	σ_x	σ_y
12	−63.624	101.720	−63.220	101.811	0.635	0.089	−63.681	101.681	0.090	0.038
14	−41.691	79.786	−41.488	79.752	0.487	0.043	−41.707	79.757	0.038	0.036
16	−27.455	65.551	−27.364	65.535	0.331	0.024	−27.467	65.529	0.044	0.034
18	−17.695	55.791	−17.665	55.792	0.170	0.002	−17.706	55.775	0.062	0.029
20	−10.714	48.810	−10.724	48.803	0.093	0.014	−10.724	48.798	0.093	0.025
22	−5.549	43.644	−5.577	43.578	0.505	0.151	−5.553	43.631	0.072	0.030
24	−1.620	39.716	−1.547	39.500	4.506	0.544	−1.569	39.695	3.148	0.053
Average relative error [%]:					0.96	0.12			0.51	0.03

4 Conclusions

This paper presents the high efficiency of inclusion of NURBS curves in the PIES method and compare obtained solutions with the existing ones. Using NURBS curves, the number of points (necessary to model the boundary) is much lower. Using 8 cubic Bézier segments in modeling circle shape, the average radius over the entire circumference is $R = 9.9944$ (expected $R = 10$). However, using one NURBS curve of the second degree, the radius is $R = 9.9999$. Therefore, using NURBS curves, the modeling accuracy increases, even with the lower number of points. Such an improvement is caused by two curve parameters: weights of points and knots. The shape modeling using NURBS in PIES is also more uniform. Till now, the curvilinear and linear segments were described using separate functions. Now, using NURBS curves, the shape will be described with one curve. The above examples present the advantages of modeling and its impact on improvement in the accuracy of obtained solutions. Therefore, using NURBS in PIES improves the process of modeling and modification of the boundary shape, significantly reduces the number of points (necessary to define the shape) and ultimately improves the accuracy of the solutions.

References

1. Boltuc, A., Zieniuk, E.: Modeling domains using Bezier surfaces in plane boundary problems defined by the Navier-Lame equation with body forces. Eng. Anal. Bound. Elem. **35**(10), 1116–1122 (2011)
2. Brebbia, C.A., Telles, J.C.F., Wrobel, L.C.: Boundary Element Techniques: Theory and Applications in Engineering. Springer, New York (1984). https://doi.org/10.1007/978-3-642-48860-3
3. Doitrand, A., Martin, E., Leguillon, D.: Numerical implementation of the coupled criterion: matched asymptotic and full finite element approaches. Finite Elem. Anal. Des. **168**, 103344 (2020)
4. Kuzelewski, A., Zieniuk, E., Kapturczak, M.: Acceleration of integration in parametric integral equation system using CUDA. Comput. Struct. **152**, 113–124 (2015)

5. Noruzi, R., Ghadai, S., Bingol, O.R., Krishnamurthy, A., Ganapathysubramanian, B.: NURBS-based microstructure design for organic photovoltaics. Comput. Aided Des. **118**, 102771 (2020)
6. Piegl, L.A., Tiller, W.: Computing the derivative of NURBS with respect to a knot. Comput. Aided Geom. Des. **15**(9), 925–934 (1998)
7. Ji, S.Y., Wang, S.L.: A coupled discrete-finite element method for the ice-induced vibrations of a conical jacket platform with a GPU-based parallel algorithm. Int. J. Comput. Methods **17**(4), 1850147 (2020)
8. Timoshenko, S.P., Goodier, J.N.: Theory of Elasticity. McGraw-Hill, Tokyo (1970)
9. Zieniuk, E., Szerszeń, K.: A separation between the boundary shape and the boundary functions in the parametric integral equation system for the 3D stokes equation. Numer. Algorithms **80**(3), 753–780 (2018). https://doi.org/10.1007/s11075-018-0505-3
10. Zieniuk, E., Kapturczak, M., Kużelewski, A.: Modification of interval arithmetic for modelling and solving uncertainly defined problems by interval parametric integral equations system. In: Shi, Y., Fu, H., Tian, Y., Krzhizhanovskaya, V.V., Lees, M.H., Dongarra, J., Sloot, P.M.A. (eds.) ICCS 2018. LNCS, vol. 10862, pp. 231–240. Springer, Cham (2018). https://doi.org/10.1007/978-3-319-93713-7_19
11. Zieniuk, E., Kapturczak, M.: Modeling the shape of boundary using NURBS curves directly in modified boundary integral equations for Laplace's equation. Comput. Appl. Math. **37**(4), 4835–4855 (2018)

Parameterizations and Lagrange Cubics for Fitting Multidimensional Data

Ryszard Kozera[1,2,3]([✉]), Lyle Noakes[2], and Magdalena Wilkołazka[3]([✉])

[1] Institute of Information Technology, Warsaw University of Life Sciences - SGGW,
Ul. Nowoursynowska 157, 02-776 Warsaw, Poland
ryszard.kozera@sggw.edu.pl, ryszard.kozera@gmail.com
[2] Faculty of Engineering and Mathematical Sciences,
The University of Western Australia, 35 Stirling Highway, Crawley,
Perth, WA 6009, Australia
lyle.noakes@uwa.edu.au
[3] Faculty of Natural and Health Sciences,
The John Paul II Catholic University of Lublin, Ul. Konstantynów 1H,
20-708 Lublin, Poland
magda.wilkolazka@gmail.com, magda8310@kul.lublin.pl

Abstract. This paper discusses the issue of interpolating data points in arbitrary Euclidean space with the aid of Lagrange cubics $\hat{\gamma}^L$ and exponential parameterization. The latter is commonly used to either fit the so-called reduced data $Q_m = \{q_i\}_{i=0}^m$ for which the associated exact interpolation knots remain unknown or to model the trajectory of the curve γ passing through Q_m. The exponential parameterization governed by a single parameter $\lambda \in [0,1]$ replaces such discrete set of unavailable knots $\{t_i\}_{i=0}^m$ ($t_i \in I$ - an internal clock) with some new values $\{\hat{t}_i\}_{i=0}^m$ ($\hat{t}_i \in \hat{I}$ - an external clock). In order to compare γ with $\hat{\gamma}^L$ the selection of some $\phi : I \to \hat{I}$ should be predetermined. For some applications and theoretical considerations the function $\phi : I \to \hat{I}$ needs to form *an injective mapping* (e.g. in length estimation of γ with any $\hat{\gamma}$ fitting Q_m). We formulate and prove two *sufficient conditions* yielding ϕ as *injective* for given Q_m and analyze their asymptotic character which forms an important question for Q_m getting sufficiently dense. The *algebraic conditions* established herein are also *geometrically visualized* in 3D plots with the aid of *Mathematica*. This work is supplemented with illustrative examples including numerical testing of the underpinning convergence rate in length estimation $d(\gamma)$ by $d(\hat{\gamma})$ (once $m \to \infty$). The *reparameterization* has potential ramifications in computer graphics and robot navigation for trajectory planning e.g. to construct a new curve $\tilde{\gamma} = \hat{\gamma} \circ \phi$ controlled by the appropriate choice of interpolation knots and of mapping ϕ (and/or possibly Q_m).

1 Introduction

Assume that $\gamma : I \to \mathbb{E}^n$ represents a smooth *regular curve* (i.e. $\dot{\gamma}(t) \neq \vec{0}$) of class C^k (usually with $k = 3, 4$) defined over a compact interval $I = [0, T]$ (with

© Springer Nature Switzerland AG 2020
V. V. Krzhizhanovskaya et al. (Eds.): ICCS 2020, LNCS 12138, pp. 124–140, 2020.
https://doi.org/10.1007/978-3-030-50417-5_10

$0 < T < \infty$). Suppose that $m + 1$ interpolation points $\{q_i\}_{i=0}^m = \{\gamma(t_i)\}_{i=0}^m$ (forming the so-called *reduced data* Q_m) belong to an arbitrary Euclidean space $\mathbb{E}^n$. Here $\mathcal{T} = \{t_i\}_{i=0}^m$ is not given (here $t_i < t_{i+1}$). We introduce now (see e.g. [1,7,12] or [19]) some preliminary notions (applicable for $m \to \infty$).

Definition 1.1. The interpolation knots $\mathcal{T}$ are *admissible* if:

$$\lim_{m \to \infty} \delta_m = 0, \text{ where } \delta_m = \max_{1 \le i \le m} \{t_i - t_{i-1} : \quad i = 1, 2, \ldots, m\}. \tag{1}$$

Definition 1.2. The interpolation knots $\mathcal{T}$ are *more-or-less uniform* if there exist constants $0 < K_l \le K_u$ such that:

$$(K_l/m) \le t_i - t_{i-1} \le (K_u/m), \tag{2}$$

for all $i = 1, 2, \ldots, m$ and any $m \in \mathbb{N}$. Alternatively, more-or-less uniformity amounts to the existence of some constant $0 < \beta \le 1$ such that $\beta \delta_m \le t_i - t_{i-1} \le \delta_m$ for all $i = 1, 2, \ldots, m$ and arbitrary $m \in \mathbb{N}$. Lastly, the subfamily $\mathcal{T}_{\beta_0}$ of more-or-less uniform samplings represents a set of β_0-*more-or-less uniform* samplings if each of its representatives satisfies $\beta_0 \le \beta \le 1$, for some $0 < \beta_0 \le 1$ fixed.

Having selected the fitting scheme $\hat{\gamma}$ of Q_m the unknown knots $\mathcal{T}$ for the interpolant $\hat{\gamma}$ must somehow be replaced by estimates $\hat{\mathcal{T}} = \{\hat{t}_i\}_{i=0}^m$ subject to $\hat{\gamma}(\hat{t}_i) = q_i$. We use here the so-called *exponential parameterization* (see e.g. [17]) which depends on a single parameter $\lambda \in [0, 1]$ according to:

$$\hat{t}_0 = 0 \quad \text{and} \quad \hat{t}_i = \hat{t}_{i-1} + \|q_i - q_{i-1}\|^\lambda, \tag{3}$$

for $i = 1, 2, \ldots, m$. It is also assumed here that $q_i \ne q_{i+1}$ so that the extra condition $\hat{t}_i < \hat{t}_{i+1}$ is preserved as stipulated generically while fitting reduced data Q_m. The case of $\lambda = 0$ in (3) gives *uniform knots* $\hat{t}_i = i$. Evidently the latter does not reflect the geometry of Q_m. On the other hand, $\lambda = 1$ yields the so-called *cumulative chord parameterization* which coincides with Euclidean distances between consecutive points q_i and q_{i+1} and as such it refers to the spread of Q_m. More information on the above topic and related issues can be found e.g. in [3,5,16,17] or [18].

The selection of the specific interpolant $\hat{\gamma} : \hat{I} = [0, \hat{T}] \to \mathbb{E}^n$ (with $\hat{T} = \hat{t}_m$) together with some knots' estimates $\hat{\mathcal{T}} \approx \mathcal{T}$ raises an important question concerning the convergence rate (if any) in approximating γ with $\hat{\gamma}$ (or its length) once $m \to \infty$. Recall first (see [1,12] or [19]):

Definition 1.3. Consider a family $\{F_{\delta_m}, \delta_m > 0\}$ of functions $F_{\delta_m} : I \to \mathbb{E}^n$. We say that F_{δ_m} is of order $O(\delta_m^\alpha)$ (denoted as $F_{\delta_m} = O(\delta_m^\alpha)$), if there is a constant $K > 0$ such that, for some $\bar{\delta} > 0$ the inequality $\|F_{\delta_m}(t)\| < K\delta_m^\alpha$ holds for all $\delta_m \in (0, \bar{\delta})$, uniformly over I.

For a given $\hat{\gamma}$ fitting dense data Q_m based on $\hat{\mathcal{T}} \approx \mathcal{T}$ (and some *a priori* selected mapping $\phi : I \to \hat{I}$) the natural question arises about the distance measurement $\|F_{\delta_m}\| = \|\gamma - \hat{\gamma} \circ \phi\|$ tending to 0 (uniformly over I), while $m \to \infty$.

Of course, by (1) proving $F_{\delta_m} = \gamma - \hat{\gamma} \circ \phi = O(\delta_m^\alpha)$ not only guarantees the latter but also establishes lower bound on convergence speed (if $\alpha > 0$). The coefficient $\alpha > 0$ appearing in Definition 1.3 is called *the convergence rate* in approximating γ by $\hat{\gamma} \circ \phi$ uniformly over $[0, T]$. If additionally such α cannot be improved (once γ and $\mathcal{T}$ are given) then α is *sharp*. The latter analogously extends to the length estimation (with $n = 1$), for which the scalar expression $F_{\delta_m} = d(\gamma) - d(\hat{\gamma}) = O(\delta_m^\beta)$ is to be considered.

For certain applications such as the analysis of the convergence rate in $d(\gamma) = \int_0^T \|\dot{\gamma}(t)\| dt \approx d(\hat{\gamma}) = \int_0^{\hat{T}} \|\hat{\gamma}'(\hat{t})\| d\hat{t}$ (see e.g. [2,5] or [15]) the mapping $\phi(t) = \hat{t}$ should be *a reparameterization* of I into $\hat{I}$ (i.e. $\dot{\phi} > 0$). In other situations such as robot's and drone path planning the extra trajectory looping of $\hat{\gamma}$ is sometimes needed (e.g. for traction line posts' inspection while making circles by drone). Of course, in many other applications robot navigation requires trajectory planning with no loops whatsoever. In that context (as well as for length estimation) one of the conditions to exclude the local looping of $\hat{\gamma} \circ \phi$ is to require ϕ to be *an injective* function (see e.g. [13]).

From now on it is assumed that $\hat{\gamma} = \hat{\gamma}^L$ which represents a piecewise-Lagrange cubic $\hat{\gamma}^L : \hat{I} = [0, \hat{T}] \to \mathbb{E}^n$ (see e.g. [1]). More precisely, the interpolant $\hat{\gamma}^L$ is defined as *a track-sum* of Lagrange cubics $\{\hat{\gamma}_{i=3k}^L\}_{k=0}^{(m-3)/3}$ with each $\hat{\gamma}_i^L : \hat{I}_i = [\hat{t}_i, \hat{t}_{i+3}] \to \mathbb{E}^n$ satisfying $q_{i+j} = \hat{\gamma}_i^L(\hat{t}_{i+j})$, for $j = 0, 1, 2, 3$. As already pointed out the unavailable knots $\mathcal{T}$ are estimated with $\hat{\mathcal{T}}$ governed by exponential parameterization (3). For simplicity we suppose that $m = 3k$, where $k \in \{1, 2, 3, \ldots\}$. In a similar fashion, one selects here $\phi = \psi^L$ defined as *a track-sum* of Lagrange cubics $\{\psi_{i=3k}^L\}_{k=0}^{(m-3)/3}$ mapping $\psi_i^L : I_i = [t_i, t_{i+3}] \to [\hat{t}_i, \hat{t}_{i+3}]$ and fulfilling $t_{i+j} = \hat{\psi}_i^L(t_{i+j})$, for $j = 0, 1, 2, 3$. Evidently if $\dot{\psi}_i^L > 0$ (as $\hat{t}_i < \hat{t}_{i+1}$) then $\psi_i^L : I_i \to \hat{I}_i = Rg(\psi_i^L)$ (here $Rg(\psi_i^L)$ denotes the range of ψ_i^L). On the other hand if ψ_i^L is not injective we may also have $\psi_i^L : I_i \to \hat{I}_i \subset Rg(\psi_i^L)$. In order to construct the composition $\hat{\gamma}_i^L \circ \psi_i^L$ as a well-defined function, each domain of $\hat{\gamma}_i^L$ is here understood as naturally extendable from $\hat{I}_i$ to $\mathbb{R}$. Such adjusted Lagrange piecewise-cubics denoted as $\check{\gamma}_i^L$ satisfy $\check{\gamma}_i^L|_{\hat{I}_i} = \hat{\gamma}_i^L$. The following result holds (see e.g. [7,9] or [19]):

Theorem 1.4. *Assume $\gamma \in C^4([0, T])$ be a regular curve in $\mathbb{E}^n$ sampled admissibly (see (1)). For $\hat{\gamma}^L$ and $\lambda = 1$ in (3) each mapping ψ_i^L is a C^∞ reparameterization of I_i into $\hat{I}_i$ and we have (uniformly over $[0, T]$):*

$$\gamma - \hat{\gamma}_i^L \circ \psi_i^L = O(\delta_m^4). \tag{4}$$

In the remaining cases of $\lambda \in [0, 1)$ from (3) let γ be sampled more-or-less uniformly (see (2)). Then for each mapping ψ_i^L combined with $\check{\gamma}_i^L$ the following holds (uniformly over $[0, T]$):

$$\gamma - \check{\gamma}_i^L \circ \psi_i^L = O(\delta_m). \tag{5}$$

Both (4) and (5) are *sharp within the class* of $\gamma \in C^4([0, T])$ and *within a given family of admitted samplings*, assumed here as either (1) or (2), respectively. By the latter we understand the existence of at least one $\gamma_0 \in C^4([0, T])$

and some admissible (or more-or-less uniform) sampling $\mathcal{T}_0$ for which $\alpha(1) = 4$ in (4) (or $\alpha(\lambda) = 1$ for $\lambda \in [0, 1)$ in (5)) are sharp according to Definition 1.3 - see also [9] or [12]. Note that ψ^L as *a track-sum* of $\{\psi^L_{i=3k}\}_{k=0}^{(m-3)/3}$ defines a piecewise C^∞ mapping of I into $\mathbb{R}$ at least continuous at $\mathcal{T}$. If ψ^L is a reparameterization (e.g. always holding asymptotically for $\lambda = 1$) then $\psi^L : I \to \hat{I}$. In particular for $\lambda = 1$ we also have $d(\gamma) - d(\hat{\gamma}^L) = O(\delta_m^4)$ - see [19]. In contrast, the injectivity of ψ^L_i and length estimation for $\lambda \in [0, 1)$ has not been so far examined.

In this paper we introduce *two sufficient conditions enforcing* each $\psi^L_i : I_i \to \hat{I}_i$ to be *injective*, for $\lambda \in [0, 1)$ governing the exponential parameterization (3). These two conditions are represented by the inequalities (6) and (7). In the next step, Theorem 2.1 is established (*the main result of this paper*) to formulate several sufficient conditions enforcing (6) and (7) to hold asymptotically. Noticeably all derived conditions stipulating asymptotically the injectivity of ψ^L are independent from γ and apply to any fixed $\lambda \in [0, 1)$ and to any preselected β_0-more-or-less-uniform samplings (i.e. to any $0 < \beta_0 < 1$ fixed *a priori*). Additionally, all re-transformed *algebraic constraints* established here are *visualized* with the aid of *3D* plots in *Mathematica* (see [22]). The conditions can also be exploited once the incomplete information about samplings is available such as *a priori* knowledge of the respective upper and lower bounds for each triples (M_{im}, N_{im}, P_{im}) characterizing $\mathcal{T}$ as specified in (8) - see also Remark 3.1. The examples illustrate Theorem 2.1 and the relevance of this work (see Example 1). The conjecture concerning the sharp convergence rate $\alpha(\lambda) = 2$ in length estimation $d(\gamma) - d(\hat{\gamma}^L) = O(\delta_m^{\alpha(\lambda)})$ (combined with (3) for all $\lambda \in [0, 1)$ yielding $\dot{\phi} > 0$) is tested numerically (see Example 2 and Remark 3.2).

2   Sufficient Conditions for Injectivity of ψ^L_i

In this section we establish and discuss the asymptotic character (i.e. applicable for m sufficiently large) of two sufficient conditions enforcing ψ^L_i to be *a genuine reparameterization* of I_i into $\hat{I}_i$ based on multidimensional reduced data Q_m.

Evidently the positivity of *the quadratic* $\dot{\psi}^L_i(t) = a_i t^2 + b_i t + c_i$ over I_i is e.g. guaranteed (for both *sparse* and *dense* data Q_m) provided if e.g. either (6) or (7) hold:

$$a_i < 0 \quad \text{and} \quad \dot{\psi}^L_i(t_i) > 0 \quad \text{and} \quad \dot{\psi}^L_i(t_{i+3}) > 0, \tag{6}$$

$$a_i > 0 \quad \text{and} \quad \dot{\psi}^L_i\left(-\frac{b_i}{2a_i}\right) > 0. \tag{7}$$

Noticeably, any admissible sampling (1) can be characterized as follows:

$$t_{i+1} - t_i = M_{im}\delta_m, \quad t_{i+2} - t_{i+1} = N_{im}\delta_m \text{ and } t_{i+3} - t_{i+2} = P_{im}\delta_m, \tag{8}$$

where $0 < M_{im}, N_{im}, P_{im} \leq 1$. *The main theoretical contribution* of this paper reads as:

Theorem 2.1. *Let* $\gamma \in C^3([0,T])$ *be sampled* β_0-*more-or-less uniformly (see Definition (1.2)) with knots* $\mathcal{T}$ *represented by* (8). *For data* Q_m *combined with exponential parameterization* (3) *(with any fixed* $\lambda \in [0,1)$) *the condition* (6) *yielding each* $\psi_i^L : I \to \hat{I}_i$ *as a reparameterization holds asymptotically, if the following three inequalities are satisfied for sufficiently large* m:

$$\frac{1}{P_{im} + N_{im} + M_{im}} \left(\frac{P_{im}^{\lambda-1} - N_{im}^{\lambda-1}}{P_{im} + N_{im}} - \frac{N_{im}^{\lambda-1} - M_{im}^{\lambda-1}}{N_{im} + M_{im}} \right) \leq \rho < 0, \qquad (9)$$

$$M_{im}^{\lambda-1} - \frac{(N_{im}^{\lambda-1} - M_{im}^{\lambda-1})M_{im}}{N_{im} + M_{im}} + \frac{(P_{im}^{\lambda-1} - N_{im}^{\lambda-1})M_{im}(N_{im} + M_{im})}{(P_{im} + N_{im})(P_{im} + N_{im} + M_{im})}$$
$$- \frac{(N_{im}^{\lambda-1} - M_{im}^{\lambda-1})M_{im}}{P_{im} + N_{im} + M_{im}} \geq \rho_1 > 0, \qquad (10)$$

$$P_{im}^{\lambda-1} - \frac{(N_{im}^{\lambda-1} - M_{im}^{\lambda-1})P_{im}(P_{im} + N_{im})}{(N_{im} + M_{im})(P_{im} + N_{im} + M_{im})} + \frac{P_{im}(P_{im}^{\lambda-1} - N_{im}^{\lambda-1})}{P_{im} + N_{im} + M_{im}}$$
$$+ \frac{P_{im}(P_{im}^{\lambda-1} - N_{im}^{\lambda-1})}{P_{im} + N_{im}} \geq \rho_2 > 0, \qquad (11)$$

with fixed $\rho < 0$, $\rho_1 > 0$ *and* $\rho_2 > 0$ *but arbitrary small. Similarly, the condition* (7) *enforcing* $\dot{\psi}_i^L > 0$ *holds asymptotically if the following two inequalities are met for sufficiently large* m:

$$\frac{1}{P_{im} + N_{im} + M_{im}} \left(\frac{P_{im}^{\lambda-1} - N_{im}^{\lambda-1}}{P_{im} + N_{im}} - \frac{N_{im}^{\lambda-1} - M_{im}^{\lambda-1}}{N_{im} + M_{im}} \right) \geq \rho_3 > 0, \qquad (12)$$

$$M_{im}^{\lambda-1} + \frac{(N_{im}^{\lambda-1} - M_{im}^{\lambda-1})(2N_{im} + M_{im})}{3(N_{im} + M_{im})} - \frac{(N_{im}^{\lambda-1} - M_{im}^{\lambda-1})^2}{3(N_{im} + M_{im})}$$
$$\cdot \frac{(P_{im} + N_{im})(P_{im} + N_{im} + M_{im})}{(P_{im}^{\lambda-1} - N_{im}^{\lambda-1})(N_{im} + M_{im}) - (N_{im}^{\lambda-1} - M_{im}^{\lambda-1})(P_{im} + N_{im})}$$
$$- \left[\frac{(P_{im}^{\lambda-1} - N_{im}^{\lambda-1})(N_{im} + M_{im}) - (N_{im}^{\lambda-1} - M_{im}^{\lambda-1})(P_{im} + N_{im})}{(N_{im} + M_{im})(P_{im} + N_{im})(P_{im} + N_{im} + M_{im})} \right]$$
$$\cdot \frac{(N_{im}^2 + N_{im}M_{im} + M_{im}^2)}{3} \geq \rho_4 > 0, \qquad (13)$$

where constants $\rho_3 > 0$ *and* $\rho_4 > 0$ *are fixed and small.*

Proof. Newton interpolation formula (see [1]) based on divided differences of ψ_i^L yields over I_i:

$$\psi_i^L(t) = \psi_i^L(t_i) + \psi_i^L[t_i, t_{i+1}](t - t_i) + \psi_i^L[t_i, t_{i+1}, t_{i+2}](t - t_i)(t - t_{i+1})$$
$$+ \psi_i^L[t_i, t_{i+1}, t_{i+2}, t_{i+3}],$$

which for each $t \in I_i$ renders $\dot{\psi}_i^L(t) =$

$$\psi_i^L[t_i, t_{i+1}] + \psi_i^L[t_i, t_{i+1}, t_{i+2}](2t - t_i - t_{i+1}) + \psi_i^L[t_i, t_{i+1}, t_{i+2}, t_{i+3}] \qquad (14)$$
$$\cdot \big((t - t_{i+1})(t - t_{i+2}) + (t - t_i)(t - t_{i+2}) + (t - t_i)(t - t_{i+1})\big).$$

We recall now the proof of (18) (see [9] or [12]) since it is vital for further arguments. As γ is regular it can be assumed to be parameterized by *arc-length* rendering $\|\dot{\gamma}(t)\| = 1$, for $t \in [0, T]$ (see [2]). The latter due to $1 \equiv \|\dot{\gamma}(t)\|^2 = \langle \dot{\gamma}(t)|\dot{\gamma}(t) \rangle$ results in $0 \equiv (\|\dot{\gamma}(t)\|^2)' = 2\langle \dot{\gamma}(t)|\ddot{\gamma}(t) \rangle$ over $t \in [0, T]$. The orthogonality of $\dot{\gamma}$ and $\ddot{\gamma}$ nullifies certain terms in the expression (for $j = i+k$ with $k = 0, 1, 2$ and any $\lambda \in [0, 1]$):

$$\hat{t}_{j+1} - \hat{t}_j = \|q_{j+1} - q_j\|^\lambda = \|\gamma(t_{j+1}) - \gamma(t_j)\|^\lambda = \langle \gamma(t_{j+1}) - \gamma(t_j)|\gamma(t_{j+1}) - \gamma(t_j) \rangle^\lambda \tag{15}$$

once Taylor expansion for $\gamma \in C^3$ is used:

$$\gamma(t_{j+1}) - \gamma(t_j) = (t_{j+1} - t_j)\dot{\gamma}(t_j) + \frac{(t_{j+1} - t_j)^2}{2}\ddot{\gamma}(t_j) + O((t_{j+1} - t_j)^2). \tag{16}$$

Indeed, upon substituting (16) into (15) and exploiting $\langle \dot{\gamma}(t)|\ddot{\gamma}(t) \rangle = 0$ one obtains:

$$\hat{t}_{j+1} - \hat{t}_j = (t_{j+1} - t_j)^\lambda \left(1 + O((t_{j+1} - t_j)^2)\right)^{\frac{\lambda}{2}}. \tag{17}$$

For any admissible samplings the constants in the term $O((t_{j+1}-t_j)^2)$ depend on the third derivative of γ which is bounded over $[0, T]$ as $\gamma \in C^3$. Again Taylor Th. applied to the function $f(x) = (1 + x)^{\frac{\lambda}{2}}$ at $x_0 = 0$ yields for all $x \in [-\varepsilon, \varepsilon] = I_\varepsilon$ (with some fixed $\varepsilon > 0$) the existence of some ξ_x satisfying $|\xi_x| < |x|$ such that $f(x) = 1 + \frac{\lambda}{2}x + \frac{\lambda}{4}(\frac{\lambda}{2} - 1)(1 + \xi_x)^{\frac{\lambda}{2}-2}$. For $0 < \varepsilon < 1$ we exclude the singularity of $\tau(\xi_x) = (1 + \xi_x)^{\frac{\lambda}{2}-2}$ at $\xi_x = -1$ (with $\lambda \in [0, 1]$) which forces τ to be bounded over I_ε. Thus for $|\xi_x| < |x| \leq \varepsilon < 1$ we have $f_1(x) = 1 + \frac{\lambda}{2}x + O(x^2)$ - the constant standing along x^2 depends now on λ (which is fixed). Take now $x = O((t_{j+1} - t_j)^2)$ determined in (17) which is asymptotically small (for m large) due to the admissibility condition (1) and thus separated from -1. Hence the second-divided differences of ψ_i^L satisfy (with $k = 0, 1, 2$):

$$\psi_i^L[t_{i+k}, t_{i+k+1}] = \frac{\hat{t}_{i+k+1} - \hat{t}_{i+k}}{t_{i+k+1} - t_{i+k}} = (t_{i+k+1} - t_{i+k})^{\lambda-1} + O((t_{i+k+1} - t_{i+k})^{1+\lambda}). \tag{18}$$

Thus, by (8) and (18) one obtains for each $\lambda \in [0, 1]$ and $k = 0, 1, 2$ the following formula for the *second divided differences* of ψ_i^L (needed also in (15)):

$$\psi_i^L[t_{i+k}, t_{i+k+1}] = R_{imk}^{\lambda-1}\delta_m^{\lambda-1} + O(\delta_m^{1+\lambda}), \tag{19}$$

with $R_{im0} = M_{im}$, $R_{im1} = N_{im}$ and $R_{im2} = P_{im}$. Furthermore still by (18) combined with $0 < (t_{i+l+1} - t_{i+l})(t_{i+2} - t_i)^{-1} \leq 1$ (for $l = 0, 1$) and telescoped $t_{i+2} - t_i = (t_{i+2} - t_{i+1}) + (t_{i+1} - t_i)$ the third-divided difference of ψ_i^L is equal to $\psi_i^L[t_i, t_{i+1}, t_{i+2}]$

$$= \frac{(t_{i+2} - t_{i+1})^{\lambda-1} - (t_{i+1} - t_i)^{\lambda-1}}{t_{i+2} - t_i} + \frac{O((t_{i+2} - t_{i+1})^{1+\lambda}) + O((t_{i+1} - t_i)^{1+\lambda})}{t_{i+2} - t_i}$$

$$= \frac{N_{im}^{\lambda-1}\delta_m^{\lambda-1} - M_{im}^{\lambda-1}\delta_m^{\lambda-1}}{(N_{im} + M_{im})\delta_m} + O\left(\frac{(t_{i+2} - t_{i+1})^{1+\lambda}}{t_{i+2} - t_i}\right) + O\left(\frac{(t_{i+1} - t_i)^{1+\lambda}}{t_{i+2} - t_i}\right)$$

$$= \frac{N_{im}^{\lambda-1} - M_{im}^{\lambda-1}}{N_{im} + M_{im}}\delta_m^{\lambda-2} + O((t_{i+2} - t_{i+1})^\lambda) + O((t_{i+1} - t_i)^\lambda). \tag{20}$$

A similar argument leads to:

$$\psi_i^L[t_{i+1}, t_{i+2}, t_{i+3}] = \frac{P_{im}^{\lambda-1} - N_{im}^{\lambda-1}}{P_{im} + N_{im}} \delta_m^{\lambda-2} + O((t_{i+3} - t_{i+2})^\lambda) + O((t_{i+2} - t_{i+1})^\lambda). \tag{21}$$

Hence by (20) and (21) (for $l = 0, 1$) *the third divided differences of* ψ_i^L (needed in (15)) read as:

$$\psi_i^L[t_{i+l}, t_{i+l+1}, t_{i+l+2}] = \frac{R_{im(l+1)}^{\lambda-1} - R_{iml}^{\lambda-1}}{R_{im(l+1)} + R_{iml}} \delta_m^{\lambda-2} + O(\delta_m^\lambda). \tag{22}$$

Coupling again (20) and (21) with telescoped $t_{i+3} - t_i = (t_{i+3} - t_{i+2}) + (t_{i+2} - t_{i+1}) + (t_{i+1} - t_i)$ and $0 < (t_{i+l+1} - t_{i+l})(t_{i+3} - t_i)^{-1} < 1$ reduces *the fourth divided difference of* ψ_i^L into:

$$\psi_i^L[t_i, t_{i+1}, t_{i+2}, t_{i+3}] = \frac{\frac{P_{im}^{\lambda-1} - N_{im}^{\lambda-1}}{P_{im} + N_{im}} - \frac{N_{im}^{\lambda-1} - M_{im}^{\lambda-1}}{N_{im} + M_{im}}}{t_{i+3} - t_i} \delta_m^{\lambda-2}$$

$$+ \sum_{l=0}^{2} O\left(\frac{(t_{i+l+1} - t_{i+l})^\lambda}{t_{i+3} - t_i}\right),$$

which ultimately yields $\psi_i^L[t_i, t_{i+1}, t_{i+2}, t_{i+3}]$

$$= \frac{1}{P_{im} + N_{im} + M_{im}} \left(\frac{P_{im}^{\lambda-1} - N_{im}^{\lambda-1}}{P_{im} + N_{im}} - \frac{N_{im}^{\lambda-1} - M_{im}^{\lambda-1}}{N_{im} + M_{im}}\right) \delta_m^{\lambda-3} + O(\delta_m^{\lambda-1}). \tag{23}$$

The proof of (23) relies on $O\left(\frac{(t_{i+l+1} - t_{i+l})^\lambda}{t_{i+3} - t_i}\right) = O((t_{i+l+1} - t_{i+l})^{\lambda-1}) = O(\delta_m^{\lambda-1})$. The second step resorts to more-or-less uniformity (3) of admitted samplings $\mathcal{T}$ for any $\lambda \in [0, 1)$ (as $\lambda - 1 < 0$). However, to keep all constants in $O(\delta_m^{\lambda-1})$ from (23) as independent from each representative of (3) from now on we admit only β_0-more-or-less uniform samplings for some fixed $0 < \beta_0 \leq 1$ (see Definition 1.3). The latter permits to exploit the inequality $|(t_{i+l+1} - t_{i+l})^{\lambda-1}| \leq \beta_0^{\lambda-1} \delta_m^{\lambda-1}$ to justify (23) with constants in $O(\delta_m^{\lambda-1})$ depending on γ and λ (but not on samplings $\mathcal{T}$).

Recalling now that $\dot{\psi}_i^L(t) = a_i t^2 + b_i t + c_i$ over I_i, by (15) we have:

$$a_i = 3\psi_i[t_i, t_{i+1}, t_{i+2}, t_{i+3}],$$
$$b_i = 2\psi_i[t_i, t_{i+1}, t_{i+2}] - 2\psi_i[t_i, t_{i+1}, t_{i+2}, t_{i+3}](t_{i+2} + t_{i+1} + t_i),$$
$$c_i = \psi_i[t_i, t_{i+1}] - \psi_i[t_i, t_{i+1} t_{i+2}](t_i + t_{i+1})$$
$$+ \psi_i[t_i, t_{i+1}, t_{i+2}, t_{i+3}](t_i t_{i+1} + t_{i+1} t_{i+2} + t_i t_{i+2}). \tag{24}$$

In the next steps both conditions (6) and (7) enforcing $\dot{\psi}_i^L > 0$ (for arbitrary m) are transformed into their *asymptotic analogues* applicable for sufficiently large m (i.e. for Q_m *sufficiently dense*). This will ultimately complete the proof of Theorem 2.1.

In doing so, both conditions (6) and (7) are reformulated into asymptotic counterparts expressed in terms of (M_{im}, N_{im}, P_{im}) (see Theorem 2.1). To save space only the first inequality from (6) i.e. $a_i < 0$ is fully addressed here (which automatically covers both *(i)* and *(iv)* - see (9) and (12)). The remaining more complicated cases *(ii)*, *(iii)* and *(v)* (listed below) are supplemented with the final asymptotic formulas (10), (11) and (13). The proof of the latter shall be given in the full journal version of this paper.

(i) By (24) the first inequality $\underline{a_i < 0}$ from (6) amounts to $\psi_i^L[t_i, t_{i+1}, t_{i+2}, t_{i+3}] < 0$ which in turn by (23) holds subject to:

$$
\left(\frac{P_{im}^{\lambda-1} - N_{im}^{\lambda-1}}{(P_{im} + N_{im})(P_{im} + N_{im} + M_{im})} - \frac{N_{im}^{\lambda-1} - M_{im}^{\lambda-1}}{(N_{im} + M_{im})(P_{im} + N_{im} + M_{im})} \right) \delta_m^{\lambda-3}
$$
$$
+ O(\delta_m^{\lambda-1}) < 0, \tag{25}
$$

for $(M_{im}, N_{im}, P_{im}) \in [\beta_0, 1]^3$. Asymptotically, for fixed $\lambda \in [0, 1)$ the slowest term determining the sign of (25) accompanies $\delta_m^{\lambda-3}$ and reads as (for all β_0-more-or-less uniform samplings):

$$
\theta_1(M_{im}, N_{im}, P_{im}) = \frac{1}{P_{im} + N_{im} + M_{im}} \left(\frac{P_{im}^{\lambda-1} - N_{im}^{\lambda-1}}{P_{im} + N_{im}} - \frac{N_{im}^{\lambda-1} - M_{im}^{\lambda-1}}{N_{im} + M_{im}} \right),
$$

provided θ_1 is not of any order $\Theta(\delta_m^{2+\varepsilon})$ with $\varepsilon \geq 0$. A possible sufficient condition guaranteeing the latter is to require:

$$
\theta_1(M_{im}, N_{im}, P_{im}) \leq \rho < 0, \tag{26}
$$

to hold for any fixed $\rho < 0$. Evidently (26) amounts to the first inequality (9) assumed to hold in Theorem 2.1 in order to enforce in turn asymptotically the first inequality in (6) (for any fixed $\lambda \in [0, 1)$).

(ii) A similar but longer argument shows that (upon combining (8), (15), (19), (22) and (23)) the asymptotic fulfillment of the second inequality from (6) i.e. $\dot{\psi}_i^L(t_i) > 0$ is met subject to (10) satisfied for any fixed, but arbitrary small $\rho_1 > 0$ and sufficiently large m.

(iii) The third inequality $\dot{\psi}_i^L(t_{i+3}) > 0$ determining (6) maps analogously into its asymptotic counterpart (11) assumed to be fulfilled for an arbitrary but fixed $\rho_2 > 0$ and m sufficiently large.

(iv) Clearly the proof of (9) yields a symmetric sufficient condition for $\underline{a_i > 0}$ (representing the first inequality in (7)) to hold asymptotically. The latter coincides with (12) stipulated to be satisfied by any fixed $\rho_3 > 0$, subject to m getting large.

(v) The reformulation of $\kappa_{im} = \dot{\psi}_i^L(\frac{-b_i}{2a_i}) > 0$ from (7) into (13) (assumed to hold for any fixed $\rho_4 > 0$ and sufficiently large m) involves a more intricate treatment (it is omitted here).

The asymptotic conditions established in Theorem 2.1 in the form of specific inequalities depend (for each i) exclusively on triples $(M_{im}, N_{im}, M_{im}) \in [\beta_0, 1]^3$

and fixed $\lambda \in [0,1)$ (not on curve γ). Consequently, they can all be also visualized geometrically in $3D$ for each $i = 3k$ and $\lambda \in [0,1)$ as well as for any regular curve γ. Several examples with $3D$ plots are presented in Sect. 3 with the aid of *Mathematica Package* [22].

We note that all asymptotic conditions from Theorem 2.1 can be extended to their $2D$ analogues (with extra argument used establishing in fact a new theorem) which in turn can be visualized in more appealing $2D$ plots. Again it is omitted here as exceeding the scope of this paper.

Recall that *uniform sampling*, for which $M_{im} = N_{im} = P_{im} = 1$ (i.e. where $\beta_0 = 1$) combined with $\lambda \in [0,1)$ or $\lambda = 1$ with (1) both yield $\psi_i^L = 1 + O(\delta_m^2) > 0$ (see [9] and [19])). Noticeably, conditions (10), (11) and (13) are met for either $\lambda = 1$ or $\mathcal{T}$ uniform and $\lambda \in [0,1)$. In contrast none of (9) or (12) (participating in either (6) or (7)) holds for the above two eventualities. A possible remedy to incorporate these two special cases in adjusted asymptotic representations of either $a_i > 0$ or $a_i < 0$ is to apply the fourth-order Taylor expansion for $\gamma \in C^4$ - see (16). The analysis (left out here) yields a modified condition for $a_i > 0$ (and thus for $a_i < 0$), this time hinging not only on triples $(M_{im}, N_{im}, P_{im}) \in [\beta_0, 1]^3$, $\lambda \in [0,1)$ but also on γ curvature $\|\ddot{\gamma}(t_i)\|^2$ along $\mathcal{T}$ (see [9] and [19]) - here $\|\dot{\gamma}(t)\| = 1$ as γ is a regular curve and as such can be assumed to be parameterized by arc-length (see [2]). The latter may not always be given in advance. Alternatively, one could rely on *a priori* imposed restrictions on curvatures of γ belonging to the prescribed family of admissible curves.

3   Experimentation and Testing

In this section first Theorem 2.1 is illustrated with some examples based on algebraic tests supported by $3D$ plots generated in *Mathematica* (see Subsect. 3.1). Next the convergence rate $\alpha(\lambda)$ for $d(\gamma) - d(\hat{\gamma}^L) = O(\delta_m^{\alpha(\lambda)})$ is numerically investigated. A special attention is given to $\lambda \in [0,1)$ yielding ψ^L as a piecewise C^∞ reparameterization of $[0, T]$ into $[0, \hat{T}]$ (see Subsect. 3.2).

In doing so, in a preliminary step, for a given fixed β_0 two families of β_0-more-or-less uniform samplings (27) and (29) are introduced. Next the fulfillment of the asymptotic sufficient conditions enforcing the injectivity of $\dot{\psi}^L > 0$ (see Theorem 2.1) is examined for various $\lambda \in [0,1)$ and both samplings (27) and (29). In particular, the inequalities (9), (10), (11), (denoted in this section by (6)*) and (12), (13) (marked here with (7)*) representing asymptotically in $3D$ both (6) and (7) are tested for different sets of triples $(M_{im}, N_{im}, P_{im}) \in [\beta_0, 1]^3$ characterizing either (27) or (29). The algebraic calculations performed herein (assuming m is sufficiently large) are supplemented by geometrical visualizations with $3D$ plots in *Mathematica*. At this point, we re-emphasize that the asymptotic conditions from Theorem 2.1 can be extended further into respective $2D$ counterparts upon some laborious calculations. In return, the latter gives some advantage in visualizing more appealing $2D$ (*versus* $3D$) plots. To save the space the relevant theory and testing concerning this extra $2D$ case are left out here.

The second example reports on tests designed to numerically evaluate $\alpha(\lambda)$ in length estimation $d(\gamma) - d(\hat{\gamma}^L) = O(\delta^{\alpha(\lambda)})$, for any $\lambda \in [0,1]$ yielding each ψ_i^L as an injective function. The conjecture concerning $\alpha(\lambda)$ is proposed in Remark 3.2 based on our numerical results.

The tests reported here are performed for $2D$ and $3D$ curves γ_{sp}, γ_S introduced in Example 2 (i.e. for $n = 2, 3$). However all established results with the accompanied experimentation are equally applicable to arbitrary multidimensional reduced data $Q_m = \{q_i\}_{i=0}^m$ with $q_i = \gamma(t_i) \in \mathbb{E}^n$.

3.1 Testing Injectivity of ψ^L

Example 1. Consider first the following family $\mathcal{T}_1$ of *more-or-less uniform sampling* (for geometrical distribution of $\{\gamma(t_i)\}_{i=0}^{15}$ with sampling (27) see also Fig. 3(a) and Fig. 4(a)):

$$t_i = \begin{cases} \frac{i}{m} + \frac{1}{2m}, & \text{for} \quad i = 4k+1, \\ \frac{i}{m} - \frac{1}{2m}, & \text{for} \quad i = 4k+3, \\ \frac{i}{m}, & \text{for} \quad i = 2k, \end{cases} \tag{27}$$

for which $K_l = \frac{1}{2}$, $K_u = \frac{3}{2}$ and $\beta_1 = \frac{1}{3}$ (see Definition 1.2). Here $0 \leq i \leq m = 3k$, where $k \in \{1, 2, \dots\}$, so that $t_0 = 0$ and $t_m = T = 1$. Upon resorting to (8) the following $3D$ compact asymptotic representation $\mathcal{T}_1^{3D}$ of $\mathcal{T}_1$ reads as (for $m = 3k$):

$$\mathcal{T}_1^{3D} = \left\{ (1, \tfrac{1}{3}, \tfrac{1}{3}), (1, 1, \tfrac{1}{3}), (\tfrac{1}{3}, 1, 1), (\tfrac{1}{3}, \tfrac{1}{3}, 1), (1, \tfrac{1}{3}, \tfrac{2}{3}), (\tfrac{1}{3}, 1, \tfrac{2}{3}) \right\}. \tag{28}$$

The last two points in (28) are generated for $m = 3k$ as $t_m = 1$. We set $\beta_0 = 0.16$ and hence as $\beta_0 \leq \beta_1$ the sampling (27) is also β_0-*more-or-less uniform*.

We also admit another β_0-*more-or-less uniform sampling* $\mathcal{T}_2$ defined according to (for geometrical spread of $\{\gamma(t_i)\}_{i=0}^{15}$ with sampling (29) see also Fig. 3(b) and Fig. 4(b)):

$$t_i = \frac{i}{m} + \frac{(-1)^{i+1}}{3m}, \tag{29}$$

with $K_l = \frac{1}{3}$, $K_u = \frac{5}{3}$ and $\beta_2 = \frac{1}{5} \geq \beta_0$ (see Definition 1.2). Again we set $t_0 = 0$ and $t_m = T = 1$ with $0 \leq i \leq m = 3k$, for $k \in \{1, 2, \dots\}$. By (8) the $3D$ asymptotic form $\mathcal{T}_2^{3D}$ of (29) reads as:

$$\mathcal{T}_2^{3D} = \left\{ (\tfrac{4}{5}, \tfrac{1}{5}, 1), (\tfrac{1}{5}, 1, \tfrac{1}{5}), (1, \tfrac{1}{5}, 1), (1, \tfrac{1}{5}, \tfrac{4}{5}), (\tfrac{1}{5}, 1, \tfrac{2}{5}) \right\}. \tag{30}$$

The last two points in (30) come for $m = 3k$ as $t_m = 1$ and the first point is due to $t_0 = 0$.

The inequalities (9), (10), (11) marked as (6)* (or (12) and (13) denoted by (7)*) enforcing asymptotically (6) (or (7)) to hold are tested over $[\beta_0, 1]^3$ for both samplings (27) and (29). The fixed parameter λ is set either to $\lambda = 0.3$ or to $\lambda = 0.9$ with $\rho = -0.001$, $\rho_1 = \rho_2 = 0.05$, $\rho_3 = 0.001$ and $\rho_4 = 0.005$ - see Table 1 and Table 2. The corresponding sets of triples $(M_{im}, N_{im}, P_{im}) \in [\beta_0, 1]^3$

Table 1. Testing conditions (6) and (7) (implied asymptotically by (6)* and (7)*) for sampling (27) (represented by (28)) and for $\lambda = 0.3$ and $\lambda = 0.9$ with $\rho = -0.001$, $\rho_1 = 0.05$, $\rho_2 = 0.05$, $\rho_3 = 0.001$ and $\rho_4 = 0.005$. Here **T** stands for *true* and **F** for *false*, respectively.

λ	$\lambda = 0.3$		$\lambda = 0.9$	
Sampling $\mathcal{T}_1^{3D}$	Conditions			
	$(6)^*$	$(7)^*$	$(6)^*$	$(7)^*$
$(1, \frac{1}{3}, \frac{1}{3})$	F	F	T	F
$(1, 1, \frac{1}{3})$	F	T	F	T
$(\frac{1}{3}, 1, 1)$	F	T	F	T
$(\frac{1}{3}, \frac{1}{3}, 1)$	F	F	T	F
$(1, \frac{1}{3}, \frac{2}{3})$	F	F	T	F
$(\frac{1}{3}, 1, \frac{2}{3})$	F	T	F	T

Table 2. Testing conditions (6) and (7) (implied asymptotically by (6)* and (7)*) for sampling (29) (represented by (30)) and for $\lambda = 0.3$ and $\lambda = 0.9$ with $\rho = -0.001$, $\rho_1 = 0.05$, $\rho_2 = 0.05$, $\rho_3 = 0.001$ and $\rho_4 = 0.005$. Here **T** stands for *true* and **F** for *false*, respectively.

λ	$\lambda = 0.3$		$\lambda = 0.9$	
Sampling $\mathcal{T}_2^{3D}$	Conditions			
	$(6)^*$	$(7)^*$	$(6)^*$	$(7)^*$
$(\frac{4}{5}, \frac{1}{5}, 1)$	F	F	T	F
$(\frac{1}{5}, 1, \frac{1}{5})$	F	T	F	T
$(1, \frac{1}{5}, 1)$	F	F	T	F
$(1, \frac{1}{5}, \frac{4}{5})$	F	F	T	F
$(\frac{1}{5}, 1, \frac{2}{5})$	F	T	F	T

satisfying either (6)* or (7)* represent the respective solids $D_{\beta_0}^{\lambda} \subset [\beta_0, 1]^3$ plotted in $3D$ by *Mathematica* as shown in Fig. 1 and Fig. 2.

Noticeably different points from $\mathcal{T}_k^{3D}$, for $k = 1, 2$ may interchangeably satisfy one of the sufficient conditions enforcing either (6) or (7) to hold asymptotically. The latter is demonstrated in Table 1 and Table 2. Indeed for $\lambda = 0.3$ all conditions from (6)* are not satisfied by both $\mathcal{T}_k^{3D}$ (for $k = 1, 2$) as we have **F** in the respective columns of both Table 1 and Table 2. Moreover, the conditions from (7)* are only fulfilled by some points (not all) from $\mathcal{T}_k^{3D}$. Consequently the injectivity of ψ_i^L for either $\mathcal{T}_1^{3D}$ or $\mathcal{T}_2^{3D}$ is not guaranteed. Geometrically both $\mathcal{T}_k^{3D}$ (for $k = 1, 2$) are not contained in the respective injectivity zones $D_{\beta_0}^{\lambda=0.3}$ (for either (6)* or (7)*). In contrast for $\lambda = 0.9$, a simple inspection of Table 1 and Table 2 reveals that all points from $\mathcal{T}_k^{3D}$ (for $k = 1, 2$) can be split into two subsets each contained in the injectivity zones $D_{\beta_0}^{\lambda=0.9}$ determined by either (6)*

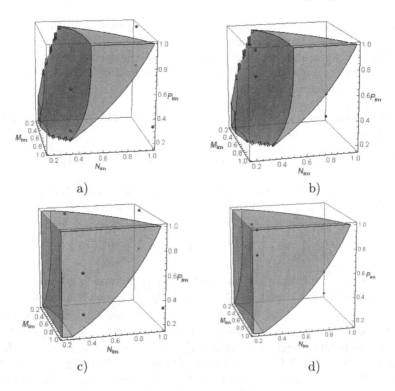

Fig. 1. Condition (6) enforced asymptotically by (6)* visualized in $3D$ plots as two solids $D_{\beta_0}^{\lambda} \subset [\beta_0, 1]^3$, for $\lambda = 0.3$ or $\lambda = 0.9$, respectively. Here $\beta_0 = 0.16$ with dotted points representing samplings: *a)* (27) mapped into (28) or *b)* (29) mapped into (30) both for $\lambda = 0.3$ and samplings: *c)* (27) mapped into (28) or *d)* (29) mapped into (30) both for $\lambda = 0.9$.

or by (7)*, respectively. Algebraically the latter yields at least one **T** in the last two columns of all rows for both Table 1 and Table 2.   □

Remark 3.1. Note that if for a given family of β_0-more-or-less uniform samplings $\mathcal{T}_{\beta_0}$ the subfamily $\mathcal{T}_{\beta_0}^{\nu} \subset \mathcal{T}_{\beta_0}$ with extra constraints $\nu_1 \leq M_{im} \leq \nu_2$, $\nu_3 \leq N_{im} \leq \nu_4$ and $\nu_5 \leq P_{im} \leq \nu_6$ (here $\nu = (\nu_1, \nu_2, \nu_3, \nu_4, \nu_5, \nu_6)$) is chosen one can also examine (for a fixed $\lambda \in [0,1)$) whether $I_\nu^{3D} \subset D_{\beta_0}^{\lambda}$, where $I_\nu^{3D} = (\nu_1, \nu_2) \times (\nu_3, \nu_4) \times (\nu_5, \nu_6)$. By Theorem 2.1, should the latter holds the entire subfamily of $\mathcal{T}_{\beta_0}^{\nu}$ yields asymptotically ψ_i^L as injective functions. The incomplete information on input samplings $\mathcal{T}$ carried by $\mathcal{T}_{\beta_0}^{\nu}$ can in certain situations accompany Q_m.

□

3.2   Numerical Testing for Length Estimation

We pass now to the experiments designed to investigate convergence rate $\alpha(\lambda)$ in length approximation by examining $d(\gamma) - d(\hat{\gamma}) = O(\delta^{\alpha(\lambda)})$ - see Definition 1.3.

The coefficient $\alpha(\lambda)$ is estimated numerically by $\tilde{\alpha}(\lambda)$ which in turn is computed using *a linear regression* on the pairs $\{(\log(m), -\log(E_m)\}_{m=m_{min}}^{m=m_{max}}$, where $E_m = |d(\gamma) - d(\hat{\gamma}^L)|$, for a given m. The slope a of the regression line $y(x) = ax + b$ found in *Mathematica* with the aid of *Normal[LinearModelFit[data]]* yields $a = \tilde{\alpha}(\lambda)$ forming a numerical estimate of $\alpha(\lambda)$.

Example 2. Consider *a 2D spiral* $\gamma_{sp} : [0,1] \to \mathbb{E}^2$ (a regular curve with $\gamma_{sp}(0) = (-0.2, 0)$ and $\gamma_{sp}(1) = (1.2, 0)$):

$$\gamma_{sp}(t) = \big((t + 0.2)\cos(\pi(1-t)), (t+0.2)\sin(\pi(1-t))\big), \tag{31}$$

and the so-called *3D Steinmetz curve* $\gamma_S : [0,1] \to \mathbb{E}^3$ (a regular closed curve with $\gamma_S(0) = \gamma_S(1) = (1, 0, 1.2)$ - see a dotted gray point in Fig. 4):

$$\gamma_S(t) = \left(\cos(2\pi t), \sin(2\pi t), \sqrt{1.2^2 - 1.0^2 \sin^2(2\pi t)}\right). \tag{32}$$

Both curves γ_{sp}, γ_S (from (31) and (32)) sampled according to either (27) or (29) are plotted in Fig. 3 and Fig. 4, respectively. The numerical results assessing the estimate $\tilde{\alpha}(\lambda)$ of $\alpha(\lambda)$ (for $d(\gamma) - d(\hat{\gamma}^L) = O(\delta_m^{\alpha(\lambda)})$) are presented in Table 3. Recall that here, a linear regression to compute $\tilde{\alpha}(\lambda)$ is applied to the collections of points $\{(\log(m), -\log(E_m)\}_{m_{min}=120}^{m_{max}=201}$, with $E_m = |d(\gamma) - d(\hat{\gamma}^L)|$ and for various $\lambda \in \{0.3, 0.7, 0.9\}$. The results from Table 3 suggest that for all $\lambda \in \{0.3, 0.7, 0.9\}$ rendering $\dot{\psi}^L > 0$ (e.g. the latter is guaranteed if Theorem 2.1 holds) one may expect $\lim_{m\to\infty} E_m = 0$ with *the quadratic convergence rate* $\alpha(\lambda) = 2 \approx \tilde{\alpha}(\lambda)$.   □

In fact the numerical results from Example 2 combined with (5) in conjunction with the argument used to prove $d(\gamma) - d(\hat{\gamma}^L) = O(\delta_m^4)$ for $\lambda = 1$ (see [7]) or [19]) lead to expect $\alpha(\lambda) = 2$ in $d(\gamma) - d(\hat{\gamma}^L) = O(\delta_m^{\alpha(\lambda)})$, for all $\lambda \in [0,1)$ yielding ψ^L as a piecewise C^∞ reparametrization. The latter forms an open problem which can be stated as:

Remark 3.2. Assume $\gamma \in C^4([0, T])$ be a regular curve in $\mathbb{E}^n$ sampled more-or-less uniformly (see Definition 1.2). For the interpolant $\hat{\gamma}^L$ and any $\lambda \in [0, 1)$ in (3) yielding each $\psi_i^L : I \to \hat{I}$ as a C^∞ genuine reparameterization Example 2 suggests *a sharp quadratic convergence rate* in:

$$d(\gamma) - d(\hat{\gamma}_i^L \circ \psi_i^L) = O(\delta_m^2). \tag{33}$$

In particular if Theorem 2.1 holds (and β_0-more-or-less uniform samplings are used) the mapping ψ^L is asymptotically a reparameterization which in turn hints to expect (33). Recall that by *sharpness* of (33) we understand the existence of at least one regular curve of class C^4 and of at least one samplings from $\mathcal{T}_{\beta_0}$ such that in (33) the convergence rate $\alpha(\lambda)$ has exactly order 2 (i.e. is not faster than quadratic).   □

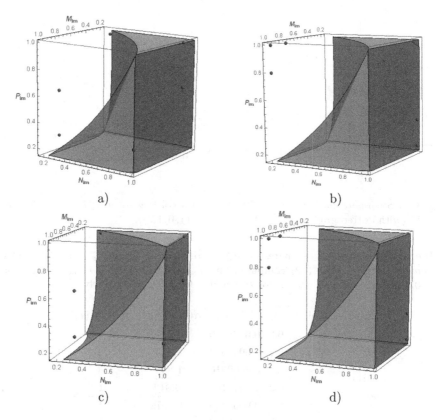

Fig. 2. Condition (7) enforced asymptotically by (7)* visualized in 3D plots as two solids $D_{\beta_0}^{\lambda} \subset [\beta_0, 1]^3$, for $\lambda = 0.3$ or $\lambda = 0.9$, respectively. Here $\beta_0 = 0.16$ with dotted points representing samplings: *a)* (27) mapped into (28) or *b)* (29) mapped into (30) both for $\lambda = 0.3$ and samplings: *c)* (27) mapped into (28) or *d)* (29) mapped into (30) both for $\lambda = 0.9$.

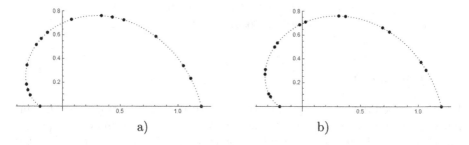

Fig. 3. *A spiral curve* γ_{sp} *from* (31) *sampled according to:* a) (27) *or* b) (29), *for* $m = 15$.

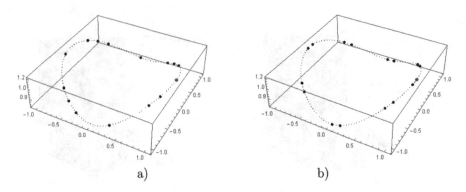

a)   b)

Fig. 4. *A Steinmetz curve* γ_S *from* (32) *sampled according to: a)* (27) *or b)* (29), *for* $m = 15$ *(with dotted gray point* $\gamma_S(0) = \gamma_S(1) = (1, 0, 1.2)$*)*.

Table 3. The numerical estimates of $\alpha(\lambda) \approx \tilde{\alpha}(\lambda)$ for *a spiral* γ_{sp} from (31) and a *Steinmetz curve* γ_S from (32) computed for $m_{min} = 120 \leq m \leq m_{max} = 201$ and $\lambda \in \{0.3, 0.7, 0.9\}$. Here **T** stands for *true* and **F** for *false*, respectively.

Curve	Sampling	λ	$E_{m=201}$	$\alpha(\lambda) \approx \tilde{\alpha}(\lambda)$	$(6)^*$ or $(7)^*$
(31)	(27)	0.3	0.0735200	0.044	F
		0.7	0.0000083	1.945	T
		0.9	0.0000016	1.885	T
	(29)	0.3	2.4619100	-0.012	F
		0.7	0.0050445	0.007	F
		0.9	0.0000319	1.989	T
(32)	(27)	0.3	0.2036000	0.033	F
		0.7	0.0000897	2.015	T
		0.9	0.0000181	2.092	T
	(29)	0.3	6.7392400	-0.009	F
		0.7	0.0132964	-0.080	F
		0.9	0.0003419	1.985	T

4   Conclusions

Fitting reduced data (see e.g. [3] or [16]) constitutes an important task in computer vision and graphics, engineering, microbiology, physics and other applications like medical image processing (e.g. for area, length and boundary estimation or trajectory planning) - see e.g. [4, 6, 8, 11, 14, 15, 17, 20] or [21].

Two sufficient conditions (6) and (7) are first formulated to ensure that the Lagrange piecewise-cubic $\psi^L : [0, T] \to [0, \hat{T}]$ (introduced in Sect. 1) is a genuine *reparameterization*. The latter applies to both sparse and dense reduced data Q_m. Here the unknown interpolation knots $\mathcal{T}$ are replaced by $\hat{\mathcal{T}}$ which in turn is determined by exponential parameterization (3) controlled by a single parameter $\lambda \in [0, 1]$ and Q_m. *The main contribution* established in Theorem 2.1 (see Sect. 2) reformulates (6) and (7) into respective asymptotic representatives valid for sufficiently large m (i.e. for Q_m getting denser). These new transformed conditions (specified in Theorem 2.1) depend exclusively on $\lambda \in [0, 1)$ and $\mathcal{T}$ characterized by (8) within the admitted class of β_0-more-or-less uniform samplings (see Definition 1.2) and apply to any regular curve $\gamma \in C^3([0, T])$ (with $0 < T < \infty$). Lastly, in Sect. 3 two illustrative examples are presented. The attached $3D$ plots generated in *Mathematica* [22] illustrate the algebraic character of the asymptotic conditions justified in Theorem 2.1 (see Example 1). In addition, the numerical examination of the convergence rate in length estimation of interpolated γ for $\lambda \in \{0.3, 0.7, 0.9\}$ are performed. Consequently, based on the latter the conjecture suggesting the quadratic convergence rate for $d(\gamma) - d(\hat{\gamma}^L) = O(\delta_m^2)$ is posed (see Example 2 and Remark 3.2), subject to the injectivity of ψ^L. At this point we remark that all asymptotic formulas from Theorem 2.1 are extendable to the corresponding inequalities expressed in (x, y)-variables. This can be achieved by converting first (with the aid of special *homogeneous mapping*) each triple (M_{im}, N_{im}, P_{im}) from (8) into a pair $(x(M_{im}, N_{im}, P_{im}), y(M_{im}, N_{im}, P_{im}))$ and then by reformulating all conditions from Theorem 2.1, accordingly in terms of (x, y). The satisfaction of such new conditions enforces (9), (10) and (11) or (12) and (13) asymptotically (and thus of (6) or (7)). It is a big advantage to reduce the illustrations from $3D$ to more appealing $2D$ analogues. We omit here the theoretical discussion and the geometrical insight of this $2D$ extension of Theorem 2.1. Similarly, recall that only items *(i)* and *(iv)* (see Sect. 2) are given here a full proof. In contrast, the final steps of proving *(ii)*, *(iii)* and *(v)* are left out as treated later exhaustively in a journal version of this work (together with the mentioned above $2D$ extension of Theorem 2.1).

Future work may include various interpolation schemes $\hat{\gamma}$ or ϕ based on Q_m combined with either (3) or with other $\hat{\mathcal{T}}$ compensating the unknown knots $\mathcal{T}$ (see e.g. [3,10,13] or [16]). Searching for alternative sufficient conditions enforcing ψ_i^L to be injective forms an interesting topic. Lastly the theoretical justification of (33) poses another open problem.

References

1. de Boor, C.: A Practical Guide to Spline. Springer, New York, 1985. https://www.researchgate.net/publication/200744645_A_Practical_Guide_to_Spline
2. do Carmo, M.P.: Differential Geometry of Curves and Surfaces. Prentice-Hall, Englewood Cliffs (1976). http://www2.ing.unipi.it/griff/files/dC.pdf
3. Epstein, M.P.: On the influence of parameterization in parametric interpolation. SIAM J. Numer. Anal. **13**(2), 261–268 (1976). https://doi.org/10.1137/0713025
4. Farin, G.: Curves and Surfaces for Computer Aided Geometric Design. Academic Press, Cambridge (1993). https://www.sciencedirect.com/book/9780122490521/curves-and-surfaces-for-computer-aided-geometric-design
5. Floater, M.S.: Chordal cubic spline interpolation is fourth-order accurate. IMA J. Numer. Anal. **25**(1), 25–33 (2005). https://doi.org/10.1093/imanum/dri022

6. Janik, M., Kozera, R., Kozioł, P.: Reduced data for curve modeling - applications in graphics, computer vision and physics. Adv. Sci. Technol. Res. J. **7**(18), 28–35 (2013). https://doi.org/10.5604/20804075.1049599

7. Kozera, R.: Curve modeling via interpolation based on multidimensional reduced data. Stud. Inform. **25**(4B(61)), 1–140 (2004). https://doi.org/10.21936/si2004_v25.n4B

8. Kozera, R., Wilkołazka, M.: A natural spline interpolation and exponential parameterization for length estimation of curves. Proc. Amer. Inst. Phys. **1863**(1), 400010-1–400010-4 (2017). https://doi.org/10.1063/1.4992579

9. Kozera, R., Wilkołazka, M.: Convergence order in trajectory estimation by piecewise cubics and exponential parameterization. Math. Model. Anal. **24**(1), 72–94 (2019). https://doi.org/10.3846/mma.2019.006

10. Kozera, R., Wilkołazka, M.: A note on modified Hermite interpolation. Math. Comput. Sci. **14**, 223–239 (2020). https://doi.org/10.1007/s11786-019-00434-3

11. Kozera, R., Noakes, L.: C^1 interpolation with cumulative chord cubics. Fundam. Inform. **61**(3), 285–301 (2004). https://dl.acm.org/doi/abs/10.5555/1031956.1031962

12. Kozera, R., Noakes, L.: Piecewise-quadratics and exponential parameterization for reduced data. Appl. Math. Comput. **221**, 620–638 (2013). https://doi.org/10.1016/j.amc.2013.06.060

13. Kozera, R., Noakes, L.: Piecewise-quadratics and reparameterizations for interpolating reduced data. In: Gerdt, V.P., Koepf, W., Seiler, W.M., Vorozhtsov, E.V. (eds.) CASC 2015. LNCS, vol. 9301, pp. 260–274. Springer, Cham (2015). https://doi.org/10.1007/978-3-319-24021-3_20

14. Kozera, R., Noakes, L., Wilkołazka, M.: A modified complete spline interpolation and exponential parameterization. In: Saeed, K., Homenda, W. (eds.) CISIM 2015. LNCS, vol. 9339, pp. 98–110. Springer, Cham (2015). https://doi.org/10.1007/978-3-319-24369-6_8

15. Kozera, R., Noakes, L., Szmielew, P.: Convergence orders in length estimation with exponential parameterization and ε-uniformly sampled reduced data. Appl. Math. Inf. Sci. **10**(1), 107–115 (2016). https://doi.org/10.18576/amis/100110

16. Kuznetsov, E.B., Yakimovich, A.Y.: The best parameterization for parametric interpolation. J. Comput. Appl. Math. **191**(2), 239–245 (2006). https://core.ac.uk/download/pdf/81959885.pdf

17. Kvasov, B.I.: Methods of Shape-Preserving Spline Approximation. World Scientific Publishing Company, Singapore (2000). https://doi.org/10.1142/4172

18. Lee, E.T.Y.: Choosing nodes in parametric curve interpolation. Comput. Aided Des. **21**(6), 363–370 (1989). https://doi.org/10.1016/0010-4485(89)90003-1

19. Noakes, L., Kozera, R.: Cumulative chords piecewise-quadratics and piecewise-cubics. In: Klette, R., Kozera, R., Noakes, L., Weickert, J. (eds.) Geometric Properties of Incomplete Data, Computational Imaging and Vision, chap. 4, vol. 31, pp. 59–75. Springer, Dordrecht (2006). https://doi.org/10.1007/1-4020-3858-8_4

20. Piegl, L., Tiller, W.: The NURBS Book. Springer, Heidelberg (1997). https://doi.org/10.1007/978-3-642-59223-2

21. Rababah, A.: High order approximation methods for curves. Comput. Aided Geom. Des. **12**(1), 89–102 (1995). https://doi.org/10.1016/0167-8396(94)00004-C

22. Wolfram, S.: The Mathematica Book. Wolfram Media Inc., Champaign (2003). https://www.wolfram.com/books/profile.cgi?id=4939

Loop Aggregation for Approximate Scientific Computing

June Sallou[1,2]($\boxtimes$), Alexandre Gauvain[2], Johann Bourcier[1]($\boxtimes$),
Benoit Combemale[3]($\boxtimes$), and Jean-Raynald de Dreuzy[2]

[1] Univ Rennes, Inria, CNRS, IRISA, Rennes, France
`{june.benvegnu-sallou,johann.bourcier}@irisa.fr`
[2] Geosciences Rennes, OSUR, Rennes, France
`{alexandre.gauvain,jean-raynald.de-dreuzy}@univ-rennes1.fr`
[3] Inria and University of Toulouse Jean Jaurès, Toulouse, France
`benoit.combemale@inria.fr`

Abstract. Trading off some accuracy for better performances in scientific computing is an appealing approach to ease the exploration of various alternatives on complex simulation models. Existing approaches involve the application of either time-consuming model reduction techniques or resource-demanding statistical approaches. Such requirements prevent any opportunistic model exploration, e.g., exploring various scenarios on environmental models. This limits the ability to analyse new models for scientists, to support trade-off analysis for decision-makers and to empower the general public towards informed environmental intelligence. In this paper, we present a new approximate computing technique, aka. `loop aggregation`, which consists in automatically reducing the main loop of a simulation model by aggregating the corresponding spatial or temporal data. We apply this approximate scientific computing approach on a geophysical model of a hydraulic simulation with various input data. The experimentation demonstrates the ability to drastically decrease the simulation time while preserving acceptable results with a minimal set-up. We obtain a median speed-up of 95.13% and up to 99.78% across all the 23 case studies.

Keywords: Approximate computing · Trade-off · Computational science

1 Introduction

There is a long-standing history on numerical analysis to provide a better accuracy in scientific computing. Research activities in the past decades result in efficient solvers to accurately simulate ever more complex models. For example, complex climate change models can nowadays be simulated to elaborate global warming scenarios and their consequences. However, such simulations come with the price of large computing and memory resources. While extremely useful to elaborate accurate results on precise scenarios, the required resources limit the

© Springer Nature Switzerland AG 2020
V. V. Krzhizhanovskaya et al. (Eds.): ICCS 2020, LNCS 12138, pp. 141–155, 2020.
https://doi.org/10.1007/978-3-030-50417-5_11

possible interactions with the models. This prevents any *interactive*, *live* and *customized* manipulation of such models, e.g., the ability to analyse new models for scientists, to support trade-off analysis for decision-makers and to empower the general public towards informed environmental intelligence.

In this context, trading some accuracy for better performance in scientific computing is an appealing approach to facilitate the exploration of various alternatives on complex simulation models. This has been initially explored through model reduction techniques [1, 11], which require specific knowledge on the model and highly-qualified analysis. Alternatively, statistical approaches are explored to automatically infer a new (surrogate) model from a large and well-qualified set of input/output simulations [9]. Although these approaches are mostly automatic and potentially accurate and efficient, they usually require a large, possibly controlled, set-up before providing acceptable results from the inferred model.

Model reduction and statistical techniques are potentially very efficient, but all require an important initial set-up (either time-consuming or resource-demanding) that prevents any *opportunistic* model exploration, e.g., exploration of various scenarios on environmental models.

In this paper, we present a new approximate computing (AC) technique, called `loop aggregation`. According to the main variable of interest, we automatically reduce the main loop of the simulation model by aggregating to a given degree the corresponding spatial or temporal data. This aggregation can be either applied as a pre-processing of the input data or by model transformation. For example, in the case of an a posteriori study of the causes of soil drying-up with all the necessary data available, it is worthwhile to use the pre-processing implementation. Or, in the case of a crisis situation and the continuous monitoring of a sudden and dangerous flooding episode, the model transformation will be able to manage the available on-going data. We apply `loop aggregation` on a geo-physical model of a hydraulic simulation with various input data from different sites and climate series. The geophysical model concerns the groundwater flow in coastal areas where sea level rise changes the distribution of saturation inland and potentially generates risks of floodings even when submersion is properly managed. Experiments are performed on 23 sites to investigate the potential reduction of computational time without significant modification of the assessment of groundwater-issued vulnerability. Results show a median speed-up of 95.13%, demonstrating the ability to drastically reduce the simulation time while preserving acceptable results with a minimal set-up. We discuss the capacities and limitations of the approach in the perspectives of further generalisation.

To follow up, Sect. 2 presents the simulation model of interest and motivates the needs for improved interactivity. After an introduction of the background related to AC in Sect. 3, Sect. 4 details the overall approach and the rationales for a specific AC technique in the context of scientific computing. Section 5 gives the details of experimenting the proposed approach on the simulation model of interest and Sect. 6 describes the evaluation conducted to validate the approach. Section 7 presents related works on statistical approaches, model reduction techniques and AC. Section 8 concludes the paper and gives the perspectives related to this work.

2   Motivating Example

Hydrologists are working to determine the impact of the sea level rise on coastal aquifers, on increased saturation levels and associated consequences on inland vulnerability. Between the current state of the aquifers and the predicted sea level rise and climate scenarios, hydrological models are expected to provide predictions.

Those models are based on the three-dimensional software Modflow [8] considered to be an international standard for simulating and predicting groundwater movements. It is based on the Darcy's law and conservation principles to represent the groundwater flows. Groundwater flows are essentially modelled by a diffusion equation with Dirichlet boundary conditions when the groundwater level reaches the surface. The resulting parabolic partial differential equation is discretised with a finite difference method and integrated with classical implicit temporal schemes [8]. The quantity of interest to assess the groundwater-issued vulnerability is derived from the depth to the groundwater level. When groundwater levels rise to some tens of centimeters to the surface, vulnerability becomes difficult to mitigate. Modflow requires both the geological and geographical settings of the studied site (inputs illustrated as *Geology* and *Land Use* in Fig. 1) and the meteorological forcing term (represented as the *Weather* input in Fig. 1) driving the infiltration and the recharge to the aquifer. This configuration does not change over the simulation period. The meteorological forcing term comes from climate scenarios available on the next century with the estimation of the different elements of the hydrological balance taken here as the input (recharge) to the aquifer. The groundwater flow model provides over the simulation period the location of the groundwater surface, more generally called water table.

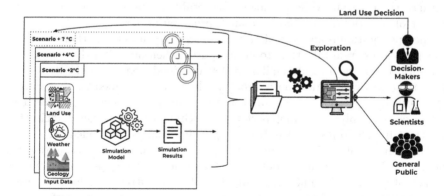

Fig. 1. Exploration of several climate scenarios simulations for various stakeholders.

As shown in Fig. 1, although the model was created by and for hydrologists, the simulation results can also be of interest to other users. Indeed, with growing awareness of climate change, general public including decision-makers increasingly ask to investigate by themselves the effect of climate change on property and land use planning. Overall, people want to explore the different future climate scenarios, and associated simulations, in an interactive way to make informed decisions or to understand their impact.

Since Modflow is a complex model, its execution can last more than a day. This simulation time is multiplied by the number of scenarios to explore which prevent effective and interactive exploration of the predictions. Making the model run faster would enable such exploration, but the predictions obtained must remain scientifically acceptable to respect the main trends and avoid any significant bias.

Thus, there is a need for finding a trade-off between accuracy and performance. A solution is to simplify the model. However, hydrologists and/or decision-makers may not have the expertise to make this trade-off through model reduction. This raises several scientific questions: (i) Can we make Modflow run faster while maintaining acceptable predictions? (ii) Can we do so without any expertise in hydrology or model simplification and any time-consuming/resource-demanding set-up? (iii) More generally, how to achieve it for scientific simulation models? In this paper, we propose to investigate AC for scientific simulation models to provide relevant trade-offs between accuracy and performance.

3   Introducing Acceptable Approximation into Models

The trade-off between accuracy and performance is a well-known concern in software engineering research. Approximate computing (AC) is one way to address this concern automatically from a computational point of view. It relies on the difference between the accuracy required by the developer or the user and the accuracy given by the execution of the software [7]. It introduces approximation into the program while producing acceptable outputs with respect to the purpose of the application thanks to an approximation strategy.

Loop perforation is an approximation strategy which assumes that iterations of loops in a program take time to be computed when not all of them would be necessary to achieve a result similar to that obtained from all of the iterations. Within the software code, the appropriate loops are modified so that only a subset of the iterations is realised. In practice, for a loop incremented by 1 at each iteration, applying the loop perforation technique would mean changing the increment from 1 to p so that only every p-th iteration is performed [12].

Thus, the number of iterations is reduced, fewer calculations are performed and a gain in performance is obtained. The choice of p is made according to the acceptability constraints made on the software outputs. Loop perforation is a technique that achieves the trade-off between accuracy and performance without knowledge of the application domain.

Another AC strategy, approximate loop unrolling [10], proposes, in addition to skipping iterations, to interpolate the results of non-computed iterations thanks to an interpolation function. It enables to better preserve accuracy.

Models, such as scientific simulation models whose consideration is producing an output within an acceptable precision range and are being by definition only an approximation of reality, can naturally benefit from a trade-off between accuracy and performance through approximate computing.

For scientific simulation models, the techniques of loop perforation or approximate unrolling cannot be straightforwardly applied. Removing iterations from a simulation leads to different case studies and non-comparable simulations. There is the exception of convergent simulations for which it is possible to safely remove iterations whose results do not provide more information. However without any information about the convergent criteria, it is not safe to assume that scientific simulations with different numbers of iteration can be comparable. In the case of our motivating example, Modflow, fully removing some iterations corresponding to time steps would alter the duration of the simulation period. Even more, meteorologic data are highly variable and cannot be easily inferred between time steps. The recharge rate (i.e. quantity of water to enter the aquifer per time unit) varies on a daily basis. Removing iterations would change the climate scenario. Thus, the model of Modflow undergoing loop perforation or loop unrolling would not be comparable to the initial model.

4   Approach

To handle the issue of non-comparable simulations, we reduce the number of computations while ensuring comparable conditions (e.g. same duration of the simulation period for Modflow). We introduce a new AC strategy adapted to scientific models, the `loop aggregation` technique (depicted in Fig. 2). This technique is similar to loop perforation and loop unrolling since it skips some loop iterations but adds specific stages to keep the results and simulation consistent with the baseline. It acts on the main loop of the simulator which is the loop iterating at the highest level on all the input data and enclosing all the processing of those data. The `loop aggregation` technique consists in three stages:

- *aggregation* (highlighted in blue): the input values of the main variable of interest are aggregated through an aggregation function. In the case of Modflow, the aggregate function consists in combining the recharge rate per day.
- *processing* (highlighted in violet): the operations within the loop are only performed on aggregated values.
- *interpolation* (highlighted in pink): the intermediate results are retrieved through an interpolation function.

As shown in Fig. 2, the simulation context guides the type of the `loop aggregation` implementation: data pre-processing or model transformation. Both implementations are equivalent as they reflect the `loop aggregation` approach. The difference is that the three stages are not carried out at the same moment. When all the input data are available before the model is run, a black-box implementation relies on the separation of the three stages (Data pre-processing). The aggregation stage acts as a pre-processing before the model execution, hence the name of the strategy. The values of the input data are

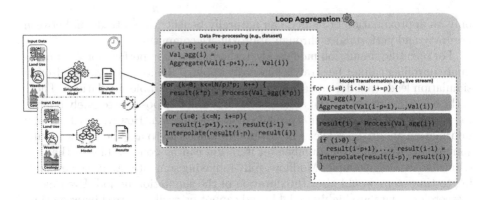

Fig. 2. The `Loop Aggregation` technique. (Color figure online)

aggregated according to an aggregation factor, p. The processing stage is carried out when the model runs with the model remaining as is. The interpolation stage is then performed as post-processing after the model execution. With a simulation context of dynamic data flows (i.e. stream data), the model transformation strategy is used. It is then necessary to perform the three stages of approximation dynamically to take data into account when they are retrieved. This implies accessing the model source code and adopting a white-box approach. All the stages and the modifications are made within the main loop of the model and the value of the iteration step is replaced by p, the value of the aggregation factor.

In essence, the `loop aggregation` technique adds an aggregation stage to the approximation process described in approximate loop unrolling to enable AC with scientific models. It enables a black-box implementation with separate stages running at different times when all data are available or a white-box implementation with a model transformation when dealing with dynamic data. The number of computations is reduced by the use of the p-factor and the approximate simulation is still comparable to the reference one. In theory, the technique can be applied to all scientific models with a main loop iterating over temporal or spatial data. There is no need for specific knowledge about the application domain of the model except for information about the use of the model.

5   Experimenting Loop Aggregation on Modflow

5.1   The Case Study of Modflow

The case study of our motivating example is based on the prediction of groundwater movements in a watershed near Lestre in Normandy in France to assess the risk of increased saturation at this site. The prediction period for the simulation is 42 years and is represented by 15340 stress periods (i.e. time steps) whose duration is set to correspond to a simulated day. The parameters of the model have been set by hydrologists. Executing Modflow with those inputs and non-aggregated data constitutes the reference simulation.

5.2   Approximating the Model with the Loop Aggregation Approach

We derive the approximate simulations using our loop aggregation technique with different aggregation p-factors to assess the variation of the simulation execution time.

In the case of Modflow, the main loop iterating over the stress periods, the computation reduction is done by removing some of them. The values of the recharge rate are aggregated as it is the variable of interest. As all the input data are available upstream, we perform our loop aggregation approach on those data following the Data Pre-processing implementation. We experiment two strategies for the aggregation stage (detailed hereafter) and linear regression for results interpolation.

Strategy with the Mean as Aggregation Function of the Recharge. The recharge data are aggregated for p being equal to 2, 7, 30, 90, 182, 365, 730 and 3652, corresponding to stress period durations of 2 days, 7 days, 1 month, 3 months, 6 months, 1 year, 2 years and 10 years. Those values of p were chosen to represent meaningful periods for hydraulic and meteorological events.

The mean is chosen as the aggregation function. The choice has been made according to the use of the recharge rate and according to the advice of experts to maintain the overall flux balance. To ensure comparable approximate simulations, the aggregation has also to impact the values of the stress period duration. Indeed, the simulation period must represent a span of 42 years. The stress period duration of the aggregated stress periods is thus changed into the value of p. For instance, with $p = 2$, the inputs are modified as shown in Fig. 3a.

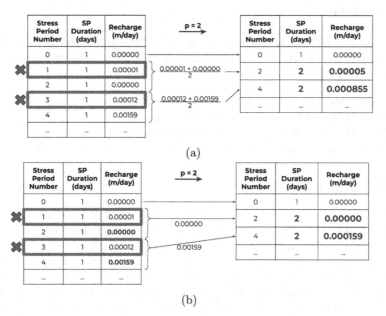

(a)

(b)

Fig. 3. Aggregation strategies with $p = 2$. a: mean as the aggregation function of the recharge. b: p-th value assigned as the aggregated value of recharge.

Strategy with Assigning the p-th Value as the Aggregated Value of Recharge. The aggregation is again carried out for p being equal to 2, 7, 30, 90, 182, 365, 730 and 3652. The stress period duration is changed into the value of p and the recharge value of the p-th stress period is assigned as the mean recharge value for the corresponding aggregated stress periods. The aggregation is carried out as presented in Fig. 3b for $p = 2$.

5.3   Conditions of the Experimentation

Modflow is run as a fortran written executable with compiled code for computing the groundwater flows in the aquifer. A wrapping software layer written in Python by hydrologists is used to configure Modflow and format the simulation inputs. The version of Modflow used is MODFLOW-NWT-SWR1, the U.S. Geological Survey modular finite-difference groundwater-flow model with Newton formulation and with the version number 1.1.4 released on 04/01/2018 associated with the SWR1 which version number is 1.04.0 released on 09/15/2016.

The experimentation is done on a single node with a Intel(R) Xeon(R) CPU E5-2650 v4 processor with 2.20 GHz. Each simulation is run on a single core with 2 threads and 8 GB RAM. The model is embedded inside a Docker image deployed on a virtual machine for each simulation. The virtual machine is a Alpine Linux 3.4.3 amd64. Through the Docker image, the memory available for each simulation is limited to 2 Gb. These measures are taken to limit variations in the experimentation environment.

6   Evaluation

In this section, we validate our ability to apply the `loop aggregation` on our motivating scenario presented in Sect. 2. The goal is to answer the following research questions:

RQ1: Is `loop aggregation` able to perform substantial performance increase while maintaining meaningful results for experts?

RQ2: Is the `loop aggregation` technique able to produce relevant trade-offs for various input data such as climate scenarios and geographical sites?

6.1   Acceptation Criterion

Domain experts have established an approximation indicator called acceptance criterion, which represents a threshold under which the indicator value should remain for the approximated results to be considered acceptable.

Simulation approximations are defined on the quantities of interest of the models. Considering the issues of coastal saturation, the relevant quantities derive from the proximity of the aquifer to the surface. When the top of the aquifer (water table) approaches the soil surface at a distance smaller than d_c, water resources, soil humidity, flooding risks and other human activities are

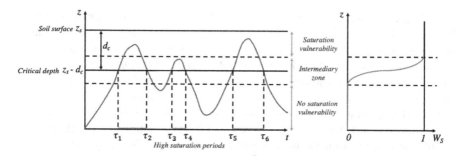

Fig. 4. Vulnerability zone and the associated representation of Ws.

impacted. The characteristic distance d_c depends on the type of human activity (e.g. agriculture or cites). It also depends on local choices of collectivities.

However, the transition is not sharp. Rather, it is a transition zone from not vulnerable conditions when the aquifer is deep enough to vulnerable close to the surface (Fig. 4). The width of the transition zone will be noted Δd_c and be taken as a linear function of d_c with $\Delta d_c\left(x\right) = \alpha d_c\left(x\right)$ where α is the proportionality factor and x is the position. With d_c typically of the order of 30 cm and α equal to $1/3$, the Δd_c the transition width is of the order of 10 cm.

Approximations on the quantities of interest will thus be weighted according to their proximity to the surface with the function $W_s\left(h\right)$, presented in Eq. 1, where h is the piezometric level and z_s is the altitude of the soil surface.

$$W_s\left(h\right) = \begin{cases} 0 & \text{if } h < z_s - \left(d_c + \frac{\Delta d_c}{2}\right) \\ sin\left(\frac{\pi}{2}\frac{h-\left(z_s-\left(d_c+\frac{\Delta d_c}{2}\right)\right)}{\Delta d_c}\right) & \text{if } z_s - \left(d_c - \frac{\Delta d_c}{2}\right) \leq h \leq z_s - \left(d_c + \frac{\Delta d_c}{2}\right) \\ 1 & \text{if } h > z_s - \left(d_c - \frac{\Delta d_c}{2}\right) \end{cases}$$
(1)

The H indicator $\|\Delta h\|_2$ on the saturation level is defined by Eq. 2. The threshold for the H indicator is set by hydrologists to 0.1 m, meaning that any approximation of the water table depth within a margin of 10 cm is acceptable. Variables issued by the reference and approximate simulations are indexed by the letters R and A respectively.

$$\|\Delta h\|_2 = \sqrt{\frac{\sum_t \sum_x max\left[W_s\left(h_R\left(x,t\right)\right), W_s\left(h_A\left(x,t\right)\right)\right] * \left(h_R\left(x,t\right) - h_A\left(x,t\right)\right)^2}{\sum_t \sum_x max\left[W_s\left(h_R\left(x,t\right)\right), W_s\left(h_A\left(x,t\right)\right)\right]}}$$
(2)

6.2 [RQ1] Performance Increase with Loop Aggregation

To answer **RQ1**, we assess the performance increase and the acceptation criterion when applying `loop aggregation` on the Modflow model (Sect. 2). We use two aggregation strategies with the same inputs (site, i.e. Lestre, and recharge series). The reference simulation is run in 34032 s, i.e. 9 h, 27 min and 12 s.

Experiments with the Strategy of the Mean as the Aggregation Function. We observe in Fig. 5a&c that the most approximated the simulation is, the fastest it is. It follows the rational idea that, for a dominantly linear model, the duration of the execution is directly linked to the number of iterations in the loop. With respect to the acceptance criterion, the approximated simulations performed in our experiment with a period of less than one year ($p = 365$) are considered to produce acceptable outputs. Within these acceptable outputs, the shortest execution time (1149 s or 19 min and 9.0 s) is obtained with the simulation of one year stress periods. The execution time is reduced by more than 29 times, a speed-up of more than 96.6%.

Test with the Strategy of the p-th Value Assigned to the Aggregated Value. Loop aggregation shows again a performance gain (Fig. 5b&d). We observe that the H indicator is higher for the same aggregation rate than for the previous strategy which is consistent with the fact that we introduce more approximation here (i.e. the recharge values of the aggregated iterations are not taken into account). The best speed-up is 87.49% with $p = 30$ (Fig. 5d).

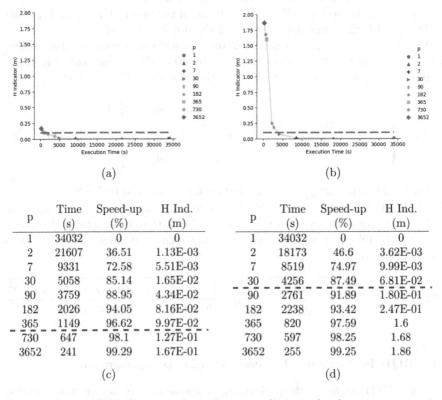

(a)   (b)

p	Time (s)	Speed-up (%)	H Ind. (m)	p	Time (s)	Speed-up (%)	H Ind. (m)
1	34032	0	0	1	34032	0	0
2	21607	36.51	1.13E-03	2	18173	46.6	3.62E-03
7	9331	72.58	5.51E-03	7	8519	74.97	9.99E-03
30	5058	85.14	1.65E-02	30	4256	87.49	6.81E-02
90	3759	88.95	4.34E-02	90	2761	91.89	1.80E-01
182	2026	94.05	8.16E-02	182	2238	93.42	2.47E-01
365	1149	96.62	9.97E-02	365	820	97.59	1.6
730	647	98.1	1.27E-01	730	597	98.25	1.68
3652	241	99.29	1.67E-01	3652	255	99.25	1.86

(c)   (d)

Fig. 5. Evolution of H indicator and speed-up according to p for the mean aggregation function strategy (a and c) and for the strategy with the p-th value as the aggregated value (b and d). The red dashed line represents the value of the acceptation criterion. c, d: H Ind. = H Indicator. (Color figure online)

Table 1. Variability of time across replicate simulations. RSE = Relative Standard Error.

p	Number of replicates	Mean (s)	Median (s)	Standard Deviation (s)	RSE (%)
1	30	3.57E+04	3.66E+04	3.01E+03	8.42
365	30	1.02E+03	9.68E+02	1.71E+02	16.77
3652	30	2.07E+02	2.00E+02	2.76E+01	13.33

Stability of the Execution Time Across Simulations. To assess the stability of the execution time obtained for the simulations, we run 30 replicates of the reference simulation and 30 replicates of the simulation with $p = 365$ and $p = 3652$. We use here the approximation with the mean aggregation function. The summary of the results is shown in Table 1. The execution times are stable enough to back the conclusion of substantial performance increase, i.e. the standard deviations and relative standard errors are significantly lower than the speed-up.

To answer **RQ1**, the `loop aggregation` provides substantial performance increase while preserving accepted results for the Modflow hydraulic simulator.

6.3   Approach Robustness

To answer **RQ2**, we experiment `loop aggregation` with other inputs such as climate scenarios (i.e. recharge series) or geographic sites. In these experiments, we use the mean aggregation strategy for the following case studies.

Another Climate Scenario. In this experiment, we use another climate scenario while the rest of the experiment inputs remain the same as in inprevious section. Again, `loop aggregation` leads to a substantial speed up 84.84% ($p = 90$), while remaining within the acceptation criterion.

Replication on Other Geographical Sites. We conduct the same experimentation on 22 other geographical sites. The cumulative execution time amounts to 24 days, 15 h, 44 min and 27 s. The speed-up between the reference simulation and the fastest acceptable approximation is illustrated in Fig. 6 across the sites.

Empirically, we find that `loop aggregation` enables an acceptable approximate simulation for all sites. The gains are not homogeneous but they are substantial. The mean and median speed-up are 91.93% and 95.13% with a minimum of 72.26% (Doville) and a maximum of 99.78% (Graye-sur-Mer).

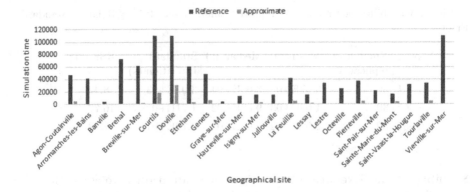

Fig. 6. Speed-up across different geographical sites.

To answer **RQ2**, the `loop aggregation` technique provides appealing speed-up while maintaining acceptable results with various inputs (i.e. recharge series and geographical sites).

6.4 Threats to Validity

Although we empirically validate that the `loop aggregation` approach gives conclusive results for the Modflow scientific simulator in several scenarios, some internal and external limitations remain. Our approach is not analytical but empirical. Our technique may not be the only answer to find trade-off between performance and acceptability with a minimal set-up. The experimentation was carried out on a specific environment. Care should be taken to ensure that the conclusion can be made with other environments. Moreover, while the indicator used to determine the acceptability of the approximated results were given by experts, it may not meet the expectations of other Modflow experts. To mitigate these limitations, the experimentations have been carried out with various inputs (several sites, another recharge series) and a second aggregation strategy. Regarding external limitations, our implementation of `loop aggregation` is limited to a single scientific simulation model based on a differential equation and to aggregating temporal data. Following works are needed to apply the technique with an aggregation on spatial data as well as including other scientific simulation models (e.g. other forms of equations).

7 Related Work

Other scientific approaches address the problem of providing trade-off between accuracy and performance. The common goals of these works is to create a surrogate model that can provide an acceptable solution while using less resources. We can classify all these various works in three different categories: the statistical approaches, model reduction approaches and AC approaches.

7.1   Statistical Approaches

We gather under the umbrella of statistical approaches all techniques that consist in finding correlation between a set of inputs and outputs of the model. It therefore encompasses most machine learning and regression techniques. These black-box techniques do not impose access to the inner model, but require both access to a large number of inputs/outputs of the model and a training period. At the end of the process, the original model is replaced by the learned model.

In [5], authors present an approach to efficiently explore architectural design spaces through the replacement of simulator by a learned artificial neural network. The artificial neural network is trained through sample inputs which were obtained by repeated execution of the initial simulator.

In [4], authors describe an approach which builds a surrogate model of a mobile network simulator. This surrogate model is then used to guide an optimisation technique. In [13], authors elaborate on a technique which leverage Kriging models to be used for global multidisciplinary design optimisation.

7.2   Model Reduction Approaches

Model reduction approaches [6] refer to a set of techniques that aim to reduce the complexity of the simulation model used to represent the natural phenomenon. The design of the reduced model is done manually and requires both a deep understanding of the natural phenomenon and model reduction techniques. At the end of the process, the original model is replaced by the reduced model.

In [6], authors review the various model reduction techniques used in fluid dynamics systems. In [2], authors present a tailored used of space and time decomposition to transform the original model into several reduced models.

7.3   Approximate Computing Approaches

AC approaches [7] refer to a set of techniques that aim to balance accuracy of computing with performance. These techniques leverage the initial software or model and automatically reduce its computation time by performing systematic approximations. These techniques are automatic and systematic and keep the structure of the original model.

In [12], authors present one of the major technique of AC: **loop perforation**. This technique, which is presented in more details in Sect. 3, consists in reducing the required computing for a software loop by performing only a subset of its iteration. In [10], authors present an **approximate loop unrolling** technique for trading performance of loop over precision. They applied their technique, on computer vision library, music synthesizer library, text search engine and a machine learning library.

Our loop approximation technique is inspired by the approximate loop unrolling technique but as mentioned in Sect. 4, applying this approach directly on a scientific computation model would radically change the case study and therefore it is not suitable.

Other works have been proposed to reduce the computation of loop iterating over matrices. In [3], authors present an approach, called randomized numerical linear algebra, which intends to remove randomly (according to a distribution law) some data from a matrix to accelerate its processing. As mentioned before, removing data is not a suitable approach since it will completely change the subject of study.

8   Conclusion and Future Work

In this paper, we propose `loop aggregation`, an approximate scientific computing technique, that enables to automatically and systematically reduce the main loop of a simulation model by aggregating the corresponding spatial or temporal data. It can either be implemented as a black-box approximation with a data pre-processing or as a white-box model transformation. Our experimentation on a hydraulic simulator shows a median 95.13% speed-up of the simulation time while preserving acceptable results for all the 23 use cases. The approach is supported with a minimal set-up as opposed to time-consuming model reduction and resource-demanding statistical techniques. The flexibility provided ensures that users can explore the simulations according to their specific constraints.

These results encourage further validation of the `loop aggregation` approach on other scientific simulation models with other forms of equations or other simulation contexts. The technique does not replace the other statistical or model reduction approaches applied to scientific models but rather complements them. Indeed, thanks to its minimal set-up, it can be used during a first approximation phase to generate input/output pairs that can later be used for more efficient statistical approaches as well as allowing a first exploration of the model to better understand it for a possible model reduction approach later on.

References

1. Chaturantabut, S., Sorensen, D.C.: Nonlinear model reduction via discrete empirical interpolation. SIAM J. Sci. Comput. **32**, 2737–2764 (2010)
2. Chinesta, F., Ladeveze, P., Cueto, E.: A short review on model order reduction based on proper generalized decomposition. Arch. Comput. Methods Eng. **18**(4), 395 (2011)
3. Drineas, P., Mahoney, M.: RandNLA. Commun. ACM, May 2016. https://doi.org/10.1145/2842602
4. Efstathiou, D., McBurney, P., Zschaler, S., Bourcier, J.: Surrogate-assisted optimisation of composite applications in mobile ad hoc networks. In: Proceedings of the 2014 Annual Conference on Genetic and Evolutionary Computation, pp. 1239–1246. ACM (2014)
5. İpek, E., McKee, S.A., Caruana, R., de Supinski, B.R., Schulz, M.: Efficiently exploring architectural design spaces via predictive modeling. In: Proceedings of the 12th International Conference on Architectural Support for Programming Languages and Operating Systems, ASPLOS XII, pp. 195–206. Association for Computing Machinery, New York (2006)

6. Lassila, T., Manzoni, A., Quarteroni, A., Rozza, G.: Model order reduction in fluid dynamics: challenges and perspectives. In: Quarteroni, A., Rozza, G. (eds.) Reduced Order Methods for Modeling and Computational Reduction. MMSA, vol. 9, pp. 235–273. Springer, Cham (2014). https://doi.org/10.1007/978-3-319-02090-7_9

7. Mittal, S.: A survey of techniques for approximate computing. ACM Comput. Surv. **48**(4), 62:1–62:33 (2016). https://doi.org/10.1145/2893356

8. Niswonger, R., Panday, S., Ibaraki, M.: MODFLOW-NWT: A Newton Formulation for MODFLOW-2005: U.S. Geological Survey Techniques and Methods 6-A37 (2011). https://pubs.usgs.gov/tm/tm6a37. Accessed 4 Dec 2019

9. Razavi, S., Tolson, B.A., Burn, D.H.: Review of surrogate modeling in water resources. Water Resour. Res. **48**(7), (2012). https://doi.org/10.1029/2011WR011527

10. Rodriguez-Cancio, M., Combemale, B., Baudry, B.: Approximate loop unrolling. In: CF 2019 - ACM International Conference on Computing Frontiers, Alghero, Sardinia, Italy, pp. 94–105. ACM (2019). https://doi.org/10.1145/3310273.3323841

11. Rowley, C.W., Dawson, S.T.M.: Model reduction for flow analysis and control. Annu. Rev. Fluid Mech. **49**(1), 387–417 (2017). https://doi.org/10.1146/annurev-fluid-010816-060042

12. Sidiroglou-Douskos, S., Misailovic, S., Hoffmann, H., Rinard, M.: Managing performance vs. accuracy trade-offs with loop perforation. In: Proceedings of the 19th ACM SIGSOFT Symposium and the 13th European Conference on Foundations of Software Engineering. p. 124–134. ESEC/FSE 2011, Association for Computing Machinery, New York, NY, USA (2011). https://doi.org/10.1145/2025113.2025133

13. Simpson, T., Mauery, T., Korte, J., Mistree, F.: Kriging models for global approximation in simulation-based multidisciplinary design optimization. AIAA J. **39**(12), 2233–2241 (2001). https://doi.org/10.2514/2.1234

Numerical Computation for a Flow Caused by a High-Speed Traveling Train and a Stationary Overpass

Shotaro Hamato[1](✉), Masashi Yamakawa[1], Yongmann M. Chung[2],
and Shinichi Asao[3]

[1] Kyoto Institute of Technology, Matsugasaki, Sakyo-ku, Kyoto 606-8585, Japan
Hamato0820@icloud.com
[2] University of Warwick, Gibbet Hill Road, Coventry CV4-7AL, UK
[3] College of Industrial Technology, Amagasaki, Hyogo 661-0047, Japan

Abstract. In the United Kingdom, old infrastructure facilities along train tracks like overpasses should be repaired as soon as possible. In order to design them more optimally, we need to know the flow field around such facilities. In this study, a traveling train and a stationary overpass were reproduced in a computational domain, and the flow field around the train and the overpass was simulated with the MCD method and the Expanded Sliding Mesh approach. In the conventional Sliding Mesh approach, sliding planes must be completely adjacent to other Sliding planes, that is, Sliding planes cannot be adjacent to both other Sliding planes and outer boundaries simultaneously. Then, the Expanded Sliding Mesh approach was proposed. The results given by using the MCD method and the Expanded Sliding Mesh approach were compared with the experimental results given by Baker, and their good qualitative and quantitative agreements were shown in almost cases. Only in the case that the overpass is very close to the train, the effects of boundary layers on laminar flow and turbulence need to be considered for more precise computation.

Keywords: Computational fluid dynamics · Compressible flow · Unstructured mesh · Moving grid · High-speed train

1 Introduction

Railway systems and trains contribute cultural and economical prosperity of cities and nations in the world. In Japan, liner motor trains which are faster than Shinkansen trains by using electrical magnet are about to use. On the other hand, the United Kingdom is the first country which starts to use railway systems. Underground was also developed and used first in the UK. However, the railway systems in the UK are older than those in any other countries, and it should be repaired or rebuild as soon as possible [1,2]. When such facilities along train tracks would be repaired, these two points should be considered:

© Springer Nature Switzerland AG 2020
V. V. Krzhizhanovskaya et al. (Eds.): ICCS 2020, LNCS 12138, pp. 156–169, 2020.
https://doi.org/10.1007/978-3-030-50417-5_12

1. Repair and rebuilding should be as cheap as possible.
2. New one is more difficult to be damaged.

Since the passage of trains is generally related to damage of facilities along tracks, it is necessary to know the flow fields and the pressure distribution around the overpass. A previous research regarding to this problem was reported by Baker [3], which was using an experimental model in 1/25 scale. However, the Baker's experiment is for obtaining data to renew Euro Code for the UK, is not for knowing the flow fields in detail.

We focused on overpasses. Overpasses are a kind of bridges for pedestrians and cars, which are fixed over train tracks. They are commonly used in countries with many plains like the UK and are necessary in people's lives. Therefore, the objectives of this study are expressing a traveling train and an overpass in computer, estimating pressure distribution on the overpass and revealing the flow fields around the traveling train and the overpass in detail. The feature of this study is that the traveling train and the stationary overpass are considered at the same time. In order to achieve the computation with objects which have different motion, the approach combining the MCD method and the Sliding Mesh approach is adopted. A previous research using the approach [4] is already reported, which showed the exactness of computation with large movement of objects or deformation of grid. In this study, the Expanded boundary conditions and the combined approach enable the computation.

2   Numerical Approach

2.1   Govening Equations

In this study, the computations are carried out within inviscid approximation to simplify calculations and shorten the calculation time. The three dimensional Euler equations for compressible flow in conservation form are used as governing equations,

$$\frac{\partial \boldsymbol{q}}{\partial t} + \frac{\partial \boldsymbol{E}}{\partial x} + \frac{\partial \boldsymbol{F}}{\partial y} + \frac{\partial \boldsymbol{G}}{\partial z} = 0 \tag{1}$$

$$\boldsymbol{q} = \begin{bmatrix} \rho \\ \rho u \\ \rho v \\ \rho w \\ e \end{bmatrix}, \boldsymbol{E} = \begin{bmatrix} \rho u \\ \rho u^2 + p \\ \rho uv \\ \rho uw \\ u(e+p) \end{bmatrix}, \boldsymbol{F} = \begin{bmatrix} \rho v \\ \rho uv \\ \rho v^2 + p \\ \rho vw \\ v(e+p) \end{bmatrix}, \boldsymbol{G} = \begin{bmatrix} \rho w \\ \rho uw \\ \rho vw \\ \rho w^2 + p \\ w(e+p) \end{bmatrix} \tag{2}$$

where $\boldsymbol{q}$ is conserved physical quantities, ρ is density and e is total energy per unit volume. Moreover, $\boldsymbol{E}, \boldsymbol{F}, \boldsymbol{G}$ are the inviscid flux functions and u, v, w are velocity components for x, y, z axes respectively. Equation of State for a Perfect gas is used to obtain pressure,

$$p = (\gamma - 1)\left\{ e - \frac{1}{2}\rho\left(u^2 + v^2 + w^2\right)\right\} \tag{3}$$

where γ is the specific heat ratio and $\gamma = 1.4$ in this study.

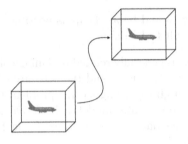

Fig. 1. Conceptual image of the MCD method

The Roe's Flux Splitting method is used for evaluating the inviscid flux functions. Physical quantities are defined at each cell center. For higher-ordering, the MUSCL method with the Venkatakrishnan's limiter is adopted. The Rational Runge-Kutta method is employed for time integration in pseudo-time steps.

2.2 MCD Method

Computation including movement and deformation of bodies is called the moving boundary problem in general. In order to solve such problems, the whole movement path and the deformation range of the bodies should be included in a computational domain. Thus, the massive computational domain is needed, and such computation will take much time and computational cost.

The MCD method (Moving Computational Domain method) [5–7], which is a kind of Finite volume method for Unstructured moving grid [8,9], has been proposed for the moving boundary problem and was adopted in this study. In the MCD method, the whole computational grid is moved according to the bodies to express the movement of bodies (Fig. 1). The main good point of the MCD method is that there is no restriction of movement. The bodies can move in infinite domain without restriction of its movement. Moreover, computational cost will be saved because computational domain do not need to include all of the movement pass and can be smaller than in conventional method. Since the four-dimensional volume in a space-time unified domain is used as control volume, the Geometric Conservation Law (GCL) [10] and the physical conservation law are satisfied automatically and exactly. Many previous researches of Racing car [11] or Tilt-Rotor plane [12] using the MCD method have been already reported. In this study, the computation of the stationary overpass and the traveling train was carried out by a new method combining the MCD method and the Expanded Sliding Mesh approach.

2.3 Sliding Mesh Approach

The moving boundary problem can be solved just by the MCD method. If the overpass and the train were included in the same computational domain, the overpass would be moved because of the movement of the computational domain

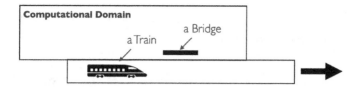

Fig. 2. Conceptual image of the Sliding Mesh approach

to express the movement of the train. Then, the Sliding Mesh approach enables to simulate objects which have different motions. In the Sliding Mesh approach, the computational domain is divided into two or three computational domains, and these domains are facing to each other and slid to express different motions (Fig. 2). Physical quantities are interpolated and communicated between these domains while sliding. The interpolation method is introduced in detail below.

Figure 3(a) shows the faces of cells on the dividing plane, which is called the Sliding plane in this paper. The face of the cell i belongs to the computational domain of the train, and the faces of the cells j belong to the computational domain of the overpass. They are adjacent to each other on the Sliding plane, that is, the cell i is adjacent to and overlapping with the cells j. The overlapping areas of the cell i and the cells j are calculated and the interpolated values are based on the each area and the each physical quantity of the cells. The interpolating formula is shown below,

$$q_{bi} = \frac{\sum_{j \in i} q_j S_{ij}}{S_i} \tag{4}$$

where q_j is each physical quantity of cells j, q_{bi} is the physical quantity of the ghost cell of the cell i, S_{ij} is the overlapping areas of the cell i and the cells j and $\sum_{j \in i}$ means the summation of the values of all the cells j which is adjacent to the cell i. Previous researches using the Sliding Mesh approach have been reported. However, in these researches, Sliding planes must be completely adjacent to other Sliding planes, that is, Sliding planes cannot be adjacent to both other Sliding planes and outer boundaries simultaneously. Then, in this study, the boundary condition on Sliding planes is improved and expanded. This Expanded Sliding Mesh approach enables to solve the problem that Sliding planes are adjacent to both other Sliding planes and outer boundaries at the same time.

Figure 3(b) shows that a sliding plane is adjacent to the other sliding plane and outer boundaries simultaneously. The part of the boundary which is adjacent to the other Sliding plane is called the Sliding boundary, and the part of the boundary which is adjacent to the outer boundary is called the Outer-Sliding boundary. At the Sliding boundary, the interpolation of the physical quantities is using the Eq. 4. On the other hand, at the Outer-Sliding boundary, the physical quantities are estimated with the MCD method as the complete outer boundary at first, and then these quantities are weighted by the outer area and added to

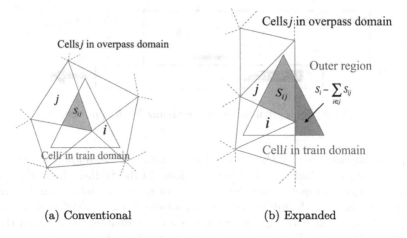

(a) Conventional   (b) Expanded

Fig. 3. Relationships of the triangles on sliding planes

the quantities computed by the Eq. 4. These computation is formulated as shown below,

$$q_{bi} = \frac{\sum_{j \in i} q_j S_{ij} + q_{oi}(S_i - \sum_{j \in i} S_{ij})}{S_i} \tag{5}$$

where q_{oi} is physical quantities computed by the MCD method as the complete outer boundary.

3   Inspection for Geometric Conservation Law

In this section, the inspection of the approach combining the MCD method and the Expanded Sliding Mesh approach is carried out with a uniform flow problem. Whether the extended boundary condition satisfies the Geometric conservation law is checked.

3.1   Computational Grid

Figure 4 shows the computational grid for the uniform flow problem. The small pink rectangular moves for positive x-axis at velocity 0.1, and the small yellow rectangular moves for negative x-axis at velocity -0.1. Unstructured grid is generated by MEGG3D [13,14]. The number of elements is 594,794.

3.2   Computational Conditions

All the cells have $\rho = 1.0, p = \frac{1}{\gamma}, u = v = w = 1.0$ as an initial condition. As boundary conditions, initial conditions are remained at outer boundary. The

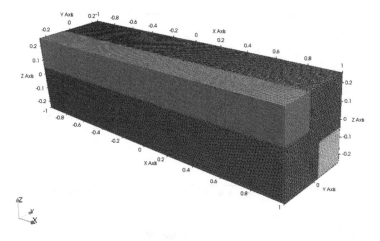

Fig. 4. Computational grid for uniform flow inspection

Sliding boundary condition is applied to the sliding boundary, and the Sliding-Outer boundary condition is applied to the Sliding-Outer boundary. Time width is 0.005 and the computation lasts for dimensionless time 16000.

3.3 Result

Density error is defined by Eq. 6. The maximum value of the density errors among all the cells is the density error at each time step.

$$\text{ERROR}_n = \max \left(\frac{|\rho_\infty - \rho_i|}{\rho_\infty} \right) \tag{6}$$

Figure 5 shows the error history of the density errors. The orders of the errors are under 10^{-12}, which is the range of mechanical error. Therefore, it is proved that the MCD method and the Expanded Sliding Mesh approach do not give any effect to the flow fields.

4   Application for a Train and an Overpass

In this section, the computation of the train passing through the stationary overpass is achieved with the MCD method and the Expanded Sliding Mesh approach. Moreover, the numerical results are compared qualitatively and quantitatively with the experimental results carried out by Baker to confirm the validity of the numerical results.

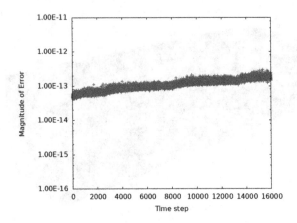

Fig. 5. Density error history

Table 1. Sizes of the experimental model of the Class390 made by Baker [3]

[m]	Real scale	1/25 scale
Length	37.35	1.48
Width	2.73	0.11
Height	3.56	0.14

4.1   Introduction of Previous Research

First, the previous experiment carried out by Baker is explained for comparing. The Baker's experiment was carried out in order to obtain a fundamental understanding of the nature of aerodynamical phenomena caused by the passages of trains and to obtain data for a variety of railway infrastructure geometries of particular relevance to the UK situation.

Experimental Model. In the experiment, 4 types of trains and 3 types of trackside facilities were adopted. In particular, the Class390 and the overpass are explained here.

First, the Class390 is the common train in the UK, which has streamline of the head carriage and the notch on the ceiling. Baker made the experimental model of the Class390 in 1/25 scale (See Fig. 6(a)). The sizes of the Class390 in real scale and in 1/25 scale are shown in Table 1.

Next, as explained in Sect. 1, overpasses are a kind of bridges for pedestrians and cars, which are fixed over train tracks. Baker made the experimental model of the overpass with an elastic plate in 1/25 scale and it is fixed over the train track with the piers, shown in Fig. 6(b). The sizes of the overpass in real scale and 1/25 scale are shown in Table 2.

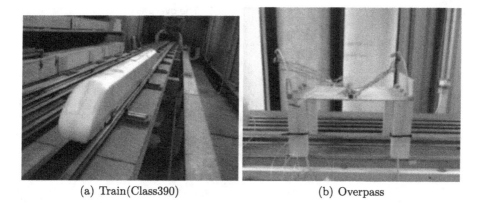

(a) Train(Class390)   (b) Overpass

Fig. 6. Experimental models made by Baker [3]

Table 2. Sizes of the experimental model of the overpass made by Baker [3]

[m]		Real scale	1/25 scale
Width		10.0	0.4
Length		5.0	0.2
Height	Case1	4.5	0.18
	Case2	5.0	0.2
	Case3	5.5	0.22
	Case4	6.0	0.24

Experimental Condition. The measurement points of pressure are explained here. The lower surface of the overpass is shown in Fig. 7. The sensors for pressure measurement are set at the stars. The sensors are set every 1.0 m in real scale (every 40 mm in 1/25 scale), they are set symmetrically for the center line of the overpass.

4.2 Evaluation of Pressure

An evaluation method for pressure measured on the lower surface of the overpass is explained in this section. In the Baker's experiment, all the results of pressure were evaluated as pressure coefficients using the Eq. 7. For comparing, our results will be evaluated as pressure coefficients using the same equation.

$$C_{\mathrm{p}} = \frac{p - p_{\infty}}{\frac{1}{2}\rho_{\infty}U^2} \tag{7}$$

where p is pressure measured at the each measurement point on the lower surface of the overpass, ρ_{∞} and p_{∞} is constant density and constant pressure respectively, and U is the traveling speed of the train.

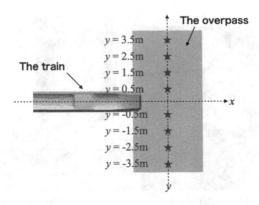

Fig. 7. Pressure measurement points on the overpass model

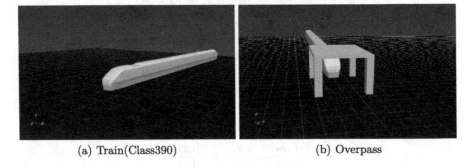

(a) Train(Class390)   (b) Overpass

Fig. 8. Computational model

4.3 Computation of the Traveling Train Under the Overpass with the Piers

First, the computations that the height of the overpass is 5.5 m and 6.0 m are carried out in order to check the validity of the code and the results qualitatively. After that, the computations that the overpass is very close to the traveling train (the height of the overpass is 4.5 m and 5.0 m) are carried out in order to obtain a variety of the results.

Computational Model and Conditions. The numerical models are shown in Fig. 8, the computational grid is shown in Fig. 9, and the overview and the sizes of the computational domains are shown in Fig. 10. The shape of the numerical models of the train and the overpass are almost the same with the experimental models. The streamline of the head carriage and the notch shape on the ceiling are reproduced. The overpass is modeled as rigid plate. The computational grid is generated by MEGG3D.

Fig. 9. Computational grid

Computation of the Wider Gap Cases. At first, the results of the cases that the height of the overpass is 5.5 m and 6.0 m are shown and discussed. These cases are called "the wider gap cases" below. Pressure distribution on the lower surface of the overpass in the wider gap cases at dimensionless time t = 45 and 56 is shown in Fig. 11. The range of the color bars is set at the same. At both time steps, pressure is higher and lower at height = 5.5 m than at height = 6.0 m, that is, pressure changes more dynamically in the case that the overpass is close to the train. It means pressure fluctuation depends on the height of the overpass. Moreover, such aerodynamical phenomenon is expressed by numerical simulation.

Difference between the maximum and the minimum pressure coefficients in the wider gap cases obtained by Baker and our numerical computation is plotted in the Fig. 12. Pressure coefficients are calculated as explained in the Sect. 4.2. As shown in Fig. 12, the experimental results and the numerical results have almost the same distribution. Thus, the qualitative and quantitative agreements of them are proved in the wider gap cases.

Computation of the Narrower Gap Cases. Next, the results of the cases that the height of the overpass is 4.5 m and 5.0 m are shown and discussed in this subsection. These cases are called "the narrower gap cases" below. Pressure distribution on the lower surface of the overpass in the narrower gap cases at dimensionless time t = 45 and 56 is shown in Fig. 13. As in the previous subsection, pressure changes more dynamically in the case that the overpass is close to the train. It is shown that the height of the overpass affects dynamical pressure fluctuation not only in the wider cases but also in the narrower cases.

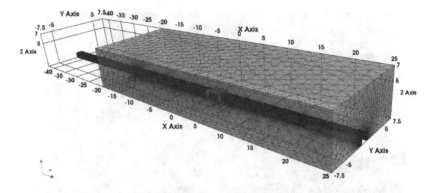

Fig. 10. Sizes of computational grid

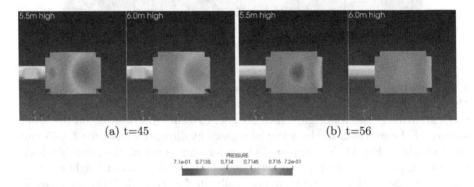

(a) t=45   (b) t=56

Fig. 11. Pressure distribution on the lower surface of the overpass (the wider gap cases)

Difference of the maximum and the minimum pressure coefficients in the narrower cases obtained by the Baker's experiment and the our numerical computation is plotted in Fig. 14. The basic tendency of pressure distribution, which is symmetrical at the center line ($y = 0$), is almost expressed also in the narrower cases. Therefore, the qualitative agreements are proved.

However, there are some quantitative disagreements in the narrower cases. First, pressure coefficients around the both ends ($y = -3.5\,\mathrm{m}$ and $3.5\,\mathrm{m}$) of the numerical results in Fig. 14 are overpredicted. Second, the maximum pressure coefficients of the numerical results are quite different from those of the experimental results. The error of the maximum pressure coefficients is about 23%.

The reasons why these differences occurred are discussed below. Ignoring viscosity and boundary layers would be one possibility. Actual fluids have viscosity. Boundary layers are appeared along the wall of moving bodies and the layers give influences to the flow fields. In our study, since it is supposed that the fluid is inviscid, the influence of the boundary layers to the flow fields is completely ignored. In the situation that the train is very close to the overpass, the influence should be considered because the gap between the train and the overpass is

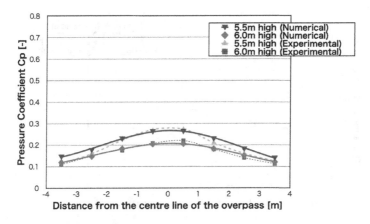

Fig. 12. The lateral peak-to-peak distribution on the lower surface of the overpass (the wider gap cases)

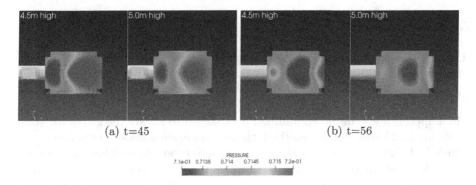

Fig. 13. Pressure distribution on the lower surface of the overpass (the narrower gap cases)

small. The order of the boundary layer based on the Prandtl's Laminar Boundary layer Equation is 1.77×10^{-3}, which is about 1% towards the gap. The boundary layer will be bigger in turbulence, and the influence of the turbulent and laminar boundary layer will be more significant. If the physical quantities in the flow fields like pressure or velocity are altered with the influence of the layers in laminar flow and turbulence, the numerical results may agree with the experimental results.

Now we have developed the numerical simulation code for viscous flow with the numerical model of the train and the overpass.

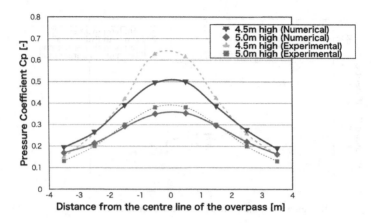

Fig. 14. The lateral peak-to-peak distribution on the lower surface of the overpass (the narrower gap cases)

5 Conclusions

The objective of this study is to simulate the flow fields driven by the train passage and to estimate the effects of them like pressure wave in order to help optimal reconstruction and rebuild of old overpasses, which are often seen in the UK. The new method combined the MCD method and the Expanded Sliding Mesh approach was proposed and had been shown to satisfy the geometric conservation law. Using the new method, the numerical simulations of the four cases which had different height of the overpass were conducted within the inviscid approximation. In the cases of the wider gap that the height of the overpass is 5.5 m and 6.0 m, the numerical results agreed almost completely with the experimental results. Therefore, it was shown to be able to reproduce the flow field around the traveling train and the overpass in computer and fully estimate pressures acting on the lower surface of the overpass even within the inviscid approximation. However, in the cases of the narrower gap that the height of the overpass is 4.5 m and 5.0 m, the pressure fluctuation of the results of the numerical simulation was under-estimated. The one of the reasons of the disagreement would be the inviscid approximation. In the real world, the flow field around the traveling train and the overpass is under the influence of laminar or turbulent boundary layer in the case that the train is very close to the overpass. But in this computation, such influences were ignored at all. Now, we have been developing the simulation code for the viscous compressible flow and computations are now running.

In the case that there is appropriate distance between the train and facilities like overpasses, it was shown to be able to estimate the effect of train passages to the facilities even within the inviscid approximation. This method combined the MCD method and the Expanded Sliding Mesh approach can be applied to computations including another facilities, for instance station platforms, or other kind of trains and also could help old facilities be reconstructed and rebuild. More accurate and detailed flow simulation considering viscosity will be required in the further study.

Acknowledgments. This publication was subsidized by JKA through its promotion funds from KEIRIN RACE.

References

1. Hori, M.: A study on railway reform in Britain. From the viewpoint of structural separation. Tour. Stud. **8**, 57 (2009). (Translated from Japanese)
2. Koyakumaru, S.: Recognition of reforms and new developments in British Railways. Transp. Econ. **72**(7), 74 (2012). (Translated from Japanese)
3. Baker, C., et al.: Transient aerodynamic pressures and forces on trackside and overhead structures due to passing trains. Part 1: Model-scale experiment. Part 2: Standards applications. J. Rail Rapid Transit **228**(1), 37–70 (2014)
4. Takii, A., Yamakawa, M., Asao, S., Tajiri, K.: Six degrees of freedom numerical simulation of tilt-rotor plane. In: Rodrigues, J.M.F., et al. (eds.) ICCS 2019. LNCS, vol. 11536, pp. 506–519. Springer, Cham (2019). https://doi.org/10.1007/978-3-030-22734-0_37
5. Yamakawa, M., Mitsunari, N., et al.: Numerical simulation of rotation of intermeshing rotors using added and eliminated mesh method. Proc. Comput. Sci. **108**, 1183–1192 (2017)
6. Asao, S., et al.: Simulations of a falling sphere with concentration in an infinite long pipe using a new moving mesh system. Appl. Therm. Eng. **72**, 29–33 (2014)
7. Asao, S., et al.: Parallel computations of incompressible flow around falling spheres in a long pipe using moving computational domain method. Comput. Fluids **88**, 850–856 (2013)
8. Yamakawa, M., Matsuno, K., et al.: Unstructured moving-grid finite-volume method for unsteady shocked flows. J. Comput. Fluids Eng. **10**(1), 24–30 (2005)
9. Yamakawa, M., Takekawa, D., Matsuno, K., et al.: Numerical simulation for a flow around a body ejection using an axisymmetric unstructured moving grid method. Comput. Therm. Sci. **4**(3), 217–223 (2012)
10. Obayashi, S., et al.: Free-stream capturing for moving coordinates in three dimensions. AAIA J. **30**, 1125–1128 (1992)
11. Watanabe, K., Matsuno, K., et al.: Moving computational domain method and its application to flow around a high-speed car passing through a hairpin curve. J. Comput. Sci. Technol. **3**(2), 449–459 (2009)
12. Yamakawa, M., Chikaguchi, S., Asao, S.: Numerical simulation of tilt-rotor plane using multi axes sliding mesh approach. In: The 27th International Symposium on Transport Phenomena, Honolulu (2016)
13. Ito, Y.: Challenges in unstructured mesh generation for practical and efficient computational fluid dynamics simulations. Comput. Fluids **85**, 47–52 (2013)
14. Ito, Y., Nakanishi, K.: Surface triangulation for polygonal models based on CAD data. Intern. J. Numer. Methods Fluids **39**(1), 75–96 (2002)

Bézier Surfaces for Modeling Inclusions in PIES

Agnieszka Bołtuć$^{(\boxtimes)}$, Eugeniusz Zieniuk , Krzysztof Szerszeń ,
and Andrzej Kużelewski

Institute of Informatics, University of Bialystok, Bialystok, Poland
{aboltuc,ezieniuk,kszerszen,akuzel}@ii.uwb.edu.pl

Abstract. The paper presents the approach for solving 2D elastic
boundary value problems defined in domains with inclusions with different material properties using the parametric integral equation system
(PIES). The main feature of the proposed strategy is using Bézier surfaces for global modeling of inclusions. Polygonal inclusions are defined
by bilinear surfaces, while others by bicubic surfaces. It is beneficial over
other numerical methods (such as FEM and BEM) due to the lack of
discretization. Integration over inclusions defined by surfaces is also performed globally without division into subareas. The considered problem
is solved iteratively in order to simulate different material properties by
applying initial stresses within the inclusion. This way of solving avoids
increasing the number of unknowns and can also be used for elasto-
plastic problems without significant changes. Some numerical tests are
presented, in which the results obtained are compared with those calculated by other numerical methods.

Keywords: PIES · Inclusions · Bézier surfaces

1 Introduction

Practical elastic problems very often require the analysis of piecewise homogeneous domains in which several regions exist, each with various material properties. Well known numerical methods such as the finite element method (FEM)
[1–3] and the boundary element method (BEM) [1,4–6] deal with such situations, however, they demand a completely different approach. Including two or
more materials in FEM is quite straightforward, because different properties can
be assigned to specific finite elements, which are always generated at the stage
of body modeling, regardless of the problem solved. It is reduced to the proper
discretization of the body and linking the right attributes with the right finite
elements. On the other hand, most problems solved by BEM require defining
only the boundary elements, which causes that elements inside the domain do
not exist. Therefore, the only way to handle piecewise homogeneous problems in
BEM is to divide the model into subregions or zones. Each zone has its own set
of material properties and they are connected along a common interface. At the

© Springer Nature Switzerland AG 2020
V. V. Krzhizhanovskaya et al. (Eds.): ICCS 2020, LNCS 12138, pp. 170–183, 2020.
https://doi.org/10.1007/978-3-030-50417-5_13

beginning, various zones are treated as separate BEM models, and finally they are combined into a single system, using so-called constraint equations. The disadvantage of this strategy lies in the additional degrees of freedom arising from boundary elements on the common interface, which have to consider different results in separate zones.

The described above multi-region approach used in BEM results in the larger system of equations, therefore in [7,8] the authors propose the different technique. The classical BEM is extended to include heterogeneous domains by introducing the volume effect. Such defined problem should be solved iteratively in order to modify the solution from the one with elastic homogeneous domain to this with presence of inclusions with different material properties. The only drawback of the proposed approach is that it requires cells for the evaluation of the domain integrals, which is an additional effort and is technically similar to discretization in FEM. Therefore, in order to overcome the need for a domain discretization, another concept is introduced in [9]. The new idea is to define subregion by two NURBS curves and a linear interpolation between them. It eliminates the need for cell generating, but the subdivision into integration regions still exists.

Taking into account mentioned above disadvantages, in this paper another approach is presented. The arising volume is not discretized into cells or defined by two curves, but is entirely modeled with one parametric surface. For subregions with linear boundary the surfaces of the first degree can be used, while for those with curvilinear edges the surfaces of the third degree are applied. Moreover, the evaluation of volume integrals is done in global manner, without dividing the area into subregions. Mentioned features (global modeling and integrating without classical discretization) are main advantages of the parametric integral equation system (PIES) [10]. This method was successfully used to solve 2D and 3D problems like potential [11], elastic [12], acoustic [13] and most recently elastoplastic [14]. As was emphasized above, PIES is characterized by no discretization and flexible way of modeling both the boundary and the domain by any curves and surfaces known from computer graphics [15,16]. This crucial advantage comes from the fact that the shape is analytically included in the mathematical formalism of the method. Moreover, such approach gives a possibility for applying various methods for approximating of boundary and domain functions, because it is separated from approximating of the shape. Mentioned feature causes simple, independent improving of the accuracy of solutions without interfering with the shape of the boundary and the domain.

The main aim of the paper is to develop the approach for solving elastic problems with inclusions using PIES and global modeling of the shape. The idea bases on treating the whole solid as single region and simulating iteratively different material properties by applying initial stresses within the inclusion. The geometry of the inclusion is defined globally by Bézier surface. The approximation of initial stresses is performed by the Lagrange polynomial with various number and arrangement of interpolation nodes. Finally, the approach is verified compared to other well-known numerical methods.

2   Parametric Integral Equation System (PIES)

The parametric integral equation system (PIES) for elastic problems with inclusions taking into account incremental initial stress formulation can be presented in the following form

$$0.5\dot{\mathbf{u}}_l(\bar{s}) = \sum_{j=1}^{n} \int_{s_{j-1}}^{s_j} \left\{ \mathbf{U}_{lj}^*(\bar{s},s)\dot{\mathbf{p}}_j(s) - \mathbf{P}_{lj}^*(\bar{s},s)\dot{\mathbf{u}}_j(s) \right\} J_j(s)ds \tag{1}$$

$$+ \int_{\Omega} \mathbf{E}_l^*(\bar{s},\mathbf{y})\dot{\boldsymbol{\sigma}}_0(\mathbf{y})d\Omega(\mathbf{y}),$$

where $\mathbf{U}_{lj}^*(\bar{s},s)$, $\mathbf{P}_{lj}^*(\bar{s},s)$, $\mathbf{E}_l^*(\bar{s},\mathbf{y})$ are fundamental solutions for the displacements, tractions and strains respectively. Functions $\dot{\mathbf{p}}_j(s)$, $\dot{\mathbf{u}}_j(s)$ are incremental forms of parametric functions corresponding to tractions and displacements on the boundary. Vector $\dot{\boldsymbol{\sigma}}_0(\mathbf{y})$ contains increments of initial stresses inside the inclusion. $s,\bar{s}$ are parameters in one-dimensional parametric reference system in which the boundary in PIES is defined and $s_{l-1} \leq \bar{s} \leq s_l$, $s_{j-1} \leq s \leq s_j$. In PIES s_{l-1} and s_{j-1} correspond to the beginning of lth and jth segments, while s_l and s_j to the end of these segments. $J_j(s)$ is the Jacobian, n is the number of boundary segments, $\mathbf{y} \in \Omega$ and $l,j = 1..n$.

The displacement fundamental solution $\mathbf{U}_{lj}^*(\bar{s},s)$ is presented explicitly in [12] by

$$\mathbf{U}_{lj}^*(\bar{s},s) = -\frac{1}{8\pi(1-\nu)\mu} \begin{bmatrix} (3-4\nu)ln(\eta) - \frac{\eta_1^2}{\eta^2} & -\frac{\eta_1\eta_2}{\eta^2} \\ -\frac{\eta_1\eta_2}{\eta^2} & (3-4\nu)ln(\eta) - \frac{\eta_2^2}{\eta^2} \end{bmatrix}, \tag{2}$$

where $\eta_1 = \Gamma_j^{(1)}(s) - \Gamma_l^{(1)}(\bar{s})$, $\eta_2 = \Gamma_j^{(2)}(s) - \Gamma_l^{(2)}(\bar{s})$, $\eta = \left[\eta_1^2 + \eta_2^2\right]^{0.5}$, ν is Poisson's ratio and μ is the shear modulus.

The traction fundamental solution $\mathbf{P}_{lj}^*(\bar{s},s)$ is given by [12]

$$\mathbf{P}_{lj}^*(\bar{s},s) = -\frac{1}{4\pi(1-\nu)\eta} \begin{bmatrix} P_{11} & P_{12} \\ P_{21} & P_{22} \end{bmatrix}, \tag{3}$$

where

$$P_{11} = \left\{(1-2\nu) + 2\frac{\eta_1^2}{\eta^2}\right\} \frac{\partial \eta}{\partial n}, \quad P_{22} = \left\{(1-2\nu) + 2\frac{\eta_2^2}{\eta^2}\right\} \frac{\partial \eta}{\partial n},$$

$$P_{21} = P_{12} = \left\{2\frac{\eta_1\eta_2}{\eta^2}\frac{\partial \eta}{\partial n} - (1-2\nu)\left[\frac{\eta_1}{\eta}n_2(s) + \frac{\eta_2}{\eta}n_1(s)\right]\right\},$$

and $\frac{\partial \eta}{\partial n} = \frac{\partial \eta_1}{\partial n}n_1(s) + \frac{\partial \eta_2}{\partial n}n_2(s)$, while $n_1(s)$ and $n_2(s)$ are direction cosines of the external normal to jth segment of the boundary.

The strains fundamental solution $\mathbf{E}_l^*(\bar{s},\mathbf{y})$ can be presented by

$$\mathbf{E}_l^*(\bar{s},\mathbf{y}) = -\frac{1}{8\pi(1-\nu)G\bar{\eta}} \begin{bmatrix} 2A\bar{\eta}_1 - \bar{\eta}_1 + 2\bar{\eta}_1^3 & -\bar{\eta}_2 + 2\bar{\eta}_1^2\bar{\eta}_2 \\ A\bar{\eta}_2 + 2\bar{\eta}_1^2\bar{\eta}_2 & A\bar{\eta}_1 + 2\bar{\eta}_1\bar{\eta}_2^2 \\ A\bar{\eta}_2 + 2\bar{\eta}_1^2\bar{\eta}_2 & A\bar{\eta}_1 + 2\bar{\eta}_1\bar{\eta}_2^2 \\ -\bar{\eta}_1 + 2\bar{\eta}_1\bar{\eta}_2^2 & 2A\bar{\eta}_2 - \bar{\eta}_2 + 2\bar{\eta}_2^3 \end{bmatrix}^T, \tag{4}$$

where $A = (1 - 2\nu)$, $\bar{\eta} = [\bar{\eta}_1^2 + \bar{\eta}_2^2]^{0.5}$, $\bar{\eta}_1 = G^{(1)}(\boldsymbol{y}) - \Gamma_l^{(1)}(\bar{s})$ and $\bar{\eta}_2 = G^{(2)}(\boldsymbol{y}) - \Gamma_l^{(2)}(\bar{s})$.

The shape of the boundary and the domain is analytically included in (2),(3) and (4). The boundary can be modeled by means of any parametric curves $\boldsymbol{\Gamma}_j(s) = \left[\Gamma_j^{(1)}(s), \Gamma_j^{(2)}(s)\right]^T$ [15,16], which are included into functions η_1 and η_2. Functions $\bar{\eta}_1$ and $\bar{\eta}_2$ contain $\boldsymbol{G}(\boldsymbol{y}) = \left[G^{(1)}(\boldsymbol{y}), G^{(2)}(\boldsymbol{y}), G^{(3)}(\boldsymbol{y})\right]^T$, which is a parametric surface known from computer graphics [15,16]. It should be emphasized that for 2D problems considered in this paper $G^{(3)}(\boldsymbol{y}) = 0$.

As can be seen formula (1) requires initial stresses inside the inclusion. They can be calculated using the strains and generalized Hooke's law. For this reason the integral identity for strains is also required.

3   Internal Results

As mentioned in the previous section, the proposed strategy requires calculating strains inside the inclusion. They can be obtained using the following integral equation

$$\dot{\boldsymbol{\varepsilon}}(\boldsymbol{x}) = \sum_{j=1}^{n} \int_{s_{j-1}}^{s_j} \left\{ \hat{\boldsymbol{D}}_j^*(\boldsymbol{x}, s)\dot{\boldsymbol{p}}_j(s) - \hat{\boldsymbol{S}}_j^*(\boldsymbol{x}, s)\dot{\boldsymbol{u}}_j(s) \right\} J_j(s)ds$$
$$+ \int_{\Omega} \hat{\boldsymbol{W}}^*(\boldsymbol{x}, \boldsymbol{y})\dot{\boldsymbol{\sigma}}_0(\boldsymbol{y})d\Omega(\boldsymbol{y}) + \hat{\boldsymbol{f}}\dot{\boldsymbol{\sigma}}_0(\boldsymbol{x}). \tag{5}$$

The integrand $\hat{\boldsymbol{S}}_j^*(\boldsymbol{x}, s)$ is presented by the following formula (in plain strain)

$$\hat{\boldsymbol{S}}_j^*(\boldsymbol{x}, s) = \frac{1}{4\pi(1 - \nu)r^2} \begin{bmatrix} S_{111} & S_{211} \\ S_{112} & S_{212} \\ S_{121} & S_{221} \\ S_{122} & S_{222} \end{bmatrix}, \tag{6}$$

$S_{111} = [2\frac{\partial r}{\partial n}[2\nu r_1 + r_1 - 4r_1^3] + (1 - 2\nu)(n_1 + 2r_1^2 n_1) + 4\nu r_1^2 n_1]$,
$S_{112} = S_{121} = [2\frac{\partial r}{\partial n}[\nu r_2 - 4r_1^2 r_2] + (1 - 2\nu)(n_2 + 2r_1 r_2 n_1) + 2\nu(r_1^2 n_2 + r_1 r_2 n_1)]$,
$S_{122} = [2\frac{\partial r}{\partial n}[r_1 - 4r_1 r_2^2] + (1 - 2\nu)(-n_1 + 2r_2^2 n_1) + 4\nu r_1 r_2 n_2]$,
$S_{211} = [2\frac{\partial r}{\partial n}[r_2 - 4r_2 r_1^2] + (1 - 2\nu)(-n_2 + 2r_1^2 n_2) + 4\nu r_1 r_2 n_2]$,
$S_{212} = S_{221} = [2\frac{\partial r}{\partial n}[\nu r_1 - 4r_2^2 r_1] + (1 - 2\nu)(n_1 + 2r_1 r_2 n_2) + 2\nu(r_2^2 n_1 + r_1 r_2 n_2)]$,
$S_{222} = [2\frac{\partial r}{\partial n}[2\nu r_2 + r_2 - 4r_2^3] + (1 - 2\nu)(n_2 + 2r_2^2 n_2) + 4\nu r_2^2 n_2]$,
where $\frac{\partial r}{\partial n} = \frac{\partial r_1}{\partial r}n_1(s) + \frac{\partial r_2}{\partial r}n_2(s)$, $r = [r_1^2 + r_2^2]^{0.5}$, $r_1 = \Gamma_j^{(1)}(s) - G^{(1)}(\boldsymbol{x})$ and $r_2 = \Gamma_j^{(2)}(s) - G^{(2)}(\boldsymbol{x})$.

The integrand $\hat{\boldsymbol{D}}_j^*(\boldsymbol{x}, s)$ can be described by (4) multiplied by -1, in which $\bar{\eta}, \bar{\eta}_1, \bar{\eta}_2$ are replaced by r, r_1, r_2.

The last integral in (5) can be presented by the following expression

$$\hat{\boldsymbol{W}}^*(\boldsymbol{x},\boldsymbol{y}) = \frac{1}{8\pi G(1-\nu)\bar{r}^2} \begin{bmatrix} W_{1111} & W_{1112} & W_{1121} & W_{1122} \\ W_{1211} & W_{1212} & W_{1221} & W_{1222} \\ W_{2111} & W_{2112} & W_{2121} & W_{2122} \\ W_{2211} & W_{2212} & W_{2221} & W_{2222} \end{bmatrix}, \tag{7}$$

$W_{1111} = [2(1-2\nu) - 1 + 8\nu\bar{r}_1^2 + 2(2\bar{r}_1^2 - 4\bar{r}_1^4)]$,
$W_{1112} = W_{1121} = W_{1211} = W_{2111} = [4\nu\bar{r}_1\bar{r}_2 + 2(\bar{r}_1\bar{r}_2 - 4\bar{r}_1^3\bar{r}_2)]$,
$W_{1122} = W_{2211} = [-1 + 2(\bar{r}_1^2 + \bar{r}_2^2 - 4\bar{r}_1^2\bar{r}_2^2)]$,
$W_{1221} = W_{1212} = W_{2112} = W_{2121} = [(1-2\nu) + 2\nu(\bar{r}_1^2 + \bar{r}_2^2) - 8\bar{r}_1^2\bar{r}_2^2]$,
$W_{1222} = W_{2122} = W_{2221} = W_{2212} = [4\nu\bar{r}_1\bar{r}_2 + 2(\bar{r}_1\bar{r}_2 - 4\bar{r}_1\bar{r}_2^3)]$,
$W_{1111} = [2(1-2\nu) - 1 + 8\nu\bar{r}_2^2 + 2(2\bar{r}_2^2 - 4\bar{r}_2^4)]$,
where $\bar{r} = [\bar{r}_1^2 + \bar{r}_2^2]^{0.5}$, $\bar{r}_1 = G^{(1)}(\boldsymbol{y}) - G^{(1)}(\boldsymbol{x})$ and $\bar{r}_2 = G^{(2)}(\boldsymbol{y}) - G^{(2)}(\boldsymbol{x})$.

The free term from (5) for the plane strain case is given by

$$\hat{\boldsymbol{f}} = -\frac{1}{16G(1-\nu)} \begin{bmatrix} 1 - 2(3-4\nu) & 0 & 0 & -(3-4\nu) \\ 0 & -(3-4\nu) & 1 & 0 \\ 0 & 1 & -(3-4\nu) & 0 \\ -(3-4\nu) & 0 & 0 & 1 - 2(3-4\nu) \end{bmatrix}. \tag{8}$$

4   Solving PIES

Solving problems without inclusions by PIES is reduced to finding functions $\dot{\boldsymbol{u}}_j(s)$ and $\dot{\boldsymbol{p}}_j(s)$. They are approximated using series with various base functions, e.g. in this paper the Lagrange polynomials are used [14]. After substituting such series into PIES, writing down the resultant equation at all interpolation (collocation) points and reordering the following system of equation is obtained

$$\boldsymbol{A}\dot{\boldsymbol{x}} = \dot{\boldsymbol{b}}, \tag{9}$$

where $\boldsymbol{A}$ is a matrix that contains mixed values of both integrands (2, 3), $\dot{\boldsymbol{x}}$ contains the unknown boundary values, while $\dot{\boldsymbol{b}}$ contains prescribed values on the boundary.

If inclusions are considered, this equation should be extended by the additional term

$$\boldsymbol{A}\dot{\boldsymbol{x}} = \dot{\boldsymbol{b}} + \dot{\boldsymbol{f}}, \tag{10}$$

where $\dot{\boldsymbol{f}}$ includes integrals (4) with the initial stresses.

Initial stresses are also approximated using series similar to those applied for boundary functions, but they depend on two variables

$$\dot{\boldsymbol{\sigma}}_0(\boldsymbol{y}) = \sum_{r=0}^{R_1-1} \sum_{w=0}^{R_2-1} \dot{\boldsymbol{\sigma}}_0^{rw}(\boldsymbol{y}) L_{rw}(\boldsymbol{y}), \tag{11}$$

where

$$L_{rw}(\boldsymbol{y}) = L_r(y_1) L_w(y_2),$$

$$L_r(y_1) = \prod_{o=0, o \neq r}^{R_1} \frac{y_1 - y_{1o}}{y_{1r} - y_{1o}}, \quad L_w(y_2) = \prod_{o=0, o \neq w}^{R_2} \frac{y_2 - y_{2o}}{y_{2w} - y_{2o}},$$

and $N = R_1 \times R_2$ is the given number of interpolation nodes, while $\dot{\sigma}_0^{rw}(\boldsymbol{y})$ is the value of initial stress at interpolation node with (y_1, y_2) coordinates.

As was mentioned in the previous section, to calculate initial stresses strains are required. The final matrix form of the formula for calculating strains can be obtained in a similar as above manner using (5)

$$\dot{\varepsilon} = -\boldsymbol{A}' \dot{\boldsymbol{x}} + \boldsymbol{b}' + (\boldsymbol{W} + \boldsymbol{F}) \dot{\boldsymbol{\sigma}}_0, \tag{12}$$

where $(\boldsymbol{W} + \boldsymbol{F})$ correspond to expressions (7,8), $\boldsymbol{A}'$ is a matrix that contains mixed values of two first integrands in (5), while $\boldsymbol{b}'$ contains prescribed values on the boundary.

Integrals required for (10) and most of integrals for (12) are regular or weakly singular, therefore they are calculated using Gauss integration and the subdivision technique applied to the surfaces [4–6]. The integral over the domain with function (7), which should be calculated in (12), is strongly singular, with singularity of order $\frac{1}{r^2}$ for 2D problems. The singularity is isolated by replacing the integral by two integrals. The first is weakly singular and is treated by the subdivision technique, while second is transformed into the boundary as presented in [17].

5   Modeling of Inclusions

As can be seen in (1) and (5), last integrals require defining the domain of inclusion. In FEM, regardless of the problem, the whole body is modeled by finite elements. Therefore, considering the inclusion requires assuming various material properties for various groups of elements (Fig. 1).

In BEM, the inclusion is divided into so-called cells, which technically resemble finite elements (Fig. 2). In both methods the number, type and arrangement of those elements influence also the accuracy of the solutions. Such approach is troublesome and often forces the use of more elements than the shape of the geometry actually requires. Moreover, integration also is performed over elements, even if mapping method is used [9], which bases on modeling the inclusion using two NURBS curves and interpolation between them.

For this reason in this paper another approach is proposed. It bases on the popular tool of computer graphics, namely on surface patches [15,16]. Unlike other methods, the inclusion area is modeled entirely using a single Bézier surface, but also other types of surfaces can be used instead. If the polygonal inclusion is considered, then bilinear surfaces should be used (Fig. 3), while for other curvilinear shapes of subdomains bicubic surfaces can be applied (Fig. 4).

As can be seen in Fig. 3 and Fig. 4, the proposed way of definition requires smaller number of points than in FEM and BEM. For example the inclusion presented in Fig. 3 is created using only 4 control points (■), which are actually

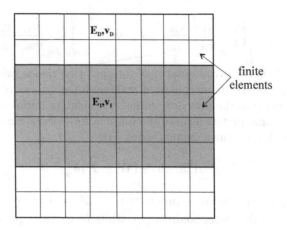

E_D,v_D - material properties of the body
E_I,v_I - material properties of the inclusion

Fig. 1. Modeling the inclusion in FEM

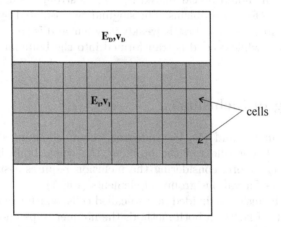

E_D,v_D - material properties of the body
E_I,v_I - material properties of the inclusion

Fig. 2. Modeling the inclusion in BEM

corner points. For comparison, the same inclusion in FEM (Fig. 1) and BEM (Fig. 2) is composed of 32 finite elements and 32 cells respectively. If we assume that those elements and cells are linear, both cases require 128 nodes for defining them, while FEM additionally needs finite elements for modeling the whole body (64 elements and 256 nodes in total). The curvilinear shape of the inclusion in Fig. 4 requires 12 control points (■), which define its boundary (other 4 are not important due to 2D nature of the problem).

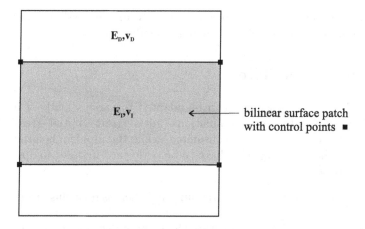

E_D, v_D - material properties of the body
E_I, v_I - material properties of the inclusion

Fig. 3. Modeling the polygonal inclusion in PIES

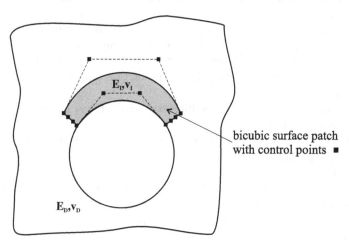

E_D, v_D - material properties of the body
E_I, v_I - material properties of the inclusion

Fig. 4. Modeling the curvilinear inclusion in PIES

Moreover, the proposed approach gives the opportunity for simple modification of the defined shape. Moving even one control point causes significant change in the shape of the inclusion. Such feature can be very useful when dealing with identification or optimization of the shape. Furthermore, modification of the geometry automatically modifies the mathematical formalism of PIES (1), because the shape is included in it. It also allows for separation of shape modeling from the approximation of solutions, which results in possibility of applying various methods for both stages of solving boundary problems.

6   Iterative Procedure

The general iterative procedure can be adapted to various kinds of inclusions (elastic, inelastic) and various formulations (initial strains, initial stress). In this paper only elastic inclusions are considered, while the problem is formulated as initial stress. The following steps have to be performed in order to approximate final results [7–9]:

a) The elastic problem is solved assuming that there is no inclusions (using (9))
$$A\dot{x}_{i=0} = \dot{b}.$$
b) The strains $\dot{\varepsilon}$ are calculated within the inclusion (using (12)).
c) The total boundary and internal results are initilized
$$x_{total} = \dot{x}_{i=0},$$
$$\varepsilon_{total} = \dot{\varepsilon}_{i=0}.$$
d) Convert the internal strains into the initial stresses using
$$\dot{\sigma}_0 = (C_D - C_I)\dot{\varepsilon},$$
where C_D is the constitutive matrix of the domain used in a) as a homogeneous, while C_I is the constitutive matrix for the inclusion.
e) Compute the last integral over the domain from (1) using kernel (4) and obtained in d) values of $\dot{\sigma}_0$. The result of that operation is the residual vector $\dot{f}_i$.
f) Check if the residual vector is sufficiently small. If yes the iterative process ends, otherwise it continues.
g) The residual vector obtained in e) is applied as the right hand side for the system of equation (9)
$$A\dot{x}_i = \dot{f}_i.$$
h) The above system of equations is solved and once again the strains inside the inclusions are calculated.
i) The final boundary and internal results are updated
$$x_i = x_{i-1} + \dot{x}_i,$$
$$\varepsilon_i = \varepsilon_{i-1} + \dot{\varepsilon}_i.$$
j) Repeat the procedure from step e).

7   Numerical Verification and Discussion

The example concerns a square plate ($2\,\mathrm{m} \times 2\,\mathrm{m}$) with a square inclusion in the center ($1\,\mathrm{m} \times 1\,\mathrm{m}$). The considered body is fixed at the bottom and loaded on the top with a constant pressure $p = 1\,\mathrm{MN/m^2}$ (Fig. 5). The plate is in plane strain conditions and it is composed of two different materials. The material of the plate (D) and the material of the inclusion (I) are characterized by the following properties: $E_D = 5000\,\mathrm{MN/m^2}$, $\nu_D = 0.3$ and $E_I = 2500\,\mathrm{MN/m^2}$, $\nu_I = 0.3$.

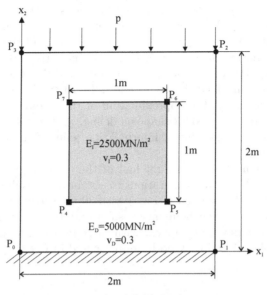

Fig. 5. Geometry and boundary conditions for the square plate with the inclusion

At first, the main feature of the proposed approach is analyzed - the way of modeling of inclusions. As can be seen in Fig. 5, the boundary of the plate in PIES is modeled by four linear Bézier segments, using two corner ● points for each of them. Such approach requires only four corner points (P_0, P_1, P_2, P_3). The inclusion is defined using single Bézier surface of the first degree, which is created also by only four corner points ■ (P_4, P_5, P_6, P_7).

For comparison, the considered body was modeled using other numerical methods. The model corresponding to FEM was designed with the help of 400 quadratic finite elements (100 of them concern the inclusion itself). It means that the proposed approach allows modeling with significantly fewer input data than classical FEM (even several hundred times). The described above FEM model was also used for numerical calculations presented later in this section.

The way of modeling of inclusions in PIES was also compared with BEM. For this purpose, models available in the literature were used. Two approaches were taken into account: first concerns classical discretization into cells [7], while second presented in [9] uses NURBS curves and a linear interpolation between them. Using classical BEM approach the plate was defined by 48 quadratic boundary elements, while the inclusion by 36 quadratic cells. In the isogeometric BEM two linear NURBS curves are used. As can be seen, the first way (classical BEM) requires defining several times more nodes than in PIES, while the second uses only 4 control points to define two curves, but approximated area should still be divided into subareas for integration.

It should be emphasized that in classical versions of the so-called element methods (FEM, BEM), the number of elements, their type and arrangement (shape approximation) are closely related to the accuracy of the obtained solutions (the approximation of solutions). In many cases, the more elements, the higher the accuracy. However, it should be remembered that the more elements, the greater the system of equations to solve. In PIES the approximation of solutions is independent of the approximation of the shape. Therefore, only a minimal amount of data is used for shape modeling, while the accuracy is steered by the number and arrangement of interpolation points (e.g. see formula (11) for initial stresses).

Vertical displacements on the right half of the upper segment of the body were obtained. It took only a few iterations to obtain such results in PIES. They are based on the model presented in Fig. 5 with 20 interpolation nodes assumed for the approximation of the boundary results and 16 for the initial stresses. Obtained displacements were compared with those returned by FEM (using the model described above) and they both are presented in Fig. 6.

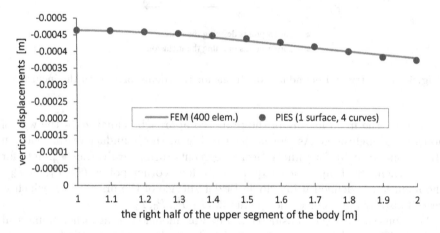

Fig. 6. Vertical displacements on the right half of the upper segment of the body obtained by FEM and PIES

In Fig. 6 very good agreement between FEM and PIES results can be noticed. It should be remembered that they were obtained with a completely different amount of data used for modeling and a completely different amount of solved equations (both numbers in favor of PIES).

The next stage of studies concerns comparison of the deformated shape of the boundary taking into account the body with and without the inclusion. Obtained by PIES boundaries are shown in Fig. 7 and Fig. 8. The way of deformation of the body with the inclusion simulated by PIES agrees with that presented in [18] obtained by BEM.

The presented in this section results confirm the high efficiency of the proposed method (efficient modeling of inclusions with the small number of data)

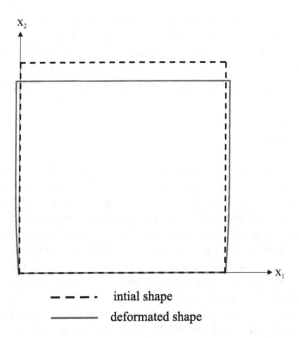

Fig. 7. The deformated boundary for the body without the inclusion

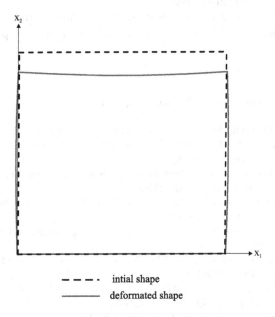

Fig. 8. The deformated boundary for the body with the inclusion

and also its accuracy (the results are consistent with another methods). The effectiveness of the method measured by the calculation time in comparison with other methods was not checked. It comes from the fact that it is not reasonable to compare the time of execution of the program created by the authors for the research being considered with the commercial product for finite element analysis.

8   Conclusions

The paper presents the PIES method for solving 2D elastic boundary value problems with inclusions. The geometry of the inclusion is defined using Bézier surface, without classical discretization. Such approach reduces the number of data required for modeling and gives possibility for easy modification of the shape. PIES separates the approximation of the shape from the approximation of the solutions, therefore for displacements, tractions and initial stresses approximation, Lagrange polynomials are used. The accuracy of solution in such case depends only on the number and arrangement of interpolation nodes.

The verification of the proposed approach was performed on the example with elastic inclusion in comparison to other numerical methods. Obtained results are in good agreement with FEM and BEM solutions. It should be mentioned that PIES is especially efficient taking into account the way of modeling of inclusions, but the accuracy is also satisfactory.

The proposed strategy requires tests on more complicated examples, especially when the inclusions with nonlinear material behavior are present.

References

1. Ameen, M.: Computational Elasticity. Alpha Science International Ltd., Harrow (2005)
2. Zienkiewicz, O.C.: The Finite Element Methods. McGraw-Hill, London (1977)
3. Liu, G.R., Quek, S.S.: The Finite Element Method: A Practical Course. Butterworth Heinemann, Oxford (2003)
4. Gao, X.W., Davies, T.G.: Boundary Element Programming in Mechanics. Cambridge University Press, Cambridge (2002)
5. Aliabadi, M.H.: The Boundary Element Method. Applications in Solids and Structures, vol. 2. Wiley, Chichester (2002)
6. Becker, A.A.: The Boundary Element Method in Engineering: A Complete Course. McGraw-Hill (1992)
7. Beer, G., Smith, I., Duenser, C.: The Boundary Element Method with Programming. Springer, Wien (2008). https://doi.org/10.1007/978-3-211-71576-5
8. Riederer, K., Duenser, C., Beer, G.: Simulation of linear inclusions with the BEM. Eng. Anal. Bound. Elem. **33**, 959–965 (2009)
9. Beer, G., Marussig, B., Zechner, J., Dünser, Ch., Friesa, T.P.: Isogeometric Boundary Element analysis with elasto-plastic inclusions. Part 1: Plane problems. Comput. Methods Appl. Mech. Eng. **308**, 552–570 (2016)
10. Zieniuk, E.: Potential problems with polygonal boundaries by a BEM with parametric linear functions. Eng. Anal. Boundary Elem. **25**(3), 185–190 (2001)

11. Zieniuk, E., Szerszeń, K.: Triangular Bézier surface patches in modeling shape of boundary geometry for potential problems in 3D. Eng. Comput. **29**(4), 517–527 (2012). https://doi.org/10.1007/s00366-012-0278-6

12. Zieniuk, E., Bołtuć, A.: Non-element method of solving 2D boundary problems defined on polygonal domains modeled by Navier equation. Int. J. Solids Struct. **43**(25–26), 7939–7958 (2006)

13. Zieniuk, E., Bołtuć, A.: Bézier curves in the modeling of boundary geometry for 2D boundary problems defined by Helmholtz equation. J. Comput. Acoust. **14**(3), 353–367 (2006)

14. Bołtuć, A.: Parametric integral equation system (PIES) for 2D elastoplastic analysis. Eng. Anal. Bound. Elem. **69**, 21–31 (2016)

15. Farin, G.: Curves and Surfaces for CAGD: A Practical Guide. Morgan Kaufmann Publishers, San Francisco (2002)

16. Salomon, D.: Curves and Surfaces for Computer Graphics. Springer, New York (2006). https://doi.org/10.1007/0-387-28452-4

17. Gao, X.W.: Evaluation of regular and singular domain integrals with boundary-only discretization - theory and Fortran code. J. Comput. Appl. Math. **175**, 265–290 (2005)

18. Riederer, K.: Modelling of ground support in tunnelling using the BEM, Ph.D. thesis. Graz University of Technology (2010)

Impact of Water on Methane Adsorption in Nanopores: A Hybrid GCMC-MD Simulation Study

Ji Zhou[1,2] ⓘ, Wenbin Jiang[1,2(✉)] ⓘ, Mian Lin[1,2(✉)] ⓘ,
Lili Ji[1,2(✉)] ⓘ, and Gaohui Cao[1] ⓘ

[1] Institute of Mechanics, Chinese Academy of Sciences, Beijing 100190,
People's Republic of China
{jiangwenbin, linmian}@imech.ac.cn
[2] School of Engineering Science, University of Chinese Academy of Sciences,
Beijing 100049, People's Republic of China

Abstract. Adsorbed methane is an important component of shale gas. Shale generally contains a certain amount of primary water, and isothermal adsorption experiments on wet samples show that water inhibits methane adsorption. Researches on methane adsorption mainly focus on the conditions of low pressure and water content. In this study, a hybrid GCMC-MD simulation method is proposed to study methane adsorption characteristics under high pressure and water content in pores of different sizes. This method can obtain the bulk pressure of the system while ensuring the simultaneous movement of methane and water molecules, and has high efficiency and reliability. It is found that the existence of water does not change the morphology of excess isotherm, and the relative decrease of adsorption capacity due to the existence of water is not sensitive to temperature. In ≤ 3 nm pores, water molecules form water clusters and partially occupy wall adsorption sites, and the adsorption amount decreases linearly with increasing water saturation. In the 5 nm wide pore with 40% water saturation, water films formed and methane adsorption is strongly suppressed. It is expected these findings could provide guidance for the evaluation of the amount of adsorbed methane with primary water.

Keywords: Shale gas · Adsorption · GCMC · MD · Water saturation

1 Introduction

Methane is the main component of natural gas in low-porosity and low-permeability shale gas reservoir. Isothermal adsorption experiments reveal that the capacity of methane adsorption in shale is considerable, and the amount of adsorbed gas could be equivalent to that of free gas, accounts for 20%–85% of the total gas content in the USA and China [1–3]. In addition, adsorbed gas will influence the behavior of gas flow and diffusion [3]. Thus, it is very important for the evaluation of shale gas content and recovery to obtain a deeper understanding of methane adsorption characteristics.

Currently, methane isothermal adsorption experiments are mainly conducted on dry shale by many departments of industry and researchers. It is found that: 1) the excess

© Springer Nature Switzerland AG 2020
V. V. Krzhizhanovskaya et al. (Eds.): ICCS 2020, LNCS 12138, pp. 184–196, 2020.
https://doi.org/10.1007/978-3-030-50417-5_14

adsorption amount firstly increases and then decreases with the increase of pressure, and the isotherm could not be fitted by the standard Langmuir model; 2) the maximum adsorption amount decreases with the increase of temperature; 3) the maximum adsorption amount is proportional to the specific surface. Through high-resolution scanning electron microscopy (SEM) imaging and low-temperature gas adsorption analysis, abundant of nanopores with high specific surface area are found in shale [4]. Based on this finding, researchers simulated the adsorption of methane molecules in nanometer channels with side walls of different compositions, such as graphite (substitute of organic matter), different types of kerogen, quartz, illite and montmorillonite [5, 6]. A high-methane-density layer was observed near the wall and the excess loading can be determined, and the excess isotherms show a similar trend as that obtained from experiments. Such consistency delineates that the van der Waals force and physical adsorption play the leading role in methane adsorption in shale. Molecular simulation is proved to be able to reveal the microscopic mechanism of adsorption.

Gas-bearing shale always contains a certain amount of primary water. In recent years, some researchers conducted isothermal adsorption experiments on shales with moisture, and found that the moisture could reduce the adsorption capacity of methane by more than 50% [7–9]. In recent years, some molecular simulations of competitive adsorption of methane and water have also been carried out. Billemont et al. [10] used grand canonical Monte Carlo (GCMC) to simulate the methane sorption with three different water contents in the graphite pore at 300 K, and observed that the preloaded water notably decreased the adsorption capacity of methane. Sui et al. [11] conducted GCMC simulations on methane adsorption in montmorillonite plates with and without Na^+ cations containing 0 mmol/cm^3, 10 mmol/cm^3 and 20 mmol/cm^3 water at 298 K and pressures of 0–20 MPa, and the adsorbed amount of methane in montmorillonite pores with or without Na^+ both significantly decrease with the increase of water molecules. A similar result is also found by Zhao et al. [12], who used GCMC to simulate the methane isothermal adsorption with 0–3% water content at temperatures of 298 K, 323 K, 348 K and pressure ranges from 0–20 MPa. In summary, molecular simulations of competitive adsorption of methane and water were primarily limited to relatively low pressure and water content, and did not cover the range of pressure and water saturation of shale reservoir deeper than 2000 m.

GCMC is widely adopted in the simulation of adsorption isotherm as the pressure is given as an input parameter. However, in high-density systems, the probability of acceptance of random molecules operations is relatively low. When considering mixed sorbates, different molecules need to be operated separately to make them move, and the computational consumption will double. MD is capable to simulate the movement of different molecules simultaneously via solving the Newtonian equations directly with high parallelism and efficiency. The limited of MD lie in that the pressure near the wall is hard to calculate, and the pressure and density of bulk phase (external system connected with the simulation cell with the same chemical potential) are difficult to determine accurately.

In this work, we proposed a hybrid GCMC-MD method to simulate competitive adsorption of methane and water in nanopores with water saturation up to 40% under pressure up to 50 MP. The rests of the paper are arranged as follows: the detail of molecular simulation method is introduced in the next section. Section 3 shows the validation of simulation method and models. The results and some discussions are addressed in Sect. 4. Finally, the conclusions of this study are drawn.

2 Simulation Models and Methods

2.1 Models

K. Lin et al. [13] found that the adsorption characteristics of methane in the pores with parallel graphite walls were similar to those in shale. Considering the focus of this work is the effect of different pore sizes and water contents on the methane adsorption, it is more convenient to perform a single factor analysis in pores with parallel graphite walls to disclose internal mechanism. A rectangular simulation cell is built with periodicities in the x, y and z directions. Four graphene sheets separated by a distance of 0.335 nm form the pore wall to meet the minimum mirror criterion. The bond length of C-C is 0.142 nm and the angle is 120°. The dimension along the x-y surface is 9.116 nm × 7.681 nm. The pore size H is determined by the length of vacuum in z direction. In this study, a united atom model is used to represent the methane molecule [5]. The water molecule is described as SPC/E model [14]. The bond length between the hydrogen atom and the oxygen atom is 0.1 nm, and the bond angle is 109.47°. The schematic representation of the slit-like graphite pore is presented in Fig. 1.

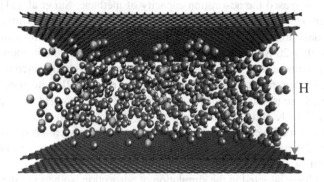

Fig. 1. Simulation models. Gray, green, red and white color represent the carbon atoms, methane molecules, oxygen atoms and hydrogen atoms, respectively. (Color figure online)

Two interactions included van der Waals force and Coulomb force are taken into account. The Lennard-Jones (L-J) potential model is used to describe van der Waals force:

$$E_{vdw} = \begin{cases} 4\varepsilon_{ij}\left[\left(\dfrac{\sigma_{ij}}{r_{ij}}\right)^{12} - \left(\dfrac{\sigma_{ij}}{r_{ij}}\right)^{6}\right], & r_{ij} \leq r_{cut} \\ 0 & r_{ij} > r_{cut} \end{cases} \tag{1}$$

while the Coulomb force is represented by the following equation

$$E_{col} = \begin{cases} \frac{q_i q_j}{4\pi\varepsilon_0 r_{ij}}, & r_{ij} \leq r_{cut} \\ \text{pppm solver}, & r_{ij} > r_{cut} \end{cases} \quad (2)$$

where r_{ij} is the distance between atom i and j; q_i and q_j are the charges of atom i and j; ε_{ij} and σ_{ij} are the L-J well depth and the L-J zero-energy separation distance, respectively; r_{cut} is the cutoff distance, in this study it is set to be $4\sigma_{max}$, and σ_{max} is the maximum value of L-J zero-energy separation distances of all types of atoms. The long-range effectively-infinite Coulombic pairwise interactions within the cutoff distance are computed directly; the interactions outside this distance are computed in reciprocal space using the particle-particle particle-mesh solver (pppm) solver with an accuracy of 10^{-4}. Table 1 lists the value of these parameters of the particles involved in the simulation. Lorentz-Berthelot mixing rule is utilized to compute the parameters of unlike particle's interaction.

Table 1. Potential parameters.

Interaction	ε(Kcal/mol)	σ(nm)	$q(e)$
C-C	0.0556	0.34	
CH_4-CH_4	0.294	0.373	
He-He	0.02	0.2556	
O-O	0.15535	0.3166	−0.8472
H-H	0	0	0.4236

2.2 Simulation and Analysis Method

In this work, we propose a simulation method (GCMC-MD) in which GCMC and MD are performed alternately to simulate the adsorption of methane with a certain water saturation. GCMC operation is applied only on methane molecules to ensure the chemical potential of the system is consistent with the target value, and the bulk phase density is known. MD computation is applied on both methane and water molecules to simulate their movements based on potential parameters and Newtonian equations. This method combines the advantages of both GCMC and MD. Comparing with pure GCMC simulation, this model reduces GCMC operations by halves as the number of water molecules is predetermined and unchanged, therefore, it is more efficient. On the other side, this method overcomes the difficulty of determining the bulk density in MD simulation through GCMC operations on methane molecules. It is implemented using LAMMPS [15]. The procedure of GCMC-MD is described as follow:

1) Performing one step of GCMC operation after every 50 fs, 100 random exchanges (insertions or deletions) of methane molecules are attempted;

2) MD simulation is performed in every time step. the *NVT* ensemble (atom number N, volume V and temperature T are constant) is applied with a time step of 2 fs. The Nose/Hoover thermostat with a relaxation time of 200 fs is adopted to maintain the system temperature;

3) Repeating 1) and 2) until the total simulation time step reaches the preset value.

The total computation time is set to be 1.0 ns in this work. For a given simulation condition with a prefix pressure, temperature, water saturation and pore width, the first 2/3 time steps are for equilibration and the data in the last 1/3 time steps are recorded for analysis. The GCMC-MD simulation flow chart is shown in Fig. 2.

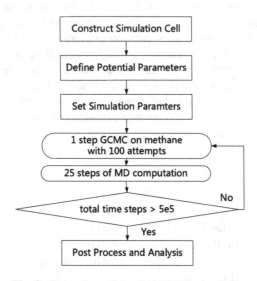

Fig. 2. Flow chart of the GCMC-MD simulation.

In general, GCMC operations include exchanges (insertions or deletions) of atoms or molecules with an imaginary reservoir and Monte Carlo (MC) moves within the simulation cell. In this method, only exchanges of atoms or molecules are performed in the GCMC step, and the following MD steps start from a sample that is generated by a successful attempt. Simulations performed with MD algorithm in the NVT ensemble are equivalent to successful MC operations. Overall, the simulated samples belong to the grand canonical ensemble.

The total number of methane molecules N_{tol} at equilibrium state can be obtained by averaging the number of methane molecules in the time steps of data recording. The mass of methane molecules in bulk phase is computed by $\rho_{free}V_{free}$ where $\rho_{free}(g/ml)$ is the bulk phase density of methane and $V_{free}(ml)$ is the volume of space that methane can enter. Then the excess adsorption amount of methane per unit surface area q_{ex} (ml/m^2) is computed following the equation.

$$q_{ex} = V_{std}\left(\frac{N_{tol}}{N_A} - \frac{\rho_{free}V_{free}}{M_{CH_4}}\right)/S \tag{3}$$

where V_{std} is the volume of 1 mol methane under the standard conditions, ml; $N_A = 6.022 \times 10^{23}\text{mol}^{-1}$ is the Avogadro constant; M_{CH_4} is the molar mass of methane, g/mol; S is the inner surface area of the simulation cell, m^2. Considering the extremely weak adsorption capacity of helium gas, it is used to measure the dead volume in the adsorption experiment. Similarly, helium is used to replace methane during the simulation in order to obtain the number of helium molecules in the system. And then the V_{free} can be calculated based on the following equation.

$$V_{free} = \frac{N_{He}}{N_A} \cdot \frac{M_{He}}{\rho_{He}} \tag{4}$$

2.3   Validation of Method and Model

We performed the bulk phase density simulations on methane, water and helium at temperature of 313 K with GCMC-MD method, respectively. The simulated densities are close to the values from the database of NIST (National Institute of Standards and Technology) (Fig. 3), indicating that the force field parameters of methane, water and helium used in the model are reasonable.

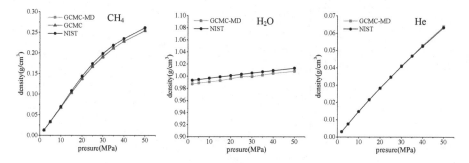

Fig. 3. Validation of bulk phase density simulation

The efficiency of GCMC-MD is also investigated. A comparison of methane bulk phase density attained from GCMC simulation and GCMC-MD simulation to investigate the difference of efficiency. The density obtained from the GCMC-MD simulation is nearly identical to that of GCMC simulation under the same simulation conditions (the interaction parameters, the time step and the size of simulation system, etc.). While the running time of GCMC is about 30 times as large as that of GCMC-MD (Table 2) on the same computer with 4 core 4 thread (i5-4690k) CPU.

Table 2. Running time of different methods

Simulation method	Running time
GCMC	12 h
GCMC-MD	0.4 h

Billemont et al. [10] conducted methane isothermal adsorption experiments on the dry and wet F400 activated carbon with 90% carbon content whose dominant pore size is 0.8 nm. We perform GCMC-MD simulation of methane adsorption in the graphite nanopore ($H = 0.8$ nm) with the same water content. In order to make the simulation procedure consistent with the experimental steps, water molecules are injected into the pore at first, and MD simulation are conducted for 0.1 ns to make water molecules reach equilibrium state. Then the methane molecules are injected into the pore to participate competitive adsorption through GCMC-MD simulation. The experimental and simulated adsorption isotherms under dry and wet condition are in good agreement with each other as depicted in Fig. 4, which indicates that the parameter setting of force field between different molecules is reasonable and the simulation process is correct and reliable.

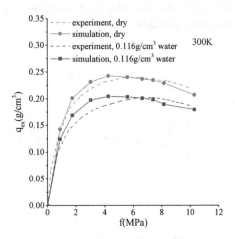

Fig. 4. Experimental and GCMC-MD simulated adsorption isotherms

3   Results and Discussion

In a pore with a size of H under given temperature (T), pressure (P) and water saturation (S_w) conditions, the excess adsorption amount $q_{ex}(S_w, P, T, H)$ can be obtained through GCMC-MD simulation. Single factor analysis is performed to investigate the influence of each factor.

3.1   Water Saturation

We simulate the methane adsorption in a 3 nm pore under 313 K, 25 MPa with $S_w = 0$–40%. The results exhibited in Fig. 5 include snapshots of the equilibrium distribution of methane (in green) and water (in red) molecules in the slit-like graphite pore, and the number density profile of methane along the vertical direction. The number density of methane molecules has a peak near the wall surface, and peaks decrease with increasing water saturation. The excess adsorption amount decline

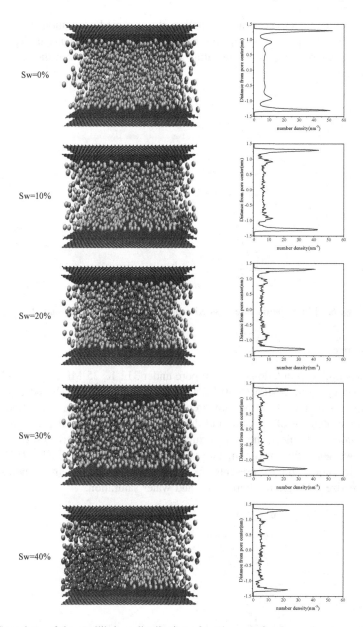

Fig. 5. Snapshots of the equilibrium distribution of methane molecules and water molecules in the slit-like graphite pore, and the number density profile of methane. (Color figure online)

linearly with the increase of water saturation (see Fig. 6). This trend is similar to the experimental result obtained by Hu et al. [8]. It can be seen from the molecular distribution in the system that the molecules occupy the wall surface in the form of water clusters, occupying the adsorption sites and causing the adsorption amount to decrease. With the increase of water saturation, the occupied adsorption sites gradually increase, so the adsorption capacity continues to decrease. Some water molecules exist around the center of the channel, resulting in zigzag fluctuations in the center of the density profile.

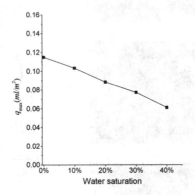

Fig. 6. The variation of excess adsorption amount with water saturation

3.2   Pressure

Methane adsorption simulations in 3 nm pore under 313 K, 25 MPa, $S_w = 0\text{--}40\%$ with pressure ranging from 2 MPa to 50 MPa are further conducted. The adsorption isotherms with different S_w show the same trend (Fig. 7) i.e. the excess adsorption amount grow with the increase of pressure and then abruptly decreasing. It means that water saturation does not affect the supercritical adsorption characteristics of methane. Therefore, we select the maximum excess adsorption amount $q_{ex,max}(S_w, T, H)$ as the characteristic value which represents the adsorption capacity of the nanopore of a certain pore size at a certain temperature and water saturation.

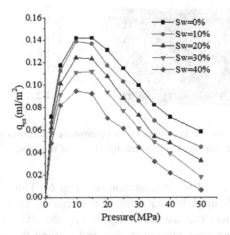

Fig. 7. The excess adsorption isotherms with different water saturation

3.3   Temperature

In the 3 nm pore, we further simulate the adsorption of methane with $S_w = 0\text{--}40\%$, $P = 2\text{--}50$ MPa at 313 K, 353 K, 393 K, and a series of excess adsorption isotherms are obtained and the maximum adsorption capacity $q_{ex,max}$ (0–40%, 313–393 K, 3 nm) can be determined. Let $\delta(S_w, T, H) = q_{ex,max}(S_w, T, H)/q_{ex,max}(0\%, T, H)$ be the relative decrease in methane adsorption capacity with different water content. The variations of $\delta(S_w, T, 3\,\text{nm})$ with S_w and T are shown in Fig. 8. At different temperatures, the relative decrease δ of adsorption capacity has the same trend with increasing water content. Under the same S_w, the relative decrease δ at different temperatures are relatively close. Because the water molecules are always liquid within the studied temperature and pressure ranges, the state of occupying the adsorption site is less affected by temperature, so the relative decrease in different temperatures is close.

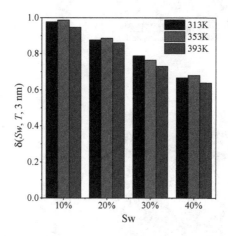

Fig. 8.   Barplot of $\delta(Sw, T, 3\,\text{nm})$

3.4   Pore Size

Methane adsorption at 25 MPa, 313 K, with 40% water saturation in pores of 1 nm, 3 nm, and 5 nm are simulated, respectively, to investigate the difference in pores with different size. The distribution of methane and water molecules in nanopores with different pore sizes and the density profile of methane along the vertical direction are shown in Fig. 9. It can be seen that in the 1 nm and 3 nm pores, water partially occupies the wall surface in the form of water clusters, and methane still has a higher density area at the wall surface; in the 5 nm pore, water molecules are completely spread out on the wall, forming a water film completely covering the wall surface, blocking the adsorption of methane, the methane density near the wall surface is lower than the center.

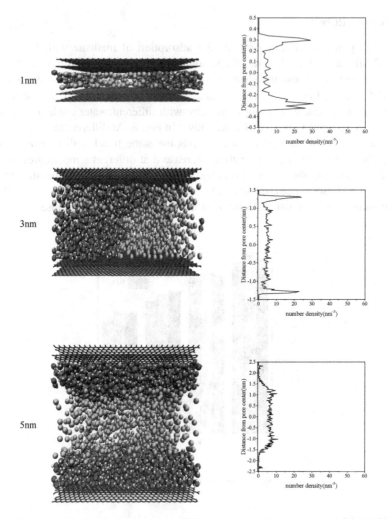

Fig. 9 Water and methane distribution at 40% water saturation in pores of different sizes

The reasons for these differences are analyzed from the perspective of inter-molecular interaction forces. In the 1 nm and 3 nm pores, the water clusters are in contact with the upper and lower wall at the same time, which results in finite wall area being occupied. In the 5 nm pores, the upper and lower walls are far away, which makes the interactions between water molecules adsorbed on the upper and lower walls are too weak to form clusters.

4 Conclusion

In order to study the characteristics of methane adsorption under the conditions of high pressure, high water content and a certain water saturation in pores with different sizes, we propose a hybrid GCMC-MD simulation method, and perform molecular simulations of methane adsorption under different temperature, pressure and water saturation conditions in pores with different pore sizes. The conclusions reached are as follows:

(1) The hybrid simulation method can obtain the bulk pressure of the system while ensuring that the methane and water molecules move under the interaction force, which has high efficiency and can generate reliable results;
(2) In ≤ 3 nm pores, water molecules form water clusters and partially occupy wall adsorption sites. As the water content increases, the adsorption amount decreases linearly. In the graphite pore with a width of 5 nm under 40% water saturation, water molecules completely cover the surface and form a water film, and the amount of methane adsorption is greatly reduced.
(3) The water-containing conditions do not change the tendency of the adsorption isotherm to rise first and then decrease with increasing pressure. The relative decrease of methane adsorption capacity due to the existence of water is not sensitive to temperature.

It is expected these findings could provide guidance for the evaluation of the amount of adsorbed methane with primary water.

Acknowledgements. This study was supported by the National Natural Science Foundation of China (No. 41690132 and 41872163), the Strategic Priority Research Program of the Chinese Academy of Sciences (XDA14010304) and the Major National Science and Technology Special Program of China (No. 2017ZX05037-001).

References

1. Curtis, J.B.: Fractured shale-gas systems. AAPG Bull. **86**(11), 1921–1938 (2002)
2. Mavor, M.: Barnett shale gas-in-place volume including sorbed and free gas volume. In: Southwest Section AAPG Convention. Fort Worth Geological Society (2003)
3. Wang, X., Gao, S., Gao, C.: Geological features of Mesozoic lacustrine shale gas in south of Ordos Basin. NW China. Petrol Explor Dev. **41**(3), 326–337 (2014)
4. Loucks, R.G., Ruppel, S.C.: Mississippian Barnett Shale: Lithofacies and depositional setting of a deep-water shale-gas succession in the Fort Worth Basin. Texas. AAPG Bull. **91**(4), 579–601 (2007)
5. Ambrose, R.J., Hartman, R.C., Diaz-Campos, M., Akkutlu, I.Y., Sondergeld, C.H.: Shale gas-in-place calculations part I: new pore-scale considerations. SPE J. **17**(01), 219–229 (2012)
6. Jiang, W., Lin, M.: Molecular dynamics investigation of conversion methods for excess adsorption amount of shale gas. J. Nat. Gas Sci. Eng. **49**, 241–249 (2018)
7. Ross, D.J., Bustin, R.M.: Shale gas potential of the lower Jurassic Gordondale member, northeastern British Columbia. Can. Bull. Can Petrol. Geol. **55**(1), 51–75 (2007)

8. Hu, Z., et al.: Influence of reservoir primary water on shale gas occurrence and flow capacity. Nat. Gas. Ind. **38**(7), 44–51 (2018). (in Chinese)
9. Chen, Z., Ning, Z., Wang, Q., Huang, L., Qi, R., Wang, J.: Experimental study on methane adsorption characteristics of hydrous shale. Fault-Block Oil Gas Field **25**(4), 510–514 (2018). (in Chinese)
10. Billemont, P., Coasne, B., De Weireld, G.: An experimental and molecular simulation study of the adsorption of carbon dioxide and methane in nanoporous carbons in the presence of water. Langmuir **27**(3), 1015–1024 (2010)
11. Sui, H., Yao, J., Zhang, L.: Molecular simulation of shale gas adsorption and diffusion in clay nanopores. Computation **3**(4), 687–700 (2015)
12. Zhao, T., et al.: Molecular simulation of methane adsorption on type II kerogen with the impact of water content. J. Petrol. Sci. Eng. **161**, 302–310 (2018)
13. Lin, K., Yuan, Q., Zhao, Y.: Using graphene to simplify the adsorption of methane on shale in MD simulations. Comput. Mater. Sci. **133**, 99–107 (2017)
14. Hermans, J., et al.: A consistent empirical potential for water–protein interactions. Biopolym. Orig. Res. Biomol. **23**(8), 1513–1518 (1984)
15. Plimpton, S.: Fast parallel algorithms for short-range molecular dynamics. J. Comput. Phys. **117**(1), 1–19 (1995)

A Stable Discontinuous Galerkin Based Isogeometric Residual Minimization for the Stokes Problem

Marcin Łoś[1], Sergio Rojas[2], Maciej Paszyński[1($\boxtimes$)], Ignacio Muga[3], and Victor M. Calo[2,4]

[1] Department of Computer Science, AGH University of Science and Technology, Krakow, Poland
paszynsk@agh.edu.pl
[2] School of Earth and Planetary Sciences, Curtin University, Bentley, Perth, WA 6102, Australia
[3] Pontificia Universidad Católica de Valparaíso, Casilla 4059, Valparaíso, Chile
[4] Mineral Resources, Commonwealth Scientific and Industrial Research Organisation (CSIRO), Kensington, Perth, WA 6152, Australia

Abstract. We investigate a residual minimization (RM) based stabilized isogeometric finite element method (IGA) for the Stokes problem. Starting from an inf-sup stable discontinuous Galerkin (DG) formulation, the method seeks for an approximation in a highly continuous trial space that minimizes the residual measured in a dual norm of the discontinuous test space. We consider two-dimensional Stokes problems with manufactured solutions and the cavity flow problem. We explore the results obtained by considering highly continuous isogeometric trial spaces, and discontinuous test spaces. We compare by the Pareto front the resulting numerical accuracy and the computational cost, expressed by the number of floating-point operations performed by the direct solver algorithm.

Keywords: Isogeometric analysis · Residual minimization · Discontinuous Galerkin · Stokes problem · Direct solvers · Computational cost

1 Introduction

The Isogeometric Analysis (IGA) [1] bridges the gap between the Computer Aided Design (CAD) and Computer Aided Engineering (CAE) communities. The idea of IGA is to apply B-spline basis functions [2] for finite element method (FEM) simulations. The ultimate goal is to perform engineering analysis directly to CAD models without expensive remeshing and recomputations. IGA has multiple applications in time-dependent simulations, including phase-field models [3,4], phase-separation simulations with application to cancer growth simulations [5,6], wind turbine aerodynamics [7], incompressible hyper-elasticity [8], turbulent flow simulations [9], transport of drugs in cardiovascular applications

© Springer Nature Switzerland AG 2020
V. V. Krzhizhanovskaya et al. (Eds.): ICCS 2020, LNCS 12138, pp. 197–211, 2020.
https://doi.org/10.1007/978-3-030-50417-5_15

[10], or the blood flow simulations and drug transport in arteries simulations [11,12].

The stability of a numerical method based on Petrov-Galerkin discretizations of a general weak form relies on the famous discrete inf-sup condition (see, e.g., [13]): "Babuška-Brezzi condition" (BBC) developed in years 1971–1974 at the same time by Ivo Babuška, and Franco Brezzi [14–16].

Let $\mathfrak{U}, \mathfrak{V}$ denote two Hilbert spaces. For a given variational formulation of the form:

$$\text{Find } \mathfrak{u} \in \mathfrak{U}, \text{ such that } \mathfrak{b}(\mathfrak{u}, \mathfrak{v}) = \mathfrak{l}(\mathfrak{v}), \quad \forall \mathfrak{v} \in \mathfrak{V}, \tag{1}$$

with $\mathfrak{b} : \mathfrak{U} \times \mathfrak{V} \to \mathbb{R}$ being a bilinear form, and $\mathfrak{l} : \mathfrak{V} \to \mathbb{R}$ being a linear form, the BBC condition states that the problem is stable if there exists a positive constant $\gamma > 0$, such that:

$$\sup_{\mathfrak{v} \in \mathfrak{V}, \mathfrak{v} \neq 0} \frac{|\mathfrak{b}(\mathfrak{u}, \mathfrak{v})|}{\|\mathfrak{v}\|_{\mathfrak{V}}} \geq \gamma \|\mathfrak{u}\|_{\mathfrak{U}}, \quad \forall \mathfrak{u} \in \mathfrak{U}. \tag{2}$$

The inf-sup condition in the above form concerns the abstract formulation where we consider all the test functions from $\mathfrak{v} \in \mathfrak{V}$ and look for solution at $\mathfrak{u} \in \mathfrak{U}$. The above condition is satisfied also if we restrict to a conforming space of trial functions $\mathfrak{U}_h \subset \mathfrak{U}$. This is,

$$\sup_{\mathfrak{v} \in \mathfrak{V}, \mathfrak{v} \neq 0} \frac{|\mathfrak{b}(\mathfrak{w}_h, \mathfrak{v})|}{\|\mathfrak{v}\|_{\mathfrak{V}}} \geq \gamma \|\mathfrak{w}_h\|_{\mathfrak{U}_h}, \quad \forall \mathfrak{w}_h \in \mathfrak{U}_h. \tag{3}$$

However, if we consider test functions from a finite dimensional test space $\mathfrak{V}_h$ (not necessarily conforming), there is not guarantee that the inf-sup condition is realized on the discrete level.

There are many methods constructing test functions providing better stability of the method for a given class of problems [17–20]. In 2010 the Discontinuous Petrov-Galerkin (DPG) method was proposed, with the modern summary of the method described in [21,22]. The key idea of the DPG method is to construct the optimal test functions "on the fly", element by element. The DPG automatically guarantee the numerical stability of difficult computational problems, thanks to the automatic selection of the optimal basis functions. The DPG method is equivalent to the residual minimization method [21]. The DPG is a practical way to implement the residual minimization method when the computational cost of the global solution is expensive (non-linear).

There is consistent literature on residual minimization methods, especially for convection-diffusion problem [23–25], where it is well known that the lack of stability is the main issue to overcome. In particular, the class of DPG methods [26,27] aim to obtain a practical approach to solve the mixed system by breaking the test spaces (at the expense of introducing a hybrid formulation).

Recently, in [28] a new stabilized finite element method based on residual minimization was introduced. The method consider first an adequate discontinuous Galerkin formulation. Then, the wanted solution is obtained by solving a residual minimization problem in terms of a dual discontinuous Galerkin norm.

As in DPG methods, the method delivers a stable approximation and an error estimator to guide the adaptivity. However, its main attractive relies in that it allows to obtain a solution in a conforming sub space with the same quality of those ones obtained with the discontinuous Galerkin formulations. Last is evidenced by the authors considering standard Lagrange FEM polynomials.

In this paper, we explore the extension of [28] to IGA. We investigate the possibility of considering highly-continuous B-splines spaces as trial and broken B-spline spaces as test. We focus on the stationary Stokes problem, that requires special stabilization effort (see [29–33]). Due to the large range of subspaces that can be considered as trial spaces, we perform experimentations considering different setups of conforming trial spaces contained in a given broken B-spline space of degree 4. We solve the global system calling the MUMPS solver [34–36], and we compare the obtained results in terms of computational cost and accuracy of the obtained solution.

2   Discontinuous Galerkin Based Isogeometric Residual Minimization (DGIRM)

In this section we briefly discuss, in an abstract setting, the main idea behind the discontinuous Galerking based residual minimization method introduced in [28] in the isogeometric context.

Assume that we want to obtain an approximation u_h, of the continuous problem (1), in a given discrete space $\mathfrak{U}_h \subset \mathfrak{U}$ (eg., a highly continuous B-spline space). The residual minimization method is constructed as follows: First, a broken B-spline polynomial space $\mathfrak{V}_h$, containing $\mathfrak{U}_h$, is considered. Next, as starting point, is considered a discontinuous Galerkin variational formulation for problem (1) of the form:

$$\text{Find } u_h^{DG} \in \mathfrak{V}_h, \text{ such that } \mathfrak{b}_h(u_h^{DG}, \mathfrak{v}_h) = \mathfrak{l}_h(\mathfrak{v}_h), \quad \forall \mathfrak{v}_h \in \mathfrak{V}_h, \qquad (4)$$

the bilinear form $\mathfrak{b}_h$ is inf-sup stable with respect to a given discrete norm $\|\cdot\|_{\mathfrak{V}_h}$ of $\mathfrak{V}_h$. This is, there exists a positive constant C_{sta}, independent of the mesh size, such that:

$$\sup_{0 \neq \mathfrak{v}_h \in \mathfrak{V}_h} \frac{\mathfrak{b}_h(\mathfrak{w}_h, \mathfrak{v}_h)}{\|\mathfrak{v}_h\|_{\mathfrak{V}_h}} \geq C_{sta} \|\mathfrak{w}_h\|_{\mathfrak{V}_h}, \quad \forall \mathfrak{v}_h \in \mathfrak{V}_h. \qquad (5)$$

Finally, instead of solving the square problem (4), the wanted solution is obtained by solving the following residual minimization problem:

$$\text{Find } u_h \in \mathfrak{U}_h, \text{ such that } u_h = \arg\min_{\mathfrak{w}_h \in \mathfrak{U}_h} \frac{1}{2} \|\mathfrak{l}_h - \mathfrak{B}_h \mathfrak{w}_h\|_{\mathfrak{V}_h'}^2, \qquad (6)$$

where $\mathfrak{V}_h'$ denotes the dual space of $\mathfrak{V}_h$, the operator $\mathfrak{B}_h : \mathfrak{V}_h \to \mathfrak{V}_h'$ is defined as:

$$< \mathfrak{B}_h \mathfrak{w}_h, \mathfrak{v}_h >_{\mathfrak{V}_h' \times \mathfrak{V}_h} := \mathfrak{b}_h(\mathfrak{w}_h, \mathfrak{v}_h), \qquad (7)$$

and, for $\phi \in \mathfrak{V}_h'$, the dual norm $\| \cdot \|_{\mathfrak{V}_h'}$ is defined as:

$$\|\phi\|_{\mathfrak{V}_h'} := \sup_{0 \neq \mathfrak{v}_h \in \mathfrak{V}_h} \frac{<\phi, \mathfrak{v}_h >_{\mathfrak{V}_h' \times \mathfrak{V}_h}}{\|\mathfrak{v}_h\|_{\mathfrak{V}_h}}. \tag{8}$$

Considering the Riesz operator:

$$R_{\mathfrak{V}_h} : \mathfrak{V}_h \ni v_h \to (\mathfrak{v}_h, .)_{\mathfrak{V}_h} \in \mathfrak{V}_h', \tag{9}$$

where $(\cdot, \cdot)_{\mathfrak{V}_h}$ denotes the inner product inducing the discrete norm $\| \cdot \|_{\mathfrak{V}_h} = (\cdot, \cdot)_{\mathfrak{V}_h}^{1/2}$, and defining the residual representative:

$$\mathfrak{r}_h := R_{\mathfrak{V}_h}^{-1} (\mathfrak{l}_h(\mathfrak{v}_h) - \mathfrak{b}_h(\mathfrak{u}_h, \mathfrak{v}_h)) = R_{\mathfrak{V}_h}^{-1} \mathfrak{b}_h(\mathfrak{u}_h^{\mathrm{DG}} - \mathfrak{u}_h, \mathfrak{v}_h), \tag{10}$$

with $R_{\mathfrak{V}_h}^{-1}$ being the inverse of the Riesz operator $R_{\mathfrak{V}_h}$, and $\mathfrak{u}_h^{\mathrm{DG}}$ being the solution of the DG problem (4), problem (6) can be equivalently written as the following saddle-point problem: Find $(\mathfrak{r}_h, \mathfrak{u}_h) \in \mathfrak{V}_h \times \mathfrak{U}_h$, such that:

$$\begin{aligned} (\mathfrak{r}_h, \mathfrak{v}_h)_{\mathfrak{V}_h} + \mathfrak{b}_h(\mathfrak{u}_h, \mathfrak{v}_h) &= \mathfrak{l}_h(\mathfrak{v}_h), \quad \forall \mathfrak{v}_h \in \mathfrak{V}_h, \\ \mathfrak{b}_h(\mathfrak{w}_h, \mathfrak{r}_h) &= 0, \qquad \forall \mathfrak{w}_h \in \mathfrak{U}_h. \end{aligned} \tag{11}$$

The main attractive of the discrete saddle-point problem (11) is that it delivers automatically a stable approximation $\mathfrak{u}_h \in \mathfrak{U}_h$ enjoying of desired properties for the solution, such as high-continuity, and a residual representation $\mathfrak{r}_h \in \mathfrak{V}_h$ that can be used as error indicator to guide an adaptive mesh refinement. Indeed, in [28] the authors proved that, under the standard assumptions for the Discontinuous Galerkin problem (4): a) inf-sup stability (see Eq. (5)), b) boundedness and c) consistency (see [37] or [28] for definitions), problem (11) is well-posed. Additionally, it delivers an approximation $\mathfrak{u}_h$ with the same quality, in terms of the norm $\mathfrak{V}_h$, of the one obtained by solving problem (4). Moreover, the residual representative $\mathfrak{r}$ is an efficient error estimator that, under an adequate saturation assumption is satisfied (see Assumptions 4 and 5 in [28]) the residual representative is also reliable.

Therefore, roughly speaking, the following two ingredients are required to perform the discontinuous Galerkin based isogeometric residual minimization:

a) A well-posed discontinuous Galerkin formulation of the form (4), satisfying the inf-sup property (5).
b) A conforming, in $\mathfrak{U}$, subspace $\mathfrak{U}_h \subset \mathfrak{V}_h$ as trial space.

3   The Stokes Problem

Let $\Omega \subset \mathbb{R}^2$ be a open bounded polygon with outer normal $\mathbf{n}$, and denote by $\partial\Omega$ its boundary. Without loss of generality, we consider $\Omega = (0, 1)^2$. The Stokes problem with no-slip boundary condition reads:

Find $\mathbf{u}, p$ such that:

$$-\Delta\mathbf{u} + \nabla p = f, \text{ in } \Omega,$$
$$\nabla \cdot \mathbf{u} = 0, \text{ in } \Omega, \tag{12}$$
$$\mathbf{u} = 0, \text{ on } \partial\Omega,$$

where $\mathbf{u} := (u_1, \ldots, u_d) : \Omega \to \mathbb{R}^2$ denotes the velocity field, $p : \Omega \to \mathbb{R}$ the pressure and $f := (f_1, \ldots, f_d) \in [L^2(\Omega)]^2$ a given forcing term. The solution of (12) is unique for the pressure p up to a constant, therefore, problem (12) is complemented with the following extra condition for p:

$$\int_\Omega p = 0. \tag{13}$$

3.1 Weak Variational Formulation

We consider the following Hilbert spaces: $L^2(\Omega) = \{v : \Omega \to \mathbb{R} : \int_\Omega v^2 < +\infty\}$, $H^1(\Omega) = \{v \in L^2(\Omega) : \nabla v \in [L^2(\Omega)]^2\}$ and $H_0^1(\Omega) = \{v \in H(\Omega) : v = 0 \text{ on } \partial\Omega\}$. Defining $U := (H_0^1(\Omega))^2$ as the space for the velocity field and, as consequence of condition (13), the space $P := L_0^2(\Omega)$, with $L_0^2(\Omega) = \{p \in L^2(\Omega) : \int_\Omega p = 0\}$ for the pressure, the weak variational formulation of the strong problem (12)–(13) reads:
Find $(\mathbf{u}, p) \in U \times P$, such that:

$$a(\mathbf{u}, \mathbf{v}) + b(\mathbf{v}, p) = (f, v)_\Omega, \forall \mathbf{v} \in U,$$
$$-b(\mathbf{u}, q) = 0, \qquad \forall q \in P, \tag{14}$$

where

$$a(\mathbf{u}, \mathbf{v}) = \int_\Omega \nabla\mathbf{u} : \nabla\mathbf{v} := \sum_{i,j=1}^2 \int_\Omega \partial_j u_i \partial_j v_i,$$
$$b(\mathbf{v}, p) = -\int_\Omega p\nabla \cdot \mathbf{v}, \tag{15}$$

and $(\cdot, \cdot)_\Omega$ denotes the inner product of $L^2(\Omega)$. It is well known that problem (14) is well-posed (see, eg. [37] or [13]) so we skip here the mathematical details.

3.2 An Equal-Order Discontinuous Galerkin Formulation

In this section we briefly introduce, in the isogeometric context, a discontinuous Galerkin formulation proposed by Cockburn et al. in [38] allowing to consider equal-order discontinuous spaces for the velocity and the pressure. A detailed discussion of alternative discontinuous Galerkin methods for the Stokes problem can be found in [37].

For a given mesh size h, denote by Ω_h a conforming isogeometric discretization of Ω [1]. Denote by F_h the set of all faces of Ω_h, and by $F_h^0 \subset F_h$ the set of internal faces. Over F_h, we define $\mathbf{n}_F$ as a predefined normal over each F being coincident with $\mathbf{n}$ when F is a boundary face. We denote by h_F the diameter of the face being the length of the edge in 2D, and equal to $2^{0.5*}$ length of the edge in 3D. We denote by $S_{c1,...c_d}^{p_1,...,p_d}$ the space, defined over Ω_h, of splines functions of degree $p_i \geq 1$, and continuity $c_i = -1 \leq p_i - 1$ in the x_i coordinate. Over F_h, for any function $v_h \in S_{c1,...c_d}^{p_1,...,p_d}$, we denote by $[v_h]$ the jump operator, and by $\{v_h\}$ the average operator, defined as follows:

$$[v_h]\,|_F = \begin{cases} v_h^- - v_h^+, & \text{if } F \in F_h^0, \\ v_h, & \text{if } F \in F_h \setminus F_h^0, \end{cases} \qquad \{v_h\}\,|_F = \begin{cases} \frac{1}{2}(v_h^- + v_h^+), & \text{if } F \in F_h^0, \\ v_h, & \text{if } F \in F_h \setminus F_h^0, \end{cases} \tag{16}$$

with v_h^- and v_h^+ denoting the left and right traces respectively, with respect to the predefined normal $\mathbf{n}_F$. Finally, for a given $p \geq 1$, define $W_h := \left[S_{-1,...,-1}^{p,...,p}\right]^2$ as the space for the discontinuous velocity, $Q_h = S_{-1,...,-1}^{p,...,p}$, and $Q_{0,h} := Q_h \cap L_0^2(\Omega)$ as the space for the discontinuous pressure. The equal-order velocity and pressure discontinous Galerkin formulation reads:

Find $(\mathbf{u}_h^{DG}, p_h^{DG}) \in W_h \times Q_{0,h}$, such that:

$$\begin{aligned} a_h(\mathbf{u}_h^{DG}, \mathbf{v}_h) + b_h(\mathbf{v}_h, p_h^{DG}) &= (f, \mathbf{v}_h)_\Omega, & \forall \mathbf{v}_h \in W_h, \\ -b_h(\mathbf{u}_h^{DG}, q_h) + s_h(p_h^{DG}, q_h) &= 0, & \forall q_h \in Q_{0,h}, \end{aligned} \tag{17}$$

with

$$\begin{aligned} a_h(\mathbf{w}_h, \mathbf{v}_h) = \sum_{i=1,...,d} \Bigg(&\sum_{K \in \Omega_h} \int_K \nabla w_{h,i} \cdot \nabla v_{h,i} - \sum_{F \in F_h} \int_F \{\nabla w_{h,i}\} \cdot \mathbf{n}_F [v_{h,i}] \\ &- \sum_{F \in F_h} \int_F [w_{h,i}] \cdot \{\nabla v_{h,i}\} \cdot \mathbf{n}_F + \sum_{F \in F_h} \int_F \frac{\eta}{h_F}[w_{h,i}][v_{h,i}] \Bigg), \end{aligned} \tag{18}$$

being the discretization of the diffusive term,

$$b_h(\mathbf{v}_h, q_h) = -\sum_{K \in \Omega_h} \int_K q_h \nabla \cdot \mathbf{v}_h + \sum_{F \in F_h} \int_F [\mathbf{v}_h] \cdot \mathbf{n}_F \{q_h\}, \tag{19}$$

is the discretization of the pressure-velocity coupling term, and

$$s_h(p_h, q_h) = \sum_{F \in F_h^0} h_F \int_F [p_h][q_h], \tag{20}$$

an extra stabilization term allowing to consider equal-order discrete spaces. In (18), $\eta > \underline{\eta}$ denotes a user-defined stabilization parameter that has to be considered large enough to guarantee the inf-sup stability (see eg. Lemma 4.12 in [37]). Notice that, by identifying $\mathfrak{V}_h = W_h \times Q_{0,h}$, $\mathfrak{u}_h^{DG} = (\mathbf{u}_h^{DG}, p_h^{DG})$, and $\mathfrak{v}_h = (\mathbf{v}_h, q_h)$, problem (17) can be equivalently written of the form (4), with $l_h(\mathfrak{v}_h) = (f, v_h)_\Omega$ and

$$\mathfrak{b}_h(\mathfrak{u}_h^{DG}, \mathfrak{v}_h) := a_h(\mathbf{u}_h^{DG}, \mathbf{v}_h) + b_h(\mathbf{v}_h, p_h^{DG}) - b_h(\mathbf{u}_h^{DG}, q_h) + s_h(p_h^{DG}, q_h). \tag{21}$$

Moreover, the bilinear form (21) satisfies the inf-sup condition (5) (see Lemma 6.13 in [37]) with the following norm:

$$
|||(\mathbf{v}_h, q_h)|||^2 = \sum_{i=1,\ldots,d} \left(\sum_{K \in T_h} \|\nabla v_{h,i}\|^2_{L^2(K)} + \sum_{F \in F_h} \frac{\eta}{h_F} \|[v_{h,i}]\|^2_{L^2(F)} \right)
$$
$$
+ \|q_h\|^2_{L^2(\Omega)} + \sum_{F \in F_h^0} h_F \|[q_h]\|^2_{L^2(F)}.
\tag{22}
$$

Remark 1 (Discarding the zero-mean value restriction). Following Remark 6.14 from [37], in practice we can ignore the zero mean-value constrain (13) in the spaces for the pressure, and call MUMPS with pivoting. Then, a zero mean-value solution can be recovered by post-processing the solution as $p = p - \frac{1}{|\Omega|} \int_\Omega p$.

3.3 Trial Spaces for the Residual Minimization Problem

The subspace condition for the trial space give a wide range of possibilities. In this paper, we focus in the two dimensional case. For a given polynomial degree p, we denote by $V_h = S^{p,p}_{-1,-1} \times S^{p,p}_{-1,-1}$ the test space for the velocity, and by $Q_h = S^{p,p}_{-1,-1}$ the test space for the pressure (see Remark 1). Denoting by $V_h \subset W_h$ the trial space for the velocity, and by $P_h \subset Q_h$ the trial space for the pressure, we consider the following conforming couples of spaces:

a) **Raviart-Thomas type:**
   $V_h := S^{p,p-1}_{c,c-1} \times S^{p-1,p}_{c-1,c}$, $P_h := S^{p-1,p-1}_{c-1,c-1}$, with $p \geq 2$ and $1 \leq c \leq p-1$.

b) **Second order Nédélec type:**
   $V_h := S^{p,p}_{c,c-1} \times S^{p,p}_{c-1,c}$, $P_h := S^{p-1,p-1}_{c-1,c-1}$, with $p \geq 2$ and $1 \leq c \leq p-1$.

c) **Taylor-Hood type:**
   $V_h := S^{p,p}_{c,c} \times S^{p,p}_{c,c}$, $P_h := S^{p-1,p-1}_{c,c}$, with $p \geq 2$ and $0 \leq c \leq p-2$.

d) **Equal-order type:**
   $V_h := S^{p,p}_{c,c} \times S^{p,p}_{c,c}$, $P_h := S^{p,p}_{c,c}$, with $p \geq 1$ and $0 \leq c \leq p-2$.

We notice that couple of spaces a), b) and c) are stable in the classical isogeometric case (see [39]), while the couple d) is not.

4   Numerical Results

In this section, we explore the results obtained considering (as starting point) the equal-order discontinuous Galerkin formulation defined in Sect. 3.2, with $W_h = S^{4,4}_{-1,-1} \times S^{4,4}_{-1,-1}$ and $Q_h = S^{4,4}_{-1,-1}$ as the discontinuous spaces for the velocity and pressure respectively, and performing the discontinuous Galerkin based residual minimization method (see Sect. 2) with the several options of conforming trial spaces defined in Sect. 3.3.

Table 1. Taylor-Hood type trial with minimum and maximum continuity respectively.

Trial spaces	$\|\mathbf{u} - \mathbf{u}_h\|_{L^2}$	$\|p - p_h\|_{L^2}$	$\|\text{div}(\mathbf{u} - \mathbf{u}_h)\|_{L^2}$	flops
$V = S_{0,0}^{4,4} \times S_{0,0}^{4,4}, P = S_{0,0}^{3,3}$	1.72e−06	5.69e−05	5.28e−06	3.96537e+12
$V = S_{2,2}^{4,4} \times S_{2,2}^{4,4}, P = S_{2,2}^{3,3}$	1.59e−05	0.000232	2.14e−05	9.52712e+10

Table 2. Raviart-Thomas type trial with minimum and maximum continuity respectively.

Trial spaces	$\|\mathbf{u} - \mathbf{u}_h\|_{L^2}$	$\|p - p_h\|_{L^2}$	$\|\text{div}(\mathbf{u} - \mathbf{u}_h)\|_{L^2}$	flops
$V = S_{1,0}^{4,3} \times S_{0,1}^{3,4}, P = S_{0,0}^{3,3}$	0.00019	0.000132	7.05e−06	4.40153e+11
$V = S_{3,2}^{4,3} \times S_{2,3}^{3,4}, P = S_{2,2}^{3,3}$	0.000493	0.000235	1.65e−06	2.7742e+10

4.1 A Smooth Analytical Solution

We consider the Stokes problem (12), defined over the 2D domain $\Omega = [0,1]^2$, and we define the source term f in such a way that the analytical solution is given by $\mathbf{u} = (u_1, u_2)$ and p, with (cf. [40]):

$$u_1(x,y) = 2e^x(-1+x)^2 x^2(y^2+y)(-1+2y),$$
$$u_2(x,y) = -e^x(-1+x)x(-2+x(3+x))(-1+y)^2 y^2,$$
$$p(x,y) = (-424 + 156e + (y^2 - y)(-456 + e^x(456 + x^2(228 - 5(y^2 - y)) +$$
$$2x(-228 + (y^2 - y)) + 2x^3(-36 + (y^2 - y)) + x^4(12 + (y^2 - y)))))).$$
$$\tag{23}$$

In Tables 1, 2, 3 and 4 we show the L^2-error in the approximation of the functions $\mathbf{u}$, p and div $\mathbf{u}$, obtained by considering a fixed mesh of size 20×20, and the extreme allowed continuities for the Taylor-Hood type, Raviart-Thomas type, second order Nédélec type, and equal-order type trial spaces respectively. We also show the number of flops required for the resolution of the corresponding saddle-point problem (see Equation (11)). As expected, all the selected trial spaces deliver good approximations for the measured quantities. Moreover, there is no a significative difference in the approximation when considering a highly-continuous trial space, while the total number of flops is reduced in almost two orders of magnitude, when compared with its C^0-trial equivalent, and the highly-continuous equal order type trial is the one that delivers a better balance between accuracy and computational cost. Last can be also appreciated in Fig. 1, where we plot the Pareto front for the previous results (see [41]), considering the the number of floating-point operations (as performed by the MUMPS direct solver) as the vertical axis, and the numerical error measured in the $|||(\cdot,\cdot)|||$-norm, defined in Equation (22), as the horizontal axis.

Finally, in Fig. 2, we plot the error $|||(\mathbf{u} - \mathbf{u}_h, p - p_h)|||$ (real), and the error of the residual estimation $|||(r_h^u, r_h^p)|||$ (estimated), where $\mathbf{r}_h^u, r_h^p$ are the residual

Table 3. Second-order Nédélec type trial with minimum and maximum continuity respectively.

Trial spaces	$\|\mathbf{u} - \mathbf{u}_h\|_{L^2}$	$\|p - p_h\|_{L^2}$	$\|\mathrm{div}(\mathbf{u} - \mathbf{u}_h)\|_{L^2}$	flops
$V = S_{1,0}^{4,4} \times S_{0,1}^{4,4}, P = S_{0,0}^{3,3}$	1.72e−06	5.69e−05	5.27e−06	9.09917e+11
$V = S_{3,2}^{4,4} \times S_{2,3}^{4,4}, P = S_{2,2}^{3,3}$	1.72e−05	0.000232	2.27e−05	5.00601e+10

Table 4. Equal-order type trial with minimum and maximum continuity respectively.

Trial spaces	$\|\mathbf{u} - \mathbf{u}_h\|_{L^2}$	$\|p - p_h\|_{L^2}$	$\|\mathrm{div}(\mathbf{u} - \mathbf{u}_h)\|_{L^2}$	flops
$V = S_{0,0}^{4,4} \times S_{0,0}^{4,4}, P = S_{0,0}^{4,4}$	1.64e−06	8.29e−05	5.27e−06	2.85826e+12
$V = S_{3,3}^{4,4} \times S_{3,3}^{4,4}, P = S_{3,3}^{4,4}$	1.79e−05	9.13e−05	2.28e−05	3.86971e+10

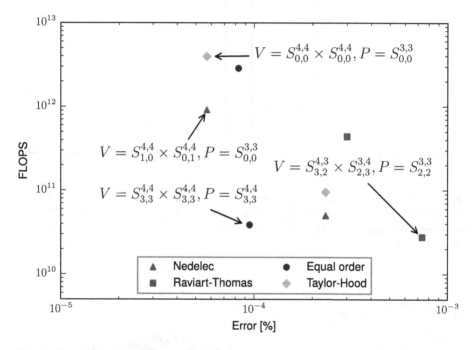

Fig. 1. Pareto front for different setups of trial spaces defined in Sect. 3.3. The vertical axis denotes the computational cost expressed in terms of the number of floating-point operations performed by MUMPS solver. The horizontal axis denotes the error in the $\||(\cdot,\cdot)\||$-norm (see (22)) for the smooth analytical problem.

associated with the velocity and pressure respectively, obtained when considering $V_h = S_{c,c}^{4,4} \times S_{c,c}^{4,4}$, $P_h = S_{c,c}^{4,4}$, with $c = 0, 1, 2, 3$ respectively, as trial spaces. As can be appreciated in the figures, increasing the continuity reduces the distance between the real and the estimated errors, implying that the error bound becomes sharper when increasing the continuity.

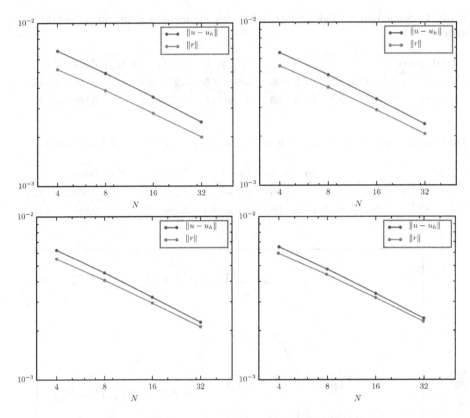

Fig. 2. Comparison of residual and error for different mesh dimensions, 4×4, 8×8, 16×16 and 32×32 considering as test the space $W_h = S^{4,4}_{-1,-1} \times S^{4,4}_{-1,-1}$, $Q_h = S^{4,4}_{-1,-1}$, and as trial the space $V_h = S^{4,4}_{c,c} \times S^{4,4}_{c,c}$, $P_h = S^{4,4}_{c,c}$, with $c = 0, 1, 2, 3$ respectively (from left to the right).

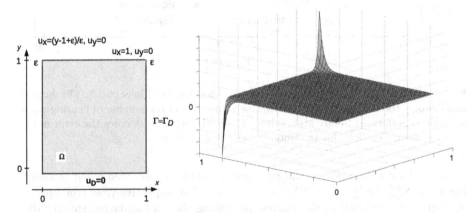

Fig. 3. Left panel: The formulas of the Dirichlet boudnary conditions for the velocity field. **Right panel:** The singularities at the pressure solution.

Table 5. Velocity (u_1, u_2) and pressure p on a series of uniformly adapted meshes.

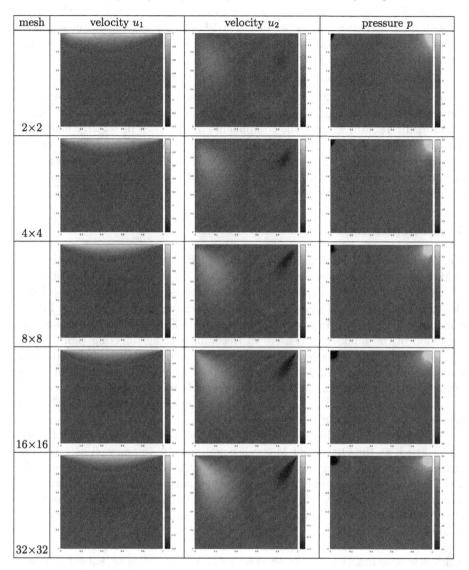

4.2   The Lid-Cavity Flow Problem

With the spirit of exploring the behavior of the method when the solution is non-smooth, as second example we consider the well-known lid-cavity flow problem (see eg. [42]).

The problem models a plane flow of an isothermal fluid in a square lid-driven cavity of size $(0, 1)^2$ (cf. [43]). The pressure solution in the problem exhibits two singularities at the corners, as presented on right panel in Fig. 3. For the numerical simulation, we enforce the Dirichlet boundary conditions for the velocity field in terms of a small parameter $\varepsilon > 0$, to obtain a solution with Dirichlet trace belonging to $H^{1/2}$ (see left panel in Fig. 3). For the pressure, we fix its value at one point, which is numerically equivalent to setting the condition (13). We set a homogeneous force $f = 0$. We consider the spaces $W_h = S^{4,4}_{-1,-1} \times S^{4,4}_{-1,-1}$ and $Q_h = S^{4,4}_{-1,-1}$ for as test for the velocity and pressure respectively, and the spaces $V_h = S^{4,4}_{3,3} \times S^{4,4}_{3,3}$, $P_h = S^{4,4}_{3,3}$ as trials for the velocity and pressure respectively, that we recall it is not stable in the classical isogeometric sense (cf. [39]). In Table 5, we plot the components of the discrete velocity field, and the discrete pressure field obtained considering several uniform meshes. As can be appreciated from the figures, also in this scenario the method delivers stable and accurate approximations, even if a highly-continuous space is chosen as trial, evidencing the performance of the method.

5   Conclusions

We investigated a Discontinuous Galerkin (DG) based residual minimization (RM) stabilization for isogeometric analysis (IGA) simulations of the stationary Stokes problem. We explore the results obtained when considering a fixed DG-type test space and several types of conforming trial spaces. The higher continuity spaces result in a lower computational effort of the solver due to the reduction of the number of degrees of freedom, without affecting significantly the approximation. Moreover, the upper error bound constant is reduced when the continuity is increased, leading to a sharper estimation of the error in terms of the analytical solution. The method is also able to capture singularities even is considering a highly-continuous trial space, as evidenced with the well-known lid-cavity flow problem in the numerical section.

As future work, we plan to extend the analysis to other kind of mixed formulations, such as the Ossen and Maxwell equations [44–46], as well as exploring parallelization techniques for the resolution of the saddle-point problem [47], and localized adaptive mesh refinement techniques based on the residual estimator (10). The future work will also involve incorporating of the DG method mixed with residual minimization formulation within adaptive finite element code [42,48].

Acknowledgments. This is supported by National Science Centre, Poland grant no. 2017/26/M/ ST1/ 00281. The CSIRO Professorial Chair in Computational Geoscience at Curtin University and the Deep Earth Imaging Enterprise Future Science Platforms of the Commonwealth Scientific Industrial Research Organisation, CSIRO, of Australia. The European Union's Horizon 2020 Research and Innovation Program of the Marie Skłodowska-Curie grant agreement No. 777778 provided additional support. At Curtin University, The Institute for Geoscience Research (TIGeR) and by the Curtin Institute for Computation, kindly provide continuing support. The work by Ignacio Muga was done in the framework of Chilean FONDECYT research project #1160774. The visit of Sergio Rojas at AGH University was partially supported by National Science Centre, Poland grant no. 2017/26/M/ ST1/ 00281.

References

1. Hughes, T.J.R., Cottrell, J.A., Bazilevs, Y.: Isogeometric analysis: CAD, finite elements, NURBS, exact geometry and mesh refinement. Computer Methods in Applied Mechanics and Engineering **39–41**, 4135–4195 (2005)
2. Piegl, L., Tiller, W.: The *NURBS* Book, 2nd edn. Springer, New York (1997). https://doi.org/10.1007/978-3-642-59223-2
3. Dedè, L., Hughes, T.J.R., Lipton, S., Calo, V.M.: Structural topology optimization with isogeometric analysis in a phase field approach. In: USNCTAM 2010, 16th US National Congress of Theoretical and Applied Mechanics (2016)
4. Dedè, L., Borden, M.J., Hughes, T.J.R.: Isogeometric analysis for topology optimization with a phase field model. ICES REPORT 11–29. The University of Texas at Austin, The Institute for Computational Engineering and Sciences (2011)
5. Gómez, H., Calo, V.M., Bazilevs, Y., Hughes, T.J.R.: Isogeometric analysis of the Cahn-Hilliard phase-field model. Comput. Methods Appl. Mech. Eng. **197**, 4333–4352 (2008)
6. Gómez, H., Hughes, T.J.R., Nogueira, X., Calo, V.M.: Isogeometric analysis of the isothermal Navier-Stokes-Korteweg equations. Comput. Methods Appl. Mech. Eng. **199**, 1828–1840 (2010)
7. Hsu, M.-C., Akkerman, I., Bazilevs, Y.: High-performance computing of wind turbine aerodynamics using isogeometric analysis. Comput. Fluids **49**(1), 93–100 (2011)
8. Duddu, R., Lavier, L., Hughes, T.J.R., Calo, V.M.: A finite strain Eulerian formulation for compressible and nearly incompressible hyper-elasticity using high-order NURBS elements. Int. J. Numer. Methods Eng. **89**(6), 762–785 (2012)
9. Chang, K., Hughes, T.J.R., Calo, V.M.: Isogeometric variational multiscale large-Eddy simulation of fully-developed turbulent flow over a wavy wall. Comput. Fluids **68**, 94–104 (2012)
10. Hossain, S., Hossainy, S.F.A., Bazilevs, Y., Calo, V.M., Hughes, T.J.R.: Mathematical modeling of coupled drug and drug-encapsulated nanoparticle transport in patient-specific coronary artery walls. Comput. Mech. (2011). https://doi.org/10.1007/s00466-011-0633-2
11. Bazilevs, Y., Calo, V.M., Cottrell, J.A., Hughes, T.J.R., Reali, A., Scovazzi, G.: Variational multiscale residual-based turbulence modeling for large Eddy simulation of incompressible flows. Comput. Methods Appl. Mech. Eng. **197**, 173–201 (2007)
12. Bazilevs, Y., Calo, V.M., Zhang, Y., Hughes, T.J.R.: Isogeometric fluid-structure interaction analysis with applications to arterial blood flow. Comput. Mech. **38**, 310–322 (2006)
13. Ern, A., Guermond, J.-L.: Theory and Practice of Finite Elements. Springer, New York (2004). https://doi.org/10.1007/978-1-4757-4355-5
14. Demkowicz, L.: Babuška<=>Brezzi??. ICES-Report 0608, The University of Texas at Austin, USA (2006). https://www.ices.utexas.edu/media/reports/2006/0608.pdf
15. Babuška, I.: Error bounds for finite element method. Numerische Mathematik **16**, 322–333 (1971)
16. Brezzi, F.: On the existence. Uniqueness and approximation of saddle-point problems arising from lagrange multipliers. ESAIM: Math. Model. Numer. Anal. - Modélisation Mathématique et Analyse Numérique **8**(R2), 129–151 (1974)

17. Hughes, T.J.R., Scovazzi, G., Tezduyar, T.E.: Stabilized methods for compressible flows. J. Sci. Comput. **43**(3), 343–368 (2010)
18. Franca, L.P., Frey, S.L., Hughes, T.J.R.: Stabilized finite element methods: I. Application to the advective-diffusive model. Comput. Methods Appl. Mech. Eng. **95**(2), 253–276 (1992)
19. Franca, L.P., Frey, S.L.: Stabilized finite element methods: II. The incompressible Navier-Stokes equations. Comput. Methods Appl. Mech. Eng. **99**(2–3), 209–233 (1992)
20. Brezzi, F., Bristeau, M.-O., Franca, L.P., Mallet, M., Rogé, G.: A relationship between stabilized finite element methods and the Galerkin method with bubble functions. Comput. Methods Appl. Mech. Eng. **96**(1), 117–129 (1992)
21. Demkowicz, L., Gopalakrishnan, J.: Recent developments in discontinuous Galerkin finite element methods for partial differential equations. In: Feng, X., Karakashian, O., Xing, Y. (eds.) IMA Volumes in Mathematics and its Applications. An Overview of the DPG Method, vol. 157, pp. 149–180 (2014)
22. Ellis, T.E., Demkowicz, L., Chan, J.L.: Locally conservative discontinuous Petrov-Galerkin finite elements for fluid problems. Comput. Math. Appl. **68**(11), 1530–1549 (2014)
23. Chan, J., Evans, J.A.: A Minimum-Residual Finite Element Method for the Convection-Diffusion Equations. ICES-Report, 13–12 (2013)
24. Broersen, D., Dahmen, W., Stevenson, R.P.: On the stability of DPG formulations of transport equations. Math. Comput. **87**, 1051–1082 (2018)
25. Broersen, D., Stevenson, R.: A robust Petrov-Galerkin discretisation of convection-diffusion equations. Comput. Math. Appl. **68**(11), 1605–1618 (2014)
26. Demkowicz, L., Heuer, N.: Robust DPG method for convection-dominated diffusion problems. SIAM J. Numer. Anal. **51**(5), 2514–2537 (2013)
27. Chan, J., Heuer, N., Bui-Thanh, T., Demkowicz, L.: A robust DPG method for convection-dominated diffusion problems II: adjoint boundary conditions and mesh-dependent test norms. Comput. Math. Appl. **67**(4), 771–795 (2014)
28. Calo, V.M., Ern, A., Muga, I., Rojas, S.: An adaptive stabilized conforming finite element method via residual minimization on dual discontinuous Galerkin norms. Comput. Methods Appl. Mech. Eng. **363**, 112891 (2020)
29. Hughes, T.J.R., Franca, L.P.: A new FEM for computational fluid dynamics: VII. The stokes problem with various well-posed boundary conditions: symmetric formulations that converge for all velocity/pressure spaces. Comput. Methods Appl. Mech. Eng. **65**, 85–96 (1987)
30. Hughes, T.J.R., Franca, L.P., Balestra, M.: A New FEM for computational fluid dynamics: V. Circumventing the Babuška-Brezzi condition: a stable Petrov-Galerkin formulation of the stokes problem accomodating equal- order interpolations. Comput. Methods Appl. Mech. Eng. **59**, 85–99 (1986)
31. Jansen, K.E., Collis, S.S., Whithing, C., Shakib, F.: A better consistency for low-order stabilized finite element methods. Comput. Methods Appl. Mech. Eng. **174**, 153–170 (1997)
32. Maniatty, A.M., Liu, L., Klaas, O., Shephard, M.S.: Stabilized finite element method for viscoplastic flow: formulation and a simple progressive solution strategy. Comput. Methods Appl. Mech. Eng. **190**, 4609–4625 (2001)
33. Matuszyk, P.J., Boryczko, K.: A parallel preconditioning for the nonlinear stokes problem. In: Wyrzykowski, R., Dongarra, J., Meyer, N., Waśniewski, J. (eds.) PPAM 2005. LNCS, vol. 3911, pp. 534–541. Springer, Heidelberg (2006). https://doi.org/10.1007/11752578_64

34. Amestoy, P.R., Duff, I.S.: Multifrontal parallel distributed symmetric and unsymmetric solvers. Comput. Methods Appl. Mech. Eng. **184**, 501–520 (2000)
35. Amestoy, P.R., Duff, I.S., Koster, J., L'Excellent, J.Y.: A fully asynchronous multifrontal solver using distributed dynamic scheduling. SIAM J. Matrix Anal. Appl. **1**(23), 15–41 (2001)
36. Amestoy, P.R., Guermouche, A., L'Excellent, J.-Y., Pralet, S.: Hybrid scheduling for the parallel solution of linear systems. Comput. Methods Appl. Mech. Eng. **2**(32), 136–156 (2001)
37. Di Pietro, D., Ern, A.: Mathematical Aspects of Discontinuous Galerkin Methods. Springer, Heidelberg (2011). https://doi.org/10.1007/978-3-642-22980-0
38. Cockburn, B., Karniadakis, G., Shu, C.-W.: Discontinuous Galerkin Methods, Theory. Lecture Notes in Computational Science and Engineering. Computation and Applications. Springer, Heidelberg (2000). https://doi.org/10.1007/978-3-642-59721-3
39. Buffa, A., de Falco, C., Sangalli, G.: Isogeometric analysis: new stable elements for the Stokes equation. Int. J. Numer. Methods Fluids **65**(11–12), 1407–1422 (2011)
40. Buffa, A., de Falco, C., Sangalli, G.: IsoGeometric analysis: stable elements for the 2D Stokes equation. Int. J. Numer. Methods Fluids **65**(11–12), 1407–1422 (2011)
41. Fudenberg, D., Tirole, J.: Game Theory, pp. 18–23. MIT Press, Cambridge (1991). Lawrence Berkeley National Laboratory, LBNL-44289 (1999). http://crd.lbl.gov/xiaoye/SuperLU/
42. Matuszyk, P., Paszyński, M.: Fully automatic HP finite element method for the Stokes problem in two dimensions. Comput. Methods Appl. Mech. Eng. **197**(51–52), 4549–4558 (2008)
43. Donea, J., Huerta, A.: Finite Element Methods for Flow Problems, 1st edn. Willey, Hoboken (2003)
44. Hochbruck, M., Jahnke, T., Schnaubelt, R.: Convergence of an ADI splitting for Maxwell's equations. Numerishe Mathematik **129**, 535–561 (2015)
45. Liping, G.: Stability and super convergence analysis of ADI-FDTD for the 2D Maxwell equations in a lossy medium. Acta Mathematica Scientia **32**(6), 2341–2368 (2012)
46. Paszyński, M., Demkowicz, L., Pardo, D.: Verification of goal-oriented HP-adaptivity. Comput. Math. Appl. **50**(8–9), 1395–1404 (2005)
47. Woźniak, M., Kuźnik, K., Paszyński, M.: Computational cost estimates for parallel shared memory isogeometric multi-frontal solvers. Comput. Math. Appl. **67**(10), 1864–1883 (2014)
48. Paszyńska, A., Paszyński, M., Grabska, E.: Graph transformations for modeling hp-adaptive finite element method with mixed triangular and rectangular elements. In: Allen, G., Nabrzyski, J., Seidel, E., van Albada, G.D., Dongarra, J., Sloot, P.M.A. (eds.) ICCS 2009. LNCS, vol. 5545, pp. 875–884. Springer, Heidelberg (2009). https://doi.org/10.1007/978-3-642-01973-9_97

Numerical Modeling of the Two-Phase Flow of Water with Ice in the Tom River

Vladislava Churuksaeva$^{(\boxtimes)}$ and Alexander Starchenko

National Research Tomsk State University, Tomsk 634050, Russia
chu.vv@mail.ru, starch@math.tsu.ru

Abstract. A new mathematical model and a numerical method were constructed for numerical investigation of a two-phase turbulent flow in an open channel. Solid particles with a density close to that of water were considered a continuous phase with effective properties. This new model is based on a continuum-mechanics approach, a hydrostatic assumption, and equations averaged by the flow depth. Turbulent closure of the equations was done with a two-parameter $k - \varepsilon$ turbulence model modified by Pourahmadi and Humphrey to account for the influence of the particles on the turbulent structure of the flow. The new numerical method is based on partial elimination algorithm for computing areas of the two-phase flow free of ice particles and uses semi-implicit approximation in time. The influence of the dynamic parameters of the dispersed phase on the structure of the flow was also investigated by computing several scenarios of the flow in an open channel with a 90-degree bend. Applications of the approach to the modeling of riverside flooding due to sudden increase in the river depth after a release of an ice jam illustrate the capabilities of the model.

Keywords: Mathematical modeling · Two-phase dispersed flow · Continuum-mechanics approach · Depth-averaged equations · Turbulent flow · Solid particles · Finite volume method · Partial elimination algorithm

1 Introduction

Dispersed two-phase flows occur in such areas of environmental modeling as the atmosphere (clouds, airborne particulate matter), sediment transport in rivers, and river flow with ice as well as in technological facilities (coolants in cooling systems, fuel combustion). The majority of the flows mentioned above are turbulent. In spite of geophysical flows involving sediment transport in rivers being among the first flows observed from the mechanical point of view, river flow with ice is much less researched than multiphase flows in technological facilities and flows with sediment transport [1, 2]. Nevertheless several hydrodynamic approaches to modeling river ice processes have been developed. One of the most important applications of these models is predicting riverside flooding caused by ice jams.

Static Models
Classical theoretical research on surface ice jams formed on a river during the breakup is based on the assumption that an ice jam is a one-dimensional static formation of ice

© Springer Nature Switzerland AG 2020
V. V. Krzhizhanovskaya et al. (Eds.): ICCS 2020, LNCS 12138, pp. 212–224, 2020.
https://doi.org/10.1007/978-3-030-50417-5_16

floes with constant porosity. River flow velocity and characteristics of ice cover are considered to be constant across the river.

This approach allows evaluating the thickness of the existing ice jam but cannot indicate the initiation of an ice jam in a particular place and the conditions in which a jam could form, because they cannot correctly describe transport of ice and ice interaction with a flow. Static models are applicable when there is much data about the object of modeling such as ice thickness, density, friction between ice particles in the jam, and some other characteristics of ice cover. Wide review and comparison of the static models based on test cases is found in [1, 3].

Dynamic Models

Dynamic models of ice processes in rivers include a hydrodynamic model of the flow and a model of moving particles. Because multiphase flow is a very complex phenomenon, dynamic models began to develop from one-dimensional spatial approximation [4–6]. One-dimensional models, however, have limited application, because the transport of ice particles is a significantly two-dimensional process due to the wall friction, bathymetry, and mutual influence of the flow and the ice jam.

An advanced approach to modeling river flow with ice involves a multidimensional hydrodynamic model of the flow and a model to describe the movement of solid particles. Some of the multidimensional dynamic models that are precise in describing a flow with ice particles are that of Shen et al. [7–9], which is built within the finite-element Eulerian approach and Lagrangian model of ice particles; the three-dimensional unsteady Eulerian two-phase model of the flow with ice particles suggested by Wang et al. [10]; the model based on two-dimensional inviscid shallow water equations and Lagrangian model for ice particles by Shlychkov et al. [11]; and the two-dimensional hydrodynamic code with DEM for ice particles by Stockstill et al. [12].

Mathematical models that are part of much current hydrological software are quite robust in mathematical description of physical processes and require high-quality input data about a hydrological object, especially its bathymetry. Therefore developing mathematical models for hydrodynamic investigation of small rivers without precise bathymetric data about them is of particular interest.

Developing effective numerical algorithms for solving hydrodynamic equations are also of particular importance, especially in cases of river breakup, floods, and flows with ice particles, because of the complexity of these problems.

In this work, both a new mathematical model and a numerical method for computing two-phase flow of water with light particles densely placed at the water surface are presented. This approach meets all the requirements that exist for an advanced, up-to-date model.

2 Mathematical Model

Two-phase isothermal flow of the mixture of water and light particles in an open channel (river bed) is considered. Thermal exchange between phases is not considered because the temperatures of the water and the environment are close and change slightly during the period modeled. Because the considered density of ice $\rho_i^0 = 910\,\text{kg/m}^3$ is less than the density of water $\rho_l^0 = 1000\,\text{kg/m}^3$, ice particles are considered to be packed in the upper layer of water and their concentration remains constant at the inlet of the channel (or the section of the river). Interactions between particles (collisions and friction) are also accounted for. Horizontal dimensions of the area of modeling are presumed to be much greater than the water depth. The size of ice particles is significantly less than the characteristic linear dimension of the channel (river bed).

Hydrostatic balance was applied because of the significant difference in the scale of the process in vertical and horizontal, and therefore all the terms in the equation for vertical velocity are negligible except ones that describe pressure and the force of gravity.

The mathematical model of the process described is based on continuum mechanics approach [13]. Light particles densely placed on the water surface are regarded as a continuum with effective properties.

Equations that describe the flow of the liquid phase (water) are [14]

$$\frac{\partial h'}{\partial t} + \nabla \cdot (h' \vec{w}_l) = 0; \tag{1}$$

$$h'\left(\frac{\partial \vec{w}_l}{\partial t} + (\vec{w}_l \cdot \nabla)\vec{w}_l\right) = -gh'\psi\nabla(z_b + h) + \nabla \cdot h'\tau_1(\vec{w}_l) - c_f|\vec{w}_l|\vec{w}_l + \vec{F}_1. \tag{2}$$

For the dispersed phase of light particles (ice)

$$\frac{\partial h''}{\partial t} + \nabla \cdot (h'' \vec{w}_i) = 0; \tag{3}$$

$$h''\left(\frac{\partial \vec{w}_i}{\partial t} + (\vec{w}_i \cdot \nabla)\vec{w}_i\right) = -\frac{\rho_l^0}{\rho_i^0}gh''\psi\nabla(z_b + h) + \nabla \cdot h''\tau_1(\vec{w}_i) - c_f^i|\vec{w}_i|\vec{w}_i + \vec{F}_i. \tag{4}$$

Here g is the gravitational acceleration; $\vec{w} = (w_1, w_2)$ is the velocity of phase; ρ_l^0 is the density of water. $z_b = z_b(x, y)$ is the function that describes the bathymetry; and h is the water depth. $\int_{h-h_i+z_b}^{h+z_b} \alpha_i dz = \bar{\alpha}_i h_i = h''$, $h' = h - h''$ where h_i is the characteristic depth of the layer of ice particles. $c_f = \frac{gn^2}{h^{0.333}}$ is the bed friction of the liquid phase, and $n > 0$ is the Manning coefficient.

The bed friction of the dispersed phase in shallow regions is computed as $c_f^i|\vec{w}_i|\vec{w}_i$, where $c_f^i = 0.0025|\vec{w}_i|^{-1}$ [15]. $\psi = \bar{\alpha}_l + \left(\frac{\rho_l^0}{\rho_b^0} - 1\right)(1 - \bar{\alpha}_l)$, where $\bar{\alpha}_l$ is the depth-averaged volume fraction of water ($0 < \bar{\alpha}_l \leq 1$).

$$\tau_{kj}(\vec{w}) = (\nu^0 + \nu')\left[\left(\frac{\partial w_k}{\partial x_j} + \frac{\partial w_j}{\partial x_k}\right) - \frac{2}{3}\delta_{kj}\text{div }\vec{w}\right] - \frac{2}{3}\delta_{kj}k, \quad (j, k = 1, 2)$$

are the components of the viscous and turbulent stress tensor for the liquid/ice particles; $v_{l,i}^0$ is the viscosity of liquid/dispersed phase; $v_{l,i}^t$ is the eddy viscosity of the flow (liquid/ice); δ_{kj} is the Kronecker delta; and k is the turbulent kinetic energy. Viscous stress tensor in the dispersed phase appears due to collisions of particles and friction between them.

2.1   Force Terms in the Momentum Equations for Phases

The force term in the momentum equation for the dispersed phase is $\vec{F}_i = \vec{F}_A + \vec{F}_\mu + \vec{F}_{VM} + \vec{F}_C$, which is the sum of the following forces [13]:

- buoyancy

$$\vec{F}_A = h'' \frac{\rho_l^0}{\rho_i^0} \left(\frac{D_l \vec{w}_1}{D_l t} - \vec{g} \right); \quad \frac{D_l \vec{w}_1}{Dt} = \frac{\partial \vec{w}_1}{\partial t} + (\vec{w}_1 \cdot \nabla) \vec{w}_1;$$

- fluid drag force [16]

$$\vec{F}_\mu = \begin{cases} \left[150 \frac{h'' \tilde{\alpha}_i (1 - \tilde{\alpha}_l) \mu_l}{\tilde{\alpha}_l \rho_i^0 (d_i f_i)^2} + 1.75 \frac{h'' \rho_l^0 |\vec{w}_1 - \vec{w}_i|}{\rho_i^0 d_i f_i} \right] (\vec{w}_1 - \vec{w}_i), & \tilde{\alpha}_l \le 0.8; \\ \dfrac{3}{4} \dfrac{h'' \rho_l^0 c_D \tilde{\alpha}_l^{-1.65}}{\rho_i^0 d_i f_i} |\vec{w}_1 - \vec{w}_i| (\vec{w}_1 - \vec{w}_i), & \tilde{\alpha}_l > 0.8; \end{cases}$$

where $c_D = \max \left[\frac{24}{Re_p} \left(1 + 0.15 R_p^{0.687} \right), 0.44 \right]$ is the dimensionless drag coefficient, and d_i is the characteristic diameter of the ice particles; $Re_p = \frac{|\vec{w}_1 - \vec{w}_i| d_i \rho_l^0 \alpha_l}{\mu_l^0}$;

- virtual mass

$$\vec{F}_{VM} = c_{VM} h'' \frac{\rho_l^0}{\rho_i^0} \left(\frac{D_l \vec{w}_1}{Dt} - \frac{D_l \vec{w}_i}{Dt} \right), \quad C_{VM} \approx 0.5 \text{ for spherical particles;}$$

- coriolis force $\vec{F}_C = h'' (\vec{w}_i \times \vec{\omega})$.

   The model is closed with the high-Reynolds $k - \varepsilon$ turbulence model for depth-averaged equations [17] with a modification suggested by Pourahmadi and Humphrey [18] to account for the influence of particles on the flow. Constants of the model are the same as in the standard high-Reynolds $\bar{k} - \bar{\varepsilon}$ model [17].

Initial and Boundary Conditions

At the initial moment $t = 0$ the following conditions are used:

$h'' = h_{ice}$, h_{ice} is a known value;
$u_i = u_{i0}$, $v_i = v_{i0}$; $h' = h - h_{ice}$; $u_l = u_{l0}$, $v_l = v_{l0}$.

At the inlet of the channel, parameters of the liquid phase are considered to be known; at the outlet normal derivatives of the parameters are set to zero. At the solid boundaries shear stresses are considered both for liquid and for particles. Both normal and tangential velocities on the wall are set to zero. In the liquid phase, near the wall the stresses and turbulent characteristics of the liquid phase were set by Launder-Spalding wall functions [19].

3   Numerical Method

The area of the flow is contained in the rectangle covered with structured mesh. The equations of the model are discretized with the finite volume method on staggered mesh (Fig. 1), that is, finite volumes for velocity components are half-cell shifted from the node P, which is the center of the cell where scalar values h', h'', $\bar{k}$, $\bar{\varepsilon}$ are defined.

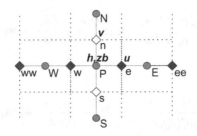

Fig. 1. Mesh stencil. Uppercase letters are for centers of the finite volumes, lowercase letters are for midpoints of their edges.

All the terms in the Eqs. (1)–(4) were approximated explicitly in time except for the drag force in the momentum Eqs. (2) and (4) which was approximated implicitly.

Convective fluxes in the momentum equations and advection-diffusion equations are approximated the MUSCL scheme [20].

To reach the third order accuracy in space in regions where the functions are monotonic, variables on the edges of the finite volumes are computed with linear interpolation from the values in the centers of the volumes (Fig. 1) [21]:

$$\Phi_e = \Phi_P + \frac{\Delta x}{2}\sigma_e^+, \quad \bar{u}_e > 0;$$

$$\Phi_e = \Phi_E - \frac{\Delta x}{2}\sigma_e^-, \quad \bar{u}_e \leq 0;$$

where $\sigma_e^+ = \frac{\delta_e}{\Delta x}\phi(\theta_e)$, $\sigma_e^- = \frac{\delta_e}{\Delta x}\phi(\theta_{ee}^{-1})$ are the slopes, limited by the function $\phi(\theta_e)$; $\delta_e = \Phi_E - \Phi_P$. Slope limiter $\phi(\theta)$ was chosen to satisfy the sufficient condition of the Harten's theorem [22] $\left(0 \leq \frac{\phi(\theta_e)}{\delta_e} \leq 2, \ 0 \leq \phi(\theta_e) \leq 2\right)$, which means that reconstructed value is within the values used for reconstruction. In this work the function $\phi(\theta) = \max\left[0, \min\left(2\theta, \frac{2+\theta}{3}, 2\right)\right]$ [20] was used as a limiter. Here $\theta_e = \frac{\delta_e}{\delta_w}$ ($\delta_w \neq 0$), and δ_e could be approximated with left, right, or central finite difference.

First order upwind scheme was used to approximate convective terms in the equations of the turbulence model.

Central finite difference scheme

$$
\left[gh\frac{\partial(z_b + h)}{\partial x} \right]_e \approx
\begin{cases}
gh_e\frac{h_E - h_P + z_{bE} - z_{bP}}{\Delta x}, & h_E \text{ and } h_P > \varepsilon_{wd} > 0 \text{ (if the cells with} \\
& \text{centers in E and P are wet),} \\
0, & h_E \text{ or } h_P < \varepsilon_{wd} \text{ (if one of the cells is dry);} \\
& \varepsilon_{wd} \text{ is the small positive value}
\end{cases}
$$

was used to approximate the source terms, which represent the influence of the bed slope in momentum equations.

The diffusive terms are discretized with the second order central-difference scheme.

In order to reduce a limitation on the time step for momentum equations, terms that express the dynamic interaction between phases (friction) was approximated with an implicit scheme and the partial elimination algorithm that allows solving equations in the areas with no dispersed phase ($h'' = 0$). A brief description of this approach is below.

Consider Eqs. (2) and (4) for the longitudinal velocity component for the inner node e of the mesh shown at Fig. 1.

$$
\frac{h'^0_e u_{le} - h'^0_e u^0_{le}}{\Delta t}\Delta x \Delta y = \Phi^0_e + \beta^0 h''^0_e(u_{ie} - u_{le}); \tag{5}
$$

$$
\frac{h''^0_e u_{ie} - h''^0_e u^0_{ie}}{\Delta t}\Delta x \Delta y = \Psi^0_e + \frac{\rho^0_l}{\rho^0_i}\beta^0 h''^0_e(u_{le} - u_{ie}). \tag{6}
$$

Φ^0_e and Ψ^0_e combine approximations of convective, diffusive, and source terms, β^0 is the coefficient of difference between velocities of the phases in the expression for the drag force. Upper index '0' is for the discrete values from the previous time step. Here and below in the section the over-bar for averaging is omitted.

In the right side of the discrete Eqs. (5)–(6), terms $\beta^0 h''^0_e(u_{ie} - u_{le})$ and $\frac{\rho^0_l}{\rho^0_i}\beta^0 h''^0_e$ $(u_{le} - u_{ie})$ describe dynamic interactions between phases and contain the differences between velocities of phases. Explicit approximation of these terms leads to a more strict convergence condition than the Courant–Friedrichs–Lewy condition

$$
\tau < \frac{0.5\,\Delta x\,\Delta y}{\max|u_e|\Delta y + \max|v_n|\Delta x + \sqrt{gh}(\Delta x + \Delta y)}
$$

in the case of flow with particles with large inertia.

In the case of $h'' = 0$ Eq. (6) becomes an identity. To avoid this, rewrite Eqs. (5)–(6):

$$
\left(h'^0_e\Delta x\Delta y + \Delta t\beta^0 h''^0_e\right)u_{le} - \Delta t\beta^0 h''^0_e u_{ie} = \Delta t\Phi^0_e + h'^0_e u^0_{le}\Delta x\Delta y; \tag{7}
$$

$$\left(\Delta x \Delta y + \frac{\rho_i^0}{\rho_i^0} \Delta t \beta \right) u_{ie} - \frac{\rho_i^0}{\rho_i^0} \Delta t \beta u_{le} = \Delta t \Psi_e^0 \Big/ h_e^{\prime\prime 0} + u_{ie}^0 \Delta x \Delta y. \qquad (8)$$

In (8), the term $\Psi_e^0 \Big/ h_e^{\prime\prime 0}$ Ψ_e^0 is equated to zero if $h_e^{\prime\prime 0} < \varepsilon$, where ε is an infinitely small positive value. The determinant of the SLAE (7)–(8) is nonzero and the system is being solved by Cramer's rule. The SLAE for $v_{l\ n}$, $v_{i\ n}$ was solved in the same way.

The wet-dry boundary treatment for an unsteady flow is of particular complexity because the solution becomes unstable due to very little water depth in the boundary cell [23, 24]. One of the simplest approaches to wet-dry boundary treatment is choosing a small positive value $\varepsilon > 0$ such that if the depth becomes less than ε, the cell is considered dry and excluded from computations. In the case of the two-phase flow considered in this article, h', which is the depth of the liquid phase, was compared to ε. When the cell became dry, the depth of the dispersed phase h'' was also equated to zero.

Several test cases of unsteady flow in open channels were computed with the model and the method presented above. The results were compared to experimental data to evaluate the model.

4   Results and Discussion

4.1   Two-Phase Flow in a 180° Bend Flume

Unsteady turbulent flow in a 180° bend flume with polypropylene particles to model ice was computed. This test case was investigated both experimentally and numerically in [25]. The depth of the flow at the inlet of the channel was set to 0.45 m, and the volumetric flow rate was 0.16 m³/s. These parameters define turbulent flow with Reynolds number $\mathrm{Re}_h = 77170$ defined by the depth of the flow. Spherical polypropylene particles ($\rho = 900$ kg/m³) with the diameter $d_i = 0.005$ m were used to model ice floes. Particles were tossed with constant rate into the steady flow at the end of the first straight section of the channel. A wire-mesh screen was placed after the bend to initiate ice jam development and support its downstream end.

Results of the computations made were compared with those of Urroz and Ettema [25] (Fig. 2).

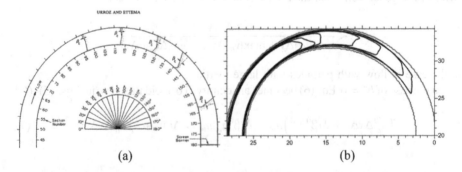

Fig. 2. Jam-head profiles around the bend (a) experiment [25], (b) computations

Moving along the flume, particles tend to accumulate near the left wall because of the centrifugal force, which also causes the increase in the velocity towards the left wall. After crossing the longitudinal axis of the channel, the flow tends to move closer to the right bank (Fig. 2).

This showed that the mathematical model and the computational method proposed accurately predicted both the velocity field and the distribution of the particles in the channel: they showed the increase in ice jam thickness both downstream and toward the inner bank of the bend as it was observed in experiments and gave a jam-head shape similar to those observed.

In [25] it was also noted that a similar shape of the front of ice particles was observed on the Iowa River during the breakup.

4.2   Two-Phase Flow in Open Channel with a 90° Bend

Two-phase flow in the channel with a 90° bend was computed. The first section of the channel was 5.555 m long and 0.86 m wide with a flat bottom, and the second section was 4.43 m long and 0.72 wide with a flat bottom. At the end of the first section there was a step with a change in the bed elevation of 0.013 m. One-phase flow in this channel was investigated in detail in [23, 26]. In the present article, the two-phase flow of the water with ice particles in the described channel is discussed. At the inlet, the longitudinal velocity of the flow was $U_0 = 0.2$ m/s, the initial depth of the flow was $h = 0.175$ m, and the depth of the dispersed phase was $h'' = 0.04$ m. Parameters of the dispersed phase are shown in Table 1. Structured mesh of 254×208 nodes was used in computations. Time step was chosen automatically from the Courant–Friedrichs–Lewy condition.

Table 1. Parameters of the test cases: 1 basic case, 2–3 parametric computations

	Particle diameter, d_i, m	Shape parameter, f_i	Effective viscosity of the liquid phase, m²/s
1	0.1	0.166 (cubic particles)	0.01
2	0.01	0.166	0.01
3	0.1	1 (spherical particles)	0.01

Computations led to the following conclusions:

1. Velocities of the smaller and the larger particles did not differ significantly near the bend. A recirculation area formed behind the bend near the right wall, and recirculating flow was more intensive for the larger particles ($d_i = 0.1$ m).

   In the corner of the channel, velocities of the phases were also different, but for cases (1) and (2) the particles did not accumulate in the corner because the particles had little inertia and wall friction to resist the dragging force of the water. It was also found that the flow with smaller particles had less turbulent kinetic energy behind the bend.

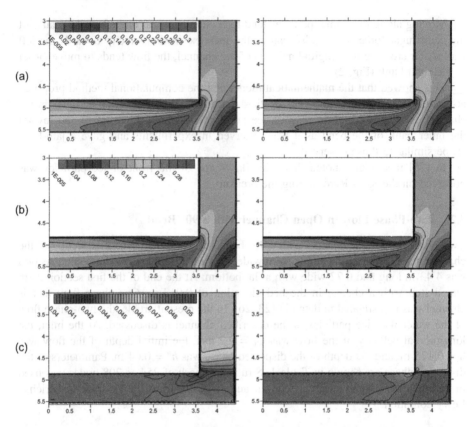

Fig. 3. Liquid phase velocity field (a), dispersed phase velocity field (b), depth of the dispersed phase layer (c) for the basic variant (1), and computation with smaller particles (2)

2. The change in bottom elevation influenced the velocities and the depth of the dispersed phase layer more than the size of the particles. Contours of the velocity magnitudes of both phases were not smooth near the step on the bottom because particles with little inertia in the same way as the liquid phase reacted to the change in the flow conditions (Fig. 3). At the same time, the distribution of h'' showed that with the increase in size of particles (and consequently their inertia) h'' increases more smoothly near the step and again near the bend.

Evaluation of the Influence of the Shape of Particles
With decrease in f_i from $f_i = 1$ to $f_i = 0.16$, resistance of the dispersed phase to the dragging force of the liquid phase increased (Fig. 4). The increase in the depth h'' in the corner of the channel was more efficient for spherical particles. In the case of the flow with spherical particles, the intensity of the recirculating flow was greater and the contours of h'' near the step were smoother. Computed distribution of the free surface in the bend was in accordance with experimental data [27].

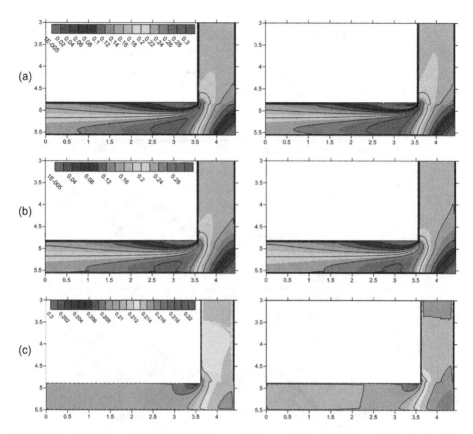

Fig. 4. Liquid phase velocity field (a), dispersed phase velocity field (b), and the depth of the dispersed phase (c) for the basic variant (1) and computation with spherical particles (3)

4.3   Modeling the Floods of the Tom River Near Tomsk, Russia

The abrupt increase in the flow rate of the river at the inlet of the section studied was computed. Such a phenomenon occurs when a sudden failure of an ice jam upstream releases large quantities of water and ice and often causes severe damage in the floodplain. As a precondition, flow with a longitudinal velocity of 1.25 m/s and free surface elevation of 69.8 m above sea level was computed for 20 h. After flow stabilization, the depth at the inlet of the section studied was increased by 2 m. The structured mesh of 221 × 180 nodes was used in computations.

With the increase in volume of fluid moving from the inlet, the depth of the flow in the section studied increased and water flooded the areas of the floodplain with the lowest heights: islands near 12,000 m and lowlands on the left bank of the river between 18,000 and 20,000 m (Fig. 5). The depth of the lake on the left bank of the river (near 6,000) increased and the lake connected to the river. Flooding of these areas is usually observed during the spring breakup.

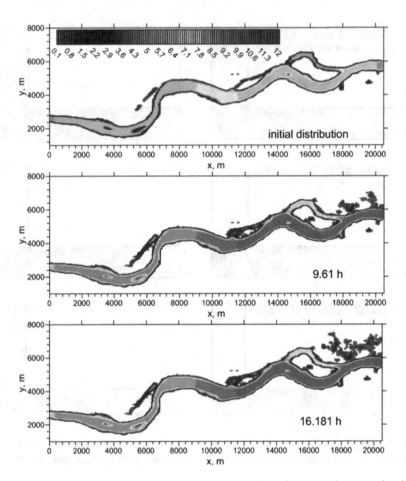

Fig. 5. Free surface level before increasing water level, 9.61 h after increasing water level, and 16.181 h after increasing the water level

5   Conclusions

A new mathematical model that is based on mechanics of interacting, interpenetrating continua was developed and tested. It is one of the first Eulerian hydrodynamic models of the two-phase flow of water with ice particles within the shallow water approach that takes into account interaction between phases, interactions between ice particles, interaction of both phases with the river bed, and the turbulence of the flow.

A computational algorithm based on finite volume method, semi-implicit discretization in time, and original partial elimination technique was developed to avoid uncertainty in the areas with no dispersed phase. The novelty of the computational method is in combining the technique for detecting the moving boundary of the river, which is of particular importance in modeling floods, with the partial elimination technique that ensures the correctness of computations throughout the domain in cases of areas with no ice particles.

The approach presented in this work was applied to numerical investigation of the Tom River during the breakup in spring near the city of Tomsk (Russia), where ice jams often form and cause localized flooding in the area behind the blockage. Two-phase river flow modeling showed that the approach developed correctly modeled changes in hydrodynamic characteristics of the river during the breakup and gave accurate results for the cases of flooding of the floodplain that involve change in the boundary of the river.

The investigation of the two-phase flow in the channel with 90° bend showed that the dispersed phase is more influenced by the bottom relief than by sharp change in flow direction. It was also shown that the presence of the ice particles in the flow increased the nonuniformity in the free surface level.

Acknowledgments. The numerical method was developed and the computations were performed by V. Churuksaeva, and this work was funded by Russian Foundation for Basic Research under research project N 18-31-00386.

The problem was stated and the mathematical model was formulated by A. Starchenko, and this work was supported by the Ministry of Science and Higher Education of the Russian Federation under project of Regional Scientific and Educational Mathematical Center of Tomsk State University.

The authors wish to thank Jean Kollantai of Tomsk State University for her assistance with the style of the paper.

References

1. Carson, R., et al.: Comparative testing of numerical models of river ice jams. Can. J. Civil Eng. **38**, 669–678 (2011). https://doi.org/10.1139/l11-036
2. Wang, J., Chen, P.: Progress in studies on ice accumulation in river bends. J. Hydrodyn. **23** (6), 737–744 (2011). https://doi.org/10.1016/S1001-6058(10)60171-0
3. Beltaos, S.: Rivers ice jams: theory, case studies, and applications. J. Hydraul. Eng. **109**(10), 1338–1359 (1983). https://doi.org/10.1061/(ASCE)0733-9429(1983)109:10(1338)
4. Daly, S.F.: Wave-propagation in ice-covered channels. J. Hydraul. Eng. **119**(2), 895–910 (1993). https://doi.org/10.1061/(ASCE)0733-9429(1993)119:8(895)
5. Shen, H.T., Shen, H.H., Tsai, S.M.: Dynamic transport of river ice. J. Hydraul. Res. **28**(9), 659–671 (1990). https://doi.org/10.1080/00221689009499017
6. Debolskaya, E.I., Derbenev, M.V., Maslikova, O.J.: Numerical modeling of ice jams. Water Resour. **31**(5), 533–539 (2004). [in Russian]
7. Shen, H.T., Su, J., Liu, L.: SPH simulation of river ice dynamics. J. Comput. Phys. **165**(2), 752–770 (2000). https://doi.org/10.1006/jcph.2000.6639
8. Shen, H.T., Liu, L.: Shokotsu river ice jam formation. Cold Reg. Sci. Technol. **37**(1), 35–49 (2003). https://doi.org/10.1016/S0165-232X(03)00034-X
9. Shen, H.T.: Mathematical modeling of river ice processes. Cold Reg. Sci. Technol. **62**, 3–13 (2010). https://doi.org/10.1016/j.coldregions.2010.02.007
10. Wang, X., Jiang, Z., Zhang, A., Zhou, Z., An, J.: Two-phase flow numerical simulation of a bend-type ice sluice in the diversion water channel of powerhouse. Cold Reg. Sci. Technol. **81**, 36–47 (2012). https://doi.org/10.1016/j.coldregions.2012.02.00410.1016/j.coldregions.2012.02.004

11. Shlychkov, V.A.: Hydrodynamic model of a river breakup for studying ice jams. Vestnik NSU. Ser. Math. Mech. Inform. **13**(2), 126–130 (2013). [in Russian]
12. Stockstill, R.L., Daly, S.F., Hopkins, M.A.: Modeling floating objects at river structures. J. Hydraul. Eng. **135**(5), 403–414 (2009). https://doi.org/10.1061/(ASCE)0733-9429(2009) 135:5(403)
13. Nigmatulin, R.I.: Dynamics of Multiphase Media, 1st edn. CRC Press, Boca Raton (1990)
14. Churuksaeva, V., Starchenko, A.: Numerical modeling of the two-phase flow of a liquid with particles in an open channel. Tomsk State Univ. J. Math. Mech. **6**(44), 88–103 (2016). https://doi.org/10.17223/19988621/44/8
15. Gidaspov, D.: Multiphase Flow and Fluidization: Continuum and Kinetic Theory Descriptions. Academic Press, Boston (1994)
16. Bubenchikov, A.M., Salomatov, V.V., Starchenko, A.V., Stropus, V.V.: Numerical investigation of the aerodynamics in the circulating fluidized bed installations. Russ. J. Eng. Thermophys. **3**(3), 257–264 (1993)
17. McGuirk, J.J., Rodi, W.: A depth-averaged mathematical model for the near field of side discharges into open channel flow. J. Fluid Mech. **88**, 761–781 (1978)
18. Pourahmadi, F., Humphrey, J.A.C.: Modelling solid-fluid turbulent flows with application to predicting erosive wear. Physico-Chem. Hydrodyn. **4**(3), 191–219 (1983). https://doi.org/10. 1016/0270-0255%2885%2990029-6
19. Launder, B.E., Spalding, D.B.: The numerical computation of turbulent flows. Comput. Methods Appl. Mech. Eng. **2**(3), 269–289 (1974)
20. Cada, M., Torrilhon, M.: Compact third-order limiter functions for finite volume methods. J. Comput. Phys. **228**, 4118–4145 (2009). https://doi.org/10.1016/j.jcp.2009.02.020
21. van Leer, B.: Towards the ultimate conservative difference scheme. Part II. Monotonicity and conservation combined in a second order scheme. J. Comput. Phys. **14**(4), 361–370 (1974). https://doi.org/10.1016/0021-9991(74)90019-9
22. Harten, A.: High resolution schemes for hyperbolic conservation laws. J. Comput. Phys. **135** (2), 260–278 (1997). https://doi.org/10.1006/jcph.1997.5713
23. Cea, L., Puertas, J., Vazquez-Cendon, M.E.: Depth averaged modelling of turbulent shallow water flow with wet-dry fronts. Arch. Comput. Methods Eng. **14**(3), 303–341 (2007). https:// doi.org/10.1007/s11831-007-9009-3
24. Hou, J., Simons, F., Mahgoub, M., Hinkelmann, R.: A robust well-balanced model on unstructured grids for shallow water flows with wetting and drying over complex topography. Comput. Methods Appl. Mech. Eng. **257**, 126–149 (2013). https://doi.org/10. 1016/j.cma.2013.01.015
25. Urroz, G.E., Ettema, R.: Bend ice jams: laboratory observations. Can. J. Civ. Eng. **19**, 855–864 (1992). https://doi.org/10.1139/l92-097
26. Churuksaeva, V., Starchenko, A.: Mathematical modeling of a river stream based on a shallow water approach. Procedia Comput. Sci. **66**, 200–209 (2015). https://doi.org/10.1016/ j.procs.2015.11.024
27. Han, S.S., Ramamurthy, A.S., Biron, P.M.: Characteristics of flow around open channel 90° bends with vanes. J. Irrig. Drain. Eng. **137**(10), 668–676 (2011). https://doi.org/10.1061/ (ASCE)IR.1943-4774.0000337

Remarks on Kaczmarz Algorithm for Solving Consistent and Inconsistent System of Linear Equations

Xinyin Huang[1], Gang Liu[2]([⊠]), and Qiang Niu[2,3]([⊠]) [iD]

[1] Department of Mathematics, King's College of London, London, UK
[2] Department of Mathematical Sciences, Xi'an Jiaotong-Liverpool University,
Suzhou 215123, Jiangsu, People's Republic of China
[3] XJTLU Laboratory for Intelligent Computation and Financial Technology,
Suzhou 215123, Jiangsu, People's Republic of China
{Gang.Liu,Qiang.Niu}@xjtlu.edu.cn

Abstract. In this paper we consider the classical Kaczmarz algorithm for solving system of linear equations. Based on the geometric relationship between the error vector and rows of the coefficient matrix, we derive the optimal strategy of selecting rows at each step of the algorithm for solving consistent system of linear equations. For solving perturbed system of linear equations, a new upper bound in the convergence rate of the randomized Kaczmarz algorithm is obtained.

Keywords: Iterative methods · Kaczmarz method · Convergence rate · Orthogonal projection · Linear systems

1 Introduction

Kaczmarz algorithm [28] is an iterative method for solving system of linear equations of the form

$$Ax = b, \tag{1}$$

where $A \in \mathcal{R}^{m \times n}$ has full column rank, $m \geq n$ and $b \in \mathcal{R}^m$. In the consistent case, the solution of (1) can be regarded as the coordinate of the common point of hyperplanes defined by each single equation in (1):

$$\mathcal{P}_i = \{x | a_i^T x = b_i\}, \tag{2}$$

where a_i^T, $i = 1, 2, \cdots, m$, denotes the ith row of A and b_i is the ith element of vector b.

This work was supported by XJTLU research enhancement fund with no. REF-18-01-04 and Key Programme Special Fund (KSF) in XJTLU with nos. KSF-E-32, KSF-P-02 and KSF-E-21. Partial of the work was supported by the Qing Lan project of Jiangsu Province.

© Springer Nature Switzerland AG 2020
V. V. Krzhizhanovskaya et al. (Eds.): ICCS 2020, LNCS 12138, pp. 225–236, 2020.
https://doi.org/10.1007/978-3-030-50417-5_17

The idea of the Kaczmarz type algorithms is to exploit the geometric structure of the problem (1), and the using a sequential of projections to seek the solution. The recursive process can be formulated as follows. Let x_0 be an initial guess to the solution of (1), then the classical Kaczmarz algorithm iteratively generates a sequence of approximate solutions x_k by the recursive formula:

$$x_{k+1} = x_k + \frac{b_i - a_i^T x_k}{||a_i||_2^2} a_i, \tag{3}$$

where $i = mod(k, m) + 1$. For a given x_k, from (3) we can see that x_{k+1} satisfies the ith equation in (1), i.e., $a_i^T x_{k+1} = b_i$. The updating formula (3) implicitly produces a solution to the following constraint optimization problem [21, 37]

$$\min_{\{x | a_i^T x_{k+1} = b\}} ||x - x_k||_2,$$

which is equivalent to finding the projection of x_k from the hyperplane $\mathcal{P}_i$. Two geometric explanations of the above process can be illustrated by Fig. 1:

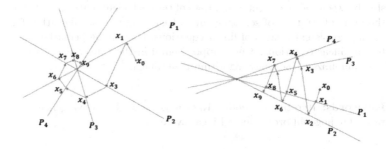

Fig. 1. Geometric illustrations of the classical Kaczmarz iterations with $m = 4$.

By comparing the projection processes displayed in Fig. 1, it is natural to have the intuition that convergence of the classical Kaczmarz algorithm highly depends on the geometric positions of the associated hyperplanes. If the normal vectors of every two successive hyperplanes keep reasonably large angles, the convergence of the classical Kaczmarz algorithm will be fast, whereas two nearly parallel consecutive hyperplanes will make the convergence slow down. The Kaczmarz algorithm can be regarded as a special application of famous von Neumann's alternating projection [35] originally distributed in 1933. The fundamental idea can even trace the history back to Schwarz [38] in 1870s.

In the past few years, the Karzmarz algorithm has been interpreted as successive projection methods [4, 7, 8, 11–13], which are also known as projection onto convex sets (POCS) [9, 17, 18, 42–44] in the optimization community. Notice that each iteration of the Kaczmarz algorithm just need $\mathcal{O}(n)$ flops and the cost is independent with the number of equations, this type of algorithms are well-suited to problems with $m \gg n$. Due to its simplicity and generality, Kaczmarz algorithms find viable applications in the area of image processing and signal process [19, 20, 24–26, 30, 36] under the name of algebraic reconstruction techniques (ART). Since 1980s, relaxation variants [11, 25, 41]

$$x_{k+1} = x_k + \lambda_k \frac{b_i - a_i^T x_k}{||a_i||_2^2} a_i, \tag{4}$$

and the block versions [3, 33, 34]

$$x_{k+1} = x_k + A_\tau^\dagger (b_\tau - A_\tau^T x_k), \; with \; A = \begin{pmatrix} A_1 \\ A_2 \\ \vdots \\ A_M \end{pmatrix}, b = \begin{pmatrix} b_1 \\ b_2 \\ \vdots \\ b_M \end{pmatrix}, \tau \in \{1, 2, \cdots, M\}, \tag{5}$$

of the Kaczmarz algorithm have been widely investigated, and some fruitful theoretical results have been obtained. In particular, for consistent linear systems, it is shown [5, 21, 31, 39] that the Kaczmarz iterations converges to the least square norm solution $x = A^\dagger b$ with any starting vector x_0 in the column space of A^T. For inconsistent linear systems, the cyclic subsequences generated by the Kaczmarz algorithm converges to a weighted least squares solution when the relaxation parameter λ_k goes to zero [12].

As indicated in Fig. 1, convergence of the classical Kaczmarz algorithm depends on the sequence of successive projections, which relies upon the ordering of the rows in the matrix A. In some real applications, it is observed [25, 30] that instead of selecting rows of the matrix A sequentially at each step of the Kaczmarz algorithm, randomly selection can often improve its convergence. Recently, in the remarkable paper [39], Strohmer and Vershynin proved the rate of convergence for the following randomized Kaczmarz algorithm

$$x_{k+1} = x_k + \frac{b_{r(i)} - a_{r(i)}^T x_k}{||a_{r(i)}||_2^2} a_{r(i)}$$

where $r(i)$ is chosen from $\{1, 2, \cdots, m\}$ with probabilities $\frac{||a_{r(i)}||_2^2}{||A||_F^2}$. In particular, the following bound on the expected rate of convergence for the randomized Kaczmarz method is proved

$$\mathbb{E}||x_k - x||_2^2 \leq (1 - \frac{1}{\kappa(A)^2})^k ||x_0 - x||_2^2, \tag{6}$$

where $\kappa(A) = ||A||_F ||A^{-1}||_2$, with $||A^{-1}||_2 = \inf\{M : M||Ax||_2 \geq ||x||_2\}$ be the scaled conditioned number of A introduced by J. Demmel [14]. Due to this pioneering work that characterized the convergence rate for the randomized Kaczmarz algorithms, the idea stimulated considerable interest in this area and various investigations [1, 2, 6, 10, 15] have been performed recently. In particular, some acceleration strategies have been proposed [6, 16, 22] and convergence analysis was performed in [21, 23, 27, 29, 31, 32]. See also [21, 23] for some comments on equivalent interpretations of the randomized Kaczmarz algorithms.

2   Optimal Row Selecting Strategy of the Kaczmarz Algorithm for Solving Consistent System of Linear Equations

In this section, we consider the case that system of linear equations (1) is consistent and x is a solution. If the ith row is selected at the $(k+1)$th iteration of the Kaczmarz algorithm, i.e.,

$$x_{k+1} = x_k + \frac{b_i - a_i^T x_k}{||a_i||_2^2} a_i,$$

then x_{k+1} can be reformulated as

$$\begin{aligned}
x_{k+1} &= x_k + \frac{b_i - a_i^T x_k}{||a_i||_2^2} a_i \\
&= x_k + \frac{b_i}{||a_i||_2^2} a_i - \frac{x_k^T a_i}{||a_i||_2^2} a_i \\
&= x_k + \frac{a_i^T x}{||a_i||_2^2} a_i - \frac{a_i^T x_k}{||a_i||_2^2} a_i \\
&= x_k + \frac{a_i^T (x - x_k)}{||a_i||_2^2} a_i \\
&= x_k + \frac{a_i a_i^T}{||a_i||_2^2} (x - x_k).
\end{aligned}$$

It follows that

$$\begin{aligned}
x - x_{k+1} &= x - x_k - \frac{a_i a_i^T}{||a_i||_2^2}(x - x_k) \\
&= (I - \frac{a_i a_i^T}{||a_i||_2^2})(x - x_k)
\end{aligned} \tag{7}$$

and thus

$$x_{k+1} - x_k = \frac{a_i a_i^T}{||a_i||_2^2}(x - x_k). \tag{8}$$

From (7) and (8), we can see that

$$x - x_{k+1} \perp x_{k+1} - x_k, \tag{9}$$

i.e.,

$$x - x_{k+1} \perp a_i. \tag{10}$$

To this end, let us make the following orthogonal direct sum decomposition $x - x_k$,

$$x - x_k = \alpha \hat{a}_i + \beta \hat{a}_i^\perp, \tag{11}$$

where $\hat{a}_i = \frac{a_i}{||a_i||_2}$ and $\hat{a}_i^\perp$ is a normalized vector orthogonal to a_i. Then coefficients α and β can be written as

$$\alpha = ||x - x_k||_2 \cos\theta_{k_i},$$

$$\beta = ||x - x_k||_2 \sin\theta_{k_i},$$

where $\theta_{k_i} = \angle(x - x_k, a_i)$ is the angle between the vectors $(x - x_k)$ and a_i.

Substituting the above decomposition (11) into (7) gives

$$x - x_{k+1} = (I - \frac{a_i a_i^T}{||a_i||_2^2})(\alpha \hat{a}_i + \beta \hat{a}_i^\perp)$$
$$= \beta \hat{a}_i^\perp \qquad (12)$$
$$= ||x - x_k||_2 \sin \theta_{k_i} \hat{a}_i^\perp .$$

It follows that

$$||x - x_{k+1}||_2 = ||x - x_k||_2 \cdot |\sin \theta_{k_i}| . \qquad (13)$$

From (13) we can see that the error norms generated by the Kaczmarz algorithm are monotonically nonincreasing. Moreover, the convergence can be optimized if $|\sin \theta_{k_i}|$ is minimized at every iteration, which is equivalent to selecting the row a_i that solves the optimization problem

$$|\sin \angle (x - x_k, a_i)| = \min_j |\sin \angle (x - x_k, a_j)| .$$

As x is the unknown solution, the above minimization problems seems unsolvable. However, noting that consistent linear system (1) implies

$$a_j^T x = b_j, \ j = 1, 2, \cdots, m$$

and x_k is fixed at the $(k+1)$th iteration. The minimization problem can be tackled by maximizing $|\cos \angle (x - x_k, a_j)|$, i.e.,

$$|\cos \angle (x - x_k, a_j)| = \frac{|a_j^T (x - x_k)|}{||x - x_k||_2 ||a_j||_2}$$
$$= \frac{|b_j - a_j^T x_k|}{||x - x_k||_2 ||a_j||_2} \qquad (14)$$
$$= \frac{|r_k(j)|}{||x - x_k||_2 ||a_j||_2} ,$$

where $r_k = b - A x_k = \left(r_k(1), r_k(2), \cdots, r_k(m) \right)^T$.

It is clear from (14) that the optimal updating strategy for the Kaczmarz algorithm is to select the row $\hat{i}$ that satisfies

$$|b_{\hat{i}} - a_{\hat{i}}^T x_k| = \max_j |b_j - a_j^T x_k| = ||b - A x_k||_\infty ,$$

i.e., the index where r_k has the largest entry in absolute value. We refer to the above row selection method as the *optimal selecting strategy*, and call the Kaczmarz algorithm with the optimal selecting strategy as the *optimal Kaczmarz algorithm*.

Next, we analyze the convergence of the optimal Kaczmarz algorithm for solving consistent system of linear equations. To simplify the analysis, we introduce two notations

$$\theta_k^{\hat{i}} = \min_j \angle (x_k - x, a_j),$$

and
$$\theta_p^{\hat{i}} = \max_k \theta_k^{\hat{i}},$$

where $1 \le \hat{i} \le m$ and $1 \le p \le k$.

Based on (13), the $(k+1)$th error can be bounded as follows

$$
\begin{aligned}
||x - x_{k+1}||_2 &= ||x - x_k||_2 \cdot |\sin \theta_k^{\hat{i}}| \\
&= ||x - x_0||_2 \cdot |\sin \theta_k^{\hat{i}}| \cdot |\sin \theta_{k-1}^{\hat{i}}| \cdots |\sin \theta_0^{\hat{i}}| \\
&\le ||x - x_0||_2 \cdot |\sin \theta_p^{\hat{i}}|^k,
\end{aligned}
\tag{15}
$$

where $1 \le p \le k$.

Notice that
$$0 \le \sin \theta_p^{\hat{i}} \le 1,$$

we can theoretically divide the convergence history of the Kaczmarz algorithm into two periods:

- when $\sin \theta_p^{\hat{i}} < 1$, the algorithm converge exponentially,
- when $\sin \theta_p^{\hat{i}} = 1$, we have

$$\max_j a_j^T (x_p - x) = 0$$

and thus,
$$a_j^T (x_p - x) = 0, \quad j = 1, 2, \cdots, m.$$

This implies that $Ax_p = b$, i.e., x_p solves the system of linear equation (1).

In summary, for solving consistent system of linear equations (1), there exists a theoretical optimal selecting strategy or optimal randomization strategy for Kaczmarz algorithm. With the strategy, the algorithm converges exponentially and will achieve convergence when

$$\max_k \min_{1 \le j \le m} \angle(x_k - x, a_j) = \frac{\pi}{2}.$$

3   Randomized Kaczmarz Algorithm for Solving Inconsistent System of Linear Equations

Suppose (1) is a consistent system of linear equations and its right hand side is perturbed with a noise vector r as follows:

$$Ax \simeq b + r,\tag{16}$$

where (16) can be either consistent or inconsistent. In this section, we give some remarks on the convergence of randomized Kaczmarz algorithm for solving (16), which was investigated by D. Needell [32].

Firstly, we recall the Lemma 2.2 in [32].

Lemma 1. *Let H_i be the affine subspaces of $\mathcal{R}^n$ consisting of the solutions to unperturbed equations, $H_i = \{x \mid \langle a_i, x \rangle = b_i\}$. Let $\tilde{H}_i$ be the solution spaces of the noisy equations, $\tilde{H}_i = \{x \mid \langle a_i, x \rangle = b_i + r_i\}$. Then*

$$\tilde{H}_i = \{w + \alpha_i a_i \mid w \in H_i\}$$

where $\alpha_i = \frac{r_i}{||a_i||_2^2}$.

Remarks: If the Lemma 1 is used to interpret the Kaczmarz algorithm for solving the perturbed and unperturbed equations, we need to introduce a vector $a_i^{\perp}$ in the as the orthogonal complement of the vector a_i, and write $\tilde{x}_i \in \tilde{H}_i$ as

$$\tilde{x}_i = x_i + \alpha_i a_i + \beta v_i$$

where x_i is a solution generated by Kaczmarz algorithm for solving the unperturbed equations, and v_i is a vector in the orthogonal complement of a_i.

Example 1. Consider the 2×2 system of linear equations

$$\begin{cases} x_1 + x_2 = 1, \\ x_1 - x_2 = 1, \end{cases}$$

and the perturbed equations

$$\begin{cases} x_1 + x_2 = 1.5, \\ x_1 - x_2 = 1.5, \end{cases}$$

i.e., $A = \begin{pmatrix} 1 & 1 \\ 1 & -1 \end{pmatrix}$, $b = \begin{pmatrix} 1 \\ 1 \end{pmatrix}$ and $r = \begin{pmatrix} 0.5 \\ 0.5 \end{pmatrix}$.

Let

$$H_i \doteq \{x \mid \langle a_i, x \rangle = b_i\}$$

and

$$\tilde{H}_i \doteq \{\tilde{x} \mid \langle a_i, \tilde{x} \rangle = b_i + r_i\}.$$

If we use $x_0 = \begin{pmatrix} 1 \\ 0 \end{pmatrix}$ as the same initial guess for the perturbed and unperturbed linear system, then

$$H_1 = \{ \begin{pmatrix} 1 \\ 0 \end{pmatrix} + \xi \begin{pmatrix} -1 \\ 1 \end{pmatrix} \mid \xi \in \mathcal{R}\}$$

and

$$\tilde{H}_1 = \{ \begin{pmatrix} 1.5 \\ 0 \end{pmatrix} + \xi \begin{pmatrix} -1 \\ 1 \end{pmatrix} \mid \xi \in \mathcal{R}\}$$

Note that $a_1 = \begin{pmatrix} 1 \\ 1 \end{pmatrix}$, $||a_1||_2^2 = 2$ and $r_1 = \frac{1}{2}$. We have

$$\begin{pmatrix} 1.5 \\ 0 \end{pmatrix} = \begin{pmatrix} 1 \\ 0 \end{pmatrix} + \frac{1}{4} \begin{pmatrix} 1 \\ 1 \end{pmatrix} + \frac{1}{4} \begin{pmatrix} 1 \\ -1 \end{pmatrix},$$

i.e.,

$$\tilde{x}_1 = x_1 + \frac{r_1}{||a_1||_2^2} a_1 + \frac{r_1}{||a_1||_2^2} a_1^{\perp}.$$

In order to derive the convergence rate of randomized Kaczmarz algorithm for solving the perturbed linear equations (16), we need to make use of the established convergence results [39] for the unperturbed linear system (1), together with the relationship between the approximate solutions generated by the Kaczmarz algorithm [39] for perturbed and unperturbed linear equations. In [32], D. Needell analyzed the convergence rate and error bound of the randomized Kaczmarz algorithm for solving the perturbed linear equations, in which the author take the approximate solution to the perturbed linear equations as the guess for the unperturbed system, which make the derivation process simplified. However, the approximate solutions generated by applying the randomized Kaczmarz algorithm to the perturbed linear system may not converge to the solution of the unperturbed linear system.

In what follows, we will consider the convergence rate of the randomized Kaczmarz algorithm for solving (16) from a different perspective. We try to bound the difference between the solution for the unperturbed linear system (1) and approximate solutions generated by applying the randomized Kaczmarz algorithm to the perturbed linear system.

In the following discussion, we use x_k and $\tilde{x}_k$ to denote the approximate solutions generated by applying the randomized Kaczmarz algorithm to (1) and (16), respectively. The recursive formulas can be written as

$$x_{k+1} = x_k + \frac{b_{i_k} - x_k^T a_{i_k}}{||a_{i_k}||_2^2} a_{i_k} \tag{17}$$

and

$$\tilde{x}_{k+1} = \tilde{x}_k + \frac{b_{i_k} + r_{i_k} - \tilde{x}_k^T a_{i_k}}{||a_{i_k}||_2^2} a_{i_k}, \tag{18}$$

where the subscript $i_k \in \{1, 2, \cdots, m\}$ is used to denote that the i_kth row is selected with probability $\frac{||a_{i_k}||_2^2}{||A||_F^2}$ at the kth iteration.

Suppose the same initial guess $x_0 = \tilde{x}_0$ is used as the starting vector. Then

$$\tilde{x}_1 = \tilde{x}_0 + \frac{b_{i_0} + r_{i_0} - \tilde{x}_0^T a_{i_0}}{||a_{i_0}||_2^2} a_{i_0}$$

and potentially

$$x_1 = x_0 + \frac{b_{i_0} - x_k^T a_{i_0}}{||a_{i_0}||_2^2} a_{i_0}.$$

It follows that

$$\tilde{x}_1 = x_1 + \frac{r_{i_0} a_{i_0}}{||a_{i_0}||_2^2}. \tag{19}$$

In the next iteration, we have

$$
\begin{aligned}
\tilde{x}_2 &= \tilde{x}_1 + \frac{b_{i_1} + r_{i_1} - \tilde{x}_1^T a_{i_1}}{||a_{i_1}||_2^2} a_{i_1} \\
&= x_1 + \frac{r_{i_0} a_{i_0}}{||a_{i_0}||_2^2} + \frac{b_{i_1} - (x_1 + \frac{r_{i_0} a_{i_0}}{||a_{r_0}||_2^2})^T a_{i_1}}{||a_{i_1}||_2^2} a_{i_1} + \frac{r_{i_1} a_{i_1}}{||a_{r_1}||_2^2} \\
&= x_1 + \underbrace{\frac{b_{i_1} - x_1^T a_{i_1}}{||a_{i_1}||_2^2} a_{i_1}}_{x_2} + \frac{r_{i_1} a_{i_1}}{||a_{i_1}||_2^2} + \underbrace{(I - \frac{a_{i_1} a_{i_1}^T}{||a_{i_1}||_2^2}) \frac{r_{i_0} a_{i_0}}{||a_{i_0}||_2^2}}_{a_{i_1}^\perp} \\
&= x_2 + \frac{r_{i_1} a_{i_1}}{||a_{i_1}||_2^2} + v_{i_1}
\end{aligned}
$$

where $v_{i_1} = (I - \frac{a_{i_1} a_{i_1}^T}{||a_{i_1}||_2^2}) \frac{r_{i_0} a_{i_0}}{||a_{i_0}||_2^2} \in span\{a_{i_1}\}^\perp$ with $||v_{i_1}||_2 = \frac{|r_{i_0}|}{||a_{i_0}||_2}$.
Continue the above process, we have

$$
\tilde{x}_k = x_k + \frac{r_{i_{k-1}}}{||a_{i_{k-1}}||_2^2} a_{i_{k-1}} + \sum_{j=1}^{k-2} v_{i_j}, \tag{20}
$$

where $v_{i_j} = (I - \frac{a_{i_j} a_{i_j}^T}{||a_{i_j}||_2^2}) \frac{r_{i_{j-1}} a_{i_{j-1}}}{||a_{i_{j-1}}||_2^2} \in span\{a_{i_j}\}^\perp$ and $||v_{i_j}||_2 = \frac{|r_{i_{j-1}}|}{||a_{i_{j-1}}||_2}$.
Subtracting x on both sides of (20) gives

$$
\tilde{x}_k - x = x_k - x + \frac{r_{i_{k-1}} a_{i_{k-1}}}{||a_{i_{k-1}}||_2^2} + \sum_{j=1}^{k-2} v_{i_j}. \tag{21}
$$

Based on Jensen's inequality and (6), we have

$$
\mathbb{E}||x_k - x||_2 \le (1 - \frac{1}{\kappa(A)^2})^{\frac{k}{2}} ||x_0 - x||_2, \tag{22}
$$

where $\kappa(A) = ||A||_F ||A^{-1}||_2$, with $||A^{-1}||_2 = \inf\{M : M||Ax||_2 \ge ||x||_2\}$.
Taking norm on both sides of (21) and using triangle inequality, we have

$$
\begin{aligned}
\mathbb{E}(||\tilde{x}_k - x||_2) &\le \mathbb{E}(||x_k - x||_2) + ||\frac{r_{i_{k-1}} a_{i_{k-1}}}{||a_{i_{k-1}}||_2^2}||_2 + \sum_{j=1}^{k-2} ||v_{i_j}||_2 \\
&\le (1 - \frac{1}{\kappa(A)^2})^{\frac{k}{2}} ||x_0 - x||_2 + \sum_{j=1}^{k-1} ||v_{i_j}||_2 \\
&= (1 - \frac{1}{\kappa(A)^2})^{\frac{k}{2}} ||x_0 - x||_2 + \sum_{j=1}^{k-1} \frac{|r_{i_j}|}{||a_{i_j}||_2} \\
&= (1 - \frac{1}{\kappa(A)^2})^{\frac{k}{2}} ||x_0 - x||_2 + (k-1)\gamma
\end{aligned}
$$

where $\gamma = \max\limits_{1 \le i \le m} \frac{|r_i|}{||a_i||_2}$.
In conclusion, we have derived the following theorem.

Theorem 1. *Let A be a matrix full column rank and assume the system $Ax = b$ is consistent. Let $\tilde{x}_k$ be the kth iterate of the noisy randomized Kaczmarz method run with $Ax \simeq b + r$, and let $a_1, \cdots, a_m$ denote the rows of A. Then we have*

$$\mathbb{E}||\tilde{x}_k - x||_2 \le (1 - \frac{1}{\kappa(A)^2})^{\frac{k}{2}}||x_0 - x||_2 + (k-1)\gamma,$$

where $\kappa(A) = ||A||_F ||A^{-1}||_2$ and $\gamma = \max\limits_{1 \le i \le m} \frac{|r_i|}{||a_i||_2}$.

4   Conclusions

In this paper, we provide a new look at the Kaczmarz algorithm for solving system of linear equations. The optimal row selecting strategy of the Kaczmarz algorithm for solving consistent system of linear equations is derived. The convergence of the randomized Kaczmarz algorithm for solving perturbed system of linear equations is analyzed and a new bound of the convergence rate is obtained from a new perspective.

References

1. Agaskar, A., Wang, C., Lu, Y.M.: Randomized Kaczmarz algorithms: exact MSE analysis and optimal sampling probabilities. In: Proceedings of the 2014 IEEE Global Conference on Signal and Information Processing (GlobalSIP), Atlanta, GA, 3–5 December, pp. 389–393 (2014)
2. Ivanov, A.A., Zhdanov, A.I.: Kaczmarz algorithm for Tikhonov regularization problem. Appl. Math. E-Notes **13**, 270–276 (2013)
3. Ivanovy, A.A., Zhdanovz, A.I.: The block Kaczmarz algorithm based on solving linear systems with arrowhead matrices. Appl. Math. E-Notes **17**, 142–156 (2017)
4. Bai, Z.-Z., Liu, X.-G.: On the Meany inequality with applications to convergence ananlysis of several row-action methods. Numer. Math. **124**, 215–236 (2013)
5. Bai, Z.-Z., Wu, W.-T.: On greedy randomized Kaczmarz method for solving large sparse linear systems. SIAM J. Sci. Comput. **40**, A592–A606 (2018)
6. Bai, Z.-Z., Wu, W.-T.: On relaxed greedy randomized Kaczmarz methods for solving large sparse linear systems. Appl. Math. Lett. **83**, 21–26 (2018)
7. Benzi, M., Meyer, C.D.: A direct projection method for sparse linear systems. SIAM J. Sci. Comput. **16**, 1159–1176 (1995)
8. Brenzinski, C.: Projection Methods for Systems of Equations. Elsevier Science B.V., Amsterdam (1997)
9. Benzi, M., Meyer, C.D.: The relaxation method of finding the common point of convex sets and its application to the solution of problems in convex programming. USSR Comput. Math. Math. Phys. **7**, 200–217 (1967)
10. Cai, J., Tang, Y.: A new randomized Kaczmarz based kernel canonical correlation analysis algorithm with applications to information retrieval. Neural Netw. **98**, 178–191 (2018)
11. Censor, Y.: Row-action methods for huge and sparse systems and their applications. SIAM Rev. **23**, 444–466 (1981)

12. Censor, Y., Eggermont, P.P.B., Gordon, D.: Strong underrelaxation in Kaczmarz's method for inconsistent systems. Numer. Math. **41**, 83–92 (1983)
13. Censor, Y., Herman, G.T., Jiang, M.: A note on the behaviour of the randomized Kaczmarz algorithm of Strohmer and Vershynin. J. Fourier Anal. Appl. **15**, 431–436 (2009)
14. Demmel, J.: The probability that a numerical analysis problem is difficult. Math. Comput. **50**, 449–480 (1988)
15. De Loera, J.A., Haddock, J., Needell, D.: A sampling Kaczmarz-Motzkin algorithm for linear feasibility. SIAM J. Sci. Comput. **39**, S66–S87 (2017)
16. Eldar, Y.C., Needell, D.: Acceleration of randomized Kaczmarzmethod via the Johnson–Lindenstrauss lemma. Numer. Algorithms **58**, 163–177 (2011)
17. Feichtinger, H.G., Cenker, C., Mayer, M., Steier, H., Strohmer, T.: New variants of the POCS method using affine subspaces of finite codimension with applications to irregular sampling. In: VCIP. SPIE, pp. 299–310 (1992)
18. Galántai, A.: On the rate of convergence of the alternating projection method in finite dimensional spaces. J. Math. Anal. Appl. **310**, 30–44 (2005)
19. Gordon, R., Bender, R., Herman, G.T.: Algebraic Reconstruction Techniques (ART) for threedimensional electron microscopy and x-ray photography. J. Theor. Biol. **29**, 471–481 (1970)
20. Gordon, R., Herman, G.T., Johnson, S.A.: Image reconstruction from projections. Sci. Am. **233**, 56–71 (1975)
21. Gower, R.M., Richtárik, P.: Randomized iterative methods for linear systems. SIAM J. Matrix Anal. **36**, 1660–1690 (2015)
22. Hanke, M., Niethammer, W.: On the acceleration of Kaczmarz's method for inconsistent linear systems. Linear Algebra Appl. **130**, 83–98 (1990)
23. Hefny, A., Needell, D., Ramdas, A.: Rows vs. columns: randomized Kaczmarz or gauss-seidel for ridge regression. SIAM J. Sci. Comput. **39**, S528–S542 (2016)
24. Herman, G.T.: Image Reconstruction from Projection, the Fundamentals of the Computerized Tomography. Academic Press, New York (1980)
25. Herman, G.T., Lent, A., Lutz, P.H.: Relaxation methods for image reconstruction. Commun. Assoc. Comput. Mach. **21**, 152–158 (1978)
26. Herman, G.T., Meyer, L.B.: Algebraic reconstruction techniques can be made computationally efficient. IEEE Trans. Med. Imaging **12**, 600–609 (1993)
27. Jiao, Y.-L., Jin, B.-T., Lu, X.-L.: Preasymptotic convergence of randomized Kaczmarz method. Inverse Prob. **33**, 125012 (2017)
28. Kaczmarz, S.: Angenäherte auflösung von systemen linearer gleichungen. Bull. Acad. Polon. Sci. Lett. A **35**, 355–357 (1933)
29. Leventhal, L., Lewis, A.S.: Randomized methods for linear constraints: convergence rates and conditioning. Math. Oper. Res. **35**, 641–654 (2010)
30. Natterer, F.: Themathematics of Computerized Tomography. SIAM (2001)
31. Ma, A., Needell, D., Ramdas, A.: Convergence properties of the randomized extended Gauss-Seidel and Kaczmarz methods. SIAM J. Matrix Anal. Appl. **36**, 1590–1604 (2015)
32. Needell, D.: Randomized Kaczmarz solver for noisy linear systems. BIT **50**, 395–403 (2010)
33. Needell, D., Zhao, R., Zouzias, A.: Randomized block Kaczmarz method with projection for solving least squares. Linear Algebra Appl. **484**, 322–343 (2015)
34. Needell, D., Tropp, J.A.: Paved with good intentions: analysis of a randomized block Kaczmarz method. Linear Algebra Appl. **441**, 199–221 (2014)

35. von Neumann, J.: The Geometry of Orthogonal Spaces, vol. 2. Princeton University Press, Princeton (1950). This is a mimeographed lecture notes, first distributed in 1933
36. Nutini, J., Sepehry, B., Laradji, I., Schmidt, M., Koepke, H., Virani, A.: Convergence rates for greedy Kaczmarz algorithms, and faster randomized Kaczmarz rules using the orthogonality graph, UAI (2016)
37. Schmidt, M.: Notes on randomized Kaczmarz, Lecture notes, 9 April 2015
38. Schwarz, H.A.: Ueber einen Grenzübergang durch alternirendes Verfahren. Vierteljahrsschrift der Naturforschenden Gessellschaft in Zurich 15, 272–286 (1870)
39. Strohmer, T., Vershynin, R.: A randomized Kaczmarz algorithm with exponential convergence. J. Fourier Anal. Appl. 15, 262–278 (2009)
40. Strohmer, T.: Comments on the randomized Kaczmarz method (2009, unpublished manuscript)
41. Tanabe, K.: Projection method for solving a singular system of linear equations and its applications. Numer. Math. 17, 203–214 (1971)
42. Trussell, H., Civanlar, M.: Signal deconvolution by projection onto convex sets. In: IEEE International Conference on Acoustics, Speech, and Signal Processing, ICASSP 1984, vol. 9, pp. 496–499 (1984)
43. Youla, D.C., Webb, H.: Image restoration by the method of convex projections: part 1 theory. IEEE Trans. Med. Imaging 1, 81–94 (1982)
44. Youla, D.C., Webb, H.: Image restoration by the method of convex projections: part 2 applications and numerical results. IEEE Trans. Med. Imaging 1, 95–101 (1982)
45. Zouzias, A., Freris, N.M.: Randomized extended Kaczmarz for solving least squares. SIAM J. Matrix Anal. 34, 773–793 (2013)

Investigating the Benefit of FP16-Enabled Mixed-Precision Solvers for Symmetric Positive Definite Matrices Using GPUs

Ahmad Abdelfattah[1]($\boxtimes$), Stan Tomov[1], and Jack Dongarra[1,2,3]

[1] University of Tennessee, Knoxville, USA
{ahmad,tomov,dongarra}@icl.utk.edu
[2] Oak Ridge National Laboratory, Oak Ridge, USA
[3] University of Manchester, Manchester, UK

Abstract. Half-precision computation refers to performing floating-point operations in a 16-bit format. While half-precision has been driven largely by machine learning applications, recent algorithmic advances in numerical linear algebra have discovered beneficial use cases for half precision in accelerating the solution of linear systems of equations at higher precisions. In this paper, we present a high-performance, mixed-precision linear solver ($Ax = b$) for symmetric positive definite systems in double-precision using graphics processing units (GPUs). The solver is based on a mixed-precision Cholesky factorization that utilizes the high-performance tensor core units in CUDA-enabled GPUs. Since the Cholesky factors are affected by the low precision, an iterative refinement (IR) solver is required to recover the solution back to double-precision accuracy. Two different types of IR solvers are discussed on a wide range of test matrices. A preprocessing step is also developed, which scales and shifts the matrix, if necessary, in order to preserve its positive-definiteness in lower precisions. Our experiments on the V100 GPU show that performance speedups are up to 4.7× against a direct double-precision solver. However, matrix properties such as the condition number and the eigenvalue distribution can affect the convergence rate, which would consequently affect the overall performance.

Keywords: Mixed-precision solvers · Half-precision · GPU computing

1 Introduction

The solution of a dense linear system of equations ($Ax = b$) is a critical component in many scientific applications. The standard way of solving such systems includes two steps: a matrix factorization step and a triangular solve step. In this paper, we discuss the specific case where the matrix $A_{N \times N}$ is dense and symmetric positive definite (SPD). It is also assumed that A, b, and x are stored in 64-bit double precision format (FP64).

© Springer Nature Switzerland AG 2020
V. V. Krzhizhanovskaya et al. (Eds.): ICCS 2020, LNCS 12138, pp. 237–250, 2020.
https://doi.org/10.1007/978-3-030-50417-5_18

The standard LAPACK software [1] provides the dposv routine for solving $Ax = b$ for SPD systems in FP64. The routine starts with a *Cholesky factorization* (dpotrf) of A, such that $A = LL^T$, where L is a lower triangular matrix. The factors are used to find the solution x using two triangular solves with respect to b (dpotrs). Throughout the paper, we assume that b is an $N \times 1$ vector, and so the triangular solve step requires $\mathcal{O}(N^2)$ floating-point operations (FLOPs). In such a case, the Cholesky factorization dominates the execution time, since it requires $\mathcal{O}(N^3)$ FLOPs. Therefore, any performance improvements for solving $Ax = b$ usually focus on improving the factorization performance.

A full FP64 factorization extracts its high performance from a blocked implementation that traverses the matrix in panels of width nb (which is often called the blocking size). A blocked design enables high performance through the compute-bound Level 3 BLAS[1] routines. Sufficiently optimized routines such as matrix multiplication (dgemm) and symmetric rank-k updates (dsyrk) would guarantee a high performance Cholesky factorization that is close to the hardware peak performance. As an example, both cuSOLVER [14] (the vendor library) and the MAGMA library [4,11] reach an asymptotic performance of $\approx$6.3 teraFLOP/s on the V100 GPU for dpotrf. This is about 90% of the dgemm peak performance, meaning that there is little room for improving the performance of the factorization. Another direction to achieve more performance is to change the algorithmic steps for solving $Ax = b$. This is where *mixed-precision iterative refinement (MP-IR) algorithms* come into play. The basic idea of MP-IR solvers is to perform the Cholesky factorization using a "reduced precision." If FP32 is used for the factorization instead of FP64, a natural 2× improvement is expected. However, we cannot use the traditional triangular solves with the low-precision factors of A. In order to recover the solution back to FP64 accuracy, an extra algorithmic component is required: *iterative refinement (IR)*. It applies iterative corrections to an initial solution vector until it converges to FP64 accuracy. Early efforts to implement such algorithms in LAPACK were introduced by Langou et al. [12], and Baboulin et al. [5]. GPU-accelerated versions of the MP-IR solver also exist in the MAGMA library [4,11].

The algorithmic structure of MP-IR solvers did not change for almost a decade. This was true until half precision (16-bit floating-point format) was introduced into commercial HPC hardware (e.g., NVIDIA GPUs). The original motivation for FP16 computation was to accelerate machine learning applications rather than scientific HPC workloads. NVIDIA GPUs support the "binary16" format which is defined by the IEEE-754 standard [2]. Intel and Google support a different format called "bfloat16". Since our study targets GPUs, we focus on the binary16 format, which we also call half precision or simply FP16. NVIDIA's Volta and Turing architectures provide hardware accelerators, called Tensor Cores (TCs), for gemm in FP16. TCs can also perform a mixed-precision gemm, by accepting operands in FP16 while accumulating the result in FP32. TCs are theoretically 4× faster than using the regular FP16 peak performance on the Volta GPU. Applications that take advantage of TCs have access to up to 125

---

[1] Basic Linear Algebra Subroutines.

teraFLOP/s of performance. The vendor library cuBLAS [13] provides a number of matrix multiplication routines that can take advantage of TCs. Some other efforts introduced open-source routines that are competitive with cuBLAS [3].

Such a high performance of half-precision has drawn the attention of the HPC community to assess its benefit for scientific HPC workloads. Originally motivated by the analysis of Carson and Higham [6,7], the work done by Haidar et al. [9] introduced a mixed-precision solver that is different in several ways from the ones introduced in [12] and [5]. **First**, the new method uses three precisions (double, single, and half) to solve $Ax = b$ up to double-precision accuracy. **Second**, the new solver uses a *mixed-precision LU factorization*, where the dominant trailing matrix updates are performed using a mixed-precision **gemm**. **Third**, the new solver uses a new IR algorithm based on the GMRES method, instead of the classic IR solver that is based on triangular solves. The GMRES-based IR uses the original matrix A preconditioned by its low-precision factors, which yields a faster convergence and thus a higher performance.

In this paper, we design a similar mixed-precision solver for SPD matrices. Technically, the LU factorization supports such matrices, but (1) its operation count is much higher than a Cholesky factorization, and (2) SPD matrices don't need pivoting, which is a plus for performance. We show that the developed solver works well with problems whose condition number $\kappa_\infty(A)$ is up to $\mathcal{O}(10^9)$. We also implement an optional preprocessing step that includes scaling and diagonal shifts. The preprocessing step, which is based on [10], protects the matrix from losing its definiteness when FP16 is used in the factorization. Therefore, it helps solve a wider range of problems. Our experiments are conducted on a Tesla V100 GPU and span a wide range of dense SPD matrices with different condition numbers and eigenvalue distributions. We show how these two properties affect the convergence rate of GMRES-based IR, which in turn affects the performance. Our results show that the developed solution can be up to 4.7× faster than a direct full FP64 solver. This work is lined up for integration into the MAGMA library [4,11].

2   Background and Related Work

Classic MP-IR solvers for SPD systems used to perform the Cholesky factorization in single precision. The refinement phase iteratively updates the solution vector $\hat{x}$ until it is accurate enough. At each refinement iteration, three main steps are performed. **First**, the residual $r = b - Ax$ is computed in FP64. **Second**, we solve for the correction vector c, such that $Ac = r$. This step uses the low precision factors of A. **Finally**, the solution vector is updated $\hat{x}_{i+1} = \hat{x}_i + c$. Convergence is achieved when the residual is small enough.

A key factor for the high performance of MP-IR solvers is the number of iterations in the refinement stage. As mentioned before, a maximum of 2× speedup is expected from the factorization stage in FP32. This performance advantage can be completely gone if too many iterations are required for convergence. Typically, an MP-IR solver (FP32→FP64) requires 2–3 iterations for a well-conditioned problem. This is considered a *best case scenario*, since the asymptotic speedup approaches 2×, meaning a minimal overhead by the IR stage. In most cases, an MP-IR solver is asymptotically 1.8× faster than a full FP64 solver.

Using half precision in legacy MP-IR algorithms was mostly unsuccessful. Performing the factorization in FP16 further worsens the quality of the factors of A, which leads to a longer convergence or even a divergence. For SPD matrices, an FP16 factorization can fail due to the loss of definiteness during the conversion to FP16. While countermeasures have been proposed by Higham et al. [10], a more practical approach for high performance is possible. Similar to [9], we adopt a mixed-precision Cholesky factorization, in which the rank-k updates are performed using a mixed-precision `gemm` (FP16→FP32), while all other steps are performed in FP32. The quality of the Cholesky factors would be better than a full FP16 factorization. We also apply a slightly modified version of the preprocessing proposed by Higham et al. [10] in order to support matrices with higher condition numbers and avoid the loss of definiteness, overflow, and possibly underflow.

Now, considering the IR step, the low quality of the produced factors leads to the likely failure of the classic IR algorithm (e.g., following the classic mixed-precision solvers' convergence theory [12]). In fact, classic IR would only work for matrices with relatively small condition numbers, as we show later in Sect. 7. An alternative approach, which further improves the numerical stability and convergence of the overall solver, is to solve the correction equation ($Ac = r$) using an iterative method, such as GMRES [16]. The solver thus uses two nested refinement loops, which are also often referred to as "inner-outer" iterative solvers [15,17]. We call the new IR algorithm IRGMRES. The recent work by Carson and Higham [6,7] analyzes this type of solvers when three precisions are used (e.g., {FP16, FP32, FP64} or {FP16, FP64, FP128} for {factorization, working precision, residual precision}, respectively). They prove that, if a preconditioned GMRES is used to solve the correction equation, then forward and backward errors in the order of $10^{-8}/10^{-16}$ are achievable if the condition number of A satisfies $\kappa_\infty(A) < 10^8/10^{12}$, respectively. The work in [9] implements a simplified version of GMRES with just two precisions, typically using the working precision as the residual precision. By preconditioning GMRES using the low-precision factors of A, FP64 accuracy can be achieved for matrices with condition numbers up to 10^5. Our study expands upon this work for SPD matrices using a mixed-precision Cholesky factorization. Successful convergence is achieved for condition numbers up to 10^9. In addition, we study the behavior of both IR and IRGMRES for a wide range of SPD matrices, and show how the condition number and the eigenvalue distribution affect the convergence of the IRGMRES solver. Finally, we show that the modified version of the preprocessing steps proposed in [10] enable our solver to support harder problems that were not solvable otherwise (i.e., without preprocessing).

3   System Setup

All the experiments reported in this paper are conducted on a system with two Intel Broadwell CPUs (Intel Xeon CPU E5-2698 v4 @ 2.20 GHz), with 20 cores per CPU. The system has 512 GB of memory. The GPU is a Tesla V100-SXM2, with 80 multiprocessors clocked at 1.53 GHz. Our solver is developed as part of the MAGMA library, which is compiled using CUDA-10.1 and MKL-2018.0.1 for the CPU workloads. The number of MKL threads is set to 40 throughout all the experiments.

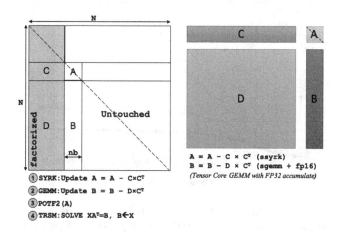

Fig. 1. Steps of a single iteration in the left-looking Cholesky factorization, as well as the mixed-precision update (syrk + gemm).

4   Mixed-Precision Cholesky Factorization

The first step in our solver is to obtain the Cholesky factorization $(A = LL^T)$. This step is expected to be much faster than a factorization in FP64 or FP32. The performance advantage obtained in this step serves as an upper bound for the speedup achieved by the whole solver. As mentioned before, we use an FP32 factorization that uses mixed-precision updates. Figure 1 shows the steps of the mixed-precision factorization. Both the potf2 and trsm steps are performed in FP32. We adopt the left-looking variant of the factorization, since it relies on gemm as the dominant operation in the update. The factorization is designed similarly to other factorizations in MAGMA. The panel step is performed on the CPU. This "hybrid execution" has the advantage of hiding the panel task on the CPU while the GPU is performing the update [18].

The sgemm updates are replaced by a call to a cuBLAS routine that performs an implicit FP32→FP16 conversion of the multiplicands, while accumulating the result in FP32. A tuning experiment was conducted to find the best blocking size

nb for the mixed-precision factorization. The details of the experiment are omitted for lack of space, but its final outcome suggests that setting $nb = 512$ achieves the best performance for the mixed-precision factorization. Figure 2 shows the performance of the mixed-precision Cholesky factorization (spotrf_fp16). The figure shows significant speedups against full-precision factorizations. In fact, the asymptotic speedup approaches $3\times$ against single precision, and $6\times$ compared to double precision. As mentioned before, we expect the IR phase to consume some of these performance gains.

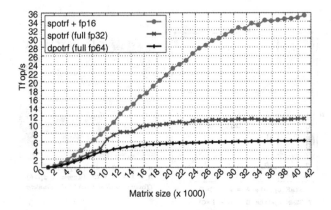

Fig. 2. Performance of the mixed-precision Cholesky factorization (spotrf_fp16) against full-precision factorizations in FP32 (spotrf) and FP64 (dpotrf). Results are shown on a Tesla V100-SXM2 GPU, and two 20-core Intel Broadwell CPUs.

5   GMRES-Based Iterative Refinement

The main difference between classic IR and GMRES-based IR is how the correction equation $Ac = r$ is solved. Classic IR solvers use a direct method using two triangular solves with respect to the Cholesky factors of A. This method works well for matrices with relatively small condition numbers. However, the quality of the correction vector is often impacted by the low-precision factors, which might lead to a long convergence. As mentioned in Sect. 2, it is important to keep the iteration count small in order to achieve an overall performance gain. The proposition by Carson and Higham [6, 7] was to use a GMRES solver to solve $Ac = r$. The solver uses the original matrix A preconditioned by its Cholesky factors. This produces a correction vector of a much higher quality than a classic IR, eventually leading to a faster convergence. As an example, Fig. 3 shows the convergence history of both the classic IR solver and GMRES-based one (IRGMRES) for two matrices of size $10k$. The matrices share the same distribution of eigenvalues, but have different condition numbers. Our observations are (1) IRGMRES usually converges faster than classic IR, and (2) IR fails to converge for relatively large condition numbers. However, the gap between IR and IRGMRES is not big for well-conditioned matrices. Both variants converge in few iterations, and so the final performance would be similar.

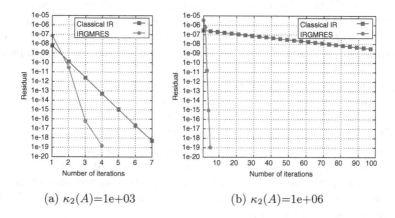

(a) $\kappa_2(A)$=1e+03   (b) $\kappa_2(A)$=1e+06

Fig. 3. Comparisons of the conversion history between IR and IRGMRES. The test matrix in both cases has a clustered distribution of eigenvalues ($\lambda_i = 1, 1, \cdots, \frac{1}{\kappa_2(A)}$).

It is worth mentioning that a conjugate gradient (CG) solver can be used instead of a GMRES solver. In fact, the study by Higham et al. [10] shows that both GMRES and CG converge within mostly similar iteration counts. However, the error analysis in [6,7] is based on the backward stability of GMRES. This means that a new error analysis is required for a CG-based IR solver, since its backward stability requires a well-conditioned matrix or a good preconditioner [8].

6   Scaling and Shifting

Higham et al. [10] proposed some countermeasures to ensure a successful factorization in FP16. The countermeasures avoid the loss of definiteness, overflow, and possibly underflow. In this study, the factorization uses two precisions (FP32 + FP16), so these countermeasures are still legitimate for our implementation. We also point out that the work done in [10] focuses only on the numerical analysis part, with no actual implementation on a high-performance hardware. Since our work focuses more on the performance, we are interested in determining the extent to which these safeguards ensure a successful factorization and convergence without too much impact on the performance. More specifically, our preprocessing works as follows:

1. **Two-sided diagonal scaling.** A lightweight GPU kernel computes the matrix $H = D^{-1} A_{fp32} D^{-1}$, where D is a diagonal matrix such that $D_{ii} = \sqrt{a_{ii}}, i = 1, \cdots, N$. This operation equilibrates the matrix rows and columns,

and reduces their range to $[0, 1]$. The multiplication by diagonal matrices can be simplified to a row-wise or a column-wise matrix scaling. Therefore, the GPU kernel is very lightweight with a nearly negligible execution time.

2. **An optional diagonal shift**. In order to avoid the loss of positive definiteness, the GPU kernel allows an optional small perturbation on the diagonal of H. Note that the diagonal of H is all ones. This step forms the matrix $G = H + cu_h I$, where u_h is the unit roundoff (machine epsilon) of FP16, and c is a constant parameter. The original proposition is to set c as a small positive integer constant. However, we show that this shift is sometimes unnecessary, and setting it anyway might affect the convergence of the GMRES solver. We also allow $c < 1$, since our shift occurs in FP32, where u_h is possibly a large shift to start with. We can shift by a fraction of u_h.

3. **Matrix scaling**. Finally, the entire matrix is scaled by μ, where $\mu = \frac{\theta x_{max}}{1 + cu_h}$. The constant x_{max} is 6.55×10^4. The constant θ is a parameter that is set to 0.1 in all of our experiments, but in general $\theta \in (0, 1)$. The purpose of this scaling operation is to make a better use of the half-precision range. This scaling step avoids overflow and reduces the chances of underflow. Further details can be found in [10].

All of these preprocessing steps are performed by one lightweight GPU kernel. The preprocessing step obviously implies modifications in other numerical steps. In an IRGMRES solver, the matrix A is preconditioned by the Cholesky factors. However, the action of the preconditioner on a vector is obtained by a triangular solve (similar to the classical IR), and then a matrix-vector multiplication with respect to A. Noting that $A = \frac{1}{\mu} DHD$, any triangular solve $(Ap = q)$ inside the GMRES solver now solves for y with respect to $D^{-1}q$ and then forms $p = \mu D^{-1}y$. Another GPU kernel that performs diagonal matrix-vector products has been developed for such a purpose.

Table 1. Eigenvalue distributions used in the test matrices.

Distribution Name	Specification ($i = 1, 2, \cdots, N$)
Arithmetic	$\lambda_i = 1 - (\frac{i-1}{N-1})(1 - \frac{1}{\kappa_2(A)})$
Clustered	$\lambda_1 = 1, \lambda_i = \frac{1}{\kappa_2(A)}$ for $i > 1$
Logarithmic	$\log(\lambda_i)$ uniform on $[\log(\frac{1}{\kappa_2(A)}), \log(1)]$
Geometric	$\lambda_i = \kappa_2(A)^{(\frac{1-i}{N-1})}$
Custom-clustered	$\lambda_i = 1$ for $i \leq \lfloor \frac{N}{10} \rfloor$, $\frac{1}{\kappa_2(A)}$ otherwise

7   Performance Results

Test Matrices and General Outlines. Our experiments use a matrix generator that is available in MAGMA, which is similar to the LAPACK routine dlatms. It generates random dense SPD matrices with (1) a specified 2-norm condition number $\kappa_2(A)$, and (2) a specified distribution of eigenvalues. The matrix is generated as the product $A = V\lambda V^T$, where λ is the diagonal matrix of eigenvalues and V is a random orthogonal matrix. Performance results are shown for matrices with different types of distributions and different condition numbers. Table 1 shows the distributions used in this paper.

Throughout this section, the performance is measured in tera FLOPs per second (teraFLOP/s). In order to have a fair comparison, a constant number of FLOPs for each matrix size is divided by the time-to-solution of each tested solver. That constant is equal to the operation count of a full FP64 solver, which is equal to $(\frac{N^3}{3} + \frac{5N^2}{2} + \frac{N}{6})$ for one right-hand side. Performance figures have the left Y-axis with a fixed maximum value of 30 teraFLOP/s. The right Y-axis displays the infinity norm condition number $(\kappa_\infty(A) = \|A\|_\infty \|A^{-1}\|_\infty)$, since this condition number is the one used in the error analysis of the IRGMRES solver [6,7]. The 2-norm condition number is constant across a single figure, and is equivalent to the ratio between the maximum and the minimum eigenvalues. We accept convergence when the residual $r = \frac{\|b - Ax\|_\infty}{N\|A\|_\infty}$ is at most $\mathcal{O}(10^{-14})$. Each performance graph features some or all of the following solvers:

- dposv: a direct solver in full double precision.
- dsposv: a classic MP-IR solver with two precisions (FP64→FP32).
- dsposv-fp16-ir : our new MP-IR solver with three precisions.
- dsposv-fp16-irgmres : our new MP-IRGMRES solver with three precisions. This solver always scales and equilibrates the matrix, but the shift is optional. The time of the these preprocessing steps is included in the final timing of the solver.

Matrices with an Arithmetic Distribution of Eigenvalues. Figure 4a shows a "best case scenario" for a small $\kappa_2(A)$. The infinity norm condition number is capped at 10^4. Both dsposv-fp16-ir and dsposv-fp16-irgmres converge within 3 iterations at most, which yields significant performance gains. The asymptotic performance reaches 28.5 teraFLOP/s, which is 4.7× faster than dposv, and 2.7× faster than dsposv. Figure 4b shows the impact of increasing the condition number. The dsposv-fp16-irgmres solver converges within 7–8 iterations in most cases, while the dsposv-fp16-ir solver converges within 6–11 iterations, leading to performance drops at some points. The increased iteration count on both sides leads to a drop in the asymptotic performance, which is now measured at 24 teraFLOP/s. This is still 4× faster than dposv and 2.3× faster than dsposv.

Matrices with a Clustered Distribution of Eigenvalues. Figure 5a shows a performance similar to the best case scenario of Fig. 4a. However, there is a slight advantage for using the dsposv-fp16-irgmres solver. It converges in 3–4 iterations, while the dsposv-fp16-ir solver requires 3–6 iterations. The dsposv-fp16-irgmres solver maintains asymptotic speedups of 4.5×/2.6× against dposv/dsposv, respectively. Now we increase $\kappa_2(A)$ to 10^8, which results in $\kappa_\infty(A)$ in the range of 10^9. No convergence was achieved except for the

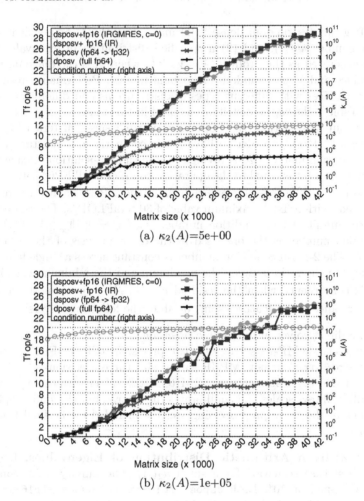

Fig. 4. Performance on matrices with an arithmetic distribution of eigenvalues.

`dsposv-fp16-irgmres` solver. This is a test case where classic IR fails in both `dsposv` and `dsposv-fp16-ir` . As Fig. 5b shows, the `dsposv-fp16-irgmres` solver requires 5 iterations for this type of matrices, leading to an asymptotic performance that is 4.4× faster than `dposv`. The result of this experiemnt also encourages using the GMRES-based IR with single-precision factorization. While this combination is not discussed this paper, the performance would be similar to `dsposv` in Fig. 4a.

Matrices with Logarithmic/Geometric Distributions of Eigenvalues.
It is clear that by trying harder-to-solve matrices, the `dsposv-fp16-irgmres` solver requires more iterations, which would impact the final performance of the solver. Figure 6 shows two example for such a case, where the benefit of using half-precision is limited only to large matrices. The condition number

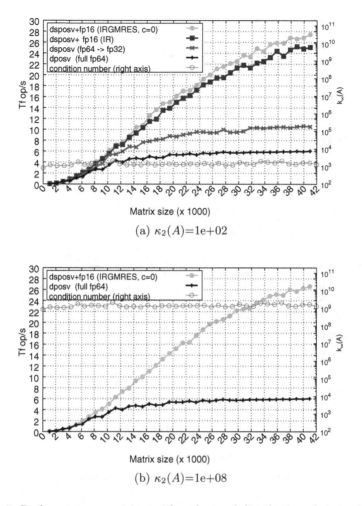

(a) $\kappa_2(A)$=1e+02

(b) $\kappa_2(A)$=1e+08

Fig. 5. Performance on matrices with a clustered distribution of eigenvalues.

$\kappa_\infty(A)$ is intentionally high to show such a behavior. Several useful observations can be taken away from these results. **First**, this is the first time we see a benefit for the matrix preprocessing stage. Both dsposv-fp16-ir and and the dsposv-fp16-irgmres (without preprocessing) fail during the factorization, meaning that the matrix loses its positive-definiteness during the mixed-precision updates. **Second**, our proposition for smaller shifts proves to achieve a better performance against limiting the constant c to an integer. **Third**, the number of iterations for the dsposv-fp16-irgmres solver ($c = 0.4$) is asymptotically measured at 27 for Fig. 6a, and at 32 for Fig. 6. Such large iteration counts consume most of the performance gains achieved in the factorization. Performance speedups are observed only for large matrices ($N \geq 27$k). Figure 6a shows

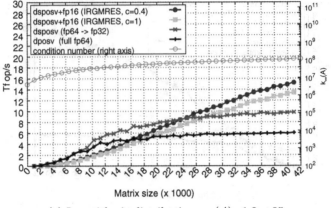

(a) Logarithmic distribution, $\kappa_2(A)=1.2e+05$

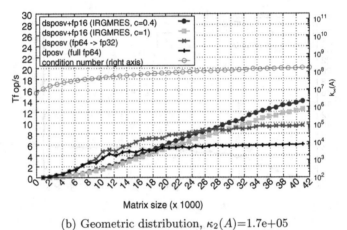

(b) Geometric distribution, $\kappa_2(A)=1.7e+05$

Fig. 6. Performance on matrices with logarithmic (a) and geometric (b) distributions of eigenvalues.

an asymptotic speedup of $2.5\times/1.56\times$ against `dposv`/`dsposv`, respectively. The respective speedups of Fig. 6b are measured at $2.3\times/1.46\times$.

Matrices with a Custom-Clustered Distributions of Eigenvalues. This distribution assigns 10% of the eigenvalues to 1, and the other 90% to $\frac{1}{\kappa_2(A)}$. Figure 7 shows the results, in which the two variants of `dsposv-fp16-irgmres` (with/without preprocessing) successfully converge. However, the preprocessed solver converges within 15–16 iterations in most cases, as opposed to at least 37 iterations without preprocessing. This means that the produced Cholesky factors without preprocessing do not form a good preconditioner for A. The performance gains for the preprocessed solver are noticeable much earlier than its regular variant. The asymptotic speedups for the preprocessed `dsposv-fp16-irgmres` are $3.3\times/1.96\times$ against `dposv`/`dsposv`, respectively.

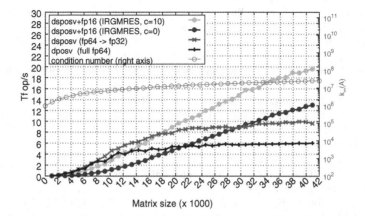

Fig. 7. Performance on matrices with a custom-clustered distribution ($\kappa_2(A) = 10^4$).

8   Conclusion and Future Work

This paper presented an FP16-accelerated dense linear solver for SPD systems. The proposed solution combines a mixed-precision Cholesky factorization with a GMRES-based iterative refinement algorithms in order to achieve double precision accuracy. Optional safeguards are developed (scaling and shifting) to ensure successful factorization and solve for matrices with relatively large condition numbers. The accelerated solver can be up to $4.7\times$ faster than a direct solve in full FP64 precision.

Future directions include integrating the GMRES-based IR solver into dual-precision solvers (i.e., FP32→FP64), which would improve their performance for matrices with higher condition numbers. It is also useful to study the impact of the preprocessing stage (especially the diagonal shift) on the convergence of the GMRES-based IR solver. As per our results, there is no single setting that works well across the board, and each matrix has to be treated separately. Another potential direction is to add support for the complex precision (Hermitian Positive Definite systems), which requires half-complex BLAS routines.

References

1. LAPACK - Linear Algebra PACKage. http://www.netlib.org/lapack/
2. IEEE standard for floating-point arithmetic. IEEE Std 754–2008, pp. 1–70, August 2008. https://doi.org/10.1109/IEEESTD.2008.4610935. https://ieeexplore.ieee.org/document/4610935

3. Abdelfattah, A., Tomov, S., Dongarra, J.J.: Fast batch matrix multiplication for small sizes using half precision arithmetic on GPUs. In: 2019 IEEE International Parallel and Distributed Processing Symposium, IPDPS 2019, Rio de Janeiro, Brazil, 20–24 May 2019, pp. 111–122 (2019)
4. Agullo, E., et al.: Numerical linear algebra on emerging architectures the PLASMA and MAGMA projects. J. Phys. Conf. Ser. **180**(1), 012937 (2009)
5. Baboulin, M., et al.: Accelerating scientific computations with mixed precision algorithms. Comput. Phys. Commun. **180**(12), 2526–2533 (2009)
6. Carson, E., Higham, N.: A new analysis of iterative refinement and its application to accurate solution of ill-conditioned sparse linear systems. SIAM J. Sci. Comput. **39**(6), A2834–A2856 (2017). https://doi.org/10.1137/17M1122918
7. Carson, E., Higham, N.: Accelerating the solution of linear systems by iterative refinement in three precisions. SIAM J. Sci. Comput. **40**(2), A817–A847 (2018). https://doi.org/10.1137/17M1140819
8. Greenbaum, A.: Estimating the attainable accuracy of recursively computed residual methods. SIAM J. Matrix Anal. Appl. **18**(3), 535–551 (1997). https://doi.org/10.1137/S0895479895284944
9. Haidar, A., Tomov, S., Dongarra, J., Higham, N.J.: Harnessing GPU tensor cores for fast FP16 arithmetic to speed up mixed-precision iterative refinement solvers. In: Proceedings of the International Conference for High Performance Computing, Networking, Storage, and Analysis (SC 2018), pp. 47:1–47:11. IEEE Press, Piscataway (2018). https://doi.org/10.1109/SC.2018.00050
10. Higham, N., Pranesh, S.: Exploiting lower precision arithmetic in solving symmetric positive definite linear systems and least squares problems. Technical report 1749–9097, November 2019. http://eprints.maths.manchester.ac.uk/2736/
11. MAGMA: Matrix Algebra on GPU and Multicore Architectures. http://icl.cs.utk.edu/magma/
12. Langou, J., Langou, J., Luszczek, P., Kurzak, J., Buttari, A., Dongarra, J.J.: Exploiting the performance of 32 bit floating point arithmetic in obtaining 64 bit accuracy (revisiting iterative refinement for linear systems). In: Proceedings of the ACM/IEEE SC2006 Conference on High Performance Networking and Computing, 11–17 November 2006, Tampa, FL, USA. p. 113 (2006). https://doi.org/10.1145/1188455.1188573
13. NVIDIA CUDA Basic Linear Algebra Subroutines (CUBLAS). https://developer.nvidia.com/cublas
14. NVIDIA cuSOLVER: A Collection of Dense and Sparse Direct Solvers. https://developer.nvidia.com/cusolver
15. Saad, Y.: A flexible inner-outer preconditioned GMRES algorithm. SIAM J. Sci. Comput. **14**(2), 461–469 (1993). https://doi.org/10.1137/0914028
16. Saad, Y., Schultz, M.H.: GMRES: a generalized minimal residual algorithm for solving nonsymmetric linear systems. SIAM J. Sci. Stat. Comput. **7**(3), 856–869 (1986). https://doi.org/10.1137/0907058
17. Simoncini, V., Szyld, D.: Flexible inner-outer Krylov subspace methods. SIAM J. Numer. Anal. **40**(6), 2219–2239 (2002). https://doi.org/10.1137/S0036142902401074
18. Tomov, S., Dongarra, J.J., Baboulin, M.: Towards dense linear algebra for hybrid GPU accelerated manycore systems. Parallel Comput. **36**(5–6), 232–240 (2010). https://doi.org/10.1016/j.parco.2009.12.005

Simulation Versus an Ordered–Fuzzy-Numbers-Driven Approach to the Multi-depot Vehicle Cyclic Routing and Scheduling Problem

Grzegorz Bocewicz[1]([⊠])[iD], Zbigniew Banaszak[1][iD],
Czeslaw Smutnicki[2][iD], Katarzyna Rudnik[3][iD], Marcin Witczak[4][iD],
and Robert Wójcik[2][iD]

[1] Faculty of Electronics and Computer Science,
Koszalin University of Technology, Koszalin, Poland
bocewicz@weii.tu.koszalin.pl,
zbigniew.banaszak@tu.koszalin.pl
[2] Faculty of Electronics, Wroclaw University of Science and Technology,
Wrocław, Poland
{czeslaw.smutnicki, robert.wojcik}@pwr.edu.pl
[3] Faculty of Production Engineering and Logistics,
Opole University of Technology, Opole, Poland
k.rudnik@po.opole.pl
[4] Institute of Control and Computation Engineering,
University of Zielona Gora, Zielona Gora, Poland
m.witczak@issi.uz.zgora.pl

Abstract. It is an undeniable fact that material handling systems aim at supplying the right materials at the right locations at the right time. This fact creates the need for the design of logistic-train-fleet-oriented, distributed and scalability-robust control policies ensuring deadlock-free operations. The paper presents a solution to a multi-item and multi-depot vehicle routing and scheduling problem subject to fuzzy pick-up and delivery transportation time constraints. Since this type of problem can be treated as a fuzzy constraint satisfaction problem, a solution to it can be determined using both computer simulation and analytical ordered-fuzzy-numbers-driven calculations. The accuracy of both approaches is verified based on the results of multiple simulations. In this context, our contribution consists of proposing an alternative approach that allows avoiding time-consuming computer simulation-based calculations of logistic train fleet schedules.

Keywords: Vehicle routing problem · Ordered fuzzy numbers

1 Introduction

To solve a Vehicle Routing Problem (VRP), one has to create a serving plan specifying how much a given fleet of vehicles should deliver and what cyclic routes the vehicles should travel to provide the required supplies on time. Since the VRP belongs to a class

© Springer Nature Switzerland AG 2020
V. V. Krzhizhanovskaya et al. (Eds.): ICCS 2020, LNCS 12138, pp. 251–266, 2020.
https://doi.org/10.1007/978-3-030-50417-5_19

of NP-hard logistic train routing and scheduling problems, various heuristic methods that return approximate solutions are used to solve it. Due to the growing interest in logistic networks based on autonomous vehicles and Milk-run systems [11, 14], there is a need to build models of these systems that take into account the uncertainty of the parameters describing them. In response to this need and deficiencies of the currently used approaches [3, 14–16], we wanted to investigate the possibility of using declarative modelling methods [2] supported by the ordered fuzzy number (OFN) framework in solutions that provide interactive decision support for prototyping congestion-free vehicle traffic in in-plant distribution systems. More specifically, we assessed computer simulation methods and ordered-fuzzy-number-driven approaches to solve multi-depot vehicle cyclic routing and scheduling problems. The present study is a continuation of our previous work that explored methods of fast prototyping of solutions to problems related to routing and scheduling of tasks typically performed in batch flow production systems, as well as problems related to the planning and control of production flow in departments of automotive companies [2]. The main contributions of this paper are summarized as follows: 1) In contrast to the usually accepted assumptions, we assume that transport processes have a deterministic nature and an uncertain course, which requires taking into account the human factor. We take into consideration the related distribution of delivery moments, which allows constructing more realistic, i.e., more accurate, models for assessing the effectiveness of prototyped route variants. 2) We formulate in detail a declarative-modelling-driven approach to the assessment of alternative routing and scheduling variants for a fleet of vehicles. The obtained ordered-fuzzy-number-driven model allows searching for congestion-free logistic train routes in terms of the Fuzzy Constraint Satisfaction Problem. 3) The proposed approach enables the replacement of the usually used computer simulation methods for route prototyping with an analytical method employing the OFN formalism. It is an outperforming approach to solving in-plant Milk-run-driven delivery problems.

The remainder of this paper is as follows: Sect. 2 presents a review of selected literature of the subject, including necessary information about OFNs. A motivation example introducing the problem under consideration is in Sect. 3. Section 4 formulates a declarative model and a Fuzzy Constraint Satisfaction Problem for planning delivery missions of a vehicle fleet. Section 5 shows how to use the model in supply-cycle-prototyping tasks. Section 6 summarizes the principal conclusions and proposes the main directions for future research.

2 Literature Review

2.1 Vehicle Routing and Scheduling

VRPs belong to a class of combinatorial optimization problems. Because VRPs are problems in which a set of vehicles have to serve a set of pick-up/delivery points and satisfy assumed constraints, while minimizing different objectives such as cost, distance, or time, they are usually NP-hard problems for which, so far, no efficient solution algorithm has been found. Different constraints, depending on the specific characteristics of the problem and the objective(s) of the decision-making process, lead to a variety of task-specific problems. Examples of such problems [4, 7, 15, 19] include Mix Fleet VRP, Multi-depot VRP, Split-up Delivery VRP, Pick-up and Delivery VRP, VRP with Time Windows, and many similar ones. The VRP can be seen as a generalization of the Traveling Salesman Problem aimed at finding the optimal set of routes for a fleet of vehicles delivering goods or services to various locations. Most of the research in the field of distribution logistics is devoted to the analysis of methods of organizing transport processes in ways that minimize the size of the fleet, the distance travelled (energy consumed), or the space occupied by a distribution system. In focusing on the search for optimal solutions, these studies implicitly assume that there exist admissible solutions, e.g., ones that ensure collision-free and/or deadlock-free (congestion-free) flow of concurrent transport processes. In practice, this kind of assumption requires either on-line updating (revision) of the routing policies used or prior (offline) planning of congestion-free vehicle routes and schedules. Studies on generating dynamic routing policies are conducted sporadically [4]; even less frequent are investigations of robust routing and scheduling of Milk-run traffic, which are, by and large, limited to Automated Guided Vehicle (AGV) systems. In a Milk-run system, routes, time schedules, and the type and number of parts to be transported are assigned to different logistic trains so that they can collect orders from different suppliers [11]. The benefits of using a system of this type include improved efficiency of the overall logistics system and substantial potential savings of environmental and human resources along with remarkable cost reductions related to inventory and transportation [7, 14]. The Congestion Avoidance Problem, which conditions the existence of admissible solutions, is an NP-hard problem [21]. Because the necessary and sufficient conditions for deadlock-free execution of concurrent processes are not known, system analysis (i.e., analysis of the states potentially leading to system deadlocks) is most frequently performed using the laborious and time-consuming computer simulation methods [4, 9]. In practical applications, congestion avoidance methods are used, in which the sufficient conditions for collision-free execution of processes are implemented. This means that the time-consuming method of analyzing distribution networks with a view to detecting situations that lead to deadlocks between concurrent transport flows can be replaced by searching for a synchronization mechanism that would guarantee cyclic execution of these flows. Methods that are most commonly employed for such purposes include those that use the formalism of max-plus algebra [17], simulations [8], graph theory [20], and constraint programming [2, 19]. It should be noted that the possibility of fast implementation of the process-synchronization mechanism comes at the expense of omitting some of the potentially possible scenarios for deadlock-free execution of the processes.

In many real situations, not all the constraints and objective functions can be valued in a precise way. The majority of models of the so-called Fuzzy VRP only assume vagueness for fuzzy demands to be collected and fuzzy service and travel times. It should be emphasized that the literature on these issues is very scarce [3, 10].

2.2   Ordered Fuzzy Number Algebra Framework

The multi-depot vehicle routing and scheduling problems developed so far have limited use due to the data uncertainty observed in practice. The values describing parameters such as transport time, loading/unloading times, depend on the human factor, which means they cannot be determined precisely. Accounting for data uncertainty by including fuzzy variables in these models is difficult due to the imperfections of the classical fuzzy numbers algebra [1]. Relations describing the relationships between fuzzy variables (variables with fuzzy values) by algebraic operations (in particular, addition and multiplication) do not meet the conditions of the Ring (among others if the condition $\forall_{A \in \mathcal{F}} A + 0 = A$ is met, then condition $\forall_{A \in \mathcal{F}} \exists!_{B \in \mathcal{F}} A + B = 0$ is not met). In addition, algebraic operations based on standard fuzzy numbers follow Zadeh's extension principle. In practice, this means that no matter what algebraic operations are used, the support of the fuzzy number, being the result, expands. Consequently, it is impossible to solve algebraic equations with fuzzy variables. In particular, this means that for any fuzzy numbers a, b, c does not hold the following implication $(a + b = c) \Rightarrow [(c - b = a) \wedge (c - a = b)]$. This makes it impossible to solve a simple equation $A + X = C$. This fact significantly hinders the use of approaches based on declarative models, in which most of the relationships between decision variables are described as linear/nonlinear equations and/or algebraic inequalities. There are various approaches in the literature that work around the above-mentioned deficiencies [1, 12], but they are quite complex.

We address these issues by proposing a declarative model of congestion-free vehicle routing and scheduling that implements the formalism of OFN algebra, which assumes the existence of a neutral element (zero) for operations such as addition and multiplication, making it possible to solve algebraic equations in the model. The concept of OFNs can be defined as follows [13]:

Definition 1. An OFN is defined as a pair of continuous real functions defined by the interval [0, 1], i.e.:

$$\hat{A} = (f_A, g_A), \text{where} : f_A, g_A : [0, 1] \to \mathbb{R}. \tag{1}$$

The functions f_A and g_A are called the up part and the down part of an OFN $\hat{A}$, respectively. They are also referred to as branches of the fuzzy number $\hat{A}$. The values of these continuous functions are limited ranges, which can be defined as the following bounded intervals: $UP_A = (l_{A0}, l_{A1})$ and $DOWN_A = (p_{A1}, p_{A0})$. Assuming that: f_A is increasing and g_A is decreasing as well as that $f_A \leq g_A$, the membership function μ_A of the OFN $\hat{A}$ is as shown in Figs. 1a) and b):

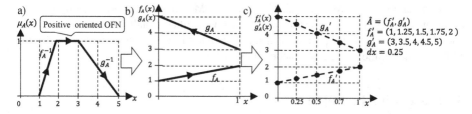

Fig. 1. a) OFN $\hat{A}$ represented as a convex fuzzy number, b) functions f_A, g_A determining $\hat{A}$ (positive orientation), c) discrete representation of $\hat{A}(dx = 0.25)$ (based on [13])

$$\mu_A(x) = \begin{cases} f_A^{-1}(x) & when\, x \in UP_A \\ g_A^{-1}(x) & when\, x \in DOWN_A \\ 1 & when\, x \in [l_{A1}, p_{A1}] \\ 0 & in\, the\, remaining\, cases \end{cases} \tag{2}$$

An additional property called orientation (direction) is defined for an OFN. There are two types of orientation: positive, when $\hat{A} = (f_A, g_A)$ the direction is consistent with the direction of the OX axis and negative, when $\hat{A} = (g_A, f_A)$ the direction is opposite to the direction of the OX axis. Assuming that the values of all fuzzy variables may have a different orientation, let us define algebraic operations that meet the listed conditions of the Ring. The definitions of algebraic operations used in the proposed model are as follows:

Definition 2. Let $\hat{A} = (f_A, g_A)$ and $\hat{B} = (f_B, g_B)$ be OFNs. $\hat{A}$ is a number equal to $\hat{B}$ $(\hat{A} = \hat{B})$, $\hat{A}$ is a number greater than $\hat{B}$ or equal to or greater than $\hat{B}$ $(\hat{A} > \hat{B}; \hat{A} \geq \hat{B})$, $\hat{A}$ is less than $\hat{B}$ or equal to or less than $\hat{B}$ $(\hat{A} < \hat{B}, \hat{A} \leq \hat{B})$ if: $\forall_{x \in [0,1]} f_A(x) * f_B(x) \wedge g_A(x) * g_B(x)$, where: the symbol $*$ stands for: $=$, $>$, $\geq$, $<$, or $\leq$.

Definition 3. Let $\hat{A} = (f_A, g_A)$, $\hat{B} = (f_B, g_B)$, and $\hat{C} = (f_C, g_C)$ be OFNs. The operations of addition $\hat{C} = \hat{A} + \hat{B}$, subtraction $\hat{C} = \hat{A} - \hat{B}$, multiplication $\hat{C} = \hat{A} \times \hat{B}$ and division $\hat{C} = \hat{A}/\hat{B}$ are defined as follows: $\forall_{x \in [0,1]} f_C(x) = f_A(x) * f_B(x) \wedge g_C(x) = g_A(x) * g_B(x)$, where: the symbol $*$ stands for $+$, $-$, $\times$, or $\div$; The operation of division is defined for $\hat{B}$ such that $|f_B| > 0$ and $|g_B| > 0$ for $x \in [0, 1]$.

In recent years, the concept of OFNs has continuously been developed and used in various practical applications. Many publications have been devoted to the analysis of the OFN model in relation to convex fuzzy sets [5, 6]. The concept of defining imprecise values as OFNs has also been used in critical path analysis. A practical implementation of OFN arithmetic in the monitoring of a crisis control centre was described in [6]. Another recently popular area of OFN's applications is multi-criteria decision making (MCDM) methods [18]. In MCDM methods, the orientation of OFNs differentiates the type of criterion used (cost vs profit). Finally, to the best of our knowledge, the approach proposed in this paper is the first attempt to use OFNs for Milk-run-like traffic routing and scheduling.

3 Illustrative Example

Let us consider graph $G = (N, E)$ modelling a distribution network composed of $|N| = \omega = 11$ pick-up/delivery points (i.e., workstations and warehouses), as shown in Fig. 2. The pick-up/delivery points (hereinafter referred to as nodes) include 2 nodes representing warehouses N_1 and N_5 and 9 nodes representing workstations N_2-N_4, N_6-N_{11}. Each node is labelled with an index which indicates the beginning moments of node occupation x_λ and node release xs_λ, as well as the time spent at the node, i.e. pick-up/delivery operation time t_λ. Nodes are cyclically supplied with goods in time windows repeated (with size $T = 2970$ s). The goods are supplemented in intervals determined by the delivery deadline dx_λ and delivery margin τ_λ, i.e. $x_\lambda + t_\lambda = y_\lambda \in [dx_\lambda - \tau_\lambda, dx_\lambda]$ (see intervals identified by grey bars in Fig. 4). In turn, each edge $(N_\beta, N_\lambda) \in E$ linking nodes N_β and N_λ is labelled with an index representing travelling time $d_{\beta,\lambda}$ between nodes N_β and N_λ and a set of indexes $K_{\beta,\lambda}$ indicating the transport zones located along the edge. It is assumed that the set of edges E model the routes travelled by logistic trains between nodes N_β and N_λ. It is also assumed that each edge (N_β, N_λ) is composed of a set of transport zones labelled by a set of indexes $K_{\beta,\lambda}$. Given is a fleet of logistic trains LT which handle deliveries in the distribution network under consideration. The routes travelled by the logistic trains LT_v are denoted by sequences of nodes: $\pi_v = (N_{v_1}, \ldots, N_{v_i}, N_{v_{i+1}}, \ldots, N_{v_\mu})$, where: $v_i \in \{1, .., ln\}$, $\forall_{v_i \neq v_j} N_{v_i} \neq N_{v_j}$, $(N_{v_i}, N_{v_{i+1}}) \in E$. To each edge $(N_{v_i}, N_{v_{i+1}}) \in E$ of the route π_v, a time period is assigned in which the edge is occupied by the logistic train: $IN_{v_i, v_{i+1}} = [xs_{v_i}, x_{v_{i+1}}]$. The set of routes π_v of the available fleet of logistic trains is marked by Π. It is assumed that nodes representing a warehouse (e.g. N_1, N_5) appear on every route, and each node representing a workstation (e.g. N_2-N_4, N_6-N_{11}) occurs only on one route from the set Π. In order to avoid collisions/blockades between trains, the following condition must also be met: given are two routes $\pi_v, \pi_w \in \Pi$. If for any pair of edges $(N_{v_i}, N_{v_{i+1}})$ belonging to π_v and $(N_{w_j}, N_{w_{j+1}})$ belonging to π_w, the following condition holds $[(K_{v_i, v_{i+1}} \cap K_{w_j, w_{j+1}} \neq \emptyset) \wedge (IN_{v_i, v_{i+1}} \cap IN_{w_j, w_{j+1}} \neq \emptyset)]$, then trains LT_v, LT_w which travel along routes π_v, π_w are collision/blockade-free. In other words, it means that two trains LT_v, LT_w travelling along routes π_v, π_w will not block each other if they do not occupy the same edge during the same period of time.

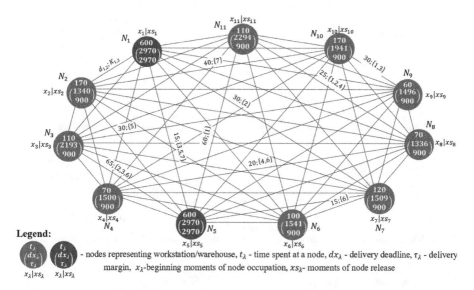

Fig. 2. Graph model of a distribution network

Taking into account the assumptions mentioned above, we are looking for a set of routes of logistic trains and the associated delivery schedules that guarantee congestion-free and timely delivery of goods to the nodes. Examples of routings that guarantee timely delivery of goods and the resulting schedule are presented in Figs. 3 and 4. The routes are the following sequences of nodes visited repetitively by LT_1 and LT_2: $\pi_1 = (N_1, N_7, N_6, N_4, N_8, N_{11}, N_5)$, $\pi_2 = (N_1, N_2, N_{10}, N_9, N_3, N_5)$. These routes guarantee collision-free and deadlock-free delivery. However, in many cases (e.g. in Milk-run systems), transport operations and loading/unloading operations are usually carried out by people, which means they are quite uncertain. The uncertainty of the duration of the operations results in uncertain moments of node occupation and release. Consequently, the actual implementation of the schedule may differ significantly from the planned one, and even minor deviations from the plan may have serious implications, such as blockages. Figure 3 illustrates a situation in which a 90-s delay (relative to the deadline resulting from the planned schedule in Fig. 4) of train LT_2 with the simultaneous acceleration of the train LT_1 by 60 s leads to blockade in edges N_{10}-N_9 and N_8-N_{11} (the condition introduced above does not hold). Therefore, there is a need to synthesize such routes, which, assuming a specific range of data uncertainty, still guarantee collision-free and deadlock-free performance of periodically repeating delivery operations.

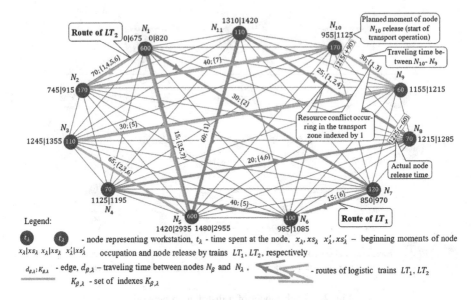

Fig. 3. Routes of trains LT_1 and LT_2 servicing the distribution network from Fig. 2

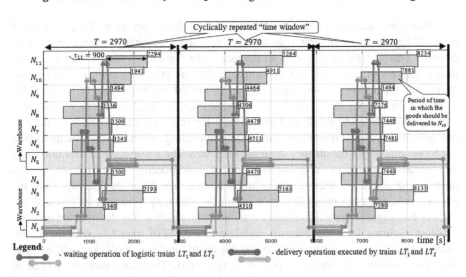

Fig. 4. Gantt chart of a multi-depot delivery schedule for the train routes from Fig. 3

4   Problem Formulation

4.1   Assumptions

The problem under consideration can be defined as follows. Assuming that:

- there is a known transportation network $G = (N, E)$, where N is a set of nodes and E is a set of edges; the set N contains the subsets of nodes representing workstations $NC \subseteq N$ and warehouses $NW \subseteq N$: $NW \cup NC = N$, $NW \cap NC = \emptyset$,

- each edge $(N_\beta, N_\lambda) \in E$ is labelled by a fuzzy value $\widehat{d_{\beta,\lambda}}$ (represented in terms of OFN) determining the travel time between nodes N_β and N_λ,
- each edge $(N_\beta, N_\lambda) \in E$ consists of sectors described by a set of indexes $K_{\beta,\lambda} \subseteq \mathbb{N}$,
- given is a fleet of logistic trains LT, in which each of the trains LT_v corresponds to a route π_v ($\pi_v \in \Pi$) described by a sequence of successively visited nodes,
- trains can only move between nodes connected by an edge,
- if for any pair of edges: $(N_{v_i}, N_{v_{i+1}})$ and $(N_{w_j}, N_{w_{j+1}})$ belonging to π_v, π_w, the following condition holds $\left[\left(K_{v_i, v_{i+1}} \cap K_{w_j, w_{j+1}} \neq \emptyset \right) \wedge \left(IN_{v_i, v_{i+1}} \cap IN_{w_j, w_{j+1}} \neq \emptyset \right) \right]$, then the trains travelling along routes π_v, π_w are congestion-free,
- each node $N_\lambda \in NC$ occurs exactly on one route of the set Π,
- each node $N_\lambda \in NW$ occurs exactly on all routes of the set Π,
- node N_λ located on route π_v is associated with the delivery operation $o_\lambda \in \mathcal{O}$,
- the duration of the delivery operation is determined by the fuzzy value $\widehat{t_\lambda}$,
- deliveries of goods take place cyclically in time windows repeated with a period $\widehat{T}$,
- goods are delivered in accordance with the fuzzy delivery deadline $\widehat{dx_\lambda}$ and fuzzy delivery margin $\widehat{\tau_\lambda}$ (represented as OFN),
- fuzzy beginning moments of node occupation $\widehat{x_\lambda}$ and node release $\widehat{xs_\lambda}$ (represented as OFN) make up the fuzzy cyclic schedule $\widehat{X}$,

the following question can be considered: Does there exist a set of routes Π operated by the given fleet LT, which ensures that a fuzzy cyclic schedule $\widehat{X}$ will guarantee timely delivery (with given deadlines $\widehat{dx_\lambda}$ and delivery margin $\widehat{\tau_\lambda}$) of goods to the nodes?

The proposed model uses decision variables whose values OFNs as defined in Definition 1. For the needs of the model, OFN $\hat{A}$ is specified by sequences f'_A and g'_A containing values of functions f_A and g_A obtained as a result of discretization of the interval $[0, 1]$, i.e.

$$f'_A = (f_A(0), f_A(dx), ..., f_A((M-1)dx), f_A(1)), \tag{3}$$

$$g'_A = (g_A(1), g_A((M-1)dx), ..., g_A(1dx), g_A(0)), dx = \frac{1}{M}, \tag{4}$$

where $(M+1)$ is the number of discrete points (Fig. 1c). The adoption of such an OFN representation allows to implement the defined operations (see Definition 2 and 3).

4.2 Declarative Model

The previously introduced terminology and symbols referring to OFN and the following notation were used in designing the Milk-run like traffic model:

Symbols:

$N_\lambda \in N$: λ-th node.

$LT_v \in LT$: v-th logistic train.

$o_\lambda \in \mathcal{O}$:   operation of delivery of materials to node N_λ on route π_v

Parameters:

Crisp parameters:

$G = (N, E)$:   graph of a transportation network: $N = \{N_1 \ldots N_\omega\}$ is a set of nodes, $E = \{(N_i, N_j) | i, j \in N, i \neq j\}$ is a set of edges, ω – the number of nodes.

ln: the number of logistic trains.

$K_{\beta,\lambda}$:   a set of indexes assigned to zones located along the edge (N_β, N_λ).

Imprecise parameters: (defined as positive-oriented OFNs and marked by ⌢):

$\widehat{d_{\beta,\lambda}}$:   time of a transport operation executed along the edge (N_β, N_λ).

$\widehat{t_\lambda}$:   time of operation o_λ.

$\widehat{dx_\lambda}$:   deadline of delivery of containers to node N_λ (see example in Fig. 3).

$\widehat{\tau_\lambda}$:   delivery margin, (see Fig. 3),

$\widehat{T}$:   window width understood as a period, repeated at regular intervals, in which deliveries must be made to all nodes (see Fig. 3).

Variables:

Crisp variables:

rb_λ: an index of the operation that precedes the operation o_λ; $rb_\lambda = 0$ means that operation o_λ, is the first one on the route.

rf_λ: an index of the operation that follows o_λ.

Imprecise variables (positive-/negative-oriented OFNs):

$\widehat{x_\lambda}$:   moment of commencement of the delivery operation o_λ on node N_λ.

$\widehat{y_\lambda}$:   moment of completion of the operation o_λ on node N_λ.

$\widehat{xs_\lambda}$:   moment of release of node N_λ by operation o_λ.

Sets and sequences:

NC: a subset of nodes representing workstations $NC \subseteq N$.

NW: a subset of nodes representing warehouses $NW \subseteq N$.

RB: a sequence of predecessor indexes of delivery operations, $RB = (rb_1, \ldots, rb_\alpha, \ldots, rb_{|NC| + ln \times |NW|})$, $rb_\alpha \in \{0, \ldots, \omega\}$.

RF: a sequence of successor indexes of delivery operations, $RF = (rf_1, \ldots, rf_\alpha, \ldots, rf_{|NC| + ln \times |NW|})$, $rf_\alpha \in \{1, \ldots, \omega\}$, e.g. RB and RF that determine routes π_1 and π_2 (see Fig. 3), and take the following form:

$$
\begin{array}{cccccccccccccc}
 & N_1 & N_2 & N_3 & N_4 & N_5 & N_6 & N_7 & N_8 & N_9 & N_{10} & N_{11} & N_1' & N_5' \\
RB = & (0, & 1', & 9, & 6, & 11, & 7, & 1, & 4, & 10, & 2, & 8, & 0, & 3) \\
RF = & (7, & 10, & 5', & 8, & 1, & 4, & 6, & 11, & 3, & 9, & 5, & 2, & 1')
\end{array}
$$

The symbol ' refers to nodes associated with the warehouses visited by train LT_2.

π_v: route of the train LT_v, $\pi_v = (N_{v_1}, \ldots, N_{v_i}, N_{v_{i+1}}, \ldots, N_{v_\mu})$, where: $v_{i+1} = rf_{v_i}$ for $i = 1, \ldots, \mu - 1$ and $v_1 = rf_{v_\mu}$.

$\widehat{X'}$:   a sequence of moments $\widehat{x_\lambda}$: $\widehat{X'} = (\widehat{x_1}, \ldots, \widehat{x_\lambda}, \ldots, \widehat{x_\omega})$.

$\widehat{Y'}$:   a sequence of moments $\widehat{y_\lambda}$: $\widehat{Y'} = (\widehat{y_1}, \ldots, \widehat{y_\lambda}, \ldots, \widehat{y_\omega})$.

$\widehat{Xs'}$:   a sequence of moments $\widehat{xs_\lambda}$: $\widehat{Xs'} = (\widehat{xs_1}, \ldots, \widehat{xs_\lambda}, \ldots, \widehat{xs_\omega})$.

$\widehat{X}$:   a fuzzy cyclic schedule: $\widehat{X} = \left(\widehat{X'}, \widehat{Y'}, \widehat{Xs'} \right)$.

Constraints:

1. constraints describing the orders of operations depending on the logistic train routes:

$$\widehat{y_\lambda} = \widehat{x_\lambda} + \widehat{t_\lambda}, \forall o_\lambda \in \mathcal{O}, \tag{5}$$

$$rb_\lambda = 0, \forall \lambda \in BS{\subseteq}BI = \{1, \ldots, \omega\}, |BS| = ln, \tag{6}$$

$$b_\lambda \neq rb_\beta, \forall \lambda, \beta \in BI \backslash BS, \lambda \neq \beta, \tag{7}$$

$$rf_\lambda \neq rf_\beta, \forall \lambda, \beta \in BI, \lambda \neq \beta, \tag{8}$$

$$(rb_\lambda = \beta) \Rightarrow (rf_\beta = \lambda), \forall b_\lambda \neq 0, \tag{9}$$

$$\widehat{xs_\lambda} \geq \widehat{y_\lambda}, \forall o_\lambda \in \mathcal{O}, \tag{10}$$

$$[(f_\lambda = \beta) \wedge (b_\beta = 0)] \Rightarrow \left(\widehat{xs_\lambda} = \widehat{x_\beta} + \widehat{T} - \widehat{d_{\lambda,\beta}} \right), \forall o_\lambda, o_\beta \in \mathcal{O}, \tag{11}$$

$$[(f_\lambda = \beta) \wedge (b_\beta \neq 0)] \Rightarrow \left(\widehat{xs_\lambda} = \widehat{x_\beta} - \widehat{d_{\lambda,\beta}} \right), \forall o_\lambda, o_\beta \in \mathcal{O}, \tag{12}$$

2. if edge (N_ε, N_β) has common sectors with the edge (N_λ, N_γ), then:

$$\left(K_{\varepsilon,\beta} \cap K_{\lambda,\gamma} \neq \emptyset \right) \Rightarrow \left[(\widehat{x_\beta} \leq \widehat{xs_\lambda}) \vee (\widehat{x_\gamma} \leq \widehat{xs_\varepsilon}) \right], \forall o_\lambda, o_\beta, o_\varepsilon, o_\gamma \in \mathcal{O}, \tag{13}$$

3. the delivery operation o_λ should be completed before the given delivery deadline dx_λ (with a margin $\widehat{\tau_\lambda}$) resulting from the production flows of an individual product:

$$\widehat{y_\lambda} + c \times \widehat{T} \leq \widehat{dx_\lambda}, \forall o_\lambda \in \mathcal{O}, \tag{14}$$

$$\widehat{y_\lambda} + c \times \widehat{T} \geq \widehat{dx_\lambda} + \widehat{\tau_\lambda}, \forall o_\lambda \in \mathcal{O}; c \in \mathbb{N}. \tag{15}$$

4.3   Fuzzy Constraint Satisfaction Problem

Our problem can be viewed as a Fuzzy Constraint Satisfaction (FCS) Problem (16):

$$\widehat{FCS} = \left((\hat{\mathcal{V}}, \hat{\mathcal{D}}), \hat{\mathcal{C}} \right), \tag{16}$$

where: $\hat{\mathcal{V}} = \{\hat{X}, \Pi\}$ – a set of decision variables, including: $\hat{X}$ – a fuzzy cyclic schedule: $\hat{X} = \left(\widehat{X'}, \widehat{Y'}, \widehat{Xs'} \right)$, Π – a set of routes determined by sequences RB, FR. $\hat{\mathcal{D}}$ – a finite set of decision variable domains: $\hat{x}_\lambda$, $\hat{y}_\lambda$, $\widehat{xs}_\lambda \in \mathcal{F}$ ($\mathcal{F}$ is a set of OFNs (1)), $rb_\lambda \in \{0, \ldots \omega\}$, $rf_\lambda \in \{1, \ldots \omega\}$, C_{RE} – a set of constraints specifying the relationships between the operations implemented in Milk-run cycles (5)–(15).

To solve $\widehat{FCS}$ (16), the values of the decision variables from the adopted set of domains for which the given constraints are satisfied must be determined. Implementation of $\widehat{FCS}$ in a constraint programming environment such as OzMozart, allows us to find the answer.

5   Computational Experiments

Consider the system layout from Fig. 2. The goal is to find congestion-free routes for the given fleet of logistic trains (i.e. the set Π). The trains cyclically supply goods to nodes N_1-N_{11} in time windows with a width of $\hat{T} = 2970$ [s] (in the case under consideration, $\hat{T}$ is defined as a singleton – an OFN with a strictly neutral direction). It is assumed that the available vehicle fleet consists of two trains $LT = \{LT_1, LT_2\}$. It is also assumed that the fuzzy times of a delivery operation $(\hat{t}_\lambda)$ and admissible fuzzy travel times $(\widehat{d}_{\beta,\lambda})$ are as shown in Figs. 5b and 5c. The answer to the following question is sought: *Does there exist a set of routes Π operated by the given logistic trains LT_1 and LT_2, which ensures that there exists a fuzzy cyclic schedule $\hat{X}$, that guarantees timely delivery of the goods to the nodes?* While searching for the answer, problem $\widehat{FCS}$ (16) was formulated, and then implemented in the constraint programming environment OzMozart (Windows 10, Intel Core Duo2 3.00 GHz, 4 GB RAM). The solution time of this scale of problems with up to 12 nodes does not exceed 2 s. The results are shown in graphical form in Figs. 6 and 7. The obtained sequences RB and RF make up the following routes (Fig. 6): $\pi_1 = (N_1, N_8, N_9, N_2, N_5)$ and $\pi_2 = (N_1, N_7, N_6, N_4, N_3, N_{10}, N_{11}, N_5)$.

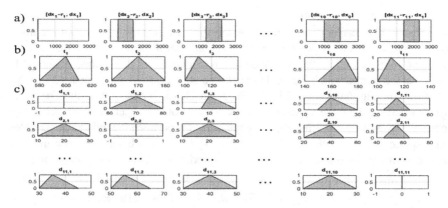

Fig. 5. Input data specifying the delivery time windows a), loading/unloading times b), periods in which a train moves between a pair of nodes c).

The fuzzy values of decision variable $\hat{X}$, and the cyclic schedule determined by them, which guarantees timely delivery of the goods, are presented in Fig. 6. In the Gantt's chart like schedule, the execution of each operation is represented as a ribbon-like "arterial road", whose increasing width represents the time of train movement resulting from the growing uncertainty of the moments of occupation and release of nodes. For example, the moment when the node N_{11} can be occupied is determined by the fuzzy variable $\widehat{x_{11}}$ (Fig. 6a), whose support is the interval [1473 s, 1750 s] (interval width of 277 s). In turn, the moment the node is released is determined by $\widehat{y_{11}}$; for which the support is the interval [1573 s, 1880 s] (interval width of 307 s).

It is worth noting that the width of the ribbon-like arterial roads increases until the next time-window begins. The uncertainty of decision variables is, however, reduced at the end of each time window as a result of the operation of trains waiting on nodes N_1, N_5. So, increasing uncertainty is not transferred to subsequent cycles of the system. Uncertainty is reduced as a result of the implementation of the OFN formalism. Fuzzy variables describing the waiting time of trains on nodes N_1, N_5 have a negative orientation (see Fig. 7 – laytimes $\widehat{w_1}$ and $\widehat{w_5}$), which means that the results of algebraic operations ($\widehat{xs_1} = \widehat{y_1} + \widehat{w_1}$ and $\widehat{xs_5} = \widehat{y_5} + \widehat{w_5}$) using these variables leads to a decrease in uncertainty. Uncertainty cannot be decreased in the same way using standard fuzzy numbers. According to Zadeh's extension principle, the uncertainty of variables would grow with each subsequent cycle of system operation until the information about their value ceased to be useful. It is worth noting that the adoption of such a schedule guarantees congestion-free movement of the logistic trains despite the uncertainty of the parameters specified in Fig. 5. In order to verify the results, we ran a simulation of the delivery of goods in the system shown in Fig. 2. In this network, two logistic trains move along routes π_1 and π_2 (see Fig. 6). The trains' travel times between nodes ($\widehat{d_{\beta,\lambda}}$) and the delivery times ($\widehat{t_\lambda}$) are assumed to be random variables given by triangular distribution probability functions whose parameters correspond to the variation ranges from Fig. 5. The results of the simulation are shown in Fig. 7. For each of the nodes N_1-N_{11}, OFNs of the starting moments ($\widehat{x_\lambda}$) and termination ($\widehat{y_\lambda}$) of the delivery operation (o_λ), as well as the corresponding histograms, are determined. It should be noted that the frames which are used to mark operations carried out on nodes N_7, N_9, N_{11} are shown in Fig. 6a, too. In Fig. 7, the green charts correspond to the operations performed along the route π_1 and the orange ones to route π_2 operations. The charts are connected by arcs representing algebraic relationships between the individual variables. For instance, for the route travelled by train LT_2 (π_2), the relations between the variables describing the operations performed on nodes N_8, N_9, N_2, N_5, N_1 are as follows: $\widehat{x_9} = \widehat{y_8} + \widehat{d_{8,9}}$ (N_9 can be serviced only after N_8 has been released), $\widehat{x_2} = \widehat{y_9} + \widehat{d_{9,2}}$, $\widehat{x_5} = \widehat{y_2} + \widehat{d_{2,5}}$, $\widehat{x_1} = \widehat{y_5} + \widehat{d_{5,1}}$, $\widehat{x_8} = \widehat{xs_1} + \widehat{d_{1,8}} - \widehat{T}$. These relations were used during the simulation. All of the histograms we obtained fall within the range of calculated OFN values (see Fig. 7). It should be underlined that in none of the simulated variants (1 000 000) did any congestion occur between the trains.

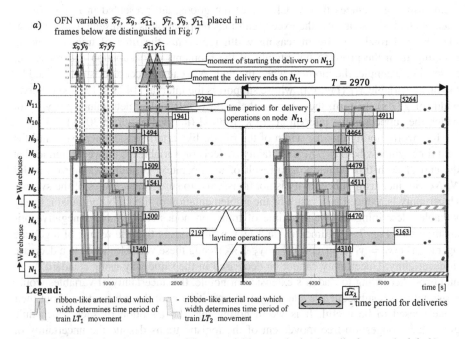

Fig. 6. Graph showing sample of fuzzy variables a), obtained cyclic fuzzy schedule b)

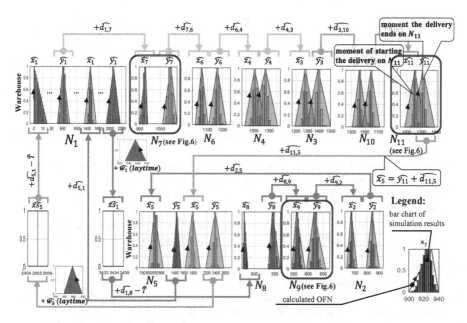

Fig. 7. Graphic summary of simulation results

6 Concluding Remarks

The results of the tests demonstrate that the proposed approach provides formal framework enabling to formulate and solve both routing and scheduling problems. In other words, it allows to more realistically model the movement of human-driven vehicles and replaces the usually used computer simulation methods of route proto-typing by analytical methods employing the OFN formalism. It is worth nothing that the proposed approach has yet another advantage of allowing to determine analytically the size of the vehicle fleet and a congestion-free routing that guarantee successful delivery of ordered goods.

Our future work is on finding sufficient conditions that would allow planners to reschedule Milk-run flows while guaranteeing smooth transition between two succes-sive cyclic steady states corresponding to the current and rescheduled logistic train fleet flows.

References

1. Bocewicz, G., Nielsen, I., Banaszak, Z.: Production flows scheduling subject to fuzzy processing time constraints. Int. J. Comput. Integr. Manuf. **29**, 1105–1127 (2016)
2. Bocewicz, G., Bożejko, W., Wójcik, R., Banaszak, Z.: Milk-run routing and scheduling subject to a trade-off between vehicle fleet size and storage capacity. Manag. Prod. Eng. Rev. **10**(3), 41–53 (2019). https://doi.org/10.24425/mper.2019.129597

3. Brito, J., Moreno-Pérez, J.A., Verdegay, J.L.: Fuzzy optimization in vehicle routing problems. In: Proceedings of the joint 2009 International Fuzzy Systems Association, World Congress and European Society of Fuzzy Logic and Technology, pp. 1547–1552 (2009)
4. Caric, T., Gold, H.: Vehicle Routing Problem. I-Tech, Vienna, Austria, p. 142, September 2008. ISBN 978-953-7619-09-1
5. Chwastyk, A., Kosiński, W.: Fuzzy calculus with applications. Mathematica Applicanda **41** (1), 47–96 (2013)
6. Czerniak, J.M., Dobrosielski, W.T., Apiecionek, Ł., Ewald, D., Paprzycki, M.: Practical application of OFN arithmetics in a crisis control center monitoring. Stud. Comput. Intell. **655**, 51–64 (2016)
7. De Moura, D.A., Botter, R.C.: Delivery and pick-up problem transportation – milk run or conventional systems. Independent J. Manag. Prod. **7**(3), 746–770 (2016)
8. Fedorko, G., Vasil, M., Bartosova, M.: Use of simulation model for measurement of MilkRun system performance. Open Eng. **9**, 600–605 (2019)
9. Güner, A.R., Chinnam, R.B.: Dynamic routing for milk-run tours with time windows in stochastic time-dependent networks. Transp. Res. Part E Logistics Transp. Rev. **97**(C), 251–267 (2017)
10. He, Y., Xu, J.: A class of random fuzzy programming model and its application to vehicle routing problem. World J. Model. Simul. **1**(1), 3–11 (2005)
11. Kilic, H.S., Durmusoglu, M.B., Baskak, M.: Classification and modeling for in-plant milk-run distribution systems. Int. J. Adv. Manuf. Technol. **62**, 1135–1146 (2012)
12. Klir, G.J.: Fuzzy arithmetic with requisite constraints. Fuzzy Sets Syst. **91**, 165–175 (1997)
13. Kosiński, W., Prokopowicz, P., Ślęzak, D.: On algebraic operations on fuzzy numbers. In: Kłopotek, M.A., Wierzchoń, S.T., Trojanowski, K. (eds.) Intelligent Information Processing and Web Mining. AINSC, vol. 22, pp. 353–362. Springer, Heidelberg (2003). https://doi. org/10.1007/978-3-540-36562-4_37
14. Meyer, A.: Milk Run Design (Definitions, Concepts and Solution Approaches). Ph.D. thesis, Institute of Technology. Fakultät für Maschinenbau, KIT Scientific Publishing (2015)
15. Mirabi, M., Shokri, N., Sadeghieh, A.: Modeling and solving the multi-depot vehicle routing problem with time window by considering the flexible end depot in each route. Int. J. Supply Oper. Manag. **3**(3), 1373–1390 (2016)
16. Nguyen, P.K., Crainic, T.G., Toulouse, M.: Multi-trip pickup and delivery problem with time windows and synchronization. Ann. Oper. Res. **253**(2), 899–934 (2017)
17. Polak, M., Majdzik, P., Banaszak, Z., Wójcik, R.: The performance evaluation tool for automated prototyping of concurrent cyclic processes. Fundamenta Informaticae. **60**(1–4), 269–289 (2004)
18. Rudnik, K., Kacprzak, D.: Fuzzy TOPSIS method with ordered fuzzy numbers for flow control in a manufacturing system. Appl. Soft Comput. J. **52**, 1020–1041 (2017)
19. Sitek, P., Wikarek, J.: Capacitated vehicle routing problem with pick-up and alternative delivery (CVRPPAD) – model and implementation using hybrid approach. Ann. Oper. Res. (2017). https://doi.org/10.1007/s10479-017-2722-x
20. Smutnicki, C.: Minimizing cycle time in manufacturing systems with additional technological constraints. In: Proceedings of the 22nd International Conference on Methods and Models in Automation & Robotics, pp. 463–470 (2017). https://doi.org/10.1109/MMAR. 2017.8046872
21. Toth, P., Vigo, D.: The Vehicle Routing Problem. SIAM, Philadelphia (2002). https://doi. org/10.1137/1.9780898718515

Epigenetic Modification of Genetic Algorithm

Kornel Chrominski$^{(\boxtimes)}$ ⓘ, Magdalena Tkacz ⓘ, and Mariusz Boryczka ⓘ

Institute of Computer Science, University of Silesia in Katowice,
Bedzinska 39, 40-080 Sosnowiec, Poland
{kornel.chrominski,magdalena.tkacz,mariusz.boryczkaz}@us.edu.pl

Abstract. The article presents a new operation in the genetic algorithm. This operation mimics the epigenetic process of cytosine methylation. The Epigenetic processes have a huge impact on the functioning of living organisms, but have not yet been reflected in the operations of genetic algorithms. In a study on the evaluation of the operation mimics epigenetics process were used genetic algorithm for Knapsack issue.

Keywords: Genetic algorithms · Epigenetics · Knapsack problem

1 Introduction

For many years man has been trying to imitate nature and use solutions seen in nature. Inspiration processes occurring in nature have also been reflected in computer science. Based on nature observation, algorithms have been developed that use operations that mimic the processes that occur in nature. Algorithms inspired by nature have found a special application in the case of optimization problems that classic deterministic algorithms cannot cope with. Optimization problems are a group of problems for which we are able to define a certain function for which iterative we choose the appropriate values so as to obtain its optimal value. A special type of algorithms inspired by nature - genetic algorithms are the subject of research in this article. They belong to the group of evolutionary algorithms. The idea of genetic algorithms is based on the theory of evolution and processes occurring in the cells of every living organism. In recent years there has been a significant progress in science in the field of biology and molecular genetics, many new mechanisms have been discovered in the process of inheritance and responsible for the evolution of organisms. These mechanisms are not closely related to direct DNA sequence modifications and have been called epigenetic processes. Epigenetic processes have a significant impact on inheritance and are responsible for the influence of external factors on the functioning of living organisms. Classic genetic algorithms were developed in the 1960s, so it was not possible to include all currently known processes regarding inheritance and variability of species. It is these relatively new discoveries in the field of molecular genetics (which are epigenetic processes) that have become

ⓒ Springer Nature Switzerland AG 2020
V. V. Krzhizhanovskaya et al. (Eds.): ICCS 2020, LNCS 12138, pp. 267–278, 2020.
https://doi.org/10.1007/978-3-030-50417-5_20

the starting point for the research presented in this article. The starting point for the introduction of new operations in the genetic algorithm (modeled on epigenetic processes) was that since they are of great importance in nature when adjusting living organisms to environmental conditions, it is possible that their proper mapping in genetic algorithms will improve their efficiency.

2   Background and Motivation

Epigenetic processes occurring in the living organism have a huge impact on their functioning. Genetic algorithms mimics the operations of living organisms, but epigenetic operations have not yet been implemented in genetic algorithm. The use of operations mimic epigenetic processes is aimed at improving the efficiency of genetic algorithms by reducing the number of new generations being created, which influence the time need to find solution by the genetic algorithm. This article presents the results of the appropriate probability of occurrence the modification mimicking the cytosine methylation process, as well as the results of studies on the impact on the reduction of the number of generations. The time of operation of the algorithm with and without operation mimics epigenetic process was also presented. The research results presented in the article are part of the research on the possibility of using processes that mimic epigenetic processes in genetic algorithms.

2.1   Genetic Algorithms

Genetic algorithms [1–5] have found application in solving optimization problems in cases where deterministic approaches did not work well. The idea of a genetic algorithm, as an algorithm inspired by the theory of evolution, was presented by John Holland in the 1960s, and then developed by David E. Goldberg in the 1980s. In genetic algorithms, the terminology was taken from biological sciences and adapted for the needs of genetic algorithms. The pseudo-code [6–8] of the classical genetic algorithm was presented as an Algorithm 1.

The first necessary step in genetic algorithms is choosing the right coding method for the individual - this has a significant impact on whether the algorithm will get the expected result. The next step in constructing the task for the genetic algorithm is to properly construct the function of assessing individuals (fitness function). The fitness function return a value that shows how good the solution is. The value of the fitness function [9–12] is calculated for all individuals in a given population, and its value affects the likelihood of an individual's participation in reproduction – creating a new population. Operations imitating biological processes are carried out on the created population of possible solutions. These operations include individual selection, crossover, and mutation of individuals. In addition to standard operations in the genetic algorithm, there may also be additional operations or modifications of standard operations.

Algorithm 1. Genetic algorithm

    Data: population of individuals;
    Result: Best individuals;
1 **begin**
2     $t = 0$;
3   Create an initial population– $Pop(0)$;
4   Evaluate individuals – calculate the value of the fitness function for each individual in the population P_0;
5   **do**
6   select individuals for the new population $Pop(t)$ – selection;
7   perform crossover operation;
8   perform the mutation of individuals;
9   evaluate individuals;
10   replace old population a new one;
11       $t = t + 1$;
12   **while** *stop condition reached*;
13 **end**

2.2 Epigenetics

Epigenetics [14–16] is a science that studies the processes of extragenic inheritance, as well as the impact of external factors on the level of gene expression. Gene expression determines the phenotypic characteristics of an individual, i.e. its adaptation to the environment, and behavior, appearance, etc. For a long time, scientists have been wondering, for example, why there is a difference in the appearance and behavior of identical twins, or why cloned individuals, despite identical sequences genes, however, have a different coloration, exhibited a different behavior. Some of these puzzles could be explained by discovering the mechanisms that are the subject of epigenetics research.

Concepts related to epigenetics emerged at the time of the discovery that some changes in the genotype of living organisms are not directly related to the structure of DNA, its changes and inheritance processes. They began to wonder what could cause these changes. The result of the research was the discovery of numerous molecules that affect the processes that occur in living organisms, affecting how the genetic code will be read. In other words - how the individual's phenotype will change without changing its DNA.

It turned out that epigenetic processes play a significant role in differentiating the population and adapting to new conditions. Epigenetic modifications are also a source of some diseases, and also affect individual traits such as, for example, perception of the world or personality traits. It can be stated that the genotype of living organisms is a place of storage of relatively static genetic information, and epigenetic processes are specific dynamic controllers (inhibitors or catalysts) responsible for activating certain information.

Epigenetic processes, through biochemical changes, affect the structure of DNA in living organisms, as well as the level of gene expression. Launching a given epigenetic process may occur as a result of changes occurring in the body (also randomly), as well as as a result of external (environmental) factors. Current research indicates that the environment in which the organism resides, components of the diet, as well as chemical compounds can be impulses to launch a given epigenetic process in the body. Scientists estimate that a thorough understanding of epigenetic processes will allow the development of effective gene therapies for many diseases. The main modifications occurring in our epigenetic genotype include:

- Modifications resulting from the interaction of free particles – in each cell, apart from the DNA strand, they find numerous short fragments resembling the structure of the DNA strand, which can attach to DNA and modify the protein encoded by a given fragment of the genotype (e.g. inheritance with prion).
- modifications at the level of DNA strand - most often they lead to silencing the expression of a particular gene, i.e. reducing its expression level (e.g. cytosine methylation process);
- modifications at the histone level - histones are small proteins around which the DNA strand is wrapped; their modification may affect the increase or decrease in the synthesis of proteins encoded in a given DNA segment (e.g. deacetylation, allelic exclusion);

Due to the fact that epigenetics is a relatively young field of science, and epigenetic processes themselves have been discovered relatively recently, they have not yet been reflected in evolutionary algorithms. Therefore, the author decided to check whether, as in the case of living organisms (where epigenetics affects faster differentiation of the population and its better adaptation to environmental conditions), also in the case of genetic algorithms, the use of operations imitating epigenetic processes will improve the operation of these algorithms (by faster finding the optimal result by a given genetic algorithm).

3   Genetic Algorithm with a Modification that Mimics Cytosine Methylation

This section presents the proposed modification that mimics the epigenetic process of cytosine methylation. In the first subsection there is information about the genetic algorithm for solving the Knapsack problem, this algorithm was used as the basis for placing operation mimics the epigenetic process. The next part provides information on the biological basis of modification, followed by the idea of modification and how to implement it in a genetic algorithm.

3.1   Genetic Algorithm for Solving the Knapsack Problem

The formal definition of the Knapsack problem [17,18] can be formulated as follows:

A backpack with the maximum load capacity V_K is available, and a set of N elements $\{x_1, ..., x_j, ..., x_N\}$, with what each element of the set has a specific weight of m_j, and the value of c_j, $j \in \mathbb{N}$.
maximize:

$$\sum_{j=1}^{N} c_j x_j \qquad (1)$$

Assuming:

$$\sum_{j=1}^{N} m_j x_j \leq V_K, \qquad x_j = 0 \ or \ 1, \ j = 1, ..., N \qquad (2)$$

The genetic algorithm used for Knapsack is based on a classic genetic algorithm. Contains standard operations for genetic algorithms such as crossing, mutation, and selection of individuals. Standard operations used in the genetic algorithm for solving Knapsack problem:

1. **Coding of individuals:** The algorithm uses binary coding of individuals. A value of 1 corresponds to the information that the item should be loaded into the backpack, a value of 0 means that the item should not be loaded into the backpack.
2. **Fitness function:** The algorithm uses the fitness function calculated on the basis of three parameters:
   - weight of the item;
   - item value;
   - maximum capacity of the backpack.
   The best adapted individual is the one for which the most valuable items will be loaded, without exceeding the maximum capacity of the backpack.
3. **Individual selection:** The algorithm used is a tournament selection method.
4. **Crossing individuals:** One-point crossing operations were used in the algorithm implementation.
5. **Mutation of individuals:** point mutations were used to change individual genes in the genotype of the individual.

3.2 Biological Basis of the Cytosine Methylation Process

The operation mimics epigenetic processes proposed in this article is based on the epigenetic process of cytosine methylation. Cytosine methylation is a process involving the attachment of methyl groups (-CH3) to the nitrogenous bases of nucleotides (the basic building block of DNA and RNA nucleic acids). The attachment of a methyl group to nucleotides reduces the expression level of genes encoded by a given DNA fragment. If a large DNA strand is methylated, the sequence fragment may be blocked so that the gene cannot be read. Cytosine methylation may also affect the transfer of genetic information by preventing the transfer of a specific gene in the process of inheritance. This process in living organisms also affects the formation of tissues (cell specialization). It is also extremely important in the overall functioning of the body, because only a small number of genes have a constant level of expression (so called housekeeping genes responsible for basic life functions). Other genes, as a result of various biological processes, have variable levels of expression, which affects the proper functioning of the body. Cytosine methylation can also help the body get rid of the external DNA code (e.g., from viruses), which protects the body against this code. Disturbance of the methylation process may lead to the development of, for example, cancer or other genetic diseases (e.g. Angelman, Prader-Willi, Beckwith-Wiedeman syndromes).

3.3   Metylation of Cytosine in Genetic Algorithm

The proposed modification simulated the process of blocking a fragment of a subject's genotype sequence. The epigenetic modification was implemented in the crossover of individuals, modifying this operation. The progeny contain genotype fragments of each parent. If a process simulating cytosine methylation occurs in the crossing operation, a particular fragment of the individual's genotype is blocked, so that he does not participate in the crossing. This is equivalent to the fact that the genome fragment undergoing cytosine methylation will not be passed on to the descendants, i.e. it will not appear in the new population.

Figure 1 shows a diagram of the process of cytosine methylation in the process of crossing individuals. The dark gray fragment of the individual's genotype has undergone the methylation process, i.e. it does not participate in the crossing process.

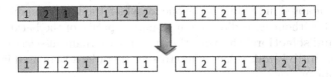

Fig. 1. Epigenetic operation imitating the process of cytosine methylation

In order to prevent the loss of the best solutions in the process of crossing, the process of blocking the sequence occurs in individuals with the lowest values of the fitness function, which aims to transfer to the descendants a larger fragment of the better-adapted individual's genotype. This means that both progeny receive only a specific genotype fragment from the parent having a better fitness function value. Thanks to this, the gene sequence with the poorer fitness function (from the less adapted individual) is "silenced".

Algorithm 2 represents the pseudo-code of the process simulating the process of cytosine methylation. The proposed epigenetic modification occurs in the process of crossing with an appropriate probability, the optimal value of which has been determined empirically.

Algorithm 2. Modified crossover algorithm including simulations of the cytosine methylation process

---

Data: individuals population, methylation probability (P_m), individuals fitness function F_p

1 select with probability P_m a pair of individuals to undergo the cross-over operation with the Epi modification;

2 **foreach** *selected pair of individuals* **do**

3 | generate a length and location of the blocked part of genotype;

4 | **if** $F_p(individual1) > F_p(individual2)$ **then**

5 | | block a part of genotype in individual2;

6 | **end**

7 | **else**

8 | | block a part of genotype in individual1;

9 | **end**

10 | perform the crossover operation without the blocked part of genotype;

11 **end**

The presented algorithm of epigenetics modification (Algorithm 2) shows the modified crossover operation at the time of modification, the rest of the algorithm operation corresponds to basic genetic algorithm. Based on the Algorithm 2, a certain group of individuals selected from the entire population undergoes a modified crossing operation, the remaining individuals undergo a standard crossing operation. In the first step, the length of the genotype to be blocked as part of epigenetic modification and the location of the blocking is randomly generated. In the next step, the values of the fitness function of individuals involved in crossing are compared. In an individual with a lower value of the fitness function, transmission of the fragment of the genotype to the descendants, whose location and length was determined in an earlier step, is blocked. Crosses are performed on individuals without the blocked genotype fragment of one of the parents.

4   Experiments

The proposed modification based on the cytosine methylation process has been tested in terms of selecting the optimal probability of occurrence of operations in the genetic algorithm. The optimal probability is the one for which it obtains the biggest reduction the number of generations. In the next step, it was checked how the introduction of the operation (with optimal probability) mimicking the process of cytosine methylation into the genetic algorithm affects the number of algorithm generations. In the last step, the operation time of the algorithms was compared with and without epigenetic modification.

Five data sets were used in the experiments, differing in the possible number of items to be packed and the maximum permissible capacity of the backpack.

Data sets are artificial collections created for the purposes of experiments. The description of the data sets for the load optimization algorithm is presented in Table 1.

Table 1. Datasets used in experiments

Dataset	Number of items to pack	Max capacity of backpack
Dataset 1	10	20
Dataset 2	20	40
Dataset 3	40	80
Dataset 4	80	160
Dataset 5	160	320

Parameters such as crossing probability, mutation and stop condition have been established for the genetic algorithm used in the study. The parameters set for the load optimization algorithm are standard values recommended in the literature and are presented in the Table 2.

The set parameters for GA for loading optimization were used both in the case of the base algorithm and algorithms modified for operations imitating epigenetic processes.

Table 2. Parameters of the genetic algorithm used in experiments

Parameters	Value
Crossover probability	90%
Probability of mutation	1%
Individual selection method	Tournament
Stop condition	Determined by the length of the set number generation without improving the value of the fitness function

In the first step, it was checked whether the occurrence of epigenetic operations affects the number of iterations needed to obtain the optimal result. The number of generations needed to obtain the best solution depending on the probability of occurrence of the proposed operation for individual test sets is presented in the Table 3. The results presented in Table 3 are the average value of the number of generations needed to obtain the expected result for 100 repeats of each algorithm.

Based on the Table 3, it can be seen that the smallest number of generations needed to obtain the best solution for most test sets occurs in the case of 40% probability of epigenetic operation. The exception is test set 3, in which the minimum number of generations was reached in the case of a modification probability of 30%.

Table 3. Number of generations needed to obtain the best result for individual in dataset depending on the probability of epigenetics operations (The smallest number of iterations for a given test set is marked in bold italics)

Probability of modification occurring	The number of generations needed to find the best solution				
	Dataset 1	Dataset 2	Dataset 3	Dataset 4	Dataset 5
0% (without modification)	154	631	973	1355	1513
5%	178	624	968	1454	1574
10%	172	614	972	1286	1530
20%	144	578	690	1103	1588
30%	126	430	*641*	1186	1591
40%	*113*	*406*	644	*889*	*1404*
50%	131	419	658	1016	1459
60%	148	425	741	1021	1589
70%	152	437	951	1165	1598
80%	159	410	963	1302	1581
90%	166	623	967	1398	1598
100%	185	645	980	1465	1597

At the Fig. 2 are presented the results of the comparison of changes in the fitness function for the genetic algorithm with epigenetic modification (with the optimal probability of modification occurring), and the algorithm without modification for individual test sets. From the graphs it is possible to read how quickly the value of the fitness function of the best individual increases with next generations of algorithms.

Based on the graphs in Fig. 2, it can be seen that in the case of the genetic algorithm with the epigenetic modification, the value of the best individual fitness function increased faster compared to the algorithm without modification. The largest differences in the increase in the value of the fitness function for genetic algorithm with epigenetic operation compared to basic genetic algorithm can be seen for the second and the fifth data sets (Figs. 2b and 2e). However, the smallest differences in the change in the value of the fitness function can be observed in the case of set 4 (Fig. 2d).

The Table 4 shows the average running time (for 100 repeats) of the algorithm with and without modification. The table also shows the standard deviation value and the percentage value of the algorithm reduction time after application the operation mimicking the cytosine methylation process.

Although the introduced epigenetic modification is an additional operation in the genetic algorithm, a reduction in the total time needed to obtain the expected result by the genetic algorithm has been observed. For most data sets, the time reduction was above 25%. The impact on the reduction of the algorithm's operation time with modification was the reduction of the number of generations in the algorithm.

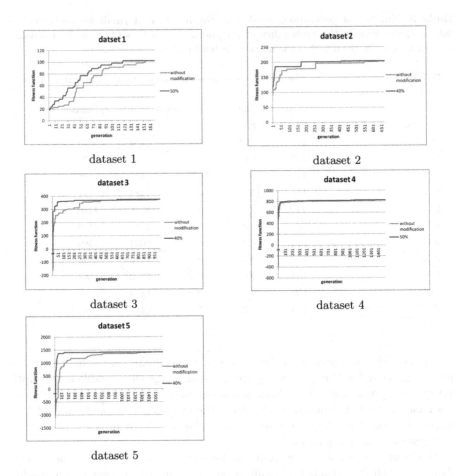

dataset 1   dataset 2

dataset 3   dataset 4

dataset 5

Fig. 2. Changes in the value of the fitness function for generation in the case of an algorithm with and without epigenetic modification

Table 4. Comparison of one generation times for a genetic algorithm with and without epigenetic operation

Dataset	Time [s]				Change %
	Epi_GA		GA		
	Mean	Sd	Mean	Sd	
Dataset 1	7.230	0.68	9.70	1.850	25.46%
Dataset 2	32.480	4.470	45.430	6.310	28.51%
Dataset 3	70.510	9.62	95.350	20.430	26.05%
Dataset 4	160.909	14.224	228.995	29.810	29.73%
Dataset 5	475.956	161.460	481.134	173.995	1.080%

Table 5. Statistics for two sample t-test - number of generations in genetic algorithm with modifications and without.

Statistics for two sample t test	
t	-3.0754
df	4
p-value	0.0185

The Table 5 shows the T test value for the number of generations for the algorithm with and without epigenetic modification.

Based on the Table 5, it can be confirmed that the difference in the number of generations for the algorithm with epigenetic modification compared to the algorithm without modification is statistically significant - a p-value below 0.05.

This section presents the impact of a modification that mimics the epigenetic process of cytosine methylation on the operation of the genetic algorithm.

5 Conclusion and Future Works

The article presents the modification of the genetic algorithm. The modification mimics the epigenetic process of cytosine methylation. Based on the conducted research, it was shown that the application of the proposed modification improves the efficiency of the genetic algorithm. The efficiency is improved by reducing the number of generations in the genetic algorithm, and thus reducing the duration of the genetic algorithm. The presented results are part of the research on the possibility of using epigenetic processes in genetic algorithms. As part of our research, it is planned to examine the assembly of other modifications that mimic other epigenetic processes in various genetic algorithms. Research is also being carried out into the development of an epigenetic algorithm.

References

1. Abdel-Magid, H.L., Dawoud, M.M.: Optimal agc tuning with genetic algorithms. Electr. Power Syst. Res. **38**(3), 231–238 (1996)
2. Arabas, J.: Wyklady z algorytmow ewolucyjnych. WNT (2004). (in Polish)
3. Goldberg, D.E.: Genetic Algorithms in Search, Optimization, and Machine Learning. Addison-Wesley Professional (1989)
4. Goldberg, D.E., Deb, K.: A comparative analysis of selection schemes used in genetic algorithms. In: Foundations of Genetic Algorithms, pp. 69–93, Elsevier (1991)
5. Hager, W.W.: Computational optimization and applications (2017)
6. Greenwell, R.N., Angus, J.E., Finck, M.: Optimal mutation probability for genetical algorithms. Math. Comput. Model. **21**(8), 1–11 (1995)
7. Abdoun, O., Abouchabaka, J.: A comparative study of adaptive crossover operators for genetic algorithms to resolve the traveling salesman problem. Int. J. Comput. Appl. **31**(11) (2012)

8. Eiben, A.E., Smit, S.K.: Evolutionary algorithm parameters and methods to tune them. In: Hamadi, Y., Monfroy, E., Saubion, F. (eds.) Autonomous Search, pp. 15–36. Springer, Heidelberg (2011). https://doi.org/10.1007/978-3-642-21434-9_2

9. Hopper, E., Turton, B.: Application of genetic algorithms to packing problems - a review. In: Chawdhry, P.K., Roy, R., Pant, R.K. (eds.) Soft Computing in Engineering Design and Manufacturing, pp. 279–288. Springer, London (1998). https://doi.org/10.1007/978-1-4471-0427-8_30

10. Hussain, A., Muhammad, Y., Nawaz, A.: Optimization through genetic algorithm with a new and efficient crossover operator. Int. J. Adv. Math. **2018**(1), 1–14 (2018)

11. Jebari, K.: Selection methods for genetic algorithms. Int. J. Emerg. Sci. **3**, 333–344 (2013)

12. Khan, G.M.: Evolutionary computation. Evolution of Artificial Neural Development. SCI, vol. 725, pp. 29–37. Springer, Cham (2018). https://doi.org/10.1007/978-3-319-67466-7_3

13. Moore, D., David, S.: The Developing Genome: An Introduction to Behavioral Epigenetics. OXFORD UNIV PR (2015)

14. Weglenski, P.: Genetyka molekularna. PWN Warszawa (2006). (in Polish)

15. Dupont, C., Armant, D., Brenner, C.: Epigenetics: definition, mechanisms and clinical perspective. Semin. Reprod. Med. **27**(05), 351–357 (2009)

16. Al-Haddad, R., et al.: Epigenetic changes in diabetes. Neurosci. Lett. **625**, 64–69 (2016)

17. Li, X., Zhao, Z., Zhang, K.: A genetic algorithm for the three-dimensional bin packing problem with heterogeneous bins. In: Industrial and Systems Engineering Research (2014)

18. Man, K., Tang, K.-S., Kwong, S.: Genetic Algorithms. Springer, London (2012). https://doi.org/10.1007/978-1-4471-0577-0

ITP-KNN: Encrypted Video Flow Identification Based on the Intermittent Traffic Pattern of Video and K-Nearest Neighbors Classification

Youting Liu[1,2,3], Shu Li[1,2,3]($\boxtimes$), Chengwei Zhang[1,2], Chao Zheng[1,2], Yong Sun[1,2], and Qingyun Liu[1,2,3]

[1] Institute of Information Engineering, Chinese Academy of Sciences, Beijing, China
`lishu@iie.ac.cn`
[2] National Engineering Laboratory of Information Security Technologies, Beijing, China
[3] School of Cyber Security, University of Chinese Academy of Sciences, Beijing, China

Abstract. As video dominates internet traffic, researchers tend to pay attention to video-related fields, such as video shaping, differentiated service, multimedia protocol tunneling detection. Some video-related fields, e.g., traffic measurement and the metrics for Quality of Experience, are based on video flow identification. However, video flow identification faces challenges. Firstly, the increasing adoption of Transport Layer Security makes payload-based methods no longer applicable. Secondly, traffic features differ when generated by different streaming protocols.

This paper proposes a video flow identification method, called ITP-KNN, which utilizes the intermittent traffic pattern-related features (ITP) and the K-nearest neighbors (KNN) algorithm. The intermittent traffic pattern is caused by fragmented transmission, which is common among video streamings generated by different streaming protocols. Therefore, the intermittent traffic pattern is useful for overcoming the above challenges and then differentiating video traffic from not-video traffic. We develop a set of features to describe the intermittent traffic pattern. Preliminary results show the promise of ITP-KNN, yielding high identification recall and precision over a range of video content and encoding qualities.

Keywords: Video streaming · Encrypted traffic · Traffic identification · Traffic pattern · Explainable machine learning · Feature selection

Supported by the Strategic Priority Research Program of the Chinese Academy of Sciences (Grant No. XDC02030600), Youth Innovation Promotion Association CAS and CAS Key Technology Talent Program.

V. V. Krzhizhanovskaya et al. (Eds.): ICCS 2020, LNCS 12138, pp. 279–293, 2020.
https://doi.org/10.1007/978-3-030-50417-5_21

1   Introduction

Video dominates the Internet: streaming video occupies 65% of traffic globally [28]. The increasing popularity of video has attracted the interest of different research groups. As a result, several video-related research directions have become hot topics, such as video traffic model [24,34], the metrics for Quality of Experience (QoE) [10], video title identification [9,13,29], multimedia protocol tunneling [2].

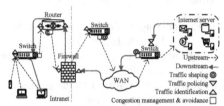

(a) An example of Quality of Service (QoS)
technology implementations .

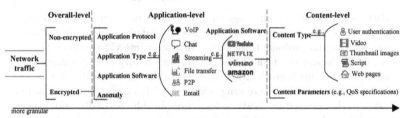

(b) Multilevel categorization of traffic identification methods. We aim to solve a content-level traffic identification problem.

Fig. 1. What is video flow identification? QoS measures the ability of a network to provide differentiated service guarantees for diverse traffic following changing and complex network conditions [15]. Furthermore, traffic identification is a basic QoS technique, and ISPs can provide differentiated services based on traffic identification. Video flow identification is a content-level identification.

Video flow identification (described in Fig. 1) is needed. There is an assumption in much research [9,13,29,32]: the flow is labeled as a video flow or not-video flow precisely. In other words, video flow identification is essential for practical applications of some video-related research, e.g., the metrics for internet video QoE. Besides, limiting or blocking video transmission also requires video flow identification. When streaming video transmission is limited, the resolution of the streaming video will switch to a lower one, in order not to interrupt the video playback. Thus, it is possible to limit video transmission to mitigate network congestion without significantly affected the Quality of Service. As an example, streaming-video services such as Netflix and YouTube began switching to standard definition to manage internet congestion during COVID-19 [26].

Two challenges of video flow identification are protocol diversity and traffic encryption.

Protocol diversity challenges video flow identification as it makes traffic features differ. Protocol diversity results from the application of kinds of streaming protocols. Streaming protocols are designed to provide online video playback without completely downloading it first. There are various popular streaming protocols, e.g., Dynamic Adaptive Streaming over HTTP (DASH), HTTP Live Streaming (HLS), or even private protocol. Protocol diversity makes traffic features differ. For example, packet size is different when different streaming protocols generate flows. The advanced method presented by Li et al. [18] requires detecting the upstream request, which depends on the packet size, so it is not universal enough to deal with other streaming protocols.

Video flow identification faces the other challenge, traffic encryption. *Sandvine*'s report states that more than 50% of global traffic is encrypted [27]. When traffic is encrypted, conventional payload-based methods, such as deep packet inspection, are no longer applicable. Fortunately, traffic analysis (TA) still work even when traffic is encrypted [23]. However, for encrypted traffic, fine-grained classification is a tough task [4], and video flow identification is a kind of fine-grained identification.

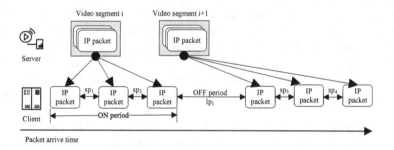

Fig. 2. Transmitting video in segments leads to the traffic showing an intermittent pattern—continuously arrive of packets during ON periods, and a suspend for transmission during OFF periods. The intermittent traffic pattern consists of lots of ON periods and OFF periods one-to-one. Furthermore, continuously arrive of packets means a set of small PAIs (e.g., sp_i), smaller than the PAI (e.g., lp_j) that reflects the suspend for transmission (OFF period). Therefore, the probability distribution image of PAIs will contain two peaks. One peak is influenced by ON periods, consisting of sp_i, and i value range is from 1 to n in increments of 1 while n is the number of small PAIs (sp_i). The other peak reflects OFF periods, consisting of lp_j, and j value range is from 1 to z in increments of 1 while z is the number of larger PAIs (lp_j).

In this paper, we propose a video flow identification method: ITP-KNN. ITP-KNN is a kind of machine learning-based traffic analysis, using intermittent traffic pattern-related features (ITP) as input and the K-nearest neighbors (KNN) algorithm as classifier. Regularly and fragmentarily transmission of video, which is caused by the application of streaming protocols, makes the traffic traces showing the intermittent pattern [21] (as shown in Fig. 2). Thus, the intermittent traffic pattern helps differentiate video traffic from not-video traffic.

We summarize our key contributions as follows:

- We develop a set of features to depict video transmission patterns: the intermittent traffic pattern.
- We interpret why the features we developed can help to identify video flows.
- We implement and evaluate ML-based TA in video flow identification: we propose a video flow identification framework, and evaluate it on five public datasets.

2 Related Work

2.1 Machine Learning (ML)-Based Traffic Analysis (TA)

TA has been proved applicable in encrypted traffic identification [4,22]. McGregor et al. [20] first apply ML algorithms to identify traffic. After that, researchers began to pay attention to the application of ML algorithms in traffic identification [7].

Manual feature selection is required when using conventional ML algorithms [35]. Encrypted traffic identification lacks public datasets to evaluate different methods. Draper-Gil et al. [8,17] released their datasets, and they used time-related features to realize VPN or Tor traffic identification. However, they did not explain why the time-related features can help identify traffic. Shi et al. [30] developed a set of network path-related features to identify the traffic source; the disadvantage is that models that use network path-related features need to retrain regularly.

In recent years, deep learning (DL) has been implemented in traffic identification with different aims [25]. Wang [37] showed that compared with conventional ML algorithms, DL had improved the result of traffic identification. However, their model is not explainable.

DL obviates the requirement of manually feature selection since the feature selection runs automatically through training [25]. It is a high cost for domain experts to select features for identifying various types of traffic. Therefore, the DL-based TA is more suitable for multiclass identification than conventional ML-based TA. Conventional ML-based TA is not inferior to the DL-based one; the problems that prevent the application of DL are that DL requires a large and representative dataset, and the result of DL lacks interpretability.

2.2 Video Streaming

Li et al. [18] presented Silhouette to identify YouTube video flows. Silhouette depends on the packet size of the upstream request, so it is not universal enough to deal with other streaming protocols. Shi et al. [32] proposed a method that can identify the video source. However, the features adopted by them are strongly related to the network condition, so regular training is needed. Casas et al. [5] achieved an application-level identification; some features used by them are computational complex. They concluded that the flows' label in the ground-truth dataset should carry what content (e.g., video, Web pages or YouTube thumbnail images) the flow carries. Garcia et al. [11] improved Ground Truth techniques by applying unsupervised clustering methods on DPI-labeled traffic. However, the method is designed for offline analysis only.

Video title identification enjoys the most attention. Advanced researches choose to establish the video fingerprints database [9,13,29]. Video fingerprint is a type of information leakage caused by DASH and Variable Bit Rate. The common flaw is that building video fingerprint is hard to cope with the growing number of video titles. Besides, the fingerprint-based method can only identify the known video, and the best result for video outside the training set is to mark as unknown.

As early as 2016, Dubin et al. [9] first identified YouTube video titles, assuming that adversaries can directly observe encrypted video flows at the network layer, yet their detectors are susceptible to noise. Schuster et al. [29] extended the attack scenario to one where direct eavesdropping was not feasible. They used a neural network to identify video titles, and there were no false positives for video outside the training set. Recently, Gu et al. [13] presented a method to build video fingerprint by using differentiated bit rate, which eliminated the impact of the switching of video resolution.

3 Preliminaries

3.1 Streaming Protocols

As the popularity of Internet video streaming services has increased, streaming protocols have developed several generations to make the utmost of advanced networking techniques. Nowadays, HTTP-based adaptive streaming protocols become the most popular because of the advantages of easy to deploy and firewall penetration [3,18,21]. Video-related companies or institutions (e.g., Netflix, Adobe, Apple, and Microsoft) have all proposed HTTP-based adaptive streaming protocols [19].

HTTP-based adaptive streaming protocols are designed to provide smooth video playback [21]. With the adoption of HTTP-based adaptive streaming protocols, the video resolution can be dynamically adjusted according to the network condition. Moreover, HTTP-based adaptive streaming protocol can avoid the waste of bandwidth resources and reduce network congestion. For example, the customer may not watch the entire video. In this case, the video with such protocols is periodically delivered in segments if the customer continues watching.

Streaming protocols have the above-mentioned advantages due to the segmented transmission of video. As shown in Fig. 2, video is divided into several video segments, and each video segment usually contains content that lasts for a few seconds. The client request one video segment at a time, and then the server will send the video segment. Therefore, the video resolution can be adjusted as soon as the network condition changes. Moreover, video transmission can be suspended immediately when the customer does not continue watching the video.

3.2 The Intermittent Traffic Pattern

As described in Sect. 3.1 and Fig. 2, video traffic exhibits the intermittent pattern because of the segmented transmission. To be transmitted over the network, one video segment will be divided into several IP packets. Consisting of many ON periods and OFF periods, the intermittent traffic pattern seem like a faucet: when the faucet is turned on (ON periods), the water flooding, and when the faucet is turned off (OFF periods), transmission suspend. During the ON period, a video segment is transmitted, resulting in the continuous arrival of packets; during the OFF period, video transmission will not be performed. Through Packet Arrival Interval (PAI), we can observe the intermittent traffic pattern of video more intuitive, as shown in Fig. 2. Moerover, the intermittent traffic pattern can be observed in most HTTP video streaming [21], such as video streaming generated by different video service providers, e.g., Youtube [30] and Netflix [1].

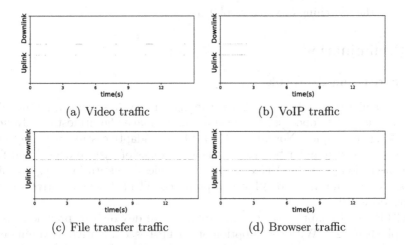

(a) Video traffic

(b) VoIP traffic

(c) File transfer traffic

(d) Browser traffic

Fig. 3. Packet arrival times of different traffic. We ignore the packet size, and all packets appear as points on the horizontal line. Compared with not-video traffic, video traffic shows an intermittent pattern (Downstream). Besides, intervals of upstream requests of video traffic are related to the duration of video segments, consistent with the traffic model proposed by Waldmann et al. [34].

We use PAI to describe the intermittent traffic pattern, as shown in Fig. 3. There is a gap between two video segments because one video segment is transmitted after the sever received a request from the client. As a result, the arrival interval of two successive packets that belong to the same video segment is smaller than the interval between two successive video segment transmission. Therefore, PAI obtained in ON periods is generally smaller than PAI obtained in OFF periods. As a consequence, the distribution of the PAI of video traffic is characteristic. The probability distribution image of PAIs will contain two peaks. The highest peak consist of smaller PAIs (ON period-related), and the other one consist of PAIs that are related to the interval between two successive video segment transmission.

3.3   State-of-the-Art Method

Silhouette. Silhouette, presented by Li et al., is a method to detect YouTube videos from network traffic dataflow. The method has several steps. First, two traffic features (i.e., average downstream payload size, data rate) are extracted. Second, Application Data Units (ADUs) are detected based on two thresholds (i.e., segment length threshold for video ADU, packet length threshold for upstream request). Third, three thresholds are used to determine a flow as a YouTube video. Thresholds are tuned from observing hundreds of YouTube video sessions.

Within encrypted traffic classification, Silhouette is a rare exception, which achieves high recall and none false positives without machine learning algorithms. Therefore, we choose it as one of the state-of-the-art methods.

Machine Learning-Based Methods

Candidate Methods. Since it was applied for traffic classification in 2004 [20], Machine Learning (ML)-based traffic analysis has been attracting much interest [4,22]. Different researchers use different methodological datasets to evaluate their methods [33]. As a result, their methods are not directly comparable [33], increasing the difficulty of selecting a method as the state-of-the-art method. Below, we describe three types of traffic features that are often used in traffic classification.

- **Time-related features.** Time-related features are a set of time-related features (e.g., packets per second, bytes per second, flow duration) that first proposed by draper et al. [8].
- **Network path-related features.** Network path-related features refer to the distribution of Packet Arrival Intervals (PAIs). The distribution of PAIs is proved related to the network path [30].
- **Raw traffic.** The first few packets of the flow are observed enough for traffic identification [25]. Besides, wang et al. use the first 784 bytes of the flow with Convolutional Neural Network (CNN) algorithms, and finally get a better result than conventional ML algorithms in encrypted traffic identification [35,36]. Therefore, we use the first 784 bytes of the flow as a type of traffic features, marked as raw traffic.

We choose KNN and CNN as two candidate algorithms, the reasons list as follows:

– KNN. KNN, representatives for conventional machine learning algorithms, has been proved effective in encrypted traffic classification [9,22,30]. Moreover, it behaves better than other algorithms (e.g., SVM, J48) in some tasks [8,30] (e.g., video source identification).
– CNN. CNN, a typical algorithm of deep learning, is good at classifying sequential data (network traffic can be seen as sequential data) [12,14]. Besides, wang et al. find CNN better than conventional ML algorithms in application-level traffic identification [35,36]. Moreover, CNN was proved to have an excellent result in video title identification [29].

Table 1. Candidate methods vs. our method (ITP-KNN).

Mark	Feature	Algorithm
Draper-Gil et al.'s method	Time-related features	KNN
Shi et al.'s method	Network path-related features	KNN
Wang et al.'s method	Raw traffic	CNN
ITP-KNN (our method)	Intermittent traffic pattern-related features	KNN

Finally, we get three Candidate methods, list in Table 1.

Table 2. The F_1 score of candidate methods (%).

	K = 1	K = 3	K = 5	K = 7	K = 9
Draper-Gil et al.'s method	96.47	96.96	96.95	97.05	97.00
Shi et al.'s method	97.62	97.95	98.24	98.39	98.49
Wang et al.'s method	4.27 (recall = 43.8%, precision = 2.24%)				

Final Battle. We select 94500 flows randomly from five public datasets [8–10, 17,18]. We use these flows as a dataset to evaluate candidate methods. 10-fold cross-validation was used when tested the performance, and all the features are extracted at the end of the flow. Results are listed in Table 2. Shi et al.'s method performs best; therefore, we choose Shi et al.'s method as one of the state-of-the-art methods to judge our methods.

4   Methods

4.1   Overview

We compare our method with two state-of-the-art methods. Therefore, we implemented three methods, as shown in Fig. 4. The three methods are described as following. First, our method, we develop a set of intermittent traffic pattern-related (ITP) features and use the K-nearest neighbors (KNN) algorithm. Thus, our method is called ITP-KNN. Second, we choose Silhouette as one of the state-of-the-art methods, since Silhouette is a training-free method with high recall and none false alarm [18]. Third, we evaluated three kinds of encrypted traffic identification methods. According to the evaluation results, we choose Shi et al.'s method (marked as Shi et al.'s method) as the other of the state-of-the-art methods. Shi et al.'s method achieves an application-level traffic identification with extremely high recall [30].

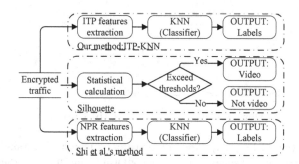

Fig. 4. Overview: our method vs. state-of-the-art methods. We develop a set of intermittent traffic pattern-related (ITP) features. We compare our method (ITP-KNN), with two state-of-the-art methods (Silhouette, and Shi et al.'s method). Firstly, our method ITP-KNN uses K-nearest neighbors (KNN) algorithm and ITP features to identify video flows in encrypted traffic. Secondly, Silhouette is a training-free method that identifies YouTube video flows based on some thresholds and heuristics [18]. Thirdly, Shi et al.'s method uses machine learning algorithms and a set of network path-related (NPR) features. As a result, it needs to retrain regularly [30].

- **ITP-KNN.** The core of this part is feature extraction. We develop a set of features to highlight the intermittent traffic pattern and identify video flows based on these features. ITP-KNN is described in Sect. 4.2.
- **Silhouette** (detailed in Sect. 3.3). Silhouette is a training-free light-weight method. The experiment results show that Silhouette can identify YouTube video flows in encrypted traffic with high recall (QUIC-based streaming: 99%) and a zero false-positive rate [18].
- **Shi et al.'s method** (detailed in Sect. 3.3). Shi et al. [30] proposed a method that can identify video traffic sources (in other words, application software)

at a client-side firewall. Furthermore, their method achieves the best average true positive rate at 94% while using the Nearest Neighbor algorithm [6]. For the sake of fairness, we modify the algorithm used in this method: we change the Nearest Neighbor algorithm to the KNN algorithm.

4.2 Our Framework: KNN that Uses ITP Features as Input

Feature Set: ITP Features. As described in Sect. 3.2, PAI can be used to describe the intermittent pattern of video traffic. There is a series of small PAIs (continuous packets arrival, ON period) followed by a large PAI (the duration of OFF period). Therefore, we can observe differences between video class and non-video class in the distribution of PAI. We randomly selected 50 video instances and 50 non-video instances and drawn their probability density function in Fig. 5.

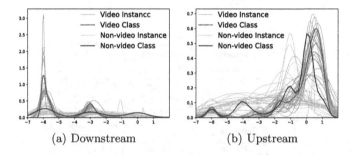

(a) Downstream   (b) Upstream

Fig. 5. Distribution of log(PAI). Non-video instances include file transfer. There are two peaks in video class (downstream), which conform to our inference that video traffic shows intermittent pattern. Moreover, the log(PAI) corresponding to the highest peak in video class (upstream) is between 0 and 1, which is related to the duration of a video segment (1 to 10 s).

Moerover, video traffic is asymmetric, especially for the amount of the transmitted data. Downstream (client-to-server direction) of the video traffic and upstream (server-to-client direction) of the video traffic both show an intermittent pattern, but, they have differences, compared as follows:

- **Downstream.** During the ON period (the duration of ON period is recorded as T_{on}), a video segment is transmitted; thus, packets transmission bursts. The OFF periods are essentially the pauses between the two segments being transmitted.
- **Upstream.** A request contains few packets, and the interval between two upstream requests (recorded as T_{req}) is related to T_{on}; therefore, the data amount and transmission rate of upstream are smaller than those of downstream. (Waldmann et al. [34] found that the distribution of T_{req} is centered precisely around the duration of a video segment (T_{seg}), and T_{on} is close to but slightly less than T_{seg}, which can be observed in Fig. 3 and 5 as well).

Table 3. Description of intermittent traffic pattern-related (ITP) features.

Feature category		Direction	Description
Quantized PAIs		Upstream	23 bin, the first one covering 0 to 5 µs, and each subsequent bin has an upper limit two times larger than the previous one
		Downstream	23 bin, the first one covering 0 to 5 µs, and each subsequent bin has an upper limit two times larger than the previous one
Summary statistics	Simple descriptive statistics	Upstream	Maximum, minimum, and mean over the PAI time series; the two largest values and their bin number of the Quantized PAIs
		Downstream	Maximum, minimum, and mean over the PAI time series; the two largest values and their bin number of the Quantized PAIs
	Higher-order statistics	Upstream	The skew, kurtosis, and coefficient of variation of the PAI time series
		Downstream	The skew, kurtosis, and coefficient of variation of the PAI time series
	Rates	Upstream	Bytes per seconds and packets per second
		Downstream	Bytes per seconds and packets per second

We use 70-dimensional features, listed in Table 3. We extract ITP features from each 15 s network trace and divide these features into two categories, as follows:

– **Quantized PAIs.** We use 23 bin with the first one covering 0 to 5 µs, and each subsequent bin has an upper limit two times larger than the previous one. The last bin covers 10 s to 20 s. ITP-KNN processes upstream packets and downstream packets separately and concatenates the resulting features together. Therefore, Quantized PAIs are 46 dimensions.

- **Summary statistics**. Encrypted traffic provides two main sources of data: a time series of PAIs, and a time series of packet lengths [2]. This kind of feature comprises a sequence of summary statistics computed over the network traces of encrypted traffic, which is a set of features that prevalent for the problem of traffic identification [8,16,22]. As for the selection of summary statistics, we compute multiple descriptive statistics for upstream/ downstream traffic individually. This feature set includes simple descriptive statistics over the PAI time series - such as maximum, minimum, and mean - as well as higher-order statistics like the skew or kurtosis of the time series.

Selected Classifiers. We describe the K-nearest neighbors (KNN) algorithm we have chosen for conducting our experiments as follows:

KNN is a supervised learning algorithm. Many researchers have utilized KNN to realize an application-level traffic classification and proved its better performance [9,22,30]. KNN performs even better than a deep learning algorithm, deep neural network, in video source identification [30,31].

5   Evaluation and Discussion

In the experiments, we use five public datasets [8–10,17,18], together with some traffic captured by ourselves using Wireshark. In totally, we collected more than 450 thousand flows (about 350G). All the traffic was encrypted (HTTPS, QUIC, VPN, Tor, Shadowsocks).

From the above traffic traces, we randomly select 20010 video flows and 282650 not-video flows as our dataset. We use these 302660 flows to evaluate our method, i.e., ML-based TA that uses PR features (marked as ITP-KNN). We compare our method, ITP-KNN, with two state-of-the-art methods, Silhouette (training-free method) and Shi et al.'s method (performed best among ML-based methods in our evaluation).

Video flow identification requires high timely ability, which means the response time of the identification method should be as short as possible. For example, ISPs need to identify video flows in time in order to provide differentiated service. The response time of the identification method, affected mainly by the feature extraction time [8], shows how timely the method is. Therefore, we set the feature extraction time (flow duration) at 15 s (shorter than Shi et al.'s method).

We used 10-fold cross-validation in the evaluation. Table 4 depicts the recall (R), precision (P), and F_1 score (F_1) obtained by using our method (ITP-KNN), and two state-of-the-art methods. Next, we present our main findings.

1. The heuristic method that does not use machine learning algorithms (Silhouette) possesses a limited capability for identifying video flows. This finding is supported by the fact that Silhouette attains a recall of 84.65% (far lower

Table 4. Our method (ITP-KNN) vs. two state-of-the-art methods (%).

	Silhouette			Shi et al.'s method			ITP-KNN		
	R	P	F_1	R	P	F_1	R	P	F_1
K = 1	84.65	99.63	91.53	92.25	92.07	92.16	97.46	97.32	97.39
K = 3				93.95	94.00	93.97	97.36	97.76	97.56
K = 5				94.60	94.13	94.36	97.36	97.90	97.63
K = 7				94.75	94.00	94.37	97.55	97.95	97.75
K = 9				95.00	94.06	94.53	97.82	98.18	98.00

than ML-based methods). We regard recall as the most important metric for the reason that high recall means that a method identified most of the video flows. For example, ISPs need to provide differentiated service guarantees, which require the ability to identify video flows as much as possible.

2. The application-level traffic identification method shows promising results for the identification of video flows (content-level). Results in Table 4 show that Shi et al.'s method performs well in this task. To sum up, these results were very encouraging, which means the present traffic identification methods can be applied to a more fine-grained identification task. However, when the identification task requires high recall and precision, the present traffic identification methods need to be improved.

3. Fine-grained traffic identification task demand for in-depth analysis of traffic. The results in Table 4 suggest this finding: our method (ITP-KNN) behaves best overall in the specific fine-grained traffic identification task (identification of video flows). Our method is based on an in-depth analysis of video traffic: we develop a set of features to stress out the unique transport pattern of video, the intermittent traffic pattern.

6   Conclusion

Experimental evaluations prove the advantages of our method; compared with baseline, our method is better overall. The results show that our method can identify video flows from encrypted traffic.

Our framework is based on one assume: most online video traffic shows the intermittent pattern. Though theoretically universal, it has a flaw that it is only applicable to identify video traffic that generated by present streaming protocols. Though better than the baseline in universality, our method still has a margin of improvement. In future work, we will apply our method to the actual engineering to realize video flows identification in high-speed and large-scale network traffic.

References

1. Ameigeiras, P., Ramos-Munoz, J.J., Navarro-Ortiz, J., Lopez-Soler, J.M.: Analysis and modelling of YouTube traffic. Transa. Emerg. Telecommun. Technol. **23**(4), 360–377 (2012)

2. Barradas, D., Santos, N., Rodrigues, L.: Effective detection of multimedia protocol tunneling using machine learning. In: 27th {USENIX} Security Symposium ({USENIX} Security 18), pp. 169–185 (2018)

3. Bitmovin: Bitmovin video developer report 2019 (2019). https://go.bitmovin.com/video-developer-report-2019. Accessed 1 Jan 2020

4. Cao, Z., Xiong, G., Zhao, Y., Li, Z., Guo, L.: A survey on encrypted traffic classification. In: Batten, L., Li, G., Niu, W., Warren, M. (eds.) ATIS 2014. CCIS, vol. 490, pp. 73–81. Springer, Heidelberg (2014). https://doi.org/10.1007/978-3-662-45670-5_8

5. Casas, P., Mazel, J., Owezarski, P.: MINETRAC: mining flows for unsupervised analysis & semi-supervised classification. In: Proceedings of the 23rd International Teletraffic Congress, pp. 87–94. International Teletraffic Congress (2011)

6. Cover, T., Hart, P.: Nearest neighbor pattern classification. IEEE Trans. Inf. Theory **13**(1), 21–27 (1967)

7. Dainotti, A., Pescape, A., Claffy, K.C.: Issues and future directions in traffic classification. IEEE Netw. **26**(1), 35–40 (2012)

8. Draper-Gil, G., Lashkari, A.H., Mamun, M.S.I., Ghorbani, A.A.: Characterization of encrypted and VPN traffic using time-related. In: Proceedings of the 2nd International Conference on Information Systems Security and Privacy (ICISSP), pp. 407–414 (2016)

9. Dubin, R., Dvir, A., Pele, O., Hadar, O.: I know what you saw last minute–encrypted HTTP adaptive video streaming title classification. IEEE Trans. Inf. Forensics Secur. **12**(12), 3039–3049 (2017)

10. Dubin, R., Dvir, A., Pele, O., Hadar, O., Richman, I., Trabelsi, O.: Real time video quality representation classification of encrypted HTTP adaptive video streaming-the case of safari. arXiv preprint arXiv:1602.00489 (2016)

11. Garcia, J., Brunstrom, A.: Clustering-based separation of media transfers in DPI-classified cellular video and VoIP traffic. In: 2018 IEEE Wireless Communications and Networking Conference (WCNC), pp. 1–6. IEEE (2018)

12. Goodfellow, I., Bengio, Y., Courville, A.: Deep Learning. MIT Press, Cambridge (2016)

13. Gu, J., Wang, J., Yu, Z., Shen, K.: Walls have ears: traffic-based side-channel attack in video streaming. In: IEEE INFOCOM 2018-IEEE Conference on Computer Communications, pp. 1538–1546. IEEE (2018)

14. Gu, J., et al.: Recent advances in convolutional neural networks. Pattern Recogn. **77**, 354–377 (2018)

15. H3C_Technologies: H3c S5120-SI series ethernet switches configuration guide-release. http://www.h3c.com.hk. Accessed 10 Dec 2019

16. Hayes, J., Danezis, G.: k-fingerprinting: a robust scalable website fingerprinting technique. In: 25th {USENIX} Security Symposium ({USENIX} Security 16), pp. 1187–1203 (2016)

17. Lashkari, A.H., Draper-Gil, G., Mamun, M.S.I., Ghorbani, A.A.: Characterization of tor traffic using time based features. In: ICISSP, pp. 253–262 (2017)

18. Li, F., Chung, J.W., Claypool, M.: Silhouette: identifying YouTube video flows from encrypted traffic. In: Proceedings of the 28th ACM SIGMM Workshop on Network and Operating Systems Support for Digital Audio and Video, pp. 19–24. ACM (2018)

19. Martin, J., Fu, Y., Wourms, N., Shaw, T.: Characterizing Netflix bandwidth consumption. In: 2013 IEEE 10th Consumer Communications and Networking Conference (CCNC), pp. 230–235. IEEE (2013)

20. McGregor, A., Hall, M., Lorier, P., Brunskill, J.: Flow clustering using machine learning techniques. In: Barakat, C., Pratt, I. (eds.) PAM 2004. LNCS, vol. 3015, pp. 205–214. Springer, Heidelberg (2004). https://doi.org/10.1007/978-3-540-24668-8_21

21. Mustafa, I.B., Uddin, M., Nadeem, T.: Understanding the intermittent traffic pattern of HTTP video streaming over wireless networks. In: 2016 14th International Symposium on Modeling and Optimization in Mobile, Ad Hoc, and Wireless Networks (WiOpt), pp. 1–8. IEEE (2016)

22. Nguyen, T.T., Armitage, G.J.: A survey of techniques for internet traffic classification using machine learning. IEEE Commun. Surv. Tutor. **10**(1–4), 56–76 (2008)

23. Raymond, J.-F.: Traffic analysis: protocols, attacks, design issues, and open problems. In: Federrath, H. (ed.) Designing Privacy Enhancing Technologies. LNCS, vol. 2009, pp. 10–29. Springer, Heidelberg (2001). https://doi.org/10.1007/3-540-44702-4_2

24. Reed, A., Aikat, J.: Modeling, identifying, and simulating dynamic adaptive streaming over HTTP. In: 2013 21st IEEE International Conference on Network Protocols (ICNP), pp. 1–2. IEEE (2013)

25. Rezaei, S., Liu, X.: Deep learning for encrypted traffic classification: an overview. IEEE Commun. Mag. **57**(5), 76–81 (2019)

26. Sandvine: COVID-19 global internet trends. https://www.sandvine.com/covid-19-trends. Accessed 3 Apr 2020

27. Sandvine: The global internet phenomena report, September 2019 (2019)

28. Sandvine: The mobile internet phenomena report, February 2020 (2020)

29. Schuster, R., Shmatikov, V., Tromer, E.: Beauty and the burst: remote identification of encrypted video streams. In: 26th {USENIX} Security Symposium ({USENIX} Security 17), pp. 1357–1374 (2017)

30. Shi, Y., Biswas, S.: Protocol-independent identification of encrypted video traffic sources using traffic analysis. In: 2016 IEEE International Conference on Communications (ICC), pp. 1–6. IEEE (2016)

31. Shi, Y., Biswas, S.: A deep-learning enabled traffic analysis engine for video source identification. In: 2019 11th International Conference on Communication Systems & Networks (COMSNETS), pp. 15–21. IEEE (2019)

32. Shi, Y., Ross, A., Biswas, S.: Source identification of encrypted video traffic in the presence of heterogeneous network traffic. Comput. Commun. **129**, 101–110 (2018)

33. Velan, P., Čermák, M., Čeleda, P., Drašar, M.: A survey of methods for encrypted traffic classification and analysis. Int. J. Netw. Manag. **25**(5), 355–374 (2015)

34. Waldmann, S., Miller, K., Wolisz, A.: Traffic model for HTTP-based adaptive streaming. In: 2017 IEEE Conference on Computer Communications Workshops (INFOCOM WKSHPS), pp. 683–688. IEEE (2017)

35. Wang, W., Zhu, M., Wang, J., Zeng, X., Yang, Z.: End-to-end encrypted traffic classification with one-dimensional convolution neural networks. In: 2017 IEEE International Conference on Intelligence and Security Informatics (ISI), pp. 43–48. IEEE (2017)

36. Wang, W., Zhu, M., Zeng, X., Ye, X., Sheng, Y.: Malware traffic classification using convolutional neural network for representation learning. In: 2017 International Conference on Information Networking (ICOIN), pp. 712–717. IEEE (2017)

37. Wang, Z.: The applications of deep learning on traffic identification. BlackHat USA **24**, 1–10 (2015)

DeepAD: A Joint Embedding Approach for Anomaly Detection on Attributed Networks

Dali Zhu[1,2], Yuchen Ma[1,2(✉)], and Yinlong Liu[1,2]

[1] Institute of Information Engineering,
Chinese Academy of Sciences, Beijing, China
{zhudali,mayuchen,liuyinlong}@iie.ac.cn
[2] School of Cyber Security,
University of Chinese Academy of Sciences, Beijing, China

Abstract. Detecting anomalies in the attributed network is a vital task that is widely used, ranging from social media, finance to cybersecurity. Recently, network embedding has proven an important approach to learn low-dimensional representations of vertexes in networks. Most of the existing approaches only focus on topological information without embedding rich nodal information due to the lack of an effective mechanism to capture the interaction between two different information modalities. To solve this problem, in this paper, we propose a novel deep attributed network embedding framework named DeepAD to differentiate anomalies whose behaviors obviously deviate from the majority. DeepAD (i) simultaneously capture both of the highly non-linear topological structure and node attributes information based on the graph convolutional network (GCN) and (ii) preserve various interaction proximities between two different information modalities to make them complement each other towards a unified representation for anomaly detection. Extensive experiments on real-world attributed networks demonstrate the effectiveness of our proposed anomaly detection approach.

Keywords: Anomaly detection · Attributed networks · Autoencoder · Graph convolutional network

1 Introduction

Networks have become an important tool in many real-world applications to represent complex information systems such as social networks, transportation networks, and communication networks. In these networks, attributed networks have become a hot topic of research. Different from traditional plain networks where only the topological structure information is utilized, attributed networks also associated with rich features or attributes, which enrich the knowledge inside representations for network analysis. For example, in social networks, in addition to friend relationships, rich profile information is also an important attribute for

© Springer Nature Switzerland AG 2020
V. V. Krzhizhanovskaya et al. (Eds.): ICCS 2020, LNCS 12138, pp. 294–307, 2020.
https://doi.org/10.1007/978-3-030-50417-5_22

describing user characteristics; in online shopping networks, purchase records associated with reviews provide valuable information. Studies from social science have revealed the influence of interaction between the attributes of nodes and their structures [21,24]. Going through these insights, we can discover deeper patterns from attributed networks.

Anomaly detection plays a vital role in many information systems to achieve secure cyberspace. It aims to identify rare instances that do not conform to the expected patterns of majority [1]. Recently, there is emerging research of anomaly detection focusing on attributed networks due to the potential rich information contained in the attributed network. However, how to model network structure information and rich semantic nodal information into a unified representation is still a challenging problem.

Conventional anomaly detection methods mainly focus on exploiting the structure of the network to find patterns and spot anomalies such as structural-based or community-based [3] methods. Besides, feature-based methods assume that complex anomalies only exist in a subset according to node features. Unfortunately, existing efforts usually focus on either topological information or attribute information without insight into the complex interaction between those two different types of modalities. Moreover, other methods employing shallow models can hardly capture the highly non-linear [27] property of the attributed network. To address the problems as mentioned above, inspired by graph convolutional network (GCN) [16], we resort to embedding the input topological structure as well as nodal attributes seamlessly into a unified representation through stacking GCN layers. Meanwhile, the proposed model enforces the learned node representation to preserve various proximities. Then we aim to spot anomalies leveraging by the reconstruction errors both from the two kinds of modalities. The contributions of this paper are listed as follow:

- We propose a novel joint embedding approach DeepAD modeled by graph autoencoder DeepAD to capture the underlying high non-linearity in both topological structure and nodal attributes and detect anomalies according to the reconstruction errors.
- DeepAD jointly preserves the first-order, high-order, and cross-modal proximities in original networks towards a unified complementary representation.
- Experimental results show that DeepAD outperforms several state-of-art methods on benchmark datasets.

2   Related Work

2.1   Graph Based Anomaly Detection

Typically, graph-based anomaly detection methods are broadly divided into three classes: (1) *Structure-based methods* (2) *Community-based methods* and (3) *Feature-based methods* [3]. Structure-based methods mainly aim to identify substructures or subgraphs in the graph that are rare structurally, therefore anomalies can be sought out as the inverse of frequent subgraphs [22]. CODA [9] is one of the attempts that simultaneously finds communities as well as spots community anomalies using Markov random fields. OddBall [2] extracts features and finds patterns based on the ego-network of the graph to spot anomalous nodes. Community-based methods aim to find dense group nodes in the graph and spot anomalies that have connections across communities. One of the types of them, LOF [5], computes the local density deviation of a given data point concerning its neighbors. The main idea behind feature-based methods is that similar graphs should share the same properties, such as degree distribution, diameter, eigenvalues [14]. Recently, residual analysis [18] shows its effectiveness for anomaly detection in a more general way. However, those shallow models failed to model the underlying high non-linearity information of attributed networks into a unified complementary representation.

2.2   Deep Network Embedding

Network embedding aims to learn low-dimensional vector representations for nodes of the network, which preserves structure information and properties of graphs. With the increasing research on deep learning, a vast amount of deep models have been proposed towards various learning tasks. For plain networks, DNGR [6] utilizes deep autoencoder to capture network's non-linearity, and SDNE [27] further preserves the first-order and second-order proximity. LANE [12] jointly combine the label, attribute, and structure information into embedding. Besides, DANE [8] adopts two deep autoencoders to train structure and attributed features separately while keeping the representation consistency and complementary. Recently, Kipf and Welling [16] propose graph convolutional network (GCN) model for attributed networks that simultaneously encode the structure and attribute information into latent space and further employ it on a variational auto-encoder architecture [17]. Our model took inspiration from these methods.

3   Problem Definition

We define the anomaly detection problem in attributed networks with first-order proximity, high-order proximity, and cross-modal proximity.

Definition 1 *(Attributed Network Embedding). An attributed network is denoted as $G = (A, X)$ with n nodes, where $A = [a_{i,j}] \in \mathbb{R}^{n \times n}$ is the adjacency matrix and $X = [x_{i,j}] \in \mathbb{R}^{n \times m}$ is the attribute matrix. Each node is associated with a nodal attributes row vector $\mathbf{x}_i \in \mathbb{R}^m (i = 1, \ldots, n)$. $a_{i,j} = 1$ represents there is a link between the ith node and the jth node. Otherwise, $a_{i,j} = 0$. The objective of network embedding is to map the topological structure and nodal attributes into a representation space $H \in \mathbb{R}^{n \times d}$ through a mapping function $f : \{A, X\} \rightarrow H$. Note that, $d \ll |A|$ is the dimension of representation space.*

Network embedding aims to preserve the intrinsic information of the network into a low-dimensional representation space. To perform the embedding appropriately for anomaly detection task, we define three proximities to preserve local proximity, global proximity and interaction proximity respectively.

Definition 2 *(First-Order Proximity). The first-order proximity describes the pairwise similarity between two nodes. For each pair of nodes, $a_{i,j} > 0$ indicates there exists first-order proximity between them. Otherwise, if no interaction is observed, the first-order proximity is 0.*

Generally speaking, the first-order proximity is the most direct expression in a network. For example, people who are friends with each other in social media tend to share a common characteristic. Because of this importance, it is necessary to preserve the first-order proximity, which can be viewed as local proximity. However, due to the sparsity and incompleteness of the real-world network, it is not sufficient only considering the first-order proximity to represent the network. Therefore, we introduce high-order proximity to characterize the global proximity of the network to compensate for this problem.

Definition 3 *(High-Order Proximity). The high-order proximity describes the neighborhood similarity between two nodes. Given an attributed network $G = (A, X)$, let $M = \left(G^1 + G^2 + \dots, +G^k\right)$ denotes the high-order proximity, where G^k is the kth-order node proximity information propagation through the embeddings. Then the high-order proximity between v_i^k and v_j^k is determined by M_i^t and M_j^t.*

Intuitively, two nodes are similar if they share similar neighbors. For example, in a citation network, documents are similar if they are surrounded by similar citations, even if they are not referencing to each other [19]. Since the topological structure and nodal attributes are two different information modalities in the same network, to make them complement each other towards a unified informative representation of the same network [11], the cross-modal proximity is essential to be preserved.

Definition 4 *(Cross-Modal Proximity). The cross-modal proximity describes the similarity between nodes according to their structure and attribute information. Given an attributed network $G = (A, X)$, the cross-modal proximity of two nodes v_i and v_j is determined by (A_i, X_i) and (A_j, X_j).*

Definition 5 *(Anomaly Detection). The task of anomaly detection is to find the node instances that are rare and significantly different from the majority of the reference nodes according to their anomalous scores.*

4   The DeepAD Model

According to the previous analysis, three challenges remain for anomaly detection on attributed networks to achieve desired results:

(1) *Network sparsity:* Many real-world networks tend to be so sparse that the utilization of limited observed node interactions severely restricts the performance of network analysis.

(2) *High non-linearity:* The underlying structure of topological structure and nodal attributes is often highly non-linear and hence cannot be accurately captured by linear models [20].

(3) *Proximity preservation:* The combination of the two information modalities describes different aspects of the network information. How to preserve complex interaction proximity and complement each other to form a unified information representation is still a thorny problem.

To address the challenges above, we present a novel deep joint model approach *DeepAD* for anomaly detection, as shown in Fig. 1. The network structure and the nodal attributes embedded through a joint framework modeled by GCN into the same representation space. In order to preserve the complex interaction between two information sources, we add constraints to refine the representation. And then, we make use of the reconstruction errors as a measure to spot anomalies. Details are introduced as follows.

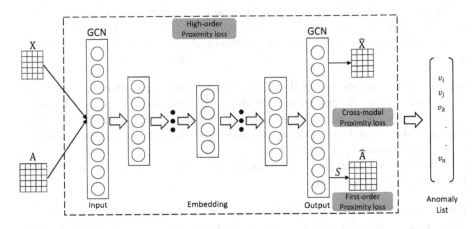

Fig. 1. The model takes the adjacency matrix A and the attribute matrix X as inputs, representing the topological structure and the nodal attributes respectively.

4.1 Framework

DeepAD embeds the input data through an autoencoder to capture the highly non-linear information simultaneously from network structure and nodal attributes. Autoencoder has proven a powerful deep learning model to learn the complex latent representation of data for various applications [13]. The primary target of autoencoder can be reduced to solving the following optimization problem:

$$\min_{\theta} \sum_{\phi \in \Phi_{tar}} \mathcal{L}\left(\psi_{dec}\left(\psi_{enc}\left(X_{\phi}\right)\right), \phi | \theta\right) \tag{1}$$

where Φ_{tar} is the target information that the embedding layers expect to preserve, and $X \in \mathbb{R}^n$ denotes the input data involved in ϕ. The encoder ψ_{enc} maps data into representation vectors, and decoder ψ_{dec} reconstructs the original data from the representation space. θ denotes the model parameters in encoders and decoders. The parameters are trained by minimizing the loss function described above, thereby preserving the desired network information Φ_{tar} in the network.

To capture the complex interaction of the topological structure and nodal attributes, inspired by the significant performance improvement of graph convolutional network (GCN) [16] in the analysis of non-Euclidean structured data such as graphs and manifolds, we use GCN layers as encoder which is defined as:

$$H^{(k)} = \sigma \left(\widetilde{D}^{-\frac{1}{2}} \widetilde{A} \widetilde{D}^{-\frac{1}{2}} H^{(k-1)} W^{(k-1)} \right) \tag{2}$$

where H^{k-1} is the input for the embedding layer $k-1$, and $H^0 = X$. $\widetilde{A} = (I + A)$ is the adjacency matrix with added self-connections. I is the identity matrix, and $\widetilde{D}$ is the diagonal matrix of $\widetilde{A}$. $\sigma(\cdot)$ denotes a non-linear activation function such as ReLU or sigmoid. W^{k-1} is a matrix of filter parameters which are shared for all input nodes. It is worth noting that unlike autoencoders that explicitly treat each node's neighbor as features to embed into a latent space separately with the nodal attributes as in SDNE and DANE, GCN implicitly applies the local neighborhood links on each encoding layer as pathways to aggregate embeddings from neighbors [10]. Given the input attribute matrix X, the convolutional layers iteratively aggregate embeddings of neighbors as well as its own to capture a higher-order node proximity information of which both the topological structure and nodal attributes are preserved. Let $A_N = \widetilde{D}^{-\frac{1}{2}} \widetilde{A} \widetilde{D}^{-\frac{1}{2}}$ denotes the normalized adjacency, the encoder can be formed as:

$$\begin{aligned} H^{(1)} &= \sigma \left(A_N H^{(0)} W^{(0)} \right) \\ Z = H^{(k)} &= \sigma \left(A_N H^{(k-1)} W^{(k-1)} \right), k = 2, ..., K \end{aligned} \tag{3}$$

Therefore, the embeddings Z is the desired low-dimensional representation of the attributed network. Correspondingly, there will be k layers in the decoder and the output $\widehat{X}$ is the reconstruction of attribute matrix. Furthermore, according to [17], the reconstructed adjacency matrix $\widehat{A}$ can be calculated as $\widehat{A} = \mathcal{S} \left(HH^{\top} \right)$ where $\mathcal{S}(x)$ is the sigmoid function. To maximize the information propagation, we choose the last H to reconstruct the adjacency matrix.

4.2 Loss Function

As aforementioned analysis, nodes with similar features are more likely to be connected in attributed networks. The **first-order proximity** can be regarded as the supervised information to restrict the similarity of a pair of nodes in the latent representations. Inspired by the idea of Laplacian eigenmaps (LE) [4], we introduce a penalty term to constrain the local proximity when similar nodes are mapped away from each other in the latent representations. The loss function is shown as follows:

$$\mathcal{L}_f = \sum_{i,j=1}^{n} \hat{a}_{i,j} \left\| \mathbf{h}_i^{(K)} - \mathbf{h}_j^{(K)} \right\|_2^2 = \sum_{i,j=1}^{n} \hat{a}_{i,j} \left\| \mathbf{h}_i - \mathbf{h}_j \right\|_2^2 \tag{4}$$

where $\mathbf{h}_i^{(K)}$ and $\mathbf{h}_j^{(K)}$ are the row vector of the hidden layer matrix $H^{(K)}$ and $\hat{a}_{i,j} \in \hat{A}$ indicates whether there exists a connection between nodes v_i and v_j. The loss function can be reformulated as the following term:

$$\mathcal{L}_f = \sum_{i,j=1}^{n} \hat{a}_{i,j} \left\| \mathbf{h}_i - \mathbf{h}_j \right\|_2^2 = 2tr \left(H^T L H \right) \tag{5}$$

where $L = D - \hat{A}$, $D \in \mathcal{R}^{n*n}$ is a diagonal matrix of $\hat{A}$, and $D_{i,i} = \sum_j \hat{a}_{i,j}$.

The **high-order proximity** refers to how similar the neighborhood information of a pair of nodes is. With the iteration of convolutions, the higher-order neighborhood information is embedded into the latent space. As SDNE proved, the constraints on reconstruction can enforce the neural network to capture the data manifold smoothly, thereby preserve the proximity among a wider range of samples. To preserve this proximity, we minimize reconstruction loss as follows:

$$L_h = R_x + \alpha R_a = \sum_{i=1}^{n} \|\hat{\mathbf{x}}_i - \mathbf{x}_i\|_2^2 + \alpha \sum_{i=1}^{n} \|\hat{\mathbf{a}}_i - \mathbf{a}_i\|_2^2 \tag{6}$$

where R_x and R_a represent the reconstruction error of the attribute matrix and adjacent matrix respectively. Specifically, if the neighborhood information is similar between two nodes, after minimizing the L_h, the learned representation H_i and H_j will also be similar. According to [7], anomalies are more difficult to reconstruct than normal nodes since their information representation does not conform to the patterns of the majority. Therefore, a larger reconstruction error indicates a higher probability of anomalies.

It's not only necessary to preserve the network proximity separately, but also essential to discover the implicit relationship since the topological structure and nodal attributes are two interdependent information modalities to describe the network. To make those two modalities complement each other, we preserve the **cross-modal proximity** by maximizing their interaction likelihood estimation as follows:

$$L_c = \prod_{i,j} p_{i,j}^{I_{i,j}} \left(1 - p_{i,j} \right)^{1 - I_{i,j}} \tag{7}$$

where $p_{i,j}$ is the joint distribution of two modalities which can be defined as $p_{ij} = \mathcal{S}(\mathbf{a}_i \mathbf{h}_j)$. Furthermore, $I_{i,j}$ indicates whether $\mathbf{a}_i$ and $\mathbf{h}_j$ are from the same node. $I_{ij} = 1$ if $i = j$. Otherwise $I_{ij} = 0$. So, the loss function can be defined in the form of the negative log-likelihood as follows:

$$L_c = - \sum_i \{\log p_{ii} - \sum_{j \neq i} \log (1 - p_{i,j})\} \tag{8}$$

The objective function of Eq. (8) constrains $\mathbf{a}_i$ and $\mathbf{h}_j$ as consistent as possible when they belong to the same node while separating them from each other when they come from different nodes, resulting in sufficient complementary interactions of two modalities. To simplify the calculation, pairwise nodes with similar first-order proximity need not be separated from each other, because representation $\mathbf{a}_i$ and $\mathbf{h}_j$ should also be similar. Therefore, the objective function can be revised as follows:

$$L_c = -\sum_i \{\log p_{ii} - \sum_{\hat{a}_{i,j}=0} \log(1 - p_{i,j})\} \tag{9}$$

As shown in Fig. 1, in order to simultaneously preserve the three proximities, we propose a semi-supervised framework which jointly combines Eq. (5), Eq. (6), and Eq. (8). The overall objective function is shown as follows:

$$\begin{aligned}
L &= L_f + L_{hx} + \alpha L_{ha} + L_c \\
&= \sum_{i,j=1}^n \hat{a}_{i,j} \|\mathbf{h}_i - \mathbf{h}_j\|_2^2 + \sum_{i=1}^n \|\hat{\mathbf{x}}_i - \mathbf{x}_i\|_2^2 + \alpha \sum_{i=1}^n \|\hat{\mathbf{a}}_i - \mathbf{a}_i\|_2^2 \\
&\quad - \sum_i \{\log p_{ii} - \sum_{\hat{a}_{i,j}=0} \log(1 - p_{i,j})\}
\end{aligned} \tag{10}$$

4.3 Anomaly Detection

By minimizing the loss function, DeepAD can iteratively learn the representations of input attributed network until the objective function converges. With a Xavier Initialization, the model parameters can be optimized by using stochastic gradient descent. After a certain number of iterations, as in [7] and [23], the reconstruction error can be directly applied to rank the abnormality of nodes. Thus the anomaly score of each node v_i can be calculated as follows:

$$socre(\mathbf{v}_i) = \|\hat{\mathbf{x}}_i - \mathbf{x}_i\|_2^2 + \alpha \|\hat{\mathbf{a}}_i - \mathbf{a}_i\|_2^2 \tag{11}$$

As a result, we can calculate the ranking of anomalies according to the corresponding abnormal scores. The higher the score, the more likely the instance is to be considered an anomaly.

4.4 Complexity Analysis

The complexity of graph convolutional network is dominated by the computation of $\tilde{D}^{-\frac{1}{2}}\tilde{A}\tilde{D}^{-\frac{1}{2}}XW$ whose complexity is $\mathcal{O}(ncdh)$ [16], where n is the number of nodes, c is the average degree of network which is usually a constant in real-world applications, d is the number of feature dimensions on the attributed network and h is the number of feature maps of W. In this way, nc represents the number of non-zero elements in A so that $\tilde{A}X$ can be efficiently calculated using sparse-dense matrix multiplications. The complexity of Eq. (5) is $\mathcal{O}(ncd)$ [27] while the complexity of Eq. (9) is $\mathcal{O}(n^2)$, thus the overall complexity of the model is $\mathcal{O}(ncd(H + n^2))$ where H is the sum of h in all layers.

5   Experiments

5.1   Datasets

We choose three benchmark datasets[1]: Cora, Citeseer, and PubMed. These three datasets are paper citation networks. The nodes and edges of each network denote documents and reference links, respectively. The attribute of each node is the bag-of-words feature vectors of each document. In order to obtain a ground truth of anomalies in the above datasets, we refer to two widely used methods [25, 26] to generate a combined set of anomalies from both the topological structure and nodal attributes perspectives for each dataset. In real-world scenarios, the small clique is a typical substructure created by anomalous activity [25]. Therefore we randomly select m nodes and connect them to each other to form a dense clique, iteratively repeat this process until n cliques are generated, and all the mn nodes in cliques are considered as anomalies. Then, we inject the same number of anomalies from the attribute perspective. Similarly, we randomly select mn nodes from the network, shuffle their attribute values to generate anomalous nodes, while the topological relationship remains unchanged. The details of dataset statistics are summarized in Table 1.

Table 1. Description of benchmark datasets

Dataset	Cora	Citeseer	PubMed
# Nodes	2780	3327	19717
# Edges	5278	4732	44338
# Attributes	1433	3703	500
# Anomalies	10%	10%	10%

5.2   Baseline Algorithms

We choose four contrast algorithms as baselines. The details are as follows:

- LOF [5] detects anomalies which have a substantially lower density and only considers attribute information.
- SCAN [28] detects anomalies at the structural level and only considers structure information.
- CODA [9] detects anomalies based on community detection within a unified probabilistic model.
- Radar [18] detects anomalies whose behavior obviously deviates from the majority according to the residuals of attribute information and its coherence with network information.

---

[1] https://github.com/kimiyoung/planetoid/tree/master/data.

5.3   Evaluation Metrics

We select *AUC*, *precision*@K and *recall*@K to evaluate the performance. Their definition is listed as follows:

- AUC: AUC (Area Under ROC Curve) is a performance measurement for classification problems. Higher the AUC, better the model is at distinguishing between normal and anomalous
- Precision@K: We evaluate the proportion of true anomalies that are discovered in the top K ranked nodes.
- Recall@K: It measures the percentage of true anomalies selected out of all the ground truth anomalies.

5.4   Parameter Settings

The architecture of our approach varies with different datasets. The dimension of each layer is summarized in Table 2. All the neural networks have three layers, and the dimension of the last encoder layer is the same.

We use ReLU as the activation function and optimize the loss function with Adam algorithm [15]. The learning rate is set to 0.025. The hyper-parameter α is tuned with grid search on each dataset. For the rest baselines, their settings are set as described in the original papers.

Table 2. Neural network structures

Dataset	# nodes in each layer
Cora	1433-200-100
Citeseer	3703-500-100
PubMed	500-200-100

5.5   Experiment Results

The experimental results in terms of AUC values are presented in Fig. 2, and the precision and recall results are reported in Table 3. The results of SCAN and CODA are not included in Table 3 since they are cluster-based methods that are incapable of providing an accurate ranking list for all nodes. From the evaluation results, we can see that the dual-modality information-based model (Radar, DeepAD) is superior to the conventional methods (LOF, SCAN, and CODA) merely based either on attribute information or structure information. However, through the comparison of the residual-based model Radar and DeepAD, we can observe that there is a significant increase in each metric. Figure 3 shows the result of five anomaly detection models on the Citeseer dataset. When the ratio of anomalies increased, our proposed DeepAD can still maintain high AUC

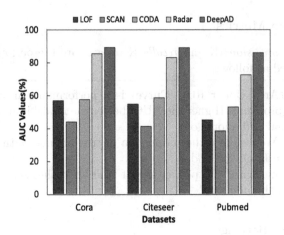

Fig. 2. Anomaly detection results by different methods.

Table 3. *precision*@K and *recall*@K on three datasets for anomaly detection.

	Cora			Citeseer			PubMed		
Precision@K									
K	50	100	200	50	100	200	50	100	200
LOF	0.480	0.375	0.314	0.525	0.462	0.410	0.080	0.075	0.053
Radar	0.786	0.770	0.756	0.780	0.765	0.726	0.575	0.583	0.560
DeepAD	**0.820**	**0.796**	**0.743**	**0.812**	**0.785**	**0.752**	**0.652**	**0.610**	**0.580**
Recall@K									
K	50	100	200	50	100	200	50	100	200
LOF	0.060	0.095	0.120	0.065	0.087	0.115	0.008	0.012	0.016
Radar	0.090	0.204	0.250	0.086	0.180	0.294	0.052	0.095	0.186
DeepAD	**0.116**	**0.235**	**0.384**	**0.095**	**0.205**	**0.265**	**0.061**	**0.105**	**0.236**

values. The main reasons may be as follows: (1) We employ a deep neural network model based on graph autoencoder, which breaks through the limitation of shallow mechanisms to handle the network sparsity issue and capture the high non-linearity information both from the topological structure and nodal attributes in attributed networks. (2) To further capture the complex interaction between two different modalities, we propose various proximities to preserve the implicit proximity make them complement each other towards a unified representation. Among the results, DeepAD outperforms other baselines on all benchmark datasets, which demonstrate the effectiveness of our proposed anomaly detection approach on attributed networks. Our future work will focus on how to develop a deep anomaly detection model robust to noise.

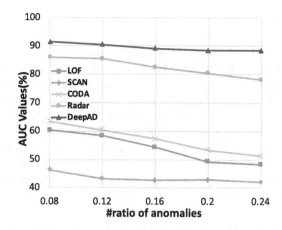

Fig. 3. AUC of anomaly detection on Citeseer dataset

5.6 Parameter Sensitivity

There are several parameters in our proposed DeepAD framework, we investigate the impact of the number of the embedding dimension and the value of hyper-parameter α on Citeseer dataset with 400 injected anomalies and report the performance variance results in Fig. 4. Figure 4(a) reports the performance of DeepAD w.r.t the dimension of the embedding layer. It can be shown that performance improves as the dimension increases. However, when the dimension continues to increase beyond 100, the performance no longer improves or even drops. The possible reason is that too large dimension of embedding will introduce noise so as to influence the latent representations. The hyper-parameter α balances the impact of three proximities on model training and anomaly scores computation. The results in Fig. 4(b) indicate that it is necessary to find a balance between those proximities to achieve better performance, and the best choice of α is 0.025.

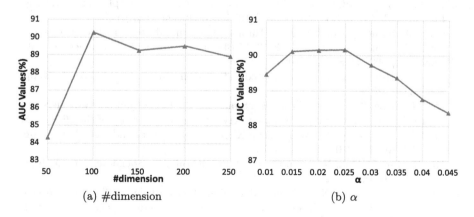

(a) #dimension

(b) α

Fig. 4. Sensitivity w.r.t dimension and the value of α

6 Conclusion

In this paper, we propose a joint embedding approach for anomaly detection on attributed networks, namely DeepAD. Specifically, to capture the highly non-linear information in network topological structure and nodal attributes. We design a graph convolutional network (GCN) based deep autoencoder model. To further address the complex interaction problem, we jointly preserve the first-order, high-order, and cross-modal proximity to make two types of information complement each other towards a unified representation. By jointly optimizing them in the deep model, the reconstruction errors are them employed to spot anomalies. The experimental results demonstrate the effectiveness of our approach to anomaly detection compared with state-of-art methods.

Acknowledgement. This work was supported by the Strategic Priority Research Program of Chinese Academy of Sciences, Grant No. XDC02040300.

References

1. Aggarwal, C.C.: Outlier analysis. In: Aggarwal, C.C., et al. (eds.) Data Mining, pp. 237–263. Springer, Cham (2015). https://doi.org/10.1007/978-3-319-14142-8_8
2. Akoglu, L., McGlohon, M., Faloutsos, C.: Oddball: spotting anomalies in weighted graphs. In: Zaki, M.J., Yu, J.X., Ravindran, B., Pudi, V. (eds.) PAKDD 2010. LNCS (LNAI), vol. 6119, pp. 410–421. Springer, Heidelberg (2010). https://doi.org/10.1007/978-3-642-13672-6_40
3. Akoglu, L., Tong, H., Koutra, D.: Graph based anomaly detection and description: a survey. Data Min. Knowl. Disc. **29**(3), 626–688 (2015). https://doi.org/10.1007/s10618-014-0365-y
4. Belkin, M., Niyogi, P.: Laplacian eigenmaps for dimensionality reduction and data representation. Neural Comput. **15**(6), 1373–1396 (2003)
5. Breunig, M.M., Kriegel, H.P., Ng, R.T., Sander, J.: LOF: identifying density-based local outliers. In: Proceedings of the 2000 ACM SIGMOD International Conference on Management of Data, pp. 93–104. ACM (2000)
6. Cao, S., Lu, W., Xu, Q.: Deep neural networks for learning graph representations. In: Thirtieth AAAI Conference on Artificial Intelligence (2016)
7. Chen, J., Sathe, S., Aggarwal, C., Turaga, D.: Outlier detection with autoencoder ensembles. In: Proceedings of the 2017 SIAM International Conference on Data Mining, pp. 90–98. SIAM (2017)
8. Gao, H., Huang, H.: Deep attributed network embedding. In: IJCAI 2018, pp. 3364–3370 (2018)
9. Gao, J., Liang, F., Fan, W., Wang, C., Sun, Y., Han, J.: On community outliers and their efficient detection in information networks. In: Proceedings of the 16th ACM SIGKDD International Conference on Knowledge Discovery and Data Mining, pp. 813–822. ACM (2010)
10. Hu, X., Tan, Q., Liu, N.: Deep representation learning for social network analysis. Front. Big Data **2**, 2 (2019)
11. Huang, X., Li, J., Hu, X.: Accelerated attributed network embedding. In: Proceedings of the 2017 SIAM International Conference on Data Mining, pp. 633–641. SIAM (2017)

12. Huang, X., Li, J., Hu, X.: Label informed attributed network embedding. In: Proceedings of the Tenth ACM International Conference on Web Search and Data Mining, pp. 731–739. ACM (2017)

13. Jiang, W., Gao, H., Chung, F.L., Huang, H.: The l2, 1-norm stacked robust autoencoders for domain adaptation. In: Thirtieth AAAI Conference on Artificial Intelligence (2016)

14. Kang, U., Papadimitriou, S., Sun, J., Tong, H.: Centralities in large networks: algorithms and observations. In: Proceedings of the 2011 SIAM International Conference on Data Mining, pp. 119–130. SIAM (2011)

15. Kingma, D.P., Ba, J.: Adam: a method for stochastic optimization. arXiv preprint arXiv:1412.6980 (2014)

16. Kipf, T.N., Welling, M.: Semi-supervised classification with graph convolutional networks. arXiv preprint arXiv:1609.02907 (2016)

17. Kipf, T.N., Welling, M.: Variational graph auto-encoders. arXiv preprint arXiv:1611.07308 (2016)

18. Li, J., Dani, H., Hu, X., Liu, H.: Radar: residual analysis for anomaly detection in attributed networks. In: IJCAI, pp. 2152–2158 (2017)

19. Liben-Nowell, D., Kleinberg, J.: The link-prediction problem for social networks. J. Am. Soc. Inf. Sci. Technol. **58**(7), 1019–1031 (2007)

20. Luo, D., Nie, F., Huang, H., Ding, C.H.: Cauchy graph embedding. In: Proceedings of the 28th International Conference on Machine Learning (ICML-11), pp. 553–560 (2011)

21. McPherson, M., Smith-Lovin, L., Cook, J.M.: Birds of a feather: homophily in social networks. Annu. Rev. Sociol. **27**(1), 415–444 (2001)

22. Noble, C.C., Cook, D.J.: Graph-based anomaly detection. In: Proceedings of the Ninth ACM SIGKDD International Conference on Knowledge Discovery and Data Mining, pp. 631–636. ACM (2003)

23. Sakurada, M., Yairi, T.: Anomaly detection using autoencoders with nonlinear dimensionality reduction. In: Proceedings of the MLSDA 2014 2nd Workshop on Machine Learning for Sensory Data Analysis, pp. 4–11 (2014)

24. Shalizi, C.R., Thomas, A.C.: Homophily and contagion are generically confounded in observational social network studies. Sociol. Methods Res. **40**(2), 211–239 (2011)

25. Skillicorn, D.B.: Detecting anomalies in graphs. In: 2007 IEEE Intelligence and Security Informatics, pp. 209–216. IEEE (2007)

26. Song, X., Wu, M., Jermaine, C., Ranka, S., et al.: Conditional anomaly detection. IEEE Trans. Knowl. Data Eng. **19**(5), 631–645 (2007)

27. Wang, D., Cui, P., Zhu, W.: Structural deep network embedding. In: Proceedings of the 22nd ACM SIGKDD International Conference on Knowledge Discovery and Data Mining, pp. 1225–1234. ACM (2016)

28. Xu, X., Yuruk, N., Feng, Z., Schweiger, T.A.: Scan: a structural clustering algorithm for networks. In: Proceedings of the 13th ACM SIGKDD International Conference on Knowledge Discovery and Data Mining, pp. 824–833. ACM (2007)

SciNER: Extracting Named Entities from Scientific Literature

Zhi Hong[1]([⊠]), Roselyne Tchoua[1], Kyle Chard[1], and Ian Foster[1,2]

[1] University of Chicago, Chicago, IL 60637, USA
{hongzhi,roselyne,chard,foster}@uchicago.edu
[2] Argonne National Lab, Lemont, USA

Abstract. The automated extraction of claims from scientific papers via computer is difficult due to the ambiguity and variability inherent in natural language. Even apparently simple tasks, such as isolating reported values for physical quantities (e.g., "the melting point of X is Y") can be complicated by such factors as domain-specific conventions about how named entities (the X in the example) are referenced. Although there are domain-specific toolkits that can handle such complications in certain areas, a generalizable, adaptable model for scientific texts is still lacking. As a first step towards automating this process, we present a generalizable neural network model, SciNER, for recognizing scientific entities in free text. Based on bidirectional LSTM networks, our model combines word embeddings, subword embeddings, and external knowledge (from DBpedia) to boost its accuracy. Experiments show that our model outperforms a leading domain-specific extraction toolkit by up to 50%, as measured by F1 score, while also being easily adapted to new domains.

Keywords: Named Entity Recognition · LSTM · Word embeddings

1 Introduction

The scholarly model has long relied on publication as a means of documenting and disseminating results. As such, scientific papers often contain the ultimate source of truth about a particular scientific entity, for example how it was produced, analyzed, and processed. Unfortunately, this approach to dissemination has obvious shortcomings, most notably that data and results are inaccessible to machines due to their esoteric encoding. Further, given the enormous number of publications—estimated to be over 2.5 million every year [31] and exponentially growing [12]—it is increasingly infeasible for individual researchers to locate important data in publications. A researcher might have to read dozens if not hundreds of papers just to get a rough idea of the state-of-the-art research in an

This work was supported in part by NIST contract 60NANB15D077, the Center for Hierarchical Materials Design, and DOE contract DE-AC02-06CH11357, and by computer resources provided by Jetstream [26].

© Springer Nature Switzerland AG 2020
V. V. Krzhizhanovskaya et al. (Eds.): ICCS 2020, LNCS 12138, pp. 308–321, 2020.
https://doi.org/10.1007/978-3-030-50417-5_23

area, and even after they have done so, there is no guarantee that they have not missed important results.

One potential solution to this publication deluge is to extract scientific facts from free text articles and then to store these facts in structured searchable databases. A researcher might then be able to issue queries such as "SELECT (*) from polymers WHERE glass_transition_temperature >= 100" to obtain data of interest when needed. However, while such databases can avoid redundant effort, building them in the first place requires extensive manual extraction and curation effort, that is furthermore complicated by the difficulty and uncertainty of extracting scientific facts from text [16].

Crowdsourcing provides one method for performing manual, human-oriented tasks in an efficient and cost-effective manner [3, 4, 7]. However, the expertise required to extract scientific facts makes crowdsourcing impractical. The few existing scientific databases and repositories, such as the Japanese Polymer Data Handbook [19], are curated by domain experts and thus costly to maintain; as a result, they quickly become out of date.

Since it is infeasible to rely solely on humans to extract scientific facts from publications, automatic approaches are needed to address the increasing rate of publication. A first problem to be tackled when extracting facts from publications is the identification of scientific entities (e.g., a chemical, sample, or anatomical region): a problem that we call scientific Named Entity Recognition (NER).

Considerable progress has been made in machine learning (ML) and natural language processing (NLP) in the last decade, with state-of-the-art models outperforming humans in various tasks [9]. However, most such efforts are centered on day-to-day language corpora, such as news articles, Twitter posts, and online product reviews. Little attention has been paid to the unique challenges associated with understanding scientific texts, such as idiosyncratic writing styles, specialized article organizations, and domain-specific vocabularies that are not common in other texts. A previous study of the biomedical literature from PubMed shows that the quality of machine learning models depends on the training corpora, model architectures, and hyper-parameters used [6]. Hence, to obtain high quality scientific NER, models must be trained on corpora from the same domain.

In this paper, we present SciNER, a NER model that is specifically designed for recognizing named entities in a scientific context. We show that this model is generalizable and can be trained on and applied to different domains. Our primary contributions are:

1. Development of SciNER using bidirectional LSTM networks and conditional random fields.
2. Integration of several word embedding models and lexicons from DBpedia as an external source of knowledge to boost learning performance.
3. Evaluation of the accuracy of SciNER on two different scientific NER problems and comparison with a domain-specific, state-of-the-art toolkit.

The paper is organized as follows. In Sect. 2 we discuss the specific problem we aim to solve and introduce the architecture of our proposed model. In Sect. 3 we describe the word embedding and lexicon features used in our model. In Sect. 4 we evaluate the accuracy of SciNER on two different scientific named entity recognition problems. Finally, we explore related work in Sect. 5 and summarize our approach in Sect. 6.

2   The SciNER Model

We focus on the task of identifying scientific named entities in scientific publications. In this section we first define our extraction problem and then outline the Bidirectional LSTM and Lexicon-infused LSTM models used in SciNER.

2.1   Problem Definition

Given a publication, comprised of sections containing natural language text, our task is to identify scientific named entities of interest. For example, in materials science publications about polymers, we want to identify polymer names, such as "*polystyrene*" in the following:

> "We measured the viscosity of unentangled, short-chain **polystyrene** films on silicon at different temperatures and found that ..." [33]

In social science publications, the task is slightly different. Here researchers explore hypotheses and make assertions based on analysis of known datasets but the dataset are often not cited like other artifacts. The authors do not always use the full names when referencing datasets in the natural language text of the paper. When given the following paragraph from a social science paper, we want to extract the boldfaced words.

> "By analyzing data from 3279 individuals who participated in the **Longitudinal Study of American Youth**, this study examines ..." [25]

While these two examples are from different domains, their named entity extraction tasks are similar: in each case, we want a model that, without altering its structure, can adapt to the task of identifying a certain class of scientific named entity (polymers and social science datasets, respectively) in scientific text.

2.2   The Basic Model: Bidirectional LSTM with CRF

The Long-Short Term Memory (LSTM) network has shown promise for various natural language processing tasks. LSTMs, like the human mind, can retain knowledge of previous tokens (i.e., words or punctuation marks) and use them to better understand the meaning of the next token in its context. SciNER aims to build upon this prior work by adapting LSTM-based approaches to the specific challenges associated with scientific NER.

Figure 1 shows an overview of the structure of a basic LSTM network model. One major advantage that LSTM has over other traditional methods is that it does not require any specific (and often manually selected) features. A common approach for applying LSTMs to NER tasks is to assign labels that indicate the entities to be extracted. For example when using beginning-inside-outside (BIO) labeling [22], a label "B," "I," or "O" is assigned to each token in the training corpus. "B" is assigned to the first word in a named entity or a single-word named entity, "I" marks a subsequent word in a multi-word entity, and all other (non-named entity) words are given the label "O."

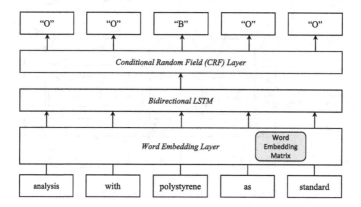

Fig. 1. Overall structure of the proposed neural network model.

For NER tasks, LSTM reads the input in one pass and assigns a BIO label to each word. However, in some cases, it is hard to tell whether a word is part of a named entity by looking at only the words *preceding* it in the sentence. For example, upon seeing the word "New" in a sentence, it is difficult to determine whether it should be given the label "B" or "O." However, if the next word is "York," then we can determine that it is likely a named entity. Bidirectional LSTM (Bi-LSTM) is used in natural language processing to address this need to exploit information about the words that come before *and* after a given word.

The Bi-LSTM network predicts a label for every word. This, however, means that the network has no awareness of the validity of the label sequence that it generates. Thus, it may output sequences (e.g., "OIO") that are invalid under the BIO labeling scheme. To penalize such invalid label sequences, we add a Conditional Random Field (CRF) layer on top of the Bi-LSTM network.

2.3   A Lexicon-Infused Bi-LSTM Network

The Bi-LSTM model can learn only about the order of the words. For example, seeing *"poly(vinyl methyl ether)"* in the training data would not indicate that the unseen text *"poly(ethylene glycol)"* is also a polymer name. However, knowledge that both *vinyl methyl ether* and *ethylene glycol* are chemical compounds and they both follow the pattern *"poly([chemical compounds])"* could be used to determine that both are in fact polymer names.

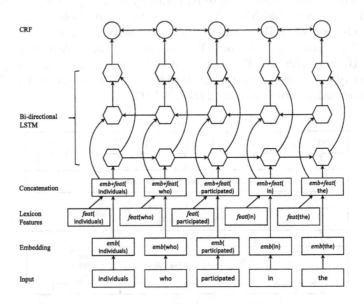

Fig. 2. The model with external lexicon knowledge.

To further improve the accuracy of our model, we introduce an external source of knowledge, by mapping words to classes obtained from DBpedia. Simply put, DBpedia is a structured version of Wikipedia, which consists of 4.2 million entries that are categorized in 774 distinct classes such as people, organization, location, and chemical compound [1]. Previous studies have explored the use of DBpedia classes for standard NER tasks such as CoNLL [6]; however, to the best of our knowledge such approaches have not been applied to scientific NER problems. Encoding lexicon features and feeding them to the LSTM gives the network more opportunities to capture the internal structure of named entities. Thus the network may be able to recognize *"poly(ethylene glycol)"* as a polymer name even when it has not been seen by the network before.

To encode the knowledge from the external lexicon, we use the same BIO labels as described above, as shown in Table 1. When overlaps occur between classes, as in this example (or within a single class, as in the case of "US Bureau of Labor Statistics" and "Bureau of Labor Statistics," both present in the Organization lexicon), we choose the longest match. We add a concatenation layer to combine these the word embeddings with the lexicon features and feed the concatenated vector to the LSTM, as illustrated in Fig. 2.

Table 1. An example showing how external knowledge from DBpedia is encoded for the Lexicon-infused Bi-LSTM Network. The first line is a sample sentence from a paper. The next two lines shows how BIO labels are assigned to the words that match entries in the Location or Organization categories in DBpedia.

	The	U.S	Bureau	of	Labor	Statistics	Industry	Injury	and	Illness	Data	reveals	that	⋯
LOC	O	B	O	O	O	O	O	O	O	O	O	O	O	⋯
ORG	O	B	I	I	I	I	O	O	O	O	O	O	O	⋯

3   Features

SciNER's LSTM models can consume input features beyond the word labels described above. To capture the context and meaning of scientific text we add various word embedding models and lexicon features as input to the models.

3.1   Word Embeddings

The LSTM models require that input sentences are represented as numerical data (i.e., vectors). The simplest way to convert words to vectors is one-hot encoding, but it is not ideal because significant syntactic and semantic properties of the word would be lost. One approach for capturing these properties is through word embeddings. In the remainder of this subsection, we introduce and compare several different word embedding models. In Sect. 4 we explore the performance of SciNER when using each of these word embedding models.

Randomly Initialized Trainable Word Embeddings. As shown in Fig. 1, the input layer is connected to the Bi-LSTM layer via the Embedding layer. By default, the weights of this layer, i.e., the word embedding matrix, is randomly initialized and is trained along with the whole neural network. No special algorithm is applied in this case, the word embedding matrix is treated in the same way as any other trainable parameters in the network.

Continuous Bag-Of-Words Model (CBOW). CBOW [17] is a popular method for training word embedding models. The core idea is that the semantic and syntactic information of a word can be determined (or represented) based on the context in which the word appears. Hence comes the idea of a fixed-sized window around the center word. Reusing the same sentence as in Fig. 1 ("... analysis with polystyrene as standard ..."), If the center word is "polystyrene" and the window size is 2, then "analysis," "with," "as," and "standard" are all in the window and are considered as context for the word "polystyrene." The context words are treated as a bag of words so the order does not matter. When given any one of these four context words, the CBOW model could predict the word "polystyrene."

Skip-Gram Model. Skip-gram [18] is another widely used method for training word embedding models. The major difference between CBOW and Skipgram is that CBOW predicts the center word based on the surrounding words within the context window, while Skip-gram does it the opposite way. When trained on the sample input "...analysis with polystyrene as standard ...", the Skip-gram model can predict any of its context words, "analysis", "with", "as", "standard" based on the center word "polystyrene."

Empirical data have suggested that CBOW is more efficient computationally, whereas Skip-gram works better when the training corpus is relatively small [11].

As pre-trained word embedding models cannot accurately represent the meaning of words in scientific contexts. We train word embeddings using CBOW or Skip-gram on the texts from the target domain, and then use them as the fixed (i.e., untrainable) word embedding matrix for the embedding layer in the neural network (grey box in Fig. 1).

FastText Word Embedding Model. The FastText model from Facebook [2] provides yet another architecture for creating word embeddings. Aside from representing a word based on its context, FastText also makes use of character n-grams. Words are mapped to character n-grams, which are then embedded in vectors. The n-gram vectors will make up a part of the embeddings for the words that do not appear frequently enough and thus do not have sufficient context in the training corpus. The addition of n-gram embeddings also greatly helps when the target word is not in the vocabulary of the pre-trained word embeddings. There is little that classic methods such as CBOW or Skip-gram can do when faced with an unknown word. They may either give it a random vector or the average of all the other vectors in the vocabulary, but, unsurprisingly, such a vector does not reflect the actual meaning of the out-of-vocabulary word. FastText, meanwhile, can capture the meaning of an unknown word better by making a word vector out of its character n-grams.

3.2   Lexicon Features

The current ontology of DBpedia has over 4.2 million entries in 774 classes. Matching all of them to the training and testing corpus is a computationally intensive task. Appending the one-hot encodings of the BIO labels for all the classes to a word vector will result in an extra $774 \times 3 = 2322$ dimensions, and for each word vector most of these dimensions will be zero. In other words, the concatenated vector will be very sparse and inefficient to compute.

To avoid diluting the dense word vector, we use only the few DBpedia classes that are relevant to the NER task at hand. For example, to identify polymers we use "Chemical Elements", "Chemical Compounds", and "Chemical Substance." When identifying social science dataset names we use "Location" and "Organization" as they are likely to be more suitable.

4   Experimental Analysis

We evaluate SciNER by applying it to two distinct scientific NER tasks: recognizing polymer names in materials science publications, and identifying dataset names in social science publications. In this section, we describe the process of obtaining and cleaning the publications, as well as how the labels were generated with minimal manual effort. We then explore the affect of using different combinations of features with SciNER to compare their influence on the extraction results. Finally, we compare SciNER against other methods for identifying scientific named entities.

4.1   Datasets

We use two distinct free-text scientific datasets to evaluate the accuracy of our models: materials science publications that contain polymer names and social science publications that contain references to datasets.

Our first dataset is a collection of 100 materials science publications from the journal *Macromolecules*. We chose this journal because we have established an agreement with its publisher, the American Chemical Society, that allows us to access the full text publications. Our second dataset comprises 6368 social science papers. We chose these papers because they are indexed by the Inter-university Consortium for Political and Social Research (ICPSR), which provides manually annotated relationships between datasets and papers in the field of social science. ICPSR has indexed over 72 000 papers. We selected a set of 6368 papers hosted by Elsevier for which we can easily download, via an API, the full text in JSON format.

4.2   Data Preparation

In order to feed our input datasets to the LSTM models we must first process the raw input publications into labeled collections of words.

Common Representation. The polymer and ICPSR datasets are represented in raw HTML or JSON formats which contain redundant information such as HTML tags, document object identifiers (DOIs) and publisher copyright statements. To remove these artifacts and create a clean format for processing, we parsed each file into a tree structure and removed any non-text related nodes in the tree. The resulting format includes only the raw text from the publication.

Tokenization. There are two steps in the tokenization process: sentence tokenization and word tokenization. First we split each paper into sentences so that each training sample consists of a sentence, not a whole passage. We do so by applying the `tokenize_sents()` function from the Python Natural Language Toolkit (NLTK). The reason why this is required is two-fold. The first is to assemble a large enough number of training examples, the second is to ensure that each training example has a reasonable length for the LSTM network to learn.

The second step is word tokenization, which converts each sentence from a string to a list of words and punctuations (tokens), so that it can be labeled in the next step.

Token Labeling. In order to be processed by our LSTM models each token in the training set must be labeled (using BIO labeling).

For the social science dataset, we have access to an ontology of named entities from ICPSR. For the materials science dataset there is no such ontology of named entities. Here we rely on domain experts to create one. However, instead of asking them to label every word in the 100 papers, we asked them to produce a list of unique polymer names from the corpus. Each paper was reviewed by two expert reviewers to label polymer names, and when disagreement arose a third more senior domain expert made the final decision. The result is a list of 495 polymer names identified by experts in the 100 papers [29].

Then, for both datasets, we then applied an automated script to search for known named entities in the unlabeled texts, and assign a label to each token according to the BIO scheme described in Sect. 2.2. To reduce the number of negative examples and create a balanced training set, we removed sentences that do not have any named entities.

Lexicon Features Labeling. We use the latest release of DBpedia [8] to associate class labels. As described in Sect. 3.2, we manually selected which classes to include for each NER task. For the social science dataset, we selected entries belonging to "Location" or "Organization." For the polymer dataset, we selected entries belonging to "Chemical Element" and "Chemical Compound" classes. We associated BIO labels automatically for each class following the same procedure as described above and encoded as additional features using one-hot encoding. The lexicon features are concatenated to the word embeddings of each word.

Splitting the Datasets. The labeled examples are then split into training, validation, and test sets.

For the 6368 social science papers, we use a 64–16–20% split, yielding 14 945, 3737, and 4699 sentences in the training, validation, and test sets, respectively.

Splitting the polymer science dataset is trickier because unlike the social science papers, multiple polymers often appear in one sentence. If we divide the sentences randomly, we may end up with many polymer names occurring in both the training and the testing set, in which case our model might learn specific polymer names rather than general concepts. To mitigate this problem, we randomly select half of the unique polymers mentioned in the 100 papers and use the sentences that mention any of those polymers for training and validation, while the rest makes up the test set. We split the first group 80–20, yielding 3676 sentences for training and 919 sentences for validation, and leaving 2497

Table 2. Experimental results when applying SciNER using different word embedding models and lexicon features to the materials science dataset. The table also includes results from the baseline ChemDataExtractor (CDE) for comparison.

#	Model and features	With lexicon features			Without lexicon features		
		Precision	Recall	F1	Precision	Recall	F1
1	SciNER w/o pre-trained word embeddings	–	–	–	93.0%	78.2%	0.850
2	SciNER with CBOW word embeddings	84.6%	71.9%	0.777	82.1%	70.9%	0.761
3	SciNER with Skip-gram word embeddings	92.3%	81.6%	0.866	85.0%	75.4%	0.799
4	SciNER with FastText word embeddings	89.6%	92.3%	**0.909**	82.3%	80.6%	0.814
5	CDE (NLP module only)	–	–	–	54.3%	58.3%	0.562
6	CDE (NLP+regex+dictionary)	–	–	–	65.1%	58.7%	0.617

sentences for testing. As a single sentence can contain more than one polymer name, polymer names can still co-occur in the training and test sets. In practice, we find that only 18.8% of polymers co-occur in this way, which we view as acceptable.

4.3 Experimental Results

We now explore the accuracy of the SciNER LSTM models using different word embedding models and lexicon features. We apply SciNER to both the materials science and social science datasets to demonstrate its effectiveness and generalizability. For the polymer name recognition task, we compare our results with a state-of-the-art domain-specific toolkit, ChemDataExtractor (CDE) [28]. For the social science dataset, in which we aim to identify dataset names, we could not identify a readily available toolkit that performs a similar tasks, so we compare our results with a basic KNN classifier.

Experiments on the Materials Science Dataset. Table 2 compares the precision, recall, and F1 scores of our LSTM model, when fed with different word embedding and lexicon features, to those achieved by CDE. As shown in the table, Tests 1–4 evaluate the effect of different word embedding models on the performance of SciNER with and without lexicon features. In Test 1, the word embedding matrix in the Embedding layer is randomly initialized as described in Sect. 3.1. In Tests 2–4, word embeddings are trained on the same materials science corpus before being fed to the model as the fixed weights in the Embedding layer. In Test 2, word embeddings are trained using the CBOW model (Sect. 3.1). Note that it produces the lowest F1 score among the first four tests, which is not surprising considering that the words used in academic papers usually follow a long tail distribution, and CBOW is not good at handling infrequent words. The model in Test 3 is fed with word vectors trained using the Skip-gram model, which is designed to better encode rare words, resulting in a 10% improvement in F1 score compared to CBOW. The fourth test uses word

Table 3. Experimental results for social science corpus

#	Model	Precision	Recall	F1 score
7	SciNER w/FastText embeddings & lexicon features	82.5%	87.0%	0.847
8	KNN classifier enhanced by rules [29]	60.0%	58.7%	0.592

embedding generated by FastText, which encodes character n-grams in addition to contextual information. It produces the best results, achieving an F1 score of 0.909.

To explore the benefits of the lexicon features we also ran Tests 2–4 without the lexicon features. The table shows that the lexicon features improves the F1 score by 11% in the best case (row 4).

Tests 5–6 show the results achieved by CDE, which is the state-of-the-art model for recognizing chemical entities [28]. When only using CDE's NLP module, we get an F1 score of 0.562. When using the entire CDE pipeline, which relies on regular expression rules and dictionary, the F1 score increases to 0.617. In either case, SciNER's F1 score exceeds CDE by approximately 50%.

Experiments on the Social Science Corpus. For the social science dataset we apply only FastText word embeddings and lexicon features, as the previous experiment demonstrated that this configuration performed best of the configurations studied. Table 3 compares the precision, recall, and F1 scores of SciNER to our previous work, in which we used a KNN classifier and many manually created rules [29]. Even in this quite different environment, SciNER achieves an F1 score of 0.847, significantly outperforming our rule-based approach that achieves an F1 score of 0.592. This result highlights the value of SciNER, as the dataset names included in social science publications are significantly different from the polymer names included in materials science publications. Each domain uses a different set of frequently used words and domain-specific jargon. Another less obvious, but more challenging, difference is that dataset names are usually much longer than polymer names. Dataset names with more than ten words are not uncommon.

5   Related Work

Researchers have explored myriad approaches to scientific NER. Most approaches rely on crowdsourcing [27,32] or rule-based systems [24]. For example, AQL is a declarative rule language used in IBM's SystemT [15]. With AQL, users can define a set of rules, which SystemT then uses to optimize and build an efficient query plan. SystemT can support complex expressions, but like all rule-based systems, still requires manual effort to define rules, and thus its accuracy is highly dependent on the proper construction of rules.

In other cases, extraction systems are dependent on rich domain-specific ontologies via which named entities can be matched directly with terms in the ontology [13,20,23]. High NER accuracy has been achieved in biomedicine [5,10], due to the availability of structured databases (e.g., Uniprot and PDB) and well-defined, unique identifiers and names (e.g., gene/protein names, diseases, organisms) that can be easily identified in free text (e.g., the string "PDB:1BFM" denotes the 1BFM protein in the PDB database, in this case a histone protein). Few other scientific communities have achieved such a high level of standardization, which is one of the reasons that we have chosen to focus on NER in domains where standard identifiers for named entities are not readily available.

Word embeddings have been shown to be effective at capturing latent information in scientific publications, including in materials science [14,30]. Prior work has shown that the embeddings can capture complex materials science concepts such as structure-property relationships and the relative positions of elements in the periodic table [30]. Those results motivated our use of unsupervised word embeddings rather than hand-curated features to represent words as input to our model.

6   Conclusion

The exponential growth in the number of academic papers has made it infeasible for researchers to manually discover important scientific facts buried deep within these free text publications.

SciNER aims to address part of this problem by automatically identifying scientific named entities in free text publications. SciNER specifiably focuses on addressing challenges associated with the rare words and terminologies used in scientific texts. By leveraging external sources of knowledge and training on scientific texts, SciNER produces more meaningful vectors than traditional word embeddings.

Our experiments demonstrate that SciNER is able to accurately identify diverse named entities from materials science and social science publications. Our best result for identifying polymer names reached an F1 score of 0.909—far exceeding the 0.617 achieved by ChemDataExtractor, the state-of-the-art domain-specific toolkit. When applied to the task of extracting social science dataset names SciNER achieved an F1 score of 0.847, significantly better than the 0.592 achieved by a KNN-based classifier.

In future work we aim to expand SciNER to more domains (e.g. biomedical research) and test its performance against widely used domain ontologies (e.g. the FDA database). We will explore the use of deep neural network-based word embeddings (e.g., BERT [9] and ELMo [21]) to improve extraction performance and design a pipeline for identifying relations between entities, of which SciNER is the first component. Our hope is that the structured data extracted from publications will benefit many applications, such as discovering new molecule pathways and enabling targeted material design.

References

1. Auer, S., Bizer, C., Kobilarov, G., Lehmann, J., Cyganiak, R., Ives, Z.: DBpedia: a nucleus for a web of open data. In: Aberer, K., et al. (eds.) ASWC/ISWC -2007. LNCS, vol. 4825, pp. 722–735. Springer, Heidelberg (2007). https://doi.org/10.1007/978-3-540-76298-0_52
2. Bojanowski, P., Grave, E., Joulin, A., Mikolov, T.: Enriching word vectors with subword information. arXiv preprint arXiv:1607.04606 (2016)
3. Bonney, R., et al.: Citizen science: a developing tool for expanding science knowledge and scientific literacy. Bioscience **59**(11), 977–984 (2009)
4. Bonney, R., et al.: Next steps for citizen science. Science **343**(6178), 1436–1437 (2014)
5. Brase, J.: DataCite-A global registration agency for research data. In: 4th International Conference on Cooperation and Promotion of Information Resources in Science and Technology, pp. 257–261. IEEE (2009)
6. Chiu, J.P., Nichols, E.: Named entity recognition with bidirectional LSTM-CNNs. Trans. Assoc. Comput. Linguist. **4**, 357–370 (2016)
7. Cohn, J.P.: Citizen science: can volunteers do real research? Bioscience **58**(3), 192–197 (2008)
8. DBpedia: DBpdia ontology (2019). https://wiki.dbpedia.org/services-resources/ontology. Accessed 11 Apr 2018
9. Devlin, J., Chang, M.W., Lee, K., Toutanova, K.: BERT: pre-training of deep bidirectional transformers for language understanding. arXiv preprint arXiv:1810.04805 (2018)
10. Duggan, M.: System and method for generating unique and persistent identifiers. US Patent App. 11/444,887, 10 January 2008
11. Enríquez, F., Troyano, J.A., López-Solaz, T.: An approach to the use of word embeddings in an opinion classification task. Expert Syst. Appl. **66**, 1–6 (2016)
12. Fortunato, S., et al.: Science of science. Science **359**(6379) (2018). https://doi.org/10.1126/science.aao0185, https://science.sciencemag.org/content/359/6379/eaao0185
13. Friedman, C., Kra, P., Yu, H., Krauthammer, M., Rzhetsky, A.: GENIES: A natural-language processing system for the extraction of molecular pathways from journal articles. In: ISMB (Supplement of Bioinformatics), pp. 74–82 (2001)
14. Isayev, O.: Text mining facilitates materials discovery. Nature **571**(7763), 42 (2019)
15. Krishnamurthy, R., Li, Y., Raghavan, S., Reiss, F., Vaithyanathan, S., Zhu, H.: SystemT: a system for declarative information extraction. ACM SIGMOD Rec. **37**(4), 7–13 (2009)
16. Mathiak, B., Boland, K.: Challenges in matching dataset citation strings to datasets in social science. D-Lib Mag. **21**(1/2), 23–28 (2015)
17. Mikolov, T., Chen, K., Corrado, G., Dean, J.: Efficient estimation of word representations in vector space. arXiv preprint arXiv:1301.3781 (2013)
18. Mikolov, T., Sutskever, I., Chen, K., Corrado, G.S., Dean, J.: Distributed representations of words and phrases and their compositionality. In: Advances in Neural Information Processing Systems, pp. 3111–3119 (2013)
19. Ohama, Y.: Handbook of Polymer-Modified Concrete and Mortars: Properties and Process Technology. William Andrew, Norwich (1995)
20. Ono, T., Hishigaki, H., Tanigami, A., Takagi, T.: Automated extraction of information on protein-protein interactions from the biological literature. Bioinformatics **17**(2), 155–161 (2001)

21. Peters, M.E., et al.: Deep contextualized word representations. arXiv preprint arXiv:1802.05365 (2018)
22. Ramshaw, L.A., Marcus, M.P.: Text chunking using transformation-based learning. In: Armstrong, S., Church, K., Isabelle, P., Manzi, S., Tzoukermann, E., Yarowsky, D. (eds.) Natural Language Processing Using Very Large Corpora. Text, Speech and Language Technology, vol. 11, pp. 157–176. Springer, Dordrecht (1999). https://doi.org/10.1007/978-94-017-2390-9_10
23. Rzhetsky, A., et al.: GeneWays: a system for extracting, analyzing, visualizing, and integrating molecular pathway data. J. Biomed. Inform. **37**(1), 43–53 (2004)
24. Shaalan, K., Raza, H.: Arabic named entity recognition from diverse text types. In: Nordström, B., Ranta, A. (eds.) GoTAL 2008. LNCS (LNAI), vol. 5221, pp. 440–451. Springer, Heidelberg (2008). https://doi.org/10.1007/978-3-540-85287-2_42
25. Sommerfeld, A.K.: Education as a collective accomplishment: how personal, peer, and parent expectations interact to promote degree attainment. Soc. Psychol. Educ. **19**(2), 345–365 (2015). https://doi.org/10.1007/s11218-015-9325-7
26. Stewart, C.A., et al.: Jetstream: a self-provisioned, scalable science and engineering cloud environment. In: XSEDE Conference (2015)
27. Sui, D., Elwood, S., Goodchild, M.: Crowdsourcing Geographic Knowledge: Volunteered Geographic Information (VGI) in Theory and Practice. Springer, Dordrecht (2012). https://doi.org/10.1007/978-94-007-4587-2
28. Swain, M.C., Cole, J.M.: ChemDataExtractor: a toolkit for automated extraction of chemical information from the scientific literature. J. Chem. Inf. Model. **56**(10), 1894–1904 (2016)
29. Tchoua, R.B., et al.: Creating training data for scientific named entity recognition with minimal human effort. In: Rodrigues, J., et al. (eds.) ICCS 2019. LNCS, vol. 11536, pp. 398–411. Springer, Cham (2019). https://doi.org/10.1007/978-3-030-22734-0_29
30. Tshitoyan, V., et al.: Unsupervised word embeddings capture latent knowledge from materials science literature. Nature **571**(7763), 95 (2019)
31. Ware, M., Mabe, M.: The STM report: an overview of scientific and scholarly journal publishing. Technical report, International Association of Scientific, Technical and Medical Publishers (2015)
32. Wiggins, A., Crowston, K.: From conservation to crowdsourcing: a typology of citizen science. In: 44th Hawaii International Conference on System Sciences, pp. 1–10. IEEE (2011)
33. Yang, Z., Fujii, Y., Lee, F.K., Lam, C.H., Tsui, O.K.: Glass transition dynamics and surface layer mobility in unentangled polystyrene films. Science **328**(5986), 1676–1679 (2010)

GPU-Embedding of kNN-Graph Representing Large and High-Dimensional Data

Bartosz Minch[✉], Mateusz Nowak, Rafał Wcisło, and Witold Dzwinel

AGH University of Science and Technology, Kraków, Poland
{minch,wcislo,dzwinel}@agh.edu.pl, mtsznowak9@gmail.com

Abstract. Interactive visual exploration of large and multidimensional data still needs more efficient $ND \rightarrow 2D$ data embedding (DE) algorithms. We claim that the visualization of very high-dimensional data is equivalent to the problem of 2D embedding of undirected kNN-graphs. We demonstrate that high quality embeddings can be produced with minimal time&memory complexity. A very efficient GPU version of IVHD (interactive visualization of high-dimensional data) algorithm is presented, and we compare it to the state-of-the-art GPU-implemented DE methods: BH-SNE-CUDA and AtSNE-CUDA. We show that memory and time requirements for IVHD-CUDA are radically lower than those for the baseline codes. For example, IVHD-CUDA is almost 30 times faster in embedding (without the procedure of kNN graph generation, which is the same for all the methods) of the largest ($M = 1.4 \cdot 10^6$) YAHOO dataset than AtSNE-CUDA. We conclude that in the expense of minor deterioration of embedding quality, compared to the baseline algorithms, IVHD well preserves the main structural properties of ND data in $2D$ for radically lower computational budget. Thus, our method can be a good candidate for a truly big data ($M = 10^{8+}$) interactive visualization.

Keywords: High-dimensional data · Data embedding · kNN graph visualization · GPU implementation

1 Introduction

In the age of data science, interactive visualization of large high-dimensional data is an essential tool in knowledge extraction. It allows for both the insight into data structure and its interactive exploration by direct manipulation on the whole or a fragment of a dataset. This way, it is possible to observe the shapes and mutual location of classes, as well as remove irrelevant data samples and identify the outliers. The multiscale structure can be explored visually by changing data embedding strategies and visualization modes (e.g., the type of the loss function), and zooming-in and out selected fragments of 2D (3D) data mappings. Summarizing, interactive visualization allows for: 1) instant verification of a number of hypotheses, 2) precise matching of data mining tools to the properties of data investigated, 3) adapting optimal parameters to machine learning algorithms, and 4) selecting the best data representation. Herein, we focus on application of data embedding (DE) methods in the interactive visualization of large ($M \sim 10^5$-10^6) and high-dimensional $N \sim 10^{2+}$ data.

© Springer Nature Switzerland AG 2020
V. V. Krzhizhanovskaya et al. (Eds.): ICCS 2020, LNCS 12138, pp. 322–336, 2020.
https://doi.org/10.1007/978-3-030-50417-5_24

Data embedding (DE) is defined as a transformation $\mathbf{B}:Y{\rightarrow}X$ of N-dimensional (ND) dataset $\mathfrak{R}^N{\ni}Y = \{y_i\}_{i=1,...M}$ into its n-dimensional (nD) representation $\mathfrak{R}^n{\ni}X = \{x_i\}_{i=1,...M}$, where $N{>>}n$ and M is the number of ND feature vectors y_i and corresponding nD embeddings x_i. The mapping $\mathbf{B}$ can be perceived as a lossy compression of data. It is performed by minimizing a loss function $E(\|Y - X\|)$, where $\|.\|$ is a measure of topological dissimilarity between Y and X. Due to the high complexity of the low-dimensional manifold, immersed in the ND feature space and occupied by data samples Y, perfect embedding of Y in the nD space is possible only for trivial cases.

In the context of high-dimensional data visualization, we assume that $n = 2$. Data embedding to 3D can be processed in a similar way. As shown in many papers (see, e.g., [9, 17]), DE of large data that is both sufficiently precise in reconstruction of ND data topology, and simultaneously, computationally affordable, is the algorithmic challenge. To preserve topological properties of Y in X both the classical MDS (multidimensional scaling) methods (e.g., [5]) and the state-of-the-art (SOTA) clones of the stochastic neighbor embedding (SNE) concept (e.g., [11, 14, 19]) require computing and storing two $O(M^2)$ arrays: (1) the dissimilarities between data vectors Y and (2) the Euclidean distances between their 2D embeddings X. That is why, the SOTA visualization algorithms, such as t-SNE [16] and its clones, suffer from high $O(M^2)$ time&memory complexity. The time complexity can be decreased to $O(M log M)$ and even $O(M)$, by using the approximated versions of t-SNE, such as BH-SNE [15] and other its variants and approximations [14, 19]. Meanwhile, the memory complexity remains $O(M^2)$ which considerably limits its use for truly big data and new parallel computer architectures. Summing up, for many of the SOTA embeddings: [1, 17, 22, 23]:

1. The time complexity of the DE procedure is dominated by the construction of the kNN-graph, which is generally $O(M log M)$ complex (for exact kNN search algorithms).
2. The computational efficiency of the DE process vastly depends on the loss function and optimization procedure applied for its minimization.
3. Calculating gradient of the loss function is $O(N{\cdot}M)$ complex, but with large proportionality coefficient, which is dominated by a relatively large value of k.

Interactive visual data exploration involves very strict time&memory performance regimes while the SOTA DE algorithms are still too time&memory consuming. The main contribution of this paper is developing a method to overcome this flaw and optimize the data embedding process. To this end, for DE, we adapt the same idea we used previously for large unweighted and undirected graphs visualization (GV) [6]. Furthermore, we clearly demonstrate that the DE of very high-dimensional data can be treated as a subproblem of GV what has not been said explicitly before (maybe except of recently published [13]). Consequently, the visualization of high-dimensional data comes down to the visualization of the kNN graphs (kNN-graphs) where each of the nodes represents a feature vector. However, unlike the other DE algorithms which also employ this trick (e.g., BH-SNE [15], LargeVis [22], UMAP [17]), we show that the value of k can be much smaller, such as $k{\sim}2, 3$ (compare it to BH-SNE [15] and LargeVis [22] where $k{\sim}10^2$). Moreover, for truly high-dimensional data ($N \gtrsim 30$), i.e., when the effects of "curse of dimensionality" on the ND space topology become evident, the floating point dissimilarities used for kNN-graph construction can be discarded. Instead, the integer indices to only a few nn nearest neighbors (instead of index k, we use nn to be consistent with the notation we use in the rest of this paper) have to be kept in the computer memory. We show here that if for each ND feature vector we:

1. store only indices of $nn < 6$ nearest neighbors ($nn = 2$ in the most cases is sufficient),
2. select very few (often just one) random neighbors rn (similar to *negative sampling* procedure) during calculations,
3. define binary distances (0 and 1, respectively) to the nearest and random neighbors,

it is possible to radically simplify the loss function and, consequently, decrease the CPU time required for its minimization. Thus, we are able to reconstruct the ND data structure in 2(3)D space in a very efficient way both in terms of computational time and storage.

Summarizing, in the paper, we demonstrate that the data embedding (DE) is equivalent to the visualization of unweighted kNN-graphs, constructed for the source Y data. The principal contribution of this paper is the essential improvement of the time&memory complexity of DE at the expense of a minor deterioration of embedding quality. Consequently, the proposed data mapping methodology allows for interactive visualization of much larger data than the state-of-the-art DE algorithms. Moreover, we demonstrate that our IVHD algorithm can be implemented in an efficient way in GPU/CUDA environment. We compare this implementation to the fastest publicly available GPU data embeddings: BH-SNE-CUDA [4] and Anchor-t-SNE [10].

2   Methods

Despite that there are many algorithms for visualization of high-dimensional data and that this topic has been extensively studied for years, to the best of our knowledge, there are only a few implementations of modern data embedding algorithms in GPU/CUDA environment [4, 10, 18]. We will focus on the publicly available ones, which generate the best embeddings of large datasets: BH-SNE-CUDA [4] and Anchor-t-SNE [10] (AtSNE). These algorithms base on the well known t-SNE (t-distributed Stochastic Neighbor Embedding) concept [16], and consist of two stages: (1) generate of a weighed kNN-graph; (2) run a proper embedding procedure which is based on: (2a) definition of a loss function and (2b) its minimization.

2.1 kNN Graph Generation

kNN-graph approximates a nD non-Cartesian manifold immersed in $\mathbb{R}^N$ sampled by the feature vectors y_i. We consider here the kNN-graph construction procedure shipped by the FAISS library [12]. Its authors claim that it is currently the fastest available kNN search algorithm implemented on GPU. The FAISS kNN-search procedure merges a very efficient and well-parallelized exact kNN algorithm and indexing structures that allow to perform approximate search. To achieve a high-efficiency search with a good accuracy we employ the IVFADC [2] indexing structure. It uses two levels of quantization combined with the vector encoding for compressing high-dimensional vectors. The main idea behind the index usage is to split the input space and divide all input samples into a number of clusters represented by their centroids. To handle a query, the algorithm compares it with the centroid centers. It picks the centroid that is the most similar to the query vector and performs an exact search within the set of samples belonging to this centroid. Apart from the efficient kNN algorithm, its CUDA implementation is extremely well optimized [12]. We use the same FAISS kNN-graph generation procedure in both the baseline and IVHD algorithms.

2.2 Loss Functions

BH-SNE-CUDA. The group of methods based on the SNE concept [16] defines the similarity of two samples i and j in terms of probabilities p_{ij} (in Y) and q_{ij} (in X), that i would pick j as its neighbor and vice versa. These probabilities are functions of distances between samples in Y and their embeddings in X, respectively. Let $\mathbf{D} = [D_{ij}]$ is the distance table in Y and D_{ij} are the distances between i and j feature vectors y_i and y_j, while $\mathbf{d} = [d_{ij}]$ is the respective distance array in X. Then, the loss function $C = E(\mathbf{D}, \mathbf{d})$ is defined by the Kullback–Leibler (KL) divergence:

$$C(.) = E(\mathbf{D}, \mathbf{d}) = \sum_i KL(P_i \| Q_i) = \sum_i \sum_j p_{ij} \log \frac{p_{ij}}{q_{ij}}, \tag{1}$$

where, for t-SNE algorithm, p_{ij} is approximated by the Gaussian $\mathcal{N}(y_i, \sigma)$, while q_{ij} is defined by the Cauchy distribution [16]. The p_{ij} and q_{ij} are defined as follows:

$$p_{ij} = \frac{exp(-D_{ij}^2/2\sigma_i^2)}{\sum_{k \neq l} exp(-D_{kl}^2/2\sigma_i^2)} \tag{2} \qquad q_{ij} = \frac{(1 + d_{ij}^2)^{-1}}{\sum_{k \neq l}(1 + d_{kl}^2)^{-1}} \tag{3}$$

The BH-SNE (Barnes-Hut t-SNE [15]) is an approximation of the t-SNE method that can reduce computational complexity of the DE from $O(M^2)$ to $O(MlogM)$ by using Barnes-Hut approximation but at the cost of increasing the algorithmic complexity and, consequently, decrease of the parallelization efficiency [15]. Its GPU version - the BH-SNE-CUDA algorithm - does not introduce any changes to the BH-SNE [15] algorithm, and just matches the instructions and data flows to the GPU architecture [4].

AtSNE-CUDA. Recently published AtSNE-CUDA (Anchor-t-SNE [10]) algorithm was created to deal with the t-SNE issues such as sensitivity to initial conditions and inadequate reconstruction of the global data structure. The Authors of the AtSNE-CUDA method claim that their method is 50% faster than BH-SNE-CUDA and has lower memory requirements. As shown in [10], it still generates good quality embeddings though some kNN quality scores are better for the classic t-SNE or the LargeVis algorithms [22].

Similar to all t-SNE clones, AtSNE minimizes the regularized KL divergence. Information about a local structure of data can be acquired from the set of approximated kNN neighbors of each y_i. Meanwhile, to reconstruct the global structure of data, x_i points receive "pulling forces" from, so called, *anchor points* generated from the original data.

This way the *anchor points* are responsible for maintaining the mutual positions and shape of classes in X. Consequently, the hierarchical embedding [10] optimizes both the positions of the *anchor points* and regular data samples.

Let A represents the set of *anchor points* in the high-dimensional space, and B its low-dimensional embedding. The probabilities $P(Y)$ and $Q(X)$ denote the high and low-dimensional distributions of the *anchor points*, respectively. To preserve both the global and local information, the following loss function are minimized:

$$E(.) = \sum_i KL(P(Y)\|Q(X)) + \sum_i KL(P(A)\|Q(B)) + \sum_i \|b_i - \frac{\sum_{y_k \in C_{b_i}} y_k}{|C_{b_i}|}\|, \quad b_i \in C_{b_i} \quad (4)$$

where C_{b_i} denotes the set of points whose K-means cluster center is b_i. The last term is a regularization term explained in details in [10].

IVHD. Unlike t-SNE based BH-SNE-CUDA and AtSNE-CUDA, IVHD-CUDA utilizes classical MDS stress function. However, the number and sort of distances IVHD employs, are radically different than those in classical MDS and the baseline algorithms.

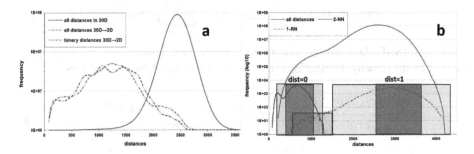

Fig. 1. The envelopes of histograms for MNIST dataset (after PCA transformation $784D \rightarrow 30D$). *Red solid line*: all **D** distances (a - linear, b - logarithmic scale); a) *dashed lines*: all **d** distances; b) *blue solid line*: **D** distances only between samples and their 2-NNs, and *red dashed line*: **D** distances only between samples and one random neighbor. (Color figure online)

In Fig. 1 we display the envelopes of histograms of both **D** and **d** distances for MNIST dataset before and after IVHD embedding. Although MNIST dataset has a varied structure, the envelope of **D** histogram in linear coordinates (Fig. 1a) is perfectly bell-shaped, while that for **d** is more deformed but still resembles the Cauchy distribution. To increase distances diversification (see Fig. 1b), instead of all $M(M - 1)/2$ floating point distances we can consider binary distances to only a few (*nn*) nearest neighbors and just one (*rn*) randomly selected neighbor. This is because, the most of real distances (95%) to the nearest and the random neighbors are located in separated and rather distant intervals (darker blue and red boxes in Fig. 1b). For higher dimensions, the random neighbors are getting almost equidistant from y_i due to the "curse of dimensionality" effect. Whereas, the overlapping region (green) contains only 0.3%

of distances. So, we can assume additionally that $O_{nn}(i) \cap O_{rn}(i) = \emptyset$. However, this assumption is superfluous because the probability of picking the nearest neighbor as a random neighbor is negligibly small for large N. As shown in Fig. 1a, for non-binarized and binarized source distances, their histograms for respective 2D embeddings are very similar. Thus, let $O_{nn}(i)$ and $O_{rn}(i)$ will be the sets of indices of nn nearest (connected) neighbors and rn (unconnected) random neighbors of a feature vector y_i in kNN-graph, respectively. We define binary dissimilarity measure as follows (see Fig. 1b):

$$\forall y_i \in Y : D_{ij} = \begin{cases} 0 \; if \; j \in O_{nn}(i) \\ 1 \; if \; j \in O_{rn}(i) \end{cases} . \tag{5}$$

Thus, unlike in the baseline algorithms, we are not interested even in an approximated ordering of $kNNs$ for each $y_i \in Y$. This is justified for small nn, because distances to a few first NN, in general, cannot differ too much (see blue plot in Fig. 1b) and the ordering of NN can result from measurement errors. We assume, that the number of nn neighbors has to fulfill two conditions. Firstly, the kNN-graph should be fully (or approximately - i.e., the size of the largest component should be comparable to the size of the full graph) connected. For example, for MNIST and FMNIST datasets (see Table 1) and $k = nn = 2$ the largest components consist of 99% of nodes, and respectively: SmallNorb ($nn = 5$ and 82%), RCV-Reuters ($nn = 3$ and 95%). Secondly, the kNN-graph augmented with approximately rn edges should be at least a minimal n-rigid graph (in 2D: 2-rigid). The term "rigidity" can be understood as a property of a nD structure made of rods (distances) and joints (data vectors) that it does not bend or flex under an applied force. The lower band of the number of connections L, required for making the augmented kNN-graph 2-rigid, is $L \sim 2 \cdot M$. Meanwhile, the augmented kNN-graph has approximately $L \sim n_{vi} \cdot M$ edges, where $n_{vi} = nn + rn > 2$ [7]. As our experience shows, the probability that the largest connected component is rigid (or approximately rigid) is very high. Summarizing, to obtain the largest connected component approximately equal to the full kNN-graph, the number of the nearest neighbors nn can be very low (mostly $nn = 2$ but for some specific datasets with very similar samples it can be a bit larger). Assuming additionally that $rn = 1$, we can obtain stable and rigid 2-D embedding of the kNN-graph.

This way, instead of the $O(M^2)$ floating point $\mathbf{D}$ matrix, we have as the input data $O(nn \cdot M)$ integers - the list of kNN-graph edges. The indices of rn random neighbors can be generated ad hoc during embedding process. Thus the embedding of high-dimensional data reduces to the embedding of the corresponding sparse kNN-graph. To this end, we minimize the following stress function:

$$E(\|\mathbf{D} - \mathbf{d}\|) = \sum_i \sum_{j \in O_{nn}(i) \cup O_{rn}(i)} \begin{cases} d_{ij}^2 & if \; j \in O_{nn}(i) \\ c \cdot (1 - d_{ij})^2 & if \; j \in O_{rn}(i) \end{cases} , \tag{6}$$

which represents the error between dissimilarities $D_{ij} \in \{0, 1\}$ and corresponding Euclidean distances d_{ij}, where: $i, j = 1, \ldots, M$, and $c \in (0, 1)$ is the scaling factor for random neighbors. Thus, IVHD uses the stress function, which is much easier to optimize than a KL-based loss function employed in the baseline algorithms.

2.3 Optimization

BH-SNE-CUDA. To minimize KL divergence, t-SNE and its clones employ the optimal, matched by the Authors, modern gradient descent optimization schemes [21]. The gradient of the loss function $C(.)$ (Eq. 1) is as follows:

$$\frac{\delta C}{\delta y_i} = 4 \sum_j (p_{ij} - q_{ij}) q_{ij} (y_i - y_j). \tag{7}$$

AtSNE-CUDA. The AtSNE-CUDA algorithm preserves the global information by minimizing the term $KL(P(A)\|Q(B))$ in Eq. 4, while the first term of the loss function is responsible for preserving the local structure of data. Because, the gradient of the loss function is more complicated than this defined by Eq. 7, AtSNE-CUDA uses the Hierarchical Optimization Algorithm [10]. This algorithm consists of two optimization layers: the global and the local ones. Optimization proceeds in a top-down manner. The main idea is to optimize the two-layer layout alternately: 1) fix the layout of the ordinary points and optimize the layout of the anchor points; 2) fix the anchor point layout and optimize the layout of ordinary points.

IVHD. A few state-of-the-art optimization schemes such as: Nesterov, Adagrad, Adadelta, RMSprop, NAG, Adam, [21] and force-directed (F-D) method [7] have been implemented to calculate minimum of the loss function (Eq. 6). As shown in Table 1, where the best GPU implementations of the optimization methods are collected, the force-directed (F-D) approach and synchronized Nestorov scheme appear to be the best in terms of accuracy (i.e., a better minimum of the loss function reached).

Table 1. Results of cf_5 and cf_{10} accuracies (see Eq. 8) achieved by IVHD-CUDA implementation for various optimizers and baseline datasets.

		MNIST			Fashion-MNIST			Small NORB			RCV		YAHOO	
		1s	3s	5s	1s	3s	5s	1s	3s	5s	15s	25s	15s	25s
F-D	cf_5	0.501	0.852	**0.89**	0.472	0.671	**0.68**	0.87	0.946	**0.954**	0.666	**0.672**	0.562	**0.618**
	cf_{10}	0.499	0.851	**0.889**	0.469	0.667	**0.676**	0.846	0.933	**0.945**	0.666	**0.671**	0.562	**0.618**
Nest.	cf_5	0.522	0.846	**0.888**	0.473	0.671	**0.688**	0.86	0.942	**0.957**	0.432	0.471	0.102	0.252
	cf_{10}	0.52	0.844	**0.887**	0.47	0.667	**0.684**	0.835	0.929	**0.948**	0.432	0.47	0.102	0.252
Adad.	cf_5	0.87	0.868	0.866	0.687	0.679	0.672	0.936	0.925	0.914	0.581	0.626	0.261	0.57
	cf_{10}	0.869	0.867	0.866	0.685	0.676	0.67	0.927	0.916	0.904	0.58	0.625	0.261	0.57
Adam	cf_5	0.64	0.872	0.867	0.551	0.687	0.686	0.835	0.958	0.951	0.37	0.437	0.102	0.145
	cf_{10}	0.638	0.871	0.866	0.548	0.684	0.682	0.807	0.947	0.941	0.369	0.436	0.102	0.145

3 GPU Implementation

The advantage of using GPU accelerator is a huge number of computational units that can perform calculations in parallel. In a typical GPU, there are a few thousands of cores. GPUs boards are Single Instruction, Multiple Data (SIMD) computers. It means that multiple threads are executed using the same instruction set on various data. In the case of the Nvidia CUDA framework, it means that all threads inside a warp will be calculated by using the same instruction set. In the case of instructions branching, the execution of two alternatives will be carried out separately and one of the two will be selected after evaluation of the *if* condition. This process is inefficient and if it occurs frequently (*branch divergence*), it might lead to a poor utilization of GPU. That is why the optimization of the code needs to develop special GPU-dedicated algorithms.

Optimization Schemes on GPU. On a CPU unit, subsequent iterations for the classical stochastic gradient descent (SGD) based optimization schemes are executed sequentially and asynchronously. Meanwhile, to achieve the best GPU performance, we had to execute the iterations synchronously, such as in the force-directed approach [7], i.e., all the updated variables of the loss function are calculated using values from the previous iteration what helps to avoid race conditions between CUDA threads. It appears that (see Table 1) synchronous version of the Nesterov method produces almost identical results as the F-D scheme.

IVHD-CUDA. In IVHD-CUDA we employ the force directed optimization scheme, similar in spirit to the synchronous *momentum* and *Nesterov* methods, described in detail in [7]. Let *connection* mean two "particles" x_i and x_j (the nearest or random neighbors) in 2D embedded X space joined by an edge. The main IVHD-CUDA implementation loop consists of the following steps: 1) calculating "forces" for each *connection* where the "force" is proportional to the gradient dependent on the two "particles'" positions and their velocities; 2) summing up the forces and 3) updating positions and velocities for all the particles simultaneously. The simplified pseudo-code is presented below as Algorithm 1.

Algorithm 1: Simplified IVHD-CUDA scheme.

Input: kNN-graph precalculated with FAISS library.

initialization;

while *running* **do**

    allocate subsets of force components to threads;

    for *thread in threads* **do**

       | calculate force components for each connection;

    end

    join threads;

    allocate subsets of samples to threads;

    for *thread in threads* **do**

       | sum force/gradient components;

       | calculate new velocities;

       | update particle positions;

    end

end

Because the steps are executed sequentially, it allows to avoid the race condition between CUDA threads. Each connection receives two different memory addresses where calculated values can be stored - one *positive* and one *negative*. If a particle *i* attracts particle *j* (*positive* value), then at the same time particle *j* attracts particle *i* (*negative* value). In the first step, each thread calculates some subset of the inter-particle force/gradient components. Simultaneously, they execute exactly the same instructions and thus, any branching is not required. However, this approach has a minor bottleneck. To save the result for each *connection*, a given thread has to execute *write* instructions to completely different memory addresses representing two samples. Meanwhile, minimizing the number of *read/write* operations is always beneficial as they are very time consuming.

In the second step, each thread is assigned with a set of particles to process. As a precaution against processing the particles that interact with different number of particles (what might cause branch divergence) we sort the particles before the first iteration by considering the number of force components to calculate. After sorting it is guaranteed that all the threads in a warp will process samples with the same number of *connections*. The percentage share of each component of DE procedure (without *k*NN graph generation), e.g., for 70.000 feature vectors with approximately three neighbors each (*nn*=2 and *rn*=1), is as follows: (1) force components calculation (78.97%), (2) positions update (20.6%), initialization (3) (0.43%). In IVHD-CUDA implementation the global, register and constant memory are used. In the constant memory, the algorithm parameters are stored. While, the *connections*, force components and positions are stored in the global memory. A register memory is used for auxiliary and temporary calculations.

4   Computational Environment

Herein, we present: the baseline datasets used in experiments, evaluation metrics and computational environment.

Datasets. The main properties of each baseline dataset are: the number of samples M, dimensionality N, and the number of classes K. Here, we consider datasets with (1) a huge number of samples and relatively low dimensionality, (2) a smaller number of samples but larger number of features, (3) highly imbalanced data (RCV-Reuters), and (4) skewed data (Small NORB). Datasets used in the experiments are described in Table 2.

Table 2. The list of baseline datasets.

Dataset	N	M	K	Short description
MNIST	784	70 000	10	Well balanced set of grayscale images of handwritten digits
Fashion-MNIST	784	70 000	10	More difficult MNIST version. Instead of handwritten digits it consists of apparel images
Small NORB	2048	48 600	5	It contains stereo image pairs of 50 uniform-colored toys under 18 azimuths, 9 elevations, and 6 lighting conditions
RCV-Reuters	30	804 409	8	Corpus of press articles preprocessed to 30D by PCA
YAHOO	100	1.4 million	10	Questions and answers from YAHOO. The answers service preprocessed with FastText [20]

Evaluation of Embedding Quality. Because of the high computational load required for computing precision/recall coefficients, to compare data separability and class purity, we define the following simple metrics:

$$cf_{nn} = \frac{\sum_{i=1}^{M} nn(i)}{nn \cdot M} \quad \text{and} \quad cf = \frac{\sum_{nn=1}^{nn_{max}} cf_{nn}}{nn_{max}}, \tag{8}$$

where $nn(i)$ is the number of the nearest neighbors of x_i in X space, which belong to the same class as y_i. The value of cf is computed for arbitrarily defined value of nn_{max} dependent on the number of feature vectors in classes. To reflect a wide range of embedding properties, we use $nn_{max}=100$. The value of $cf \sim 1$ for well separated and pure classes, while $cf \sim 1/K$ for random points from K classes. These simple metrics allow for evaluating the quality of the embeddings by calculating several cf_{nn} values for small, medium, and greater number of nn. The differences in this criterion for confronted methods allow for inferring their embedding qualities for very local ($nn=2$), local ($nn=10$) and medium ($nn=100$) reconstruction depth. The stability of cf_{nn} for increasing nn means more compact and circular shape of classes. For elongated and mixed classes, the values of cf_{nn} decrease faster with nn.

Hardware. All GPU/CUDA implementations of the baseline embedding methods were executed on the separated remote server with: CPU Intel Xeon E5-2620 v3, GPU Nvidia GeForce GTX 1070 (1920 CUDA cores, 8GB GDDR5), 252 GB RAM, OS: Ubuntu 18.04.3, architecture x86, 64. The codes were compiled using GCC-7.4 and CUDA Toolkit 10.0. IVHD-CUDA code was tested on the GPU device with capability 2.7 and CUDA toolkit V8.0, and no issues were observed. The source code used in this paper can be found at https://github.com/mtsznowak/ivhd-cuda.

5   Results and Comparisons

In this section we compare our IVHD-CUDA implementation to the most efficient and robust publicly available GPU-implemented data embedding codes: BH-SNE-CUDA and AtSNE-CUDA. The most important parameters of the methods are collected in Table 3. For the baseline methods we use default parameters (appx. 15 parameters) proposed in [10, 15] and submitted to the GitHub repositories with respective GPU codes [3,4]. IVHD parameters are also selected by default, e.g., $rn = 1$ while nn is fitted automatically as a minimal value producing the largest connected component which approximates (with a given high accuracy) kNN-graph. The value of $c \in [0.01, 0.1]$ from Eq. 6 is the only parameter, which can be adjusted interactively.

Table 3. The glossary of parameters of DE algorithms used in experiments. All IVHD parameters are displayed. The baseline algorithms require matching a considerably higher amount of parameters (about 10) [3,4]), so due to the brevity only the most important ones, i.e., *perplexity* and *nn* are given.

Method	Dataset	Perplexity	nn	rn	c	Metric
IVHD-CUDA	MNIST	-	2	1	0.01	Euclidean
IVHD-CUDA	Fashion-MNIST	-	2	1	0.01	Cosine
IVHD-CUDA	Small Norb	-	5	1	0.01	Cosine
IVHD-CUDA	RCV-Reuters	-	3	1	0.1	Cosine
IVHD-CUDA	YAHOO	-	2	1	0.1	Cosine
BH-SNE-CUDA	All	50	32	-	-	Euclidean
AtSNE-CUDA	All	50	100	-	-	Euclidean

Based on visualizations presented in Figs. 2, 3 one can make the following observations:

1. In Fig. 2a, IVHD clearly reflects the global structure of the MNIST classes and their separation. However, due to the "crowding effect", the classes become very dense in the centers, and sparse between them. On the other hand, the results of graph visualization presented in [6], demonstrate that the "crowding effect" can be controlled by tuning IVHD-CUDA parameters. Consequently, even the fine-grained neighborhood can be preserved with an amazingly high accuracy (see [8]).

2. For the Small NORB dataset (see Fig. 2b), IVHD-CUDA was able to visualize clearly separable three big clusters with fine-grained data structure. For BH-SNE-CUDA and AtSNE-CUDA the quality of embeddings is much worse and more fragmented. The changes of the parameter values of the baseline algorithms (*perplexity*) do not improve the visualization quality.

3. IVHD applied to the Fashion-MNIST (see Fig. 2c) creates separate clusters of various, mainly elongated shapes. Moreover, the generated mapping is fuzzy. It is clearly reflected by decreasing values of cf_{nn} particularly for $nn=100$. On the other hand, both AtSNE-CUDA and BH-SNE-CUDA are much better creating oblate and clearly separated clusters. As shown in Table 4, the values cf_{nn} are more stable than those for the IVHD method. Nevertheless, unlike the IVHD result, some classes reproduced by the baseline methods are mixed and fragmented.

4. Similar conclusions can be drawn by visualizing RCV and YAHOO 2D embeddings (see Fig. 3a, 3b, respectively). IVHD-CUDA generates more fuzzy output than AtSNE-CUDA but the samples from the same class are closer together than those generated by the baseline method. Meanwhile, AtSNE-CUDA is able to separate clearly visible but fragmented clusters. BH-SNE-CUDA was too slow to visualize these big datasets in a reasonable time budget.

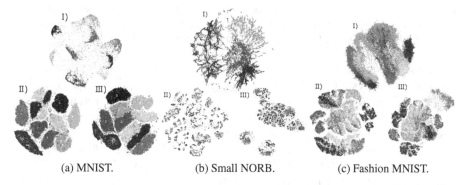

(a) MNIST.	(b) Small NORB.	(c) Fashion MNIST.

Fig. 2. Visualization of datasets using: I) IVHD-CUDA, II) BH-SNE-CUDA, III) AtSNE.

As shown in Table 4, IVHD-CUDA is the fastest method for all the baseline datasets. Unfortunately, we were not able to control the run-times of AtSNE-CUDA and BH-SNE-CUDA implementations, thus the comparisons for various time budgets are not possible. As shown above, the quality and fidelity of embeddings strongly depends on the structure of a specific dataset, user expectations, and their data visualization requirements. In general, the local structure of the *source* data is better reconstructed by the two baseline SNE algorithms. This is not a surprise because the IVHD algorithm does not use the correct ordering of kNN neighbors and original distances between a sample y_i and its kNN neighbors. Moreover, the value of k (*nn* in this paper) is extremely small compared to AtSNE-CUDA and BH-SNE. Nevertheless, despite of this drastic approximation, IVHD properly preserves the class structure and its relative locations. For fine-grained structures of classes (such as in the Small NORB dataset), IVHD outperforms its competitors in both the efficiency and - slightly - in the *cf* accuracy. Moreover, unlike AtSNE-CUDA and BH-SNE-CUDA, IVHD is able to visualize the separated and not fragmented classes. The same can be observed for highly imbalanced dataset RCV. For MNIST and Fashion MNIST datasets, the dominance of t-SNE based methods in reproducing the local - "microscopic" - data structure is visible, mainly due to the strong "crowding effect" seen for IVHD embeddings. However, the "macroscopic" view of the classes is more convincing for IVHD visualization, though due to the high *perplexity* parameter (*perplexity* = 50), the baseline algorithms are tuned also for better coarse-grained visualization.

(a) YAHOO	(b) RCV-Reuters

Fig. 3. Visualization of datasets using: I) IVHD-CUDA, II) AtSNE-CUDA.

Table 4. The values of cf_{nn} accuracy for various datasets and embedding methods. The timings show: the overall embedding time (*time*), kNN-graph generation time ($time_{gg}$), net embedding time ($time_{emb}$). The results are the averages over 10 simulations.

Dataset	Algorithm	*time* [s]	$time_{gg}$ [s]	$time_{emb}$ [s]	cf_2	cf_{10}	cf_{100}
MNIST	BH-SNE-CUDA	32.588	**5.813**	26.775	0.94	0.938	0.933
	AtSNE	15.980		10.167	0.944	**0.943**	**0.938**
	IVHD-CUDA	**7.326**		**1.261**	**0.946**	0.936	0.924
FMNIST	BH-SNE-CUDA	32.913	**6.734**	26.179	0.757	0.755	**0.738**
	AtSNE	17.453		10.719	0.76	**0.757**	0.737
	IVHD-CUDA	**8.177**		**1.443**	**0.767**	0.726	0.670
Small NORB	BH-SNE-CUDA	38.673	**15.517**	23.161	0.944	0.919	0.745
	AtSNE	20.521		5.009	**0.97**	**0.94**	0.73
	IVHD-CUDA	**16.151**		**0.634**	0.936	0.921	**0.828**
RCV-Reuters	BH-SNE-CUDA	-	**45.302**	-	-	-	-
	AtSNE	220.39		175.088	0.82	0.82	**0.818**
	IVHD-CUDA	**60.72**		**15.418**	**0.835**	**0.828**	0.803
YAHOO	BH-SNE-CUDA	-	**52.930**	-	-	-	-
	AtSNE	628.63		575.7	**0.686**	**0.686**	**0.686**
	IVHD-CUDA	**70.12**		**18.930**	0.668	0.662	0.653

The main advantage of IVHD is that after storing the kNN-graph in the disc cache, and neglecting the computational time required for its generation, IVHD embedding can be more than one order of magnitude faster than the baseline methods. This allows for a very detailed interactive exploration of multi-scale data structure by employing broad spectrum of parameter values and various versions of the stress function without a need for the kNN-graph recalculation.

6   Conclusions

In this paper, we have compared GPU/CUDA implementations of data embedding algorithms in the context of interactive visualization of large and high-dimensional data. We demonstrate that in comparison to the baseline algorithms and their GPU implementations, IVHD-CUDA data embedding is the fastest and the most storage saving GPU-implemented DE algorithm. We can demonstrate that it allows for immediate response on interactive requests during exploration of large datasets.

Surprisingly, despite radical simplifications, IVHD still properly reconstructs the main structural properties of large ND datasets at the cost of rather minor, and incomparable to the scale of these radical approximations deterioration of embedding quality. In our opinion, it is the most important result of this research, which shows how robust is the "backbone" of high-dimensional data represented by its very sparse (k is small) kNN-graph.

It is interesting, how robustness of this "backbone" is correlated to the complexity of a low-dimensional manifold occupied by data samples and embedded in a very high-dimensional feature space. The algorithms based on t-SNE still assume that Euclidean distances are responsible for the structure of data. But this is not true at all for very complex manifolds resulting from strong feature interdependence. As a result, they too often produce very fragmented visualizations. Just the sparse kNN-graph (small k) is the most appropriate structure, which is able to approximate such manifolds [17]. Therefore, in the future work, we would like to concentrate on the robustness of the IVHD algorithm depending on data complexity and also on the noise and errors in data (e.g. erroneous labels). The method is planned to be tested on really big datasets $M = 10^{7+}$, too demanding computationally for the SOTA DE algorithms. To this end, the most powerful multi-GPU boards will be used what involves substantial revision of GPU-code. Summarizing, we believe that our method would be very helpful for visualization of truly big and complex data, where low storage and high computational speed of DE algorithm are the crucial issues.

Acknowledgments. The research presented in this paper was supported by the funds assigned to AGH University of Science and Technology by the Polish Ministry of Science and Higher Education. The authors used PL-Grid Infrastructure and computing resources of ACK Cyfronet.

References

1. Amid, E., Warmuth, M.K.: TriMap: Large-scale Dimensionality Reduction Using Triplets. arXiv preprint arXiv:1910.00204 (2019)
2. Cofaru, C.: Inverted file system with asymmetric distance computation for billion-scale approximate nearest neighbor search (2019). https://github.com/zgornel/IVFADC.jl/. Accessed 20 Nov 2019
3. Fu, C., Yonghui Zhang, D.C., Ren, X.: Anchor-t-SNE for large-scale and high-dimension vector visualization. (2019). https://github.com/ZJULearning/AtSNE. Accessed 20 Nov 2019
4. Chan, D.M., Roshan Rao, F.H., Canny, J.F.: GPU accelerated t-distributed stochastic neighbor embedding. J. Parallel Distrib. Comput. **131**, 1–13 (2019). https://github.com/CannyLab/tsne-cuda/. Accessed 20 Nov 2019
5. Dzwinel, W., Blasiak, J.: Method of particles in visual clustering of multi-dimensional and large data sets. Future Gener. Comp. Syst. **15**, 365–379 (1999)
6. Dzwinel, W., Wcisło, R., Czech, W.: ivga: a fast force-directed method for interactive visualization of complex networks. J. Comput. Sci. **21**, 448–459 (2017)
7. Dzwinel, W., Wcislo, R., Matwin, S.: 2-d embedding of large and high-dimensional data with minimal memory and computational time requirements. arXiv preprint arXiv:1902.01108 (2019)
8. Dzwinel, W., Wcislo, R., Strzoda, M.: ivga: visualization of the network of historical events. In: Proceedings of the 1st International Conference on Internet of Things and Machine Learning. ACM (2017)
9. France, S.L., Carroll, J.D.: Two-way multidimensional scaling: a review. IEEE Trans. Syst. Man Cybern. **41**(5), 644–661 (2011)
10. Fu, C., Zhang, Y., Cai, D., Ren, X.: AtSNE: efficient and robust visualization on GPU through hierarchical optimization. In: Proceedings of the 25th ACM SIGKDD International Conference on Knowledge Discovery & Data Mining, pp. 176–186 (2019)
11. Hinton, G.E., Roweis, S.T.: Stochastic neighbor embedding. In: Advances in Neural Information Processing Systems, pp. 857–864 (2003)
12. Johnson, J., Douze, M., Jégou, H.: Billion-scale similarity search with gpus. arXiv preprint arXiv:1702.08734 (2017)

13. Kumari, N., Rupela, A., Gupta, P., Krishnamurthy, B.: Shapevis: High-dimensional data visu-alization at scale. arXiv preprint arXiv:2001.05166 (2020)
14. Linderman, G., Rachh, M., Hoskins, J.: Efficient algorithms for t-distributed stochastic neighborhood embedding. arXiv preprint arXiv:1712.09005 (2017)
15. van der Maaten, L.: Accelerating t-SNE using tree-based algorithms. J. Mach. Learn. Res. **15**, 3221–3245 (2014)
16. van der Maaten, L., Hinton, G.: Visualizing data using t-SNE. J. Mach. Learn. Res. **9**, 2579–2605 (2008)
17. McInnes, L., Healy, J., Melville, J.: Umap: Uniform manifold approximation and projection for dimension reduction. arXiv preprint arXiv:1802.03426 (2018)
18. Pawliczek, P., Dzwinel, W., Yuen, D.: Visual exploration of data by using multidimensional scaling on multicore CPU, GPU, and MPI cluster. Concurrency Comput. Pract. Exp. **3**, 1–21 (2014)
19. Pezzotti, N., Höllt, T., Lelieveldt, B., Eisemann, E., Vilanova, A.: Hierarchical stochastic neighbor embedding. Comput. Graph. Forum **35**, 21–30 (2016)
20. Facebook AI Research: Library for fast text representation and classification. (2018). Accessed 20 Nov 2019. https://github.com/facebookresearch/fastText
21. Ruder, S.: An overview of gradient descent optimization algorithms. arXiv preprint arXiv:1609.04747 (2017)
22. Tang, J., Liu, J., Zhang, M., Mei, Q.: Visualizing large-scale and high-dimensional data. In: Proceedings of the 25th International Conference on World Wide Web, pp. 287–297 (2016)
23. Wang, R., Zhang, X.: Capacity preserving mapping for high-dimensional data visualization. arXiv preprint arXiv:1909.13322 (2019)

Evolving Long Short-Term Memory Networks

Vicente Coelho Lobo Neto📵, Leandro Aparecido Passos📵,
and João Paulo Papa(✉)📵

Recogna Laboratory, School of Sciences, São Paulo State University, Bauru, Brazil
{vicente.lobo,leandro.passos,joao.papa}@unesp.br
http://www.recogna.tech/

Abstract. Machine learning techniques have been massively employed in the last years over a wide variety of applications, especially those based on deep learning, which obtained state-of-the-art results in several research fields. Despite the success, such techniques still suffer from some shortcomings, such as the sensitivity to their hyperparameters, whose proper selection is context-dependent, i.e., the model may perform better over each dataset when using a specific set of hyperparameters. Therefore, we propose an approach based on evolutionary optimization techniques for fine-tuning Long Short-Term Memory networks. Experiments were conducted over three public word-processing datasets for part-of-speech tagging. The results showed the robustness of the proposed approach for the aforementioned task.

Keywords: Long Short-Term Memory · Part-of-Speech tagging · Metaheuristic optimization · Evolutionary algorithms

1 Introduction

Machine learning techniques achieved outstanding outcomes in the last years, mostly due to the notable results obtained using deep learning techniques in a wide variety of applications [1], ranging from medicine [2] to route obstruction detection [3]. In this context, a specific kind of model, the so-called Recurrent Neural Network (RNN), has been extensively employed to model temporal-dependent data, such as stock market forecasting and applications that involve text and speech processing.

Such networks are time-dependent models whose neurons receive recurrent feedback from others, in contrast to densely connected models (e.g., Multilayer Perceptron). In short, the model's current output at time t depends on its input as well as the output obtained at time $t-1$. Due to this intrinsic characteristic, this kind of network deals very well with data that has temporally-related information, such as videos, audio, and texts.

The authors are grateful to FAPESP grants #2013/07375-0, #2014/12236-1, #2017/25908-6, #2018/10100-6, #2019/07665-4, as well as CNPq grants #307066/2017-7 and #427968/2018-6.

V. V. Krzhizhanovskaya et al. (Eds.): ICCS 2020, LNCS 12138, pp. 337–350, 2020.
https://doi.org/10.1007/978-3-030-50417-5_25

Among a wide variety of RNNs, such as the Hopfield Neural Network [4] and the Gated Recurrent Unit Networks [5], the Long Short-Term Memory network, also known as LSTM [6] has drawn considerable attention in the last years. LSTM is an RNN architecture designed to model temporal sequences and their long-range dependencies more accurately than conventional RNNs. It obtained impressive results over a variety of time-series problems, such as named entity recognition [7], chunking [8], acoustic modelling [9], and Part-of-Speech (PoS) tagging [7], to cite a few. However, the method still suffers from a well-known problem related to neural network-based approaches: the proper selection of their hyperparameters. A few years ago, Greff et al. [10] used a random search to handle such a problem. Later on, Reimers e Gurevych [11] studied how these hyperparameters affect the performance of LSTM networks, demonstrating that their optimization is crucial for satisfactory outcomes.

Recently, many works addressed the problem of hyperparameter fine-tuning in neural networks using nature-inspired metaheuristic optimization techniques. Fedorovici et al. [12], for instance, employed the Gravitational Search Algorithm [13] to optimize Convolutional Neural Networks (CNN) [14], while Rosa et al. [15] proposed a similar approach using Harmonic Search [16]. Later, Rosa et al. [17] used the Firefly algorithm [18] to fine-tune Deep Belief Networks [19], as well as Passos et al. [20,21] applied a similar approach in the context of Deep Boltzmann Machines [22].

However, only a very few works employed evolutionary optimization techniques to fine-tune LSTM networks to date [23,24], but none concerning the PoS tagging task. Therefore, the main contribution of this paper is to fill up this gap by introducing evolutionary techniques for LSTM hyperparameter fine-tuning in the context of Part-of-Speech tagging. For such a task, we compared four evolutionary-based optimization approaches: Genetic Algorithm (GA) [25], Genetic Programming (GP) [26], Geometric Semantic Genetic Programming (GSGP) [27], and the Stack-based Genetic Programming (SGP) [28], as well as a random search as the baseline.

The remainder of this paper is organized as follows. Section 2 briefly introduces the main concepts of LSTM and PoS tagging, while Sect. 3 describes the approach considered for LSTM hyperparameter fine-tuning. Further, the methodology is detailed in Sect. 4, while the results are presented in Sect. 5. Finally, Sect. 6 states conclusions and future works.

2   Theoretical Background

This section briefly introduces the main concepts regarding the Part-of-Speech tagging, as well as the Long Short-Term Memory Networks.

2.1   Part-of-Speech Tagging

Generally speaking, a sentence is a composition of words joined according to some rules, whose intention is to express a complete idea. In such a context, each word in a sentence performs a specific function, e.g., a verb indicates an action, an adjective adds some attribute to a noun, and the latter is used to name any being.

Natural languages, such as English, Portuguese, and any other, possess a considerable amount of words. Therefore, it is essential to note that many of them may admit ambiguous meanings, i.e., the same word can hold two or more different connotations. To illustrate the idea, suppose the word "seed": it may stand for a noun, referring to the reproductive unity of a plant, as well as denote a verb, representing the act of sowing the soil with seeds.

To handle such ambiguous meanings, one can make use of the Part-of-Speech tagging process. The method aims at, given a sentence, identifying each compounding word's grammatical function inside it, as depicted in Fig. 1.

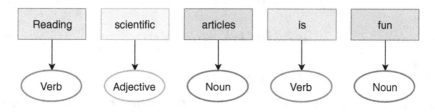

Fig. 1. Example of PoS tagging.

Understanding the true meaning of a word in a sentence is essential to comprehend the message conveyed through it thoroughly. Though it stands for a trivial task for human beings, automatic PoS tagging presents itself as quite challenging for computer systems. Among the various ways of solving this task, one can cite rule-based algorithms [29], as well as Hidden Markov Models [30].

Finally, one should notice the importance of contextual information in order to assign a tag to a word correctly. One approach that yields good results while addressing such context employs machine learning techniques through Recurrent Neural Networks. Therefore, the next section introduces the LSTM, the current state-of-art approach for natural language processing.

2.2   Long Short-Term Memory Network

Although LSTMs present an architecture similar to traditional RNNs methods, they employ a different function for computing their hidden states. Such modification is essential for avoiding problems such as the well-known vanishing gradient, which can occur while relating current information to long-term network knowledge, known as long-term dependency. LSTM addresses this problem by using structures called **cells**, which receive as input their previous states as well as the current input. Internally, these cells decide what should be kept and forwarded to the next iterations, and what should be erased from memory.

Therefore, they are capable of combining input information with previous states, as well as the current memory.

A cell is composed of states, as well as three gates: (i) the *input gate*, which is responsible for deciding what information is relevant and how much of it will be updated or deleted; (ii) the *forget gate*, used when the network needs to erase any previous information present in the cell's memory in order to store a new document; and (iii) the *output gate*, that is employed to compute the cell's state [31]. Besides, every gate is activated by a sigmoid function, and the cell's output is subject to a combination of the gates' outputs, which will decide whether information should be forgotten or kept in the cell's memory.

Let f^t, i^t, $o^t \in \Re^{m \times 1}$ be the outputs of the forget, input, and output gates at time step t, respectively, which can be computed as follows:

$$f^t = \sigma(W_f^t x^t + U_f^t h^{t-1} + b_f^t), \tag{1}$$

$$i^t = \sigma(W_i^t x^t + U_i^t h^{t-1} + b_i^t), \tag{2}$$

and

$$o^t = \sigma(W_o^t x^t + U_o^t h^{t-1} + b_o^t), \tag{3}$$

where $x^t \in \mathbb{R}^{n \times 1}$ denotes the input vector at time step t, $h^{t-1} \in \mathbb{R}^{m \times 1}$ stands for the output of the previous cell, and $W_f^t, W_i^t, W_o^t \in \mathbb{R}^{m \times n}$ and $U_f^t, U_i^t, U_o^t \in \mathbb{R}^{m \times m}$ denote the weight matrices at time step t. Finally, $b_f^t, b_i^t, b_o^t \in \mathbb{R}^{m \times 1}$ correspond to the biases of each gate at time step t.

Further, one can compute the cell's state $c^t \in \mathbb{R}^{m \times 1}$ at time step t considering the forget and input gates' output as follows:

$$c^t = f^t \otimes c^{t-1} + i^t \otimes tanh(W_c^t x^t + U_c^t h^{t-1} + b_c^t), \tag{4}$$

where $\otimes$ stands for the Hadamard product, and $W_c \in \mathbb{R}^{m \times n}$ and $U_c \in \mathbb{R}^{m \times m}$ are the weight matrices, and $b_c \in \mathbb{R}^{m \times 1}$ corresponds to the bias of the cell's state. Therefore, one can compute the output of a cell at time step t as follows:

$$h^t = o^t \otimes tanh(c^t). \tag{5}$$

Finally, the weights of an LSTM network are updated using a gradient-based backward pass as follows:

$$^z W_f^{t+1} = {}^z W_f^t - \alpha(^z \delta_f^t \otimes h^t), \tag{6}$$

$$^z W_i^{t+1} = {}^z W_i^t - \alpha(^z \delta_i^t \otimes h^t), \tag{7}$$

and

$$^z W_o^{t+1} = {}^z W_o^t - \alpha(^z \delta_o^t \otimes h^t), \tag{8}$$

where $\alpha \in \mathbb{R}$ is the learning rate and $^z W_f^t$, $^z W_i^t$, and $^z W_o^t$ stand for the zth column of matrices W_f^t, W_i^t, and W_o^t, respectively. Notice that $\delta_f^t, \delta_i^t, \delta_o^t \in \mathbb{R}^{m \times n}$

denote the partial derivatives at time step t, and $^z\delta_f^t$, $^z\delta_i^t$, and $^z\delta_o^t$ correspond to the zth column of matrices $^z\delta_f^t$, $^z\delta_i^t$, and $^z\delta_o^t$, respectively. Further, the bias of each gate is also updated in a similar fashion, and the gradient is propagated to the former layers using the backpropagation chain's rule. Finally, the weight matrices U_f^t, U_i^t, and U_o^t are updated similarly to their respective counterparts, i.e., Eqs. 6, 7, and 8.

3   LSTM Fine-Tuning as an Optimization Problem

In the context of parameterized machine learning techniques, we must refer to two different denominations: (i) *parameters* and (ii) *hyperparameters*. Typically, the first term represents low-level parameters that are not controlled by the user, such as connection weights in neural networks, for example. The other term refers to high-level parameters that can be adjusted and chosen by the user, such as the learning rate and the number of layers, among others. Both terms are of crucial importance for improving the performance of neural models.

Regarding LSTM parameter optimization, the well known back-propagated using gradient descent present itself as a suitable method for the task. However, a proper selection of its hyperparameters poses a more challenging task since it requires the user a previous knowledge of the problem. Hence, an automatic method to select the model most relevant hyperparameters is strongly desirable, i.e., the *learning rate*, which denotes the step-size towards the gradient direction, the *embedding layer size*, which is responsible for the representation of words and their relative meanings, and the *LSTM size* itself, which denotes the number of LSTM's cells composing the model.

Therefore, this work proposes employing evolutionary algorithms to fine-tune such hyperparameters. These methods are capable of automatically selecting a good set of hyperparameters by randomly initializing a set of candidate solutions that evolve over time. In this context, the candidate solutions are represented by a vector, where each position denotes a hyperparameter to be fine-tuned. In this work, for instance, it is considered a 3-dimensional representation, denoting the learning rate, the embedding layer size, and the LSTM size.

Further, every solution is evaluated after each evolution cycle. This evaluation consists of training the LSTM network using both the candidate solution set of hyperparameters together with the training set. Afterward, the trained model is employed to classify the evaluation set, and the loss function obtained over this procedure, denoted fitness function, is used to evaluate and update the solutions towards a minimum (ideally, the global minimum). At the end of the process, it is expected that the model learns a reasonably good set of hyperparameters. Figure 2 depicts the model adopted in the work.

3.1   Evolutionary Optimization Algorithms

This section briefly describes the four metaheuristic optimization techniques employed in this work concerning the task of LSTM hyperparameter fine-tuning.

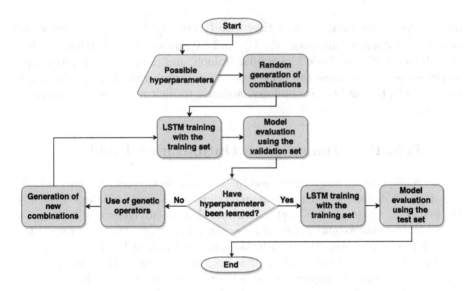

Fig. 2. Pipeline employed in the work.

3.1.1   Genetic Algorithm

GA is a specific case of evolutionary algorithms that employ concepts and expressions from natural genetics that models each possible solution as a chromosome. Among a variety of implementations of the model, this work represents each individual as an array composed of the hyperparameter to be optimized, letting aside the binary version. Moreover, the following evolution operators were employed:

1. **Elitism**: it retains the individuals with the best fitness values of the current population using elitism to accelerate the convergence;
2. **Random selection**: it adds some random elements of the current population to the new one to promote greater genetic diversity;
3. **Mutation**: it modifies a random position of individuals of the new population generating values within the allowed limits; and
4. **Crossover**: it randomly selects two individuals from the population and creates two new individuals through the single-point crossover.

3.1.2   Genetic Programming

GP employs the very same idea behind the GA, except for data structure, which is represented by an expression tree. In this model, each tree's leaf node contains an array, similar to the ones applied in GA, representing the hyperparameters to be optimized. Further, inner-nodes stand for mathematical operators, such as addition, multiplication, and logarithm, among others.

3.1.3   Geometric Semantic Genetic Programming

Despite the advantages of GP, since it uses purely syntactic operators, instead of the values of the individuals' decision variables while applying the operators, it may perform some very abrupt changes in the values of the variables as the evolution process occurs, which may adversely affect the convergence.

Therefore, GSGP considers the evolution of individuals by selecting semantically close solutions [27]. The procedure is performed by introducing some restrictions in the mutation and crossover operators.

3.1.4   Stack-Based Genetic Programming

The SGP is a variation of GP that, instead of using a tree data structure, it employs two stacks: one for data input and one for the output. Thus, the SGP input stack is composed of a set of operations and operands responsible for describing a mathematical expression, similar to GP, while the output stack contains the expression evaluation result [28].

Since there are no syntactic constraints in this technique, individuals can be represented as vectors. Besides, it is possible to use the same genetic operators as the standard Genetic Algorithm for mutation and crossover. Thence, the technique combines the great individuals' variability of GP with the simplicity of GA's genetic operators, thus providing better results than GP, in general, depending on the implementation and the problem addressed.

4   Methodology

This section presents the methodology employed in the experiments, i.e., the datasets, experimental setup, and network modeling concerning the PoS tagging problem.

4.1   Datasets

This work employs three well-known public datasets for the task of Part-Of-Speech tagging. Such datasets are composed of natural language text fragments, and they are available at the Natural Language Toolkit (NLTK) library [32]:

- *Brown Corpus of Standard American English* (**Brown**): approximately one million English text sentences divided into 15 different categories, including newspapers, scientific articles, and other sources.
- *Floresta Sintá(c)tica* (**Floresta**): sentences in Brazilian Portuguese extracted from interviews' transcriptions, parliamentary discussions, literary texts, and articles from the Brazilian newspaper Folha de São Paulo compose this corpus.
- *Penn Treebank* (**Treebank**): covering a variety of subjects, the sentences of this corpus are in English and were extracted from three years of The Wall Street Journal.

4.2   Modeling the PoS Tagging Problem with LSTM

Since LSTM networks require a vector composed of real numbers as input, the datasets were pre-processed such that the words were encoded into a numerical dictionary, where each word corresponds to a non-zero number. Besides, the input is composed of an n-dimensional vector whose features should fit in, i.e., all sentences should have the same size. Therefore, n was selected based on the longest sentence's size. Finally, a padding approach was employed to fulfill all the non-occupied spaces with zeros.

Further, each sentence is represented by an array of tuples using the pattern (word, PoS tag). Later, these tuples are split, and the tags are converted into their categorical (one-hot) representation to match the LSTM output format.

Therefore, the sequential architecture of the LSTM employed in this work for the task of PoS tagging is composed of the following layers:

1. **Input**: Receives input data. It outputs data with (n, d) shape, where d and n stand for the number of inputs and the number of input features, respectively;
2. **Embedding**: It provides a dense representation of words and their relative meanings, converting the indices into dense vectors. Output data has (n, d, E) shape, where E stands for the embedding layer size;
3. **Bidirectional LSTM layer**: is the core of the network, comprising the LSTM units. The bidirectional version was selected instead of the unidirectional due to the superior results obtained by the model. Output shape is $(n, d, 2L)$, where L represents the number of LSTM units. As the bidirectional version was used, $2L$ units must be employed;
4. **Time-distributed Dense**: A standard densely connected layer with a wrapper to handle the input data temporal distribution. Considering that the tags were converted to the categorical representation, the output shape of this layer is (n, d, j), with j standing for the number of available tags.

Figure 3 depicts the model mentioned above.

Fig. 3. Model representation with output shapes.

Further, the Softmax is then employed as the activation function. Concerning the loss function, the model aims to minimize the categorical cross-entropy, which is the most commonly applied approach for text classification tasks.

4.3   Experimental Setup

This work compares the performance of four well-known evolutionary optimization algorithms, i.e., GA, GP, GSGP, and SGP, as well as a random search, to the task of LSTM network hyperparameter fine-tuning in the context of PoS tagging. The experiments were performed over 400 randomly selected sentences from three public datasets commonly employed for the task: Brown, Floresta, and Treebank. Further, these sentences were partitioned into three sets: training, validation, and testing using the percentages of 60%, 20% and 20%, respectively.

This work focuses on finding the set of hyperparameters that minimizes the categorical cross-entropy loss function, thus maximizing the LSTM performance for the task of PoS tagging using metaheuristic evolutive algorithms. Therefore, the problem was designed to search for three hyperparameters: (i) the layer size L and (ii) the embedding layer size E, both selected from a closed set of powers of 2 ranging from 2^1 to 2^{10}, i.e., $\{2, 4, 8, 16, 32, 64, 128, 256, 512, 1024\}$, and (iii) the learning rate $\alpha \in [0.001, 0.1]$. Since the metaheuristic techniques work with real numbers only, one should convert them to the nearest power of 2 when dealing with L and E.

Concerning the optimization procedure, each technique employed 10 individuals evolving during 20 iterations and considered a retention rate of 20%, a random selection rate of 5%, and 1% of mutation rate. Moreover, GP and GSGP employed expression trees with a maximum tree height of 5, while the SGP employed arrays with 10 elements. Also, they implemented the addition, subtraction, multiplication, division, modulo, negation, logarithm, and square root operators. Notice the latter two were not employed in the GP since they adversely affected convergence[1]. Finally, these techniques were compared against a random search, which stands for the experiments' baseline.

Afterward, the best hyperparameters obtained during the optimization process are employed to train the network for 200 iterations for further classification. Besides, each experiment was performed during 20 runs for statistical analysis through the Wilcoxon signed-rank test [33] with a significance of 5%. Finally, the experiments were conducted using both the Keras open-source library [34] and Tensorflow [35].

5   Experimental Results

This section presents the experimental results. Table 1 provides the mean accuracy and the standard deviation concerning the Brown Corpus of Standard American English, Floresta Sintá(c)tica, and the Penn Treebank datasets. Notice that the best results, according to the Wilcoxon signed-rank test, are presented in bold.

From these results, it is possible to extract two main conclusions: (i) metaheuristic optimization techniques performed better than a random search concerning the three datasets, which validates the employment of such methods for the task; (ii) the tree-based algorithms, i.e., GP, GSGP, and SGP, performed better than GA in two out of three datasets. Such behavior is expected since these techniques employ more robust approaches to solve the optimization problem.

---

[1] Notice that all these parameters and configurations were set up empirically.

Table 1. Mean accuracy results over Brown, Floresta and Treebank datasets.

	GA	GP	GSGP	SGP	Random
Brown	**0.734 ± 0.043**	**0.740 ± 0.092**	**0.726 ± 0.100**	**0.673 ± 0.115**	0.566 ± 0.229
Floresta	0.656 ± 0.121	**0.706 ± 0.025**	0.676 ± 0.038	**0.603 ± 0.207**	0.470 ± 0.271
Treebank	0.748 ± 0.052	**0.685 ± 0.202**	**0.782 ± 0.024**	**0.751 ± 0.098**	0.658 ± 0.206

5.1   Training Evaluation

Considering the evaluation of the learned models, Fig. 4 depicts the mean loss obtained by each model considering the three test sets. One can observe that minimum values do not necessarily reflect in the best accuracy, as depicted in Brown, where GA obtained the minimum loss, and GP obtained the higher accuracy; and the Treebank dataset, where SGP obtained the minimum loss and GSGP obtained the higher accuracies. Nevertheless, all techniques still present mean loss values considerably lower than the random search.

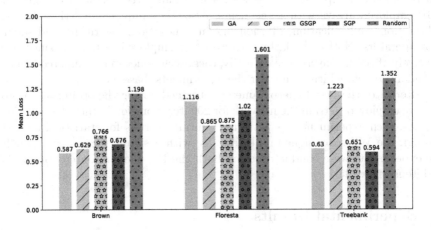

Fig. 4. Mean loss per data set using evolutionary optimization and random selection for LSTM training purposes.

5.2   Optimization Progress

Figure 5 depicts each model's convergence, considering the loss over the validation set, during the training procedure. The figure clearly shows that GA obtained the best values regarding the first iterations considering both Brown and Floresta datasets, depicted in Figs. 5(a) and (b), respectively. Additionally, concerning the Treebank dataset, although GSGP obtained the best values in the initial iterations, it is outperformed by GA in the third iteration, whose convergence keeps improving until reaching the stop criteria of 20 iterations. Moreover, it can be observed that GSGP and SGP do not present significant progress during the iterations, getting stuck on local optima and even get worst results than the ones obtained in previous iterations. Finally, considering the GP algorithm, despite the slightly positive convergence, it was not capable of outperforming GA in any of the three datasets.

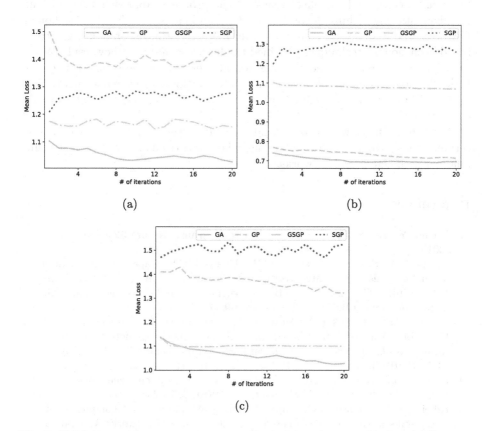

Fig. 5. Validation set loss convergence over the training steps considering: (a) Brown, (b) Floresta, and (c) Treebank datasets.

It is interesting to notice that, although GA outperformed the tree-based methods concerning the validation set over a training composed of 20 iterations when considering the same hyperparameters over the testing set considering 200 iterations for training, the tree-based methods obtained the best results. Such a phenomenon may suggest the model becomes overfitted when trained considering the GA set of hyperparameters.

6   Conclusions and Future Works

In this paper, we addressed the problem of LSTM hyperparameter fine-tuning through evolutionary metaheuristic optimization algorithms in the context of

PoS tagging. For this task, four techniques were compared, i.e., the Genetic Algorithm, the Genetic Programming, the Geometric Semantic Genetic Programming, and the Stack-based Genetic Programming, as well as a random search, employed as the baseline. Experiments conducted over three well-known public datasets confirmed the effectiveness of the proposed approach since all four metaheuristic algorithms outperformed the random search. Further, although GA presented a smoother convergence during the optimization steps, the tree-based evolutionary techniques obtained the best loss and accuracy results when considering a more extended training period over the testing sets. Such results highlight that those complex mathematical operations, like the ones performed by GP, GSGP, and SGP, contribute to the convergence of the model in this context.

Regarding future works, we intend to validate the proposed approach concerning other contexts than PoS tagging, whether textual or not, as well as using different artificial neural networks.

References

1. LeCun, Y., Bengio, Y., Hinton, G.E.: Deep learning. Nature **521**(7553), 436–444 (2015)
2. Passos, L.A., Santos, C., Pereira, C.R., Afonso, L.C.S., Papa, J.P.: A hybrid approach for breast mass categorization. In: Tavares, J.M.R.S., Natal Jorge, R.M. (eds.) VipIMAGE 2019. LNCVB, vol. 34, pp. 159–168. Springer, Cham (2019). https://doi.org/10.1007/978-3-030-32040-9_17
3. Santana, M.C., Passos, L.A., Moreira, T., Colombo, D., De Albuquerque, V.H.C., Papa, J.P.: A novel siamese-based approach for scene change detection with applications to obstructed routes in hazardous environments. IEEE Intell. Syst. **35**, 44–53 (2019). https://doi.org/10.1109/MIS.2019.2949984
4. Hopfield, J.J.: Neural networks and physical systems with emergent collective computational abilities. Proc. Natl. Acad. Sci. **79**(8), 2554–2558 (1982)
5. Ribeiro, L.C., Afonso, L.C., Papa, J.P.: Bag of samplings for computer-assisted Parkinson's disease diagnosis based on recurrent neural networks. Comput. Biol. Med. **115**, 103477 (2019)
6. Hochreiter, S., Schmidhuber, J.: Long short-term memory. Neural Comput. **9**(8), 1735–1780 (1997)
7. Ma, X., Hovy, E.H.: End-to-end sequence labeling via bi-directional LSTM-CNNs-CRF, CoRR abs/1603.01354
8. Komninos, A., Manandhar, S.: Dependency based embeddings for sentence classification tasks. In: Proceedings of the 2016 Conference of the North American Chapter of the Association for Computational Linguistics: Human Language Technologies, pp. 1490–1500 (2016)
9. Sak, H., Senior, A., Beaufays, F.: Long short-term memory recurrent neural network architectures for large scale acoustic modeling. In: Fifteenth Annual Conference of the International Speech Communication Association, pp. 338–342 (2014)
10. Greff, K., Srivastava, R.K., Koutník, J., Steunebrink, B.R., Schmidhuber, J.: LSTM: a search space odyssey, CoRR abs/1503.04069
11. Reimers, N., Gurevych, I.: Optimal hyperparameters for deep LSTM-networks for sequence labeling tasks. CoRR abs/1707.06799

12. Fedorovici, L., Precup, R., Dragan, F., David, R., Purcaru, C.: Embedding gravitational search algorithms in convolutional neural networks for OCR applications. In: 7th IEEE International Symposium on Applied Computational Intelligence and Informatics. SACI 2012, pp. 125–130 (2012). https://doi.org/10.1109/SACI.2012. 6249989
13. Rashedi, E., Nezamabadi-pour, H., Saryazdi, S.: GSA: a gravitational search algorithm. Inf. Sci. **179**(13), 2232–2248 (2009)
14. Lecun, Y., Bottou, L., Bengio, Y., Haffner, P.: Gradient-based learning applied to document recognition. Proc. IEEE **86**(11), 2278–2324 (1998)
15. Rosa, G., Papa, J., Marana, A., Scheirer, W., Cox, D.: Fine-tuning convolutional neural networks using harmony search. CIARP 2015. LNCS, vol. 9423, pp. 683–690. Springer, Cham (2015). https://doi.org/10.1007/978-3-319-25751-8_82
16. Geem, Z.W.: Music-Inspired Harmony Search Algorithm: Theory and Applications, 1st edn. Springer, Heidelberg (2009). https://doi.org/10.1007/978-3-642-00185-7
17. Rosa, G., Papa, J., Costa, K., Passos, L., Pereira, C., Yang, X.-S.: Learning parameters in deep belief networks through firefly algorithm. In: Schwenker, F., Abbas, H.M., El Gayar, N., Trentin, E. (eds.) ANNPR 2016. LNCS (LNAI), vol. 9896, pp. 138–149. Springer, Cham (2016). https://doi.org/10.1007/978-3-319-46182-3_12
18. Yang, X.-S.: Firefly algorithm, stochastic test functions and design optimisation. Int. J. Bio-Inspired Comput. **2**(2), 78–84 (2010)
19. Hinton, G.E., Osindero, S., Teh, Y.-W.: A fast learning algorithm for deep belief nets. Neural Comput. **18**(7), 1527–1554 (2006)
20. Passos, L.A., Rodrigues, D.R., Papa, J.P., Fine tuning deep Boltzmann machines through meta-heuristic approaches. In: 2018 IEEE 12th International Symposium on Applied Computational Intelligence and Informatics (SACI), pp. 000419–000424. IEEE (2018)
21. Passos, L.A., Papa, J.P.: Temperature-based deep Boltzmann machines. Neural Process. Lett. **48**(1), 95–107 (2018). https://doi.org/10.1007/s11063-017-9707-2
22. Salakhutdinov, R., Hinton, G.E.: An efficient learning procedure for deep Boltzmann machines. Neural Comput. **24**(8), 1967–2006 (2012)
23. Chung, H., Shin, K.-S.: Genetic algorithm-optimized long short-term memory network for stock market prediction. Sustainability **10**(10), 3765 (2018)
24. Kim, T., Cho, S.: Particle swarm optimization-based CNN-LSTM networks for forecasting energy consumption. In: 2019 IEEE Congress on Evolutionary Computation, pp. 1510–1516 (2019)
25. Goldberg, D.E., Holland, J.H.: Genetic algorithms and machine learning. Mach. Learn. **3**, 95–99 (1988)
26. Koza, J.R., Koza, J.R.: Genetic Programming: On the Programming of Computers by Means of Natural Selection, vol. 1. MIT Press, Cambridge (1992)
27. Moraglio, A., Krawiec, K., Johnson, C.G.: Geometric semantic genetic programming. In: Coello, C.A.C., Cutello, V., Deb, K., Forrest, S., Nicosia, G., Pavone, M. (eds.) PPSN 2012. LNCS, vol. 7491, pp. 21–31. Springer, Heidelberg (2012). https://doi.org/10.1007/978-3-642-32937-1_3
28. Perkis, T.: Stack-based genetic programming. In: Proceedings of the First IEEE Conference on Evolutionary Computation. IEEE World Congress on Computational Intelligence, vol. 1, pp. 148–153 (1994). https://doi.org/10.1109/ICEC.1994. 350025
29. Brill, E.: A simple rule-based part of speech tagger. In: Proceedings of the Third Conference on Applied Natural Language Processing, pp. 152–155. Association for Computational Linguistics (1992)

30. Kupiec, J.: Robust part-of-speech tagging using a hidden Markov model. Comput. Speech Lang. **6**(3), 225–242 (1992)
31. Gers, F.A., Schmidhuber, J., Cummins, F.: Learning to forget: continual prediction with LSTM. In: 9th International Conference on Artificial Neural Networks: ICANN 1999, pp. 850–855. Istituto Dalle Molle Di Studi Sull Intelligenza Artificiale (1999)
32. Loper, E., Bird, S.: Proceedings of the ACL-02 Workshop on Effective Tools and Methodologies for Teaching Natural Language Processing and Computational Linguistics - Volume 1. ETMTNLP 2002, pp. 63–70. Association for Computational Linguistics, Stroudsburg (2002). https://doi.org/10.3115/1118108.1118117
33. Wilcoxon, F.: Individual comparisons by ranking methods. Biom. Bull. **1**(6), 80–83 (1945)
34. Chollet, F., et al.: Keras. GitHub (2015). https://github.com/fchollet/keras
35. Abadi, M., et al.: TensorFlow: large-scale machine learning on heterogeneous systems, software available from tensorflow.org (2015). http://tensorflow.org/

Personality Recognition from Source Code Based on Lexical, Syntactic and Semantic Features

Mikołaj Biel, Marcin Kuta(✉)[iD], and Jacek Kitowski[iD]

Department of Computer Science, Faculty of Computer Science,
Electronics and Telecommunications, AGH University of Science and Technology,
Al. Mickiewicza 30, 30-059 Krakow, Poland
mkuta@agh.edu.pl

Abstract. Automatic personality recognition from source code is a scarcely explored problem. We propose personality recognition with handcrafted features, based on lexical, syntactic and semantic properties of source code. Out of 35 proposed features, 22 features are completely novel. We also show that n-gram features are simple but surprisingly good predictors of personality and present results arising from joint usage of both handcrafted and baseline features. Additionally we compare our results with scores obtained within the Personality Recognition in SOurce COde track during Forum for Information Retrieval Evaluation 2016 and set up state-of-the-art results for conscientiousness and neuroticism traits.

Keywords: Automatic personality recognition · Personality traits · Source code · Feature engineering

1 Introduction

Personality influences many aspects of human behaviour, e.g. made decisions, propensity for communication with other people, way of writing or listened music [22]. In the context of computer science, personality may influence organization of created source code, or a choice of a software project a person takes part in.

While automatic personality recognition from text attained remarkable attention [37], personality recognition from source code is still a scarcely explored problem.

Automatic personality recognition can be useful to customize learning process or to assess cultural fit in a company. Each company has a different culture [26] – there are places where programmers are supposed to often contact clients and conversely where they talk only to their supervisors. Firms may also differ in workplace organization – is it an open plan office, a small room, or remote work. Cultural fit is whether an employee is satisfied with these and some other matters in his workplace. If a person fits in his company, he is more involved in what he is doing, more satisfied with things he has accomplished during his work time

© Springer Nature Switzerland AG 2020
V. V. Krzhizhanovskaya et al. (Eds.): ICCS 2020, LNCS 12138, pp. 351–363, 2020.
https://doi.org/10.1007/978-3-030-50417-5_26

and more productive, which is beneficial both for him and his employer. Cultural fit depends on one's personality, thus an automatic personality recognition system that detects whether a person fits into the company's environment based on programming assessment completed during a recruitment phase could save both employee's and employer's time and stress. Both in academia and industry, psychological and sociological predispositions of programmers could be examined to better recognize their soft skills and choose job.

This paper proposes personality recognition from source code with random tree forest on the basis of 35 handcrafted features based on lexical, syntactic and semantic properties of source code. Out of these 35 features, 22 features are novel and have not been used earlier in personality recognition from source code. We compare above features with n-gram features serving as baseline features and present results arising from joint usage of both handcrafted and baseline features. Finally, we compare our results with scores obtained within *Personality Recognition in SOurce COde* (PR-SOCO) track during Forum for Information Retrieval Evaluation (PAN@FIRE 2016).

As a model of personality, we adopt Big Five – a five-factor model of personality [27,28]. The Big Five is a widely accepted model, being a result of long-time research, and there is a consensus that its five traits concisely describe independent personality dimensions [5]. The Big Five model assumes that personality can be described by the following five personality traits:

- Conscientiousness (C) - consistency, persistence, good organizational skills.
- Agreeableness (A) - attitude towards others (whether a person is suspicious or trustful, modest, willing to compromise).
- Neuroticism (N) - impulsiveness, susceptibility to stress and anxiety.
- Openness to experience (O) - intellectual curiosity, willingness to explore, rich imagination of examined person, searching original solutions rather than following in someone's footsteps.
- Extroversion (E) - assertiveness, building relationship at ease.

Each personality trait can be divided further into six facets, but facets are out of scope of our work.

2   Related Work

Deep learning personality predictors require no feature engineering, no preprocessing, scanning, nor parsing of source codes. An example of such an approach is an LSTM neural network which reads source code byte by byte [12]. Low amount of learning data is, however, especially problematic for this approach, as not only the correct predictor (a classifier or a regressor), but also the relevant features should be learned from data.

Features designed for personality recognition from source code were based mainly on source code, but also on structure of the project, content of comments, or code complexity. In the PR-SOCO task, the following features were taken into account [33]:

- number of files submitted by each programmer,
- mean number of lines in programs,
- mean length of variables,
- mean number of classes,
- mean length of classes (computed on the basis of the number of lines of code),
- mean number of attributes, methods in a class,
- number of programs implementing the same class,
- number of errors,
- Halstead complexity measures (e.g. difficulty and time needed for implementation and understanding),
- duplicated fragments of source code,
- cyclomatic complexity,
- frequency of occurrence of comments and their length,
- occurrence of comments written exclusively in capital letters,
- number of comments in classes,
- number of words inside comments,
- usage of punctuation marks inside comments,
- number of lines with missing white characters inside arithmetic expressions,
- number of `import` declarations, which import the whole content of module (usage of * instead of concrete classes),
- used white characters,
- ways of indentation and formatting used by the programmer,
- number of empty lines between methods, blocks of code and number of white characters between parentheses,
- occurrence of digits, capital and small letters and symbol _ in names, as well as length of names.

In [3], frequency distribution of different types of nodes in an abstract syntax tree was examined, yielding however low results, little above baseline approaches.

Another type of features are character n-grams – versatile, easy to implement features, which are language independent and have a wide range of applications in classification tasks, including authorship attribution [14,35], author profiling [2], authorship verification [9,20] and plagiarism detection [23]. They may also provide convenient features for a baseline solution of PR-SOCO. In the context of personality recognition from source code, character n-grams were used in [17,32].

The choice of a predictor (a regressor or a classifier) is a more standard procedure and includes mainly: linear regression [16,25,32], support vector regression [7,11], decision trees [11], nearest neighbours [25,36] and neural networks [12,36].

The research concerning personality recognition from source code is scarce and extraction of novel features will likely extend possiblities of distinguishing traits.

3   Proposed Features

Table 1 shows proposed handcrafted lexical, syntactic and semantic features for automatic personality recognition from source code. Consistecy in using curly brackets around one-line branches of code is implemented in two variants so it gives rise to two features. Number of consecutive lines with aligned characters represents four features, as it is computed separately for four groups of characters. Thus, in total there are 35 proposed features.

Proposed features are grounded in the extension of lexical hypothesis to programming languages. Lexical hypothesis [1] says that the most important differences in personality are reflected in used natural language, vocabulary. According to [21], the more important the difference, the more likely it will be reflected in a single word. We suppose that in the domain of programming code a conscientious person will likely apply consistent indentation throught the code; a person high in openness might use richer vocabulary while an extrovert might use longer names for variables, methods and classes. Additionaly, corellation between personality traits and programming style has been found in [10], according to which persons high in openness prefer breadth-first programming style, while persons low in openness prefer depth-first programming style.

We describe in detail three features, most complicated due to their involved implementation: the number of code duplications, length of comments in characters and the level of indentation.

Detection of *code duplication* is quite a complex task, which could be even cast as another machine learning problem, provided suitable learning data would be available or generated [24]. We adopted a simpler solution consuming less computing resources – syntax tree rewriting [30]. Two pieces of code, one being a duplicate of another, exhibit the same structure but differ in names of constants, variables or methods.

The syntax tree generated with the javalang parser is transformed to a topologically equivalent syntax tree, where tree nodes are simplified, to only reflect the structure of the code and discard irrelevant data. For instance, a name of declared method has been discarded, but structure of its body, type of formal parameters and returned type have been retained. For blocks of instructions, information about entrance conditions has been discarded. Detection of code duplication in one block is performed on the basis of such a simplified tree. A list of all subtrees in the block is created and subtrees which serialize to the same expression are treated as duplications.

Computing *length of comment*, otherwise simple, requires detection whether a comment contains parts of source code. Parsing a comment with a parser of Java would end up with a failure, as programmers usually comment a few lines of code or methods rather than entire programs. To solve this problem, besides the main parser of the whole program, parsers of smaller grammatical units of a program are used.

As white characters are discarded during lexical analysis and even less information is passed to the parser, the *level of indentation* feature was implemented as a state machine (separate from the used parser), which reads tokens, one by one, and tracks the level of indentation. One difficulty in implementing this feature lies in distinguishing between a correct and wrong indentation after a sequence of empty lines of code. Although based only on finite automata

Table 1. Features of source code used as predictors of personality traits. Novel features are marked with *

Feature	Range	Type	Extraction
length of lines	$\mathbb{R}^+$	lexical	average, 80th percentile
length of variables	$\mathbb{R}^+$	lexical	average, 80th percentile
length of methods (in lines)	$\mathbb{R}^+$	lexical	average, maximum, 80th percentile
* length of comments (in characters, not taking into account code which was commented out)	$\mathbb{R}^+$	lexical	average, maximum, 80th percentile
length of comments (in lines) in ratio to length of code (in lines)	$[0; 1]$	lexical	normalization
* length of code (in lines) which was commented out	$[0; 1]$	lexical	normalization
* number of lines with more than one instruction	$[0; 1]$	lexical	normalization
* number of consecutive lines with aligned characters, e.g. (or = (4 features)	$[0; 1]$	lexical	normalization
number of white characters	$[0; 1]$	lexical	normalization
* number of occurrences of -1 (special value)	$[0; 1]$	lexical	normalization
* ratio of the number of words in English to the number of words in languages other than English in names of variables, methods and classes	$[0; 1]$	lexical	
* preserving naming convention (e.g. *PascalCase* or *snake_case*)	$[0; 1]$	lexical	
* consistency in using curly brackets in the next line or at the end of a line containing a method or class declaration (2 features)	$[0; 1]$	lexical	
* consistent application of curly brackets around one-line branches of code	$[0; 1]$	lexical	
* level of indentation - whether it increases with the beginning of a new block of code (and only then)	$[0; 1]$	lexical	normalization
* degree of exploitation of language syntax (e.g. using various syntax of `for` loop, using lambdas)	$\mathbb{N}$	syntactic	
* depth of references to fields and methods of fields or results of methods of objects (e.g. `Cubiculos.get(i).Casilleros.get(j)`)	$\mathbb{R}^+$	syntactic	maximum, 80th percentile

(continued)

Table 1. (continued)

Feature	Range	Type	Extraction
number of methods in a class	$\mathbb{R}^+$	semantic	average, maximum, 80th percentile
* number of used `switch` instructions	$[0;1]$	semantic	normalization
number of separated logic blocks of code within methods	$[0;1]$	semantic	normalization
* number of code duplications	$[0;1]$	semantic	normalization
* maximum nesting depth of instructions	$\mathbb{N}$	semantic	

formalism, the state machine has to roughly understand the syntax of Java – it tracks the number of opening parentheses or curly brackets, closing an open block at the correct indentation level, reopening a block of code at a wrong indentation level. The state machine also knows which instructions require indentation. Additional difficulty arises from one-line bodies of `if` and `for` instructions, where curly brackets are not required. This seemingly simple task becomes a complex programming problem due to the great number of cases which should be considered.

Due to above difficulties, the implementation of the discussed feature ignores checking the level of indentation in conditional instructions and loops whose bodies contain only one line of code; and in switch instructions. For the switch instruction, it is even impossible to determine which notation is correct, as the flat form was used by programmers mainly in the past, while switch instruction in the indented form is predominant currently.

3.1 Choice of a Parser

As many proposed features were based on the syntactic structure of source code, choice of a parser of Java was an important part of the feature engineering. Three parsers were considered due to their established popularity: ANTLR[1] generated parser, JavaParser[2] and javalang[3]. Table 2 presents measured time of parsing source code with above parsers for the Hello world program (Listing 1.1) and the PR-SOCO corpus. The parser generated by ANTLR was incorrect as it was not able to parse all source code from the training corpus. It was also very slow. Although JavaParser turned out faster than javalang on the PR-SOCO corpus, we chose the latter, as it was implemented in Python, which was the language of the whole project.

---

[1] https://www.antlr.org.
[2] http://javaparser.org.
[3] https://pypi.org/project/javalang.

```
class Hello {
    public static void main(String args[]) {
        System.out.println("Hello world!");
    }
}
```

Listing 1.1. Program Hello world

Table 2. Parsing time (wall time) with ANTLR, JavaParser and javalang parsers, (secs). Results has been rounded up to the nearest integer

Parser	Hello world	PR-SOCO corpus
ANTLR	10	×
JavaParser	1	16
javalang	1	54

4   Data

As a learning and evaluation data we used the corpus of source codes, released for the PR-SOCO track [33], which accompanied PAN@FIRE 2016. The track was aimed at automatic personality recognition of programmers on the basis of Java source codes they authored. In the PR-SOCO corpus, personality was modelled with Big Five, and each trait was given a value from [20, 80]. The corpus contains 2492 source code programs written in Java by 70 students of computer science along with values of their personality traits. Values of personality traits were found on the basis of 25-item BFI questionnaire called *Big Five locator* which was completed by students. The students made their code submission through a web-based online judge for grading. The judge system does not have tools for style correction. However, it is not known whether students used an IDE before the submission or not. The training and test set contain source codes of 49 and 21 programmers, respectively. During the PR-SOCO contest, personalities of 21 persons from the test set were concealed from participating teams. Each team was allowed to submit 6 trial solutions (shots). A single solution predicted five traits for each of 21 persons from the test set.

Figure 1 presents distribution of values taken by each of five traits. Values from range [0, 20) and [80, 100] are never taken by any trait.

We followed the PR-SOCO track and used two measures to assess our solution and compare with existing personality predictors: Root Mean Square Error (*RMSE*) and Pearson Product-Moment Correlation coefficient (*PCC*).

RMSE measures the effectiveness of a regressor. For each personality trait t, root mean square error is defined as:

$$RMSE = \sqrt{\frac{1}{N} \sum_{i=1}^{N} (y_i - x_i)^2} \ , \tag{1}$$

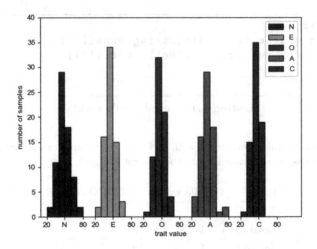

Fig. 1. Distribution of values taken by each of five traits in the PR-SOCO corpus

where x_i denotes true value of trait t for i-th instance (programmer), y_i is a value of trait t predicted by a personality predictor, and N is the number of instances (programmers). The lower $RMSE$ the better.

Pearson Product-Moment Correlation is defined as:

$$r = \frac{\sum_{i=1}^{N}(x_i - \bar{x})(y_i - \bar{y})}{\sqrt{\sum_{i=1}^{N}(x_i - \bar{x})^2}\sqrt{\sum_{i=1}^{N}(y_i - \bar{y})^2}} \ , \tag{2}$$

with $\bar{x}, \bar{y}$ denoting mean values of samples $(x_i)_{i=1}^{N}$ and $(y_i)_{i=1}^{N}$, respectively. PCC indicates whether obtained $RMSE$ is a random artifact or there is a correlation between actual and guessed values of traits. The larger the absolute value of PCC the better.

5 Experiments and Results

In the experiments we took the profile-based approach, i.e., all source codes of a programmer were treated as one learning instance. Since personality traits take continuous values, personality recognition was cast as a regression problem with random forest regression [4] from scikit-learn package [31] as the prediction module. Random forest regressors were trained on 85% of the original training set, remaining 15% of the training set was reserved for the model selection procedure. We examined the random forest regression with the number of decision trees varying from 64 to 128 and their depth varying from 2 to 6. Optimal values of the above hyperparameters were selected separately for each personality trait with grid search [8]. Mean Square Error (MSE) was used as the function measuring the quality of a split.

Beside regression with 35 handcrafted features, we used $N = 1500$ n-gram features as our baseline: $N_1 = 1000$ most frequent character trigrams (n-grams with $n = 3$) and $N_2 = 500$ most frequent token trigrams. By tokens we mean lexical units returned by the Java scanner. Finally, we tried personality recognition with 1535 features, both handcrafted and n-gram features.

5.1 Results

Table 3 presents results of personality recognition we obtained with 3 sets of features: proposed handcrafted features, n-grams and handcrafted features together with n-grams. For comparison, best results, medians and mean results of FIRE competitors are given in Table 4 (summary of FIRE competition [33] shows also first, second and third quantiles, all extreme values, and detailed results of all participating teams). Additionally we computed confidence intervals with the pairs bootstrap method [13].

For conscientiousness personality trait, the model with handcrafted features obtained *RMSE* equal 8.17 (with 95% confidence interval [6.00, 9.98]) which is lower than the minimum error achieved in the competition. Obtained value of *PCC* is 0.33 (with 95% confidence interval [−0.02, 0.65]) and equals to the corresponding maximum *PCC* value of PR-SOCO competitors.

Additionally, for openness we obtained *RMSE* lower than the median and close to the best score of PR-SOCO. For all personality traits absolute values of obtained *PCC*s were higher than the median values of PR-SOCO.

N-gram features turned out surprisingly good predictors of personality. For all traits, achieved *RMSE* and *PCC* values were better or equal than median values of PR-SOCO competitors. For neuroticism trait, n-gram features set up a new state-of-the-art result both for *RMSE* and *PCC*: *RMSE* was 9.65 with 95% confidence interval [6.03, 12.97] and *PCC* was 0.40 with 95% confidence interval [0.03, 0.73].

Table 3. Performance measures (*RMSE* and *PCC*) of automatic personality recognition with different sets of features

Trait	Handcrafted features		n-gram features		All features	
	RMSE	*PCC*	*RMSE*	*PCC*	*RMSE*	*PCC*
neuroticism	11.79	−0.13	**9.65**	**0.40**	9.68	0.37
extroversion	10.16	−0.19	9.45	−0.12	9.32	−0.07
openness	7.25	0.38	7.86	−0.08	7.78	−0.05
agreeableness	10.09	0.04	9.50	−0.03	9.20	0.12
conscientiousness	**8.17**	**0.33**	8.33	0.23	8.45	0.15

For the state-of-the-art results, we inspected the random forest regressors and found the features with the highest importance. For the model predicting conscientiousness with handcrafted features, the following features were the most important (more important features come first):

Table 4. Summary of personality recognition results obtained during PR-SOCO track of FIRE 2016 competition

Trait	Best		Median		Mean	
	RMSE	*PCC*	*RMSE*	*PCC*	*RMSE*	*PCC*
neuroticism	9.78	0.36	10.77	0.05	12.75	0.04
extroversion	8.60	0.47	9.55	0.08	12.27	0.06
openness	6.95	0.62	8.14	0.07	10.49	0.09
agreeableness	8.79	0.38	9.71	−0.03	12.07	−0.01
conscientiousness	8.38	0.33	8.99	−0.01	10.74	−0.01

- using conventional indentation
- average length of method
- average length of comments
- number of methods in a class (80th percentile)
- length of method (80th percentile)
- average length of names
- ratio of words in English to words in other languages
- length of names (80th percentile)
- number of white characters
- maximum length of comments.

The model predicting neuroticism with n-gram features benefitted the most from the following character or token trigrams (token trigrams are denoted as $\langle \cdot, \cdot, \cdot \rangle$, space characters in character trigrans are denoted as ␣):

//␣	$\langle \rangle$, ;, int$\rangle$	)␣+	cur	se␣
Com	.ne	{␣/	␣fa	;␣t
$\langle$0,], =$\rangle$	␣")	␣el	er.	nar
␣ma	1])	$\langle$., length, ;$\rangle$	$\langle$array, ., length$\rangle$	par

The effect of joint usage of handcrafted features and n-grams is the reduced error (in comparison to usage of only one type of features) for extroversion and agreeableness, although it does not set up new state-of-the-art results.

Finally, we examined statistical significance of obtained trait predictions (*RMSEs*). Statistical tests, conducted on the STAC platform [34], were computed for 14 algorithms (11 solutions from the PR-SOCO task and our three solutions: with handcrafted features, n-grams and all features) and five datasets (predictions for each of five traits were counted as a separate dataset). For solution from the PR-SOCO task we always chose the best shot. As the omnibus test we used Friedman F-test [15] for testing hypothesis H_0 that the means of the results of two or more algorithms are the same, followed by Nemenyi test [29] as the post-hoc test for pairwise comparison of predictors. At the significance level $\alpha = 0.05$, hypothesis H_0 should be rejected but pairwise comparison revealed no pair of algorithms with a statistical difference in results.

6   Conclusions

In this work we proposed new features for automatic personality recognition from source code. Handcrafted features turned out to be most useful for predicting openness and conscientiousness, traits (together with extroversion) connected with programming aptitude [19]. These features, despite their low number, achieved the state-of-the-art-results for conscientiousness. The lowest error in conscientiousness prediction is in line with the fact, that conscientiousness (and extroversion) are easily inferred from even slices of behaviour [6,18].

N-gram features are surprisingly good predictors of personality, at the same time they are easy to implement and language independent.

While the programmers' personalities may be connected with the code they write, we could not capture the relation between them. The results we achieved in neuroticism and conscientiousness recognition are state-of-the-art in personality recognition from source code, yet still insufficient to state that such a correlation exists.

Large confidence intervals of *RMSEs* and *PCCs*, and conducted statistical tests prove that larger datasets are needed to increase statistical strength of our results as well as other methods proposed so far. New datasets should take into account more programming languages and programmers, including professional programmers.

Acknowledgments. The research presented in this paper was supported by the funds assigned to AGH University of Science and Technology by the Polish Ministry of Science and Higher Education. Paolo Rosso, Francisco Rangel and Felipe Restrepo-Calle are acknowledged for making the PR-SOCO corpus available for our research and information about its construction.

References

1. Allport, G.W., Odbert, H.: Trait names: a psycho-lexical study. Psychol. Monogr. **47**(1), i–171 (1936). https://doi.org/10.1037/h0093360
2. Basile, A., Dwyer, G., Medvedeva, M., Rawee, J., Haagsma, H., Nissim, M.: N-gram: new groningen author-profiling model. In: Cappellato, L., Ferro, N., Goeuriot, L., Mandl, T. (eds.) Working Notes of CLEF 2017 - Conference and Labs of the Evaluation Forum (2017)
3. Bilan, I., Saller, E., Roth, B., Krytchak, M.: CAPS-PRC: a system for personality recognition in programming code. In: Majumder, P., Mitra, M., Mehta, P., Sankhavara, J., Ghosh, K. (eds.) Working Notes of FIRE 2016 - Forum for Information Retrieval Evaluation, pp. 21–24 (2016)
4. Breiman, L.: Random forests. Mach. Learn. **45**(1), 5–32 (2001). https://doi.org/10.1023/A:1010933404324
5. Calefato, F., Lanubile, F., Vasilescu, B.: A large-scale, in-depth analysis of developers' personalities in the Apache ecosystem. Inf. Softw. Technol. **114**, 1–20 (2019)
6. Carney, D., Colvin, R., Hall, J.: A thin slice perspective on the accuracy of first impressions. J. Res. Pers. **41**, 1054–1072 (2007)

7. Castellanos, H.A.: Personality recognition applying machine learning techniques on source code metrics. In: Majumder, P., Mitra, M., Mehta, P., Sankhavara, J., Ghosh, K. (eds.) Working Notes of FIRE 2016 - Forum for Information Retrieval Evaluation, pp. 25–29 (2016)

8. Claesen, M., Moor, B.D.: Hyperparameter search in machine learning. CoRR abs/1502.02127 (2015)

9. van Dam, M.: A basic character N-gram approach to authorship verification notebook for PAN at CLEF 2013. In: Forner, P., Navigli, R., Tufis, D., Ferro, N. (eds.) Working Notes for CLEF 2013 Conference (2013)

10. Dehkordi, Z.K., Baraani-Dastjerdi, A., Ghasem-Aghaee, N., Wagner, S.: Links between the personalities, styles and performance in computer programming. J. Syst. Softw. **111**, 228–241 (2016)

11. Delair, R., Mahajan, R.: A supervised approach for personality recognition in source code using code analysis tool at FIRE 2016. In: Majumder, P., Mitra, M., Mehta, P., Sankhavara, J., Ghosh, K. (eds.) Working Notes of FIRE 2016 - Forum for Information Retrieval Evaluation, pp. 30–32 (2016)

12. Doval, Y., Gómez-Rodríguez, C., Vilares, J.: Shallow recurrent neural network for personality recognition in source code. In: Majumder, P., Mitra, M., Mehta, P., Sankhavara, J., Ghosh, K. (eds.) Working Notes of FIRE 2016 - Forum for Information Retrieval Evaluation, pp. 33–37 (2016)

13. Efron, B., Tibshirani, R.J.: An Introduction to the Bootstrap. No. 57 in Monographs on Statistics and Applied Probability. Chapman & Hall/CRC, Boca Raton (1993)

14. Escalante, H.J., Solorio, T., Montes-y-Gómez, M.: Local histograms of character N-grams for authorship attribution. In: Lin, D., Matsumoto, Y., Mihalcea, R. (eds.) The 49th Annual Meeting of the Association for Computational Linguistics: Human Language Technologies, pp. 288–298 (2011)

15. Friedman, M.: The use of ranks to avoid the assumption of normality implicit in the analysis of variance. J. Am. Stat. Assoc. **32**(200), 675–701 (1937)

16. Ghosh, K., Parui, S.K.: Indian Statistical Institute Kolkata at PR-SOCO 2016: a simple linear regression based approach. In: Majumder, P., Mitra, M., Mehta, P., Sankhavara, J., Ghosh, K. (eds.) Working Notes of FIRE 2016 - Forum for Information Retrieval Evaluation, pp. 48–51 (2016)

17. Giménez, M., Paredes, R.: PRHLT at PR-SOCO: a regression model for predicting personality traits from source code. In: Majumder, P., Mitra, M., Mehta, P., Sankhavara, J., Ghosh, K. (eds.) Working Notes of FIRE 2016 - Forum for Information Retrieval Evaluation, pp. 38–42 (2016)

18. Gnambs, T.: The elusive general factor of personality: the acquaintance effect. Eur. J. Pers. **27**(5), 507–520 (2013)

19. Gnambs, T.: What makes a computer wiz? Linking personality traits and programming aptitude. J. Res. Pers. **58**, 31–34 (2015)

20. Houvardas, J., Stamatatos, E.: N-gram feature selection for authorship identification. In: Euzenat, J., Domingue, J. (eds.) 12th International Conference on Artificial Intelligence: Methodology, Systems, and Applications, AIMSA 2006. LNCS (LNAI), vol. 4183, pp. 77–86. Springer, Heidelberg (2006). https://doi.org/10.1007/11861461_10

21. John, O.P., Srivastava, S.: The big five trait taxonomy: history, measurement, and theoretical perspectives, pp. 102–138. Guilford Press (1999)

22. Kleć, M.: The influence of listener personality on music choices. Comput. Sci. (AGH) **18**(2), 163–178 (2017)

23. Kuta, M., Kitowski, J.: Optimisation of character n-gram profiles method for intrinsic plagiarism detection. In: Rutkowski, L., Korytkowski, M., Scherer, R., Tadeusiewicz, R., Zadeh, L.A., Zurada, J.M. (eds.) ICAISC 2014. LNCS (LNAI), vol. 8468, pp. 500–511. Springer, Cham (2014). https://doi.org/10.1007/978-3-319-07176-3_44

24. Li, L., Feng, H., Zhuang, W., Meng, N., Ryder, B.G.: CCLearner: a deep learning-based clone detection approach. In: 2017 IEEE International Conference on Software Maintenance and Evolution, ICSME 2017, pp. 249–260 (2017)

25. Liebeck, M., Modaresi, P., Askinadze, A., Conrad, S.: Pisco: a computational approach to predict personality types from Java source code. In: Majumder, P., Mitra, M., Mehta, P., Sankhavara, J., Ghosh, K. (eds.) Working Notes of FIRE 2016 - Forum for Information Retrieval Evaluation, pp. 43–47 (2016)

26. Martin, J.: Organizational Culture: Mapping the Terrain. Sage Publications, Thousand Oaks (2002)

27. McCrae, R., Costa, P.: Validation of the five-factor model of personality across instruments and observers. J. Pers. Soc. Psychol. **52**(1), 81–90 (1987)

28. McCrae, R., John, O.: An introduction to the five-factor model and its applications. J. Pers. **60**(2), 175–215 (1992)

29. Nemenyi, P.: Distribution-free multiple comparisons. Ph.D. thesis, Princeton University (1963)

30. Parr, T.: Language Implementation Patterns: Create Your Own Domain-Specific and General Programming Languages. Pragmatic Bookshelf, Raleigh (2009)

31. Pedregosa, F., et al.: Scikit-learn: machine learning in python. J. Mach. Learn. Res. **12**, 2825–2830 (2011)

32. Phani, S., Lahiri, S., Biswas, A.: Personality recognition in source code working note: team BESUMich. In: Majumder, P., Mitra, M., Mehta, P., Sankhavara, J., Ghosh, K. (eds.) Working Notes of FIRE 2016 - Forum for Information Retrieval Evaluation, pp. 16–20 (2016)

33. Rangel, F., González, F., Restrepo, F., Montes, M., Rosso, P.: PAN@FIRE: overview of the PR-SOCO track on personality recognition in SOurce COde. In: Majumder, P., Mitra, M., Mehta, P., Sankhavara, J. (eds.) FIRE 2016. LNCS, vol. 10478, pp. 1–19. Springer, Cham (2018). https://doi.org/10.1007/978-3-319-73606-8_1

34. Rodríguez-Fdez, I., Canosa, A., Mucientes, M., Bugarín, A.: STAC: a web platform for the comparison of algorithms using statistical tests. In: 2015 IEEE International Conference on Fuzzy Systems, FUZZ-IEEE, pp. 1–8 (2015). https://doi.org/10.1109/FUZZ-IEEE.2015.7337889

35. Sapkota, U., Bethard, S., Montes-y-Gómez, M., Solorio, T.: Not all character N-grams are created equal: a study in authorship attribution. In: Mihalcea, R., Chai, J.Y., Sarkar, A. (eds.) NAACL HLT 2015, The 2015 Conference of the North American Chapter of the Association for Computational Linguistics: Human Language Technologies, pp. 93–102 (2015)

36. Vázquez, E.V., et al.: UAEMex system for identifying personality traits from source code. In: Majumder, P., Mitra, M., Mehta, P., Sankhavara, J., Ghosh, K. (eds.) Working Notes of FIRE 2016 - Forum for Information Retrieval Evaluation, pp. 52–55 (2016)

37. Vinciarelli, A., Mohammadi, G.: A survey of personality computing. IEEE Trans. Affect. Comput. **5**(3), 273–291 (2014)

Data Fitting by Exponential Sums with Equal Weights

Petr Chunaev$^{(\boxtimes)}$ and Ildar Safiullin

National Center for Cognitive Technologies, ITMO University,
Saint Petersburg 199034, Russia
chunaev@itmo.ru, ild.safiullin2013@gmail.com

Abstract. In this paper, we introduce a Prony-type data fitting problem consisting in interpolating the table $\{m, g(m)\}_{m=0}^{M}$ with $g(0) \neq 0$ in the sense of least squares by exponential sums with equal weights. We further study how to choose the parameters of the sums properly to solve the problem. Moreover, we show that the sums have some advantages in data fitting over the classical Prony exponential sums. Namely, we prove that the parameters of our sums are a priori well-controlled and thus can be found via a stable numerical framework, in contrast to those of the Prony ones. In several numerical experiments, we also compare the behaviour of both the sums and illustrate the above-mentioned advantages.

Keywords: Exponential sum · Data fitting · Prony interpolation

1 Introduction and Main Results

Interpolation (or data fitting) of a complex-valued function $g = g(z)$, $z \in \mathbb{C}$, by a linear combination of exponents, e.g. by the *Prony exponential sums*

$$\mathcal{H}_K(z) = \sum_{k=1}^{K} \mu_k \exp(\lambda_k z), \quad \text{where } \mu_k \in \mathbb{C} \text{ and } \lambda_k \in \overline{\mathbb{C}} \text{ are parameters,} \quad (1)$$

appears in many contexts in Science and Engineering, in particular, in time series analysis, physical phenomena modelling and signal processing. There exist numerous state-of-the-art analytical and numerical methods for solving the corresponding data fitting problems, see the overview in [2,5,7,9–12]. One of the main challenges in the area is that the exponential data fitting is ill-conditioned in many cases [10]. Among other things, this can be connected with the behaviour of exponential sum parameters (such as μ_k and λ_k in (1)). In particular, one can find in [6] a family of g whose corresponding parameters of (1) tend to infinity thus making the interpolation process divergent.

The research was financially supported by Russian Science Foundation, Agreement 19-71-10078.

To address this disadvantage, we consider in this paper a modified version of the exponential sums (1) and introduce and solve the corresponding Prony-type data fitting problem. Moreover, we prove that the parameters in our exponential sums are a priori well-controlled and thus can be found via a stable numerical framework, in contrast to those in the sums (1). Let us move to the formal description of our approach.

We introduce the following Prony-type data fitting problem: numerically interpolate the table

$$\{m, g(m)\}_{m=0}^{M}, \qquad g(0) \neq 0, \tag{2}$$

by the following *exponential sums with equal weights*

$$H_K(z) = \frac{\mu}{K} \sum_{k=1}^{K} \exp(\lambda_k z), \qquad \mu \in \mathbb{C}, \quad \lambda_k \in \overline{\mathbb{C}}, \quad K \leqslant M. \tag{3}$$

The fitting should be carried out by a proper choice of the parameters μ and $\{\lambda_k\}_{k=1}^{K}$ which depend on K, M and g. The fitting quality has to be evaluated in the sense of least squares via a numerical optimization method so that

$$\sum_{m=0}^{M} |g(m) - H_K(m)|^2 = \delta^2 \qquad \text{for some } \delta \geqslant 0. \tag{4}$$

The main result of the paper in the above-mentioned settings is about the construction and the a priori control of the parameters of H_K and is as follows.

Theorem 1. *Given a table (2), the parameters of H_K are as follows: $\mu = g(0)$ and the set $\{\lambda_k\}_{k=1}^{K}$ is the solution to the Newton-type moment problem*

$$\sum_{m=1}^{M} \left| g(m) - \frac{g(0)}{K} \sum_{k=1}^{K} l_k^m \right|^2 = \delta^2, \qquad \text{where} \quad l_k := \exp(\lambda_k). \tag{5}$$

Furthermore, if $|g(m)| \leqslant \frac{|g(0)|}{K} a^m$ for all $m = 1, \ldots, M$ and some $a \geqslant 1$, then[1]

$$\max_{k=1,\ldots,K} |l_k| \leqslant \kappa a \quad \Rightarrow \quad -\infty \leqslant \mathrm{Re}\,(\lambda_k) \leqslant \ln(\kappa a), \qquad \text{where } \kappa := 3 + \frac{2K\delta}{|g(0)|},$$

and moreover $|H_K(x)| \leqslant |g(0)| \, (\kappa a)^x$ for any $x \in \mathbb{R}$.

Thus the parameters of H_K are well-controlled by g in (2) and by δ in (4).

The paper is organised as follows. Section 2 provides necessary facts on and the theoretical comparison of the sums H_K and $\mathcal{H}_K$. Section 3 is devoted to the proof of Theorem 1. Section 4 contains several experiments with H_K and $\mathcal{H}_K$ illustrating the difference of their behaviour in numerical processes.

---

[1] Due to the periodicity of the exponential function, one can think that $|\mathrm{Im}\,\lambda_k| \leqslant \pi$.

2   Necessary Facts on and the Comparison of H_K and Classical Prony Exponential Sums

The analytical construction of (3) with $K = M$ for the table (2) was introduced in [1]. It was also shown that the problem (4) then can be solved analytically, with $\delta = 0$. Moreover, if the corresponding μ and $\{\lambda_k\}_{k=1}^M$ are known precisely and determined via the so-called Newton-Prony method, then $g(z) = H_M(z)$ for $z \in \{m\}_{m=0}^M$. Now we briefly describe the approach from [1].

Let $L_M := \{l_k\}_{k=1}^M$, where $l_k \in \mathbb{C}$. Consider the *power sums*

$$S_m := S_m(L_M) = \sum_{k=1}^M l_k^m, \qquad m = 1, 2, \dots. \tag{6}$$

One can find the set L_M from the *Newton moment problem*

$$S_m = s_m, \quad m = 1, \dots, M, \qquad \text{where } s_m \in \mathbb{C} \text{ are given}, \tag{7}$$

via the *elementary symmetric polynomials*

$$\sigma_m = \sigma_m(L_M) := \sum_{1 \leqslant j_1 < \dots < j_m \leqslant M} l_{j_1} \cdots l_{j_m}, \qquad m = 1, \dots, M. \tag{8}$$

The connection between the power sums (6) and the polynomials (8) is expressed by the well-known *Newton-Girard formulas* [13, Section 3.1]:

$$\sigma_1 = S_1, \quad \sigma_m = \frac{(-1)^{m+1}}{m} \left(S_m + \sum_{j=1}^{m-1} (-1)^j S_{m-j} \sigma_j \right), \quad m = 2, \dots, M. \tag{9}$$

Moreover, L_M is then the set of M roots of the *unitary* polynomial

$$P_M(l) := l^M - \sigma_1 l^{M-1} + \sigma_2 l^{M-2} + \dots + (-1)^M \sigma_M. \tag{10}$$

Consequently, given any s_m, one can solve (7) using (9) and (10) and the *solution* L_M *always exists and is unique*. What is more, it is shown in [1] that the elements of L_M are well-controlled by the table (2) under few natural assumption on g. Namely, the following result is a consequence of what is proved in [1]:

For $M \geqslant 2$, $\gamma > 0$ and $a \geqslant 1$, if $|s_m| \leqslant \gamma a^m$ in (7) for $m = 1, \dots, M$, then

$$\max_{k=1,\dots,K} |l_k| \leqslant (1 + 2\gamma) a. \tag{11}$$

Now we turn our attention to the Prony exponential sums (1). There exist a classical analytical Prony method for interpolating the table (2) with odd M (so that $M + 1$ is even) by the sums (1) with $K = \tilde{M} := (M + 1)/2$, see e.g. [3,6,8,10]. The main idea behind it is to separate calculating the parameters μ_k

and λ_k. Namely, by supposing $l_k := \exp(\lambda_k)$ and $s_m = g(m)$ one comes to the following *Prony moment problem* that is a weighted version of (7):

$$S_m^* = s_m, \qquad \text{where} \qquad S_m^* := \sum_{k=1}^{\tilde{M}} \mu_k l_k^m, \qquad m = 0, \dots, M. \qquad (12)$$

If this problem has a unique solution, then $L_{\tilde{M}} = \{l_k\}_{k=1}^{\tilde{M}}$ is the set of $\tilde{M}$ roots of the (possibly *non-unitary*) polynomial

$$P_{\tilde{M}}^*(l) := \sum_{m=0}^{\tilde{M}} \sigma_m^* l^m = \begin{vmatrix} 1 & l & l^2 & \cdots & l^{\tilde{M}} \\ s_0 & s_1 & s_2 & \cdots & s_{\tilde{M}} \\ s_1 & s_2 & s_3 & \cdots & s_{\tilde{M}+1} \\ \cdots & \cdots & \cdots & \cdots & \cdots \\ s_{\tilde{M}-1} & s_{\tilde{M}} & s_{\tilde{M}+1} & \cdots & s_{2\tilde{M}-1} \end{vmatrix}, \qquad (13)$$

which is an analogue of (9) and (10). Note that $\{\mu_k, \lambda_k\}_{k=1}^{\tilde{M}}$ can be uniquely determined (and thus (1) constructed) if and only if the polynomial (13) for a given set $\{s_m\}_{m=0}^{M}$ is exactly of degree $\tilde{M}$ and all its roots are pairwise distinct [6]. In general, the problem (12) can be unsolvable or have multiple solutions [6].

Let us mention that the system (12) is not only related to (1) but plays an important role in different areas of approximation theory and is closely related to Hankel matrices, Gauss quadratures, moment problems and classical Padé rational fractions (a survey can be found e.g. in [6, Section 2] or [8]).

Due to the importance of (1) and (12) in approximation theory and applied data fitting (see Introduction), multiple numerical approaches for data fitting by the sums (1) with $K \leqslant \tilde{M}$ have been proposed [2,4,5,11,12]. The fitting quality evaluation is usually performed in the sense of least squares:

$$\sum_{m=0}^{M} |g(m) - \mathcal{H}_K(m)|^2 \to \min. \qquad (14)$$

However, in sharp contrast to H_K within (4) and (7), there are no more or less general estimates for the parameters of $\mathcal{H}_K$ similar to those in Theorem 1 and the theorems from [1], as seen from the huge bibliography on the Prony exponential sums ([2,3,5,6,8,11,12] and references therein). Moreover, numerical processes for finding solutions to (14) and (12) can be unstable and divergent (see [2,6]).

To finish the comparison of H_K and $\mathcal{H}_K$, let us also observe that computing (3) requires less arithmetic operations than that of (1) for each fixed z and known μ, μ_k and λ_k, see [1].

3   The Proof of Theorem 1

The identity for μ follows directly from (3) and (4) for $m = 0$. The optimization problem (5) is then deduced from (4) by the exchange $l_k := \exp(\lambda_k)$.

Now let $|g(m)| \leqslant \frac{|g(0)|}{K} a^m$ in (2) for all $m = 1, \ldots, M$ and some $a \geqslant 1$. Once the solution $\{l_k\}_{k=1}^K$ to (5) and a certain δ are found by a numerical optimization method, we obtain by the triangle inequality that

$$\left| \frac{g(0)}{K} \sum_{k=1}^K l_k^m \right| \leqslant |g(m)| + \delta \leqslant \left(\frac{|g(0)|}{K} + \delta \right) a^m, \quad m = 1, \ldots, K.$$

Note that m here is from 1 to K, not M. Consequently, we come to the system (7) with K unknowns and K equations:

$$|S_m| = \left| \sum_{k=1}^K l_k^m \right| = |s_m| \leqslant \left(1 + \frac{K\delta}{|g(0)|} \right) a^m, \quad m = 1, \ldots, K.$$

Furthermore, by (11) we easily obtain the required inequalities for $\max_{k=1,\ldots,K} |l_k|$ and $\mathrm{Re}\,\lambda_k$. What is more, $|H_K(x)| \leqslant \frac{|g(0)|}{K} \sum_{k=1}^K (\kappa a)^x \leqslant |g(0)|(\kappa a)^x$ for $x \in \mathbb{R}$.

4   Numerical Experiments

Experiments below show the difference in the behaviour of H_K and $\mathcal{H}_K$ in numerical processes and illustrate Theorem 1. In fact, one can use any suitable numerical method for finding the parameters of H_K and $\mathcal{H}_K$ as the examples below

Table 1. The results of experiments for different ε (five digits are provided).

ε	μ in H_3	$\lambda_{1,2,3}$ in H_3	$\mu_{1,2}$ in $\mathcal{H}_2$	$\lambda_{1,2}$ in $\mathcal{H}_2$
10^{-1}	$1.00000 + 10^{-1}$	-0.01641 $-0.06788 + 1.96110i$ $-0.06788 - 1.96110i$	-0.00151 1.10150	2.12972 -2.53519
10^{-6}	$1.00000 + 10^{-6}$	$-1.66666 \cdot 10^{-7}$ $0.00002 + 2.09440i$ $0.00002 - 2.09440i$	$-1.00000 \cdot 10^{-18}$ $1.00000 + 10^{-6}$	13.81550 -13.81551
10^{-7}	$1.00000 + 10^{-7}$	$-1.70000 \cdot 10^{-8}$ $-9.17072 \cdot 10^{-8} + 2.09439i$ $-9.17072 \cdot 10^{-8} - 2.09439i$	$-1.00000 \cdot 10^{-21}$ $1.00000 + 10^{-7}$	16.11809 -16.11809
10^{-8}	$1.00000 + 10^{-8}$	$-2.00000 \cdot 10^{-9}$ $-9.28556 \cdot 10^{-9} + 2.09439i$ $-9.28556 \cdot 10^{-9} - 2.09439i$	$-1.00000 \cdot 10^{-24}$ $1.00000 + 10^{-8}$	18.42068 -18.42068
0	1	0 $2.09440i$ $-2.09440i$	0 1	$+\infty$ $-\infty$

are constructed to be independent of the method (see [2,5,11,12] for the methods suitable for $\mathcal{H}_K$; any numerical method for non-linear least squares can be adapted for H_K). We produced calculations using Maple 2019 tools with 200 significant digits and same environment for both H_K and $\mathcal{H}_K$, to avoid the influence of a particular method and its precision.

In order to have the same number of free parameters in the exponential sums and to establish the connection with the problems (7) and (12), we solve the data fitting problems (4) and (14) with $K = M$ and $K = (M+1)/2$, correspondingly.

We consider the following parametric table of the form (2) with $M = 3$:

$$(0, 1 + \varepsilon), \ (1, 0 + \varepsilon), \ (2, 0 - \varepsilon), \ (3, 1 - \varepsilon), \qquad 0 < \varepsilon \leqslant 10^{-1}. \tag{15}$$

Note that (15) can be easily complemented to contain arbitrarily many elements. However, since the data (15) should be still fitted, the same effects will be observed for the corresponding exponential sums H_3 and $\mathcal{H}_2$ (see [6, Section 7]).

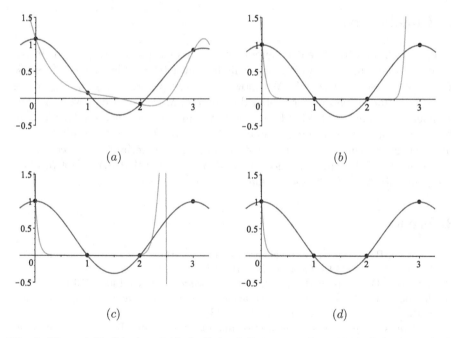

Fig. 1. Plots of H_3 (blue) and $\mathcal{H}_2$ (red) for different ε in the table (15) (black points): (a) $\varepsilon = 10^{-1}$, (b) $\varepsilon = 10^{-6}$, (c) $\varepsilon = 10^{-7}$, (d) $\varepsilon = 10^{-8}$ (Color figure online).

It directly follows from (13) that (15) leads to a unique solution to (12) for $0 < \varepsilon \leqslant 10^{-1}$ (note that (7) always has such a solution) and therefore a proper numerical method should converge to this solution. Below we choose $\delta = 10^{-10}$ in (4) and (14). Table 1 and Fig. 1 contain the results for different $\varepsilon > 0$ and the extreme case $\varepsilon = 0$ (this case for $\mathcal{H}_2$ was studied in [6, Section 7]).

It can be observed that for $10^{-5} \leqslant \varepsilon \leqslant 10^{-1}$ both H_3 and $\mathcal{H}_2$ solve the data fitting problem with good quality (see Fig. 1(a)). For $\varepsilon < 10^{-5}$, the precision of $\mathcal{H}_2$ however shrinks significantly and the data is not fitted properly, while the fitting quality of H_3 is still good (see Fig. 1(b), (c), (d)). This can be explained as follows.

The table (15) with $\varepsilon \to 0$ leads to an unsolvable problem (12). Thus one has a divergent numerical process for $\mathcal{H}_2$ (with $|\lambda_{1,2}| \to \infty$ and $|\mu_1| \to 0$) as $\varepsilon \to 0$. The corresponding computational errors significantly grow (200 significant digits for the parameters of $\mathcal{H}_2$ are not already enough for a reasonable precision of $\mathcal{H}_2$ itself) and this leads to the incorrect data fitting by $\mathcal{H}_2$ (especially for $m = 3$) represented in Fig. 1 (b), (c), (d).

Oppositely, since the parameters of H_3 are well-controlled (in particular, Theorem 1 implies for the table (15) that $\mu = 1 + \varepsilon$, $\max_{k=1,\ldots,3} |l_k| \leqslant 3.1(1+\varepsilon)$ and $-\infty \leqslant \mathrm{Re}\,(\lambda_k) \leqslant \varepsilon + \ln 3.1)$, 200 significant digits stay enough for rather accurate calculation of H_3 and its representation in Fig. 1 (b), (c), (d).

5   Conclusions

In this paper, we considered the exponential sums with equal weights (3) in the context of the Prony-type problem of fitting data (the table (2)) in the sense of non-linear least squares (4). We showed how the parameters μ and λ_k of the sums (3) should be chosen for proper data fitting. Moreover, we proved that the parameters are well-controlled by (2) and by δ in (4) (see Theorem 1) that is in sharp contrast to the case of parameters in the well-known Prony exponential sums (1). Furthermore, in several special numerical experiments, we compared the behaviour of the sums (3) and (1) and demonstrated that (3) do not produce divergent data fitting processes while (1) do in some cases.

References

1. Chunaev, P.: Interpolation by generalized exponential sums with equal weights. J. Approx. Theory **254** (2020). Article: 105397
2. Batenkov, D.: Decimated generalized Prony systems. arXiv:1308.0753 (2013)
3. Batenkov, D., Yomdin, Y.: On the accuracy of solving confluent Prony systems. SIAM J. Appl. Math. **73**(1), 134–154 (2013)
4. Beylkin, G., Monzón, L.: On approximation of functions by exponential sums. Appl. Comput. Harmonic Anal. **19**(1), 17–48 (2005)
5. Beylkin, G., Monzón, L.: Approximation by exponential sums revisited. Appl. Comput. Harmonic Anal. **28**, 131–149 (2010)
6. Chunaev, P., Danchenko, V.: Approximation by amplitude and frequency operators. J. Approx. Theory **207**, 1–31 (2016)
7. Keller, I., Plonka, G.: Modifications of Prony's method for the recovery and sparse approximation of generalized exponential sums. arXiv:2001.03651 (2020)
8. Lyubich, Y.I.: The Sylvester-Ramanujan system of equations and the complex power moment problem. Ramanujan J. **8**, 23–45 (2004). https://doi.org/10.1023/B:RAMA.0000027196.19661.b7

9. Mourrain, B.: Polynomial-exponential decomposition from moments. Found. Comput. Math. **18**, 1435–1492 (2018). https://doi.org/10.1007/s10208-017-9372-x
10. Pereyra, V., Scherer, G.: Exponential Data Fitting and Its Applications. Bentham Science Publishers, Sharjah (2018)
11. Plonka, G., Tasche, M.: Prony methods for recovery of structured functions. GAMM-Mitt. **37**(2), 239–258 (2014)
12. Potts, D., Tasche, M.: Parameter estimation for nonincreasing exponential sums by Prony-like methods. Linear Algebra Appl. **439**(4), 1024–1039 (2013)
13. Prasolov, V.V.: Polynomials. Algorithms and Computation in Mathematics, vol. 11. Springer, Heidelberg (2004). https://doi.org/10.1007/978-3-642-03980-5

A Combination of Moment Descriptors, Fourier Transform and Matching Measures for Action Recognition Based on Shape

Katarzyna Gościewska[(✉)] and Dariusz Frejlichowski

Faculty of Computer Science and Information Technology, West Pomeranian University of Technology, Szczecin, Żołnierska 52, 71-210 Szczecin, Poland
{kgosciewska,dfrejlichowski}@wi.zut.edu.pl

Abstract. This paper presents an approach for human action recognition based on shape analysis. The purpose of the approach is to classify simple actions by applying shape descriptors to sequences of binary silhouettes. The recognition process consists of several main stages: shape representation, action sequence representation and action sequence classification. Firstly, each shape is represented using a selected shape descriptor. Secondly, shape descriptors of each sequence are matched, matching values are put into a vector and transformed into final action representation—we employ Fourier transform-based methods to obtain action representations equal in size. A classification into eight classes is performed using leave-one-out cross-validation and template matching approaches. We present results of the experiments on classification accuracy using moment-based shape descriptors (Zernike Moments, Moment Invariants and Contour Sequence Moments) and three matching measures (Euclidean distance, correlation coefficient and C1 correlation). Different combinations of the above-mentioned algorithms are examined in order to indicate the most effective one. The experiments show that satisfactory results are obtained when low-order Zernike Moments are used for shape representation and absolute values of Fourier transform are applied to represent action sequences. Moreover, the selection of matching technique strongly influences final classification results.

Keywords: Action recognition · Silhouette sequences · Shape descriptors

1 Introduction

An automatic recognition of human movements has gained popularity in recent years due to its wide range of applications, especially related to surveillance systems and human-computer interaction. Other applications include quality-of-life improvement for elderly care, sports analytics, and video retrieval and annotation. This implies a diversity of data and a need for different solutions.

© Springer Nature Switzerland AG 2020
V. V. Krzhizhanovskaya et al. (Eds.): ICCS 2020, LNCS 12138, pp. 372–386, 2020.
https://doi.org/10.1007/978-3-030-50417-5_28

Human action can be defined as a sequence of elementary movements that is clearly identifiable by the observer. Combinations of elementary movements can create single (e.g. bending) or periodic (e.g. running) motion patterns [5]. An action is also defined as an activity composed of multiple gestures organized in time, and a gesture is an elementary movement of the body part [20]. To perform action recognition it is common to apply low-level features such as shape which is considered as a distinctive feature supporting accurate classification. Additionally, the order and repeatability of individual silhouettes can help distinguish between actions. Despite a few characteristic elements the recognition process is still a challenging task because of the variations in motion performance, personal differences, speed or duration of individual actions [17].

In this paper we propose an original combination of well-known methods and algorithms aimed to recognize actions based on information contained in a binary foreground masks that were extracted from consecutive video frames representing people performing simple actions. The novelty is accomplished by creating a synthesis of some already approved methods and by testing existing knowledge in a different manner. The proposed approach is applied on coarsely classified sequences. Then the recognition is performed in each subgroup separately using the same procedure composed of three main steps: single shape representation, single action representation and action classification.

The rest of the paper is organized as follows: Sect. 2 presents several related works on action classification based on shape features. The proposed approach is explained in detail in Sect. 3 and some methods are presented in Sect. 4. Section 5 defines experimental conditions and presents the results of the experiments carried out with the use of three moment-based shape descriptors, namely Zernike Moments, Moment Invariants and Contour Sequence Moments. Section 6 summarizes the paper.

2   Related Works

This section describes several methods that are similar to our approach due to the use of shape features and similar input data. We focus on a shape-based action recognition that is classified in [20] as a non-hierarchical approach. This category covers the recognition of short and primitive actions. To recognize such actions, we can use solutions based on space-time volume, like this presented in [4]. The proposed approach generates motion energy images (MEI) to show where the movement is, and motion history images (MHI) to show how the object is moving. Then Hu moments are extracted from MEI and MHI, and the resultant action descriptors are matched using Mahalanobis distance. Hu moments are statistical descriptors which are scale and translation invariant, and allow for good shape discrimination. Another popular space-time volume technique is proposed in [10]. It accumulates silhouettes into space-time cubes (3D representation) and employs a Poisson equation to extract features of human actions, among which are local space-time saliency, action dynamics, shape structure and orientation. Space-time volume is also a global approach—the localized foreground region of interest is encoded as a whole and much of the information is carried. Popular holistic representations are based on silhouettes, edges or optical flow [17]. Among these a silhouette is our most interest. In [12,14] shape features are calculated for each object separately and objects are not accumulated. The authors of [12] introduced new feature extraction techniques based on Trace transform, namely History Trace Templates and History Triple Features. In the first method, Trace transform is applied to binary silhouettes. The resultant transforms are composed into final history template that represents the whole action sequence and contains much of the spatial-temporal features. In the second method, Trace transform is used to construct a set of features that are invariant to translation, rotation and scaling, as well as robust to noise. Features are calculated for every video frame separately. Ultimately, LDA is applied to reduce dimensionality of final representations. In turn, in [14] every silhouette is transformed into time series and each of these is converted into the symbolic vector—a SAX representation. A set of all vectors represents an action and is called a SAX-Shape.

Action recognition can be performed using only some silhouettes extracted from selected video frames, so called key poses, e.g. [2,7,16]. The authors of [2] introduce a shape representation and matching technique that represents each key pose as a collection of line-pairs and can estimate similarity between two frames. A k-medoids clustering algorithm and learning algorithm are used to extract candidate key poses. During the classification process every frame is compared with all key poses in order to assign a label. Then majority voting is used to classify action sequences. Another solution using key poses is presented in [16]. The authors proposed extensive pyramidal features (EPFs) to describe poses. EPFs include Gabor, Gaussian and wavelet pyramids. AdaBoost algorithm is used to learn a subset of discriminative poses. Actions are classified with a new classifier—weighted local naive Bayes nearest neighbour. In [7] the proposed method uses the distance between all contour points and silhouette's centre of gravity to represent a pose. Then, K-means clustering with Euclidean distance is applied to learn key poses and Dynamic Time Warping is used to classify sequences of key poses.

Another solution for action recognition is a fusion of multiple features. The authors of [1] proposed a new algorithm based on Aligned Motion Images (AMIs), where each AMI is a single image that represents the motion of all frames of a single video. Two features are combined—Derivatives of Chord-Distance Signature based on contour and Histogram of Oriented Gradients which capture visual components of a silhouette's region. Action classification is performed using K-Nearest Neighbour and Support Vector Machine (SVM). Another approach based on contour and shape features is presented in [21]. It combines information obtained from the R-transform and averaged energy silhouette images which are used to generate feature vectors based on edge distribution of gradients and directional pixels. Classification is carried out with the use of multi-class SVM.

3   The Proposed Approach

The proposed approach is composed of selected methods and algorithms, among which are: shape description algorithms based on moments, signal processing algorithms based on Fourier transform as well as distance and correlation-based matching measures. The selection of methods results from the continuation of the works presented in [11], where we have also tested moment-based descriptors using following procedure: each silhouette was represented using selected shape descriptor, shape representations were matched using Euclidean distance and normalized matching values were put into a vector called distance vector. To obtain final sequence representations all distance vectors were transformed using Fast Fourier Transform and periodogram. Classification process was performed iteratively using template matching approach and k-fold cross-validation. In each iteration the database was divided into templates (class representatives) and test objects. Each test object was matched with all templates to indicate the most probable class. Final classification accuracy was an average of all iterations.

In this paper the results of new experiments are given. When compared with previous work there are some significant differences. Firstly, the experimental database consists of eight instead of five classes, four classes for each subgroup. Matching process is performed using three various measures instead of one. This applies to both comparison of shape descriptors and classification of final action representations. Our previous experiments have shown that the accuracy depends on applied matching measure. Secondly, final action representations are prepared using three various methods instead of one. In case of classification process, the leave-one-out cross-validation with template matching approach is applied instead of k-fold cross-validation technique. This is done to avoid a situation in which a set of class representatives affects classification accuracy. Moreover, a coarse classification step has been added.

Here we focus on testing various combinations of several algorithms in order to select relevant features for action description. Therefore, the proposed approach has a form of a general procedure composed of consecutive data processing steps which are:

Step 1. Data preparation
The proposed approach bases on binary silhouettes. We use the Weizmann [3] database which is composed of action sequences—one action sequence is represented by a set of frames from which foreground binary masks are extracted. Each foreground mask contains one silhouette. The dataset is divided into two subgroups based on the centroid trajectory—actions performed in place (a trajectory is very short) and actions with changing location of a silhouette (longer trajectory). Then the approach is applied in each subgroup separately. Let us denote each input action sequence as a set of binary masks $BM_i = \{bm_1, bm_2, ..., bm_n\}$, where n is the number of frames in a particular sequence.

Step 2. Single shape representation
In the next step, we take each bm_i and represent it using selected shape description algorithm. Various methods can be applied and here we examine Zernike

Moments, Moments Invariants and Contour Sequence Moments (see Sect. 4.1). In result, we obtain a set of shape descriptors for each action sequence which can be denoted as $SD_i = \{sd_1, sd_2, ..., sd_n\}$. The number of descriptors equals the number of frames. A sd_i can be a matrix or a vector, depending on the applied shape descriptor.

Step 3. Single action representation

Action representation is based on the calculation of similarity or dissimilarity measures for each SD_i separately. We can use various solutions, such as Euclidean distance, correlation coefficient and C1 correlation (see Sect. 4.2). The shape descriptor of a first frame sd_1 is matched with the rest of descriptors and matching values are put into a vector $MD_i = \{md_1, md_2, ..., md_{n-1}\}$. A sd_1 is not matched with itself therefore we obtain one element less. Here, for instance, md_1 is a matching value calculated using sd_1 and sd_2. All MD vectors are normalized and transformed into frequency domain using periodogram or Fast Fourier Transform algorithm (a magnitude is taken). Each transformed vector creates one-dimensional descriptor of a sequence—a final action representation AR. The transformation into frequency domain makes all representations equal in size—we use a predefined number of elements which exceeds the number of frames in the longest video sequences. Moreover, the resultant transforms reveal some hidden periodicities in the data.

Step 4. Classification

AR vectors are classified based on the leave-one-out cross-validation process and template matching technique. Here template matching is understood as a process that compares each test object with all templates and indicates the most similar one, which corresponds to the probable class of a test object, e.g. we take AR_1 and match it with the rest of AR vectors using methods explained in Sect. 4.2. The percentage of correctly classified actions gives classification accuracy.

4   Shape Description and Matching

4.1   Shape Description Algorithms Based on Moments

The Zernike Moments are derived using Zernike orthogonal polynomials and the formula below [22]:

$$V_{nm}(x, y) = V_{nm}(r \cos\theta, \sin\theta) = R_{nm}(r) \exp(jm\theta), \tag{1}$$

where $R_{nm}(r)$ is the orthogonal radial polynomial [22]:

$$R_{nm}(r) = \sum_{s=0}^{(n-|m|)/2} (-1)^s \frac{(n-s)!}{s! \times \left(\frac{n-2s+|m|}{2}\right)! \left(\frac{n-2s-|m|}{2}\right)!} r^{n-2s}, \tag{2}$$

where $n = 0, 1, 2, \ldots$; $0 \le |m| \le n$; $n - |m|$ is even.

The Zernike Moments of order n and repetition m of a region shape $f(x, y)$ are calculated by means of this formula [22]:

$$Z_{nm} = \frac{n+1}{\pi} \sum_r \sum_\theta f(r \cos\theta, r \sin\theta) \cdot R_{nm}(r) \cdot \exp(jm\theta), \; r \le 1. \tag{3}$$

According to [8,13,18], to obtain Moment Invariants, the general geometrical moments are firstly calculated using the following formula:

$$m_{pq} = \sum_x \sum_y x^p y^q f(x, y). \tag{4}$$

The $f(x, y)$ function value is equal to 1 for pixels belonging to an object and 0 for background pixels. The representation is invariant to translation thanks to the use of centroid, which is calculated as follows:

$$x_c = \frac{m_{10}}{m_{00}}, \quad y_c = \frac{m_{01}}{m_{00}}. \tag{5}$$

Then, Central Moments are calculated using the centroid:

$$\mu_{pq} = \sum_x \sum_y (x - x_c)^p (y - y_c)^q f(x, y). \tag{6}$$

In turn, the invariance to scaling is obtained by central normalized moments:

$$\eta_{pq} = \frac{\mu_{pq}}{\mu_{00}^{\frac{p+q+2}{2}}}. \tag{7}$$

Finally, Moment Invariants are derived (seven first values):

$$\phi_1 = \eta_{20} + \eta_{02}$$

$$\phi_2 = (\eta_{20} + \eta_{02})^2 + 4\eta_{11}^2$$

$$\phi_3 = (\eta_{30} - 3\eta_{12})^2 + (3\eta_{21} - \eta_{03})^2$$

$$\phi_4 = (\eta_{30} + \eta_{12})^2 + (\eta_{21} + \eta_{03})^2$$

$$\phi_5 = (\eta_{30} - 3\eta_{12})(\eta_{30} + \eta_{12})[(\eta_{30} + \eta_{12})^2 - 3(\eta_{03} + \eta_{21})^2] \\ + (3\eta_{21} - \eta_{03})(\eta_{03} + \eta_{21})[3(\eta_{30} + \eta_{12})^2 - (\eta_{03} + \eta_{21})^2] \tag{8}$$

$$\phi_6 = (\eta_{20} - \eta_{02})[(\eta_{30} + \eta_{12})^2 - (\eta_{21} + \eta_{03}^2] \\ + 4\eta_{11}(\eta_{30} + \eta_{12})(\eta_{03} + \eta_{21})$$

$$\phi_7 = (3\eta_{21} - \eta_{03})(\eta_{30} + \eta_{12})[(\eta_{30} + \eta_{12})^2 - 3(\eta_{03} + \eta_{21})^2] \\ - (\eta_{30} - 3\eta_{12})(\eta_{03} + \eta_{21})[3(\eta_{30} + \eta_{12})^2 - (\eta_{03} + \eta_{21})^2]$$

Based on [19], the calculation of Contour Sequence Moments starts from representing a contour as ordered sequence $z(i)$ which elements are the Euclidean distances from the centroid to N contour points. Then, one-dimensional normalized contour sequence moments are derived as follows:

$$m_r = \frac{1}{N} \sum_{i=1}^N [z(i)]^r, \quad \mu_r = \frac{1}{N} \sum_{i=1}^N [z(i) - m_1]^r. \tag{9}$$

The r-th normalized contour sequence moment and normalized central sequence moment are calculated using the following formulas:

$$\bar{m}_r = \frac{m_r}{(\mu_2)^{r/2}}, \quad \bar{\mu}_r = \frac{\mu_r}{(\mu_2)^{r/2}}. \tag{10}$$

The final shape description consists of four values:

$$F_1 = \frac{(\mu_2)^{1/2}}{m_1}, \quad F_2 = \frac{\mu_3}{(\mu_2)^{3/2}}, \quad F_3 = \frac{\mu_4}{(\mu_2)^2}, \quad F_4 = \bar{\mu}_5. \tag{11}$$

4.2   Similarity and Dissimilarity Measures

For matching we have selected standard Euclidean distance as a dissimilarity measure and two correlations measuring similarity—correlation coefficient based on Pearson's correlation and C1 correlation based on L1-norm (introduced in [6]).

Let us take two exemplary vectors $V_A(a_1, a_2, \ldots, A_N)$ and $V_B(b_1, b_2, \ldots, B_N)$ which represent object A and object B in a N-dimensional feature space. The Euclidean distance d_E between these two vectors is defined by means of the following formula [15]:

$$d_E(V_A, V_B) = \sqrt{\sum_{i=1}^{N}(a_i - b_i)^2}. \tag{12}$$

The correlation coefficient may be calculated both for the matrix and vector representations of a shape. The correlation between two matrices can be derived using the formula [9]:

$$c_c = \frac{\sum_m \sum_n (A_{nm} - \bar{A})(B_{nm} - \bar{B})}{\sqrt{\left(\sum_m \sum_n (A_{nm} - \bar{A})^2\right)\left(\sum_m \sum_n (B_{nm} - \bar{B})^2\right)}}, \tag{13}$$

where:

A_{mn}, B_{mn}—pixel value with coordinates (m, n), respectively in image A and B,
$\bar{A}$, $\bar{B}$—average value of all pixels, respectively in image A and B.

The C1 correlation is also a similarity measure based on shape correlation. It is obtained by means of the following formula [6]:

$$c_1(A, B) = 1 - \frac{\sum_{i=1}^{H} \sum_{j=1}^{W} |a_{ij} - b_{ij}|}{\sum_{i=1}^{H} \sum_{j=1}^{W} (|a_{ij}| - |b_{ij}|)}, \tag{14}$$

where:

A, B—matched shape representations,
H, W—height and width of the representation.

5   Experiments and Results

5.1   Data and Conditions

The experiments were carried out with the use of the Weizmann dataset [3]. The original database consists of 90 video sequences (144 × 180 px) recorded at 50 fps. The video sequences are very short, each lasting up to several seconds and differing in the number of frames. We have selected eight action types: 'bend', 'jumping jack', 'jump forward on two legs', 'jump in place on two legs', 'run', 'skip', 'walk' and 'wave one hand'. Exemplary frames from selected video sequences representing actions performed in place are depicted in Fig. 1. Figure 2 shows exemplary frames representing actions with changing location of a silhouette. Binary masks corresponding to all video sequences in the database are available and were used as input data (see Fig. 3 for examples).

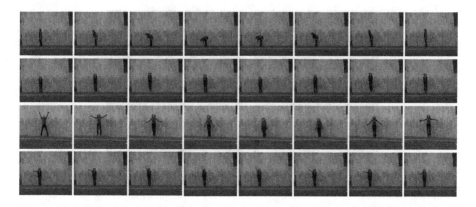

Fig. 1. Exemplary video frames representing actions performed in place (in rows): 'bend', 'jump in place', 'jumping jack' and 'wave one hand' respectively (based on [3]).

The aim of the experiments was to indicate the best result by means of the highest classification accuracy. Action sequences are coarsely classified into two subgroups: actions performed in place ('bend', 'jumping jack', 'jump in place on two legs', 'wave one hand') and actions with changing location of a silhouette ('jump forward on two legs', 'run', 'skip', 'walk'). The following procedure is performed in each subgroup separately. As a single experiment we assume the use of our approach for one shape description algorithm within several tests employing various action representations as well as different matching measures. Thanks to this we can indicate which methods should be used for a specific shape descriptor. Here shapes are represented using Zernike Moments, Moment Invariants or Contour Sequence Moments. Vectors with matching values are transformed into sequence representations using periodogram or Fast Fourier Transform. Both shape descriptors and sequence representations are matched using Euclidean distance, correlation coefficient or C1 correlation. The classification step is based on

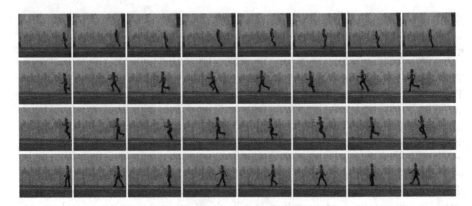

Fig. 2. Exemplary video frames representing actions with changing location of a silhouette (in rows): 'jump', 'run', 'skip' and 'walk' respectively (based on [3]).

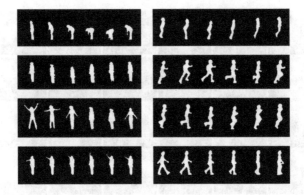

Fig. 3. Exemplary binary masks from the database—left column corresponds to actions presented in Fig. 1 (in rows): 'bend', 'jump in place', 'jumping jack' and 'wave one hand' respectively, and right column corresponds to actions presented in Fig. 2 (in rows): 'jump', 'run', 'skip' and 'walk' respectively (based on [3]).

the leave-one-out cross-validation and template matching. We take each action representation and match it with the rest of representations. Then we select the most similar one which indicates the probable action class. The percentage of correct classifications, averaged for both subgroups, gives final accuracy.

Additionally, for some shape description algorithms it is possible to calculate shape representations of different size, e.g. Zernike Moments of orders from 1st to 12th with representation size varying from 2 to 49 values. We have performed a set of experiments to select the order of Zernike Moments that gives the highest averaged accuracy. The results for orders from 1st to 12th are as follows: 71%, 71%, 73.04%, 63.31%, 64.97%, 62.59%, 64.06%, 60.03%, 66.25%, 62.22%, 70.10% and 61.84% respectively. The highest averaged accuracy can be obtained using moments of 3rd order. Moreover, Zernike Moments of 1st and 2nd order also give good results. Therefore, only these orders are considered for shape representation during experiments.

5.2 Results

The experimental results were grouped according to the applied shape description algorithm. Therefore, we can indicate which combination of techniques is the most effective when a specific shape representation is employed. Table 1 presents the averaged results for the experiment using Zernike Moments of 3rd order. The highest accuracy is 73.04% and is obtained when silhouette descriptors are matched using Euclidean distance, sequence representation is prepared using Fast Fourier Transform and final representations are matched using C1 correlation.

Table 1. Averaged classification accuracy for Zernike Moments of the 3rd order.

Zernike Moments		Silhouettes matched by:		
		Euclidean distance	Correlation Coefficient	C1 correlation
Sequences matched by:				
FFT magnitude	+ Correlation Coefficient	46.42%	44.76%	61.84%
and periodogram	+ C1 correlation	48.83%	37.97%	44.23%
	+ Euclidean distance	44.80%	35.60%	45.51%
periodogram	+ Correlation Coefficient	36.16%	46.04%	49.36%
only	+ C1 correlation	58.18%	23.45%	59.28%
	+ Euclidean distance	44.42%	25.11%	53.39%
FFT magnitude	+ Correlation Coefficient	53.96%	43.67%	47.89%
only	+ C1 correlation	**73.04%**	37.41%	50.64%
	+ Euclidean distance	63.88%	36.12%	52.11%

In case of Moment Invariants (see Table 2) the averaged accuracy reached 65.35% in the experiment using Euclidean distance for shape matching, FFT for action representation and correlation coefficient for action matching. The use of Contour Sequence Moments (see Table 3) did not exceed 52%.

Table 2. Averaged classification accuracy for Moment Invariants.

Moment Invariants		Silhouettes matched by:		
		Euclidean distance	Correlation Coefficient	C1 correlation
Sequences matched by:				
FFT magnitude	+ Correlation Coefficient	**65.52%**	53.96%	55.43%
and periodogram	+ C1 correlation	48.45%	35.97%	51.92%
	+ Euclidean distance	45.70%	44.61%	56.33%
periodogram	+ Correlation Coefficient	50.11%	49.36%	50.83%
only	+ C1 correlation	46.98%	44.42%	41.48%
	+ Euclidean distance	36.35%	40.57%	40.20%
FFT magnitude	+ Correlation Coefficient	65.35%	51.40%	52.49%
only	+ C1 correlation	61.31%	51.40%	53.77%
	+ Euclidean distance	59.84%	52.68%	49.74%

Table 3. Averaged classification accuracy for Contour Sequence Moments.

Contour Sequence Moments		Silhouettes matched by:		
		Euclidean distance	Correlation Coefficient	C1 correlation
Sequences matched by:				
FFT magnitude	+ Correlation Coefficient	44.04%	39.48%	45.32%
and periodogram	+ C1 correlation	45.51%	33.41%	41.86%
	+ Euclidean distance	34.88%	36.16%	34.50%
periodogram	+ Correlation Coefficient	41.67%	44.04%	42.95%
only	+ C1 correlation	45.51%	42.38%	47.17%
	+ Euclidean distance	45.70%	38.35%	49.92%
FFT magnitude	+ Correlation Coefficient	**51.58%**	42.57%	45.70%
only	+ C1 correlation	40.01%	42.38%	48.64%
	+ Euclidean distance	47.17%	38.35%	47.36%

We should take a closer look at the results in subgroups separately. It turned out that it is advised to apply different algorithms in each subgroup. Table 4 contains results for several experiments in which classification accuracy exceeds

Table 4. Results for the experiments with accuracy exceeding 70%.

Shape descriptor	Shape matching	Action representation	Action matching	Averaged accuracy	Actions with changing location	Actions performed in place
Moment invariants	Euclidean distance	FFT magnitude only	Correlation Coefficient	65.35%	51.28%	**79.41%**
Zernike Moments (3rd order)	Euclidean distance	FFT magnitude only	C1 Correlation	**73.04%**	66.67%	**79.41%**
Zernike Moments (1st order)	Euclidean distance	FFT magnitude only	C1 Correlation	71.00%	**74.36%**	67.65%

Table 5. Classification quality measures for the best experiments.

Shape descriptor	Subgroup	Class	Precision	Recall
Zernike Moments (1st order)	Actions with changing location	'jump forward'	0.71	0.56
		'run'	0.50	0.70
		'skip'	0.67	0.60
		'walk'	1.00	0.90
Zernike Moments (3rd order)	Actions performed in place	'bend'	0.90	1.00
		'jumping jack'	0.67	0.75
		'jump in place'	0.64	0.78
		'wave one hand'	1.00	0.5

70%. Considering as small shape representation as possible we can indicate the use of Zernike Moments of 3rd order for actions performed in place and Zernike Moments of 1st order for the other subgroup. In addition, we present classification quality measures for these two best experiments (see Table 5), including standard precision and recall for each class.

In Sect. 3 we have listed several changes introduced in our approach, and based on the experimental results we can conclude that the selection of matching measure strongly affects the accuracy. Secondly, an additional coarse classification step increased overall classification quality despite more action classes.

Thirdly, the application of a new experimental procedure helped to avoid a situation in which the results depend on the set of templates (class representatives). However, there are some classes which are not classified precisely. For instance, running is often confused with jumping or skipping. To improve the results, using a different shape descriptor can be considered or adding another feature that will distinguish between problematic actions.

6   Conclusions

The paper covered the topic of action recognition based on shape features. The presented approach uses binary silhouettes, shape description algorithms and matching techniques to classify action sequences. We represent each silhouette using selected shape descriptor, match all descriptors of a single sequence and put matching values into a vector. Then we transform each vector into frequency domain and classify. We use additional step of coarse classification based on centroid location and perform experiments in each subgroup separately. We have experimentally tested various combinations of the following: shape description algorithms (Zernike Moments, Moment Invariants and Contour Sequence Moments), matching measures (Euclidean distance, correlation coefficient and C1 correlation) and frequency domain techniques (Fast Fourier Transform and periodogram).

The best results are obtained when we use a combination of Zernike Moments for shape representation, Euclidean distance for shape matching, Fast Fourier Transform for action representation and C1 correlation for action classification. The highest averaged accuracy was 73.04% for Zernike Moments of 3rd order— 79.41% for actions performed in place and 66.67% for actions with changing location. Moreover, for the second subgroup better results can be obtained by using the Zernike Moments of 1st order instead of 3rd order. Then accuracy equals 74.36% and shape representation is smaller.

References

1. Al-Ali, S., Milanova, M., Al-Rizzo, H., Fox, V.L.: Human action recognition: contour-based and silhouette-based approaches. In: Favorskaya, M.N., Jain, L.C. (eds.) Computer Vision in Control Systems-2. ISRL, vol. 75, pp. 11–47. Springer, Cham (2015). https://doi.org/10.1007/978-3-319-11430-9_2
2. Baysal, S., Kurt, M.C., Duygulu, P.: Recognizing human actions using key poses. In: 2010 20th International Conference on Pattern Recognition, pp. 1727–1730 (2010). https://doi.org/10.1109/ICPR.2010.427
3. Blank, M., Gorelick, L., Shechtman, E., Irani, M., Basri, R.: Actions as space-time shapes. In: Proceedings of the Tenth IEEE International Conference on Computer Vision, ICCV 2005, vol. 2, pp. 1395–1402. IEEE Computer Society, Washington, DC (2005). https://doi.org/10.1109/ICCV.2005.28

4. Bobick, A.F., Davis, J.W.: The recognition of human movement using temporal templates. IEEE Trans. Pattern Anal. Mach. Intell. **23**(3), 257–267 (2001). https://doi.org/10.1109/34.910878

5. Borges, P.V.K., Conci, N., Cavallaro, A.: Video-based human behavior understanding: a survey. IEEE Trans. Circ. Syst. Video Technol. **23**(11), 1993–2008 (2013). https://doi.org/10.1109/TCSVT.2013.2270402

6. Brunelli, R., Messelodi, S.: Robust estimation of correlation with applications to computer vision. Pattern Recogn. **28**(6), 833–841 (1995). https://doi.org/10.1016/0031-3203(94)00170-Q

7. Chaaraoui, A.A., Climent-Pérez, P., Flórez-Revuelta, F.: Silhouette-based human action recognition using sequences of key poses. Pattern Recogn. Lett. **34**(15), 1799–1807 (2013). https://doi.org/10.1016/j.patrec.2013.01.021

8. Liu, C.-B., Ahuja, N.: Vision based fire detection. In: 2004 Proceedings of the 17th International Conference on Pattern Recognition, ICPR 2004, vol. 4, pp. 134–137 (2004). https://doi.org/10.1109/ICPR.2004.1333722

9. Chwastek, T., Mikrut, S.: The problem of automatic measurement of fiducial mark on air images (in Polish). Arch. Photogramm. Cartogr. Remote Sens. **16**, 125–133 (2006)

10. Gorelick, L., Blank, M., Shechtman, E., Irani, M., Basri, R.: Actions as space-time shapes. IEEE Trans. Pattern Anal. Mach. Intell. **29**(12), 2247–2253 (2007). https://doi.org/10.1109/TPAMI.2007.70711

11. Gościewska, K., Frejlichowski, D.: Moment shape descriptors applied for action recognition in video sequences. In: Nguyen, N.T., Tojo, S., Nguyen, L.M., Trawiński, B. (eds.) ACIIDS 2017. LNCS (LNAI), vol. 10192, pp. 197–206. Springer, Cham (2017). https://doi.org/10.1007/978-3-319-54430-4_19

12. Goudelis, G., Karpouzis, K., Kollias, S.: Exploring trace transform for robust human action recognition. Pattern Recogn. **46**(12), 3238–3248 (2013). https://doi.org/10.1016/j.patcog.2013.06.006

13. Hupkens, T., de Clippeleir, J.: Noise and intensity invariant moments. Pattern Recogn. Lett. **16**(4), 371–376 (1995). https://doi.org/10.1016/0167-8655(94)00110-O

14. Junejo, I.N., Junejo, K.N., Aghbari, Z.A.: Silhouette-based human action recognition using SAX-Shapes. Vis. Comput. **30**(3), 259–269 (2013). https://doi.org/10.1007/s00371-013-0842-0

15. Kpalma, K., Ronsin, J.: An overview of advances of pattern recognition systems in computer vision. In: Obinata, G., Dutta, A. (eds.) Vision Systems, Chap. 10. IntechOpen, Rijeka (2007). https://doi.org/10.5772/4960

16. Liu, L., Shao, L., Zhen, X., Li, X.: Learning discriminative key poses for action recognition. IEEE Trans. Cybern. **43**(6), 1860–1870 (2013). https://doi.org/10.1109/TSMCB.2012.2231959

17. Poppe, R.: A survey on vision-based human action recognition. Image Vis. Comput. **28**(6), 976–990 (2010). https://doi.org/10.1016/j.imavis.2009.11.014

18. Rothe, I., Susse, H., Voss, K.: The method of normalization to determine invariants. IEEE Trans. Pattern Anal. Mach. Intell. **18**(4), 366–376 (1996). https://doi.org/10.1109/34.491618

19. Sonka, M., Hlavac, V., Boyle, R.: Image Processing, Analysis, and Machine Vision. Thomson-Engineering (2007)

20. Vishwakarma, S., Agrawal, A.: A survey on activity recognition and behavior understanding in video surveillance. Vis. Comput. **29**(10), 983–1009 (2013). https://doi.org/10.1007/s00371-012-0752-6
21. Vishwakarma, D., Dhiman, A., Maheshwari, R., Kapoor, R.: Human motion analysis by fusion of silhouette orientation and shape features. Procedia Comput. Sci. **57**, 438–447 (2015). https://doi.org/10.1016/j.procs.2015.07.515
22. Zhang, D., Lu, G.: Shape-based image retrieval using generic Fourier descriptor. Sig. Process. Image Commun. **17**(10), 825–848 (2002). https://doi.org/10.1016/S0923-5965(02)00084-X

Improving Accuracy and Speeding Up Document Image Classification Through Parallel Systems

Javier Ferrando[1]([envelope]), Juan Luis Domínguez[1], Jordi Torres[1,2], Raúl García[1], David García[1], Daniel Garrido[3], Jordi Cortada[4], and Mateo Valero[1,2]

[1] Barcelona Supercomputing Center - Centro Nacional de Supercomputación, Barcelona, Spain
{javier.ferrando,juan.dominguez,jordi.torres,raul.garcia,david.garcia2, mateo.valero}@bsc.es
[2] Universitat Politècnica de Catalunya, UPC-BarcelonaTech, Barcelona, Spain
daniel_garrido@serimagmedia.com
[3] Serimag Media - TAAD, Barcelona, Spain
[4] CaixaBank, Valencia, Spain
jorge.cortada@caixabank.com

Abstract. This paper presents a study showing the benefits of the EfficientNet models compared with heavier Convolutional Neural Networks (CNNs) in the Document Classification task, essential problem in the digitalization process of institutions. We show in the RVL-CDIP dataset that we can improve previous results with a much lighter model and present its transfer learning capabilities on a smaller in-domain dataset such as Tobacco3482. Moreover, we present an ensemble pipeline which is able to boost solely image input by combining image model predictions with the ones generated by BERT model on extracted text by OCR. We also show that the batch size can be effectively increased without hindering its accuracy so that the training process can be sped up by parallelizing throughout multiple GPUs, decreasing the computational time needed. Lastly, we expose the training performance differences between PyTorch and Tensorflow Deep Learning frameworks.

Keywords: Document image classification · Deep learning · Parallel systems · EfficientNet · BERT · Scalability · TensorFlow · PyTorch

1 Introduction

Document digitization has become a common practice in a wide variety of industries that deal with vast amounts of archives. Document classification is a task to face when trying to automate their document processes, but high intra-class and low inter-class variability between documents have made this a challenging problem.

© Springer Nature Switzerland AG 2020
V. V. Krzhizhanovskaya et al. (Eds.): ICCS 2020, LNCS 12138, pp. 387–400, 2020.
https://doi.org/10.1007/978-3-030-50417-5_29

First attempts focused on structural similarity between documents [40] and on feature extraction [12,24,30] to differentiate characteristics of each class. The combination of both approaches has also been tested [14].

Several classic machine learning techniques have been applied to these problem, i. e. K-Nearest Neighbor approach [7], Hidden Markov Model [19] and Random Forest Classifier [24,29] while using SURF local descriptors before the Convolutional Neural Networks (CNNs) came into scene.

With the rise of Deep Learning, researchers have tried deep neural networks to improve the accuracy of their classifiers. CNNs have been proposed in past works, initially in 2014 by Kang *et al.* [26] who started with a simple 4-layer CNN trained from scratch. Then, transfer learning was demonstrated to work effectively [1,21] by using a network pre-trained on ImageNet [17]. And latest models have become increasingly heavier (greater number of parameters) [2,16, 46] as shown in Table 1, with the speed and computational resources drawback this entails.

Recently, textual information has been used by itself or as a combination together with visual features extracted by the previously mentioned models. Although Optical Character Recognition (OCR) is prone to errors, particularly when dealing with handwritten documents, the use of modern Natural Language Processing (NLP) techniques have demonstrated a boost in the classifiers performance [5,6,35].

The contributions of this paper can be summarized in two main topics:

- Algorithmic performance: we propose a model and a training procedure to deal with images and text that outperforms the state-of-the- art in several settings and is lighter than any previous neural network used to classify the BigTobacco dataset, the most popular benchmark for Document Image Classification (Table 1).
- Training process speed up: we demonstrate the ability of these models to maintain their performance while saving a large amount of time by parallelizing over several GPUs. We also show the performance differences between the two most popular Deep Learning frameworks (TensorFlow and Pytorch), when using their own libraries dedicated to this task.

2   Document Image Classification

Document Image Classification task tries to predict the class which a document belongs to by means of analyzing its image representation. This challenge can be tackled in two ways, as an image classification problem and as a text classification problem. The former tries to look for patterns in the pixels of the image to find elements such as shapes or textures that can be associated to a certain class. The latter tries to understand the language written in the document and relate this to the different classes.

2.1   Datasets

As mentioned earlier, in this work we make use of two publicly available datasets containing samples of images from scanned documents from USA Tobacco companies, published by Legacy Tobacco Industry Documents and created by the University of California San Francisco (UCSF). We find these datasets a good representation of what enterprises and institutions may face with, based on the quality and type of classes. Furthermore, they have been go-to datasets in this research field since 2014 with which we can compare results.

RVL-CDIP (Ryerson Vision Lab Complex Document Information Processing) is a 400.000 document sample (BigTobacco from now onwards) presented in [21] for document classification tasks. This dataset contains the first page of each of the documents, which are labeled in 16 different classes with equal number of elements per class. A smaller sample containing 3482 images was proposed in [24] as Tobacco3482 (SmallTobacco henceforth). This dataset is formed by documents belonging to 10 classes not uniformly distributed.

Table 1. Parameters of the CNNs architectures used in BigTobacco.

Model	#Params
AlexNet	60.97M
VGG-16	138.36M
ResNet-50	25.56M
Inception-V3	23.83M
EfficientNet-B2	**9.2M**
EfficientNet-B0	**5.3M**

2.2   Deep Learning

The proposed methods in this work are based on supervised Deep Learning, where each document is associated to a class (label) so that the algorithms are trained by minimizing the error between the predictions and the truth. Deep Learning is a branch of machine learning that deals with deep neural networks, where each of the layers is trained to extract higher level representations of the previous ones. These models are trained by solving iteratively an unconstrained optimization problem. In each iteration, a random batch of the training data is fed into the model to compute the loss function value. Then, the gradient of the loss function with respect to the weights of the network is computed (backpropagation) and an update of the weights in the negative direction of the gradient is done. These networks are trained until they converge into a loss function minimum.

2.3   Computer Vision

The field where machines try to get an understanding of visual data is known as Computer Vision (CV). One of the most well-known tasks in CV is image classification. In 2010 The ImageNet Large Scale Visual Recognition Challenge (ILSVRC) was introduced, a competition that dealt with a 1.2 million images dataset belonging to 1000 classes. In 2012 the first CNN-based model significantly reduced the error rate, setting the beginning of the explosion of deep neural networks. From then onwards, deeper networks have become the norm.

The most used architecture in Computer Vision have been CNN-based networks. Their main operation is the convolution one, which consists on a succession of dot products between the vector representations of both the input space $(L_q \times B_q \times d_q)$ and the filters $(F_q \times F_q \times d_q)$. We slide each filter around the input volume getting an *activation map* of dimension $L_{q+1} = (L_q - F_q + 1)$ and $B_{q+1} = (B_q - F_q + 1)$. The output volume then has a dimension of $L_{q+1} \times B_{q+1} \times d_{q+1}$, where d_{q+1} refers to the number of filters used. We refer to [3] (we used the same notation for simplicity) to a more detailed explanation. Usually, each convolution layer is associated to an activation layer, where an activation function is applied to the whole output volume. To reduce the number of parameters of the network, a pooling layer is typically located between convolution operations. The pooling layer takes a region $P_q \times P_q$ in each of the d_q activation maps and performs an arithmetic operation. The most used pooling layer is the max-pool, which returns the maximum value of the aforementioned region.

2.4   Natural Language Processing

The features learned from the OCR output are achieved by means of Natural Language Processing techniques. NLP is the field that deals with the understanding of human language by computers, which captures underlying meanings and relationships between words.

The way machines deal with words is by means of a real values vector representation. Word2Vec [34] showed that a vector could represent semantic and syntactic relationships between words. CoVe [32] introduced the concept of context-based embeddings, where the same word can have a different vector representation depending on the surrounding text. ELMo [36] followed Cove but with a different training approach, by predicting the next word in a text sequence (Language Modelling), which made it possible to train on large available text corpus. Depending on the task (such as text classification, named entity recognition...) the output of the model can be treated in different ways. Moreover, custom layers can be added to the features extracted by these NLP models. For instance, ULM-Fit [23] introduced a language model and a fine-tuning strategy to effectively adapt the model to various downstream tasks, which pushed transfer learning in the NLP field. Lately, the Transformer architecture [47] has dominated the scene, being the bidirectional Transformer encoder (BERT) [18] the one who established recently state-of-the-art results over several downstream tasks.

3   Related Work

Several ways of measuring models have been shown in the past years regarding document classification on the Legacy Tobacco Industry Documents [31]. Some authors have tested their models on a large-scale sample BigTobacco. Others tried on a smaller version named SmallTobacco, which could be seen as a more realistic scale of annotated data that users might be able to find. Lastly, transfer learning from in-domain datasets has been tested by using BigTobacco to pre-train the models to finally fine-tune on SmallTobacco. Table 2 summarizes the results of previous works in the different categories over time.

First results in the Deep Learning era have been mainly based on CNNs using transfer learning techniques. Multiple networks were trained on specific sections of the documents [21] to learn region-based high dimensional features later compressed via Principal Component Analysis (PCA). The use of multiple Deep Learning models was also exploited by Das *et al.* by using an ensemble as a meta-classifier [16]. A VGG-16 [41] stack of networks using 5 different classifiers has been proposed, one of them trained on the full document and the others specifically over the header, footer, left body and right body. The Multi Layer Perceptron (MLP) was the ensemble that performed the better. A committee of models but with a SVM as the ensemble was also proposed [37].

Table 2. Previous results comparison (accuracy in %).

Author	BigTobacco	SmallTobacco				
		BigTobacco pre-training		No pre-training		
	Image	Image	Image + Text	Image	Image + Text	
Kumar et al. [24]				43.8		
Kang et al. [26]				65.37		
Afzal et al. [1]				77.6		
Harley et al. [21]	89.8			79.9		
Csurka et al. [15]	90.7					
Noce et al. [35]					79.8	
Afzal et al. [2]	90.97	91.13				
Tensmeyer et al. [46]	90.8					
Das et al. [16]	92.21					
Audebert et al. [6]				84.5	87.8	
Asim et al. [5]		93.2[a]	95.8[b]			
Proposed work (2020)	**92.31**	**94.04**	94.9	**85.99**	**89.47**	

[a] Accuracy obtained in 9 classes that overlap in BigTobacco
[b] Evaluation method not specified

The addition of content-based information has been investigated on Small-Tobacco by extracting text through OCR and embedding the obtained features into the original document images as a previous phase to the training process [35]. Lately, a MobilenetV2 architecture [38] together with a CNN 2D [27,49] taking as input FastText embeddings [9,25] have achieved the best results in SmallTobacco [6].

A study of several CNNs was carried out [2], where VGG-16 architecture was found optimal. Afzal *et al.* also demonstrated that transfer learning from in-domain datataset like BigTobacco increases by a large margin the results in SmallTobacco. This was further investigated by adding content-based information with CNN 2D with ranking textual features (ACC2) to the OCR extracted.

As far as we are concerned, there is no study about the use of multiple GPUs in the training process for the task of Document Image Classification. However, parallelizing a computer vision task has been shown to work properly using ResNet-50, which is a widely used network that usually gives good results despite its low complexity architecture. Several training procedures are demonstrated to work effectively with this model [4,20]. A learning rate value proportional to the batch size, warmup learning rate behaviour, batch normalization, SGD to RMSProp optimizer transition are some of the techniques exposed in these works. A study of the distributed training methods using ResNet-50 architecture on a HPC cluster is shown in [10,11]. To know more about the algorithms used in this field we refer to [8].

4   Proposed Approach

In this section we present the models used and a brief explanation of them. We also show the training procedure used in both BigTobacco and SmallTobacco and the pipeline of our approach to the problem.

4.1   Image Model

EfficientNets [45] are a set of light CNNs designed to scale up in a structured manner. The network's width (w), depth (d) and resolution (r) are defined as: $w = \alpha^\phi$, $d = \beta^\phi$ and $r = \gamma^\phi$, where ϕ is the scaling compound coefficient. The optimization problem is set by constraining $\alpha \cdot \beta^2 \cdot \gamma^2 \approx 2$ and $\alpha \geq 1, \beta \geq 1, \gamma \geq 1$.

By means of a grid search of α, β, γ with AutoML MNAS framework [44] and fixing $\phi = 1$, a baseline model (B0) is generated optimizing FLOPs and accuracy. Then, the baseline network is scaled up uniformly fixing α, β, γ and increasing ϕ. We find that scaling the resolution parameter as proposed in [45] does not improve the accuracy obtained. In our experiments in Sect. 5 we proceed with an input image size of 384×384, which corresponds to a resolution $r = 1.71$, as proposed by Tensmeyer *et al.* in [46] with AlexNet architecture [28].

The main block of the EfficientNets is the mobile inverted bottleneck convolution [38,44]. This block is formed by two linear bottlenecks connected through both a shortcut connection and an intermediate expansion layer with a depthwise separable convolution (3×3) [13]. Probabilities $P(class|FC)$ are obtained by applying the softmax function on top of the fully connected layer FC of the EfficientNet model.

Pre-training on BigTobacco. We train EfficientNets (pre-trained previously on ImageNet) on BigTobacco using Stochastic Gradient Descent for 20 epochs with *Learning Rate Warmup* strategy [22], specifically we follow *STLR* (Slanted Triangular Learning Rate) [23] which linearly increases the learning rate at the beginning of the training process and linearly decreases it after a certain number of iterations. We chose the *reference learning rate* η following the formula proposed in [20] and used in [4] and [22]. Specifically, we set $\eta = 0.2 \cdot \frac{nk}{256}$, where k denotes the number of workers (GPUs) and n the number of samples per worker. Figure 1 shows the multi-GPU training procedure to get EfficientNet$_{BigTobacco}$, which represents EfficientNet model pre-trained on BigTobacco. EfficientNet is loaded with ImageNet weights (EfficientNet$_{ImageNet}$) and then located in different GPUs within the same node.

Fine-Tuning on SmallTobacco. We fine-tune on SmallTobacco the pre-trained models by freezing the entire network but the last softmax layer. Just 5 epochs are enough to get the peak of accuracy. *STLR* is used this time with $\eta = 0.8 \cdot \frac{nk}{256}$. Since only the last layer is trained, we reduce the risk of catastrophic forgetting [33]. Final fine-tuned model is represented as EfficientNet$_{BigTobacco}$ in Fig. 1.

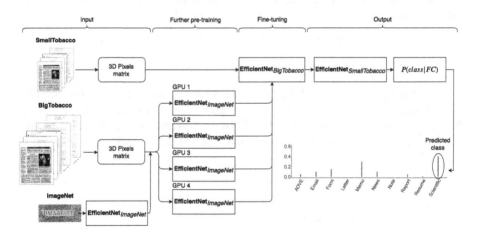

Fig. 1. Pipeline of the different stages of the pre-training of EfficientNet over multiple GPUs.

4.2 Text Model

Predictions from OCR Tesseract [42] are obtained by means of the BERT model [18]. BERT is a multi-layer bidirectional Transformer encoder model pre-trained on a large corpus. In this work we use a modification of the original pre-trained BERT$_{BASE}$ version. In our case, we reduce to 6 the number of BERT layers since we find less variance in the final results and faster training/inference times. The output vector size is kept to 768. The maximum length of the input sequence is set to 512 tokens. The first token of the sequence is defined as $[CLS]$, while $[SEP]$ is the token used at the end of each sequence.

A fully connected layer is added to the final hidden state of the $[CLS]$ token $h_{[CLS]}$ of the BERT model, which is a representation of the whole sequence. Then, a softmax operation is performed giving $P(class|h_{[CLS]})$ the probabilities of the output vector $h_{[CLS]}$, i.e the whole input sequence, pertaining to a certain *class*.

The training strategies used in this paper are similar to the ones proposed in [43,48]. We use a learning rate $\eta_B = 3e^{-5}$ for the embedding, pooling and encoder layers while a custom learning rate $\eta_C = 1e^{-6}$ for the layers on top of the BERT model. A decay factor $\xi = 1e^{-8}$ is used to reduce gradually the learning rate along the layers, $\eta^l = \xi \cdot \eta^{l-1}$. ADAM optimizer with $\beta_1 = 0.9$ and $\beta_2 = 0.999$ and L_2-weight decay factor of 0.01 is used. The dropout probability is set at 0.2. Just 5 epochs are enough to find the peak of accuracy with a batch size of 6, the maximum we could use due to memory constraints.

4.3 Image and Text Ensemble

In order to get the final enhanced prediction of the combination of both text and image model we use a simple ensemble as in [5].

$$P(class|out_{image}, out_{text}) = w_1 \cdot P(class|h_{[CLS]}) + w_2 \cdot P(class|FC)$$

$$Predicted\ Class = \arg \max_{class}(P(class|out_{image}, out_{text}))$$

In this work $w_1, w_2 = 0.5$ are found optimal. These parameters could be found by a grid search where $\sum_{i=1}^{N} w_i = 1$, being N the number of models. This procedure shows to be an effective solution when both models have a similar accuracy and it allows us to avoid another training phase [6]. In Fig. 2 this whole process is depicted.

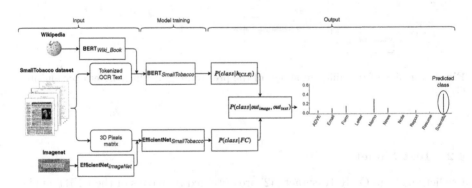

Fig. 2. Pipeline of the proposed multimodal approach.

5 Results

In this section we compare the performance of the different EfficientNets in SmallTobacco and BigTobacco as showed in Table 2 and demonstrate the benefits of the multiple GPU training. Experiments have been carried out using GPUs clusters Power-CTE[1] of the Barcelona Supercomputing Center - Centro Nacional de Supercomputación[2], each one composed by: 2 IBM Power9 8335-GTGH at 2.40 GHz (20 cores and 4 threads/core), 512 GB of main memory distributed in 16 dimms × 32 GB at 2666 MHz and 4 GPU NVIDIA V100 (Volta) with 16 GB HBM2.

The operating system is RedHat Linux 7.4. The models and their training are implemented with PyTorch[3] version 1.0 running on CUDA 10.1 and using cuDNN 7.6.4.

The only modification done to the images is a resize to 384×384 as explained in Sect. 4.1 and, in order to avoid overfitting, a shear transformation of an angle $\theta \in [-5°, 5°]$ [46] which is randomly applied in the training phase. No other modifications are used in our experiments. Source code is at https://javiferran. github.io/document-classification.

5.1 Evaluation

In order to compare with previous results in SmallTobacco dataset, we divide the dataset following the procedure in [24]. Documents are split in training, test and validation sets, containing 800, 2482 and 200 samples each one. 10 different splits of the dataset are created by randomly sampling from the 3482 documents, so that 100 samples per class are guaranteed between train and validation sets. In the Fig. 4 we give the accuracy on SmallTobacco as the median over the 10 dataset splits to compare with previous results. Accuracy on BigTobacco is shown as the one achieved on the test set. BigTobacco dataset used in Sect. 5.3 is slightly modified, where overlapping documents with SmallTobacco are extracted. Top performing model's accuracies are written down in Table 2.

5.2 Results on BigTobacco

We show in Fig. 3 the time it takes to train the different networks while using 1, 2, 3 or 4 GPUs in a single node. In order to take advantage of the multiple GPUs we use data parallelism, which consists of placing a copy of the model in each of them. Since every GPU share parameters, it is equivalent to having a single GPU with a larger batch size.

The time reduction to complete the entire training process with B0 variant is ≈61.14% lower when compared with B4 (4 GPUs). Time reduction by using multiple GPUs is clearly showed in the left plot of Fig. 3. For instance,

---

[1] https://www.bsc.es/support/POWER_CTE-ug.pdf.

[2] https://www.bsc.es.

[3] https://pytorch.org/.

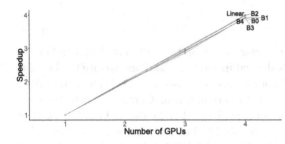

Model \ GPUs	1	2	3	4
B0	13.33	6.81	4.58	3.4
B1	19.44	9.81	6.81	4.94
B2	20.55	10.22	6.92	5.16
B3	25.64	12.94	8.78	6.55
B4	34.28	17.36	11.69	8.75

Fig. 3. Left: speedup of the training process when parallelizing. Right: total time (hours) to train each model on different number of GPUs.

EfficientNet-B0 benefits from a ≈75.4% time reduction after parallelizing over 4 GPUs. The total training time of the EfficientNets on the different number of GPUs is showed in the right side of Fig. 3. The best performing model in BigTobacco dataset is EfficienNet-B4 with 92.31% accuracy in the test set.

5.3   Results on SmallTobacco

Accuracies of the EfficientNets pre-trained on BigTobacco and finally fine-tuned on SmallTobacco are depicted in the left plot of Fig. 4. Simpler models perform with less variability between the 10 random splits than the heavier ones. The best performing model is the EfficientNet-B1, achieving a new state-of-art accuracy of 94.04% median over 10 splits.

In this work, we also wanted to test the potential of light EfficientNet models on a small dataset such as SmallTobacco without the use of transfer learning from in-domain dataset, and compared it with the previous state-of-the-art. Results given by our proposed method described in Sect. 4.3 are shown in the right plot of Fig. 4. Although we perform the tests over 10 different random splits to give

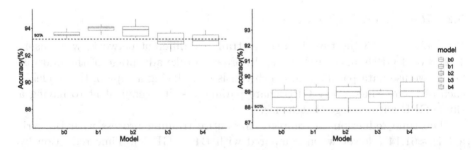

Fig. 4. Accuracy obtained in SmallTobacco by models pre-trainined on BigTobacco (Left) and without BigTobacco pre-training (Right). Previous state-of-the-art (SOTA) results are shown with a horizontal dashed line.

a wider view of how these models work, in order to compare with *Audebert et al.* [6] we calculate the average over 3 random splits, which gives us a 89.47% accuracy.

Every ensemble model achieves better accuracy than previous results, and again, there is almost no difference between different EfficientNets results.

5.4   Parallel Platforms

Single GPU training requires a huge amount of time, especially when dealing with heavy architectures like in the case of the EfficientNet-B4, which takes almost two days to complete the whole training phase. For this reason, experimenting with several workers is crucial to minimize the amount of time spent on this tasks. We test the same model and training procedure with two of the main used frameworks to train Deep Learning models, PyTorch and Tensorflow[4]. In both cases we use their own APIs for making a synchronous distributed training in several GPUs by means of data parallelism, where training on each GPU is done in its own process. We use PyTorch's DistributedDataParallel and Tensorflow's tf.distribute.Strategy (tf.distribute.MirroredStrategy). In both libraries data is loaded from the disk to page-locked memory in each host, and from there to each GPU in a parallel fashion by means of multiple workers. Each GPU is ensured to get a minibatch with non overlapping data. Every GPU has an identical copy of the model and each one does its own forward pass. Finally, NCCL is utilized as a backend to run the all-reduce algorithm to compute the gradients in parallel between GPUs, before updating the model parameters. Since we have not been able to apply the shear transformation efficiently in Tensorflow, we show the results of both frameworks without that preprocess. For this experiment we use the B0, B2 and B4 EfficientNets models. The time it takes to train each model is showed on the left side of Fig. 5. PyTorch training is faster and the speedup more linear than in the case of TensorFlow. Some of this difference

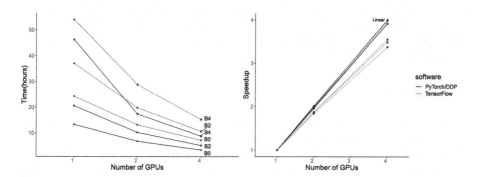

Fig. 5. Left: the time to complete a whole training process. Right: speedup curves of TensorFlow and PyTorch.

---

[4] https://www.tensorflow.org/.

could be due to the data loading process, which we have not fully optimized in TensorFlow framework.

6   Conclusion

In this paper we have presented the use of EfficientNets for the Document Image Classification task and their scaling capabilities through several GPUs. By means of two versions of the Legacy Tobacco Industry Documents, a huge and a small dataset, we demonstrated the training process to obtain high accuracy in both of them. We have compared the different versions of the EfficientNets and raised the state-of-the-art classification accuracy to 92.31% in BigTobacco and 94.04% when fine-tuned in SmallTobacco. We can consider the B0 the best choice when considering limited computational resources. We have also presented an ensemble method by adding the content extracted by OCR. A reduced version of the BERT model is trained and both models predictions are combined to achieve a new state-of-the-art accuracy of 89.47%.

Finally, we have tested the same image models and training procedures in Tensorflow and PyTorch, where we have observed similar speedup values exploiting their libraries for distributed training. We have also tried distributed training in several GPU nodes by means Horovod framework [39], however the stack of software in our IBM Power 9 cluster is still in its early stages and we have not been able to obtain desired results. Nevertheless, future work may focus in testing this approach.

Future work may also evaluate the use of different OCR engines, as we suspect this could have a great impact on the quality of the text model predictions.

With this work we also want to provide to researchers a benchmark in the Document Image Classification task, which can serve as a reference point to effortlessly test parallel systems in both PyTorch and TensorFlow.

Acknowledgements. This work was partially supported by the Spanish Ministry of Science and Innovation and the European Regional Development Fund under contract TIN2015-65316-P, by the BSC-CNS Severo Ochoa program SEV-2015-0493, and grant 2017-SGR-1414 by Generalitat de Catalunya and by the research agreement CaixaBank-BSC 2016–2021.

References

1. Afzal, M.Z., et al.: DeepDocClassifier: document classification with deep convolutional neural network. In: ICDAR, p. 1273–1278 (2015)
2. Afzal, M.Z., Kölsch, A., Liwicki, M., Ahmed, S.: Cutting the error by half: investigation of very deep CNN and advanced training strategies for document image classification. In: ICDAR (2017)
3. Aggarwal, C.C.: Neural Networks and Deep Learning: A Textbook. Springer, Cham (2018). https://doi.org/10.1007/978-3-319-94463-0
4. Akiba, T., Suzuki, S., Fukuda, K.: Extremely large minibatch SGD: Training ResNet-50 on ImageNet in 15 minutes. arXiv preprint arXiv:1711.04325 (2017)

5. Asim, M.N., Khan, M.U.G., Malik, M.I., Razzaque, K., Dengel, A., Ahmed, S.: Two stream deep network for document image classification. In: ICDAR (2019)
6. Audebert, N., Herold, C., Slimani, K., Vidal, C.: Multimodal deep networks for text and image-based document classification. In: APIA (2019)
7. Baldi, S., Marinai, S., Soda, G.: Using tree-grammars for training set expansion in page classification. In: ICDAR (2003)
8. Ben-Nun, T., Hoefler, T.: Demystifying parallel and distributed deep learning: an in-depth concurrency analysis. ACM Comput. Surv. **12** (2019). https://dl.acm.org/doi/abs/10.1145/3320060. Article no 65
9. Bojanowski, P., Grave, E., Joulin, A., Mikolov, T.: Enriching word vectors with subword information. Trans. Assoc. Comput. Linguist. (TACL) **5**, 135–146 (2017)
10. Campos, V., Sastre, F., Yagues, M., Torres, J., Giró-i-Nieto, X.: Scaling a convolutional neural network for classification of adjective noun pairs with TensorFlow on GPU clusters. In: CCGRID, pp. 677–682 (2017)
11. Campos, V., Sastre, F., Yagues, M., Torres, J., Bellver, M., Giró-i-Nieto, X.: Distributed training strategies for a computer vision deep learning training algorithm on a distributed GPU cluster. In: ICCS, pp. 315–324 (2017)
12. Chen, S., He, Y., Sun, J., Naoi, S.: Structured document classification by matching local salient features. In: ICPR, pp. 1558–1561 (2012)
13. Chollet, F.: Xception: deep learning with depthwise separable convolutions. In: CVPR (2017)
14. Collins-Thompson, K., Nickolov, R.: A clustering-based algorithm for automatic document separation. In: SIGIR, pp. 1–8 (2002)
15. Csurka, G., Larlus, D., Gordo, A., Almazan, J.: What is the right way to represent document images? arXiv preprint arXiv:1603.01076 (2016)
16. Das, A., Roy, S., Bhattacharya, U., Parui, S.K.: Document image classification with intra-domain transfer learning and stacked generalization of deep convolutional neural networks. In: ICDAR (2018)
17. Deng, J., Dong, W., Socher, R., Li, L.J., Li, K., Li, F.F.: ImageNet: a large-scale hierarchical image database. In: CVPR, pp. 248–255, June 2009
18. Devlin, J., Chang, M.W., Lee, K., Toutanova, K.: BERT: pre-training of deep bidirectional transformers for language understanding. In: NAACL (2019)
19. Diligenti, M., Frasconi, P., Gori, M.: Hidden tree Markov models for document image classification. Trans. Pattern Anal. Mach. Intell. (TPAMI) **25**(4), 519–523 (2003)
20. Goyal, P., et al.: Accurate, large minibatch SGD: Training ImageNet in 1 hour. CoRR, abs/1706.02677 (2017)
21. Harley, A.W., Ufkes, A., Derpanis, K.G.: Evaluation of deep convolutional nets for document image classification and retrieval. In: Proceedings of the ICDAR 2015, pp. 991–995. IEEE (2015)
22. He, T., Zhang, Z., Zhang, H., Zhang, Z., Xie, J., Li, M.: Bag of tricks for image classification with convolutional neural networks. arXiv preprint arXiv:1812.01187 (2018)
23. Howard, J., Ruder, S.: Universal language model fine-tuning for text classification. Assoc. Comput. Linguist. **1**, 328–339 (2018)
24. Kumar, J., Ye, P., Doermann, D.: Structural similarity for document image classification and retrieval. Pattern Recogn. Lett. **43**, 119–126 (2014)
25. Joulin, A., Grave, E., Bojanowski, P., Mikolov, T.: Bag of tricks for efficient text classification. arXiv preprint arXiv:1607.01759 (2016)
26. Kang, L., Kumar, J., Ye, P., Li, Y., Doermann, D.: Convolutional neural networks for document image classification. In: ICPR, pp. 3168–3172 (2014)

27. Kim, Y.: Convolutional neural networks for sentence classification. In: EMNLP (2014)
28. Krizhevsky, A., Sutskever, I., Hinton, G.E.: ImageNet classification with deep convolutional neural networks. In: Advances in Neural Information Processing Systems (2012)
29. Kumar, J., Doermann, D.S.: Unsupervised classification of structurally similar document images. In: ICDAR, pp. 1225–1229 (2013)
30. Kumar, J., Ye, P., Doermann, D.S.: Learning document structure for retrieval and classification. In: ICPR, pp. 653–656 (2012)
31. Lewis, D., Agam, G., Argamon, S., Frieder, O., Grossman, D., Heard., J.: Building a test collection for complex document information processing. In: SIGIR, pp. 665–666 (2006)
32. McCann, B., Bradbury, J., Xiong, C., Socher, R.: Learned in translation: contextualized word vectors. In: NeurIPS, pp. 6297–6308 (2017)
33. McCloskey, M., Cohen, N.J.: Catastrophic interference in connectionist networks: the sequential learning problem. Psychol. Learn. Motiv. **24**, 109–165 (1989)
34. Mikolov, T., Chen, K., Corrado, G., Dean, J.: Efficient estimation of word representations in vector space. In: ICLR Workshop Papers (2013)
35. Noce, L., Gallo, I., Zamberletti, A., Calefati, A.: Embedded textual content for document image classification with convolutional neural networks. In: Proceedings of the 2016 ACM Symposium on Document Engineering (DocEng 2016) (2016)
36. Peters, M.E., et al.: Deep contextualized word representations. In: Proceedings of NAACL (2018)
37. Roy, S., Das, A., Bhattacharya, U.: Generalized stacking of layerwise-trained deep convolutional neural networks for document image classification. In: 23rd International Conference on Pattern Recognition (ICPR), pp. 1273–1278 (2016)
38. Sandler, M., Howard, A., Zhu, M., Zhmoginov, A., Chen, L.C.: MobileNetV2: inverted residuals and linear bottlenecks. In: CVPR, pp. 4510–4520 (2018)
39. Sergeev, A., Balso, M.D.: Horovod: fast and easy distributed deep learning in TensorFlow. arXiv preprint arXiv:1802.05799 (2018)
40. Shin, C., Doermann, D.S.: Document image retrieval based on layout structural similarity. In: IPCV, pp. 606–612 (2006)
41. Simonyan, K., Zisserman, A.: Very deep convolutional networks for large-scale image recognition. CoRR, abs/1409.1556 (2014)
42. Smith, R.: An overview of the tesseract OCR engine. In: International Conference on Document Analysis and Recognition (ICDAR) (2007)
43. Sun, C., Qiu, X., Xu, Y., Huang, X.: How to fine-tune BERT for text classification? arXiv preprint arXiv:1905.05583 (2019)
44. Tan, M., et al.: MnasNet: platform-aware neural architecture search for mobile. In: CVPR (2019)
45. Tan, M., Le, Q.V.: EfficientNet: rethinking model scaling for convolutional neural networks. In: International Conference on Machine Learning (2019)
46. Tensmeyer, C., Martinez, T.: Analysis of convolutional neural networks for document image classification. In: ICDAR (2017)
47. Vaswani, A., et al.: Attention is all you need. In: Advances in Neural Information Processing Systems, vol. 30, pp. 6000–6010 (2017)
48. Wang, R., Su, H., Wang, C., Ji, K., Ding, J.: To tune or not to tune? How about the best of both worlds? arXiv preprint arXiv:1907.05338 (2019)
49. Zhang, Y., Wallace, B.C.: A sensitivity analysis of (and practitioners' guide to) convolutional neural networks for sentence classification. arXiv preprint arXiv:1510.03820 (2015)

Computation of the Airborne Contaminant Transport in Urban Area by the Artificial Neural Network

Anna Wawrzynczak[1,2(✉)] ⓘ and Monika Berendt-Marchel[1] ⓘ

[1] Faculty of Exact and Natural Sciences, Institute of Computer Sciences,
Siedlce University, Siedlce, Poland
anna.wawrzynczak-szaban@uph.edu.pl
[2] National Centre for Nuclear Research, Swierk-Otwock, Poland

Abstract. Providing the real-time working system able to localize the dangerous contaminant source is one of the main challenges of the cities emergency response groups. Unfortunately, all proposed up to now frameworks capable of estimating the contamination source localization based on recorded by the sensors network the substance concentrations are not able to work in real-time. The reason is the significant computational time required by the applied dispersion models. In such reconstruction systems, the parameters of the given dispersion model are sampled to fit the model output to the registrations; thus, the dispersion model is run tens of thousands of times. In this paper, we test the possibility of training an artificial neural network (ANN) to effectively simulate the atmospheric toxin transport in the highly urbanized area. The use of a fast neural network in place of computationally costly dispersion models in systems localizing the source of contamination can enable its fast response time. As a training domain, we have chosen the center of London, as it was used in the DAPPLE field experiment. The training dataset is generated by the Quick Urban & Industrial Complex (QUIC) Dispersion Modeling System. To achieve the ANN capable of estimating the contaminant concentration, we tested various ANN structures, i.e., numbers of ANN layers, neurons, and activation functions. The performed tests confirm that trained ANN has the potential to replace the dispersion model in the contaminant source localization systems.

Keywords: Machine learning · Neural networks · Airborne contaminant transport computation

1 Work Motivation

The main task of the emergency response groups existing in all cities is a quick reaction to any threats to people and the environment. The primary factor determining the success or failure of a given action is the response time. Nowadays, the chemicals are used in most areas of the industry, making the transport and

© Springer Nature Switzerland AG 2020
V. V. Krzhizhanovskaya et al. (Eds.): ICCS 2020, LNCS 12138, pp. 401–413, 2020.
https://doi.org/10.1007/978-3-030-50417-5_30

storage of the toxic materials pose a constant risk of releasing it into the atmosphere. In the cases when the source of the failure resulting in releasing the toxin into the atmosphere is known, the emergency responders can quickly undertake all necessary actions to minimize the consequences of such release. The more challenging are situations when the sensors, distributed over a city, report the non-zero concentration of the dangerous substance, which source is not known. In such cases, important is to have a system able to, in a real-time estimate the most probable location of the contamination source based solely on the concentration data reported from the sensors network. The algorithms that can cope with the task can be divided into two categories. First are based on the backward approach, but those are dedicated to the open areas or a continental-scale problem. Second, are based on the forward approach. In this case, the appropriate dispersion model parameters are sampled (among them source location) to chose the one giving the smallest distance measure between the model outputs and sensors measurement in considered spatial domain.

Such an inverse problem has no unique analytical solution but might be analyzed with probabilistic frameworks, as the Bayesian approach, where all searched quantities are modeled as random variables. Bayesian approach transforms the inverse mentioned above problem into searching for a posterior distribution based on the sampling of an ensemble of simulations using a priori knowledge and observed data. Stochastic reconstruction of the contamination source consists of two principal mechanisms. One is the dispersion model suitable for modeling of the airborne contaminant in considered terrain, and the second is the sampling algorithm able to find the optimal dispersion model parameters based on the model output comparison and the contaminant registrations. Regarding the efficiency of the applied parameter scanning algorithm, each reconstruction requires multiple runs of the dispersion model. The reconstruction in urban terrain, which is of interest in this paper, requires advanced dispersion models taking into account the turbulence of the wind field around the buildings. The most reliable and exact are the computational fluid dynamics models (CFD), but those are very computationally demanding. We must be aware of the fact that to find the most probable contamination source, the dispersion model has to be run tens of thousands of times. So, the applied dispersion model has to be fast to be applied in a real-time working emergency system.

The first reconstructions in urban scales using building models was reported in [1] and [2]. In [1], authors used an adjoint representation of the source-receptor relationship and applied a Bayesian inference methodology in conjunction with Markov Chain Monte Carlo sampling procedures. In [2] authors applied the methodology presented in [3] to the reconstruction of the flow around an isolated building and the flow during IOP3 and IOP9 of the Joint Urban 2003 Oklahoma City experiment. In this reconstruction, the FEM3MP [4] model was applied to predict the atmospheric dispersion of the released substance.

In [5] authors applied the approximate Bayesian computation algorithm (ABC) to localize the source of contamination in the highly urbanized terrain of the center of London utilizing the real field experiment data from DAPPLE experiment [6]. As the forward dispersion model, the Quick Urban Industrial Complex (QUIC) Dispersion Modeling System was applied [7]. The successful estimation of the release source required over 10000 runs of the dispersion model. Even though the QUIC model is able to simulate the airborne contaminant transport in the city relatively quickly, a single simulation over the $800\,\mathrm{m} \times 800\,\mathrm{m}$ domain takes as minimum 2 min. Thus, reconstruction on a single computer requires over 330 h. Computation time can be shortened by using a distributed system, but it is still impossible to apply it in the real-time working localization system when the answer time is crucial. Moreover, the required simulation time will increase with the enlargement of the considered terrain, e.g., for the whole city.

Even though in [5] the fast convergence of the ABC algorithm was proven, the whole framework cannot be implemented in the real-time emergency system due to the long computational time required by the dispersion model in urbanized terrain. This conclusion was an inspiration for the study presented in this paper. The idea is to train the artificial neural network (ANN) to be efficient in the simulation of the airborne contaminant transport in the urbanized terrain. If it succeeds, the ANN might work as the forward model in the system localizing the contamination source in real-time. Of course, the ANN has to be trained on the fixed city topology using the real wind conditions. This process requires lots of simulations serving as the training data-sets for the ANN. The process of training the ANN is computationally expensive, but ones trained, the ANN would be a high-speed tool for estimation the point-concentrations for a given contamination source.

2   ANN Training City Domain

Training ANN requires a large representative, reliable set of data. In this case, it should be measurements of the contaminant being a result of various release rates under different meteorological conditions. In this paper, we decided to check the possibility to train the ANN to simulate the airborne toxin transport in the area of central London where the DAPPLE experiment [6] was conducted (the main crossroad is of Marylebone Road and Gloucester Place, 51.5218N 0.1597W). The ideal situation would be if we could train the ANN on the real data. Unfortunately, it is not possible to obtain a set of data from real gas releases in urban areas that will be large enough to be a reliable set to train the ANN. Even though the city domain considered in this paper was the place of carrying out the large real field experiment DAPPLE the data available from its Trials are very limited. From four Trials, we have concentrations at 15 receptor positions for 30 min with 3-min intervals. This gives us, in sum, about 600 point-concentrations for four various source positions and release rates. This number of data is not enough to properly train the ANN. The only solution is to use the verified and well-recognized dispersion model to generate the data-set utilized to train, test, and validate the ANN. For that reason, we have used the QUIC Dispersion Modeling System. QUIC is intended for applications where the dispersion of air pollutants released near buildings must be computed relatively quickly [7]. The effectiveness of the QUIC model as the forward model in the reconstruction of the contaminant source based on the field experiment DAPPLE data was proven in [5].

The QUIC system comprises of a wind model QUIC-URB, a dispersion model QUIC-PLUME, and a graphical user interface. The modeling strategy adopted in QUIC-URB was originally developed by Röckle [8] and uses a 3D mass-consistent wind model to combine properly resolved time-averaged wind fields around buildings [9]. The code has been tested for both idealized and real-world cases (e.g., [7,10]).

To test the possibility of applying the ANN to simulate the airborne contaminant dispersion in the urbanized terrain, we have prepared the domain of size 752 m × 652 m × 80 m in which we have placed representations of the original buildings. The average building height in the area is 21.6 m (range 10 to 64 m). The whole considered domain and the estimated by the QUIC-URB sample wind field around the buildings are presented in Figs. 1 and 2.

In this domain, we have set the simulations of an ideal gas continuous release and registered its concentration for thirty minutes. To reflect the real measurement conditions we have randomly drawn the 600 contamination source locations, release rate within the interval $Q \in <10Mg; 500Mg>$ and its duration within interval $<2\,\text{min}, 30\,\text{min}>$ and 100 registration points (representing the sensor locations) per single release. The registered concentrations were normalized and logarithmized with an added background Gaussian noise at the level of $10^{-5}\,\text{g/m}^3$. The sample simulated by the QUIC model propagation of the released 250 Mg of gas during the first 30 min within the domain is presented in Fig. 3.

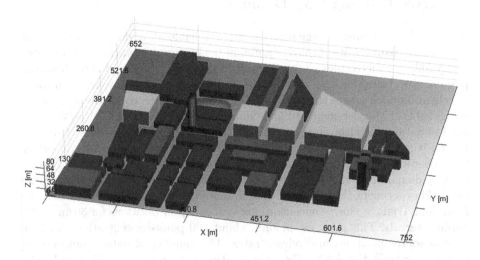

Fig. 1. The domain representing the area of central London assumed during the preparation of the ANN testing data-set (the main crossroad is of Marylebone Road and Gloucester Place, 51.5218N 0.1597W).

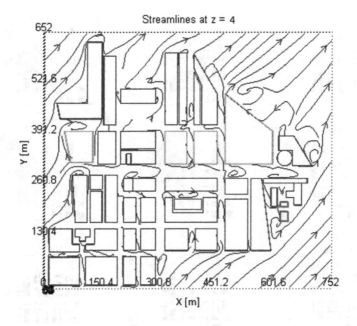

Fig. 2. The sample streamlines of the wind in the area of central London assumed during the preparation of the ANN testing data-set.

3   The Selected ANN Topology

Artificial neural networks (ANN) are computational models that consist of inter-connected elements called neurons. They are modeled on the construction of natural neurons and synapses connecting them [11]. ANN is capable of learning from training samples without knowing any laws or equations. There are several types of neural networks. In this paper, we used one of the simplest and widely used ANNs, the feed-forward neural network, e.g., [12]. This network has a one-way structure, i.e., the signal flows only in one direction from input nodes to output nodes. Feedforward neural networks were successfully used to predict the transport of pollutants in open areas, e.g., [13–15].

The ANN contaminant dispersion model is considered as a system that receives information from n distinct sets of inputs $X_i (i = 1, \ldots, n)$, namely contaminant source parameters and sensor location, and produces a specific output, in our case the concentration of the gas in the passed as the input location. No prior knowledge about the relationship between input and output variables is assumed. The input variables should be independent of each other, and each one is represented by its own input neuron $i = 1, \ldots, n$. Each neuron calculates a linear combination of the weighted inputs ω_{ij}, including a bias term b_i, from the links feeding into it and the corresponding summed value $C_j = \sum_i \omega_{ij} X_i + b_i$

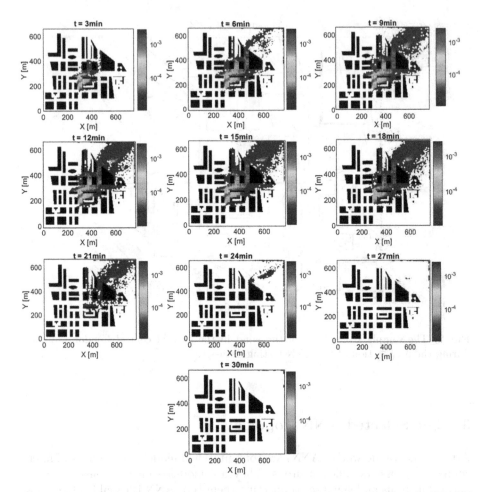

Fig. 3. The normalized concentration of the gas during thirty minutes after the 15 min release of 250 Mg of gas from the source located at $x = 300\,\mathrm{m}$, $y = 240\,\mathrm{m}$, $z = 7\,\mathrm{m}$ within the considered domain as simulated by the QUIC model.

is transformed using a function f, either linear or non-linear for example log-sigmoid or hyperbolic tangent. The bias term is included in order to allow the activation functions to be offset from zero, and it can be set randomly or to the desired value. The output obtained is then passed as a new input $\tilde{X}_j = f(C_j)$ to other nodes in the following layer, usually named hidden layer. Though one is allowed to use several neurons in this hidden layer, it is generally advantageous to somehow minimize the number of hidden neurons, in order to improve the generalization capabilities of the model and also to avoid over-fitting. Having such a framework of input variables and sets of functions, the ANN has to be trained in order to obtain the best estimate for each weight ω. The weight values are determined by an optimization procedure, the so-called learning

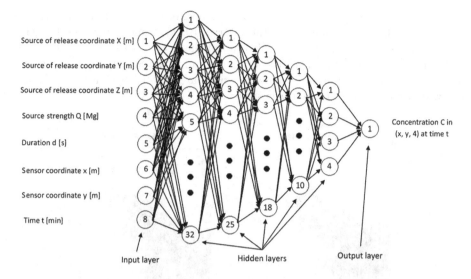

Fig. 4. The topology of the seven-layered feed-forward applied ANN with listed characteristics of input and output neurons.

algorithm [16]. The result of the neural network is compared with the target in order to calculate a predefined value of the error function. The error is sent over the network, and the algorithm adjusts the weights of each connection, respectively, to reduce the value of the error function. This repeated process corresponds to the number of training iterations that causes the network result to coincide with a state where the error between the output and the target is minimal.

During the ANN setup phase, numerous tests were performed with different combinations of the hidden layers, neuron number, learning rates, and activation functions. Figure 4 illustrates the selected topology. In the input layer, we introduce eight neurons representing the coordinates of the contamination source, release rate, and its duration and the coordinates of the registration point (sensor) within the domain and registration time after the initiation of the release. In the output layer, the tracer concentration at a given point and time will be given. It appeared that the ANN performed the best when the five hidden layers were introduced, with 32, 25, 18, 10, and 4 neurons in subsequent layers. The Levenberg-Marquardt learning method was used, which minimizes an error function in "damped" procedures, i.e., select steps proportional to the gradient of the error function. We have tested various activation functions, and the best results were achieved using the hyperbolic tangent *tanh* function in all hidden layers, and linear function in the output layer. The crucial for ANN better performance occurred scaling of all input parameters to be in the interval $(0, 1)$. Scaling allowed escaping from the problem of different scales and model instability. Additionally, the output concentrations were logarithmized. Logarithmization improved the ANN learning process because it allowed narrowing the range of concentration values. In consequence, to escape from the logarithm of the zero, the scaling was set to the interval $(0, 1)$, while hyperbolic tangent gives outputs in the range $(-1, 1)$. The input data set described in Sect. 2 was divided to training data set - 70%, validation and testing data-sets 15% each. The objective function describing the mean squared error between the real concentration and network output was set to reach $1e - 04$ value.

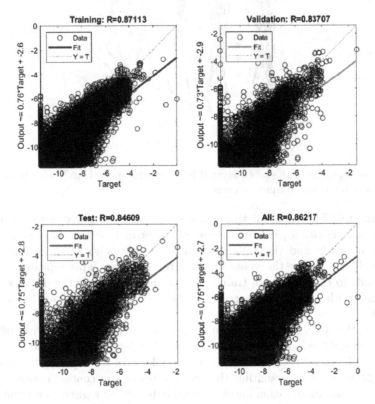

Fig. 5. The scatter plots representing results of training, testing, and validation process of the ANN. The dashed line represents the ideal fit.

4   Results

The results of the ANN training are presented in Fig. 5. Each point represents the single-point concentration as predicted by the trained ANN versus the input concentration from the QUIC model included in the training, validation, and test datasets. The ANN was trained under the Deep Learning Toolbox of the Matlab software. Taking into account the complexity of the transport of the airborne contaminant in the turbulent wind around the buildings, the quality of

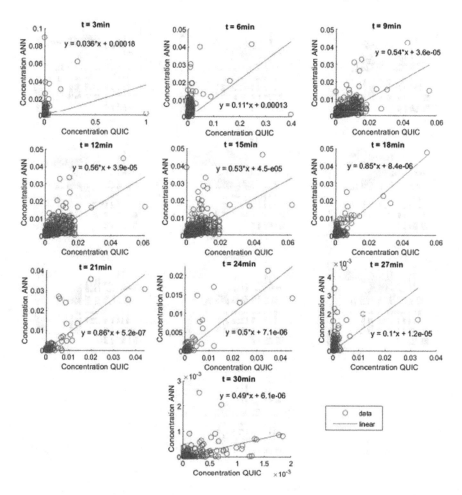

Fig. 6. The scatter plots representing results of training ANN versus the QUIC training-set data broken down by time. The line represents the linear fit.

the trained ANN is quite good. The R-value for training is 0.87, and together with test and validation data equals 0.86. These values indicate a significant relationship between the outputs and targets. The regression lines show that ANN slightly underestimates the higher concentrations. A more detailed analysis of the ANN performance is shown in Fig. 6. Figure 6 presents the comparison of the concentrations predicted by the trained ANN versus the concentrations from the QUIC model, taking into account the time dependence of concentration at the given registration point. The concentrations were sampled every three-minutes.

The scatter plots display that the best agreement is achieved after eighteen minutes from starting of the release. With time the ANN starts to underpre-dicts the concentrations. After nine minutes, the correlation coefficient increases

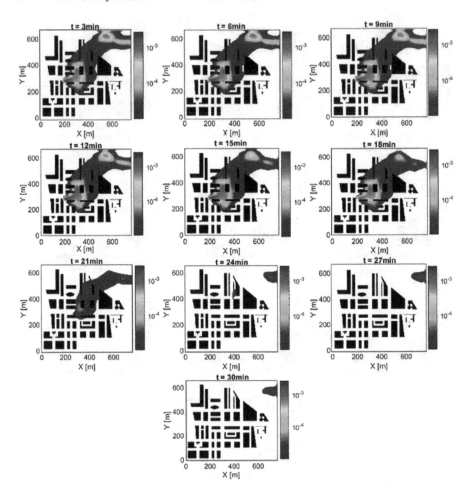

Fig. 7. The dispersion of the contaminant simulated by the ANN during consecutive thirty minutes after the 15 min release of 250 Mg of gas from the source located at $x = 300$ m, $y = 240$ m, $z = 7$ m within the considered domain.

from 0.54 up to 0.86 in the 21st minute. Moreover, it is visible that ANN has a tendency to underpredicts the smaller concentrations, while for higher concentrations level of agreement increases. A possible reason is that in the training dataset, more scenarios are leading to smaller concentrations and fewer favorable to increased concentrations. Nevertheless, the crucial question is, does the ANN learned the physics standing behind the gas dispersion over the highly urbanized area? Figs. 7 and 8 presents the simulated by the ANN contaminant transport for thirty minutes after two release scenarios. In the simulation of the gas by the trained ANN we have assumed the source location at the position with coordinates x = 300 m, y = 240 m, z = 7 m within the considered domain. The concentrations predicted by the ANN were sampled homogenously every

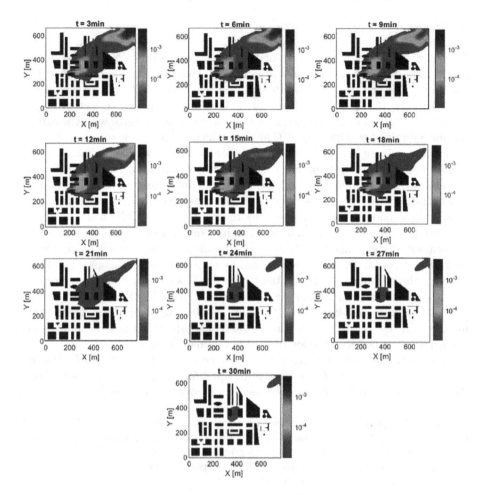

Fig. 8. The dispersion of the contaminant simulated by the ANN during consecutive thirty minutes after the 15 min release of 400 Mg of gas from the source located at $x = 300$ m, $y = 240$ m, $z = 7$ m within the considered domain.

4 m within the domain with 3 min time span. One can see that the simulated by the ANN dispersion of the contaminant gas agrees with the QUIC simulation presented in Fig. 3. The gas is spread in the wind direction set to 225°. As a result, the concentration of gas slowly decreases. Comparison of Fig. 7 where the released mass equals 250 Mg and Fig. 8 for released mass equal to 400 Mg confirms the correct estimation of gas concentrations by ANN. The concentrations predicted by ANN for release of 400 Mg of gas are greater than in the case of 250 Mg release. The regions of higher concentrations stay longer in the vicinity of buildings. We can conclude that simulated by the trained ANN transport of contaminants in the vicinity of the center of London agrees with the simulations performed by the QUIC model (Fig. 3).

5   Summary and Future Work

Presented results confirm that trained ANN can sufficiently simulate the turbulent transport of airborne toxins in the highly urbanized area. Such a result has not been published before. Even though we do not obtain the one-to-one agreement between the QUIC and ANN model concentrations, the trajectory of gas particles and gradient of concentrations predicted by ANN agree with the expectations. Obtained results suggest that the trained ANN can be successfully used in the contaminant source localization system as the forward dispersion model. In such systems, the contaminant source is estimated based on sampling the dispersion model set guided by minimizing the distance measure between the real concentrations from the sensors network and concentrations expected from the forward dispersion model. Therefore, more crucial is that ANN should correctly estimate the concentration gradients than its exact values. The main aim of the application of the trained ANN in such a localization system was to enable its operation in real-time. The time required by the presented in this paper ANN to estimate thirty-minute gas concentrations in a 196 000 sensor-points, as required by the simulations presented in Fig. 7 was equal to 3 s, while for the QUIC model it is estimated as at least 300 s, this gives us 100 times speed up. Taking this into account the reconstruction time in the real accidental situation can be short, resulting in the fast localization of the contaminant source.

The continuation of the presented research results will be the use of a trained neural network in place of the dispersion model for reconstruction based on real data from the DAPPLE field experiment, as it was presented in [5].

Acknowledgement. We thank Michael Brown and Los Alamos National Laboratory for the possibility to use the Quick Urban & Industrial Complex Dispersion Modeling System. Special thanks to Piotr Kopka for help in preparing the QUIC simulation. The authors acknowledge the partial financial support by the Polish National Science Centre, grant no. 2016/22/E/HS5/00406.

References

1. Keats, A., Yee, E., Lien, F.-S.: Bayesian inference for source determination with applications to a complex urban environment. Atmos. Environ. **41**(3), 465–479 (2007)
2. Chow, F.-K., Kosovia, B., Chan, S.: Source inversion for contaminant plume dispersion in urban environments using building-resolving simulations. J. Appl. Meteorol. Climatol. **47**(6), 1553–1572 (2008)
3. Johannesson, G., Hanley, B., Nitao, J.: Dynamic Bayesian models via Monte Carlo-an introduction with examples (No. UCRL-TR-207173). Lawrence Livermore National Lab., Livermore, CA, US (2004)
4. Chan, S.-T., Leach, M.-J.: A validation of FEM3MP with joint urban 2003 data. J. Appl. Meteorol. Climatol. **46**(12), 2127–2146 (2007)
5. Kopka, P., Wawrzynczak, A.: Framework for stochastic identification of atmospheric contamination source in an urban area. Atmos. Environ. **195**, 63–77 (2018)

6. Wood, C.-R., et al.: Dispersion experiments in central London: the 2007 DAPPLE project. Bull. Am. Meteorol. Soc. **90**(7), 955–970 (2009)
7. Williams, M.-D., Brown, M.-J., Singh, B., Boswell, D.: QUIC-PLUME theory guide, vol. 43. Los Alamos National Laboratory (2004)
8. Röckle, R.: Bestimmung der Strömungsverhältnisse im Bereich komplexer Bebauungsstrukturen. na (1990)
9. Sherman, C.-A.: A mass-consistent model for wind fields over complex terrain. J. Appl. Meteorol. **17**(3), 312–319 (1978)
10. Gowardhan, A.-A., Brown, M.-J., Pardyjak, E.-R.: Evaluation of a fast response pressure solver for flow around an isolated cube. Environ. Fluid Mech. **10**(3), 311–328 (2010). https://doi.org/10.1007/s10652-009-9152-5
11. Bishop, C.-M.: Neural Networks for Pattern Recognition. Oxford University Press, Oxford (1995)
12. Hecht-Nielsen, R.: Theory of the backpropagation neural network. In: Neural Networks for Perception, pp. 65–93. Academic Press (1992). ISBN 9780127412528
13. Hossain, K.: Predictive ability of improved neural network models to simulate pollutant dispersion. Int. J. Atmos. Sci. **2014**, 141923 (2014). https://doi.org/10.1155/2014/141923
14. Ma, D., Zhang, Z.: Contaminant dispersion prediction and source estimation with integrated Gaussian-machine learning network model for point source emission in atmosphere. J. Hazard. Mater. **311**, 237–245 (2016)
15. Wawrzynczak, A., Berendt-Marchel, M.: Application of the artificial neural network in the forecasting of the airborne contaminant. J. Phys: Conf. Ser. **1391**, 012092 (2019)
16. Haykin, S.: Neural networks: A Comprehensive Foundation. Prentice Hall PTR, Upper Saddle River (1994)

Exploring Musical Structure Using Tonnetz Lattice Geometry and LSTMs

Manuchehr Aminian[ID], Eric Kehoe[✉][ID], Xiaofeng Ma[ID], Amy Peterson[ID], and Michael Kirby

Colorado State University, Fort Collins, CO 80523, USA
{manuchehr.aminian, eric.kehoe, xiaofeng.ma, amy.peterson, michael.kirby}@colostate.edu

Abstract. We study the use of Long Short-Term Memory neural networks to the modeling and prediction of music. Approaches to applying machine learning in modeling and prediction of music often apply little, if any, music theory as part of their algorithms. In contrast, we propose an approach which employs minimal music theory to embed the relationships between notes and chord structure explicitly. We extend the Tonnetz lattice, originally developed by Euler to introduce a metric between notes, in order to induce a metric between chords. Multidimensional scaling is employed to embed chords in twenty dimensions while best preserving this music-theoretic metric. We then demonstrate the utility of this embedding in the prediction of the next chord in a musical piece, having observed a short sequence of previous chords. Applying a standard training, test, and validation methodology to a dataset of Bach chorales, we achieve an accuracy rate of 50.4% on validation data, compared to an expected rate of 0.2% when guessing the chord randomly. This suggests that using Euler's Tonnetz for embedding provides a framework in which machine learning tools can excel in classification and prediction tasks with musical data.

Keywords: Long short-term memory · LSTM · Music theory · Multidimensional scaling · Euler's Tonnetz

1 Introduction

Long Short-Term Memory (LSTM) neural networks are well known and well utilized neural networks for prediction and classification tasks [1,3,4]. In particular they have been used to create polyphonic music models—models that can predict or create music by learning from existing musical pieces. Usually one classifies or predicts on collections of notes called chords. Such a predictive model can aid

This research is partially supported by the National Science Foundation under Grant No. DMS-1322508. Any opinions, findings, and conclusions or recommendations expressed in this material are those of the authors and do not necessarily reflect the views of the National Science Foundation.

© Springer Nature Switzerland AG 2020
V. V. Krzhizhanovskaya et al. (Eds.): ICCS 2020, LNCS 12138, pp. 414–424, 2020.
https://doi.org/10.1007/978-3-030-50417-5_31

in composing musical scores which capture the harmonic stylings of a particular musical artist or genre. Learning the large scale harmonic structure, such as harmonic cadences, from the level of individual to an entire musical tradition is of great interest to the musical community. In particular, we will use music theory to construct an embedding of chords as a foundation for an LSTM neural network to predict in a musical piece the chord following its previous measure.

Prior work has been done in applying machine learning to the prediction, classification, and composition of music in various contexts, and in particular in the utility of LSTMs to learn long sequences of patterns in ordered data. There is a focus on music composition, generation, and improvisation using LSTMs [4–6,8] which involve future prediction on different time scales and styles using many approaches. Some work has been done on an even broader scale; trying to incorporate large scale musical structure [7] and classification of musical genre [16].

Similarly, there have been approaches to incorporating music theory; see [13] as an example which relates information about major and their relative minor keys, among other things. Analysis has been done on networks which explicitly encode information beyond harmonic structure, including pitch and meter [14], which suffers (as they suggest in their paper title) from an extremely large state and hyperparameter spaces.

Towards the analysis of LSTM network structure in music prediction, the authors in [1] use the JSB chorales dataset to predict the next set of notes in a given chorale. They use this to test various LSTM architectures to quantify the use of gates and peephole connections in an LSTM structure.

Finally, approaches to embedding chords via a "chord2vec" (in analogy with the famous word2vec in natural language processing) has been explored in [3] which embeds the data via the word2vec methodology then studies LSTM performance, among a variety of null models.

Taking a different approach, we choose to train our LSTMs on chord types, so collections of notes modulo an octave. Chords take into account the information of pitch class, e.g. there is a difference between the notes $C4$ and $C5$. In our setting the chord $(C4, E5, G4)$ is identified with $(C5, E3, G5)$, both of which are called a C-major triad. We then can measure harmonic distance between chord types by measuring how far apart they are from one another on the Tonnetz lattice—a lattice structure originally developed by Euler to represent musical harmony, see Fig. 1 for Eulers original picture. Multidimensional scaling can then be used to embed chords as vectors in a relatively low dimensional space using the Tonnetz distance. This embedding uses musical information to create the feature space of the chords rather than trying to train a neural network directly on one-hot encoded chords or the use of other embedding methods such as chord2vec [3].

A description of the training and the setup of the model are in the Sect. 2 below along with an explanation of the dataset on which we train our model. We also reference our visualization illustrating the process of a song being classified by chords [17].

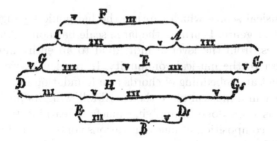

Fig. 1. The original Tonnetz depicted first in Euler's 1739 *Tentamen novae theoriae musicae ex certissismis harmoniae principiis dilucide expositae.*

In the following sections, we investigate the size of the hidden layers in their ability to predict chords. The details of this model are also discussed in Sect. 2. We look at the results of two different types of prediction and various hidden layer sizes in Sect. 3.

2   Evaluation

For the setup of our experiments, that will be discussed in this section, we focus on two important aspects: the embedding method and the structure of the LSTMs. First, we describe the embedding method of chords which takes into account the structure of chords utilizing Euler's Tonnetz lattice and the dataset that will be used. Second, we give the details of the LSTM structure and how it is applied to the embedded chords.

2.1   Data Set and Embedding

We train, test, and validate our model on a set of 382 Bach chorales, JSB Chorales [9]. This collection of Bach chorales is pre-partitioned into sub-collections of training, testing, and validation songs. The format of the songs is given in MIDI or Musical Instrument Digital Interface. Unlike other music formats such as .wav or .mp3, MIDI allows one to access individual notes and beats within a song. Musical software can then be used to play the music similar to how a player piano plays a piano roll. The advantage of the data being represented in this format is that one can work with music on the level of individual notes and beats to create discretizations of songs which preserve their harmonic or melodic structure.

We preprocess the MIDI files using the python package music21 [10]. In particular, we process MIDI files by partitioning their time axis into 8th note bins and snapping all the notes in a particular bin to that bin's left time point. This process is called quantization. Once the music has been quantized all the musical notes at each eighth note are grouped into a collection of notes called chords. Of course in some time bins there may be no notes present, this is a musical rest. To preserve the time structure of the song we replace the rest with the previous non-empty chord. A chord $\mathbf{C}$ in a song $\mathbf{S}$ can then be represented as a subset of musicals pitches e.g. $\mathbf{C} = \{C_3, E_4, G_3\}$. Here the subscript 3 indicates the specific pitch of the note type C. For instance, middle C on a piano is given by C_3 while C_4 is an octave above. We then quotient the set of chords by their pitch class, thereby identifying C_3 with C_4 and so on. By identifying the pitch classes $\mathbf{N} = \{C, C\sharp/D\flat, D, D\sharp, \ldots, A\sharp/B\flat, B\}$ with $\mathbb{Z}_{12}$, we define the set of all chord types as the power set

$$\mathbf{Ch} = 2^{\mathbb{Z}_{12}}.$$

This set of chord types is quite large, so we restrict ourselves to only those chord types which are subsets of 7-chords which are common amongst many generations of music. We build 7-chords in the following manner. First define the sets:

1. Set of fifths $\underline{\mathbf{5}} = \{6, 7, 8\}$ (diminished, perfect, augmented)
2. Set of thirds $\underline{\mathbf{3}} = \{2, 3, 4, 5\}$ (diminished, minor, major, augmented)
3. Set of sevenths $\underline{\mathbf{7}} = \{9, 10, 11\}$ (diminished, minor, major)

We then build the set of C 7-chords as

$$\mathbf{Ch}_7(C) = \{\{0, x, y, z\} \mid x \in \underline{\mathbf{5}}, y \in \underline{\mathbf{3}}, z \in \underline{\mathbf{7}}\}.$$

Now the set of all 7-chords is the union over all translates of $\mathbf{Ch}_7(C)$

$$\mathbf{Ch}_7 = \bigcup_{k \in \mathbb{Z}_{12}} (\mathbf{Ch}_7(C) + k).$$

Hence the collection of all sub-chords of 7-chords is the union of powersets

$$\mathbf{Ch}_{\leq 7} = \bigcup_{C \in \mathbf{Ch}_7} 2^C.$$

The collection $\mathbf{Ch}_{\leq 7}$ has exactly 529 chord types consisting of single notes, dyads, triads, and tetrachords subsetted from 7-chord types.

Given a chord $\mathbf{C}$ in the song $\mathbf{S}$ we use music21's root, fifth, third, and seventh functions to project $\mathbf{C}$ to a subchord $\tilde{\mathbf{C}}$ lying in $\mathbf{Ch}_{\leq 7}$. From this point of view we project a song to a lower dimensional harmonic space where prediction becomes a more feasible task. We then embed these projected chords into a weighted version of the well-known Tonnetz lattice $\mathbf{T}$ originally developed by Euler, see Fig. 2 for a modern unrolled display. The Tonnetz lattice is built using the following construction:

1. Let the underlying node set of $\mathbf{T}$ be $\mathbb{Z}_{12}$
2. For each node $k \in \mathbf{T}$:
   (a) Join node k to the node $k + 7$ by an edge of weight 3
   (b) Join node k to the nodes $k + 3$ and $k + 4$ by edges of weight 5

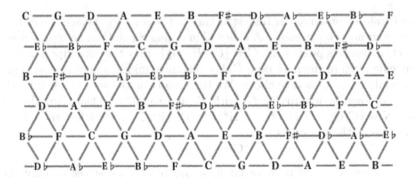

Fig. 2. The unrolled Tonnetz lattice formed by connecting notes to their neighboring fifth, major third, and minor third. Image from [11]

The significance of the weights comes from the harmonic distance between notes. For example, if we strike a key on the piano we would produce a sound wave whose displacement can be expanded into a Fourier series given by

$$\phi(t) = \sum_{n=1}^{\infty} a_n \cos(n\omega t) + b_n \sin(n\omega t).$$

Here the frequency term ω is called the *fundamental* and for $n \geq 2$ the terms $n\omega$ indicate its *overtones*. The first three overtones are the octave, fifth, and a major third respectively. A fifth occurs by tripling the frequency of the fundamental, this is represented in the Tonnetz with an edge of weight 3. Similarly a major third occurs by quintupling the fundamental frequency—so an edge weight of 5 in the Tonnetz. In order to have symmetry across major and minor modes we treat a minor third the same as a major third in harmonic distance. We then define the weighted Tonnetz metric $d_{\mathbf{T}}$ on $\mathbf{T}$ as the shortest-path-distance in $\mathbf{T}$. We can view $d_{\mathbf{T}}$ as a 2nd order approximation of the true harmonic distance between notes. For those who wish to explore the structure of the Tonnetz further, we refer are readers to [11].

We use the Tonnetz metric $d_{\mathbf{T}}$ to induce a metric $d_{\mathbf{T}}^{\text{total}}$ on $\mathbf{Ch}_{\leq 7}$ given by its *total-average*. Explicitly for $\mathbf{C}, \mathbf{D} \in \mathbf{Ch}_{\leq 7}$ we define

$$d_{\mathbf{T}}^{\text{total}}(\mathbf{C}, \mathbf{D}) = \frac{1}{|\mathbf{C}||\mathbf{D}|} \sum_{\substack{x \in \mathbf{C} \\ y \in \mathbf{D}}} d_{\mathbf{T}}(x, y).$$

It can be verified that this defines a metric. One can also use the well-known Hausdorff metric between subsets of a metric space but there will less variability in distances between chords. The total-average distance intuitively measures how every note in the first chord changes with respect to every note in the latter. We then use metric multidimensional scaling (mMDS) via python's `sklearn` package to approximately embed the metric space $\mathbf{Ch}_{\leq 7}$ into $\mathbb{R}^n$ as chord-vectors. If we order the space $\mathbf{Ch}_{\leq 7}$ we can represent $d_{\mathbf{T}}^{\text{total}}$ as a matrix D with entries δ_{ij}

giving the distance $d_{\mathbf{T}}^{\text{total}}(x_i, x_j)$ between the ith and jth chord respectively. If we let $m = 529$ denote the number of chords in $\mathbf{Ch}_{\leq 7}$ then for a given integer n mMDS solves the convex minimization problem

$$\min_{X \in \mathbb{R}^{m \times n}} \sum_{i,j} (\delta_{ij} - d_{ij}(X))^2$$

where $d_{ij}(X)$ denotes the Euclidean distance between the rows $X^{(i)}$ and $X^{(j)}$ of X. The above objective function is called the *stress* of the embedding X. In particular, the package finds a solution by applying an iterative algorithm SMA-COF to find the fixed point to the so-called Guttman transform—providing the embedding. See [15] for in-depth analysis. To estimate the optimal embedding dimension for $\mathbf{Ch}_{\leq 7}$ we plot the stress of the mMDS embedding as a function of the embedding dimension and locate where the stress first stabilizes, see Fig. 3. We estimate an embedding dimension of $n = 20$ using this method. We conjecture that $d_{\mathbf{T}}^{\text{total}}$ cannot be exactly embedded into Euclidean space. This is supported by the non-zero leveling off of the stress with dimension. Essentially points in $\mathbf{Ch}_{\leq 7}$ which are far apart will always have an error in their embedded distance due to the "curvature" of the space.

We now have a map `Tonnetz2vec : Chords` $\rightarrow \mathbb{R}^{20}$. Using this map we can convert a song in MIDI format to a time-series of vectors in $\mathbb{R}^{20}$. We then train an LSTM to predict harmonic changes in music and test the validity of the Tonnetz lattice as a fundamental space for learning musical harmony.

2.2 Model Architecture and Training

We use a network with 2 uni-directional LSTM hidden layers and a linear output layer to train the embedded Bach chorales data set to predict the next chord in a song given the previous measure of chords. This is implemented in Python using the Pytorch library [18]. For each element(chord) in the input sequence (chords series) each layer computes the following functions:

$$i_t = \sigma(W^{ix}x[t] + W^{ia}a[t-1] + b_i) \qquad \textit{Input gate}$$
$$f_t = \sigma(W^{fx}x[t] + W^{fa}a[t-1] + b_f) \qquad \textit{Forget gate}$$
$$o_t = \sigma(W^{ox}x[t] + W^{oa}a[t-1] + b_o) \qquad \textit{Output gate}$$
$$\tilde{c}[t] = \tanh(W^{gx}x[t] + W^{ga}a[t-1] + b_g) \qquad \textit{Candidate cell state}$$
$$a[t] = o_t \odot \tanh(c[t]) \qquad \textit{Hidden state update}$$
$$c[t] = f_t \odot c[t-1] + i_t \odot \tilde{c}[t] \qquad \textit{Cell state update}$$

where $a[t]$ is the hidden state at time t, $c[t]$ is the cell state at time t and $x[t]$ is input at time t for $t = 1, 2, \cdots, s$. The output of the model is an affine transformation of the hidden state in the last time step $t = s$, i.e.,

$$y[s] = W^{ya}a[s] + b_y.$$

Let n be the input dimension and m be the hidden layer dimension. Now the weight matrices are $W^{ix}, W^{fx}, W^{ox}, W^{gx} \in \mathbb{R}^{m \times n}$, and $W^{ia}, W^{fa}, W^{oa}, W^{ga} \in$

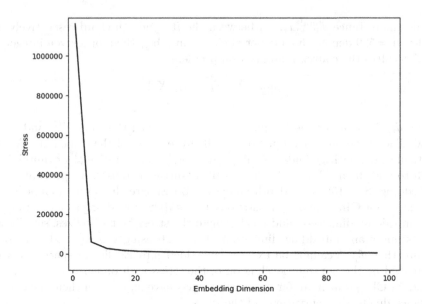

Fig. 3. mMDS stress of the Tonnetz embedding as a function of embedding dimension.

$\mathbb{R}^{m \times m}$. The bias terms are $b_i, b_f, b_o, b_g \in \mathbb{R}^m$ and for the output layer we have the weight matrix $W^{ya} \in \mathbb{R}^{n \times m}$ and the bias vector $b_y \in \mathbb{R}^n$. We parse the embedded Bach chorales data set into input-output pairs such that the input time series consists of the previous 8 chords $\{x[t-7], x[t-6], \cdots, x[t]\}$ and the output being the next chord $x[t+1]$. Each song is parsed using moving window of size 9 (8 input chords and 1 output chord). For the training phase, we train all the input-output pairs occuring in 229 Bach chorales in one batch and backpropagate with Adam optimization algorithm [12] using for the loss function the mean squared error (MSE) in the chord embedding space.

3   Results

In this section we apply the methodology described above to the JSB Chorales dataset. We first study how performance on test data varies depending on the use of simple accuracy (exact chord matches) compared to a relaxed notion of a match (at least 3 notes agree with the true chord). We find that performance is improved with the relaxation, but qualitative behavior and location of local maxima/minima is not noticeably different. Hence, model selection to avoid overfitting is not impacted by the use of different metrics.

In the second part of this study, we vary the size of the hidden layers in our LSTM model and study its effect on prediction accuracy. We see a monotone increase in accuracy on the validation data set with a maximum of 0.504 with a hidden layer size of 512 (the maximum value studied), though the values begin to plateau with a hidden layer size of 128.

3.1 Bach Chorales

The publicly available Bach chorale data set comes preprocessed and divided into training, testing, and validation sets; we use this subdivision here. Chords are mapped to 20 dimensions as described in Subsect. 2.2. Mean squared error is used as a loss function for training, and test set performance is evaluated either as an exact match "Accuracy" or as "3-note accuracy" which is a relaxation which counts a prediction as correct if at least three notes in the true and predicted chord agree. From the perspective of live music, two chords agreeing on three notes often gives a similar harmonic impression. From a machine learning perspective, this relaxes the classification task to achieving 3-note accuracy. In Fig. 4, we visualize an experiment with an LSTM trained on 200 chorales and the accuracy on test data evaluated on 77 chorales. We observe nearly identical

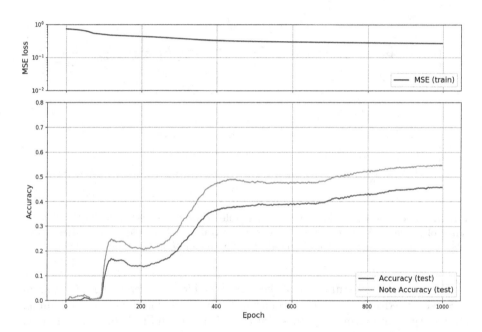

Fig. 4. Value of loss function on training data (top) and two types of accuracy on test data (bottom) for an LSTM with a hidden layer size of 128 over 1000 epochs of training. Similar qualitative behavior is observed using the two types of accuracy, making decisions for model selection the same for both.

Table 1. Training and test set performance for the JSB chorale dataset for various choices of the size of the LSTM hidden layer.

Hidden layer size	4	8	16	32	64	128	256	512
Validation accuracy	0.000	0.000	0.201	0.321	0.412	0.477	0.495	0.504
Validation accuracy (3-note)	0.100	0.073	0.247	0.376	0.467	0.550	0.596	0.614

qualitative trends in both accuracy and note accuracy, with approximately a difference of ten percentage points in reported. While a 3-note accuracy may be of further interest as more fine-tuned models are developed, for the purposes of model selection we find it sufficient to restrict to simple accuracy (exact matches).

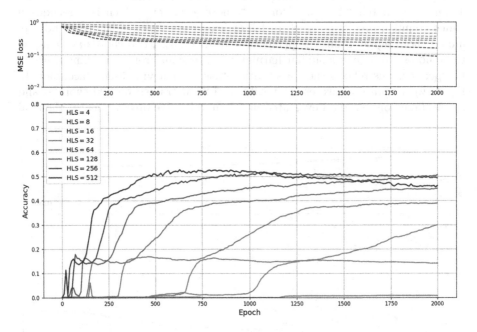

Fig. 5. Value of loss functions (top) and simple accuracy on test data (bottom) for a range of hidden layer sizes over 2000 epochs of training. For hidden layer sizes above 128, maximum accuracies are observed at different points in training.

We now study the influence of the size of the hidden layer of the LSTM in an otherwise fixed model training scenario. Decay of the loss function on training data as well as performance on test data are visualized in Fig. 5 for various hidden layer sizes. A model is chosen based on optimal performance on test data during training to reduce the likelihood of overfitting. While there is non-monotonic trends in qualitative behavior as the size of the hidden layer increases, we see a general increase in performance on test data. An optimal model is selected based on the epoch achieving maximum accuracy, then the performance is evaluated on the validation data set.

The resulting validation accuracy is in Table 1. For each hidden layer size, the performance of the optimal model on the validation data tracks that of the optimal test data performance quite closely. When the hidden layers are significant bottlenecks, of sizes 4 or 8, training is quite slow and the respective performance on validation is extremely poor. For sizes 16 upwards, we see a monotonic improvement in prediction which decelerates at a hidden layer size of 128. This gives strong evidence that successful chord prediction can be done without an extremely large neural network. Such small networks are useful for the purposes of interpretability and detailed analysis, especially if one seeks to later extend such a model for harmonic structure to also incorporate, for example, pitch and tempo.

4   Conclusion

We have developed a framework for machine learning on music which incorporates music theory via the Tonnetz lattice. This is distinguished from a one-hot encoding of notes in the chord which provides little intuition and has many challenges associated with it in the translation from output layer to identifying predicted notes or chord(s). By contrast, this embedding allows one to work directly with chords, which are often more intuitive for musicians in understanding broader musical structure. There are a broad range of directions to investigate in optimizing performance and predictions made by an LSTM model using this embedding. However, the prediction accuracy results of this paper are promising for learning the harmonic structure within a musical genre. This suggests a path forward to predict music by learning its harmonic, melodic, and rhythmic structures separately. From the machine learning perspective, understanding the performance and behavior of predictions using this embedding is of great interest. Analysis of the model in terms of well-known chord transitions and its behavior on different genres is also of note. Admittedly, our study in this paper focuses on the Western musical tradition in a genre of music with rigorous rules; towards this we are very interested in exploring how such models behave when trained in other musical traditions, and what such a model can do when asked to make predictions across culture.

References

1. Greff, K., et al.: LSTM: a search space odyssesy. IEEE Trans. Neural Netw. Learn. Syst. **28**, 2222–2232 (2017)
2. Hochreiter, S., et al.: Long short-term memory. Neural Comput. **9**(8), 1735–80 (1997)
3. Madjiheurem, S., et al. Chord2Vec: learning musical chord embeddings (2016). https://doi.org/10.13140/RG.2.2.15031.93608
4. Eck, D., Schmidhuber J.: A first look at music composition using LSTM recurrent neural networks. Technical report, Istituto Dalle Molle Di Studi Sull Intelligenza Artificiale (2002)
5. Eck, D., Schmidhuber, J.: Finding temporal structure in music: blues improvisation with LSTM recurrent networks. In: Proceedings of the 12th IEEE Workshop on Neural Networks for Signal Processing, pp. 747–756 (2002)
6. Lyu, Q., Wu, Z., et al.: Modelling high-dimensional sequences with LSTM-RTRBM: application to polyphonic music generation. In: Twenty-Fourth International Joint Conference on Artificial Intelligence (2015)

424   M. Aminian et al.

7. Eck, D., Lapalme, J.: Learning musical structure directly from sequences of music. University of Montreal, Department of Computer Science, CP 6128 48 (2008)

8. Coca, A., Coorêa, D.C., et al.: Computer-aided music composition with LSTM neural network and chaotic inspiration. In: The 2013 International Joint Conference on Neural Networks (IJCNN), pp. 1–7 (2013)

9. Boulanger-Lewandowski, N., et al.: Modeling temporal dependencies in high-dimensional sequences: application to polyphonic music generation and transcription (2012). http://www-etud.iro.umontreal.ca/~boulanni/icml2012

10. Cuthbert, M., Ariza, C.: music21: a toolkit for computer-aided musicology and symbolic music data. In: International Society for Music Information Retrieval (2010). https://web.mit.edu/music21/

11. Bergstrom, T., Karahalios, K., Hart, J: Isochords: visualizing structure in music. In: Proceedings of the Graphics Interface 2007 Conference, pp. 297–304 (2007). https://doi.org/10.1145/1268517.1268565

12. Kingma, D., Ba, J.: Adam: a method for stochastic optimization. In: 3rd International Conference for Learning Representations, San Diego (2015)

13. Brunner, G., Wang, Y., et al.: JamBot: music theory aware chord based generation of polyphonic music with LSTMs. In: IEEE 29th International Conference on Tools with Artificial Intelligence (ICTAI), pp. 519–526 (2017)

14. Sturm, B.: What do these 5,599,881 parameters mean?: an analysis of a specific LSTM music transcription model, starting with the 70,281 parameters of its softmax layer. In: International Conference on Computational Creativity (2018)

15. De Leeuw, J.: Applications of convex analysis to multidimensional scaling in recent developments in statistics. In: Barra, J.R., Brodeau, F., Romier, G., Van Cutsem, B. (eds.) pp. 133–145, North Holland Publishing Company, Amsterdam (1977). http://deleeuwpdx.net/janspubs/1977/chapters/deleeuw_C_77.pdf

16. Tang, C., Chui, K., et al.: Music genre classification using a hierarchical long short term memory (LSTM) model. In: Third International Workshop on Pattern Recognition, vol. 10828 (2018)

17. Tonnetz2vec. https://github.com/ekehoe32/Tonnetz. Accessed 7 Feb 2020

18. Paszke, A., Gross, S., et al.: PyTorch: an imperative style, high-performance deep learning library. In: Advances in Neural Information Processing Systems, pp. 8024–8035 (2019). https://pytorch.org/

Modeling of Anti-tracking Network Based on Convex-Polytope Topology

Changbo Tian[1,2], Yongzheng Zhang[1,2(✉)], and Tao Yin[1,2]

[1] Institute of Information Engineering, Chinese Academy of Sciences,
Beijing 100093, China
[2] School of Cyber Security, University of Chinese Academy of Sciences,
Beijing 100049, China
{tianchangbo,zhangyongzheng,yintao}@iie.ac.cn

Abstract. Anti-tracking network plays an important role in protection of network users' identities and communication privacy. Confronted with the frequent network attacks or infiltration to anti-tracking network, a robust and destroy-resistant network topology is an important prerequisite to maintain the stability and security of anti-tracking network. From the aspects of network stability, network resilience and destroy-resistance, we propose the convex-polytope topology (CPT) applied in the anti-tracking network. CPT has three main advantages: (1) CPT can easily avoid the threat of key nodes and cut vertices to network structure; (2) Even the nodes could randomly join in or quit the network, CPT can easily keep the network topology in stable structure without the global view of network; (3) CPT can easily achieve the self-optimization of network topology. Anti-tracking network based on CPT can achieve the self-maintenance and self-optimization of its network topology. We compare CPT with other methods of topology optimization. From the experimental results, CPT has better robustness, resilience and destroy-resistance confronted with dynamically changed topology, and performs better in the efficiency of network self-optimization.

Keywords: Anti-tracking network · Topology management · Network topology · Self-organization · Cyber security

1 Introduction

The rapid growth of Internet access has made communication privacy an increasingly important security requirement. And the low threshold and convenience of the techniques about network monitoring and tracing also have posed a great threat on the online privacy network users [1–3]. Anti-tracking network is proposed as the countermeasure to fight against the network monitoring and censorship [4–6]. So, the attacks on anti-tracking network increase rapidly in recent

Supported by the national natural science foundation of China under grant No. U1736218 and No.61572496.

V. V. Krzhizhanovskaya et al. (Eds.): ICCS 2020, LNCS 12138, pp. 425–438, 2020.
https://doi.org/10.1007/978-3-030-50417-5_32

years, such as network paralysis, network infiltration, communication tracking and so on. Especially, for the P2P-based anti-tracking network which takes the advantage of the wide distribution of nodes for tracking-resistant communication, it is important to keep a stable, secure and destroy-resistant topology structure. Also, considered that the nodes in P2P-based anti-tracking network can join in or quit the network freely and randomly, it increases the uncertainty and insecurity for anti-tracking network to maintain a robust network topology.

Much research has been done regarding the management or optimization of network topology. But as for the aspect of anti-tracking network, current research have some limitations which are concluded as follows:

(1) **The cost of network maintenance.** Some of current research needs the global view of network to optimize the network topology [7–9]. Confronted with dynamically changed network topology, the cost of network maintenance would be very high.

(2) **Weak destroy-resistance.** P2P-based anti-tracking network confronts not only the problem of communication tracing, but also the threats of network attacks [10–12]. Especially, the attack to cut vertices and key nodes will destroy the network structure tremendously.

(3) **Weak tracking-resistance.** Network infiltration is another threat on the security of anti-tracking network [13–15]. Malicious nodes can be deployed in the network to measure the network topology and traceback the communication.

To address the above problems, we propose the convex-polytope topology (CPT) which can be applied in anti-tracking network. The anti-tracking network based on CPT achieves the self-optimization of its network. The novelty and advantages of our proposal are summarized as follows:

(1) We apply the convex-polytope topology in the anti-tracking network. The anti-tracking network based on CPT has better performance in network stability and resilience.

(2) We propose a topology maintenance mechanism based on CPT. Each node in network only needs to maintain its local topology to keep the whole network in stable convex-polytope structure.

(3) We propose a network self-optimization mechanism based on CPT. With the self-optimization mechanism, anti-tracking network can optimize its network topology as the network topology changes.

2 Convex-Polytope Topology

Convex-polytope topology (CPT) is a structured topology in which all nodes are constructed into the shape of convex-polytope, as illustrated in Fig. 1. CPT has the following advantages to maintain the robust and destroy-resistant topology of anti-tracking network.

- Stability: The structure of CPT has the advantage in avoiding the cut vertices, except the extreme cases in which the connectivity of CPT is too sparse, such as ring structure. So, the maintenance of network topology is just to keep the convex-polytope structure.
- Resilience: Confronted with the dynamically changed topology, each node in CPT only needs to maintain its local topology to accord with the properties of convex-polytope, then the whole network will be in the convex-polytope structure.
- Self-optimization: The network stability, robustness and destroy-resistance benefit from the balanced distribution of nodes' connectivity. CPT can achieve the self-optimization of the whole network through the optimization of each node's local topology.

Fig. 1. An example of convex-polytope topology.

2.1   The CPT Model

Consider the topology structure of CPT illustrated in Fig. 1, the whole network is constructed in a convex-polytope structure logically. To maintain this structure, each node has to record both of the neighbouring nodes and the corresponding surfaces in the convex-polytope.

Let v denotes a node, CN denotes its neighbouring node collection, and CS denotes its surface collection. We take the example of node v shown in Fig. 1, then $CN = \{v_1, v_2, v_3\}$ and $CS = \{(v_1, v_2), (v_1, v_3), (v_2, v_3)\}$. So, if one neighbouring node of v disconnects, then node v knows which surfaces are affected by the lost neighbouring node, and adjusts its local topology accordingly.

As we can see from the Fig. 1, each surface of convex-polytope can be different polygons. For a fixed number of nodes, if each surface has more edges, the connectivity of the whole network becomes more sparse. On the contrary, each surface has less edges, the connectivity of the whole network becomes more dense. If all surfaces have three edges, then the connectivity of the network reaches the maximum. So, in our work, we construct CPT of which each surface is triangle to provide better stability and resilience of anti-tracking network.

Assume there are n nodes to construct CPT with triangle surfaces, then the number of surfaces and edges in this kind of CPT is fixed. Theorem 1 gives the details about the calculation of the number of this CPT's surfaces and edges.

Theorem 1. *If a convex-polytope has n nodes, and each surface is triangle, then the number of surfaces and edges of this convex-polytope is fixed. The number of edges is: $l = 3 \times (n-2)$, the number of surfaces is: $m = 2 \times (n-2)$.*

Proof. Because each surface of convex-polytope with n nodes is triangle, and each edge is shared by two surfaces, assume the number of edges is l, and the number of surfaces is m, then $3 \times m = 2 \times l$. According to Euler Theorem of Polyhedron, we have the equation $n + m - l = 2$.

$$\begin{cases} 3 \times m = 2 \times l \\ n + m - l = 2 \end{cases} \Rightarrow \begin{cases} l = 3 \times (n-2) \\ m = 2 \times (n-2) \end{cases} \tag{1}$$

2.2   The Construction of CPT

Even though CPT has the advantages in the maintenance of network topology, in some extreme cases shown in Fig. 2, CPT still has the potential threat in topology structure, such as key nodes. Key nodes would have big influence on the stability, robustness and security of network topology.

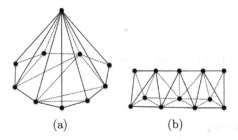

(a)   (b)

Fig. 2. Two examples of extreme convex-polytope topology.

In the construction process of CPT, the balanced distribution of each node's connectivity is the priority to construct a stable and robust anti-tracking network. We propose the construction algorithm to construct a CPT with balanced distribution of nodes' connectivity. The construction algorithm takes two main steps shown as follows:

(1) **Constructing a circuit:** All nodes construct a circuit.
(2) **Adding edges iteratively:** Each node takes turns at selecting a node which has the span of 2 and allow new connections along with one direction of the circuit, and adding an edge with it to generate a triangle surface. If all surfaces of one node are triangles, it will refuse new connection with it in order to keep the whole topology in convex-polytope structure. After all nodes have triangle surfaces, the CPT with triangle surfaces has been constructed.

Algorithm 1. Construction Algorithm

Input: $C = [v_1, v_2, ..., v_n]$: node collection.

```
 1: BuildNodeCircuit(C).                          ▷ All nodes construct a circuit.
 2: N ← n.
 3: while True do
 4:     v ← GetOneNode(C).
 5:     u ← GetLinkNode(v)        ▷ Get a node u which has the span of 2 with v.
 6:     Connect(v, u).
 7:     if The surfaces of v are all triangles then
 8:         N ← N − 1.
 9:     end if
10:     if N ≤ 0 then
11:         Break.
12:     end if
13: end while
```

Algorithm 1 gives the pseudocode for the construction algorithm of CPT. The time complexity of the construction algorithm mainly depends on the executions of the process of building node circuit (line 1) and while loop (line 3∼6). The time complexity of building node circuit is $O(n)$. Because the edges of CPT is fixed which has been proved in Theorem 1, and the node circuit has formed n edges. Then, the while loop in Algorithm 1 only needs to be executed $(2 \times n - 6)$ times. So, the time complexity of Algorithm 1 is $O(n)$.

To make straightforward sense of CPT construction algorithm, Fig. 3 illustrates the construction process of CPT with 9 nodes, labeled $v_1, v_2, ..., v_9$. Firstly, these 9 nodes construct a circuit. We take v_1 as the first node to add an edge with v_3, then v_1 has a triangle surface (v_1, v_2, v_3). Iteratively, v_2 adds an edge with v_4, and v_3 adds an edge with v_5, untill the surfaces of all nodes are triangles. In Fig. 3(f), v_2 needs to select a node with the span of 2 to add an edge, there are three nodes, v_5, v_6, v_8. Because all the surfaces of v_5 and v_8 are triangles, the extra edges may break the structure of convex-polytope. So, v_2 can only add an edge with v_6.

2.3 The Maintenance of CPT

As the nodes can join in or quit the anti-tracking network freely, the frequent change of network topology inevitably has a big influence on the stability and security of network structure. So, an effective maintenance mechanism of network topology guanrantees the robustness and usability of anti-tracking network. In this section, we will discuss the maintenance mechanism in which the relevant nodes only update its local topology to keep the anti-tracking network in convex-polytope structure when confronted with the dynamically changed topology.

In the maintenance mechanism of CPT, two cases are taken into consideration: that of the nodes join in the network and that of the nodes quit the network.

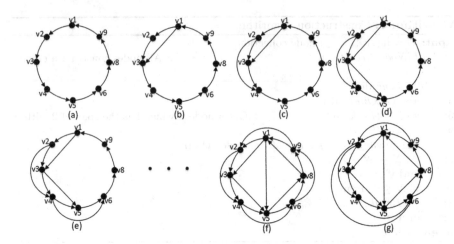

Fig. 3. Construction process of CPT

Nodes Join in CPT. All surfaces of CPT are triangles. If new nodes want to join in CPT, they only need to select one surface of CPT and connect each node of this surface.

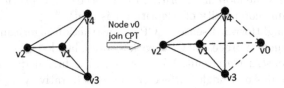

Fig. 4. Node v_0 join the network with CPT

As illustrated in Fig. 4, node v_1 has three surfaces, (v_2, v_3), (v_2, v_4) and (v_3, v_4). A new node v_0 connects v_1 and selects the surface (v_3, v_4) to join in the network. Because of the join of v_0, v_1 breaks the surface (v_3, v_4) to generates two new surfaces (v_0, v_3) and (v_0, v_4). The new node v_0 generates three surfaces, (v_1, v_3), (v_1, v_4) and (v_3, v_4). So, the node join of CPT only changes the topology of one surface. The whole topology can be easily maintained in the convex-polytope structure.

Nodes Quit CPT. If a node quits CPT, all the surfaces of this node will merge into one surface. If the quitting node has a high degree, CPT will generate a large surface which is detrimental to the stable and robust network topology.

For example, Fig. 5(a) shows a node with degree of 3 quits CPT. When v_1 quits CPT, its three surfaces (v_2, v_3), (v_2, v_4) and (v_3, v_4) merge into one surface which is still a triangle. In this case, CPT need no additional actions to maintain the topology.

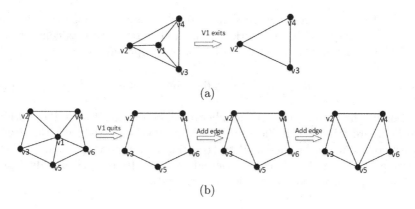

Fig. 5. CPT maintains its structure when some nodes exit

But in Fig. 5(b), node v_1 with degree of 5 quits CPT, the surface generated by quitting v_1 is pentagon. In this case, the relevant nodes have to update its local topology to keep convex-polytope structure with all triangle surfaces. The maintenance process only exists in the nodes of the generated pentagon surface. Through the connection between these nodes, the pentagon surface can be divided into different triangle surfaces.

The maintenance of the generated pentagon surface is similiar with the construction algorithm of CPT. If all surfaces of a node are triangles, it will refuse any connections. Each node in the pentagon surface connects with a node which has the span of 2 and allow new connection iteratively until all nodes in this surface has triangle surfaces. For example shown in Fig. 5(b), after v_1 quits CPT, v_2 firstly establishes a connection with v_5 to generate a triangle surface (v_3, v_5). Then v_3 has all triangle surfaces and refuse any connection with it. Next, v_5 establishes a connection with v_4 to generate a triangle. Then v_6 has all triangle surfaces and refuse any connections. At last, the maintenance of CPT is finished.

3   Self-optimization Mechanism Based on CPT

As we have discussed above, CPT can keep the network topology in the stable and resilient convex-polytope structure. But the convex-polytope structure still has some extreme cases shown in Fig. 2 which may pose a potential threat on the structure stability and communication efficiency of anti-tracking network because of the key nodes. According to the properties of CPT, we can achieve the self-optimization of anti-tracking network to balance the distribution of nodes' connectivity.

As we have mentioned in Theorem 1 that the number of edges in convex-polytope of which all surfaces are triangles is: $l = 3 \times (n - 2)$, and n denotes the number of the nodes. Then, we can calculate the average degree of CPT as shown in Eq. 2. When n is very large, the average degree of CPT $\bar{d}$ is close to 6. So, we usually set the upper limit of nodes' degree as 10 in CPT. The nodes of which the degree is bigger than 10 should adjust its local topology to reduce its connectivity. Every node in network has the ability to control its connectivity, then CPT has the ability to optimize the network topology.

$$\bar{d} = \lim_{n \to \infty} \frac{2 \times l}{n} = 6 - \lim_{n \to \infty} \frac{12}{n} = 6 \tag{2}$$

Theorem 1 has proved that the number of edges are fixed for a CPT of which all surfaces are triangles, if the degree of some nodes are reduced, the degree of other nodes should be increased to keep the convex-polytope structure. So, the self-optimization of CPT can be viewed as a transfer process of node degree from the high-degree nodes to the low-degree nodes.

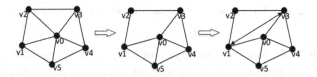

Fig. 6. The self-optimization process of CPT

Assume that node v_0 has n neighbouring nodes which are labeled in sequence from v_1 to v_n. Because all the neighbouring nodes of v_0 form a circuit, we use v_{i-1} and v_{i+1} respectively represent the adjacent nodes of v_i in the node circuit. If node v_0 needs to reduce its degree, firstly node v_0 has to request for the degree d_i of its each neighbouring node $v_i(1 \leq i \leq n)$. According to the degree of its neighbouring nodes, node v_0 selects a suitable neighbouring node to disconnect. To maintain the convex-polytope structure, the corresponding two neighbouring nodes of node v_0 have to connect with each other. As illustrated in Fig. 6, node v_0 disconnects with node v_2, a quadrilateral surface (v_0, v_1, v_2, v_3) is formed. After v_1 connects with v_3, the degree of node v_0 and v_2 is reduced and the degree of v_1 and v_3 is increased.

To balance the distribution of nodes' connectivity, it is better to guarantee that two disconnected nodes are with high-degree and the two connected nodes are with low-degree. So, before node v_0 decides the disconnected node, it needs to calculate which node is more suitable to disconnect according to the degree of each neighbouring node.

For each neighbouring node v_i of node v_0, we calculate the fitness value c_i of node v_i as shown in Eq. 3. If the degree of node v_i is smaller than one of its two adjacent nodes, the fitness value of node v_i is 0. Because after node v_0 disconnects with this candidate node, the degree of its adjacent nodes will become bigger. If the degree of node v_i is bigger than both of its adjacent nodes, node v_0 calculate the fitness value by the multiplication of $(d_i - d_{i-1})$ and $(d_i - d_{i+1})$. At last, node v_0 selects the candidate node with the biggest fitness value to disconnect. Then, node v_0 instructs the two adjacent nodes of candidate node to connect with each other to maintain the topology stable.

Algorithm 2. Self-optimization Algorithm

Input: v_0: current node; d_0: the degree of v_0.

1: **while** $d_0 > DegreeLimit$ **do**
2:     $(d_1, d_2, ..., d_n) \leftarrow GetDegree()$.   ▷ Get the degree of each neighbouring node of node v_0.
3:     $(c_1, c_2, ..., c_n) \leftarrow CalculateFitnessValue()$.   ▷ Calculate the fitness value for each neighbouring node.
4:     $v_i \leftarrow GetMaxValueNode()$.   ▷ Get the node with the highest fitness value.
5:     $Disconnect(v_0, v_i)$.
6:     $Connect(v_{i-1}, v_{i+1})$.
7:     $Update(v_0)$.   ▷ Update the edges, surfaces of node v_0.
8: **end while**

$$c_i = \begin{cases} 0 & if \ d_i \leq d_{i-1} \ or \ d_i \leq d_{i+1} \\ (d_i - d_{i-1}) \times (d_i - d_{i+1}) & if \ d_i > d_{i-1} \ and \ d_i > d_{i+1} \end{cases} \tag{3}$$

Algorithm 2 gives the pseudocode for the self-optimization algorithm of each node. Once the degree of each node is bigger than the upper limit of degree (We usually set the upper limit of degree is 10), the current node begins the self-optimization process to adjust its local topology. And the basic idea of self-optimization is to transfer the degree from the high-degree node to low-degree node. With the collaboration of all nodes in anti-tracking network, CPT can achieve the self-optimization of network topology.

4   Experiments and Analysis

In this section, we compare CPT with two self-organizing methods of network topology, one is a based on the neural network (NN) [16] and the other is based on distributed hash tables (DHT) [17]. NN achieves the self-optimization of topology by the neural network algorithm deployed in each node. DHT achieves the self-organizing network based on the hierarchical aggregation structure. We evaluate the three methods from the following aspects: (1)network robustness, (2)maintenance efficiency, and (3)communication efficiency.

4.1   Evaluation of Network Robustness

Firstly, we seperately simulate three networks with 5000 nodes by CPT, NN and DHT, calculate the distribution of nodes' degree D_d and the distribution of nodes' density D_s to compare the difference of the three network topologies. We use the number of nodes which have the span not more than 2 with v_i to represent the density of v_i.

As illustrated in Fig. 7(a), D_d of CPT is mainly distributed in the interval $[3, 10]$, because of the convex-polytope structure limitation. D_d of NN is mainly distributed in the interval $[3, 15]$. D_d of DHT is distributed widely. The distribution of nodes' density has the similar tendency with the distribution of

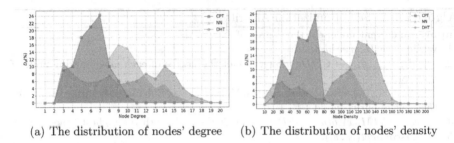

(a) The distribution of nodes' degree   (b) The distribution of nodes' density

Fig. 7. The evaluation of the distribution of nodes' degree and density

nodes' degree. In general, the local density of nodes is in direct proportion to their degree, and the nodes with high density becomes the key nodes easily. In Fig. 7(b), D_s of CPT is mainly distributed in the low density area which means the topology structure of CPT is more stable than NN and DHT.

To evaluate the network resilience, we firstly give a metric to quantify the performance of network resilience. As shown in Eq. 4, we use a graph G to represent the original network, $G(p)$ denotes the subgraph after p percent of nodes is removed from G, $MCS(G(p))$ denotes the maximum connected subgraph of $G(p)$, $Num(g)$ denotes the node number of a graph g, N denotes the node number of G. The metric β measures the maximum connectivity of the network after some nodes are removed from the network. The β is higher, the network resilience is better.

$$\beta = \frac{Num(MCS(G(p)))}{N} \tag{4}$$

Based on the above simulated networks constructed by CPT, NN and DHT, we remove p percent nodes from the three networks each time until all nodes are removed. In each round, we calculate the β to evaluate network resilience. In the experiments, we use two different ways to remove nodes shown as follows.

- **Random-p Removal.** In each round of nodes removal, we remove p percent of nodes from the network randomly.
- **Top-p Removal.** In each round of nodes removal, we remove p percent of nodes with the highest degree.

From the experimental results shown in Fig. 8, CPT has a better performance in network resilience and β keeps higher than NN and DHT in both random-p removal and top-p removal. In random-p removal, even β of CPT is always higher than NN and DHT, the difference of the three methods is not very high. The performance of the three methods in network resilience begins to decrease when β is less than 40%.

But in top-p removal, the performance of DHT degrades sharply. The performance of NN in network resilience begins decrease when β is less than 20%. But CPT still maintains good performance in network resilience under top-p removal. The experiments show that top-p removal causes bigger damage to the

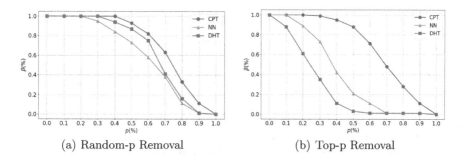

(a) Random-p Removal   (b) Top-p Removal

Fig. 8. The evaluation of network resilience

network of NN and DHT than random-p removal. But CPT maintains good performance of network resilience in both random-p removal and top-p removal.

4.2   Evaluation of Maintenance Efficiency

The real network environment is complex and changeable, the maintenance efficiency has a direct influence on the performance of network self-optimization. If the network maintenance process costs too much time or computing power, it is ineffective in the practical application. In this section, we evaluate the maintenance efficiency of CPT, NN and DHT from two aspects.

- **Efficiency of network construction.** According to the netword construction methods of CPT, NN and DHT, we construct a network in simulation environment and count the time requirement for comparison.
- **Efficiency of network self-optimization.** For each network constructed by CPT, NN and DHT, we randomly remove p percent of nodes and count the time requirement of network self-optimization as the metric. By increasing the percentage of removed nodes, we compare the change of time requirement to evaluate the efficiency of network self-optimization.

Figure 9(a) illustrates the time requirement of network construction with CPT, NN and DHT. The value of X-axis denotes the node number of the constructing network. With the increase of the network node number, the time requirement of CPT increases slower than NN and DHT. Limited by the computation of neural network, the time requirement of NN increases sharply. DHT relays too much on the agent nodes to construct and manage the network topology. Benefited from the convex-polytope structure, the whole network of CPT keeps the balanced distribution of nodes' connectivity. Without the too much computation like NN and too many connections like DHT, CPT can achieve better efficiency of network construction.

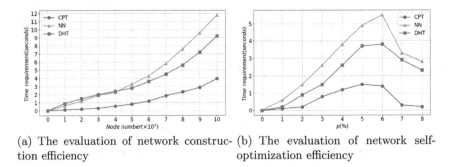

(a) The evaluation of network construc- (b) The evaluation of network self-
tion efficiency   optimization efficiency

Fig. 9. The evaluation of network maintenance efficiency

Figure 9(b) illustrates the time requirement of network self-optimization after
p percent of nodes are removed from the network randomly. When more than 50%
nodes are removed randomly, the network will be split into different subgraphs.
In this case, we need to connect the different subgraphs into one whole network,
then continue the network self-optimization process and count the time requirement. As we can see in Fig. 9(b), the curves of three methods increase at the
beginning, and then decrease when p is more than 50%. Because along with the
increase of the percentage of removed nodes, the network size becomes smaller.
So, when the network size reduces to a certain degree, the time requirement of
network self-optimization will decrease.

In this experiment, CPT performs better in both network construction and
network self-optimization. The simple structured topology of CPT improves the
efficiency of network construction, maintenance and self-optimization.

4.3   Evaluation of Communication Efficiency

To evaluate the communication efficiency, we randomly select two nodes to com-
municate in each test and count the average time requirement after 100 round
tests. With the increase of network size, we compare the communication efficiency of CPT, NN and DHT through the average time requirement.

The network diameter and average path length increase with the increase of
network size. The randomly selected two nodes would spend more time to com-
municate in each network. As illustrated in Fig. 10, the average time requirement
of CPT is always higher than NN and DHT. But, the difference of communication
efficiency between CPT, NN and DHT is not very big and it is also acceptable
for anti-tracking network.

The reason of CPT's low communication efficiency blames for the convex-
polytope structure. To keep the convex-polytope structure, CPT improves the
network robustness and destroy-resistance at the cost of communication efficiency. And the research has proved that anti-tracking network can not achieve
all the three properties: strong tracking-resistance, low bandwidth overhead,
and low latency overhead [18]. But from the tracking-resistant standpoint, the

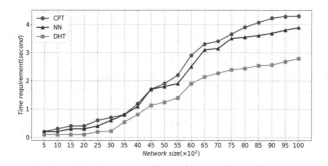

Fig. 10. Evaluation of communication efficiency between CPT, NN and DHT

weakness in communication efficiency does not impact the overall efficiency of anti-tracking network. And also, we can add extra connections between different nodes along the diagonal inside the convex-polytope topology to reduce the network diameter, then the communication efficiency of CPT would be improved greatly.

5   Conclusion

In this paper, we propose convex-polytope topology (CPT) which can be applied in anti-tracking network. We construct the anti-tracking network according to the properties of convex-polytope, and maintain the network topology in the convex-polytope structure. When the node join in or quit the network, CPT can still maintain the convex-polytope topology to keep a stable and resilient network. Based on the convex-polytope topology, we design a self-optimization mechanism for anti-tracking network. The experimental results show that CPT has better performance in network robustness, maintenance efficiency than current works.

Acknowledgments. We thank the anonymous reviewers for their insightful comments. This research was supported in part by the national natural science foundation of China under grant No. U1736218 and No.61572496. Yongzheng Zhang is the corresponding author.

References

1. Fatemeh, S., Milivoj, S., Rizwan, A.M., et al.: A survey on routing in anonymous communication protocols. ACM Comput. Surv. **51**(3), 1–39 (2018)
2. Sun, Y., Edmundson, A., Vanbever, L., et al.: RAPTOR: routing attacks on privacy in tor. In: 24th USENIX Security Symposium (USENIX Security 2015), pp. 271–286 (2015)
3. Buchanan, T., Paine, C., Joinson, A.N., et al.: Development of measures of online privacy concern and protection for use on the Internet. J. Am. Soc. Inf. Sci. Technol. **58**(2), 157–165 (2007)

4. Tian, C., Zhang, Y., Yin, T., et al.: Achieving dynamic communication path for anti-tracking network. In: 2019 IEEE Global Communications Conference (GLOBECOM), pp. 1–6, IEEE (2019)

5. Jan, M.A., Nanda, P., He, X., et al.: A Sybil attack detection scheme for a forest wildfire monitoring application. Future Gener. Comput. Syst. **80**, 613–626 (2018)

6. Tiwari, R., Saxena, T.: A review on Sybil and Sinkhole of service attack in VANET. Recent Trends Electron. Commun. Syst. **5**(1), 7–11 (2018)

7. Khan, S.A., Engelbrecht, A.P.: A fuzzy particle swarm optimization algorithm for computer communication network topology design. Appl. Intell. **36**(1), 161–177 (2012)

8. Auvinen, A., Keltanen, T., Vapa, M.: Topology management in unstructured P2P networks using neural networks. In: 2007 IEEE Congress on Evolutionary Computation, pp. 2358–2365. IEEE (2007)

9. Nakano, T., Suda, T.: Self-organizing network services with evolutionary adaptation. IEEE Trans. Neural Netw. **16**(5), 1269–1278 (2005)

10. Tian, C., Zhang, Y.Z., Yin, T., Tuo, Y., Ge, R.: A loss-tolerant mechanism of message segmentation and reconstruction in multi-path communication of anti-tracking network. In: Chen, S., Choo, K.-K.R., Fu, X., Lou, W., Mohaisen, A. (eds.) SecureComm 2019. LNICST, vol. 304, pp. 490–508. Springer, Cham (2019). https://doi.org/10.1007/978-3-030-37228-6_24

11. Yang, M., Luo, J., Ling, Z., et al.: De-anonymizing and countermeasures in anonymous communication networks. IEEE Commun. Mag. **53**(4), 60–66 (2015)

12. Yang, M., Luo, J., Zhang, L., et al.: How to block Tor's hidden bridges: detecting methods and countermeasures. J. Supercomput. **66**(3), 1285–1305 (2013)

13. Mittal, P., Borisov, N.: Information leaks in structured peer-to-peer anonymous communication systems. ACM Trans. Inf. Syst. Secur. **15**(1), 1–28 (2012)

14. Zang, W., Zhang, P., Wang, X., et al.: Detecting Sybil nodes in anonymous communication systems. Procedia Compu. Sci. **17**, 861–869 (2013)

15. Wang, J., Wang, T., Yang, Z., et al.: SEINA: a stealthy and effective internal attack in Hadoop systems. In: 2017 International Conference on Computing, Networking and Communications (ICNC), pp. 525–530. IEEE (2017)

16. Tian, C., Zhang, Y.Z., Yin, T., Tuo, Y., Ge, R.: A smart topology construction method for anti-tracking network based on the neural network. In: Wang, X., Gao, H., Iqbal, M., Min, G. (eds.) CollaborateCom 2019. LNICST, vol. 292, pp. 439–454. Springer, Cham (2019). https://doi.org/10.1007/978-3-030-30146-0_31

17. Li, Z.P., Huang, J.H., Tang, H.: A P2P computing based self-organizing network routing model. J. Softw. **16**(5), 916–930 (2005)

18. Das, D., Meiser, S., Mohammadi, E., Kate, A.: Anonymity trilemma: strong anonymity, low bandwidth overhead, low latency-choose two. In: 2018 IEEE Symposium on Security and Privacy (SP), pp. 108–126. IEEE (2018)

A Workload Division Differential Privacy Algorithm to Improve the Accuracy for Linear Computations

Jun Li[1,2], Huan Ma[3], Guangjun Wu[1(✉)], Yanqin Zhang[4], Bingnan Ma[3], Zhen Hui[3], Lei Zhang[1], and Bingqing Zhu[1,2]

[1] Institute of Information Engineering, Chinese Academy of Sciences, Beijing, China
{lijun,wuguangjun,zhanglei,zhubingqing}@iie.ac.cn
[2] School of Cyber Security, University of Chinese Academy of Sciences,
Beijing, China
[3] National Computer Network Emergency Response Technical Team/Coordination,
Center of China (CNCERT/CC), Beijing, China
{mahuang,huizhen}@cert.org.cn, mabingnan90@gmail.com
[4] China Petroleum Engineering and Construction CORP. Beijing Design Company,
Beijing, China
zhangyanqin.cpe@cnpc.com.cn

Abstract. Differential privacy algorithm is an effective technology to protect data privacy, and there are many pieces of research about differential privacy and some practical applications from the Internet companies, such as Apple and Google, etc. By differential privacy technology, the data organizations can allow external data scientists to explore their sensitive datasets, and the data owners can be ensured provable privacy guarantees meanwhile. It is inevitable that the query results that will cause the error, as a consequence that the differential privacy algorithm would disturb the data, and some differential privacy algorithms are aimed to reduce the introduced noise. However, those algorithms just adopt to the simple or relative uniform data, when the data distribution is complex, some algorithms will lose efficiency. In this paper, we propose a new simple ε-differential privacy algorithm. Our approach includes two key points: Firstly, we used Laplace-based noise to disturb answer to reduce the error of the linear computation queries under intensive data items by workload-aware noise; Secondly, we propose an optimized workload division method. We divide the queries recursively to reduce the added noise, which can reduce computation error when there exists query hot spot in the workload. We conduct extensive evaluation over six real-world datasets to examine the performance of our approach. The experimental results show that our approach can reduce nearly 40% computation error for linear computation when compared with MWEM, DAWA, and Identity. Meanwhile, our approach can achieve better response time to answer the query cases compared with the start-of-the-art algorithms.

This work was supported by the National Natural Science Foundation of China (No.61931019).

V. V. Krzhizhanovskaya et al. (Eds.): ICCS 2020, LNCS 12138, pp. 439–452, 2020.
https://doi.org/10.1007/978-3-030-50417-5_33

Keywords: Differential privacy · Privacy-preserving · Data security

1   Introduction

Among the data privacy protection technologies, existing research is based on the solution from the following perspectives: anonymity-based methods, encryption-based methods, noise-based method, and differential privacy-based method. There have been many reliable encryption-based method technology, such as DES [7], 3DES, Blowfish [23], RC5 [21], IDEA, RSA, etc. The advance of the encryption technology is their security. However, the analyzability will be lost due to the encryption. The anonymity-based methods to protect data privacy can keep the data's analyzability, the mainly anonymity-based technologies are k-anonymity [24], L-diversity [3] and T-closeness [18]. However anonymity-based methods have fatal weaknesses, and the anonymous data might suffer anti-anonymity. For the data organizers, there exist security and privacy problems on data collection and publishing. Among the data privacy attacks, differential attack is a way that the attacker infers private data through statistical information over two homogeneous datasets. For example, an attacker can infer a person's specific shopping goods by differential attacks via different queries. To explore whether a person bought an *object*, the attacker can conduct two queries, and one query obtains the count of persons that have bought the *object*, and another query the count on the data set that excludes the person by the quasi-identifier, such as timestamp, gender, region, age, etc. By the two query results, the attack can infer whether the person bought the *object*.

To solve the differential attack, many differential privacy algorithms can be used, such as matrix mechanism [17], DAWA algorithm [16], MWEM [13], and RAPPOR [10], etc. The differential privacy technology can be used in many fields [5,6,11,20,26]. Differential privacy was first defined by Dwork et al. [8,9], and it protects the individual data by injecting noise to the results according to the privacy budget. A number of ε-differential privacy algorithms have been proposed [2,13,15–17], and some of them workload-aware and data-dependent [2,13,16,17]. From the method of disturbing results view, ε-differential adopts three ways: Laplace Mechanism [8], Exponential Mechanism [19], and Randomized Response [25]. Random response mechanism is an effective way to protect the privacy of the frequency vector. The random response mechanism has been used in privacy protection of collecting sensitive data since the 1960s. RAPPOR [10] is ε-differential privacy technology that Google company has already used in the browser, and it adopts the random response. MWEM [13] is classical ε-differential privacy, and it is based on a combination of the Mechanism Exponential Mechanism with the Multiplicative Weights update rule. The MWEM algorithm selects and poses queries using the Exponential and Laplace Mechanisms, and it improves the approximation using the Multiplicative Weights update rule. DAWA [16] is a data-dependent and workload-aware algorithm, and it adds noise according to the input data and the given workload and it is a two-stage mechanism for answering range queries under ε-differential privacy. In

2016, Michael Hay et al. propose an evaluation framework for standardized evaluation of privacy, called DPBENCH [14]. In 2018, Dan Zhang et al. [27] propose a programming framework and system called ϵktelo to implement the existing algorithms. For the task of answering linear counting queries, ϵktelo allows both privacy novices and experts to easily design algorithms, and the APEx [12] is a novel differential privacy system that allows data analysts to pose adaptively chosen sequences of queries along with required accuracy bounds.

Most of the algorithms are related to data distribution, especially when the data items are sparse, i.e., there are a large number of items are empty, these algorithms can effectively reduce the introduced errors. The same conclusion can be reached in the paper [14,16]. While, these algorithms are not suitable for all data situations, as in the situation the data items are intensive and the data has complex distribution, and the conclusion is also shown in [14,16]. Current ε-differential privacy algorithms will cause computation error for linear computations over the intensive data domain. Inspired by the partition of the data domain, we propose a novel ε-differential privacy algorithm via Laplace-based noise and optimized workload division to decrease the computation error in complex data situation. We make the following contributions:

(1) We propose a novel ε-differential privacy algorithm in complex data situation. We used Laplace-based noise to disturb the query results. This disturbation can reduce the error of the linear computation queries under intensive data items by workload-aware noise.
(2) We propose an optimized workload division method. We divide the queries recursively to reduce the added noise. This division can effectively reduce computation error when there exists a hot spot, i.e., some domain is frequently queried in the workload.
(3) We conduct extensive experiments on six real-world datasets and conduct a comparison with differential privacy algorithms (MWEM, DAWA, and Identity). The evaluation results show that the proposed algorithm can effectively reduce the computation error and has better efficiency relatively.

2   Approach Overview

We propose a ε-differential privacy algorithm for the linear computation queries. The algorithm aims to reduce the results error in the case that the sensitivity of workload is high and there exists frequency queried dom($\mathbb{B}$) item due to the hot issue or statistical attack queries and the frequency count x is complex. In the algorithm, we adopt Laplace Mechanism to disturb the query results. To reduce the random added noise, we propose a novel perspective that the added noise might be reduced by dividing the queries into several clusters and add Laplace-base noise respectively. Furthermore, based on the Laplace division, we propose a simple and effective recursion division for the query workloads and the privacy budget. The method recursively divides the queries workload and privacy budget into two parts when the expected noise is less than that before dividing. To sum up, the algorithm can solve three problems: (1) The current ε-algorithms can reduce the error limitly, meanwhile, those algorithms will cost much computation resources. (2) When the sensitivity of a query workload is large, the current algorithms can't reduce the noise obviously. This can be shown in [16]. (3) In the situation that the data distribution is intensive, the current algorithms cannot fit it and will cause much error for the answer to query workload.

The method we propose satisfies ε-differential privacy rigorously, and ε-differential privacy is the privacy protection mechanism proposed by Dwork in 2006 and regulates privacy protection. We will define ε-differential privacy formally.

Definition 1 (ε-differential privacy). *An algorithm M is a ε-differential privacy algorithm if for any neighboring database I and I' ($|I - I'| \leq 1$), and any subset of output S satisfies the following formula:*

$$Pr[M(I) \in S] \leq \exp(\varepsilon) \times Pr[M(I') \in S]$$

The ε-differential privacy algorithm protects privacy data by disturbing the answer and the attackers cannot distinguish the results over the neighboring database I and I', and the parameter ε is the privacy budget and it determines the privacy-preserving capacity. If the privacy budget is lower, the differential algorithm will protect privacy more effectively. For the random algorithm M, if the results over the two adjacent datasets I and I' are close to each other, and it is difficult to infer whether a data item exists by $M(X)$ and $M(Y)$. ε-differential privacy has the following three primary properties.

Property 1. For the random algorithm ε_i-difference privacy M_1, and function $M(X)$ is an arbitrary deterministic function: $R \to R'$. Then $M_1(M(X))$ still satisfies ε differential privacy.

Property 2. For the random algorithm M_i and it satisfies ε_i-difference privacy. Defining a random function M that it is a process of a random sequence of M_i. The random function M satisfies $\sum_{i=1}^{k} \varepsilon_i$-difference privacy.

Property 3. The data set X make up of k data sets $\{X_1, ...Xi, ...X_k\}$, and $M_i(X_i)$ satisfies ε-differential privacy, respectively. $M(X) = \{M_1(X_1), ..., X_k(M_k)\}$ satisfies $\max_{\varepsilon_i} \varepsilon_i$-differential privacy.

Our algorithm reduces the results error when answering the linear computation query under the ε-differential privacy, and we will define the linear computation query. For a database instance I whose relational schema attributes $\mathbb{A} = \{A_1, A_2, ..., A_l\}$. In $\mathbb{A}$, each attribute data can be discrete or continuous. For the continuous data, the data can be treated as discrete in the data domain as well. The *workload* means a set of queries over the attributes $\mathbb{B} = \{B_1, B_2, ..., B_k\}, \mathbb{B} \in \mathbb{A}$. For example, if the *workload* queries in a subset of three-dimensional range query over attributes A_1, A_2, and A_3, $\mathbb{B} = \{A_1, A_2, A_3\}$. We then present a frequency vector x, and $x_i \in dom(\mathbb{B})$. For example, $dom(\mathbb{B}) = \{(1, 1, 1), (1, 1, 2), ...\}$ and for each $dom(\mathbb{B})_i$, x_i is the frequency of tuples values

$dom(\mathbb{B})_i$. A linear computation query computes a linear combination of the frequency in x, as described the following SQL query and we define the linear computation query as follows definition formally.

Select count(*) from R Where $dom(\mathbb{B}) = dom(\mathbb{B})_i$ or... $dom(\mathbb{B}) = dom(\mathbb{B})_k$

$$W = \begin{bmatrix} 1 & 1 & 0 & 0 & 0 & 0 \\ 1 & 1 & 1 & 0 & 0 & 0 \\ 0 & 1 & 1 & 0 & 0 & 0 \\ 0 & 0 & 1 & 0 & 0 & 0 \\ 0 & 0 & 1 & 0 & 0 & 0 \\ 0 & 0 & 0 & 1 & 0 & 0 \\ 1 & 1 & 0 & 0 & 0 & 0 \\ 0 & 0 & 0 & 0 & 0 & 1 \end{bmatrix}, x_t = \begin{bmatrix} 2 \\ 3 \\ 4 \\ 1 \\ 0 \\ 9 \end{bmatrix}, S = W \cdot x^t = \begin{bmatrix} 5 \\ 10 \\ 7 \\ 3 \\ 4 \\ 1 \\ 0 \\ 9 \end{bmatrix} \qquad (1)$$

Fig. 1. A sample of linear computation query workload, frequency vector, and answer to the workload.

Definition 2 (Linear computation query). *A linear computation query is a length-n vector $q = [q_1, q_2, ..., q_n]$, each $q_i \in \{0,1\}$. The answer to a linear query q on x is the vector product $q \cdot x = q_1 x_1 + q_2 x_2 + ..., + q_n x_n$.*

The linear computation can be called range count query, linear count query, and point count query when the query q can be marked as range, length-n vector, or a position in x.

In the data collection situation, calculating the frequency in x can be done by the data organizers. And the data organizers has the capability to answer the linear computation query over the frequency vector x. The workload W makes up of a set of linear computation queries. If W is an $m \times n$ matrix, it means m length-n linear computation queries and the query results can be computed as the matrix product $W \cdot x$. The linear computation query is one of the most important and common queries in data mining and data analysis. The linear computation can help the analyst understand the distribution information of data and to make intelligent decisions and data prediction. Figure 1 shows a workload W, frequency vector x, and the answer to W over x.

3 Laplace-Based Disturbation

Our algorithm adopts Laplace Mechanism to add noise, and we transform the Laplace Machainsim [8] to fit the query workload and data distribution. To ensure our algorithm satisfy ε-differential privacy, the algorithm adds random noise rigorously conform to the Laplace distribution.

3.1 Laplace Mechanism

The Laplace mechanism [8] is proposed by Dwork, and the key method of Laplace Mechanism is to add noise that randomly generated through the Laplace distribution to the query results. The probability density function of the Laplace distribution is described as following:

$$Lab\,(x; a, b) = \frac{1}{2b} \exp\left(\frac{|x - a|}{b}\right) \tag{2}$$

The variance of a random variable that satisfies the Laplace distribution is $\sigma^2 = 2b^2$. To make the algorithm satisfy ε-differential privacy, we can add random noise from the $Lap(x; a, 0)$, and we denote the Laplace distribution random variable as $Lap(a)$ in the following section. For different query or query workload, the Laplace Mechanism adds noise differs against the sensitivity of the query or workload.

Definition 3 (Sensitivity). *Given a query q and the frequency vector x and* x'*, the sensitivity of the query q is:*

$$\Delta_q = \max \|q\,(x) - q\,(x')\|_1 \quad (\|x - x'\|_1 \leq 1)$$

It can be seen that the sensitivity of a query is the maximum change of the answer to a query on the neighboring frequency vectors. When the sensitivity of a query is high, the privacy data has a high probability to be attacked, and the reason is that the presence or absence of certain data can greatly change the result of the query, and it is more calculable to infer the certain sensitive data. For a query workload W, we use an $m \times n$ matrix to represent W, as shown in Fig. 2. According to the sensitivity of a query, the sensitivity of the query workload W can be defined as the following:

$$\Delta_W = max \| Wx^t - Wx'^t \| = \max_j \| \sum_{i=1}^{i=m} |W_{ij}|, (\| x - x' \|_1 \leq 1)$$

Given the definition of Laplace distribution and sensitivity, we can define the Laplace mechanism as following formally.

Definition 4 (Laplace Mechanism). *Given a workload W and a frequency vector x, $M_L(x, W, \varepsilon)$ is ε-differential privacy, if it satisfies the following condition:*

$$M_L(x, W, \varepsilon) = W \cdot x^t + (Y_1, \ldots, Y_k))$$

The random variable Y_i is generated by $Lap\left(\Delta_W \cdot \frac{1}{\varepsilon}\right)$. The proof is presented as the following, where database I and I differ at most one record, $P_I(s)$ is the probability that the output for the query database I is s.

$$\frac{p_I\left(s\right)}{p_{I'}\left(s\right)} = \prod_{i=1}^{k}\left(\frac{\exp\left(-\frac{\varepsilon\left|q(I)_i - s_i\right|}{\nabla q}\right)}{\exp\left(-\frac{\varepsilon\left|q(I')_i - z_i\right|}{\nabla q}\right)}\right)$$

$$= \prod_{i=1}^{k}\left(\frac{\exp\left(-\frac{\varepsilon\left|q(I)_i - s_i\right|}{\nabla q}\right)}{\exp\left(-\frac{\varepsilon\left|q(I')_i - s_i\right|}{\nabla q}\right)}\right) \quad (3)$$

$$= \prod_{i=1}^{k}\left(\exp\left(\frac{\varepsilon\left(\left|q\left(I'\right)_i - s_i\right| - \left|f\left(I\right)_i - s_i\right|\right)}{\nabla q}\right)\right)$$

$$\leq \prod_{i=1}^{k}\left(\exp\left(\frac{\varepsilon\left(\left|q\left(I'\right)_i - q\left(i\right)_i\right|\right)}{\nabla q}\right)\right)$$

$$\leq \exp\left(\varepsilon\right)$$

3.2 Workload-Aware Noise

To reduce the noise, we will divide the queries in workload into several workloads. Meanwhile, the privacy budget ε will be divided into the same number of privacy budgets. After dividing, different workloads will add corresponding noise according to the divided privacy budget. Formally, for the workload W, we divide it as $\{W_1, W_2, ..., W_m\}$, For each divided workload W_i, the privacy budget is also divided into $\varepsilon = \{\varepsilon_1, \varepsilon_2, ..., \varepsilon_m\}$, and add random noise from the distribution $Lap(\Delta_{W_i} 1/\varepsilon_i)$. That is, the answer to the workload is $S = [W_1, W_2, ..., W_m] \cdot x_t + [Lap(\Delta_{W_1} 1/\varepsilon_1), \ Lap(\Delta_{W_2} 1/\varepsilon_2), ... \ Lap(\Delta_{W_m} 1/\varepsilon_m)]$. It can be proved that the algorithm satisfies ε-differential privacy as the *property2*, and we can also prove it by the following process. For the neighboring database I and I', that is, $\| I - I' \|_1 \leq 1$. Let $p_I(s)$ represent the distribution probability of query x on W, and $s \in R^k$:

$$\frac{p_I\left(s\right)}{p_{I'}\left(s\right)} = \prod_{i=1}^{k}\left(\frac{\exp\left(-\frac{\varepsilon_i\left|q(I)_i - s_i\right|}{\nabla q_i}\right)}{\exp\left(-\frac{\varepsilon_i\left|q(I')_i - z_i\right|}{\nabla q_i}\right)}\right)$$

$$= \prod_{j=1}^{m}\prod_{i\in W_j}\left(\frac{\exp\left(-\frac{\varepsilon_i\left|q(I)_i - s_i\right|}{\nabla q_i}\right)}{\exp\left(-\frac{\varepsilon_i\left|q(I')_i - s_i\right|}{\nabla q_i}\right)}\right) \quad (4)$$

$$\leq \prod_{j=1}^{m}\exp\left(\varepsilon_j\right)$$

$$= \exp\left(\varepsilon\right)$$

We discuss the error change by the dividing for workload $W = \{W_1, W_2, ..., W_m\}$, and privacy budget $\varepsilon = \{\varepsilon_1, \varepsilon_2, ..., \varepsilon_m\}$. We calculate the average L_1 error for the answer to the workload. The excepted L_1 error of the answer before dividing and after dividing the workload is as follows:

$$E\left(\mid Lap(\Delta_W \cdot 1/\varepsilon) \mid\right) = \Delta_{W_i} \cdot \frac{1}{\varepsilon_i}$$

$$E\left(\frac{1}{k}\sum_{i=1}^{m} \mid Lap(\Delta_{W_i} \cdot 1/\varepsilon_i) \mid \cdot \mid W_i \mid\right) = \frac{1}{k}\sum_{i=1}^{m} \Delta_{W_i} \cdot \frac{1}{\varepsilon_i} \cdot \mid W_i \mid$$

4 Optimized Workload Division

Taking the above workload in Fig. 2 for example, the original workload and divided workloads as following. When the workload W adopts 1-differential privacy by the Laplace Mechanism. The excepted L_1 error is $\Delta_W/\varepsilon = 4$, and after dividing the workload into W_1 and W_2, the privacy budget into $\varepsilon_1 = 0.58$ and $\varepsilon_2 = 0.42$, the L_1 error will be 3.4275.

$$W = \begin{bmatrix} 1&1&0&0&0&0 \\ 1&1&1&0&0&0 \\ 0&1&1&0&0&0 \\ 0&0&1&0&0&0 \\ 0&0&1&0&0&0 \\ 0&0&0&1&0&0 \\ 1&1&0&0&0&0 \\ 0&0&0&0&0&1 \end{bmatrix}, W_1 = \begin{bmatrix} 1&1&0&0&0&0 \\ 1&1&1&0&0&0 \\ 0&1&1&0&0&0 \end{bmatrix}, W_2 = \begin{bmatrix} 0&0&1&0&0&0 \\ 0&0&1&0&0&0 \\ 0&0&0&1&0&0 \\ 1&1&0&0&0&0 \\ 0&0&0&0&0&1 \end{bmatrix} \quad (5)$$

Fig. 2. A sample of dividing for workload.

Basing on the dividing for workload and privacy budget, we propose a specific division in the data situation that the data is relatively large and the distribution of data is complex. We take the mean square error of the frequency vector x to discriminate the data distribution complexity. And in query workload, there exist data domain queried with high frequency. To reduce the added Laplace-based noise, we divide the privacy budget into two equal parts iteratively, and the workload is divided according to the sensitivity, and the process can be described in Algorithm 1.

The dividing in the algorithm will continue until the recursion finished. We will discuss the rationality of the division. The dichotomy is used as the reason that the query workload and privacy budget are divided into two parts W_1, W_2, ε_1, ε_2, and for $E(L_1) = \frac{\nabla_{W_1}}{\varepsilon_1} * |W_1| + \frac{\nabla_{W_2}}{\varepsilon_2} * |W_2|$, $\varepsilon_1 = \varepsilon_2$ is the minimum extreme point of the function. As described in Algorithm 1, at first, we set the

---

Algorithm 1. Workload dividing

1: **procedure** DIVIDEWORKLOAD(W, ε)
2:     $\varepsilon_1 = \varepsilon_2 = \varepsilon/2.0$
3:     get the most frequent item in x as x_i,
4:     $|W_1| = (|W| + \nabla_W + g)/4$
5:     select randomly $|W_1|$ queries as W_1 from W where x_i is queried and the rest
       queries as W_2
6:     $noise = |W| * \nabla_W$, $noise\_divided = |W_1| * \nabla_{W_1} + (|W| - |W_1|) \cdot \nabla_{W_2}$
7:     **if** $noise \geq noise\_divided$ **then**   ▷ Stop dividing while noise doesn't reduce
8:         **return** ($DIVIDEWORKLOAD(W_1, \varepsilon/2)$, divideWorkload($W_1, \varepsilon/2$))
9:     **return** W ▷ return W while it is unnecessary to divide

---

data domain queried by high frequency as high-frequency items. For workload W, its division is W_1 and W_2 and supposing that the sensitivity of W is the sum of W_1 and W_2. The total expected noise under the ε-differential privacy is

$$E(noise(\varepsilon_1, \varepsilon_2, W_1, W_2)) = \frac{\Delta_{W_1}}{\varepsilon_1} * |W_1| + \frac{\Delta_W - \Delta_{W_1}}{\varepsilon_2} * (|W| - |W_1|) \quad (6)$$

We can infer that $(\varepsilon/2, \varepsilon/2, W_1, W_2)$ is a point of minimum, so we adopt $\varepsilon_1 = \varepsilon_2 = 2/\varepsilon$. To get $min\ E(noise(\varepsilon_1, \varepsilon_2, W_1, W_2))$, we set $\Delta_{W_1} = |W_1|$, and we can compute that when $|W_1| = (W + \Delta_W)/4$, the $E(noise\varepsilon_1, \varepsilon_2, W_1, W2)$ will be a minimal value. The parameter g can optimize the result as a consequence of that for a workload W and its divisions W_1, W_2, $\Delta_W > (\Delta_{W_1} + \Delta_{W_2})$, which is not in accordance with our assumption. Therefore, we introduce the parameter to regulate the result and the g can be estimation by the specific workload.

5   Experimental Evaluation

We now evaluate the performance of our approach on multiple datasets and workloads and compare our algorithm with state-of-the-art differential privacy algorithms. The main metric is average error, and we evaluate the metric on differential datasets and workloads (Fig. 3).

In follow evaluation, we test our algorithm with the metric average L_1 error per query result of the given workload. The workloads we use are generated randomly and the data set is from the real public database. To make the result more convincing, we run 5 trials for each evaluation. Furthermore, we test the time efficiency of our algorithm. The synthetic workload also is used by all the comparison algorithms. In the experiment, we set the privacy budget varies in $\{10.0, 5.0, 1.0, 0.5, 0.1\}$. In the following sections, we describe the datasets, workload, and the parameters in the experiment. In the above section, we have described the properties of liner count query. When there are multiple attributes in datasets, we can still use one-dimensional frequency vector x to represent the datasets. In the experiment, we use one-dimensional data sets. We use six real data sets. Adult comes from American statistical data [4]. The frequency vector

x is built on the attribute "capital loss", which is also used in the experiment [13]. The Adult is sparse, and many frequency counts in x are zero. Income is from IPUMS American community survey data from 2001–2011, and frequency vector x is the count of personal Income [22], and Income is also used in DAWA [16]. Patent is a citation network among a subset of US patents [16]. Fidaeta is from census of fatal occupational injuries in the United States of American labor statistics [1], and both Nw and Nw2 are from a survey of compensation in the United States of American labor statistics [1], and they are the frequency vector by setting unit as 1 and 2 in the continuous value attribute. We take the length of x of the five datasets as 4096. The overview of datasets is described in Table 1.

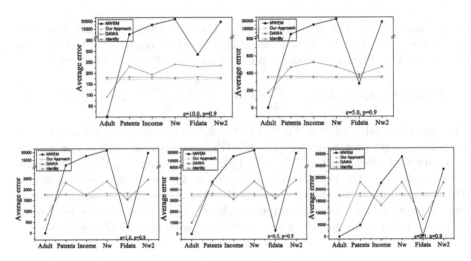

Fig. 3. Average error on the workload that the frequency count item is queried with probability as $p = 0.9$

Table 1. Overview of the datasets in the experiments.

Datasets name	Scale	% Zero Count	Mean value	Variance
Adult	17665	97.998	4.31274	263.04404
Patents	27948226	6.20118	6823.29736	3532.42422
Income	20787122	44.971	5074.98095	47859.49063
Nw	32287151	0.268	7882.60522	60262.21603
Fidata	3519442	58.178	859.23880	18942.96715
Nw2	32678757	0.0	7978.21216	84866.75896

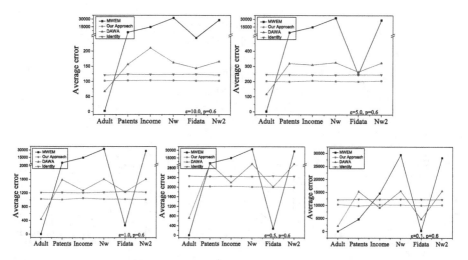

Fig. 4. Average error on the workload that the frequency count item is queried with probability as $p = 0.6$

For the query workload, we conduct the experiment on eight synthetic query workloads W. For the frequently queried item in frequency vector x, we set probability of being queried $p = \{0.2, 0.3, 0.4, 0.5, 0.6, 0.7, 0.8, 0.9\}$. A workload has 2000 queries, and each query $q \in W$ randomly selected a center cluster, and the frequency counts in x are randomly generated via the normal distribution with c as the center and 10 as the variance. Furthermore, we compare three algorithms with our algorithm. The Identity [8] algorithm adopts Laplace Mechanism that the answer results are directly added Laplace distribution noise for disturbance. MWEM [13] achieves differential privacy technology by obtaining an estimate of x through Laplace Mechanism and Exponential Mechanism. DAWA [16] algorithm adopts the partitioning method to achieve the differential privacy for range count workload and linear count workload.

Among the experimental datasets, Adult is a "sparse" data set. The data distribution is relatively even-distributed as shown in Table 1, and zero accounts for more than 97% in the frequency vector x. The other four experimental data sets are "complex" data sets with a large scale and complex data distribution. Figure 4, 5, and 6 show the L_1 average error for the parameter p as 0,9, 0.6 and 0.2. It can be seen that MWEM [13] and DAWA [16] will add more noise than the Identity [8] algorithms, meanwhile, our algorithm always adds less noise than the Identity. The results figures show that MWEM [13] and DAWA [16] algorithm are datasets-aware and when facing different datasets, both algorithms perform differently over the same workload. The MWEM is most erratic, and when the data sets are simple or approximately even-distributed, the algorithm can add less noise than the other algorithms, but not for the complex data. In Fig. 6, we compare the discount of L_1 average error by comparing it with the Identity [8]. In the experiment, we compare the different perforation with different parameter

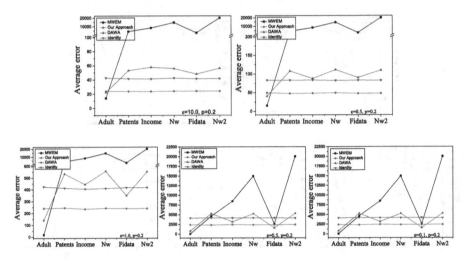

Fig. 5. Average error on the workload that the frequency count item is queried with probability as $p = 0.2$

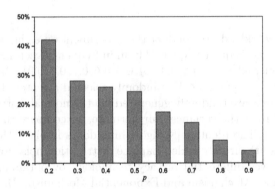

Fig. 6. The decrement of the average error by comparing our method with Identity.

p, which represents frequency of a certain $dom(\mathbb{B})_i$ in x. Figure 5 shows that in the experiment sets, when $p = 0.2$, the algorithm can reduce more than 40% the L_1 error than the Identity.

6 Conclusions

The ε-differential privacy is an effect privacy-preserving technology for linear computation. It can prompt data organizers to provide a secure third-party interface for statistical query. In this paper, we propose a novel ε-differential privacy algorithm, which uses Laplace-based noise and optimized workload division to decrease the computation error in complex data distribution for linear computations. The evaluation results show that our approach can reduce nearly

40% computation error when compared with the start-of-the-art differential privacy algorithms MWEM, DAWA, and Identity. As further work, we plan to extend our approach by optimizing the proposed work load division to reduce the introduced error.

References

1. https://www.bls.gov
2. Acs, G., Castelluccia, C., Chen, R.: Differentially private histogram publishing through lossy compression. In: 2012 IEEE 12th International Conference on Data Mining, pp. 1–10. IEEE (2012)
3. Ashwin, M., Daniel, K., Johannes, G., Muthuramakrishnan, V.: L-diversity: privacy beyond k-anonymity. ACM Trans. Knowl. Discov. Data 1(1), 1–52 (2007)
4. Bache, K., Lichman, M.: UCI machine learning repository 2013 (2013). http://archive.ics.uci.edu/ml5
5. Backes, M., Meiser, S.: Differentially private smart metering with battery recharging. In: Garcia-Alfaro, J., Lioudakis, G., Cuppens-Boulahia, N., Foley, S., Fitzgerald, W.M. (eds.) DPM/SETOP -2013. LNCS, vol. 8247, pp. 194–212. Springer, Heidelberg (2014). https://doi.org/10.1007/978-3-642-54568-9_13
6. Blocki, J., Blum, A., Datta, A., Sheffet, O.: Differentially private data analysis of social networks via restricted sensitivity. In: Proceedings of the 4th Conference on Innovations in Theoretical Computer Science, pp. 87–96. ACM (2013)
7. Diffie, W., Hellman, M.E.: Special feature exhaustive cryptanalysis of the NBS data encryption standard. Computer 10(6), 74–84 (1977)
8. Dwork, C., McSherry, F., Nissim, K., Smith, A.: Calibrating noise to sensitivity in private data analysis. In: Halevi, S., Rabin, T. (eds.) TCC 2006. LNCS, vol. 3876, pp. 265–284. Springer, Heidelberg (2006). https://doi.org/10.1007/11681878_14
9. Dwork, C., Roth, A., et al.: The algorithmic foundations of differential privacy. Found. Trends® Theor. Comput. Sci. 9(3–4), 211–407 (2014)
10. Erlingsson, Ú., Pihur, V., Korolova, A.: Rappor: randomized aggregatable privacy-preserving ordinal response. In: Proceedings of the 2014 ACM SIGSAC Conference on Computer and Communications Security, pp. 1054–1067. ACM (2014)
11. Feng, P., Zhu, H., Liu, Y., Chen, Y., Zheng, Q.: Differential privacy protection recommendation algorithm based on student learning behavior. In: 2018 IEEE 15th International Conference on e-Business Engineering (ICEBE) (2018)
12. Ge, C., He, X., Ilyas, I.F., Machanavajjhala, A.: APEx: accuracy-aware differentially private data exploration. In: Proceedings of the 2019 International Conference on Management of Data, pp. 177–194. ACM (2019)
13. Hardt, M., Ligett, K., McSherry, F.: A simple and practical algorithm for differentially private data release. In: Advances in Neural Information Processing Systems, pp. 2339–2347 (2012)
14. Hay, M., Machanavajjhala, A., Miklau, G., Chen, Y., Zhang, D.: Principled evaluation of differentially private algorithms using DPBench. In: Proceedings of the 2016 International Conference on Management of Data, pp. 139–154. ACM (2016)
15. Lee, J., Clifton, C.W.: Top-k frequent itemsets via differentially private FP-trees. In: Proceedings of the 20th ACM SIGKDD International Conference on Knowledge Discovery and Data Mining, pp. 931–940. ACM (2014)
16. Li, C., Hay, M., Miklau, G., Wang, Y.: A data-and workload-aware algorithm for range queries under differential privacy. Proc. VLDB Endow. 7(5), 341–352 (2014)

17. Li, C., Hay, M., Rastogi, V., Miklau, G., McGregor, A.: Optimizing linear counting queries under differential privacy. In: Proceedings of the Twenty-Ninth ACM SIGMOD-SIGACT-SIGART Symposium on Principles of Database Systems, pp. 123–134. ACM (2010)
18. Li, N., Li, T., Venkatasubramanian, S.: t-closeness: privacy beyond k-anonymity and l-diversity. In: 2007 IEEE 23rd International Conference on Data Engineering, pp. 106–115. IEEE (2007)
19. McSherry, F., Talwar, K.: Mechanism design via differential privacy. In: FOCS, vol. 7, pp. 94–103 (2007)
20. Pham, A.T., Raich, R.: Differential privacy for positive and unlabeled learning with known class priors. In: 2018 IEEE Statistical Signal Processing Workshop (SSP) (2018)
21. Rivest, R.L.: The RC5 encryption algorithm. In: Preneel, B. (ed.) FSE 1994. LNCS, vol. 1008, pp. 86–96. Springer, Heidelberg (1995). https://doi.org/10.1007/3-540-60590-8_7
22. Ruggles, S., Alexander, J.T., Genadek, K., Goeken, R., Schroeder, M.B., Sobek, M.: Integrated public use microdata series: Version 5.0, 2010, Minnesota Population Center, Minneapolis, MN (2015)
23. Schneier, B.: Description of a new variable-length key, 64-bit block cipher (Blowfish). In: Anderson, R. (ed.) FSE 1993. LNCS, vol. 809, pp. 191–204. Springer, Heidelberg (1994). https://doi.org/10.1007/3-540-58108-1_24
24. Sweeney, L.: k-anonymity: a model for protecting privacy. Int. J. Uncertainty, Fuzziness and Knowl.-Based Syst. **10**(05), 557–570 (2002)
25. Warner, S.L.: Randomized response: a survey technique for eliminating evasive answer bias. J. Am. Stat. Assoc. **60**(309), 63–69 (1965)
26. Xu, J., Zhang, Z., Xiao, X., Yang, Y., Yu, G., Winslett, M.: Differentially private histogram publication. VLDB J. **22**(6), 797–822 (2013). https://doi.org/10.1007/s00778-013-0309-y
27. Zhang, D., McKenna, R., Kotsogiannis, I., Hay, M., Machanavajjhala, A., Miklau, G.: Ektelo: a framework for defining differentially-private computations. In: Proceedings of the 2018 International Conference on Management of Data, pp. 115–130. ACM (2018)

On the Automated Assessment of Open-Source Cyber Threat Intelligence Sources

Andrea Tundis[1]([✉]), Samuel Ruppert[2], and Max Mühlhäuser[1]

[1] Department of Computer Science, Technische Universität Darmstadt (TUDA),
Hochschulstrasse 10, 64289 Darmstadt, Germany
{tundis,max}@tk.tu-darmstadt.de
[2] Deutsche Bahn AG, Frankfurt am Main, Germany
samuel.ruppert@deutschebahn.com

Abstract. Global malware campaigns and large-scale data breaches show how everyday life can be impacted when the defensive measures fail to protect computer systems from cyber threats. Understanding the threat landscape and the adversaries' attack tactics to perform it represent key factors for enabling an efficient defense against threats over the time. Of particular importance is the acquisition of timely and accurate information from threats intelligence sources available on the web which can provide additional intelligence on emerging threats even before they can be observed as actual attacks. In this paper, an approach to automate the assessment of cyber threat intelligence sources and predict a relevance score for each source is proposed. Specifically, a model based on meta-data and word embedding is defined and experimented by training regression models to predict the relevance score of sources on Twitter. The results evaluation show that the assigned score allows to reduce the waiting time for intelligence verification, on the basis of its relevance, thus improving the time advantage of early threat detection.

Keywords: Open source cyber threat intelligence · Cybersecurity · Machine learning · Feature engineering · Twitter

1 Introduction

Emerging vulnerabilities in computer systems can lead to far reaching impacts due to the high number of possibly affected systems. Cyber Threat Intelligence (CTI) is an emerging field whose main mission is to research and analyze trends and technical developments related to Cybercrime, Hactivism and Cyberespionage, based on the collection of intelligence using open source intelligence (OSINT), social media intelligence, human intelligence. Current research directions are exploring OSINT as a means to proactively gather CTI from individuals and organizations that share relevant information (e.g. vulnerabilities, zero-day exploits) publicly on the web, sometime just spread, sometime to openly recruit

© Springer Nature Switzerland AG 2020
V. V. Krzhizhanovskaya et al. (Eds.): ICCS 2020, LNCS 12138, pp. 453–467, 2020.
https://doi.org/10.1007/978-3-030-50417-5_34

groups (so called "hacktivists") for an imminent attack campaign [13]. This scenario, and the fact that timeliness is essential in security, emphasizes the need to determine the relevance of such information not only based on whether it is already widely spread but also on the quality and informativeness of the source itself [17]. Different publishers, security professionals, vendors and researchers provide cyber threat-related information on vulnerabilities and even hackers post information about ongoing attack campaigns or new vulnerabilities on social media like Twitter, as well as forums and marketplaces in the darkweb. Obviously, this information varies strongly with regards to credibility, timeliness and level of detail, and it is difficult to acquire and assess it in an automated manner since the sources do not only vary content-wise but also regarding their structure and syntax. To understand these evolving threats, it is essential for security experts to illuminate the threat landscape including adversaries, their tools and techniques [9]. To deal with this need, it is simply not practical to implement counter-measures in a timely and economical manner for all possible attacks, but learning about the details of cyber threats relevant sources and, prioritizing them is a vital step in defending computer systems.

For this reason the extraction of CTI from such open sources, i.e. publicly accessible data on the internet, has been the target of recent research in the field of OSINT (see Sect. 2). Dalziel et al. [2] define CTI as: *Information that has been refined, analyzed or processed such that it is relevant, actionable and valuable with regards to an organization's security objectives.* In this context, the term is used to describe threat-related information which allows cyber security experts to investigate on a certain threat, e.g. the name of a malware, adversary or vulnerability. Additionally, it is considered actionable, if it is obtained in a timely manner meaning in due time to adapt the defensive measures to the threat in question before it hits in the form of an attack. Automating the collection of CTI can improve the defense capabilities against cyber threats but itself requires to face with the selection of the most relevant sources, the balancing between precision and timeliness that lead to an earlier generation of threat alerts, which in turn provides the security experts more time to prepare against potential upcoming attacks. Relying on the intelligence alone for an emerging threat is not assumed to be sufficient, and waiting for the occurrence of additional information to confirm the threat reduces the time advantage [14].

In this direction, this paper proposes an approach for the automated assessment of the OSINT sources themselves as an additional criterion for the relevance of CTI. In particular, an upstream assessment of the publishing source itself is taken into account, both when generating intelligence-based alerts and to decide whether a source should be used for CTI collection or not. In particular, a specific OSINT source has been selected based on a survey conducted among cyber security professionals and academic researchers who are working in the field of threat intelligence. Then two feature sets, that characterize the OSINT source have been defined. A scoring function to quantify the relevance of an OSINT source with regards to CTI in particular consideration of the timeliness has been proposed. The experimentation was conducted by training 5 regression models on both feature sets to predict the relevance score for OSINT sources, by focusing on Twitter, and compared with related approaches.

The rest of the paper is organized as follows: in Sect. 2, the most related works are discussed. Section 3 elaborates the overall proposal in order to achieve the aforementioned objectives. The implementation details, the evaluation approach along with the gathered results are presented in Sect. 4, whereas Sect. 5 concludes this work.

2   Related Work

The growing interest in cyber threat intelligence (CTI) with regards to open sources (OSINT) is shown by the increasing research efforts in this field.

In [14], a ranking mechanism, to automate the evaluation of CTI sources and by selecting a subset of sources for CTI collection, is proposed. It deals with vulnerabilities disclosure in Twitter, by examining tweets which contain a Common Vulnerabilities and Exposures (CVE) ID. The authors showed that monitoring a subset of users on Twitter can be sufficient to retrieve most of the vulnerability-related information that is available on the microblogging platform. However, no ranking or scoring of the actual sources and their relevance is provided and they did not considered the detection of emerging malware and zero-day attacks. In [16], instead, the need for a quantitative evaluation of CTI sources is discussed and then an adaptive methodology for a weighted evaluation of such sources is proposed. The methodology introduces six evaluation categories on the basis of intelligence source aspects: (i) type of information, (ii) provider classification, (iii) licensing options, (iv) interoperability, (v) advanced API support and (vi) context applicability. The use of only structured data represents a limit in their methodology, furthermore, as the authors stated, other information such as the timeliness based on the time passed was not considered. In [6], a system, called Sec-Buzzer, for the detection of emerging topics related to cyber threats from expert communities on Twitter, is presented. It automatically identifies new experts on Twitter and adds them to a list of OSINT sources. In particular, the *activeness* of new candidates (i.e. potential experts) is evaluated on the basis of the number of tweets within a specified time period. The most active users are then further assessed according to their topic-relevance by examining the number of times they were mentioned in tweets and retweets by the most active existing experts. The main lack of this approach is that the user's activeness, as initial selection criterion, considers users with a high frequency of tweeting as experts. Even among cybersecurity-related Twitter accounts the number of tweets within a given time frame might not necessarily characterize a valuable threat intelligence source. In [15] instead, a system called DISCOVER is presented. It crawls both Twitter accounts of 69 international researchers and security analysts as well as a manually compiled list of 290 security blogs to discover emerging terms in the context of cyber threats. A natural language processing technique is used to preprocess the textual data as long as a list of terms related to emerging cyber threats is defined. They achieved 84% precision for warnings based on data from Twitter and 59% for the security blogs. Another research effort, called CyberTwitter [10], aimed to discover and analyze cybersecurity intelligence on Twitter, collected in real-time by using Twitter API. The considered relevant information on cyber threats was then extracted on the basis of the Security Vulnerability Concept Extractor (SVCE), that is basically a Named Entity Recognizer (NER) specialized for cyber security terms. The automatic identification and generation of warnings was based on a set of properties, such as, the maximum time period for which intelligence is considered relevant. It showed that 57.2% of all inspected entities extracted by the SVCE were marked correctly and 33.2% were partially correct. From a total of 37 relevant intelligence entries the system generated 15 warnings, 13 of them were assessed as "useful" and the 2 remaining were "maybe useful". Then, 300 discarded tweets were manually examined by obtaining 85% recall. [11] extends [10] by introducing (i) National Vulnerability Databases (NVD), security blogs, Reddit and darkweb forums as additional OSINT sources along with Twitter, (ii) as well as a hybrid structure, called VKG, which combines knowledge graphs and word embeddings in a vector space. The approach was evaluated by manually annotating 60 alerts from which 49 were marked correct with a Precision of 81.6%. Furthermore, the SPARQL query engine was evaluated by searching for concepts that were marked "similar" by the annotators. Best results were reached for word embeddings with a dimensionality of 1500 and term frequency 2 which lead to a mean average precision of 69%. In [5], the authors tried to identify cyber threat-related tweets and gather CTI by linking mentioned vulnerabilities with their associated Common Vulnerabilities and Exposures (CVE). The proposed Centroid and the One-class Support Vector Machine (OCSVM) were compared to typical SVM, MLP, CNN, by showing that the Centroid novelty classifier using the cosine similarity distance performed better than the OCSVM with 85.1% Precision and 51.7% Recall. In [18], articles related to OSINT sources were examined, to gather insight into the semantics of malicious campaigns and the stages of malware distribution. The system extracts indicators of compromise (IOC) from security articles using regular expression, since they usually have fixed formats, e.g. IP address or hashsum. During the evaluation 91.9% Precision and 97.8% Recall for the IOC detection was reached and the stage classification through word embeddings resulted in an average Precision of 78.2%. In a survey reported in [17], emerged that cybersecurity experts are still unsatisfied with regard to the timeliness of many approaches that are currently used to collect CTI. The above presented research efforts and others [7,14] aimed to achieve earlier detection of cyber threats, by confirming the importance of such requirement.

From this review, important findings emerged, that were taken into consideration to narrow down the scope of this work. The main lack is due to the limited inspection to the textual data by neglecting the sources themselves for automated threat detection and warning generation. In light of such conducted analysis, the next Section elaborates the proposed approach in order to answer the following questions: (i) How to select relevant OSINT sources to be monitored, with high potential of publishing CTI, in order to avoid a large part of unreliable or outdated intelligence? (ii) How to automatically assessed the threat intelligence's quality and credibility in order to issue a reliable warning for emerging threats?

3   Automated Assessment of OSINT Sources Driven by Features

In this Section, the process for automating the assessment of an OSINT source, for cyber threat intelligence, is described. The methodology, which is depicted in Fig. 1, can be organized in three main phases: *OSINT Sources Identification, Feature Selection* and *Score Definition*, that are elaborated in the following.

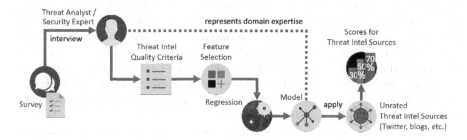

Fig. 1. Research method

3.1   OSINT Sources Identification

In the field of open-source intelligence a variety of public web sources, such as openly accessible web (e.g. vendor websites, Social network accounts, blogs) as well as forums and marketplaces in the darkweb, could be used to collect different types of threat intelligence. To deal with this challenge regarding the selection of relevant OSINT sources, an empirical study was conducted through an interview with 30 experts (ie. cyber security professionals and academic researchers) in the field of threat intelligence. The survey, which is used to establish the scope of this work but not the validity results, was based on the following questions:

1. What type of cyber threat intelligence is already being collected today?
2. How do experts rate the demand for improved CTI collection?
3. Which OSINT source are being utilized in today's CTI practice?
4. What are the most important criteria to be used to evaluate these sources?
5. How do experts rate certain sources with regards to their quality?
6. What features do the experts consider when evaluating the selected sources?

It aimed to retrieve information about (i) the type of CTI looked for in OSINT sources, such Zero-day vulnerabilities, CVE, IOC, upcoming malware, adversaries, etc. (ii) the characteristics to look for in a considered credible and qualitatively suitable source, such as technical details, code samples, author name, outgoing links, google ranking, etc.; (iii) whether a set of OSINT sources are already being used or there are new one and how they would be rated with regard to quality, credibility; (iv) OSINT sources that are planned to be examined in the future or that might be worth to be examined by motivating that; (v) how often and how new OSINT sources are looked for, for example word of mouth, links found in specialized websites, search engines; (vi) how some provided CTI sources would be rated with regards to quality, credibility, TI domain and effort, when a manual searching and processing information is conducted.

Furthermore, the selected OSINT source types were quantified with regards to 4 different characteristics that are typical for threat intelligence, that is, (i) *Level of detail*: the source provides in-depth information about a threat, (ii) *Credibility*: the source provides credible intelligence (high true positive rate); (iii) *Timeliness*: the source provides intelligence in good time to act on it, (iv) *Actionable*: the source provides intelligence which can be used directly to support an organization's security objectives. Each criterion was rated on a scale from 0 (poor) to 5 (good) depending on whether the source usually provides intelligence with low or high quality for this criterion.

The first insight, according to the domain experts, was that the most important criteria for the evaluation of OSINT sources are both the *credibility* and the *timeliness* with which a source provides intelligence. The second one was that, among the top-5 types of cyber threat intelligence, as it is shown in Fig. 2-(a), 2 of them emerged (*vulnerability* and *malware*). In particular, by using the average value of the obtained values as the threshold, the demand for intelligence on "vulnerabilities and exploits" as well as "maleware" resulted particularly higher. In addition, the participants were asked to rate the most common OSINT source types from the related work: (i) public threat feeds, (ii) third-party websites and blogs, (iii) darkweb forums and marketplaces, (iv) Twitter, (v) Reddit, (vi) Pastebin and similar text & code storage websites, as it is depicted in Fig. 2-(b).

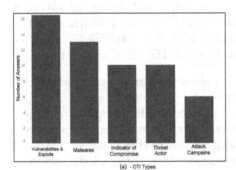 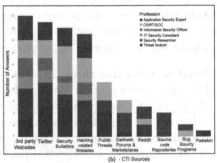

Fig. 2. CTI types and CTI sources

Since the types (iii–v) comprise many sources (i.e. user accounts) for which the same meta data (i.e. features) is available, the experts were asked to select the features that they considered promising or highly typical for valuable OSINT sources. They were also able to name additional features that they use when evaluating sources (see Sect. 3.2). On the basis of such insights, that third-party blogs, websites and Twitter emerged as the preferred sources for intelligence on new vulnerabilities and malwares. In particular, the CTI source was chosen by considering two main factors: (i) the popularity of the source in the context of threat intelligence, (ii) the type of available data that can be retrieved for supporting further analysis on them. Furthermore, even if Third-party website were rated higher with regards to the level of detail, Twitter is seen as a much more timely source type. Combining these findings and the fact that Twitter provides unified metadata on each user, which allows for better assessment and comparison, the author decided to investigate on Twitter as the OSINT source.

3.2 Feature Selection

From the analysis of related work, resulted that all existing methodologies aimed to identify cyber threat intelligence in different forms and qualities using natural language processing and machine learning techniques. Only few of them examined aspects of the source but none of them apply to the sources theirselves a feature-driven machine learning approach.

On the other hand, various research efforts have been conducted to examine the role and characteristics of influencers on Twitter, such as, users who are considered authoritative within a certain topical domain, as well as metrics to quantify the credibility of tweets and Twitter users. These approaches are often based on features extracted from profile meta data, the social graph and textual data from Tweets.

Based on such information, the first set of features, centered on meta-data listed in Table 1, has been selected by considering 3 aspects: (i) *Profile related features:* these are characteristics of a Twitter profile that are directly associated with the user profile (e.g. *registration date, the user's specified location, number of followers* and so on). (ii) *Social graph related features:* this features are related to the connections (edges) among certain users (i.e. nodes) and allows to inspect the relations between them within a group or community of connected profiles. In particular: *followed/following, retweets* and *mentioned/mentions*, where in-degree and out-degree values of each node can be computed and compared. (iii) *Tweet related features:* other features, which are specifically associated to a single Tweet, that provide additional information on the user's behaviour with regards to the published Tweets. The second feature set is based on the *word embedding* technique, that is adopted to examine only textual content of the Tweets. It is based on "doc2vec" algorithm, with a 50-dimensional word embeddings as in [12], that strives when determining the similarity between different textual data.

Table 1. Selected features based on related works and centered on Twitter meta-data

Feature	Description
num_mentions_community	The out-degree of the user in the mentions
num_hashtags	Total number of hashtags used in the observed time
ratio_retweets_replies	Ratio between retweets made by the user and replies received
num_mentioned_community	In-degree of the user in the mentions' monitored CTI social graph
num_retweets	Total number of retweets for a user
mean_mentions	Average number of mentions over all Tweets in the observed time period
num_tweets	Total number of tweets by a user
num_media	Total number of tweets containing media, for example images
verified	Whether the account has the 'verified' status by Twitter
num_likes	Total number of likes (favorites) received
num_following	Total number of friends, i.e. accounts that are followed by this user
days_since_join	Number of days since registration
mean_time_between_tweets	Average time between tweets during the observed time period in seconds
length_bio	Length of the user's description (biography)
mean_hashtags	Average number of hashtags per Tweet in the observed time period
num_followers	Total number of followers
length_username	Length of the displayed username
has_url	Whether the user profile has a website specified
length_url	Length of the website URL
mean_retweets	Average number of retweets made in the observed time period
num_mentions	Total number of mentions made by the user
mean_replies	Average number of replies received by the user
ratio_followers_following	Ratio between number of followers and following (friends)
mean_likes	Average number of likes (favorites) the user received
has_location	Whether the user profile has a location specified
num_replies	Total number of replies received by the user

3.3 Score Definition

In order to support the evaluation of the relevance of a threat intelligence source, a score function is proposed. It assigns a score R_I, between 0 and 1, to each threat intelligence source I on the basis of the weighted count of all true published intelligence $r_i \in I$. The proposed decay function, for calculating the score for a single CTI term r_i, is represented through Eq. (1).

$$r_i = score(t_i) = \begin{cases} 1 - 0.5 \cdot \left(\frac{t}{C-1}\right)^2 & s \cdot (c-1)^{1.25} \leq t \leq s \cdot c^{1.25}, \ 0 < c \leq C, \ c \in N \\ 0.5 \cdot 0.5^{\frac{t}{s}} & s \cdot C < t \end{cases} \quad (1)$$

To include the timeliness of intelligence the weighting uses the time span that passed since a CTI term has been observed for the first time within the monitored community and the moment it is mentioned again by one of the other sources. In particular, this time delta t, which is determined in seconds, is then used as an input to the function which calculates the actual weight. Additionally, for a chosen number of intervals C the score is calculated as a step function such that slight time differences during the first few minutes or hours after the first occurrence of some threat intelligence do not influence the score. This was done because users considered intelligence sufficiently timely during an initial time period after the first occurrence and wanted a decrease in the score to indicate larger time differences, i.e. change in intervals. The value C = 5 has been empirically determined, and the size of the first interval was set to s = 86,400 which corresponds to the number of seconds in a full day. For intelligence which was observed exactly after the initial time intervals s*C, the score is score(s*C) = 0.5 and intelligence mentioned later than this point of time gets a score below 0.5 assigned through the exponential decay function.

Then, all the r_i are aggregated per source I in order to assign a single relevance score to each source R_I according to Eq. (2).

$$cti\_relevance\_score(R_I) = \frac{1}{|R_I|} \sum_{i=1}^{|R_I|} r_i \cdot \frac{log(|R_I|)}{log(|R|)} \tag{2}$$

In particular, the arithmetic mean is calculated over all single relevance scores $r_i = score(t_i)$ of a source $R_I = \{r_1, r_2, ..., r_I\}$ and weighted by the logarithmically normalized number of threat-related terms that were observed for this source, where R represents the full set of all scores and R_I the scores for intelligence shared by source I. After all sources are assigned a CTI Relevance Score, they are used to train a model to predict the relevance (see Sect. 4), measured through a value between [0,1], for other sources on the basis of their features.

4   Implementation and Conducted Experiments

In this Section, first data collection, the used regressor models and evaluation metrics are described and then the experimental results are reported.

4.1   Data Collection, Regression Models and Evaluation Criteria

The data collection focused on Tweets and Twitter profiles, including metadata, related to the field of cybersecurity as a starting point to generate sets of training and testing data later on. In particular, an initial list of cyber security and cyber threat-related hashtags was manually compiled *(e.g. infosec, cybersecurity, security, threatintel, hacking, malware)*, as result of the survey.

This initial list of hashtags was then extended using the official Twitter API and third-party web services to find a more complete list of hashtags that are commonly being used in combination with one of the initial hashtags and therefore are assumed to be relevant to the field of cyber threat intelligence *(i.e. bugbounty, cve, cvss, cyberattack, cybercrime, cybercriminals, cybersec, databreach, dataleak, exploit, exploits, hacker, hackers, itsec, itsecurity, privacy, ransomware, redteam, threat intelligence, virus, vuln, vulnerabilities, vulnerability)*. This procedure was repeated on a daily basis from the 1st until the 31st of May 2019.

The official Twitter API was queried to retrieve the suggested hashtags listed under "Related Search" as well as three third-party web services, namely *keyhole.co*, *RiteKit* and *Hashtagify*. From each of these sources and for each of the hashtags in the current list, the top 3 hashtags, that is, those with the highest co-occurrence were retrieved and added to the list if they were not yet part of it. Each hashtag was used to also query the official Twitter API and retrieve the top 20 entries in the list of user accounts suggested by Twitter that recently used this hashtag. New suggestions were then added to a list of relevant Twitter users. After removal of duplicates 156 Twitter profiles remained and then they were merged with additional 16 Twitter profiles used in [5], by reaching a total of 172 profiles that represent the reference community on Twitter related to cyber threats and security. To be able to compare the features of users within this community against outside users that are not focused on cyber security, another list of Twitter profiles was retrieved from the Twitter API using the hashtags *technology, windows, linux, computer and internet of things* while making sure that they were not in the list of suggested users of any of the cyber security related hashtags from above. This was done to ensure these users are related to the domain of technology and used similar vocabulary but are not focused on cyber threat intelligence. The full list of 230 Twitter users includes 172 (75%), who are considered the CTI community and 58 users from the technology domain who have no prominent relation to the cyber security domain. Finally, after the full list of sources was compiled the meta data of these 230 Twitter profiles as well as all 1,217,213 available Tweets from the time period of 3 years (from the 1st of Jan. 2016 until 31st of Dec. 2018) were collected using the official Twitter API.

Furthermore, 5 regression algorithms were evaluated and compared to the related works, by considering that no regression approaches were adopted in the context of CTI and for Twitter as an OSINT source. Specifically, the following one have been chosen as the no-regression version is typically applied in the relate work (i) SVM Regressor (SVR) by applying the Gaussian Radial Basis Function (RBF) kernel; (ii) Random Forest Regressor (RFR) have been used to establish a baseline for comparison; (iii) a Gradient Boosting Tree regression (GBTR) model, (iv) the Extra Trees Regressor(ETR), which is less susceptible to overfitting; and (v) a Multi-Layer Perceptron regressor (MLPR). The implementation of the regression models was based on "scikit-learn" Python library [1], and the configuration parameters are reported in Table 2.

Whereas, the implemented evaluation criteria have been based on the following metrics, which are used to evaluate the performance of regression models: *Mean Squared Error (MSE)*: which is computed as the arithmetic mean of all squared errors that were made during prediction of a numeric value; (ii) *Coefficient of Determination (R^2)*: it represents the proportion of the variance in the dependent variable that is predictable from the features that the model was trained on [8]. It is used to assess how well a regression model fits the data set. In the following subsection, the results are presented and discussed.

Table 2. Description regarding the regression models configuration

Regressor	Parameter configuration description
SVR	The Gaussian Radial Basis Function (RBF) kernel has been used with the implementations default parameters, according to [12]
RFR, ETR	Maximum number of features considered for the best split is $\sqrt{(26)} \approx 5$, for the full source metadata, and $\sqrt{(50)} \approx 7$ for word embedding [3]
GBTR	500 boosting stages during training optimizing the least squares loss function and limiting the maximum depth to 5 nodes per tree, as in [4]
MLPR	Hidden layer size of 50 for the 50-dimensional word embedding features and a hidden layer size of 26 for the source meta data features, as in [12]

4.2 Results Discussion

All five regression models were trained and evaluated on the collected data set, which was split into training and testing set, by using a 10-fold cross-validation strategy according to [14,18]. The experiments exploited the list of 659 CTI terms that were found across all 230 selected sources (i.e. Twitter accounts).

Figure 3-(a) visualizes the [0, 1]-normalized scores for all sources that are sorted on the horizontal axis according to their true intelligence count (baseline). Whereas Fig. 3-(b) shows the Absolute Error between the real value and the predicted ones. It worth noticing that the MLPR performed worst of all models and its predictions have errors beyond the range of $[-0.6, 0.6]$ and are therefore not depicted.

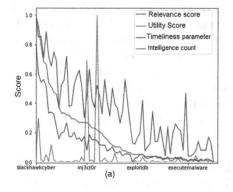
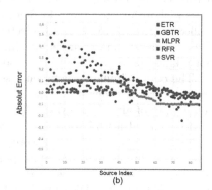

Fig. 3. CTI-Relevance-Score (a) and Absolute Error (b)

The R^2 value shows that best model for the prediction of the CTI Relevance Score on the source meta data feature set is the GBTR with an average value of $R^2 = 0.975$. The result evaluation is reported in Figure 4-(a). The same regression algorithms used for the source metadata feature set were trained on the word embedding model, that provides a single feature vector per Twitter source, by using identical parameters and metrics for training and evaluation. Figure 4-(b) displays a slight improvement in the R^2 for all models and even a large improvement for the MLPR model when using the word embedding features instead of the source meta data features. This first results indicate that the CTI Relevance Score can be predicted from CTI source features using the presented regression models.

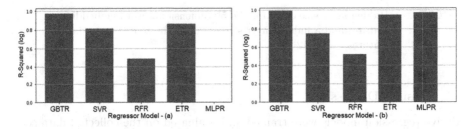

Fig. 4. Score prediction based on (a) meta-data feature set and (b) word embeddings

The other question is about, whether such a score can be used to increase the timeliness of alert generation. Similar to the method described in [5], since each source can be represented by features derived from its metadata or a word embedding vector, both types of feature vectors were used to calculate three different centroids representing the community of CTI sources and the Tweets containing true intelligence, respectively: (i) Centroid based on the meta data features of all top sources from the CTI community, i.e. the top 30% of Twitter users with respect to their CTI Relevance Score; (ii) Centroid based on the word embeddings of all top sources selected analogous to the previous centroid; (iii) Centroid based on the word embeddings of all Tweets containing true intelligence not taking the source into account, to improve the identification of CTI Tweets.

The cosine-similarity between a source and the centroid is then interpreted as the score that quantifies the source relevance, i.e. a source similar to the community of already relevant CTI sources is thereby relevant as well. In order to determine if a CTI source is relevant a threshold t needs to be established such that only sources with a score above t are classified relevant. Figure 5-(a) shows how the precision varies for possible thresholds t between [0, 1]. The red baseline indicates the precision $P_{base} = 45\%$ achieved on this data set using the count-based rule from DISCOVER [5] which only alerts on intelligence after their second occurrence. All scores reached higher precision for varying thresholds. The cosine-similarity to the third centroid (orange) reaches the highest precision but only for a rather high threshold which corresponds to a lower recall meaning that no alerts are issued for some intelligence. Considering a trade-off between

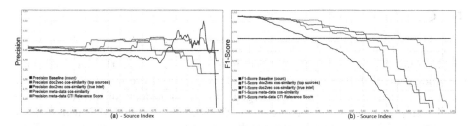

Fig. 5. Precision and F1-Score used to quantify the relevance of the predicted scores (Color figure online)

a low threshold, i.e. high recall, and a high precision the F1-Score is calculated and showed in Fig. 5-(b).

This shows that the cosine-similarity to the third centroid (orange) is actually performing worse than all other scores. The cosine-similarity for the second centroid (brown) shows a slightly better F1-Score as the predicted CTI Relevance Score on the source meta data features (blue). Interestingly, the cosine-similarity for the first centroid (green) has a F1-Score above the baseline for all thresholds up to t = 0.752. Through visual examination of the green graph a threshold of t = 0.4 is chosen to analyze the time advantage gained when using the cosine-similarity for the centroid of the source meta data features. This means that for any emerging CTI that is published by a source with a cosine-similarity above the selected threshold, an immediate alert is generated instead of waiting for a second occurrence of that intelligence from a different source. This time delay in hours is calculated for each instance in the dataset and visualized in Fig. 6. It shows not only the number of alerts that could be issued earlier but also the average time advantage to be gained: Half of all alerts could have been issued at least 32 h earlier than other count-based systems like DISCOVER.

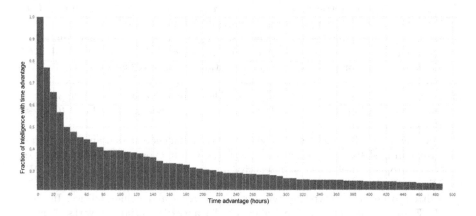

Fig. 6. The time advantage in hours gained when using the relevance score

5   Conclusion

This paper focused on the relevance assessment of OSINT sources as a cyber threat-related source. Two feature sets were engineered from the acquired data set and to quantify their relevance a CTI Relevance Score was formalized and compared with other scores. It emerged that the relevance of an open source on Twitter for CTI could be predicted through an automated feature-driven assessment of the source. As the results showed, half of all alerts could have been issued at least 32 h earlier, meaning the time advantage of preventive cyber threat detection can be increased when using the quantified source relevance as a decisive factor for automated alert generation in existing systems.

Acknowledgment. This work was performed in the context of the CHAMPIONs project, which receives funding from the EU Internal Security Fund - Police, grant no. 823705.

References

1. Scikit-learn: machine learning in python. https://scikit-learn.org/stable/
2. Dalziel, H., Olson, E., Carnall, J.: How to Define and Build an Effective Cyber Threat Intelligence Capability. Syngress is an imprint of Elsevier. http://www.books24x7.com/marc.asp?bookid=78688, OCLC: 910537102
3. Devore, J.L.: Probability and Statistics for Engineering and the Sciences, 8th edn. Brooks/Cole, Cengage Learning, California (2012)
4. Ke, G., et al.: LightGBM: a highly efficient gradient boosting decision tree. In: Advances in Neural Information Processing Systems, vol. 30, pp. 3146–3154. Curran Associates Inc. (2017)
5. Le, B.D., Wang, G., Nasim, M., Babar, A.: Gathering cyber threat intelligence from twitter using novelty classification (2019). http://arxiv.org/abs/1907.01755
6. Lee, K.-C., et al.: Sec-Buzzer: cyber security emerging topic mining with open threat intelligence retrieval and timeline event annotation. Soft Comput. **21**(11), 2883–2896 (2016). https://doi.org/10.1007/s00500-016-2265-0
7. Liao, X., et al.: Acing the IOC game: toward automatic discovery and analysis of open-source cyber threat intelligence. In: Proceedings of the 2016 ACM SIGSAC Conference on Computer and Communications Security-CCS 2016, pp. 755–766 (2016)
8. Marsland, S.: Machine Learning: An Algorithmic Perspective. Chapman & Hall/CRC Machine Learning & Pattern Recognition Series, 2nd edn. CRC Press, Boca Raton (2015)
9. Mavroeidis, V., Bromander, S.: Cyber threat intelligence model: an evaluation of taxonomies, sharing standards, and ontologies within cyber threat intelligence. In: 2017 European Intelligence and Security Informatics Conference, pp. 91–98 (2017)
10. Mittal, S., Das, P.K., Mulwad, V., Joshi, A., Finin, T.: CyberTwitter: using twitter to generate alerts for cybersecurity threats and vulnerabilities. In: 2016 IEEE/ACM International Conference on Advances in Social Networks Analysis and Mining, pp. 860–867 (2016)
11. Mittal, S., Joshi, A., Finin, T.: Thinking, fast and slow: combining vector spaces and knowledge graphs (2017). http://arxiv.org/abs/1708.03310

12. Nebot, V., Rangel, F., Berlanga, R., Rosso, P.: Identifying and classifying influencers in twitter only with textual information. In: Silberztein, M., Atigui, F., Kornyshova, E., Métais, E., Meziane, F. (eds.) NLDB 2018. LNCS, vol. 10859, pp. 28–39. Springer, Cham (2018). https://doi.org/10.1007/978-3-319-91947-8_3

13. Robertson, J.: Darkweb Cyber Threat Intelligence Mining. Cambridge University Press, New York (2017)

14. Sabottke, C., Suciu, O., Dumitras, T.: Vulnerability disclosure in the age of social media: exploiting twitter for predicting real-world exploits. In: 24th USENIX Security Symposium (USENIX Security 15), pp. 1041–1056 (2015)

15. Sapienza, A., Ernala, S.K., Bessi, A., Lerman, K., Ferrara, E.: DISCOVER: mining online chatter for emerging cyber threats. In: Companion of the The Web Conference 2018 on the Web Conference 2018 - WWW 2018, pp. 983–990 (2018)

16. Schaberreiter, T., et al.: A quantitative evaluation of trust in the quality of cyber threat intelligence sources. In: Proceedings of the 14th International Conference on Availability, Reliability and Security - ARES 2019, pp. 1–10. ACM Press (2019)

17. Tounsi, W., Rais, H.: A survey on technical threat intelligence in the age of sophisticated cyber attacks. Comput. Secur. **72**, 212–233 (2018)

18. Zhu, Z., Dumitras, T.: ChainSmith: automatically learning the semantics of malicious campaigns by mining threat intelligence reports. In: IEEE European Symposium on Security and Privacy (EuroS&P), pp. 458–472 (2018)

Malicious Domain Detection Based on K-means and SMOTE

Qing Wang[1,2], Linyu Li[1,2], Bo Jiang[1(✉)], Zhigang Lu[1,2], Junrong Liu[1],
and Shijie Jian[1,2]

[1] Institute of Information Engineering, Chinese Academy of Sciences,
Beijing 100093, China
[2] School of Cyber Security, University of Chinese Academy of Sciences,
Beijing 100029, China
{wangqing,lilinyu,jiangbo,luzhigang,liujunrong,jianshijie}@iie.ac.cn

Abstract. The Domain Name System (DNS) as the foundation of Internet, has been widely used by cybercriminals. A lot of malicious domain detection methods have received significant success in the past decades. However, existing detection methods usually use classification-based and association-based representations, which are not capable of dealing with the imbalanced problem between malicious and benign domains. To solve the problem, we propose a novel domain detection system named KSDom. KSDom designs a data collector to collect a large number of DNS traffic data and rich external DNS-related data, then employs K-means and SMOTE method to handle the imbalanced data. Finally, KSDom uses Categorical Boosting (CatBoost) algorithm to identify malicious domains. Comprehensive experimental results clearly show the effectiveness of our KSDom system and prove its good robustness in imbalanced datasets with different ratios. KSDom still has high accuracy even in extremely imbalanced DNS traffic.

Keywords: Malware domain detection · Data imbalance · K-means · SMOTE · CatBoost

1 Introduction

On the Internet, malicious activities are everywhere. As the basis of the Internet, the Domain Name System implements domain-to-IP address mapping, now is widely used by cybercriminals for malicious activities. Attackers use malicious domains for phishing, distributing malware, controlling botnets and other malicious activities. Effective detection of malicious domain names on the Internet is critical.

Malicious activities leave traces in DNS traffic, so DNS traffic can be used to detect malicious domains. Traditional detection systems analyse various DNS traffic and extract features from DNS traffic which can effectively distinguish between malicious and benign domains, such as the number of domain resolution

© Springer Nature Switzerland AG 2020
V. V. Krzhizhanovskaya et al. (Eds.): ICCS 2020, LNCS 12138, pp. 468–481, 2020.
https://doi.org/10.1007/978-3-030-50417-5_35

IP addresses [1], the standard deviation of TTL (Time-To-Live) values [2], then build a classifier through these features. This classifier can be used to detect unlabeled domains after being trained by labeled datasets.

Some researchers use an association-based approach to detect malicious domains. This method is based on the intuition that the domain associated with a malicious domain is likely to be malicious as well. By establishing association rules between domains, DNS-related data is modeled as domain-IP bipartite graphs [3] or domain-client bipartite graphs [4] or more complexity graphs [5], which is constructed to discover more malicious domains.

These existing methods have achieved excellent results to a certain extent, but rarely consider the problem of data imbalance. In the real network environment, the number of malicious domains is much smaller than benign domains, and the DNS traffic between benign domains and malicious domains is extremely imbalanced. Imbalanced datasets have a significant impact on the learning of the classifier and affect the effect of actual use.

In order to solve this problem, we propose a malicious domains detection system named KSDom, which can effectively detect malicious domains in imbalanced DNS traffic and has significant performance.

In this paper, we make the following contributions:

- We integrate active DNS data with rich external data, select powerful features that can be used to distinguish malicious domains to build a detection system based on CatBoost algorithm, which can effectively classify malicious domains.
- We propose a malicious domain detection system named KSDom based on CatBoost algorithm, which can detect malicious domains from the imbalanced DNS traffic. KSDom uses K-means and SMOTE method to handle the imbalanced DNS traffic, then uses our classifier based on CatBoost algorithm to detect malicious domains.
- The experimental results demonstrate our system is effective, competitive and can achieve state-of-the-art performance.

The rest of this paper is organized as follows. We present related work in Sect. 2 and describe our system framework and each component of it in Sect. 3. Section 4 presents the experimental results and the results of comparison experiments of several models. Section 5 presents the discussion.

2   Related Work

2.1   Malicious Domain Detection

The detection methods in the field of malicious domains can mainly be divided into two categories: classification-based methods and association-based methods. In the classification-based methods, [1] established a reputation system that dynamically assigns reputation scores to domains based on the features of the network and DNS zones and assigns low reputation scores if the domain name involves malicious activity. [6] presented a detection technology for detecting DGA domains, which is based on the following principle: domains of the same DGA algorithm will generate similar Non-Existing Domains (NXDomain) traffic. The technology uses a combination of aggregation and classification algorithms to cluster NXDomain responses to detect the relevant DGA domains. [7] used a combination of IP and string-based features to detect DGA-generated domains. [2] extracted fifteen features based on passive DNS traffic to detect different types of malicious domains, not just detects a particular type of domain. [8] used only time-of-registration features to identify malware domains when they are registered. [9] used passive DNS traffic to record domain name query history in the real network environment to generate eighteen features to detect Fast-Flux domains.

In the association-based method, systems use the association between domains and domain-related IPs (IP address) or clients to form domain graphs, then execute the inference algorithm on the domain graphs to discover more malicious domains. [10] built a client-domain bipartite graph based on the clients and the domain which it queries, effectively tracking the malware-controlled domains in the ISP network and discovering new malware control domains. [4] formed the domain-IP bipartite graph through the association between domains and IPs, then used a path-based algorithm to discover potential malicious domains. The above systems all use bipartite graphs, which only represents two types of nodes and one type of relationship. [5] constructed graphs by using multiple types of nodes and relationship and the client, domain, IP address and their relationship are modeled as HIN model. The similarity of the combined domains created on multiple meta paths represents the information contained in the DNS data.

2.2 Imbalanced Data

In recent years, more and more research begins to pay attention to the importance of imbalanced data and apply it to various fields [11], such as fraud detection, text classification and medical diagnosis in banks [12]. The extremely imbalanced data leads to poor performance of the trained classifier.

In general, there are two ways to solve the imbalanced data problem. The first one is from the algorithm layer, by modifying the bias of the algorithm on the datasets, so that the decision plane can tend to minority class samples, improving the recognition rate of minority class samples. This method is based on cost-sensitive learning [13], and the representing algorithm is AdaCost. The second method is from the data layer, by resampling imbalanced data. Resampling includes two methods: undersampling and oversampling. Undersampling, that is, by extracting some samples from majority class samples, but this may lead to the loss of classification information. The improved undersampling method is EasyEnsemble [14] and BalanceCascade [15]. Oversampling, that is, by increasing the number of samples of minority class samples, the original data can be well preserved, but it is easy to produce overfitting. The improved oversampling algorithm includes SMOTE (Synthetic Minority Oversampling Technique) [16]. There are also some hybrid methods. [17] proposed a novel hybrid learning framework which combines data-level and algorithm-level methods to deal with data imbalanced problem. [18] combined oversampling and undersampling techniques and SVM approach to handle the imbalanced data classification problem.

Unlike algorithm layer methods that need to be combined with specific classifiers, data layer methods are universally applicable, so most research methods are aimed at considering the processing of imbalanced datasets from a data perspective. In terms of network security, [19] combined ensemble learning with undersampling, using an improved EasyEnsemble method to learn imbalanced DNS traffic data and generate a classification model which can detect malicious domains. Illuminated by this research which shows resampling usefulness in malicious domain detection.

The classification model generated by using ensemble learning combined with undersampling is prone to noise, and the Decision Tree algorithm ignores the correlation between features, resulting in poor model performance. We propose KMSMOTE method, which can deal with imbalanced datasets and avoid noise well and combine with the CatBoost algorithm, which can deal with the correlation between features and avoid data overfitting.

3   Proposed KSDom System

The goal of KSDom is to detect malicious domains on the Internet. KSDom is based on the following intuitions: (i) when malicious domains perform malicious activities, the generated DNS traffic is different with the DNS traffic generated by benign domains, and attackers have to consider the costs of network resources, they prefer to choose domains with lower registration costs and tend to reuse network resources. (ii) In the real Internet, the number of benign domains is much bigger than malicious domains, DNS traffic is extremely imbalanced between benign domains and malicious domains, the classifiers trained by balanced datasets have poor performance in the real network.

Therefore, our system collects a large number of DNS traffic on the Internet, analyzes DNS traffic and selects useful features from it, oversamples the imbalanced training samples, and trains the processed samples to generate a malicious domain detection model to detect unlabeled domains. As shown in Fig. 1, KSDom consists of four main components: Data Collector, Feature Selection, Imbalanced Data Processor and Classifier component.

The operational workflow of KSDom is as follows. Data Collector aggregates imbalanced Active DNS traffic data from the local network and expands rich DNS-related data (step 1), Feature Selection component selects features from DNS traffic data to generate training samples (step 2), put the imbalanced training samples into the third component: Imbalanced Data Processor, processes imbalanced training samples with our processor (step 3), the output is fed into the fourth component: Classifier. This module trains labeled datasets based on CatBoost algorithm to build a malicious domain detection classifier (step 4), the generated classifier categorizes unlabeled domains (step 5), we analyze the classification results and add labeled domains to our domain list (step 6).

In the following, we will introduce each component in detail.

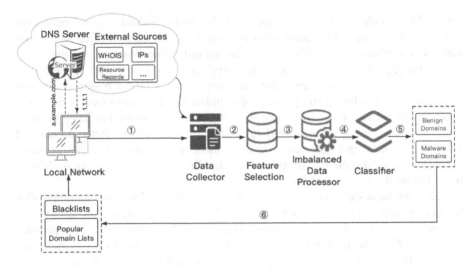

Fig. 1. Overview of KSDom

3.1   Data Collector

In the Data Collector module, we first describe the type of DNS traffic data. We divide DNS traffic into Active DNS traffic data and Passive DNS traffic data according to the collection method.

Active DNS data refers to DNS data obtained by periodically proactively parsing a large number of domains [20], which are collected from multiple public sources, such as the Alexa Top Sites [21], public whitelists, various public black-lists, etc. Active DNS data mainly captures the DNS records of domains, such as the records of domain resolved IP addresses, the information of name servers, etc. Active DNS data does not have the information of users query domains, so it does not involve any privacy issues.

Passive DNS data is obtained by deploying sensors on the DNS servers or accessing the DNS server logs to obtain real DNS queries and response informa-tion. Passive DNS data provides a summarized view of domain queries. Exper-iments have shown that active DNS data provides more kinds of records, and passive DNS data provides a tighter connection graph [22]. Passive DNS can provide richer information than active DNS, but due to privacy issues and the location of the deployed sensors, the collected data has certain limitations.

Compare to passive DNS data, the collection method of active DNS is flexible, and data collector can easily query domains of the collection lists without any private issues, and active DNS data can discover potentially malicious domains that are newly registered but not yet used [20].

In this research, we use active DNS traffic data and DNS-related external data for data enrichment. Our data collector collects the following data:

- Active DNS traffic data. We use the local network to actively query the domains. After the DNS server receives the query requests, it returns the

Table 1. Selected domain feature set

Feature set	ID	Feature name	Importance
Domain name-based feature	1	Length of domain names	0.01306
	2	Distribution of vowel characters	0.04002
	3	Distribution of consonant characters	0.17824
	4	Conversion frequency of vowels and consonants	0.04263
	5	Number of numeric characters	0.00412
	6	Ratio of numeric characters	0.00561
	7	Conversion frequency of numeric and alphabetic characters	0.10680
	8	Number of special characters	0.04600
DNS answers-based feature	9	Number of distinct IP addresses	0.06591
	10	Number of MX records	0.09031
	11	Number of domains that IP reversely resolves	0.06737
	12	Number of distinct NSs	0.07227
	13	Similarity of NS domain names	0.03003
Contextual features	14	Number of labels of the website's contents	0.11687
	15	Number of malicious IP addresses	0.19452
	16	Validity period of a domain	0.05746

corresponding response results. We collect response data as input to the data collector.

– External data related to DNS. We combine collected active DNS traffic data with DNS-related external data to enrich our datasets, such as the registration records of domains, the WHOIS information, the public IP blacklists information and other information from the associated Resource Records (RRs).

3.2 Feature Selection

This component is used to process the features we select in the DNS traffic data. In order to select features that can effectively differentiate malicious and benign domains, large amounts of DNS traffic need to be observed. By analyzing and summarizing the large amount of data obtained, we extract 16 composite features to build malicious detection classifier, as shown in Table 1. We will describe these features and explain why they can distinguish malicious domains.

DNS traffic features can be divided into internal features and contextual features. Internal features can be extracted directly from Resource Records, such as the record of domain resolution IP addresses, the character distribution of domain names. Contextual features are built from DNS-related external data information and associated Resource Records, such as Whois information for the domain, reverse mapping of domain resolved IP addresses, etc. These features can provide more information that may contain malicious behaviors than internal features, such as [23] found that domains associated with botnets typically do not have MX records.

We classify the features into three categories, based on their sources. (1) domain name-based features. These features are based on the domain name string itself, such as the character distribution of domain names. (2) DNS answer-based features. These features are based on DNS response data, such as Resource Records and IP information. (3) Contextual features. Contextual features come from external sources of data, such as WHOIS records and registration information.

Domain Name-Based Features. Benign domain applicants choose domain names that are easily remembered or beneficial to people's pronunciation, and attackers often choose characters that are difficult to remember when constructing malicious domains. Through research on a large number of malicious domains, we find that malicious domains contain more numbers, and there are a large number of numbers and letters. Therefore, the malicious domain has a more confusing combination of numbers and words, and the proportion of consonants in the random malicious domain is larger, and the conversion frequency between the vowels and the consonants is higher. Therefore, the following characteristics are constructed: the length of domain name (Feature 1), the distribution of vowel characters (Feature 2) which consists of three atomic features: the number of vowels, the ratio of vowels and the number of vowel characters in the string without a domain name suffix. The distribution of consonant characters (Feature 3) as well as a composite feature which consists of 2 atomic features: the number of consonant characters and the number of consonants in the string without a domain name suffix. The conversion frequency of vowels and consonants (Feature 4), the number of numeric characters (Feature 5), the ratio of numeric characters (Feature 6), the conversion frequency of numeric and alphabetic characters (Feature 7), the number of special characters (Feature 8).

DNS Answer-Based Features. The DNS answer records of malicious domains are very different from that of benign domains. Malicious domains usually have more address records, because malicious domain holders usually choose domains that map many IP addresses, and malicious domains usually contain fewer MX (Mail Exchanger) records and more NS (Name Server) records. Based on this information, we choose the following characteristics: the number of distinct IP addresses (Feature 9), the number of MX records (Feature 10), the number of domains that IP reversely resolves (Feature 11), the number of distinct NSs (Feature 12), and the similarity of NS domain names (Feature 13).

Contextual Features. Attackers typically do not set the website contents for the malicious domains, or just copy the specific templates to fake the phishing websites. We extract the following feature: the number of labels of the website's contents (Feature 14). Attackers usually reuse resources, and IP addresses associated with the malicious domains typically are also malicious. Query the public IP blacklists to gain how many of the IPs associated with a domain is malicious. We extract the following feature: the number of malicious IP addresses (Feature 15). Most malicious domains are used for short-lived malicious activities, and it will be soon to be recorded into blacklists, so registrants usually apply for malicious domains with shorter lifetimes. A malicious domain usually has a short validity period. The validity period of a domain can be determined by calculating the expiration date and registration date. We extract the following feature: the validity period of a domain (Feature 16).

3.3 Imbalance Data Processor

Our goal is to build a classifier that can detect real networks, but in real-world environments, the ratio of malicious and benign domains is extremely imbalanced. Therefore, we build a module in our system to oversample the imbalanced samples in the training sets to generate balanced samples.

SMOTE is one of the classic methods for solving imbalanced datasets, it is an improved method based on random oversampling. SMOTE constructs new sample points by nearest neighbors of sample points, adding sample points information, which can effectively prevent data over-fitting. It performs the following three steps: Firstly, it chooses a random sample $\vec{a}$ from minority class. Secondly, choose $\vec{b}$ from its nearest minority class neighbors. Finally, randomly interpolate $\vec{x}$ and $\vec{b}$ to generate the new sample $\vec{x}$ through this formula $\vec{x} = \vec{a} + \vec{w} \times (\vec{b} - \vec{a})$, where $\vec{w}$ is a random weight in [0,1]. But SMOTE cannot overcome the problem of data distribution, it can easily cause samples distribution marginalization problem and generates noise.

We propose an improved SMOTE method called KMSMOTE that combines K-means clustering algorithm and SMOTE oversampling algorithm. It counters small separation problems by increasing the number of sparse minority class samples, which can avoid noise by oversampling only in the "safe" region. KMSMOTE method consists of three steps.

1. In the clustering step, use K-means clusters the input samples into k groups.
2. In the filtering step, calculate the *imbalanceRatio* by $imbalanceRatio(f) \leftarrow \frac{majorityCount(c)+1}{minorityCount(c)+1}$ of each cluster, if the value of *imbalanceRatio* of cluster less than the imbalance ratio threshold, then put it into *filteredClusters*. By this step, we get *filteredClusters* containing a large number of minority class samples are reserved. Select each cluster f from *filteredClusters* to calculate *densityFactor* with formula $densityFactor(f) \leftarrow \frac{minority\ count(f)}{average\ minority\ distance(f)^{de}}$. Then we get the *sparsityFactor* by $sparsityFactor(f) \leftarrow \frac{1}{densityFactor(f)}$. We can get the *samplingWeight* by $samplingWeight(f) \leftarrow \frac{sparsityFactor(f)}{sparsitySum}$ which *sparsitySum* is the sum of all filtered clusters' *sparsityFactor*.

3. In the oversampling step, apply SMOTE to the selected clusters for over-sampling by applying formula $SMOTE(f, numberOfSamples, knn)$ where $numberOfSamples$ is computed by $samplingWeight$.

In this algorithm, it is important to find the appropriate k value to ensure the validity of KMSMOTE. According to our experimental datasets, we set k to 10 in the experiment.

3.4   Domains Classifier

Our classifier is based on CatBoost algorithm [24]. CatBoost algorithm is an implementation of Gradient Boosting. The two main features in CatBoost algorithm are: it derives ordered boosting, which is an improvement to the standard Gradient Boosting algorithm, which can avoid target leakage; it is a novel algorithm to deal with categorical features. These features are designed to solve prediction shifts, which are common in gradient boosting algorithms. Classification and regression trees cannot handle discrete features with categorical values, but CatBoost can support categorical features. An effective way to deal with categorical features i is to replace the category x_k^i of k-th training example with some Target Statistic (TS) $\hat{x}_k^i$. This value TS is calculated as follows: $\hat{x}_k^i = \mathbb{E}(y \mid x^i = x_k^i)$. CatBoost smoothes low-frequency categories through some prior p, and the estimates of these low-frequency categories are noisy. It first randomly sorts all samples and then converts the values of the categorical features into numerical values using the formula

$$x_k^i \hat{x}_k^i = \frac{\sum_{j=1}^{p-1} [x_{\sigma_j,k} = x_{\sigma_p,k}] Y_{\sigma_j} + a \cdot P}{\sum_{j=1}^{p-1} [x_{\sigma_j,k} = x_{\sigma_p,k}] + a} \tag{1}$$

where p is the priori term, which is usually be setted as the average target value in the dataset [25] and a is usually a weighting factor greater than zero. In order to solve the prediction shift that occurs in each step of gradient boosting, CatBoost proposes ordered Boosting. The method is as follows: first randomly generate a $1 - n$ permutation named σ and maintain n different supporting models $M_1, \ldots, M_n, M_i$ is the training model obtained using only the first i samples in the permutation. At each step of the iteration, in order to obtain the estimated value of the j sample residual, use the model M_{j-1} to estimate.

4   Experiments

4.1   Data

Active DNS Traffic Data. We build a domain list by ground truth which includes the benign domain lists and malicious domain lists. The lists of benign domains we select are the Second Level Domains of popular domains, such as the Alexa Top Sites [21] and the Cisco Umbrella list, and domains must survive on these lists for more than one year.

Malicious domains lists are extracted from public blacklists, such as Malware-domainlist, etc. These malicious domain lists contain various malicious activities, and we choose Second Level Domains from multiple blacklists to construct our blacklists. And through McAffee Site Advisor, and Google Safe Browsing and other reputation systems for secondary confirmation to ensure the accuracy of the domains label of our list. We obtain active DNS traffic data by sending domain requests to domains of our domain list.

DNS Related External Data. We collect DNS related external data from corresponding databases, such as obtain domain name registration records by the Whois protocol; query rich resource record database to obtain rich related resource records, such as MX, NS, PTR, etc; query the maliciousness of IPs from public IP blacklists.

We build an initial domain list contains 10000 benign domains and 6000 malicious domains and obtain an imbalanced DNS dataset by collecting Active DNS traffic data and external DNS-related data.

4.2 Malicious Domains Detection Performance

We randomly select a dataset consisting of DNS data of 10,000 benign domains and 1000 malicious domains from our imbalanced dataset, the proportion of imbalance of this dataset is large, which is 10. We then perform 10-fold cross-validation on this experimental dataset. The training data is initially divided into 10-fold. In each experiment, 9-fold data is used for training and 1-fold data is used for testing. We calculate the average of the ten results of experiments and showed them in Table 2.

As we can see in Table 2, KSDom obtains good performance, the average accuracy is 0.9842, the average F1-Score is 0.9838, and the average precision obtains 0.9926. We then use the trained classifier to classify unlabeled domains and add labeled domains to our domain list to better train the model. We extract rich features to build our classifier. These features represent the domains from different perspectives and have different effects on the generation of a classifier. We test the detection performance of each feature by selecting 70% samples for training and 30% samples for testing from our initial dataset and repeat the procedure for 10 times. Catboost provides a feature importance algorithm, and we can use get_feature_importance() method to get the importance of the features we selected, the results are shown in Table 1. From Table 1, we can see that features such as the number of malicious IP addresses, the distribution of consonant characters and the number of labels of the website's contents are all of great importance.

4.3 Performance Comparison with Different Sparsity

To test the effectiveness of KSDom, in this section, we compare KSDom with three classic classification methods based on the same dataset: Support Vector

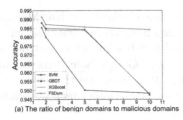
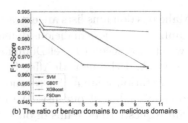

Fig. 2. Accuracy and F1-Score in different ratio of benign domains to malicious domains

Machine (SVM), Gradient Boosting Decision Tree (GBDT) and eXtreme Gradient Boosting (XGBoost). Before using these three classification methods for training, first perform one-hot encoding on categorical features, CatBoost does not need to encode them because it can process categorical features.

We perform 10-fold cross-validation to the 4 experiments to compare the performance of KSDom with other classifiers in different ratios of malicious domains and benign domains. In these 4 experiments, we set the number of benign domains in the training set is 10000, and the number of malicious domains is 6000, 4000, 2000 and 1000 (the imbalanced ratio of benign domains to malicious domains is 1.67, 2.5, 5, 10). Figure 2 shows the average of 10-fold cross-validation results.

From Fig. 2, we can see that as the imbalance ratio increases, the accuracy and F1-Score of other classifiers are extremely reduced, but KSDom still maintains good performance.

We can get further insight into the comparison of KSDom and other classification methods from Table 2. Table 2 shows the accuracy, precision, F1-Score and the running time of each method. It can be seen that the accuracy, precision and F1-Score of KSDom are higher than other classification methods, and the running time is shorter than others. For example, compared to SVM, GBDT, XGBoost, the accuracy of KSDom is increased by approximately 3.58%, 3.64% and 3.61%. As we can see, KSDom still maintains relatively good performance when the imbalanced ratio increases. It yields accuracy: 0.9915, F1-Score: 0.9907 when the imbalanced ratio is 1.67 and still obtains accuracy: 0.9842, F1-Score: 0.9838 when the imbalanced ratio increases to 10. As for the other classification methods, they can get excellent performance with a small imbalanced ratio, yet

Table 2. Detection results with different methods

Metrics	SVM	GBDT	XGBoost	KSDom
Accuracy	0.9484	0.9478	0.9481	**0.9842**
Precision	0.9508	0.9507	0.9507	**0.9926**
F1-Score	0.9641	0.9637	0.9639	**0.9838**
Running time	174.564 s	94.912 s	89.074 s	**52.518 s**

the accuracy and F1-score both decrease significantly as the imbalanced ratio increases. The reason behind this is that KSDom has the KMSMOTE algorithm, which can effectively handle imbalanced data sets and make its trained models have higher performance. The classifier used by KSDOM is Catboost, which has excellent performance. Catboost has an ordered boosting algorithm and a module that can process classified features, which improve its performance. CatBoost algorithm also implements a symmetric tree that can deal with the correlation between features, which helps avoid data overfitting and greatly accelerates predictions.

4.4 Impact of Oversampling Algorithm

In the solution of imbalanced datasets, the two classic methods in the resampling method are EasyEnsemble in undersampling and SMOTE in oversampling. Since EasyEnsemble combines undersampling and ensemble learning, we use SMOTE and CatBoost algorithm combination to compare with our KSDom.

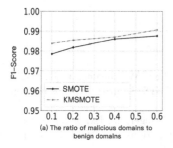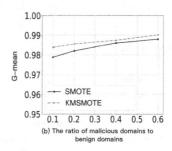

Fig. 3. F1-Score and G-mean in different ratio of malicious domains to benign domains

To prove that the K-means of our SMOTE method is better than SMOTE, we compare the two methods with the following datasets: we establish five datasets, each of which contains 10000 benign domains and 1000, 2000, 4000, 6000 malicious domains (the imbalance ratio between malicious domain and benign domain is 0.1, 0.2, 0.4 and 0.6). The experimental results are shown in Fig. 3.

Traditional evaluation metrics used to evaluate classifier performance are not applicable to imbalanced samples, such as accuracy, precision, etc. There are a number of indicators that are specifically used to assess imbalanced data, such as the F1-Score, G-mean. In this experiment, we use F1-Score to measure the classification performance of the minority class, while G-means are used to measure the overall classification performance of the datasets.

As shown in Fig. 3, our KMSMOTE method performs better than SMOTE in different ratios datasets. Especially when dealing with extremely imbalanced datasets, such as when the ratio is 0.1, the F1-Score and G-mean of KMSMOTE are both higher than SMOTE. Therefore, we can conclude that our KMSOMTE method is superior to SMOTE in dealing with imbalanced data.

5   Conclusion

In this paper, we propose a system for detecting malicious domains named KSDom, which can effectively detect the malicious domains from imbalanced DNS traffic. KSDom collects DNS traffic from the data collector and employs a KMSMOTE method to process the imbalanced DNS traffic, then uses our classifier based on CatBoost algorithm to detect malicious domains. The comparative experimental results show that KSDom has higher accuracy than other classifiers. It performs well even when the datasets imbalance ratio reaches 10:1. In our extensive evaluation, we verified KSDom's good performance and robustness.

Acknowledgment. This research is supported by National Key Research and Development Program of China (No. 2019QY1303, No. 2019QY1302, No. 2018YFB0803602), and the Strategic Priority Research Program of the Chinese Academy of Sciences (No. XDC02040100), and National Natural Science Foundation of China (No. 61802404). This work is also supported by the Program of Key Laboratory of Network Assessment Technology, the Chinese Academy of Sciences; Program of Beijing Key Laboratory of Network Security and Protection Technology.

References

1. Antonakakis, M., Perdisci, R., Dagon, D., Lee, W., Feamster, N.: Building a dynamic reputation system for DNS. In: USENIX Security Symposium, pp. 273–290 (2010)
2. Bilge, L., Sen, S., Balzarotti, D., Kirda, E., Kruegel, C.: Exposure: a passive dns analysis service to detect and report malicious domains. ACM Trans. Inf. Syst. Secur. (TISSEC) **16**(4), 14 (2014)
3. Khalil, I.M., Guan, B., Nabeel, M., Yu, T.: A domain is only as good as its buddies: detecting stealthy malicious domains via graph inference. In: Proceedings of the Eighth ACM Conference on Data and Application Security and Privacy, pp. 330–341. ACM (2018)
4. Khalil, I., Yu, T., Guan, B.: Discovering malicious domains through passive DNS data graph analysis. In: Proceedings of the 11th ACM on Asia Conference on Computer and Communications Security, pp. 663–674. ACM (2016)
5. Sun, X., Tong, M., Yang, J., Liu, X., Liu, H.: HinDom: a robust malicious domain detection system based on heterogeneous information network with transductive classification. In: RAID 2019, pp. 399–412 (2019)
6. Antonakakis, M., et al.: From throw-away traffic to bots: Detecting the rise of DGA-based malware. In: USENIX Security Symposium, vol. 12 (2012)
7. Dietrich, S. (ed.): DIMVA 2014. LNCS, vol. 8550. Springer, Cham (2014). https://doi.org/10.1007/978-3-319-08509-8
8. Hao, S., Kantchelian, A., Miller, B., Paxson, V., Feamster, N.: PREDATOR: proactive recognition and elimination of domain abuse at time-of-registration. In: Proceedings of the ACM SIGSAC Conference on Computer and Communications Security, pp. 1568–1579 (2016)
9. Zhou, C., Chen, K., Gong, X., Chen, P., Ma, H.: Detection of fast-flux domains based on passive DNS analysis. Acta Scientiarum Naturalium Universitatis Pekinensis **52**(3), 396–402 (2016)

10. Rahbarinia, B., Perdisci, R., Antonakakis, M.: Segugio: efficient behavior-based tracking of malware-control domains in large ISP networks. In: 2015 45th Annual IEEE/IFIP International Conference on Dependable Systems and Networks (DSN), pp. 403–414. IEEE (2015)
11. Chawla, N.V., Bowyer, K.W., Hall, L.O., Kegelmeyer, W.P.: SMOTE: synthetic minority over-sampling technique. J. Artif. Intell. Res. **16**(1), 321–357 (2002)
12. Maldonado, S., Weber, R., Famili, F.: Feature selection for high-dimensional class-imbalanced data sets using support vector machines. Inf. Sci. **286**, 228–246 (2014)
13. Nikolaou, N.: Cost-Sensitive Boosting: A Unified Approach. The University of Manchester (2016)
14. Liu, X.-Y., Wu, J., Zhou, Z.-H.: Exploratory undersampling for class-imbalance learning. IEEE Trans. Syst. Man Cybern. Part B Cybern. **39**(2), 539–550 (2009)
15. Liu, X.Y.: An empirical study of boosting methods on severely imbalanced data. In: Applied Mechanics and Materials, vol. 513–517, pp. 2510–2513 (2014)
16. Chawla, N.V., Bowyer, K.W., Hall, L.O., Kegelmeyer, W.P.: SMOTE: synthetic minority over-sampling technique. J. Artif. Intell. Res. **16**, 321–357 (2002)
17. Zhang, W., Wang, J.: A hybrid learning framework for imbalanced stream classification. In: IEEE International Congress on Big Data. IEEE (2017)
18. Wang, Q.: A hybrid sampling SVM approach to imbalanced data classification. Abstract Appl. Anal. **2014**(5), 1–7 (2014)
19. Zhenyan, L., Yifei, Z., Pengfei, Z., et al.: An imbalanced malicious domains detection method based on passive DNS traffic analysis. Secur. Commun. Netw. **2018**, 1–7 (2018)
20. Zhauniarovich, Y., Khalil, I., Yu, T., Dacier, M.: A survey on malicious domains detection through DNS data analysis. ACM Comput. Surv. **51**(4) 67 (2018). Article 67, 36 p
21. Amazon Web Services, Inc.: AWS — Alexa Top Sites - Up- to-date lists of the top sites on the web (2019). https://aws.amazon.com/alexa-top-sites/
22. Kountouras, A., et al.: Enabling network security through active DNS datasets. In Proceedings of the International Symposium on Research in Attacks, Intrusions, and Defenses, pp. 188–208 (2016)
23. Prieto, I., Magaña, E., Morató, D., Izal, M.: Botnet detection based on DNS records and active probing. In: Proceedings of the International Conference on Security and Cryptography, pp. 307–316 (2011)
24. Prokhorenkova, L., Gusev, G., Vorobev, A., et al.: CatBoost: unbiased boosting with categorical features (2017)
25. Micci-Barreca, D.: A preprocessing scheme for high-cardinality categorical attributes in classification and prediction problems. ACM SIGKDD Explor. Newslett. **3**(1), 27–32 (2001)

Microservice Disaster Crash Recovery: A Weak Global Referential Integrity Management

Maude Manouvrier[1]([✉]), Cesare Pautasso[2], and Marta Rukoz[1,3]

[1] Université Paris-Dauphine, PSL Research University CNRS UMR [7243]
LAMSADE, Paris, France
`maude.manouvrier@dauphine.fr`
[2] Software Institute, Faculty of Informatics, USI, Lugano, Switzerland
[3] Université Paris-Nanterre, Nanterre, France

Abstract. Microservices which use polyglot persistence (using multiple data storage techniques) cannot be recovered in a consistent state from backups taken independently. As a consequence, references across microservice boundaries may break after disaster recovery. In this paper, we give a weak global consistency definition for microservice architectures and present a recovery protocol which takes advantage of cached referenced data to reduce the amnesia interval for the recovered microservice, i.e., the time interval after the most recent backup, during which state changes may have been lost.

Keywords: Microservices · Referential integrity · Backup · Weak global consistency

1 Introduction

Microservices are small autonomous services, deployed independently, that implement a single, generally limited, business functionality [6,14,21,23]. Microservices may need to store data. Different data storage pattern exist for microservices [21]. In the Database per Service pattern, defined in [19]: each microservice stores its persistent data in a private database. Each microservice has full control of a private database, persistent data being accessible to other services only via an API [24]. The invocation of a service API will result in transactions which only involve its database.

Relationships between related entities of an application based on a microservice architecture are represented by links: the state of a microservice can include links to other entities found on other microservice APIs [18]. Following the hypermedia design principle of the REST architectural style, these links can be expressed with Uniform Resource Identifiers (URIs) which globally address the referenced entities.

Since microservices are autonomous, not only do they use the most appropriate database technology for persistent storage of their state, but they also

© Springer Nature Switzerland AG 2020
V. V. Krzhizhanovskaya et al. (Eds.): ICCS 2020, LNCS 12138, pp. 482–495, 2020.
https://doi.org/10.1007/978-3-030-50417-5_36

operate following an independent lifecycle, when their database is periodically backed up. For an entire microservice architecture, in practice, it is not very feasible to take an atomic snapshot of the state of all microservices. Thus, in case of one microservice crashes, which then needs to be recovered from its backup, the overall state of the microservice architecture may become inconsistent after recovery [18]. After recovery, such inconsistency may manifest itself as broken links between different microservices.

This paper presents a solution to ensure that the links between different entities managed by different microservices remain valid and intact even in the case of a database crash. The solution assumes that microservices referring to entities managed by other microservices will not only store the corresponding link, but also conserve a cached representation of the most recent known values. We present a recovery protocol when the crashed microservice can merge its own possibly stale backup with the possibly more recent cached representations obtained from other microservices. Thus, we revisit the definition of weak referential integrity across distributed microservice architectures.

2   Background and Related Work

2.1   Database Consistency, Durability, Backup and Disaster Crash Recovery

A database has a state, which is a value for each of its elements. The state of a database is consistent if it satisfies all the constraints [22]. Among constraints that ensure database consistency, referential integrity [8] is a database constraint that ensures that references between data are indeed valid and intact [4]. In a relational database, the referential integrity constraint states that a tuple/row in one relation referring, using a foreign key, to another relation, must refer to an existing tuple/row in that relation [11]. When a reference is defined, i.e. a value is assigned to a foreign key, the validity of the reference is checked, i.e. the referenced tuple should exist. In case of deletion, depending on the foreign key definition, the deletion of a tuple is forbidden if there are dependent foreign-key records, or the deletion of a record may cause the deletion of corresponding foreign-key records, or the corresponding foreign keys are set to null. Referential integrity is really broader and encompasses databases in general and not only relational ones [4].

Durability means that once a transaction, i.e. a set of update operations on the data, is committed, it cannot be abrogated. In the centralized databases systems, checkpoint and log are normally used to recover the state of the database in case of a system failure (e.g. the contents of main memory disappear due to a power loss and the content of a broken disk becoming illegible) [22]. Checkpoint is the point of synchronization between the database and transaction log file when all buffers are force-written to secondary storage [7]. For this kind of failure, the database can be reconstructed only if:

- the log has been stored on another disk, separately from the failure one(s),
- the log has been kept after a checkpoint, and

– the log provides information to redo changes performed by transactions before the failure, and after the latest checkpoint.

To protect the database against media failures an up-to-date backup of the database, i.e. a copy of the database separate from the database itself, is used [22]. A backup of the database and its log can be periodically copied onto offline storage media [7]. In case of database corruption or device failure, the database can be restored from one of the backup copies, typically the most recent one [3]. In this case, the recovery is carried out using the backup and the log – see [15], for more details.

A database has a Disaster Crash when the main memory and the log, or a part of the log, are lost. Therefore, to recover the database, an old, maybe obsolete, backup of the database is used. Data which was not part of the backup will be lost. In case of a disaster crash, the system cannot guarantee the durability property. However, in a centralized database, recovery from a backup provides a database which has a consistent state.

2.2   Microservices as a Federated Multidatabase

Each database of a microservice can be seen as a centralized database. Seen across an entire microservice architecture, the microservice databases represent a distributed database system. A multidatabase is a distributed database system in which each site maintains complete autonomy. Federated multidatabase is a hybrid between distributed and centralized databases. It is a distributed system for global users and a centralized one for local users [7]. According to the definitions above, stateful microservice architectures can therefore be seen as a federated multidatabase.

A microservice database can store either a snapshot of the current state of the data, containing the most recent value of data, or an event log, i.e. the current state of the data can be rebuilt by replaying the log entries, which record the changes to the microservice state in the database transaction log. Let's consider an example of a microservice managing orders. Using the snapshot architecture, the current state of an order can be stored in a row of a relational table *Order*. When using the event sourcing (log) [17], the application persists each order as a sequence of events e.g., listing the creation of the order, its update with customer details and the addition of each line item.

Each microservice ensures the durability and the consistency of its database, like in centralized databases. In the microservice context, each microservice manages its own database and stores independent backup of its own database, in order to permit disaster recovery from backup. However, while managing consistent backup is simple in a centralized database, maintaining consistent backups with distributed persistence in a federated multidatabase is challenging, as shown in the survey of [13]. So a model providing global consistent backup is necessary for microservices.

Microservice architecture deals with breaking foreign key relationships [16]. Each microservice can refer to other microservices data through loosely coupled references (i.e., URLs or links), which can be dereferenced using the API provided by the microservice managing the referenced data. Microservices are independent and the managing reference integrity between them is challenging. As for the World Wide Web [9,12], there is no guarantee that a link retrieved from a microservice points to a valid URL [18]. In the following section, we propose a model providing global reference integrity for microservices.

2.3 Microservice Disaster Recovery

In [18], the authors have addressed the problem of backing up an entire microservice architecture and recovering it in case of a disaster crash affecting one microservice. They defined the BAC theorem, inspired from the CAP Theorem [5], which states that when backing up an entire microservice architecture, it is not possible to have both availability and consistency.

Let us consider the microservice architecture defined in [18], where each microservice manages its own database and can refer to other microservices data through loosely coupled references. Each microservice does independent backup of its own database for the purpose of allowing disaster recovery from backup.

Figure 1 presents an example of two microservices with their independent backup, data of the microservice *Order* referring data of microservice *Customer*. Database of each microservice is represented in gray and data in black. Each database contains three entities. Entities C/i ($i \in \{1, 2, 3\}$) correspond to customers, described by a name, and are managed by microservice *Customer*. Entities O/i ($i \in \{1, 2, 3\}$) correspond to orders and are managed by microservice *Order*. Each order O/i refers to a customer C/i. Backups of the database are represented in blue. The backup of microservice *Customer* only contains a copy of customers $C/1$ and $C/2$. The backup and the database of microservice *Order* are, on the contrary, synchronized.

As explained in [18], in case of disaster crash, independent backup may lead to broken link (see Fig. 2): no more customer $C/3$ exists after *Customer* recovery, then $O/3$ has a broken link.

A solution to avoid broken link is to synchronize the backup of all microservices, leading to limited autonomy of microservices and loss of data. In Fig. 3, both order and customer $C/3$ and $O/3$ are lost after the recovery.

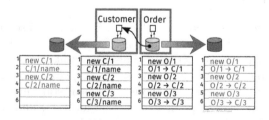

Fig. 1. An example of microservice architecture with independent backup (Color figure online)

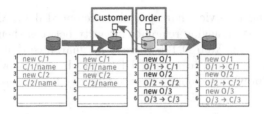

Fig. 2. The link from the *Order* microservice to entity *C/3* is broken after the recovery of *Customer* microservice from an old backup

Please note that broken link can also appear when a referenced data is deleted, e.g. when a customer is deleted in the local database of microservice *Customer*. In this case, the referential integrity is not respected.

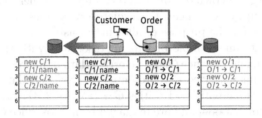

Fig. 3. Synchronized backup of an entire microservice architecture

As aforementioned, several approaches indicate that microservice architecture implies some challenging problems of data integrity and consistency management [2,18], as well as the difficulty of managing consistent backups due to distributed persistence [13]. However, as far as we know, no approach proposes a solution to such problems. In the following, we present a solution that can bypass such referential integrity violation and broken links.

3 Our Solution: A Weak Referential Integrity Management

In this work, we focus on referential integrity. We present a solution to help the user in the recovery of the system referential integrity in case of a disaster crash. We define the global consistency as a time-dependent property. We propose a new global consistency definition, called *the weak global referential consistency*. Our solution provides information about the global state in case of a disaster crash that the users can pinpoint exactly the location, and time interval, of missing data which needs to be manually repaired.

In the following, we first present the context and assumptions (Subsect. 3.1), without taking disaster crash into consideration. Then, we introduce our definition of global consistency (Subsect. 3.2). Based on this definition, we show the

Table 1. Table of symbols

Symbol	Description
μs	Microservice
D	Database of microservice μs
e	Entity of a database D
URI^e	Uniform Resource Identifier of an entity e
t_j	Date of the last update of an entity e in D
d^e	Dependency counter associated with entity e
E_k	Epoch identity, k being a timestamp
$t_{(i,k)}$	ith timestamp related to epoch E_k
B	Backup of the local database of microservice μs
$]t_{(i,k)}, t_{(j,k')}]$	Amnesia interval of an entity e

method of recovery from a disaster crash affecting one microservice. All symbols used in this article can be found in Table 1 above.

3.1   Context and Assumptions

In this article, microservice follows the pattern called Database per Service (defined in [19]), where each microservice has full control of a private database, persistent data being accessible to other services only via an API. Each microservice also use an event-driven architecture, such as the one defined in [20], consuming and publishing a sequence of state-changing events.

The following are our assumptions:

- Microservices are part of the same application.
- All microservices of an application trust each other.
- Each microservice μs has a database D storing a set of entities.
- Each entity $e \in D$ can be either a RESTful API resource, a relational tuple, a key-value record, a document or graph database item.
- Each entity e has a Uniform Resource Identifier, URI^e, that identifies the entity.
- The state of each entity e is read, updated and deleted using standard HTTP protocol primitives (GET, PUT and DELETE). In addition, we introduce two additional operations: `getReference`, `deleteReference`.
- Each microservice μs ensures the consistency and the durability of its own database D.

Taking into consideration the following ; handling the references between the different microservices and ensuring that the system is reliable when no failure occurs:

- Each microservice has its own clock. The clock of different microservices are not necessarily synchronized.
- An entity e', managed by a microservice $\mu s'$, can refer to another entity e managed by a microservice μs.
- The reference from microservice $\mu s'$ to an entity e, managed by a microservice μs, is the couple (URI^e, t_j) with the timestamp t_j marking the date of the last update of entity e in D as it is known by the microservice $\mu s'$, i.e. exactly when $\mu s'$ queries the microservice μs, using the clock of microservice μs.

There are 2 cases as far as reference storage is concerned:

1. the minimalist case consists in just storing the reference and the most recent modified timestamp, i.e. couple (URI^e, t_j);
2. the eager/self-contained backup case consists in storing a copy of the referenced entity state, that can be cached by $\mu s'$. When microservice $\mu s'$ stores a copy of the referenced entity in its cache, this former copy is considered as detached, identical to detached entity in object-relational mapping using JPA specification [10]. Detached means that the copy is not managed by μs, microservice $\mu s'$ being responsible for keeping its cache up-to-date. Cached representation is only a representation of the original entity state, thus it may only contain a projection. For our solution, we assume that it is possible to reconstruct the original entity state from its cached representation.

3.2   Global Consistency

In case of no disaster crash, the global consistency can be defined as follows:

Definition 1. The global consistency
A global state is consistent if:

- *(local database consistency) each local database is locally consistent in the traditional sense of a database, i.e. all its integrity constraints are satisfied.*

and

- *(**global referential integrity**) the timestamp value associated with each reference is less than or equal to the timestamp value of the corresponding referenced entity.*
  Formally: for each couple (URI^e, t_j) associated with an entity e' referencing another entity e of μs, $t_j \leq t_i$, with t_i the most recent update timestamp of e in μs.

Case of Snapshot Data Storage Pattern. When using the snapshot data storage pattern, each local database contains the current state of its microservice. In order to guarantee the referential integrity in the microservice architecture, a microservice μs, cannot delete an entity e, if there is an entity e' managed by another microservice $\mu s'$ that refers to the entity e. We suggest a referential integrity mechanism based on dependency counters, as follows:

- Each entity e managed by a microservice μs is associated with a dependency counter d^e. This counter indicates how many other entities managed by other microservices refer to entity e. It is initially set at 0.
- When a microservice $\mu s'$ wants to create the entity e' that refers an entity e managed by μs, it sends a `getReference` message to microservice μs. The corresponding dependency counter d^e is incremented. Then, μs sends the couple (URI^e, t_j) back to $\mu s'$, with t_j, the date of the most recent update of entity e.
- When microservice $\mu s'$ receives the information about e, it creates the entity e'.
- When microservice $\mu s'$ deletes an entity e' that refers e, it sends a message `deleteReference` to microservice μs, indicating that the reference to e does not exist any more. d^e is therefore decremented.
- Microservice μs cannot delete an entity e if its dependency counter is $d^e > 0$. It retains the most recent value of entity e with its most recent update time, t_j, and flags the entity by $\perp$ indicating that e must be deleted when its dependency counter reaches the value of 0.

According to Definition 1, a reliable microservice system using the referential integrity mechanism based on dependency counters, will always be globally consistent.

Case of Event Sourcing Data Storage Pattern. When choosing event sourcing as data storage pattern [20], each local database contains an event log, which records all changes of the microservice state. Thus, it is possible to rebuild the current state of the data by replaying the event log. In this case, we propose the following referential integrity mechanism:

- When a microservice $\mu s'$ wants to create the entity e' that refers an entity e managed by μs, it sends a `getReference` message to microservice μs. When $\mu s'$ receives the information about e, it creates the entity e' and a creation event, associated with the corresponding reference (URI^e, t_j), is stored in the log of $\mu s'$.
- When an entity e of microservice μs must be deleted, instead of deleting it, the microservice μs flags it by $\perp$, and a deletion event, associated with the related timestamp, is stored in its log, representing the most recent valid value of entity e.

Thus, it is easy to prove that global consistency state can be obtained from the logs. For each couple (URI^e, t_j) of a referenced entity e, timestamp t_j must appear in the event log of microservice μs. Moreover the most recent record associated with entity e, corresponding to an update or deletion of e, in the event log of μs, has a timestamp $t_i \geq t_j$, with t_j, any timestamp appearing in any reference couple (URI^e, t_j) stored in the event log of any other microservice referencing e.

3.3 Fault Tolerant Management of Microservice Referential Integrity

As explained in Sect. 2.3, disaster crash can occur in microservice architectures. In the following, we consider disaster crash affecting only one microservice.

To protect the local database from media-failure, each microservice stores an up-to-date backup of its database, i.e. a copy of the database separate from the database itself. Each microservice individually manages the backup of its database. The way in which microservices independently manage their backup is out of the scope of this paper.

A disaster crash of a microservice μs means that its local database and its log are lost and we have to recover the database from a past backup. The backup provides a consistent state of the local database. However, as the database has been recovered from a past backup, data could have been lost. In order to provide a state of the local database as close as possible to the one of the database before the failure, data cached by other microservices can be used. When a microservice $\mu s'$ refers to an entity managed by another microservice μs, it can store a detached copy of the referenced entity. Therefore, these detached copies can be used to update the state of the database obtained after recovery from the backup.

In the following, we present the concepts used to manage disaster crash, our recovery protocol, how to optimize it and we define the Weak Global Referential Integrity.

Backup and Recovery, Amnesia Interval and New Epoch. To manage disaster crash, our assumptions are:

- Each entity of the local database of microservice μs is associated with an epoch identity E_k. An epoch is a new period after a disaster crash recovery. A new epoch E_k begins at the first access of an entity after recovery. Therefore, a timestamp $t_{(i,k)}$ associated with an entity e represents the ith timestamp related to epoch E_k. $k = 0$ when no crash has occurred, $k > 0$ otherwise. The value of k always increases, being associated with time.
- When a backup B of the local database of microservice μs is done, operation BCK, the backup is associated with clock epoch identity $E_{k'}$, and with a creation timestamp $t_{(i,k')}$, such that: all entities e, stored in backup B, have an updated timestamp $t_{(j,k')} \leq t_{(i,k')}$. Epoch $E_{k'}$ associated with backup and epoch E_k associated with the local database are such that: $k' \leq k$.
- As long as there is no disaster crash, the local database and the backup are associated with the same epoch identity.
- In case of disaster crash of microservice μs, when the local database is locally recovered from an past obsolete backup created at time $t_{(c,k')}$, it is known that local database has an amnesia interval starting from $t_{(c,k')}$. This amnesia interval is associated with all entities saved in the backup and lasts until such entities are accessed again (see Definition 2).

- Each entity of the recovered database is associated with a timestamp related to epoch E'_k of the backup. This timestamp remains as long as no updates have been carried out. A new epoch E_k begins at the first reading or written access of an entity, k containing the current date. A written operation, PUT, overwrites whatever value was recovered. However, epoch should also be updated after a reading operation, GET. Any other microservice reading from the state of the recovered entity will establish a causal dependency, which would be in conflict with further more recent recovered values from the previous epoch (see [1] for more details).

Definition 2. Amnesia Interval
An amnesia interval of microservice μs is a time interval indicating that a disaster crash has occurred for the local database of μs. This interval is associated with each entity managed by μs. An amnesia interval $]t_{(i,k')}, t_{(j,k)}]$ of an entity e means that:

- *Epoch $E_{k'}$ is the epoch associated with the backup used for the database recovery.*
- *Timestamp $t_{(i,k')}$ corresponds to the time of most recent known update of e. It is either the timestamp associated with the backup used for recovery, or the timestamp of a cached copy of e stored in a microservice referring entity e.*
- *Timestamp $t_{(j,k)}$ corresponds to the first reading or written operation on e from another microservice $\mu s'$, after $t_{(i,k')}$ $(k' < k)$.*

Weak Global Referential Integrity. After a crash recovery, data can be lost, so we define a weak global referential integrity of the microservice architecture. Weak means that either the global referential integrity has been checked, verifying Definition 1, or an amnesia is discovered; data has been lost as well as the interval of time when the data was lost. This makes it possible to focus on the manual data recovery and reconstruction effort within the amnesia interval.

Definition 3. Weak global consistency
After a disaster crash recovery of a microservice μs, the system checks a weak global consistency iff:

- *(local database consistency) each local database is locally consistent in the traditional sense of a database, i.e. all its integrity constraints are satisfied.*

and

- *(weak global referential integrity) the timestamp value associated with each reference is either less than or equal to the timestamp value of the corresponding referenced entity or included in an amnesia interval.*
  Formally: for each couple $(URI^e, t_{(\ell,\kappa)})$ associated with an entity of $\mu s'$ referencing another entity e of μs:
    - *either $t_{(\ell,\kappa)} \leq t_{(i,k')}$, with $t_{(i,k')}$ the most recent update timestamp of e in μs, epochs κ and k' being comparable $(\kappa \leq k')$;*

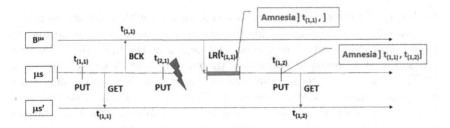

Fig. 4. Example of a scenario with 2 microservices, without cached data. PUT represents a state change of the referenced entity. BCK indicates when a backup snapshot is taken. LR shows when the microservice is locally recovered from the backup. (Color figure online)

- or $t_{(\ell,\kappa)} \in \,]t_{(i,k')}, t_{(j,k)}]$, with $]t_{(i,k')}, t_{(j,k)}]$ the amnesia interval associated with the referenced entity e, after a disaster crash of μs that manages e.

Consider a scenario of two microservices μs and $\mu s'$, $\mu s'$ referencing an entity managed by μs (see Fig. 4), but without storing any cache of the referenced entity. In figures, only timestamps of one entity of microservice μs are considered, timestamps and epoch identities being only represented by numbers. In Fig. 4, at time $t_{(1,1)}$ an entity is created by microservice μs ; operation PUT. A backup B is made, storing entities of microservice μs created before $t_{(2,1)}$; operation BCK. Microservice $\mu s'$ refers the entity of μs created at time $t_{(1,1)}$; operation GET. An update of the entity is carried out by the microservice μs at time $t_{(2,1)}$; operation PUT. When disaster crash appears to μs (see red flash), μs must recover using the backup, update of time $t_{(2,1)}$ is lost, therefore it has amnesia that begins from $t_{(1,1)}$. An update is done to the entity, then a new epoch 2 begins and timestamp $t_{(1,2)}$ is associated to the updated value. The amnesia interval is then updated to $]t_{(1,1)}, t_{(1,2)}]$. If $\mu s'$ does another GET to refresh the referenced value, the up-to-date timestamp $t_{(1,2)}$ is sent by μs.

Recovery Protocol. When a disaster crash occurs to a microservice μs, μs informs all other services of its recovery. Moreover, when microservices stored copies of the entities they refer to, in their cache, the amnesia interval associated with each recovered entity of μs can be reduced using cached replicas. In order to do so, the steps are:

- After the recovery of μs, an event indicating that there is amnesia is sent, or broadcast, to other microservices.
- When a microservice $\mu s'$ receives an amnesia event from microservice μs, managing an entity e it refers to; if it has stored a replica of e in its cache, then $\mu s'$ sends the replica of the entity it refers to, to μs.
- When microservice μs receives replies carrying information from $\mu s'$ about its entity e, it compares the value of e, associated with timestamp $t_{(i,k')}$, with the value, associated with timestamp $t_{(j,k)}$, it stored, if epochs k and k'

are comparable. Then, it retains the more up-to-date value and shrinks the amnesia interval associated with e if necessary.
- Once a read operation or an update operation is done on e, a new epoch begins, and the first timestamp associated with this new epoch represents the end of the amnesia interval.
- The beginning of the amnesia interval can still be shifted if more up-to-date values are received from belated replies from other cached replicas.

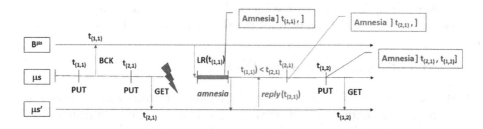

Fig. 5. Recovery scenario using cached data more recent than the backup.

In Fig. 5, $\mu s'$ stores an up-to-date value of the referenced entity, after the backup of μs ; operation GET. After the recovery from the past backup of time $t_{(1,1)}$, μs sends an event about its amnesia, associated with interval $]t_{(1,1)},]$. After receiving this amnesia event, $\mu s'$ sends its up-to-date value, associated with $t_{(2,1)} > t_{(1,1)}$, to μs ; event reply. μs stores this up-to-date value, associated with timestamp $t_{(2,1)}$, and updates the amnesia interval to $]t_{(2,1)},]$. After a update is done to the entity, a new epoch 2 begins and timestamp $t_{(1,2)}$ is associated to the updated value ; operation PUT. The amnesia interval is then updated to $]t_{(2,1)}, t_{(1,2)}]$.

Availability vs Consistency. After a disaster crash of μs: either μs is immediately available after its local recovery, or it expects information sent by other microservices that refer its entities before the disaster crash, to provide a more recent database snapshot than the past used backup, updating the value stored in the backup with the copy stored in the cache of the other microservices.

If we are uncertain that all microservices will answer the amnesia event or if μs ignores or partially knows which microservices refer to (case 1), μs can wait for a defined timeout.

If we assume that all microservices are available and will answer to the amnesia event (case 2), μs waits until all microservices have sent their reply to the amnesia event.

After the timeout (case 1) or the reception of all responses (case 2), the recovery is ended and μs is available.

When dependency counters are used (see Sect. 3.2) and if we are sure that the identity of all microservices that refer to μs is known after the disaster crash: an optimization of the recovery process can be used. In this case, an amnesia event is sent only to all microservices referring to μs, instead of broadcast, and μs waits until all the aforementioned microservices reply. To do so, the address of each microservice referring to μs should be stored by μs, when the dependency counter is updated. Each referencing microservice can either send the value it stores in its cache, or a message indicating that it is no longer concerned by the amnesia, because it currently does not refer to any entity of μs.

The choice between the aforementioned steps depends on the focus on availability (μs is available as soon as possible after its disaster crash, with a large amnesia interval) or on consistency (we prefer to wait in order to provide a more recent snapshot than the one used for the recovery before making μs available).

4   Conclusions and Future Work

In this paper, we have focused on preserving referential integrity within microservice architecture during disaster recovery. We have introduced a definition of weak global referential consistency and a recovery protocol taking advantage of replicas found in microservice caches. These are merged with local backup to reduce the amnesia interval of the recovered microservice. The approach has been validated under several assumptions: direct references to simple entities, single crashes and no concurrent recovery of more than one failed microservice.

In this paper we focused on reliability aspects, whereas as part of future work we plan to assess the performance implications of our approach in depth. We will also address more complex relationships between microservices, e.g., transitive or circular dependencies, which may span across multiple microservices. While microservice architecture is known for its ability to isolate failures, which should not cascade across multiple microservices, it remains an open question how to apply our approach to perform the concurrent recovery of multiple microservices which may have failed independently over an overlapping period of time.

Acknowledgements. The authors would like to thank Guy Pardon, Eirlys Da Costa Seixas and the referees of the article for their insightful feedback.

References

1. Ahamad, M., Neiger, G., Burns, J.E., Kohli, P., Hutto, P.W.: Causal memory: definitions, implementation, and programming. Distrib. Comput. **9**(1), 37–49 (1995). https://doi.org/10.1007/BF01784241
2. Baresi, L., Garriga, M.: Microservices: the evolution and extinction of web services? Microservices, pp. 3–28. Springer, Cham (2020). https://doi.org/10.1007/978-3-030-31646-4_1
3. Bhattacharya, S., Mohan, C., Brannon, K.W., Narang, I., Hsiao, H.I., Subramanian, M.: Coordinating backup/recovery and data consistency between database and file systems. In: ACM SIGMOD International Conference on Management of data, pp. 500–511. ACM (2002)

4. Blaha, M.: Referential integrity is important for databases. Modelsoft Consulting Corp. (2005)
5. Brewer, E.: CAP twelve years later: how the "rules" have changed. Computer **45**(2), 23–29 (2012)
6. Bucchiarone, A., Dragoni, N., Dustdar, S., Larsen, S.T., Mazzara, M.: From monolithic to microservices: an experience report from the banking domain. IEEE Softw. **35**(3), 50–55 (2018)
7. Connoly, T., Begg, C.: Database Systems. ke-3. Addison-Wesley, England (1998)
8. Date, C.J.: Referential integrity. In: 7th International Conference on Very Large Data Bases (VLDB), pp. 2–12 (1981)
9. Davis, H.C.: Referential integrity of links in open hypermedia systems. In: 9th ACM Conference on Hypertext and Hypermedia: Links, Objects, Time and Space, pp. 207–216 (1998)
10. DeMichiel, L., Keith, M.: Java persistence API. JSR 220 (2006)
11. Elmasri, R., Navathe, S.: Fundamentals of Database Systems. Addison-Wesley, Boston (2010)
12. Ingham, D., Caughey, S., Little, M.: Fixing the "broken-link" problem: the W3objects approach. Comput. Netw. ISDN Syst. **28**(7–11), 1255–1268 (1996)
13. Knoche, H., Hasselbring, W.: Drivers and barriers for microservice adoption-a survey among professionals in Germany. Enterp. Model. Inf. Syst. Architect. (EMISAJ) **14**, 1–1 (2019)
14. Lewis, J., Fowler, M.: Microservices a definition of this new architectural term (2014). http://martinfowler.com/articles/microservices.html
15. Mohan, C., Haderle, D., Lindsay, B., Pirahesh, H., Schwarz, P.: ARIES: a transaction recovery method supporting fine-granularity locking and partial rollbacks using write-ahead logging. ACM Trans. Database Syst. (TODS) **17**(1), 94–162 (1992)
16. Newman, S.: Building Microservices: Designing Fine-Grained Systems. O'Reilly, Newton (2015)
17. Overeem, M., Spoor, M., Jansen, S.: The dark side of event sourcing: managing data conversion. In: 24th International Conference on Software Analysis, Evolution and Reengineering (SANER), pp. 193–204. IEEE (2017)
18. Pardon, G., Pautasso, C., Zimmermann, O.: Consistent disaster recovery for microservices: the BAC theorem. IEEE Cloud Comput. **5**(1), 49–59 (2018)
19. Richardson, C.: Pattern: database per service (2018). https://microservices.io/patterns/data/database-per-service.html. Accessed 02 Apr 2020
20. Richardson, C.: Pattern: event sourcing (2018). https://microservices.io/patterns/data/event-sourcing.html. Accessed 01 Apr 2019
21. Taibi, D., Lenarduzzi, V., Pahl, C.: Architectural patterns for microservices: a systematic mapping study. In: CLOSER, pp. 221–232 (2018)
22. Ullman, J.D., Garcia-Molina, H., Widom, J.: Database Systems: The Complete Book, 1st edn. Prentice Hall, Upper Saddle River (2001)
23. Zimmermann, O.: Microservices tenets. Comput. Sci. Res. Dev. **32**, 301–310 (2016). https://doi.org/10.1007/s00450-016-0337-0
24. Zimmermann, O., Stocker, M., Lübke, D., Pautasso, C., Zdun, U.: Introduction to microservice API patterns (MAP). In: Joint Post-Proceedings of the First and Second International Conference on Microservices (Microservices 2017/2019). OpenAccess Series in Informatics (OASIcs), vol. 78, pp. 4:1–4:17 (2020)

Hashing Based Prediction for Large-Scale Kernel Machine

Lijing Lu[1,2], Rong Yin[1,2], Yong Liu[1(✉)], and Weiping Wang[1]

[1] Institute of Information Engineering,
Chinese Academy of Sciences, Beijing, China
[2] School of Cyber Security, University of Chinese Academy of Sciences,
Beijing, China
{lulijing,yinrong,liuyong,wangweiping}@iie.ac.cn

Abstract. Kernel Machines, such as Kernel Ridge Regression, provide an effective way to construct non-linear, nonparametric models by projecting data into high-dimensional space and play an important role in machine learning. However, when dealing with large-scale problems, high computational cost in the prediction stage limits their use in real-world applications. In this paper, we propose hashing based prediction, a fast kernel prediction algorithm leveraging hash technique. The algorithm samples a small subset from the input dataset through the locality-sensitive hashing method and computes prediction value approximately using the subset. Hashing based prediction has the minimum time complexity compared to the state-of-art kernel machine prediction approaches. We further present a theoretical analysis of the proposed algorithm showing that it can keep comparable accuracy. Experiment results on most commonly used large-scale datasets, even with million-level data points, show that the proposed algorithm outperforms the state-of-art kernel prediction methods in time cost while maintaining satisfactory accuracy.

Keywords: Kernel machine · Locality-sensitive hashing · Kernel prediction

1 Introduction

Kernel methods have been widely implemented in practice, allowing one to discover non-linear structure by mapping input data points into a feature space, where all pairwise inner products can be computed via a nonlinear kernel function. Kernel machines, such as Kernel Ridge Regression (KRR), have attracted a lot of attention as they can effectively approximate any function or decision boundary with enough training data [17–19].

Despite excellent theoretical properties, they have been limited applications in large scale learning because computing the decision function for new test samples is extremely expensive. As the scale of data increases, not only the training time will become longer, but the prediction time will also increase. However,

© Springer Nature Switzerland AG 2020
V. V. Krzhizhanovskaya et al. (Eds.): ICCS 2020, LNCS 12138, pp. 496–509, 2020.
https://doi.org/10.1007/978-3-030-50417-5_37

more and more improvement methods devoted to reducing training complexity have been proposed, while fewer algorithms have been designed to improve the performance of the prediction stage [12]. As is known to us all, the Nyström and random feature methods are devoted to reaching faster training and prediction speed by constructing low-rank approximation of kernel matrix. Nyström method [4,14,16] construct a small scale subset of landmark data points by sampling to approximate the raw kernel matrix. Random feature [17,20] maps data into a relative low-dimensional randomized feature space to improve both training and prediction speed. Random sketch [15,23] is another family of techniques that projects the kernel matrix into a small matrix to reduce the computational requirement. Although these methods perform well and are applied into practice widely, they still need huge computational requirements in the prediction stage when faced with large-scale datasets. Based on the innovation of further reducing the computational costs when dealing with large-scale datasets, we attempt to develop a fast prediction algorithm for kernel methods.

Hashing is an efficient algorithm to solve the approximate large-scale nearest Neighbor search problem. The main idea of the hashing method is to construct a family of hash functions to map the data points into a binary feature vector such that the produced hash code preserves the structure of the original space [9]. In recent years, a large number of effective hashing methods have emerged, but they are rarely used in the prediction of kernel machines. Charikar and Siminelakis presented a hashing-based framework for kernel density estimate problem in [5]. Inspired by this paper, we consider applying the hashing algorithm to the prediction stage of kernel machine and proposed the hashing based prediction (HBP) algorithm. The algorithm leverages locality-sensitive hashing (LSH) [6,8] method to search the nearest neighbors of the test data point to approximately compute the decision value in the kernel prediction problem.

Specifically, our HBP algorithm consists of two stages: the pre-processing stage and the query stage. LSH was used to find the neighbors of the test point as the sampled subset in the pre-processing stage. And in the query stage, the samples are used to approximately compute the decision value. We provide a theoretical analysis that we can compute the decision function in $\mathcal{O}(\frac{log(n/\epsilon\tau)}{\tau^{0.5}\epsilon^{2.5}})$ time cost with accuracy guarantees $\|\hat{y} - y^*\|_p \leq \epsilon \cdot (3\tau n^{1/p} + \|y^*\|_p)$, where $\epsilon, \tau \in (0,1)$. It is novel to use the hash method directly in the KRR problem which is an attempt to break through the traditional method. As will be demonstrated later, experiment results on most commonly used large-scale datasets, even with million-level data points, show that the proposed HBP algorithm outperforms the state-of-art kernel prediction methods in time cost while maintaining satisfactory accuracy.

The remainder of the paper is organized as below: We start with presenting related work in Sect. 2, and the background material is given in Sect. 3. In Sect. 4, we introduce our hashing-based prediction algorithm for the prediction of kernel machine, while in Sect. 5, we discuss the theoretical analysis. Section 6 presents the experimental results on real-world datasets. The following section is conclusions and the last section presents proof of theoretical results.

2   Related Work

In this section, we will introduce several important works on reducing time complexity in the prediction stage of KRR, and most of the algorithms can be used to other kernel machines [11]. Practical methods are proposed to overcome the computational and memory bottleneck of kernel machines. One popular family of techniques is based on low-rank approximation of the kernel matrix to improve both training and prediction speed. Nyström [13, 14, 22] method is one of the most well-known representation methods. Nyström samples a subset C of m columns to approximate the kernel matrix $G \in \mathbb{R}^{n \times n}$. Typically, the subset of columns is randomly selected by uniform sampling without replacement [13, 22]. Recently, more and more outstanding extension methods of Nyström are proposed to solve the KRR problem. For example, an accurate and scalable Nyström scheme has been proposed in [14] which first samples a large column subset from the input matrix, but then only performs an approximate SVD on the inner submatrix by using the recent randomized low-rank matrix approximation algorithms. [16] present a new Nyström algorithm based on recursive leverage score sampling. Based on Nyström method, a fast prediction method called DC-Pred++ was proposed in [11]. However, the scale of the subset in Nyström method can't be too small to keep the accuracy, and as a result, the lower bound of computational complexity is limited. Random features [17, 20] project the data into a relative low-dimensional feature space where the inner product between a pair of input points approximates their kernel evaluation so as to reduce the computation requirement of training and prediction stage. Another family of techniques is random sketches which improving the computational speed by projecting the kernel matrix into a small matrix [15, 23]. As shown in [23], a simple hash method was applied to generate the randomized sketch matrix.

Hashing algorithms [6, 8, 9] have been a continuously "hot topic" since it birth because of its superior performance in the Approximate Nearest Neighbor Search. However, hashing algorithms have not widely been used in kernel prediction problems. Inspired by the idea of "sample" of the Nyström method, we consider employing Locality-sensitive hashing (LSH) to sample the subset and approximately compute the result using the subset. Compared with the currently optimal and most popular Nyström methods, our HBP method can further greatly reduce the number of samples while ensuring accuracy. HBP reduces computational efficiency from $\mathcal{O}(n)$ to $\mathcal{O}(\frac{log(n/\epsilon\tau)}{\tau^{0.5}\epsilon^{2.5}})$ with less accuracy loss $\|\hat{y} - y^*\|_p \leq \epsilon \cdot (3\tau n^{1/p} + \|y^*\|_p)$.

3   Preliminaries

This paper focuses on the typical kernel machines: KRR. [7, 10, 21]. Given dataset $\{x_i, y_i\}_{i=1}^n$, $x_i \in \mathbb{R}^d$, $k(x_i, x_j)$ denotes the kernel function value of the two points x_i, x_j, $\hat{\alpha} \in \mathbb{R}^n$ in the formula is generated in the training process by solving the

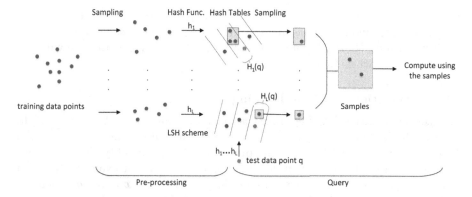

Fig. 1. Schematic diagram of algorithm structure. Given the training data points and a test point q, the HBP algorithm samples some hash functions and construct a separate hash table for each hash function leveraging LSH in the pre-processing stage. In the query stage, for each hash table, we sample a point randomly from the hash bucket that the test point maps to.

below optimization problem:

$$\hat{\alpha} \leftarrow \underset{\alpha}{argmin} \ \alpha^T K \alpha + \lambda \alpha^T \alpha - 2\alpha^T y, \tag{1}$$

where $K \in \mathbb{R}^{n \times n}$ is the kernel matrix with $K_{ij} = k(x_i, x_j)$, $y = [y_1, ..., y_n]^T \in \mathbb{R}^n$ is the response vector, $\lambda > 0$ is the regularization parameter.

Our goal in the prediction process is to compute the decision value of a testing data $\bar{x}$, following the equation:

$$y^* = \sum_{i=1}^{n} \hat{\alpha}_i k(x_i, \bar{x}), \tag{2}$$

whose time cost is $\mathcal{O}(n)$. The problem to solve Eq. (2) is challenging as n increases. To improve the speed of kernel prediction, we proposed the following HBP algorithm.

4 Hashing Based Prediction

The proposed HBP method is an importance sampling algorithm. The algorithm solves the problem of designing a model that given a set of data points $X = \{x_i, y_i\}_{i=1}^{n} \subset \mathbb{R}^d$ and a kernel function k, returns the approximation of the decision value of a test point $\bar{x}$: $\sum_{i=1}^{n} \hat{\alpha}_i k(x_i, \bar{x})$. Given a hashing based data structure, HBP performs a two-stage computation to approximate the prediction result. The architecture of HBP is illustrated in Fig. 1. In the pre-processing stage, we select data points which close to the test data point as the sampling points. In the query stage, we use the sampling points to estimate the response value. As shown in Fig. 1, the hashing technique plays an important role in the proposed method. The hashing technique we leverage is called Localization-sensitive-hashing (LSH). Before explaining the algorithm in detail, we will first introduce the LSH method.

4.1 Localization-Sensitive Hashing

Localization-sensitive hashing (LSH) is an approximate neighbor search technology. The key idea of LSH is that two adjacent data points in the original data space are mapped by the same projection, the probability that these two data points are still adjacent in new data space is high. For example, the LSH method employed by this paper map a d dimension vector $\boldsymbol{x}$ into a ρ dimension vector consisting entirely of 0 and 1, where $\rho < d$. The vector of ρ dimension is the hash value. If two data points get the same hash value through the hashing technique, it is said that the two data points fall into the same hash bucket. The construction goal of the hash scheme is to make the adjacent points in the original data space fall into the same hash bucket through the hashing function. So the hash functions need to meet two conditions shown in definition 2. We will first introduce the definition of collision probability which indicates the probability that two data points are mapped to the same hash bucket through the given hash function.

Definition 1. *Given a hashing scheme $\mathcal{H}$, the collision probability between two elements $x, \bar{x} \in \mathbb{R}^d$ is defined by $p(x, \bar{x}) := \Pr_{h \sim \mathcal{H}} [h(x) = h(\bar{x})]$, where $h \sim \mathcal{H}$ denote a hash function h is sampled from the hashing scheme $\mathcal{H}$.*

The collision probability of two data points x and $\bar{x}$ is closely related to the distance between the two points. The larger the distance, the smaller the collision probability.

Let $D(\cdot, \cdot)$ be a distance function of elements from a dataset S, and for any $x \in S$, let $\mathcal{B}(x, r)$ denote the set of elements from S within the distance r from x.

Definition 2. *A family of functions $\mathcal{H} = \{h : S \to U\}$ is called (r_1, r_2, p_1, p_2)-sensitive for $D(\cdot, \cdot)$ if for any $x, \bar{x} \in S$*

- *if $\bar{x} \in \mathcal{B}(x, r_1)$ then $p(x, \bar{x}) \geq p_1$,*
- *if $\bar{x} \notin \mathcal{B}(x, r_2)$ then $p(x, \bar{x}) \leq p_2$.*

In order to guarantee the hash functions in a locality-sensitive family to be useful, it has to satisfy that $p_1 > p_2$ and $r_1 < r_2$. The hash methods employed by this paper are presented in Algorithm 1.

Proposition 1. *For every $x, \bar{x} \in \mathbb{R}^n$, the LSH family $\mathcal{H}$ constructed by Algorithm 1 satisfies:*

$$p(x, \bar{x}) = e^{-\|x - \bar{x}\|_1 / (2\sigma)} = k(x, \bar{x})^{\frac{1}{2}}. \tag{3}$$

The proof is shown in Sect. 8.1.

After all input data points have been hashed into hash buckets, one can determine neighbors of query point by hashing the query point and searching elements in the bucket containing that query point.

Algorithm 1. Localization-sensitive hashing.

Require: Dataset $\{x_i, y_i\}_{i=1}^n$; Dimension of data d; Bandwidth σ.
Ensure: Hash function h.
1: Sample $\rho \sim Poisson(d/(2\sigma))$.
2: Sample ρ dimensions from d dimensions $\zeta_1.....\zeta_\rho \in \{1.....d\}$ at random.
3: Sample ρ reference values $\xi_1.....\xi_\rho \in [0, 1]$ at random.
4: Given a data point x, for every $i = 1.....\rho$, set $b_i = 1$ if $x_{\zeta_i} > \xi_{\zeta_i}$ and $b_i = 0$ otherwise.
5: The hash value is the concatenation of $b_1.....b_\rho$.

4.2 Hashing Based Prediction

As mentioned before, HBP uses LSH to create a two-stage structure to compute the decision value approximately. Now, we will explain the main algorithm in detail.

Pre-processing. In the pre-processing stage, training data points are hashed to different buckets according to the hash function.

1) Randomly construct L hash functions $h_1...h_L$ from hash scheme $\mathcal{H}$.
2) For each hash function h_j, sample a subset from dataset $X_j \in X$, every point in X_j is selected with probability $\delta = \frac{L}{n}$. In order to reduce computational complexity, we only run the hash process on a small subset rather than the whole training set. The result of our theorem proves that the method can still guarantee accuracy.
3) For each hash function, each data point in X_j is mapped to a hash value through the hash function. All data points in X_j of the same hash value constitute a hash bucket. And all hash buckets form a hash table corresponding to the hash function.
4) L hash functions can produce L hash tables.

Query. In the query stage, the sampling points were used to estimate the response value.

1) Compute the hash value of the test point $\bar{x}$ through each hash function, and map the test point $\bar{x}$ to a hash bucket for each hash table.
2) For each hash table, randomly select a data point $x^{(j)}, j \in (1, L)$ from the hash bucket to which the test point is mapped if that hash bucket is not empty except for the test point $\bar{x}$. The L hash tables allow us to produce at most L independent samples because if the hash bucket where the test point is mapped is empty except for the test point, we can't get a sample for this hash table.
3) Compute the estimated values using every sample point:

$$Z_j = \frac{\hat{\alpha}_j}{\sum_{j=1}^n \hat{\alpha}_j} \frac{k(x^{(j)}, \bar{x}) \cdot |b_j(\bar{x})|}{\delta \cdot p(x^{(j)}, \bar{x})}, \tag{4}$$

where $b_j(\bar{x}) = \{x \in X'_j : h_j(x) = h_j(\bar{x})\}$ represents the set of the elements in the hash bucket where the point $\bar{x}$ mapped, and $|b_j(\bar{x})|$ represents the number of elements in $b_j(\bar{x})$.

4) Compute the accurate prediction value by averaging all of the samples produced by hash tables:

$$\hat{y} = \frac{1}{L} \sum_{j=1}^{L} Z_j. \tag{5}$$

5) The approximate prediction value can achieve the accuracy of $\|\hat{y} - y^*\|_p \le \epsilon \cdot (3\tau n^{1/p} + \|y^*\|_p)$ with the time and space cost of $\mathcal{O}(\frac{\log(n/\epsilon\tau)}{\tau^{0.5}\epsilon^{2.5}})$.

Algorithm 2. Hashing Based Kernel Prediction.

Require: Dataset $\{x_i, y_i\}_{i=1}^n$; test data point $\bar{x} \in \mathbb{R}^d$; kernel $k(\cdot, \cdot)$; LSH family $\mathcal{H}$; interger $1 \le L \le n$; the result of training stage $\hat{\alpha}$;

Ensure: Estimated response value $\hat{y}$.

1: **Pre-processing:**
2: For $j = 1.....L$:
3: Sample a random hash function h_j from $\mathcal{H}$.
4: Get sampling dataset $X_j \subset X$, each point of X_j was selected with independent probability $\delta = \frac{L}{n}$.
5: For every $x \in X_j$, compute the hash value $h_j(x)$.
6: **Query:**
7: For $j = 1.....L$:
8: Compute the hash value of test data $h_j(\bar{x})$.
9: Sample a random point $x^{(j)}$ from $b_j(\bar{x}) = \{x \in X'_j : h_j(x) = h_j(\bar{x})\}$.
10: Let $Z_j \longleftarrow \frac{\hat{\alpha}_j}{\sum_{i=1}^n \hat{\alpha}_j} \cdot \frac{k(x^{(j)}, \bar{x}) \cdot |b_j(\bar{x})|}{\delta \cdot p(x^{(j)}, \bar{x})}$.
11: The prediction value $\hat{y} = \frac{1}{L} \sum_{j=1}^{L} Z_j$.

---

5   Theoretical Assessments

The problem of our proposed algorithm aims to solve is to obtain an approximation $\hat{y}$ to $y^* = \sum_{i=1}^n \hat{\alpha}_i k(x_i, \bar{x})$. In this section, we introduce the theoretical bound of the proposed algorithm.

Theorem 1. *Given a kernel k, if there exists a distribution $\mathcal{H}$ of hash functions and $M \ge 1$ such that for every $x, \bar{x} \in \mathbb{R}^d$,*

$$M^{-1} \cdot k(x, \bar{x})^{1/2} \le \Pr_{h \sim \mathcal{H}}[h(x) = h(\bar{x})] \le M \cdot k(x, \bar{x})^{1/2}, \tag{6}$$

then we can compute an approximate vector $\hat{y}$ for kernel prediction in time $\mathcal{O}(\frac{\log(n/\epsilon\tau)}{\tau^{0.5}\epsilon^{2.5}})$ such that with probability at least $1 - n^{-1}$ for all $i \in [n]$ it holds $|\hat{y}_i - y_i^*| \le 3\epsilon\tau + \epsilon |y_i^*|$ and

$$\|\hat{y} - y^*\|_p \le \epsilon \cdot (3\tau n^{1/p} + \|y^*\|_p). \tag{7}$$

Table 1. Datasets used in this paper.

Dataset	Instance	Feature	Type	Bandwidth
covtype	581,012	54	Multi-classification	0.1
SUSY	1,000,000	18	Bi-classification	0.1
Census	2,458,285	68	Regression	0.05

The proof of Theorem 1 is in Sect. 8.2. $\tau \in (0,1)$ in Theorem 1 denotes a lower bound of $\frac{1}{n}\sum_i k\,(x_i, \bar{x})$. As shown in Eq. (3) in Sect. 4.2, the hashing scheme constructed by HBP algorithm satisfies the condition in Theorem 1. The above result shows the upper bound of error and the time complexity of the HBP method. As discussed before, the computational complexity of kernel prediction is $\mathcal{O}(n)$, while we need only $\mathcal{O}(\frac{log(n/\epsilon\tau)}{\tau^{0.5}\epsilon^{2.5}})$ with accuracy guarantees $\|\hat{y} - y^*\|_p \leq \epsilon \cdot (3\tau n^{1/p} + \|y^*\|_p)$.

6 Experiments

We evaluate the efficiency and effectiveness of the proposed algorithm by experiments on 3 large-scale datasets. The experiments with these datasets use the Laplacian kernel

$$k(x, y) = e^{-\|x-y\|_1/\sigma}. \tag{8}$$

All of the experiments are conducted on a server with 2.40 GHZ Inter(R) Xeon(R) E5-2630 v3 CPU and 32 GB of RAM in Matlab.

6.1 Datasets Preparation

The performance of our proposed algorithm is presented on three large-scale real-world datasets: covtype[1], SUSY[2] and Census[3], which are generally used in the field of kernel machines [18,19]. The details of the datasets are shown in Table 1.

We randomly selected 2.5×10^5 data points on each dataset. The features of datasets have been normalized. 70% of instances are used for training experiment and the rest for prediction. We measure the error by calculating root-mean-square error (RMSE) for regression problems and calculating the classification error for classification problems.

6.2 Performance of Hashing Based Prediction

Our experiment contains two parts. Every experiment is repeated 10 times to avoid contingency. The final result is the average value of 10 experiments.

---

[1] https://www.csie.ntu.edu.tw/~cjlin/libsvmtools/datasets/.
[2] https://www.csie.ntu.edu.tw/~cjlin/libsvmtools/datasets/.
[3] https://archive.ics.uci.edu/ml/datasets/US+Census+Data+(1990).

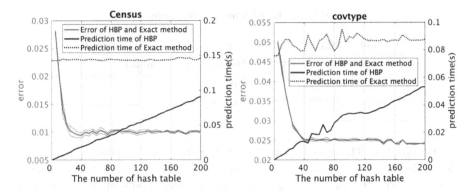

Fig. 2. Validate error between the HBP algorithm and the exact prediction method as well as the average prediction time for each test point of the two methods with respect to the number of hash table L on 2 large scale datasets: Census and covtype.

To focus on the effectiveness of our algorithm in the prediction stage, we consider the model $\hat{\alpha}$ for kernel machine is given in the first part of our experiment. We randomly generate $\hat{\alpha}$ and compare the error between the results of our approximation method and the exact value without using any approximation algorithm as well as the time complexity of the two methods.

Figure 2 shows the error and prediction time concerning the number of hash table L. The horizontal coordinate represents the number of hash table L. The vertical coordinate in the left represents the average errors of the prediction value between our algorithm and the exact algorithm, and the vertical coordinate in the right represent the prediction time of that two algorithms. With the increase of L, the error decreases at the beginning, and the rate of decrease become slow when L increases to a value. The prediction time of HBP increases with the increase of L. Our algorithm has a significant advantage over the accurate algorithm in prediction time.

In the second part of our experiment, we also compare our algorithm with 2 representative methods. Our algorithm employs KRR to finish the training process. For ensuring fairness, we use the same way to tune parameters σ in $2^{[-2:0.5:5]}$ and λ in $2^{[-40:3:8]}$ on each dataset and algorithm. The selected parameters are sufficient to achieve satisfactory results, although they may not be optimal.

The general introduction of the methods used in the experiment is as follows:

1) HBP: Our proposed algorithm, which uses Locality-Sensitive Hashing (LSH) for importance sampling to generate a fast prediction method for kernel machines.
2) Nyström [14]: An accurate and scalable Nyström scheme that made large-scale Nyström approximation possible. The algorithm first samples a large column subset from the input matrix, then only performs an approximate SVD on the inner submatrix.

3) Recursive RLS-Nyström [16]: A recently Nyström scheme using leverage score sampling.

Table 2. Comparison of prediction time and test error in solving KRR problem between HBP, Nyström and RLS-Nyström on covtype, SUSY and Census datasets. We bold the numbers of the best algorithm.

Dataset	Metric	HBP	Nyström	RLS-Nyström
covtype	Time (s) error	**4.409**	73.154	73.311
		0.2874 ± 0.00003	1.4919 ± 0.00263	0.7713 ± 0.00890
		$L = 50$	$m = 2500$	$m = 2500$
SUSY	Time (s) error	**2.974**	24.130	24.211
		0.6471 ± 0.0000	**0.6388 ± 0.00169**	0.6552 ± 0.00036
		$L = 50$	$m = 2500$	$m = 2500$
Census	Time (s) error	**5.354**	87.970	87.683
		0.2867 ± 0.00079	0.2900 ± 0.00163	1.2297 ± 0.00712
		$L = 50$	$m = 2500$	$m = 2500$

Table 2 shows the experiment result of our algorithm and other methods mentioned before. Our algorithm is at a faster prediction speed than other methods. The larger the scale of data, the more obvious the time advantage of the proposed algorithm is. Simultaneously, HBP keeps the optimal value or just a little gap with the optimal, which validates the effectiveness of our algorithm.

7 Conclusions

We propose an algorithm HBP to effectively reduce the prediction cost of kernel machines on large-scale datasets. The algorithm opens the door of solving the kernel prediction problem by the hashing method. By using Locality-sensitive hashing (LSH) for importance sampling to achieve approximate calculation, the HBP algorithm reduces computational complexity and storage cost with less prediction accuracy loss. The experimental analysis on large-scale datasets showed that our HBP algorithm outperforms previous state-of-art solutions.

Since LSH is the data-independent hash scheme, the result of the approximate search is not extremely accurate. In the future, we intend to replace the LSH with the existing data-dependent hash algorithm to further reduce the prediction errors. For example, the data-dependent hashing schemes designed by [1] and [2] can be attempted to apply to our hashing based prediction method.

8 Proof

In this section, we will provide proofs of some theorems in this paper. Section 8.1 presents the proof of Proposition 1 which references [3]. In Sect. 8.2, we present the proof of Theorem 1 which is our main theoretical result.

8.1   The Proof of Proposition 1

Proof. First assume that the data point $x, \bar{x}$ is one-dimensional. For the sake of clarity, we will describe LSH families $\mathcal{H}$ such that $\Pr_{h \sim \mathcal{H}} [h(x) = h(\bar{x})] = e^{-\|x-\bar{x}\|_1/\sigma}$. The Proposition 1 then follows simply by doubling the bandwidth σ. As shown in the Algorithm 1, we randomly sample $\xi \in [0, 1]$ and set the hash value $b(x) = 1$, if $x > \xi$, else $b(x) = 0$.

$$\Pr[b(x) = b(\bar{x})] = 1 - |x - \bar{x}|. \tag{9}$$

Then we consider the case that $x, \bar{x} \in [0, 1]^d$, applying this to a random dimension $\zeta \in \{1.....d\}$, we get:

$$\Pr[b(x) = b(\bar{x})] = \frac{1}{d} \sum_{\zeta=1}^{d} (1 - |x_\zeta - \bar{x}_\zeta|) = 1 - \frac{1}{d} \|x - \bar{x}\|_1. \tag{10}$$

We repeat ρ times independently, then,

$$\Pr_{h \sim \mathcal{H}} [h(x) = h(\bar{x})] = (1 - \frac{1}{d} \|x - \bar{x}\|_1)^\rho. \tag{11}$$

At last, we sample $\rho \sim Possion(d/\sigma)$,

$$\Pr_{h \sim \mathcal{H}} [h(x) = h(\bar{x})] = \sum_{\rho=0}^{\infty} \frac{e^{-d/\sigma} \cdot (d/\sigma)^\rho}{\rho!} \cdot (1 - \frac{1}{d} \|x - \bar{x}\|_1)^\rho = e^{-\|x-\bar{x}\|_1/\sigma}. \tag{12}$$

8.2   The Proof of Theorem 1

Note 1. Given a dataset $X = \{x_1, ..., x_n\} \subset \mathbb{R}^d$ and a test data point $\bar{x} \in \mathbb{R}^d$. $\omega_i = k(x_i, \bar{x})$ denoted the kernel value of data point x_i and test point $\bar{x}$. Let $\mathcal{H}$ be a family of hash function. For every x_i, $p_i = \Pr_{h \sim \mathcal{H}} [h(x_i) = h(\bar{x})]$ denoted the collision probability with $\bar{x}$. $b_h(\bar{x}) = \{i : h(x_i) = h(\bar{x})\}$ is the set of points with the same hash value as $\bar{x}$.

Proof. In our algorithm, we hash each point only with probability $\delta = 1/(n\tau^{1-\beta})$, where $\tau \leq \frac{1}{n} \sum_i \omega_i$. Set $r_1, ..., r_n$ be Bernoulli random variables with $\Pr[r_i] = \delta$. $b'_h(\bar{x})$ is sparsified counterpart of $b_h(\bar{x})$,

$$b'_h(\bar{x}) = \{i : h(x_i) = h(\bar{x}) \text{ and } r_i = 1\}. \tag{13}$$

We may assume that all elements of $\hat{\alpha}$ are positive otherwise we apply our algorithm to $\hat{\alpha}_+$ and $\hat{\alpha}_-$ separately. The vector $\hat{\alpha}$ can be geometrically divided into $S_1, ..., S_T$ such that all elements in each group differ by at most a factor of two, where $T = log_2(n/\epsilon\tau)$. Therefore, our problem can be expressed as T subproblem.

On the tth subproblem, our estimated response value is:

$$Z_{h,t} = \frac{\hat{\alpha}_I}{A_t} \frac{k(\bar{x}, x_I)}{\delta p_I} |b_h(\bar{x})|, \tag{14}$$

where I is a random index from $H(x) \subseteq S_t$, $A_t = \sum_{i \in S_t} \hat{\alpha}_i$.

Now we bound the variance:

$$
\begin{aligned}
\operatorname{Var}\left[Z_{h,t}\right] \leq \mathbb{E}\left[\left(Z_{h,t}\right)^2\right] &= \frac{1}{\delta^2} \mathbb{E}\left[\frac{\hat{\alpha}_i^2 \omega_i^2}{A_t^2 p_i^2 / |b_h'(\bar{x})|^2}\right] \\
&= \frac{1}{\delta^2} \underset{h \sim \mathcal{H}}{\mathbb{E}} \underset{i \in b_h'(\bar{x})}{\mathbb{E}}\left[|b_h'(\bar{x})|^2 \frac{\hat{\alpha}_i^2 \omega_i^2}{A_t^2 p_i^2}\right] \\
&== \frac{1}{\delta^2} \underset{h \sim \mathcal{H}}{\mathbb{E}}\left[|b_h'(\bar{x})| \sum_{i \in b_h'(\bar{x})} \frac{\hat{\alpha}_i^2 \omega_i^2}{A_t^2 p_i^2}\right] \\
&= \frac{1}{\delta^2} \underset{h \sim \mathcal{H}}{\mathbb{E}}\left[\sum_j [j \in b_h'(\bar{x})] \sum_i [i \in b_h'(\bar{x})] \frac{\hat{\alpha}_i^2 \omega_i^2}{A_t^2 p_i^2}\right] \\
&= \frac{1}{\delta^2} \sum_i \frac{\hat{\alpha}_i^2 \omega_i^2}{A_t^2 p_i^2} \sum_j \underset{h \sim \mathcal{H}}{\mathbb{E}}\left[[j \in b_h'(\bar{x})][i \in b_h'(\bar{x})]\right] \\
&= \frac{1}{\delta^2} \sum_i \frac{\hat{\alpha}_i^2 \omega_i^2}{A_t^2 p_i^2} \sum_j \underset{h \sim \mathcal{H}}{\operatorname{Pr}}\left[j \in b_h'(\bar{x}) \& i \in b_h'(\bar{x})\right].
\end{aligned}
\tag{15}
$$

The last term can be split into two expressions:

$$
\frac{1}{\delta^2} \sum_i \frac{\hat{\alpha}_i^2 \omega_i^2}{A_t^2 p_i^2} \sum_{j:j \neq i} \underset{h \sim \mathcal{H}}{\operatorname{Pr}}\left[j \in b_h'(\bar{x}) \& i \in b_h'(\bar{x})\right]
\tag{16}
$$

and

$$
\frac{1}{\delta^2} \sum_i \frac{\hat{\alpha}_i^2 \omega_i^2}{A_t^2 p_i^2} \underset{h \sim \mathcal{H}}{\mathbb{E}}\left[i \in b_h'(\bar{x})\right].
\tag{17}
$$

Since $j \neq i$ in Eq. (16), we have:

$$
\underset{h \sim \mathcal{H}}{\operatorname{Pr}}\left[j \in b_h'(\bar{x}) \& i \in b_h'(\bar{x})\right] = \delta^2 \underset{h \sim \mathcal{H}}{\operatorname{Pr}}\left[j \in b_h(\bar{x}) \& i \in b_h(\bar{x})\right] \leq \delta^2 p_j.
$$

Therefore, Eq. (16) is upper bounded by $\sum_i \frac{\hat{\alpha}_i^2 \omega_i^2}{A_t^2 p_i^2} \sum_j p_j$.

As $0 \leq 2 - 2\beta \leq 1$ and $0 \leq \beta \leq 1$, the inequalities $\frac{1}{n} \sum_i \omega_i^{2-2\beta} \leq \left(\frac{1}{n} \sum_i \omega_i\right)^{2-2\beta}$ and $\frac{1}{n} \sum_j \omega_j^\beta \leq \left(\frac{1}{n} \sum_i \omega_i\right)^\beta$ hold.

Using the inequalities and the definition in the theorem $\frac{\omega_i^\beta}{M} \leq p_i \leq M \omega_i^\beta$, we have:

$$
\sum_i \frac{\hat{\alpha}_i^2 \omega_i^2}{A_t^2 p_i^2} \sum_j p_j \leq \frac{4M^3}{n^2} \sum_i \omega_i^{2-2\beta} \sum_j \omega_j^\beta \leq 4M^3 \left(\frac{1}{n} \sum_i \omega_i\right)^{2-\beta}.
\tag{18}
$$

Let $\frac{1}{n} \sum_i \omega_i = \mu$, Eq. (16) is upper bounded by $4M^3 \mu^{2-\beta}$.

We observe that

$$
\underset{h \sim \mathcal{H}}{\mathbb{E}}\left[i \in b_h'(\bar{x})\right] = p_i \delta,
\tag{19}
$$

and therefore Eq. (17) is upper bounded by

$$\frac{1}{\delta}\sum_i \frac{\hat{\alpha}_i^2 \omega_i^2}{A_t^2 p_i} \leq \frac{4M}{n^2\delta}\sum_i \omega^{2-\beta}. \tag{20}$$

Since $1 \leq 2 - \beta$ and $\omega_i \leq 1$, it is easy to get:

$$\sum_i \omega_i^{2-\beta} \leq \sum_i \omega_i - n\mu. \tag{21}$$

And we have known that $\delta = \frac{1}{n\tau^{1-\beta}} \geq \frac{1}{n\mu^{1-\beta}}$, so we get the upper bound of Eq. (17) is $4M\mu^{2-\beta}$.

Equation (16) + Eq. (17)

$$\mathrm{Var}\left[Z'\right] \leq 4(M^3 + M)\mu^{2-\beta}. \tag{22}$$

It is obvious that the variance of our estimator is at most 4 times larger than the variance bound of kernel density estimate using Hashing-based-estimate method which is $(M^3 + M) \cdot \mu^{2-\beta}$. To derive Theorem 1, set $\beta = 1/2$. It is sufficient to obtain a $(1+\epsilon)$-approximation for the tth subproblem over $\mathcal{O}(\frac{1}{\tau^{0.5}\epsilon^{2.5}})$ independent samples. For all $t \in T$, the overhead of the whole process is at most a multiplicative factor $T = log(n/\epsilon\tau)$ compared to the case that we were creating a single data-structure for the same problem. Combining with the conclusion of [5], we obtain a straightforward analysis of the estimation error of the algorithm that for all $i \in [n]$ with probability at least $1 - n^{-1}$, it holds $|\hat{y}_i - y_i^*| \leq 3\epsilon\tau + \epsilon|y_i^*|$. Summing over all indices and using triangle inequality gives $\|\hat{y} - y^*\|_p \leq \epsilon \cdot (3\tau n^{1/p} + \|y^*\|_p)$.

Acknowledgements. This work was supported in part by the National Natural Science Foundation of China (No. 61703396, No. 61673293), the CCF-Tencent Open Fund, the Youth Innovation Promotion Association CAS, the Excellent Talent Introduction of Institute of Information Engineering of CAS (No. Y7Z0111107), the Beijing Municipal Science and Technology Project (No. Z191100007119002), and the Key Research Program of Frontier Sciences, CAS (No. ZDBS-LY-7024).

References

1. Andoni, A., Laarhoven, T., Razenshteyn, I., Waingarten, E.: Optimal hashing-based time-space trade-offs for approximate near neighbors. In: Proceedings of the Twenty-Eighth Annual ACM-SIAM Symposium on Discrete Algorithms, pp. 47–66. Society for Industrial and Applied Mathematics (2017)
2. Andoni, A., Razenshteyn, I.: Optimal data-dependent hashing for approximate near neighbors. In: Proceedings of the Forty-Seventh Annual ACM Symposium on Theory of Computing, pp. 793–801. ACM (2015)
3. Backurs, A., Indyk, P., Wagner, T.: Space and time efficient kernel density estimation in high dimensions. In: Advances in Neural Information Processing Systems, pp. 15773–15782 (2019)

4. Camoriano, R., Angles, T., Rudi, A., Rosasco, L.: NYTRO: when subsampling meets early stopping. In: Artificial Intelligence and Statistics, pp. 1403–1411 (2016)
5. Charikar, M., Siminelakis, P.: Hashing-based-estimators for kernel density in high dimensions. In: 2017 IEEE 58th Annual Symposium on Foundations of Computer Science (FOCS), pp. 1032–1043. IEEE (2017)
6. Datar, M., Immorlica, N., Indyk, P., Mirrokni, V.S.: Locality-sensitive hashing scheme based on p-stable distributions. In: Proceedings of the Twentieth Annual symposium on Computational Geometry, pp. 253–262. ACM (2004)
7. Van de Geer, S.A.: Applications of Empirical Process Theory, vol. 91. Cambridge University Press, Cambridge (2000)
8. Gionis, A., Indyk, P., Motwani, R., et al.: Similarity search in high dimensions via hashing. VLDB **99**, 518–529 (1999)
9. Gui, J., Liu, T., Sun, Z., Tao, D., Tan, T.: Fast supervised discrete hashing. IEEE Trans. Pattern Anal. Mach. Intell. **40**(2), 490–496 (2017)
10. Hastie, T., Tibshirani, R., Friedman, J.: The Elements of Statistical Learning: Data Mining, Inference, and Prediction. Springer, New York (2009). https://doi.org/10.1007/978-0-387-84858-7
11. Hsieh, C.J., Si, S., Dhillon, I.S.: Fast prediction for large-scale kernel machines. In: Advances in Neural Information Processing Systems, pp. 3689–3697 (2014)
12. Ju, X., Wang, T.: A hash based method for large scale nonparallel support vector machines prediction. Proc. Comput. Sci. **108**, 1281–1291 (2017)
13. Kumar, S., Mohri, M., Talwalkar, A.: Sampling techniques for the Nystrom method. In: Artificial Intelligence and Statistics, pp. 304–311 (2009)
14. Li, M., Kwok, J.T., Lu, B.L.: Making large-scale Nyström approximation possible. In: ICML, pp. 631–638 (2010)
15. Liu, M., Shang, Z., Cheng, G.: Sharp theoretical analysis for nonparametric testing under random projection. In: Conference on Learning Theory, pp. 2175–2209 (2019)
16. Musco, C., Musco, C.: Recursive sampling for the Nystrom method. In: Advances in Neural Information Processing Systems, pp. 3833–3845 (2017)
17. Rahimi, A., Recht, B.: Random features for large-scale kernel machines. In: Advances in Neural Information Processing Systems, pp. 1177–1184 (2008)
18. Rudi, A., Camoriano, R., Rosasco, L.: Less is more: Nyström computational regularization. In: Advances in Neural Information Processing Systems, pp. 1657–1665 (2015)
19. Rudi, A., Carratino, L., Rosasco, L.: FALKON: an optimal large scale kernel method. In: Advances in Neural Information Processing Systems, pp. 3888–3898 (2017)
20. Rudi, A., Rosasco, L.: Generalization properties of learning with random features. In: Advances in Neural Information Processing Systems, pp. 3215–3225 (2017)
21. Shawe-Taylor, J., Cristianini, N., et al.: Kernel Methods for Pattern Analysis. Cambridge University Press, Cambridge (2004)
22. Williams, C.K., Seeger, M.: Using the Nyström method to speed up kernel machines. In: Advances in Neural Information Processing Systems, pp. 682–688 (2001)
23. Zhang, X., Liao, S.: Incremental randomized sketching for online kernel learning. In: International Conference on Machine Learning, pp. 7394–7403 (2019)

Picking Peaches or Squeezing Lemons: Selecting Crowdsourcing Workers for Reducing Cost of Redundancy

Paulina Adamska$^{(\boxtimes)}$, Marta Juźwin , and Adam Wierzbicki

Polish-Japanese Academy of Information Technology, Warsaw, Poland
{tiia,marta.juzwin,adamw}@pja.edu.pl

Abstract. Crowdsourcing (CS) platforms are constantly gaining attention from both researchers and companies, due to the offered possibility of utilizing the "wisdom of crowds" in order to solve a great variety of problems. Despite the obvious advantages of such mechanisms, there are also numerous concerns regarding the quality assurance of work results produced by a large group of anonymous workers. In this work, we use data gathered from a real CS platform in order to study the performance of various approaches to worker selection, including a novel approach that utilizes automatic real-time monitoring of the produced results. We compare the performance of these mechanisms with respect to both overall cost and the accuracy of the final results to benchmark algorithms that aggregate results from a group of workers without pre-selection, relying solely on the "wisdom of crowd". We find that our novel approach is capable of reducing the cost of obtaining high-quality aggregated results by a factor of four, without sacrificing quality.

Keywords: Crowdsourcing · Quality · Effort · Skill · Aggregation · Algorithm

1 Introduction

Crowdsourcing (CS) platforms offer companies a possibility to delegate some tasks to a potentially large group of anonymous workers, using the "wisdom of crowd" to solve a great variety of problems. Despite numerous advantages of this model, one major concern of each potential requester is quality control. New requesters are often forced to pay third parties for the information regarding former performance of CS workers, or apply custom techniques such as: (1) utilizing custom skill tests in order to verify worker qualifications, (2) providing "benchmark" tasks, with already known solution, which are later on utilized to monitor worker reliability and attitude to solving "real" tasks, (3) requesting a certain amount of redundant solutions hoping to reduce the impact

This work is supported by the Polish National Science Centre grant 2015/19/B/ST6/03179.

V. V. Krzhizhanovskaya et al. (Eds.): ICCS 2020, LNCS 12138, pp. 510–523, 2020.
https://doi.org/10.1007/978-3-030-50417-5_38

of low quality submissions on the aggregated final result. In this work we analyze the impact of various approaches to worker selection on the accuracy of the final outcome, and the overall requester costs. We also propose and evaluate an experimental real-time worker performance monitoring technique that involves machine-learning-based solution quality estimation, and prevents further interaction with an unreliable worker. In order to obtain results that are as realistic as possible, the performance of all the previously mentioned techniques was verified using a dataset gathered from a popular CS platform.

2 Related Work

There have been many suggestions regarding quality control of the content produced by CS workers. Most of them could be assigned to one of two groups of approaches. The first group consists of research concentrated on improving initial worker selection process [5,12,14,20]. The second group of approaches is more similar to our own, as these mechanisms attempt to reduce the amount of tasks solved by unreliable CS workers. For example Hirth et al. [7] attempted to use validation tasks in order to detect cheaters among CS workers. On the other hand, our research aims to detect spammers using only one validation task, which in turn helps to reduce costs. Moreover our model does not assume that only cheaters can provide low-quality solutions. Hirth et al. compared his method to a simple majority voting approach, while our algorithm is validated using a state-of-the-art EM-based aggregation algorithm. Hirth et al. [6] also proposed to detect spammers using application level information gathering mechanism that is similar to the one used in our experiment as it gathers data regarding worker interaction with the application UI. Our proposal, however, uses a different set of features in order to estimate solution accuracy. Moreover, our classifier provides the actual accuracy estimation, not a binary classification ("qualifie" or "non-qualified"), which allows the requester to control the required accuracy threshold. Additionally, the approach proposed by Hirth et al. was verified on a significantly smaller sample of CS workers (behavioral traces from 215 workers were distributed between training and test dataset).

Liu et al. [11] attempted to answer the question of how many control tasks are required in order to identify reliable workers. The proposed Gaussian model assumes utilization of such variables as bias or label variance, which can be learned from the results of the control tasks and remain constant. In our experiment, we concentrate more on user expertise (which is assumed to be constant) and effort level (which may change over time). Feldman et al. [2] proposed a quality control mechanism that attempted to monitor user interaction with UI components (such as mouse speed acceleration, scrolling speed, etc.) to automatically detect a decrease in work quality in real-time. This approach differs from our proposal in terms of features used for performance prediction. Moreover, authors did not provide any cost analysis. Rangi and Franceschetti [16] modeled task assignment in a budget-limited setting as a Bounded Knapsack Problem. This solution however would require some major modifications in the way that requesters and workers interact with each other. Shanshan et al. [18] proposed an algorithm, that computed a proper ordering of workers' submissions based on the estimated quality of the solutions. Unlike our approach, this mechanism assumed availability of requester's feedback for all the solutions provided by the workers in the past. This assumption is realistic for the "creative" tasks, for which the algorithm was designed (like brand logo design), however may not be feasible for a typical CS task. Moreover, for such tasks it may be difficult to objectively evaluate the real quality of the solutions.

3   Experiment Design

In order to gather all data required for our analysis, we decided to study the behavior of CS workers using a popular general-purpose crowdsourcing platform. An experiment has been carried out on Amazon Mechanical Turk.[1] Our analysis is based on 8100 work results provided by 810 different CS workers - each experiment participant was asked to provide solutions to 10 Human Intelligence Tasks (HITs).

Our experimental CS task had a very high level of redundancy, as each job could have up to 810 solutions. This unique experimental approach allowed us to repeatedly sample our dataset to simulate a situation when requesters interact with different, smaller groups of workers. In a realistic case, a requester would require a much smaller redundancy (perhaps 5 solutions for each job) and would interact with a much smaller population of workers. Every time a requester uses Amazon Mechanical Turk, she may encounter a different group of workers. This means that every time, the requester has to fit workers models from scratch, encountering a cold-start problem. Our approach allows to test robustness of models to the composition of a training set of workers.

We wanted our worker sample to reflect the population of workers on Amazon Mechanical Turk as faithfully as possible. The only worker filtering method that we have applied before performing the experiment has been based on worker location - workers from outside of U.S. were not allowed to participate. This decision was made due to the nature of the chosen task, which required language-related skills. Experiment participants were asked to identify and mark misspelled words in English texts in order to help the requester evaluate the performance of different algorithms designed for automatic text processing. All the texts were excerpts of books selected from the Gutenberg project website.[2] The spelling errors were added manually by replacing one word with a similar one that was grammatically or semantically incorrect in a particular context. For example, the word "cut" was replaced by "cat", etc. The general concept behind this approach was to mimic spelling errors that could be caused by a poorly designed OCR tool, and more importantly - to eliminate the possibility of using spell check tools without reading and understanding the entire text. Our goal was to create a task of medium difficulty that allowed us to precisely measure both

---

[1] https://www.mturk.com/mturk/welcome.
[2] https://www.gutenberg.org/.

CS workers expertise, and the accuracy of the provided solutions. Each experiment participant was asked to solve 10 HITs. Additionally, workers were asked to solve a time-limited skill test (before solving the first HIT). All the stages of the experiment are described in detail in the following sections.

3.1   Skill Test

The main goal of this part of the experiment was to determine the real expertise level of each worker regarding this particular task type. In order to achieve this goal, the experiment participants were asked to mark all the words that contained spelling errors in the presented text. The total number of words was 127, and 9 of them were misspelled. The time limit for completing the test was set to one minute, and after this time workers were automatically redirected to the main task without the possibility of reviewing their solutions. For each worker we have recorded the time required to submit the solution along with the computed skill (interpreted as the accuracy of the results with respect to our 'gold solution').

3.2   Human Intelligence Task

In the main task, CS workers were asked to compare a pair of texts in terms of spelling errors and to provide two types of feedback. The first one was to indicate which of the texts contained more misspelled words. The second one was to mark the misspelled words in each text. Such task design not only allowed us to gather the required data, but also seemed to be a typical type of work for a CS setting. For example, Good et al. [4] used word tagging to capture mentions of disease in PubMed abstracts, while Finin et al. [19] applied the same technique to collect named entities annotations for Twitter status updates. Klebanov et al. [10] asked CS workers to mark words that most contributed to the overall sentiment of a sentence in order to construct subjectivity lexicon for recognizing sentiment polarity in essays. Similarly Hsueh et al. [9] utilized word tagging in order to consider the problem of classifying sentiment in political blog snippets. On the other hand, Filatova [3] performed a CS experiment that involved text tagging, aiming to create a corpus that could be used for identifying irony and sarcasm, and Prabhakaran et al. [15] attempted to collect textual data, that could be used as a training set for an automatic modality tagger.

Figure 1 shows how a pair of texts for comparison was presented to the workers. In order to gather at least some additional information about the worker effort, we divided each text into 5 parts that could be displayed by clicking the "Next" button. We have recorded the timestamp for almost every user interaction with our application, and thus we could find out whether the worker at least had a chance to see the entire text before submitting the answer, and for how long each part was displayed on the screen. The average number of words in the evaluated texts (considering all of the 10 available HITs) was approximately 120 (the shortest text consisted of 95, and the longest one of 130 words). Each pair of texts contained 10 misspelled words, and the distribution of errors was designed in such a manner that the number of the read text parts affected the quality of

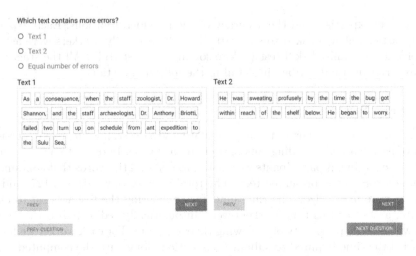

Fig. 1. A sample text comparison task UI

the final solution. Experiment participants were allowed to review the solutions submitted earlier and correct them as many times, as they wanted. The average time of completing all 10 HITs was about 30 min, which was approximately equal to the expected completion time.

4   Worker Characteristics and Solution Quality

Existing theoretical models often assume that the skill level is strongly correlated with the probability of providing a high-quality answer [8,13]. The results gathered in our experiment indicated that the correlation between skill and accuracy in the real life setting is not particularly high. We believe that this is caused by the fact that worker skill is not the only factor contributing to the accuracy of the submitted solutions. Table 1 presents inconsistencies between the skill-based and observed-accuracy-based classifications.

Note that in each group identified during the skill test (skill column), there are CS workers, whose observed behavior while working on the real tasks was incompatible with the initial skill-based prediction. Particularly, among the workers who turned out to provide the best solutions for almost all tasks, most have initially been identified as regular workers. Moreover, about 8% of the workers who have been assigned to the group of experts actually turned out to be spammers in real tasks. This phenomenon can be associated with the presence of smart adversaries in the population. Such workers put more effort into a task that is officially referred to as a "skill test", but their performance degrades significantly for ordinary tasks. Both observations lead to the conclusion that solely using a skill test may not be the best strategy to recognize the most reliable workers and thus achieve the best results. One additional observation can be made based on the data presented in Table 1. The differences in average task completion time for the group of spammers identified by the skill test suggest

Table 1. Skill-based and observer-accuracy-based classification

Label		No. of workers	% of workers	Time (average)
Skill	Observed			
expert (*skill > 0.8*)	expert*	39	5	20
	regular◊	27	3	18
	spammer†	6	1	4
regular (*0.3 ⩽ skill ⩽ 0.8*)	expert*	157	19	27
	regular◊	328	40	24
	spammer†	135	17	9
spammer (*skill < 0.3*)	expert*	7	1	25
	regular◊	49	6	27
	spammer†	62	8	11

* within 25% of workers with the highest accuracy in more than 7 tasks
◊ within 25% of workers with the highest accuracy in 1 to 7 tasks
† never in the group of 25% of workers with the highest accuracy
(all the differencies are statistically significant according to the Kruskal-Wallis test)

that about 44% of workers who did not perform well during the time-limited skill test were actually willing to compensate for the lack of skill with a certain degree of effort, achieving at least reasonable accuracy while solving the real tasks. To sum up, the presented results indicate that there is no single, simple feature that would allow to initially select a group of reliable CS workers, simply by setting up a fixed threshold. Therefore, the requesters should probably consider approaches that not only assess worker skill, but also monitor the amount of time spent on solving the task.

5   Analyzed State-of-the-Art Approaches

Apart from proposing an experimental approach to improving the quality of the information produced by CS workers, in our study we also analyzed three alternative state-of-the-art approaches. These mechanisms are described in the following sections.

5.1   Squeezing Lemons

The first and simplest state-of-the-art approach introduces some redundancy level with no initial worker selection. Removing the initial selection process allows to gather the results as fast as possible, due to the lack of additional requirements regarding worker skills. On the other hand, having several solutions to the same task allows the requester to use aggregation algorithms to generate a single output from multiple solutions, hopefully reducing the impact of potential spammers or workers who do not have a desired skill level. In this article, we are using a state-of-the-art EM-based algorithm to aggregate all solutions into a single final one. The algorithm was proposed by Raykar et al. [17] and is a generalization of the approach proposed by Dawid and Skene [1]. Note that the algorithm is not only able to estimate the ground truth label for each word, but also simultaneously estimates features of the CS workers defined by Formula 1 and 2 respectively:

$$\alpha^j := P[y_i^j = 1 \mid y_i = 1] \sim Beta(2.078, 1.843) \tag{1}$$

$$\beta^j := P[y_i^j = 0 \mid y_i = 0] \sim Beta(2.078, 1.843) \tag{2}$$

where y_i is the true label of word i, y_i^j is the label assigned to word i by worker j. The parameters of both distributions were computed using the information about the accuracy of skill test results for all the workers.

The aggregation algorithm described in this section is used in all cases in our research; however, in the "squeezing lemons" scenario, the algorithm aggregates results from workers who have not undergone any pre-selection, while in the "peach picking" scenario we apply various approaches of worker selection before we aggregate results from selected workers.

5.2   Peach Picking

Skill-Test-Based Selection Only. This technique is a typical, well-known approach to improve the quality of the redundant solutions provided by CS workers. The main idea behind this concept is that the reason for poor quality of some solutions in a CS setting is the lack of sufficient skills. Therefore, the workers are asked to provide solution to a skill test before they are offered to solve an ordinary task. If the skill of worker w denoted by $skill_w$ is lower than the threshold denoted by $threshold_{skill}$, then the worker does not have the required expertise level to provide valuable contributions. The number of workers selected depends on the specified redundancy level.

In our setting, skill is interpreted as the solution accuracy achieved in the skill test. The accuracy evaluation in our experiment is in many ways similar to binary classifier evaluation, as we have a set of words, which are supposed to be labelled as correct or misspelled by different workers. Therefore we define the following variables: the number of false positives (FP) - the number of correct words which were marked as misspelled; the number of true positives (TP) - the number of misspelled words, which were marked as misspelled; the number of false negatives (FN) - the number of misspelled words that were not marked as misspelled; the number of true negatives (TN) - the number of correct words that were not marked as misspelled. Based on these variables, we calculate the F1 score as a measure of correctness of a worker's solutions.

First-Task-Based Selection Only. The second of the studied "peach picking" techniques is based on the idea, that skill-test-based initial filtering may not be the best selection method, as workers may apply a different approach to such task. In our experiment, CS workers were not implicitly informed, whether the skill test results will or will not have any impact on their future work opportunities, however the task was time-limited, and it was called a "skill test", thus potential spammers were aware of the fact, that the main goal of this task was to identify them as reliable, or unreliable workers. In the first-task-based filtering setting, CS workers are not able to distinguish between the "skill test" and the ordinary tasks, as they both look exactly the same. The selection algorithm itself is the same as in the approach described in Sect. 5.2. The requester defines a threshold $threshold_{first}$. If the accuracy of the first task solution is at least at the $threshold_{first}$ level, the author of this solution is offered to solve all the ordinary tasks. The number of workers that are selected depends on the specified redundancy level.

6 Experimental Real-Time Monitoring Approach

There is one main disadvantage of the peach picking techniques mentioned in Sect. 5.2. Such approaches do not attempt to verify whether the quality of the solutions provided by CS workers is not changing. Changes in work results quality may occur due to various reasons. One of them may be CS workers deliberately putting more effort into a skill test, and then switching to a spammer-like behavior while solving ordinary tasks. Second reason may be CS workers not performing particularly well when solving a time-limited skill test, but being honest enough, to put as much time and effort as is required for them to accurately solve ordinary tasks. In such situations inflated skill thresholds used in conjunction with initial worker selection techniques automatically reject such workers, thus unnecessarily increasing selection costs. Another possible reason for accuracy changes may be tiredness.

As mentioned in Sect. 4, we seem to have observed all of the previously mentioned phenomena in our experiment. Approximately 8% of workers, identified as experts solely based on their skill test result, behaved like spammers when solving ordinary tasks. On the other hand, there were much more workers that behaved like experts among the skill-test-based identified regular workers in comparison to the group of skill-test-based identified experts. When the requester applies extremely inflated initial filtering threshold, all of these workers do not even get the chance to showcase their real potential and gain reputation they deserve. We have also noticed, that there were more low quality solutions for the last few tasks, which does not have to, but may indicate tiredness.

In order to address these issues we have decided to propose a novel automatic real-time solution accuracy monitoring in our experimental approach. This technique allows the requester to decide after each task, whether to further interact with this particular CS worker. Moreover it does not require to solve the exploration vs exploitation problem, unlike the mechanisms that use multiple "gold standard" tasks, which are presented to the workers among plain tasks in order to verify their current performance. The proposed approach consists of two main components, namely (1) initial worker selection mechanism, and (2) real-time solution accuracy estimation mechanism. Both of them are described in the following sections.

Algorithm 1. Real-time monitoring

T: a set of ordinary tasks
W: a set of selected workers
for all $w \in W$ **do**
  for all $t \in T$ **do**
    $solution_t^w \leftarrow w.solve(t)$
    $accuracy_t^w \leftarrow solution_w^t.estimateAccuracy(solution_w^t)$
    if $accuracy_w^t \geq threshold_{estimated}$ **then**
      $solution_t^w.setAccepted(true)$
    else
      $solution_t^w.setAccepted(false)$
      $w.setSuspended(true)$
      break
    end if
  end for
end for

Initial Worker Selection. Initial worker selection mechanism is based on the accuracy of the first task solution, as described in Sect. 5.2. We have chosen this approach as it turned out to outperform skill-test-based filtering both in terms of cost and accuracy in most cases. The number of workers, that are going to be selected is determined by the redundancy level specified by the requester. Even though skill is not considered in the initial filtering phase, each worker is asked to solve a short skill test as well. This information will be used by the real-time solution accuracy estimator.

Real-Time Solution Accuracy Estimation. Each worker $w \in W$, that has made it through the initial selection process, is offered to solve the first ordinary task. When the task is solved, a dedicated mechanism estimates the accuracy of the solution. If the solution accuracy exceeds the threshold $threshold_{estimated}$ specified by the requester, it is accepted, and CS worker is offered to solve the next task. Otherwise the solution is not accepted, and further interaction with this worker is suspended. Note that the initial filtering threshold $threshold_{first}$ described in Sect. 5.2 and the accuracy threshold $threshold_{estimated}$ used in the real-time evaluation process may be assigned different values.

We have utilized a random forest regression model to provide the accuracy estimation functionality. The model was trained based on the results of the first 405 HITs, while the simulation of our approach was done on the remaining 405 HITs. The optimal model has been selected using the RMSE metric during the 5 fold cross-validation process (grid search method). The RMSE values for the training and test datasets are 0.1447529, and 0.171867 respectively. The estimated accuracy is computed using the following features of each CS worker: skill, first task solution accuracy, first task completion time, current task no. in the set of all tasks provided by the requester, current task completion time, difference between the completion time of the first task and completion time of the

---

Algorithm 2. Acquiring solutions to the remaining tasks

T: a set of all ordinary tasks
$T_{unsolved}$: a set of ordinary tasks without any accepted solution
W: a set of selected workers
$S < T, W >$: a map of solutions provided by workers to all tasks
for all $w \in W$ **do**
    for all $t \in T_{unsolved}$ **do**
        $solution_t^w \leftarrow S < t, w >$
        if $\neg(solution_t^w \in S < T, W >)$ **then**
            $solution_t^w \leftarrow w.solve(t)$
        end if
        $solution_t^w.setAccepted(true)$
    end for
end for

---

current task, percentage difference between the completion time of the first task and completion time of the current task. The overview of the entire procedure is presented in Algorithm 1. Due to the probabilistic nature of the automatic evaluation process, CS worker receives monetary reward for submitting a solution regardless the accuracy estimation result.

If all of the selected workers have already been suspended based on the real-time accuracy estimation and there are still some unsolved tasks, then for each unsolved task the following procedure is executed: (1) if there are some solutions to this task that have already been provided and have not been accepted they are automatically marked as accepted, (2) all the suspended workers, that have not already solved the task are offered to submit their solutions, which are later on automatically accepted. The overview of the entire procedure is presented in Algorithm 2. The main idea behind this approach is the assumption, that at this point none of the selected workers can still be considered credible, as each of them has been suspended at some point of interaction. Therefore we need to maximize the redundancy level for this particular task.

7   Picking Peaches or Squeezing Lemons?

In order to compare the accuracy and cost of the techniques described in Sects. 5.1 and 5.2, the entire dataset containing solutions to 10 HITs provided by 810 workers has been divided in two parts. The first 405 workers have been assigned to a training dataset which has only been used to train the worker accuracy classifier for our experimental approach. The remaining 405 workers were assigned to the test dataset which has been used in order to verify the performance of all the studied approaches. For every approach a simulator has been used to generate the data representing a single scenario of interaction between the requester and the CS workers (by scenario we mean requester submitting the tasks and recruiting workers to provide solutions to all of them at a specified redundancy level). At each simulation run workers were chosen one at a time, at

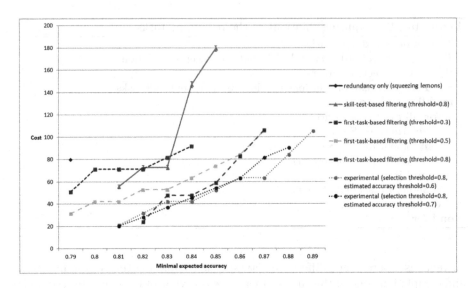

Fig. 2. Comparison of the analyzed approaches in terms of average minimum cost required to achieve a required accuracy of aggregated results. Confidence intervals' width for cost on the figure does not exceed: 5.5 for skill-test-based filtering, 1.4 for first-task-based filtering, and 1.65 for the experimental approach.

random from the entire test dataset, ensuring that none of the workers could be chosen multiple times within a single simulation run. If the chosen worker had made it through the initial filtering phase, he was offered to solve ordinary tasks according to the rules of a particular tested approach. The workers were paid $1 for every submitted solution, regardless of whether it has been accepted or not. At the end of each simulation run the following actions were taken:

1. All results available at the specified redundancy level were aggregated using the state-of-the art aggregation algorithm chosen as a benchmark in this article [17] (see Sect. 5.1).
2. The accuracy of the solution was computed using the 100% correct result as a reference.
3. The total costs were summed up and recorded.

For all the approaches apart from the experimental one, involving real-time monitoring feature, 300 simulation runs were used. For the more complex experimental approach described in Sect. 6, 1000 simulation runs were used. For each of the analyzed approaches we compute the average accuracy of all aggregated HIT solutions from all simulation runs, and average costs from all the simulation runs. We also computed confidence intervals.

The general comparison of all the analyzed approaches is depicted in Fig. 2. The figure shows the performance of all approaches for various levels of redundancy chosen so that the approach reaches a threshold of accuracy. The average minimum cost of each approach for reaching a certain accuracy is plotted on the figure.

It turned out, that for our task, the best accuracy that was achieved by the "squeezing lemons" (aka redundancy-only-based) approach was 79%. The minimum redundancy level required for this was 8 solutions per task, which translated to the minimum cost of 80.

Our study of the state-of-the-art "picking peaches" approaches shows, that the initial worker filtering based on skill test does improve the average quality of the aggregated final result assuming, that the $threshold_{skill}$ value specified by the requester is set to 0.8. For this technique higher accuracy (81%) can be achieved at a redundancy level set to only 3 solutions per task which translates to the cost of approximately 56. The drawback of this approach is that the cost associated with the worker selection process is noticeably higher, which causes the costs to dramatically increase for higher redundancy levels.

On the other hand, filtering based on the accuracy of the first task seems to significantly improve the quality of the aggregated task solutions even when $threshold_{first}$ is set to as low value as 0.3. This threshold setting can be thought of as an attempt to remove spammers from the worker pool. When the requester sets $threshold_{first}$ to 0.8, a major quality improvement of the average aggregated outcome can be achieved at a noticeably lower cost than for the skill-based filtering with the same $threshold_{skill}$ value. This "picking peaches" technique allows to achieve the accuracy of 82%, when the redundancy level is set to only 2 solutions per task, which translates to a cost of approximately 24.

For the experimental approach, we have found the following parameter settings to be optimal: (1) $threshold_{first}$ for the first task accuracy (initial worker selection phase): 0.8, $threshold_{estimated}$ for estimated ordinary task solution accuracy (real-time quality control): 0.6 and (2) $threshold_{first}$ (initial worker selection phase): 0.8, $threshold_{estimated}$ (real-time quality control): 0.7. For both settings of parameters, the experimental approach outperforms first-task-accuracy-based filtering with the same $threshold_{first}$ value in terms of average costs. An accuracy of 81% is achieved by the experimental approach with a cost of 20. The experimental approach consistently provides similar or better results in terms of accuracy as compared to other approaches. The benefits of applying real-time monitoring become particularly apparent for higher redundancy levels that achieve higher accuracy. Pre-selecting workers based on the accuracy of the first task allows to achieve a maximum accuracy of 87% at a cost of 105, compared to the benchmark accuracy of 79% at a cost of 80 achieved by the "squeezing lemons" approach without worker selection. On the other hand, the experimental approach achieves an accuracy of 87% at a cost of 61, and exceeds the benchmark accuracy of 80% at the cost of 20 (as compared to the benchmark cost of 80).

8   Conclusions

The majority of research on crowdsourcing has focused on aggregation algorithms, following the seminal work of Dawid and Skene [1]. A possible reason for this trend is the widespread belief in the "Wisdom of Crowds". Our work confirms the fact that increasing redundancy can increase quality of results, but we propose to filter workers who will contribute results that will be the basis of final aggregation. While this idea has been studied before, our research is one of the first systematic and large-scale studies of the subject.

We have conducted an experimental CS task on Amazon Mechanical Turk, mimicing typical CS task design. Our experimental task had a very high level of redundancy: 810 workers solved all our tasks. This approach allowed us to conduct data-driven simulation studies that compared various approaches of workers selection with respect to quality of aggregated results and cost of obtaining the required work results. As a benchmark, we have used a state-of-the-art aggregation algorithm [17] without worker filtering.

Our results indicate that worker pre-selection based on the results of the first real task can outperform aggregation of results without worker filtering: picking peaches is indeed better than squeezing lemons. However, not all peach picking methods work equally well. In particular, we have shown that CS workers can manipulate simple skill tests, and pre-selecting workers based on results of first real task is more effective than using skill-test results. Furthermore, we have proposed a novel, real-time dynamic filtering algorithm that significantly increases the cost-effectiveness of crowdsourcing. Our algorithm improves on the cost of the benchmark algorithm by a factor of four without decreasing quality. The algorithm also has good results for higher redundancy levels that achieve higher quality, outperforming simple worker pre-selection by a factor of two.

One of the limitations of our study is that we have considered only 10 HITs for each worker. This choice was motivated by the belief that more HITs would not change our results. However, a longitudinal study of CS workers (perhaps across multiple tasks) could possibly allow to create more in-depth models of CS workers that would further improve the ability of filtering workers for crowdsourcing tasks.

References

1. Dawid, A.P., Skene, A.M.: Maximum likelihood estimation of observer error-rates using the EM algorithm. Appl. Stat. **28**(1), 20–28 (1979)
2. Feldman, M., Bernstein, A.: Behavior-based quality assurance in crowdsourcing markets. In: Conference on Human Computation & Crowdsourcing 2014, Pittsburgh, USA, November 2014
3. Filatova, E.: Irony and sarcasm: Corpus generation and analysis using crowdsourcing. In: Language Resources and Evaluation Conference, LREC 2012 (2012)
4. Good, B.M., Nanis, M., Su, A.I.: Microtask crowdsourcing for disease mention annotation in PubMed abstracts. CoRR abs/1408.1928 (2014)

5. Hassan, U., Curry, E.: A capability requirements approach for predicting worker performance in crowdsourcing. In: 9th IEEE International Conference on Collaborative Computing: Networking, Applications and Worksharing, pp. 429–437 (2013)
6. Hirth, M., et al.: Predicting result quality in crowdsourcing using application layer monitoring. In: 2014 IEEE Fifth International Conference on Communications and Electronics (ICCE), pp. 510–515, July 2014
7. Hirth, M., Hoßfeld, T., Tran-Gia, P.: Analyzing costs and accuracy of validation mechanisms for crowdsourcing platforms. Math. Comput. Model. **57**(11–12), 2918–2932 (2013)
8. Ho, C., Jabbari, S., Vaughan, J.: Adaptive task assignment for crowdsourced classification. In: Proceedeeings of the 30th International Conference on Machine Learning - Volume 28, ICML 2013, pp. I-534–I-542. JMLR.org (2013)
9. Hsueh, P.Y., Melville, P., Sindhwani, V.: Data quality from crowdsourcing: a study of annotation selection criteria. In: Proceedings of the NAACL HLT 2009 Workshop on Active Learning for Natural Language Processing, HLT 2009, pp. 27–35. Association for Computational Linguistics, Stroudsburg (2009)
10. Klebanov, B.B., Burstein, J., Madnani, N., Faulkner, A., Tetreault, J.R.: Building subjectivity lexicon(s) from scratch for essay data. In: Proceedings of the Computational Linguistics and Intelligent Text Processing - 13th International Conference, CICLing 2012, Part I, New Delhi, India, 11–17 March 2012, pp. 591–602 (2012)
11. Liu, Q., Ihler, A.T., Steyvers, M.: Scoring workers in crowdsourcing: how many control questions are enough? In: Burges, C.J.C., Bottou, L., Ghahramani, Z., Weinberger, K.Q. (eds.) NIPS, pp. 1914–1922 (2013)
12. Liu, Q., Li, H.: Cheaper and better: selecting good workers for crowdsourcing. In: HCOMP, pp. 20–21 (2015)
13. Mavridis, P., Gross-Amblard, D., Miklós, Z.: Using hierarchical skills for optimized task assignment in knowledge-intensive crowdsourcing. In: Proceedings of the 25th International Conference on World Wide Web, WWW 2016, pp. 843–853. International World Wide Web Conferences Steering Committee, Republic and Canton of Geneva, Switzerland (2016)
14. Moayedikia, A., Yeoh, W., Ong, K.L., Boo, Y.L.: Improving accuracy and lowering cost in crowdsourcing through an unsupervised expertise estimation approach. Decis. Supp. Syst. **122**, 113065 (2019)
15. Prabhakaran, V., et al.: Statistical modality tagging from rule-based annotations and crowdsourcing. In: Proceedings of the Workshop on Extra-Propositional Aspects of Meaning in Computational Linguistics, ExProM 2012, pp. 57–64. Association for Computational Linguistics, Stroudsburg (2012)
16. Rangi, A., Franceschetti, M.: Multi-armed bandit algorithms for crowdsourcing systems with online estimation of workers' ability. In: Proceedings of the 17th International Conference on Autonomous Agents and MultiAgent Systems, AAMAS 2018, pp. 1345–1352. International Foundation for Autonomous Agents and Multiagent Systems, Richland (2018)
17. Raykar, V.C., et al.: Learning from crowds. J. Mach. Learn. Res. **11**, 1297–1322 (2010)
18. Lyu, S.: Learning representations for quality estimation of crowdsourced submissions. Inf. Process. Manag. **56**(4), 1484–1493 (2019)
19. Finin, T., et al.: Annotating named entities in Twitter data with crowdsourcing. In: Proceedings of the NAACL HLT 2010 Workshop on Creating Speech and Language Data with Amazon's Mechanical Turk, CSLDAMT 2010, pp. 80–88 (2010)
20. Yang, P., et al.: Identifying the most valuable workers in fog-assisted spatial crowdsourcing. IEEE Internet Things J. **4**, 1193–1203 (2017)

Are n-gram Categories Helpful in Text Classification?

Jakub Kruczek, Paulina Kruczek, and Marcin Kuta[(✉)][iD]

Department of Computer Science,
Faculty of Computer Science, Electronics and Telecommunications,
AGH University of Science and Technology,
Al. Mickiewicza 30, 30-059 Krakow, Poland
mkuta@agh.edu.pl

Abstract. Character n-grams are widely used in text categorization problems and are the single most successful type of feature in authorship attribution. Their primary advantage is language independence, as they can be applied to a new language with no additional effort. Typed character n-grams reflect information about their content and context. According to previous research, typed character n-grams improve the accuracy of authorship attribution. This paper examines their effectiveness in three domains: authorship attribution, author profiling and sentiment analysis. The problem of a very high number of features is tackled with distributed Apache Spark processing.

Keywords: Character n-grams · Typed n-grams · Authorship attribution · Author profiling · Sentiment analysis

1 Introduction

Character n-grams are handcrafted features which widely serve as discriminative features in text categorization [2], authorship attribution [3] authorship verification [5], plagiarism detection [9,19], spam filtering [6], native language identification of text author [8], discriminating language variety [11], and many other applications.

They also help in generating good word embeddings for unknown words, thus improving classification performance in tasks based on informal texts, where a large percentage of unknown words occurs, e.g., in sentiment analysis [1,21]. Finally, character n-grams gave notion to character n-gram graphs [4], which found applications in topic categorization of news, blog, and twitter data, but also in automatic evaluation of document summaries.

The primary advantage of character n-grams is language independence [12], i.e., the effort of porting a feature extractor and a classifier from one language to another is negligible.

Character n-grams are recognized for their surprising degree of effectiveness in authorship attribution, outperforming content words on blog data and nearly

© Springer Nature Switzerland AG 2020
V. V. Krzhizhanovskaya et al. (Eds.): ICCS 2020, LNCS 12138, pp. 524–537, 2020.
https://doi.org/10.1007/978-3-030-50417-5_39

reaching their effectiveness on email and classic literature corpora [7]. Character n-grams have also proven to be the single most effective type of feature in authorship attribution [7]. Moreover, introduction of typed character n-grams, categories and supercategories of character n-grams have contributed to improvements in authorship attribution, compared to traditional n-grams [16].

The aim of the paper is to extend research [16] and answer the question of whether typed n-grams may be effective features in author profiling and sentiment analysis as they are in authorship attribution.

Classification on the basis of character n-grams, either typed or untyped, typically introduces a very high number of features. The solution to this problem is their distributed processing, e.g., experiments in author profiling with a large number of word n-grams as features were performed in the framework of MapReduce [10]. In [18] documents from the English language Wikipedia corpus were classified according to their topic with the newer Apache Spark framework. While the authors claim their experiments to be the first implementation of a text categorization system on Apache Spark in Python using the NLTK framework, our experiments are performed with Spark on six corpora, including approximately 150 times larger PAN-AP-13 corpus [13] with up to 8 464 237 features and The Blog Authorship Corpus with up to 11 334 188 features.

By comparison, the largest work on author profiling [17] considered larger amount of data involving 15.4 million messages and 700 million instances of words, phrases, etc.

Thus, we also examine whether the distribution of preprocessing and profile classification into smaller subtasks executed on many cores and nodes is an efficient scheme in a scenario with a high number of features, larger corpora and with the application of Spark.

2   Typed n-grams

We briefly recall the notion of typed character n-grams (in short, typed n-grams) [16]. The category and supercategory of an n-gram depends on its content and position within a word or sentence. We can distinguish between *affix*, *word* and *punct* supercategories, reflecting morpho-syntax, document topic, and author's style, respectively. Within each supercategory, we can further distinguish fine-grained categories. Within the *affix* supercategory, *prefix* and *suffix* categories denote n-grams as being the proper prefixes and proper suffixes of words, while the *space-prefix* and *space-suffix* categories denote n-grams beginning and ending with a space, respectively. Categories in the *word* supercategory (*whole-word, mid-word, multi-word*) are assigned to n-grams covering an entire word, the non-affix part of a word, or spanning multiple words, respectively. The specific category of the *punct* supercategory (*beg-punct, mid-punct, end-punct*) is assigned to n-grams containing one or more punctuation characters. Examples of some of typed n-grams araising from the sentence *The actors wanted to see if the pact seemed like an old-fashioned one.* are shown in Table 1 – their detailed description can be found in [16].

Table 1. Examples of typed character n-grams of different categories for $n = 3$. Character n-grams are in red. Remaining characters (in black) denote their context. Character ␣ denotes space. Based on examples from [16]

Supercategory	Category	Examples
affix	prefix	actors wanted seemed
	suffix	pact actors wanted like
	space-prefix	␣actors ␣wanted ␣see ␣like
	space-suffix	actors␣ pact␣ wanted␣ see␣
word	whole-word	one see the
	mid-word	actors actors old-fashioned
	multi-word	actors␣wanted see␣if
punct	beg-punct	old-fashioned
	mid-punct	old-fashioned
	end-punct	old-fashioned

3 Datasets

In our experiments with n-grams we examined three problems on six datasets: authorship attribution (CCAT_50), author profiling (PAN-AP-13, Blog author gender classification data set, The Blog Authorship Corpus) and sentiment analysis (Sentiment scale dataset v1.0, Stanford Sentiment Treebank). Table 2 briefly characterizes evaluated datasets.

Table 2. Comparison of evaluated datasets

Dataset	#texts	#authors	#classes	Balanced
PAN-AP-13 (English)	500 965	283 240	6	no[a]
PAN-AP-13 (Spanish)	151 008	90 860	6	no[a]
CCAT_50	5000	50	50	yes
Blog author gender classification data set	3227	2946	2	yes
The Blog Authorship Corpus	681 288	19 320	6	no
Sentiment scale dataset v1.0	5006	4	3 and 4	no[b]
Stanford Sentiment Treebank	215 154	–	5	no

[a]Corpus is balanced by sex but imbalanced by age group
[b]Gaussian-like distribution

Figure 1 shows the proportions of categories of typed n-grams in the English part of PAN-AP-13 corpus. We can observe that together, n-grams with *multi-word* and *mid-punct* categories constitute more than half of all typed n-grams in PAN-AP-13. Figure 2 presents the number of different ngrams depending on the n-gram length. By comparison, the number of n-gram tokens in the training, validation and test sets was approximately 1 030 960 000, 58 760 000 and 77 190 000, respectively.

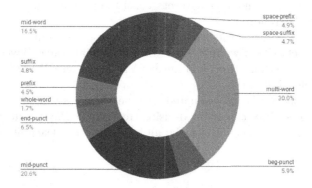

Fig. 1. Proportions of n-gram categories in the English part of PAN-AP-13 corpus

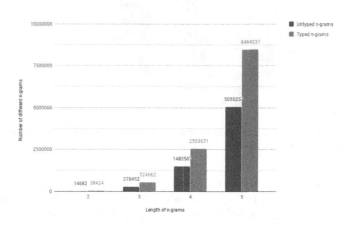

Fig. 2. Number of different character n-grams in the English part of PAN-AP-13 corpus

4   Experiments and Results

In the experiments with PAN-AP-13, corpus preprocessing involved rejecting only a few texts due to unrecognized encoding, and removing html tags and superfluous white spaces. Unknown tokens in the validation or test set were omitted.

CCAT_50 preprocessing followed the procedure from [16] and consisted of removal of citations and authors' signatures at the end of articles. Typed n-grams occurring at least five times in the dataset were taken into account as features.

Preprocessing of remaining datasets consisted of removing spurious white characters and URL addresses.

For PAN-AP-13 we adopted the predefined split into training, validation and test sets. Two classifiers were compared: multinomial Naïve Bayes (with and without feature normalization) and linear SVM based on OWLQN solver, both from Apache Spark library.

Remaining datasets were evaluated with nested cross-validation with $k = 5$ [14]. We compared three classifiers: decision trees, Naïve Bayes (multinomial and complement versions) and linear SVM, all from the scikit-learn library.

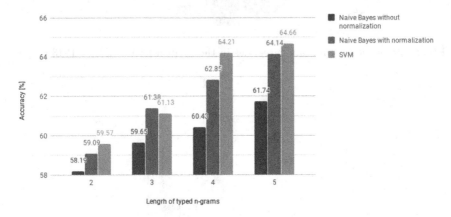

Fig. 3. Accuracy of age interval recognition depending on the length of typed n-grams and obtained on the PAN-AP-13 validation set

Table 3 presents accuracy of author profile predictions for age, sex and joint profile, evaluated on the PAN-AP-13 validation set. Parameter C denotes the regularization weight in the SVM cost function, k denotes the maximal number of iterations of the SVM solver and α is the smoothing parameter in the Naïve Bayes classification. Naïve Bayes was used with n-gram normalization.

Table 4 shows corresponding accuracies of author profiling obtained on the PAN-AP-13 test set. The obtained results outperform all solutions within the PAN-AP'13 task, which often used sophisticated features of various kinds. It is interesting to compare our outcomes with the results obtained in [10]. On the same corpus, their Naïve Bayes classifier with word n-gram features achieved a profiling accuracy of 42.57%, while using conventional character n-grams as features gave only 31.20% accuracy.

Figures 3, 4 and 5 present the accuracy of age, sex and joint recognition using typed n-grams as features, as a function of the length of used n-grams. Typed n-gram features of all categories were included in classification.

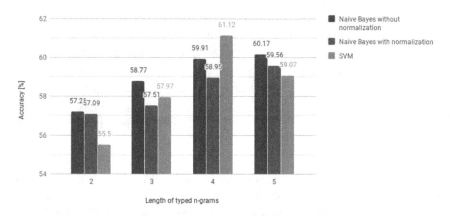

Fig. 4. Accuracy of sex recognition depending on the length of typed n-grams and obtained on the PAN-AP-13 validation set

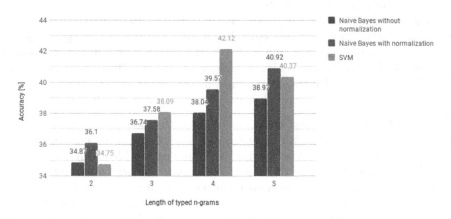

Fig. 5. Accuracy of joint profile recognition depending on the length of typed n-grams and obtained on the PAN-AP-13 validation set

Table 3. Prediction accuracy of sex and age of author on the PAN-AP-13 validation set, [%]

Classifier	N-gram length	Parameters	Age	Sex	Joint profile
SVM	4-grams	C: 500, k: 5	64.21	**61.12**	**42.12**
SVM	4-grams	C: 1000, k: 1	64.44	60.68	41.59
SVM	4-grams	C: 500, k: 1	**65.11**	58.08	41.24
Naïve Bayes	5-grams	α: 1.0	64.14	59.56	40.92
Random	–	–	33.33	50.00	16.67

Table 4. Accuracy of best models on the PAN-AP-13 test set, [%]

Classifier	N-gram length	Parameters	Age	Sex	Joint profile
SVM	4-grams	C: 500, k: 5	64.03	**60.32**	40.76
SVM	4-grams	C: 1000, k: 1	65.32	59.97	**41.02**
SVM	4-grams	C: 500, k: 1	**65.67**	57.41	40.26
SVM	4-grams	C: 0.1, k: 5	62.60	59.69	39.63
Naïve Bayes	5-gram	$\alpha = 1.0$	64.78	59.07	40.35

Usually, n-grams with $n = 3$ are considered in literature [16]. Our studies show that it is beneficial to consider longer n-grams with $n = 4$ or even $n = 5$. Using vargrams (e.g., 2-grams and 3-grams as one feature, not shown in figures) is not beneficial as they gave averaged results over n-grams with fixed n.

If time is not an issue, the choice of SVM over Naïve Bayes is preferred – this stays consistent with [20], advising SVM for classification of longer texts and Naïve Bayes for shorter texts.

The impact of feature normalization on Naïve Bayes is not clear; thus, no recommendation can be formulated. While it improves accuracy of age and joint profile classification, its effect on sex classification is negative. For feature scaling with SVM, standardization is always preferred over normalization [15], and it is the way in which SVM implementation from MLLib works.

Impact of n-gram Categories. Results in this subsection are reported for multinomial Naïve Bayes with feature normalization and size 5 n-grams. Naïve Bayes was chosen due to its better time performance over SVM. The first experiment in this part examined the impact of n-gram categories on profiling accuracy. Figures 6 and 7 shows accuracies for each of 10 categories. Additionally, classification results are shown for n-grams with no distinguished categories (*no categories*, i.e. traditional, untyped n-grams) and for features, where n-grams of *all categories* are taken into account. We observe that compared to untyped ngrams, using whole context (all categories) increases accuracy, but the increase is tiny – 40.92% for typed n-grams vs 40.43% for untyped n-grams. Typed n-grams of any single category are worse profile predictors than untyped n-grams.

The next experiment, shown in Fig. 8, looked into the discriminative power of supercategories. Profiling accuracies obtained for *all supercategories* and *all categories* features are similar. The experiment confirms findings for categories: compared to using a single supercategory, accuracy gain achieved with *all supercategories* is tiny.

Because no single n-gram category outperformed untyped n-grams and n-grams of all categories achieved the highest accuracy, in the third experiment we considered custom categories (Fig. 9). The first custom category bundled the four most discriminative categories and the second custom category bundled

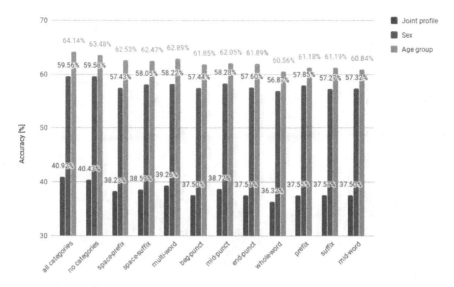

Fig. 6. Impact of n-gram categories on profiling accuracy obtained on the PAN-AP-13 validation set (English)

the nine most discriminative categories (i.e., all 10 categories but *whole-word*). Bundling more categories successively increases accuracy.

Impact of Hyperparameters. Figure 10 presents the impact of SVM hyperparameters on author profiling accuracy. Forty-five evaluations of the SVM classifier for different settings of C and k were performed. We observe that the choice of hyperparameters may impact profiling accuracy dramatically and accuracy varies from 42.12% for ($C = 5$, $k = 5$) to 21.07% for ($C = 15$, $k = 1000$). Choosing a good set of hyperparameters is much more important than the choice between typed and untyped n-grams in the case of the SVM classifier.

4.1   Further Experiments

We performed further experiments on five datasets from Table 2. First, we performed authorship attribution experiments on CCAT_50 following setup defined in [16] (Table 5).

Table 6 presents classification accuracy on five datasets performed with untyped n-grams and all-categories typed n-grams for $n = 4$ and $n = 5$.

Throughout all datasets, in most cases typed character n-grams improve classification accuracy in comparison to untyped character n-grams. The accuracy gain is however tiny – from 0.75% to 1.48%.

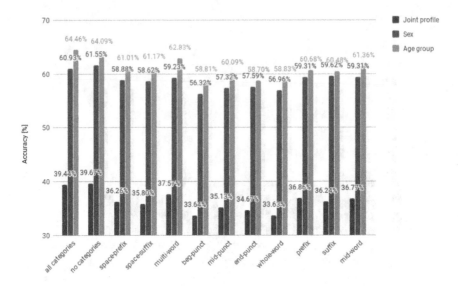

Fig. 7. Impact of n-gram categories on profiling accuracy obtained on the PAN-AP-13 validation set (Spanish)

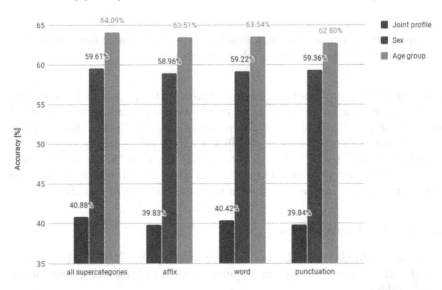

Fig. 8. Impact of n-gram supercategories on profiling accuracy obtained on the PAN-AP-13 validation set

The choice of the classifier is significant for classification with character n-grams. For all examined problems and datasets, SVM achieved higher accuracy than Naïve Bayes, with the accuracy gap up to 18%.

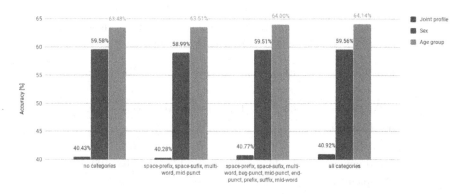

Fig. 9. Impact of custom categories of n-grams on profiling accuracy obtained on the PAN-AP-13 validation set

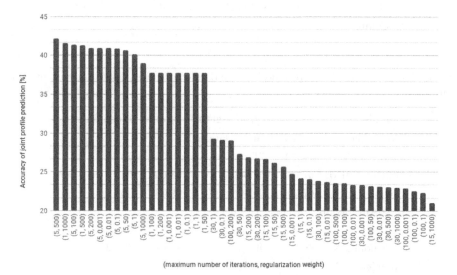

Fig. 10. Impact of SVM hyperparameters on author profiling accuracy for PAN-AP-13

We examined single-category and multiple-category n-grams. Single-category typed character n-grams differ in their predictive power w.r.t. category. Statistical tests on the Blog author gender dataset revealed that differences in accuracy are statistically significant for some pairs of categories but deeper research is needed in this area to confirm them and detect potential patterns.

Bundling more categories into typed n-grams usually results in increased accuracy. The exception was the Blog author gender classification data set, with the best results for affix+punct supercategory. Our experiments showed that information about document target label is distributed among character n-grams and their categories.

Table 5. Accuracy of authorship attribution on the CCAT_50 set, depending on used 3-gram features, [%], *acc* denotes accuracy, *N* is the number of features.

Classifier	untyped		typed		affix+punct	
	acc	*N*	*acc*	*N*	*acc*	*N*
SVM (Weka) [16]	69.20	14 461	69.10	17 062	69.30	9966
no tf-idf weighting						
SVM (libsvm)	84.30		84.72		82.98	
MultinomialNB	79.46	14689	80.06	17294	79.08	10084
ComplementNB	71.72		70.88		70.34	
with tf-idf weighting						
SVM	84.74		85.30		85.04	
MultinomialNB	78.26	14689	79.32	17294	77.84	10084
ComplementNB	73.44		73.92		72.68	

Table 6. Accuracy of untyped n-grams and all-categories typed n-grams on five datasets

Classifier	4-grams	typed 4-grams	5-grams	typed 5-grams
Blog author gender classification dataset				
SVM	**71.51**	71.23	70.06	70.80
MultinomialNB	67.05	67.64	68.60	68.79
ComplementNB	67.70	69.10	69.53	70.15
Blog Authorship Corpus				
SVM	62.50	62.98	63.54	**64.29**
MultinomialNB	46.59	46.58	47.47	47.02
ComplementNB	43.62	43.91	45.48	45.24
Sentiment scale dataset (3 classes)				
SVM	67.00	67.84	67.00	**68.48**
MultinomialNB	49.26	50.06	50.68	50.94
ComplementNB	51.56	50.56	49.46	49.24
Sentiment scale dataset (4 classes)				
SVM	58.09	58.37	59.19	**60.79**
MultinomialNB	44.35	43.61	43.53	44.51
ComplementNB	40.51	41.87	42.87	43.53
Stanford Sentiment Treebank				
SVM	59.43	60.39	60.70	**61.35**
MultinomialNB	60.19	60.10	60.72	60.00
ComplementNB	53.18	53.93	54.40	55.54

The length of n-grams affects classification results and depends on the dataset and used classifier. The highest accuracy for CCAT_50 used in authorship attribution was achieved with typed 4-grams. For all remaining datasets, the best accuracy was achieved with typed 5-grams In particular, for the Blog author gender classification dataset the highest accuracy, 71.60%, was for typed 5-grams of affix+punct supercategory (not shown in Table 6). These findings are in line with results obtained for the PAN-AP-13 corpus. Our findings clearly contradict those of [16], where authors state: *We chose $n = 3$ since our preliminary experiments found character 3-grams to be more effective than other higher level character n-grams.*

When considering typed n-grams, the highest accuracy was achieved when bundling all categories, i.e., for all-categories typed n-grams. The only exception was the Blog author gender classification data set, where affix+punct typed n-grams achieved the highest accuracy. For all datasets, using typed character n-grams of single category results in an accuracy drop in comparison to untyped character n-grams. Except for the Blog author gender classification data set, using single-supercategory n-grams resulted in lower accuracy. The best results were achieved for categories space-prefix, space-suffix, prefix, and for supercategories affix and affix+punct.

Our experiments on the Blog author gender classification dataset show that character n-grams (whether typed or untyped) give higher accuracy than word n-grams by 1%–1.15%. The downside is a larger number of arising character n-gram features than word n-gram features.

Tf-idf weighting raises classification accuracy with n-grams from 2% to 4%. The exception is authorship attribution on the CCAT_50 dataset, where accuracy increased for n-grams with $n = 2$ and $n = 3$ while there was an accuracy drop for $n = 4$ and $n = 5$.

There is no clear pattern for impact of feature normalization on accuracy. The best results were obtained with normalization according to the L_2 norm[1]. With the remaining methods - StandardScaler and MaxAbsScaler we observed suboptimal accuracy or even accuracy worse than with no normalization.

Finally, we performed qualitative analysis and looked for the most important n-grams by inspecting weights of SVM classifier. First, we analysed author profiling on the Blog author gender classification data set. For men, identified n-grams referred to wife, other men (guys) and games. The most important n-grams used by women are related to family (love, husband, mum). Found best n-grams do not suggest that text style (e.g. punctuation) is important for a classifier. Next, we analysed authorship attribution on one particular author chosen from CCAT_50: the identified n-grams were name fragments of cites, states or companies.

---

[1] Normalizer from the scikit-learn library.

5   Conclusions

The paper has shown in three domains: authorship attribution, author profiling and sentiment analysis that the choice of typed n-grams results in only a tiny increase of classification accuracy over traditional n-grams. Information about the author profile is distributed throughout all n-gram categories. No single category can be advised for classification It is worth putting much more effort into effective hyperparameter optimization and model selection than to switching from n-grams to typed n-grams or a particular category of typed n-grams.

Apache Spark allows for efficient classification with a very high number of features on large text corpora. The memory footprint is the most prohibitive aspect of such classification, which precludes experiments with n-grams longer than 5.

Acknowledgements. The research presented in this paper was supported by the funds assigned to AGH University of Science and Technology by the Polish Ministry of Science and Higher Education. This research was supported in part by the PL-Grid Infrastructure.

References

1. Bojanowski, P., Grave, E., Joulin, A., Mikolov, T.: Enriching word vectors with subword information. Trans. Assoc. Comput. Linguist. **5**, 135–146 (2017)
2. Cavnar, W.B., Trenkle, J.M.: N-gram-based text categorization. In: Proceedings of 3rd Annual Symposium on Document Analysis and Information Retrieval, SDAIR-94, pp. 161–175 (1994)
3. Escalante, H.J., Solorio, T., Montes-y-Gómez, M.: Local histograms of character n-grams for authorship attribution. In: Lin, D., Matsumoto, Y., Mihalcea, R. (eds.) The 49th Annual Meeting of the Association for Computational Linguistics: Human Language Technologies, pp. 288–298 (2011)
4. Giannakopoulos, G., Karkaletsis, V.: N-gram graphs: representing documents and document sets in summary system evaluation. In: Proceedings of the Second Text Analysis Conference, TAC 2009. NIST (2009)
5. Jankowska, M., Milios, E.E., Keselj, V.: Author verification using common n-gram profiles of text documents. In: Hajic, J., Tsujii, J. (eds.) 25th International Conference on Computational Linguistics, COLING 2014, pp. 387–397 (2014)
6. Kanaris, I., Kanaris, K., Houvardas, I., Stamatatos, E.: Words versus character n-grams for anti-spam filtering. Int. J. Artif. Intell. Tools **16**(6), 1047–1067 (2007)
7. Koppel, M., Schler, J., Argamon, S.: Computational methods in authorship attribution. J. Am. Soc. Inf. Sci. Technol. **60**(1), 9–26 (2009)
8. Koppel, M., Schler, J., Zigdon, K.: Automatically determining an anonymous author's native language. In: Kantor, P.B., et al. (eds.) Intelligence and Security Informatics, IEEE International Conference on Intelligence and Security Informatics, ISI 2005, pp. 209–217 (2005)
9. Kuta, M., Kitowski, J.: Optimisation of character n-gram profiles method for intrinsic plagiarism detection. In: Rutkowski, L., Korytkowski, M., Scherer, R., Tadeusiewicz, R., Zadeh, L.A., Zurada, J.M. (eds.) ICAISC 2014. LNCS (LNAI), vol. 8468, pp. 500–511. Springer, Cham (2014). https://doi.org/10.1007/978-3-319-07176-3_44

10. Maharjan, S., Shrestha, P., Solorio, T., Hasan, R.: A straightforward author profiling approach in MapReduce. In: Bazzan, A.L.C., Pichara, K. (eds.) IBERAMIA
2014. LNCS (LNAI), vol. 8864, pp. 95–107. Springer, Cham (2014). https://doi.
org/10.1007/978-3-319-12027-0_8

11. Malmasi, S., Dras, M.: Language identification using classifier ensembles. In:
Nakov, P., Zampieri, M., Osenova, P., Tan, L., Vertan, C., Ljubešić, N., Tiede-
mann, J. (eds.) Proceedings of the Joint Workshop on Language Technology for
Closely Related Languages, Varieties and Dialects, pp. 35–43. Association for Computational Linguistics (2015)

12. Peng, F., Schuurmans, D., Keselj, V., Wang, S.: Language independent authorship
attribution with character level n-grams. In: 10th Conference of the European
Chapter of the Association for Computational Linguistics, EACL 2003, pp. 267–
274 (2003)

13. Rangel, F., Rosso, P., Koppel, M., Stamatatos, E., Inches, G.: Overview of the
author profiling task at PAN 2013. In: Forner, P., Navigli, R., Tufis, D., Ferro, N.
(eds.) Working Notes for CLEF 2013 Conference, vol. 1179 (2013)

14. Raschka, S.: Model evaluation, model selection, and algorithm selection in machine
learning. CoRR abs/1811.12808 (2018)

15. Raschka, S., Mirjalili, V.: Python Machine Learning, 2nd edn. Packt Publishing,
Birmingham (2017)

16. Sapkota, U., Bethard, S., Montes-y-Gómez, M., Solorio, T.: Not all character n-
grams are created equal: a study in authorship attribution. In: Mihalcea, R., Chai,
J.Y., Sarkar, A. (eds.) NAACL HLT 2015, The 2015 Conference of the North American Chapter of the Association for Computational Linguistics: Human Language
Technologies, pp. 93–102 (2015)

17. Schwartz, H.A., et al.: Personality, gender, and age in the language of social media:
the open-vocabulary approach. PLoS ONE **8**(9), 1–16 (2013). https://doi.org/10.
1371/journal.pone.0073791

18. Semberecki, P., Maciejewski, H.: Distributed classification of text documents
on Apache Spark platform. In: Rutkowski, L., Korytkowski, M., Scherer, R.,
Tadeusiewicz, R., Zadeh, L.A., Zurada, J.M. (eds.) ICAISC 2016. LNCS (LNAI),
vol. 9692, pp. 621–630. Springer, Cham (2016). https://doi.org/10.1007/978-3-319-
39378-0_53

19. Stamatatos, E.: Intrinsic plagiarism detection using character n-gram profiles. In:
Stein, B., Rosso, P., Stamatatos, E., Koppel, M., Agirre, E. (eds.) SEPLN 2009
Workshop on Uncovering Plagiarism, Authorship, and Social Software Misuse,
PAN 2009, pp. 38–46 (2009)

20. Wang, S.I., Manning, C.D.: Baselines and bigrams: simple, good sentiment and
topic classification. In: Li, H., Lin, C.Y., Osborne, M., Lee, G.G., Park, J.C. (eds.)
50th Annual Meeting of the Association for Computational Linguistics, pp. 90–94
(2012)

21. Wieting, J., Bansal, M., Gimpel, K., Livescu, K.: Charagram: embedding words and
sentences via character n-grams. In: Su, J., Carreras, X., Duh, K. (eds.) Proceedings
of the 2016 Conference on Empirical Methods in Natural Language Processing,
EMNLP 2016, pp. 1504–1515 (2016)

Calculating Reactive Power Compensation for Large-Scale Street Lighting

Sebastian Ernst[1], Leszek Kotulski[1], Tomasz Lerch[2], Michał Rad[2],
Adam Sędziwy[1], and Igor Wojnicki[1(✉)]

[1] Department of Applied Computer Science, AGH University of Science
and Technology, al. Mickiewicza 30, 30-059 Kraków, Poland
wojnicki@agh.edu.pl
[2] Department of Power Electronics and Energy Control Systems, AGH University
of Science and Technology, al. Mickiewicza 30, 30-059 Kraków, Poland

Abstract. LED-based street lighting installations generate reactive
power, particularly when they are dynamically dimmed. It contributes
to power loss and efficiency reduction of the grid. The reactive power can
be compensated by installing additional dynamically connected inductors in lighting control cabinets. However such an approach significantly
increases the cost of the lighting infrastructure. The goal of this paper is
to propose another, low cost approach to reactive power compensation
for dynamically dimmed lighting installations. It is based on connecting fixed settings inductors at lighting control cabinets. The inductors
settings are calculated by the proposed algorithm for city-scale lighting systems. Its objective is to completely eliminate capacitive reactive
power and to keep inductive reactive power within acceptable limits.

Keywords: Street lighting · Reactive power · LED lighting · Reactive
power compensation

1 Introduction

In the last years we are witnessing a dynamic growth of usage of the solid state
lighting technology. This is caused and stimulated by several factors, such as
the demand of energy saving; physical properties of solid state light sources:
neglectable onset time, dimming capabilities, high expected lifespan; continuously dropping prices of LED fixtures.

The annual global energy usage related to outdoor lighting is assessed to
have 12–15% share in the total energy consumption [10]. In this context even a
small improvement in lighting energy efficiency, e.g., of the order of 1%, yields
significant total savings, due to the effect of scale. The above reasoning has
caused development of a broad scope of methods of power usage reduction [12]
and sustainable maintenance of public lighting, beginning from the well suited

© Springer Nature Switzerland AG 2020
V. V. Krzhizhanovskaya et al. (Eds.): ICCS 2020, LNCS 12138, pp. 538–550, 2020.
https://doi.org/10.1007/978-3-030-50417-5_40

lighting projects [6,9] relying on GIS-based inventory data [20,21], application of control systems [5,18,22], to sophisticated methods of tunnel illumination [16,17] or improving reflective properties of road surface [19].

One of the basic benefits of retrofitting lighting installations with LED sources is a radical drop of energy usage of the order up to 60% [1,25]. Additional savings can be achieved by adjusting luminous fluxes of luminaires when road and environment conditions change [24]. However, while dimming a luminaire to adjust lighting levels to particular needs saves energy, there is a side effect. It is a growth of *capacitive reactive power* in the power network, expressed interchangeably by means of the trigonometric values $\tan \varphi$ and $\cos \varphi$ known as *power factor*. As long as the power driver does not operate at its full capacity it dims the luminaire and introduces reactive power.

Reactive power is charged for both electric energy producers and consumers. A detailed insight into tariffs and pricing options for energy producers but also transmission providers can be found in [8]. In the case of customers it also depends on a regulatory framework [15].

Reactive power can be compensated by additional hardware components either attached to particular lamps [3,13] or introduced to a power grid. Although compensation as such is an easy task it becomes nontrivial in the context of a large (i.e., containing tens of thousands of light points), dynamically dimmed, LED-based lighting installation, powered by multiple, independent control cabinets, where the reactive power level is not constant but changes in an unpredictable manner.

The main goal of this paper is to introduce an algorithm that provides settings for static reactive power compensation for street lighting with adaptive control at each of the lighting control cabinets.

1.1 Capacitive Reactive Power Compensation

Let $u(t)$ and $i(t)$ denote respectively the voltage and current in an alternating current electric circuit, at a given time t. Moreover let us assume that both waveforms are sinusoids with the period T (hence the angular frequency for both is $\omega = \frac{2\pi}{T}$). A presence of a capacitor and/or an inductive coil in a circuit may cause a phenomenon of a phase shift between current and voltage which manifests as an additional term φ in the waveform $i(t) = i_0 \sin(\omega t + \varphi)$ (or $u(t) = u_0 \sin(\omega t - \varphi)$).

An RMS (Root-Mean-Squared) voltage value is calculated according to the following definition:

$$U = \sqrt{\frac{1}{T} \int_0^T u^2(t) dt}. \tag{1}$$

An RMS current value is given by

$$I = \sqrt{\frac{1}{T} \int_0^T i^2(t) dt}. \tag{2}$$

To obtain an **active power** we compute the integral

$$P = \frac{1}{T} \int_0^T u(t)i(t)dt. \tag{3}$$

In turn, an **apparent power** is given by

$$S = UI. \tag{4}$$

For sinusoidal waveforms of voltage and current, the power equation is met:

$$S^2 = P^2 + Q^2. \tag{5}$$

Using the Eq. 5 one can obtain straightforwardly a **reactive power**:

$$Q^2 = \sqrt{S^2 - P^2}. \tag{6}$$

The **power factor** $(\cos\varphi)$ and $\tan\varphi$ values are calculated as

$$\cos\varphi = \frac{P}{S}, \quad \tan\varphi = \frac{Q}{P}. \tag{7}$$

It can be easily implied form Eq. (7) that for $\varphi = 0$ no reactive power is produced $(P = S)$. It has to be emphasized that the power factor (PF) equals to $\cos\varphi$ when there exists only the fundamental harmonic of current or if higher harmonics are neglectable. Otherwise, one has to use the following formula:

$$PF = \frac{1}{\sqrt{1 + \text{THD}_I^2}} \cos\varphi = \frac{1}{\sqrt{1 + \left(\frac{I_{1,\text{RMS}}}{I_{\text{RMS}}}\right)^2}} \cos\varphi, \tag{8}$$

where I_{RMS} is the total current and $I_{1,\text{RMS}}$ denotes the fundamental component of current. RMS subscript means that both are computed as root mean square-values.

For our further considerations we choose $\tan\varphi$ as more handy for expressing the phase shift. It is due to a sign of the $\tan\varphi$ function which reflects a type of reactive power. A phase shift can be either negative or positive. In the former case reactive power is referred to as a *capacitive* and in the latter – as an *inductive* one.

In the case of deformed waveforms [4, 23], in order to correctly determine the values of reactive power, the definition of reactive power according to Budeanu's theory has to be used, then:

$$Q_B = \sum_{h=1}^{\infty} u_h i_h \sin\varphi_h, \tag{9}$$

where the sum iterates over all current and voltage harmonics.

In the considered case we deal with dimmed LED fixtures for which dimming implies changes of the current components $\{i_h\}$ only while the voltage remains unchanged, i.e., all harmonics except the fundamental are 0: ($u_h \approx 0$ for $h > 1$). Thus Eq. (9) reduces to the fundamental component only: $Q_B = u_1 i_1 \sin \varphi_1$.

We focus on the compensation model in which we try to fit with $\tan \varphi$ within the range of $[0, \tan \varphi_0]$. CRP compensation ($Q_{cap} < 0$) is achieved by increasing inductive reactive power ($Q_{ind} > 0$) so that a resultant $\tan \varphi$ is at least non-negative and it is less than $\tan \varphi_0$. It can be written by means of Eq. (7):

$$\tan \varphi = \frac{Q_{cap} + Q_{ind}}{P}. \tag{10}$$

In our analysis we consider $Q_{cap} = Q_{\text{fix}} + Q_{\text{power line}}$ being a sum of two negative components: the first component corresponds to a LED fixture and the second one is associated with a power line which acts as a capacitor. Although for short distances it is neglectable, for longer ones, however, it can yield capacity which has to be taken into account to avoid a further inaccurate compensation.

For a fixture working with some established dimming level and its power line, a fixed source of the inductive reactive power, Q_{ind}, compensating system with a constant reactance is employed. A reactance, X_C, of a fixture and its power line can be calculated according to the formula:

$$X_C = \frac{1}{\frac{|Q_{\text{fix}}|}{U^2} + \omega c l}, \tag{11}$$

where $U = 230\,\text{V}$ (in Europe) is a voltage value, $\omega = 2\pi f$ (the frequency $f = 50\,\text{Hz}$ in Europe), Q_{fix} stands for a CRP of a fixture, c denotes a cable capacity per unit length and l is a cable length. From the practical point of view, the capacitive reactive power compensation can be most easily achieved by attaching a parallel choke with an inductive reactance X_L to a fixture, such that $X_L = X_C$. Since an inductive reactance is given by the formula:

$$L = \frac{X_L}{\omega}, \tag{12}$$

where L is an inductance, one can obtain the desired value of the latter by combining Eqs. (11), (12) and assuming equality of reactances X_C and X_L:

$$L = \frac{1}{\frac{\omega |Q_{\text{fix}}|}{U^2} + \omega^2 c l}. \tag{13}$$

It should be remarked that such an approach allows to transform the highly penalized CRP into the inductive one with an acceptable $\tan \varphi < \tan \varphi_0$.

2   Single Luminaire Example

As a test case we have examined an Ampera Midi 106 W manufactured by Schréder. It is controlled using a 0–10 V dimmer. For the control signal s in the range 7.5–10 V the active power P is practically constant. In turn, P becomes linearly depending on s when $s \in [0.5\,\text{V}, 7.5\,\text{V}]$. This behavior can be seen in Fig. 1. Since the fixture's luminous flux intensity λ is proportional to the active power ($\lambda \approx \frac{P}{P_{\max}} \times 100\%$) being proportional to the control signal, if $s \leq 7.5\,\text{V}$, we can identify the control signal voltage with λ when the former is within the range 0–7.5 V.

Hereafter, every time when we talk about "setting/adjusting λ" we mean such setting of s so that a desired value of $\lambda(s)$ is obtained.

The Ampera Midi 106 W fixture measurements are given in Table 1. Besides the active power and $\cos\varphi$, calculated values of $\tan\varphi$ are also shown. Their negative sign is implied by the capacitive nature of the reactive power for which $\varphi < 0$.

Table 1. The characteristics of the Ampera Midi 106 W fixture, gathered remotely for a series of control signals, P denotes an active power

Ctrl [V]	0.5	1.0	1.5	2.0	2.5	3.0	3.5	4.0	4.5	5.0	5.5	6.0	6.5	7.0	7.5	8.0	8.5	9.0	9.5	10
P [W]	20,8	26,3	31,9	39,4	44,9	52,5	58,3	65,7	71,8	79,9	85,9	91,8	97,5	105,9	111,7	111,6	110,7	111,4	111,1	111,1
$\cos\varphi$	0,52	0,59	0,69	0,76	0,80	0,84	0,85	0,88	0,89	0,90	0,91	0,92	0,92	0,93	0,94	0,94	0,94	0,94	0,94	0,94
$\tan\varphi$	-1,64	-1,37	-1,05	-0,86	-0,75	-0,65	-0,62	-0,54	-0,51	-0,48	-0,46	-0,43	-0,43	-0,40	-0.36	-0.36	-0.36	-0.36	-0.36	-0.36

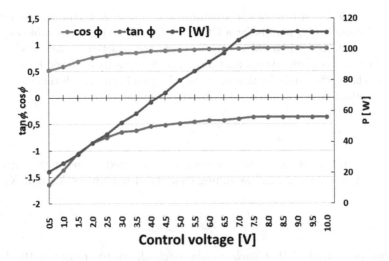

Fig. 1. The performance chart of the Ampera Midi 106 W fixture for data presented in Table 1

Having the above data we find the minimum and maximum values of the fixture's CRP $Q_{\text{fix}}(\lambda) = P(\lambda)\tan(\lambda)$: $Q_{\text{fix}}^{\min} = -41.85\,\text{Var}$ and $Q_{\text{fix}}^{\max} = -33.46\,\text{Var}$, reached for λ_1 and λ_2 respectively. The capacitive reactive power component contributed by the power line is constant. To correct a CRP of the fixture a choke generating an inductive reactive power $Q_{ind} = -Q_{\text{fix}}^{\min}$ has to be used. Thus we get

$$X_L = \frac{U^2}{Q_{ind}}. \tag{14}$$

After adding the above inductance to the circuit the vales $\tan\varphi(\lambda_1)$ and $\tan\varphi(\lambda_2)$ change. From the Eq. (10) we have (hereafter, a primed φ will refer to a phase shift after compensation) $\tan\varphi(\lambda_1) \rightarrow \tan\varphi'(\lambda_1) = 0$ and

$$\tan\varphi(\lambda_2) \rightarrow \tan\varphi'(\lambda_2) = \frac{Q_{cap}(\lambda_2) + Q_{ind}}{P(\lambda_2)} = 0.37.$$

For this case $\tan\varphi'(\lambda_2) < \tan\varphi_0$. Otherwise (the case referred to as *overcompensation*), lamp should not be dimmed to the values which violate this constraint unless a business analysis points that paying given charges for overcompensation is still more profitable.

3 Capacitive Reactive Power Compensation in Large Lighting Installations

In this section we begin with the basic case of compensation, made for a single control cabinet. Next we proceed to the case of multiple cabinets for which the issue of high analytical complexity of the system arises.

3.1 Case 1. Control Cabinet

Compensation for a series of lamps connected to a control cabinet is made similarly as in the above example. We compensate the total CRP being a sum over all fixtures: $\mathbf{Q}_{cap} = \sum_{i=1}^{n}(Q_{\text{fix},i} + Q_{\text{power line},i})$, where $Q_{\text{fix},i} = Q_{\text{fix},i}(\lambda_i)$ is a CRP of i-th fixture (depending on its luminous flux ratio λ_i) and $Q_{\text{power line},i}$ is a CRP contributed by the corresponding power line.

Getting the extreme values of $\mathbf{Q}_{cap}$ is not straightforward, however, as the fixtures are not assumed to be identical, dimmed equally and in the same time moments. It should be noted that for a dynamically controlled lighting system both minimum and maximum, $\mathbf{Q}_{cap}^{\min}, \mathbf{Q}_{cap}^{\max}$, are calculated over some time period, e.g., operating hours. Namely, with such assumptions $\lambda_i = \lambda_i(t)$ is a function of time and $\mathbf{Q}_{cap}(t) = \mathbf{Q}_{cap}(\lambda_1, \lambda_2, \ldots, \lambda_n)$ so is. In practice both the time variable and λ_i's are discretized so

$$\mathbf{Q}_{cap}^{\min} = \min_t \mathbf{Q}_{cap}(t), \quad \mathbf{Q}_{cap}^{\max} = \max_t \mathbf{Q}_{cap}(t)$$

can be calculated programatically.

It should be remarked that time discretization is possible because changes of λ's are triggered by varying environment conditions being evaluated as time averages (for example, an increased average traffic flow). λ adjustments are always made in compliance with mandatory lighting standards [7].

That step is important because due to functional specificity of particular lamps and the control patterns usually $\mathbf{Q}_{cap}^{min} \neq \sum_{i=1}^{n}(Q_{fix,i}^{min} + Q_{power\,line,i})$ which means that usually minima of the CRP of particular lamps do not meet in the one moment. For this reason $\mathbf{Q}_{cap}^{min}$ (but also $\mathbf{Q}_{cap}^{max}$) cannot be easily assessed.

The value of inductance L is derived starting from the minimum CRP reachable by the installation, i.e., $\mathbf{Q}_{cap}^{min}$. We set a required value of the inductive reactive power to be $Q_{ind} = -\mathbf{Q}_{cap}^{min}$. Having that one obtains L using (12) and (14): $L = U^2/\omega Q_{ind}$.

In the case of several circuits connected to a cabinet (e.g., for a 3-phase power line) each of them is considered separately. In particular each circuit will be attached with a dedicated compensator.

3.2   Case 2. Multiple Cabinets

The scheme presented in the previous subsection applies also to multiple cabinets (for brevity referred to as a *lighting grid*) which are compensated separately as discussed above. If a control cabinet has several circuits then we can divide it logically, for computational purposes, into virtual, single-circuit cabinets.

The difference between a single cabinet and a lighting grid is significantly higher complexity of the latter. This complexity refers to the lighting infrastructure topology and functional relations as seen from the perspective of control (adjusting λ's for particular fixtures) in terms of relations among control cabinets, circuits, lamps and streets segments.

A value of $\tan \varphi_G$ for a non-corrected lighting grid is computed as:

$$\tan \varphi_G(\lambda_1, \lambda_2, \ldots \lambda_n) = \frac{Q^* + \sum_i Q_{fix,i}(\lambda_i)}{\sum_i P_i(\lambda_i)}, \tag{15}$$

where Q^* is a total CRP of power lines and both sums iterate over all fixtures. As previously, λ_i's are time dependent. The general compensation scheme remains unchanged, i.e., we correct $\tan \varphi_G$ by adding such inductive reactive powers to particular circuits that their total value is $Q_{ind} = -(Q^* + \min_t \sum_i Q_{fix,i})$, where $\min_t$ denotes a minimum over the time. Thus, new $\tan \varphi_G$ is given as:

$$\tan \varphi'_G(\lambda_1, \lambda_2, \ldots, \lambda_n) = \frac{Q_{ind} + Q^* + \sum_i Q_{fix,i}(\lambda_i)}{\sum_i P_i(\lambda_i)}. \tag{16}$$

The major difficulty for the case of multiple cabinets is finding an optimal configuration of compensators in a reasonable time rather than a compensation as such, particularly for dynamically controlled lighting systems. An alternative to the static compensation could be using a dynamic VAR compensator. It generates, however, significantly higher cost, especially when multiplied by the number of cabinets. For this reason we focus on building an effective computational approach which enables planning configuration of compensators for a large lighting grid.

4   Algorithmic Approach to Compensation Planning

In this section we address problem of compensation planning by introducing the formal representation of a lighting installation and the algorithm which finds optimal configuration of static compensators.

4.1   Problem Overview

The luminaires illuminating public spaces (grouped into sets denoted by $\mathcal{L}$ in Fig. 2), also referred to as the light points, are organized in the hierarchical manner. At the lowest level we have single light points which are grouped in circuits. One or more circuits, dependently on a local specificity, are connected to a control cabinet (denoted by C).

A roadway network is assumed to be logically divided into the segments (denoted by S in Fig. 2). The partition is made in such a way that a given segment is a uniform street geometry assigned with a specific lighting class depending on such factors as a function of a relevant area (parking zone, road junction, freeway, residential area etc.), traffic flow and others [2,7,11,14]. The important fact related to the lighting class assignment is that standards admit changing a class if local conditions change, e.g, when the traffic flow level increases (then some more restrictive class, implying higher luminance, can be selected) or decreases (the class resulting in lower luminance required for this area is assigned). Changing an actual class allows decreasing luminous fluxes of relevant fixtures (i.e., decreasing corresponding λ's, dimming them) and thus obtain energy savings.

Our goal is to set such compensators in a lighting grid so that (see Eq. (16)): $0 \leq \tan \varphi'_G(\lambda_1, \lambda_2, \ldots, \lambda_n) < \tan \varphi_0$. This task is not straightforward because the following factors have to be taken into account:

- λ's of all relevant fixtures. Note that a value of λ depends on a fixture model but also on a public area (segment) being illuminated: a fixture's luminous flux in a residential area is usually lower than in a road junction zone. For this reason two luminaires equipped with the same fixture model can be set to two different λ's and thus produce two different capacitive reactive powers. It implies that we cannot simply multiply a number of fixtures of a given type by a unit reactive power, when computing $\sum_i Q_{\text{fix},i}$.
- An i-th fixture, f_i, installed by a segment S can be dimmed due to the lighting class reduction so one deals with the time-dependent (and, in the general case, unpredictable) profiles of the active and reactive power of f_i: $P_i(t), Q_{\text{fix},i}(t)$ (for the simplicity we can assume that t changes in a 24-h period). Obviously, the same applies to other fixtures illuminating S and to the remaining segments. Thus we have to analyze the time coincidences of the particular power profiles. Note that it may easily happen that the simultaneous dimming of all fixtures never occurs: the maxima and minima of the real and reactive powers never overlap so we are unable to make a priori any rough estimation of $\tan \varphi'_G$.

Having in mind the above one has to find $\tan \varphi'_G$ iterating over entire lighting installation covering all cabinets and segments, and over its all accessible states (dimming levels). This task cannot be achieved "manually" for its high computational complexity and the computer aided solution finding has to be applied. Prior to this, however, we need to build a formal data model to express the problem in terms applicable for a computer system. The following subsections contain both problem formalization and the compensation algorithm.

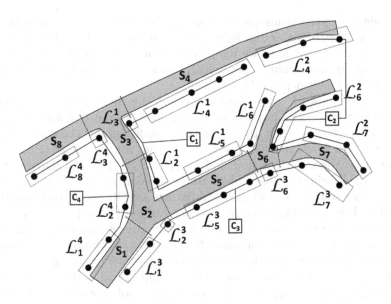

Fig. 2. The sample road and installation layout. $S_1, \ldots, S_8$ denote road segments (delimited by dashed lines); $C_1, \ldots, C_4$ are control cabinets; bold dots denote particular luminaires; $\mathcal{L}_i^k$ is a list of lamps (bounded by red dotted lines): a lower index (i) refers to a relevant segment number; an upper (k) stands for a number of cabinet to which luminaires from $\mathcal{L}_i^k$ are attached. (Color figure online)

4.2   Formal Representation

As mentioned above, we divide a system of roadways, walkways and squares being illuminated into segments $S_1, S_2, \ldots S_m$ such that in a given moment entire segment S_i is assigned with a single lighting class. Additionally it is assumed that we consider only the segments being illuminated by at least one luminaire. A subset of luminaires powered by C_k ($k = 1, \ldots, n$) and illuminating a segment S_i will be denoted as $\mathcal{L}_i^k$. It should be noted that in general case it can occur that $\mathcal{L}_i^k \cap \mathcal{L}_i^{k'} \neq \emptyset$ for $k \neq k'$ what means that some luminaires can illuminate more than one segment. It can be easily checked that $\sum_k |\mathcal{L}_i^k|$ is a number of all luminaires illuminating an i-th segment. In turn, $|\bigcup_i \mathcal{L}_i^k|$ is a total number of luminaires attached to a k-th cabinet.

To enclose the above model within a single structure passed to an algorithm we define the notion of a *structure matrix*:

$$\mathbf{M} = \{\mathcal{L}_i^k\}_{i=1,2,\ldots,m}^{k=1,2,\ldots,n}, \tag{17}$$

where n and m are numbers of cabinets and segments respectively. If no luminaire powered by C_k illuminates S_i then we denote $\mathcal{L}_i^k = \emptyset$.

For the sample layout shown in Fig. 2 we have the matrix $\mathbf{M}$ which aggregated form is shown in Table 2: particular lists, $\mathcal{L}_i^k$, are replaced by their lengths (numbers of contained luminaires).

Table 2. Aggregated matrix representation of the layout shown in Fig. 2. For better readability all cells for which $|\mathcal{L}_i^k| = 0$ are left blank.

		Segment							
		S_1	S_2	S_3	S_4	S_5	S_6	S_7	S_8
Cabinet	C_1		2	1	4	2	2		
	C_2				3		4	3	
	C_3	2	1			3	1	2	
	C_4	2	2	1					2

4.3 Compensation Strategy – Algorithmic Approach

The goal of the algorithm (see Algorithm 1) is to determine a list of compensators' inductances such as a sum (denoted as Δ) of charges corresponding to exceeding the $\tan \varphi_0$ threshold and the annual power usage expenses, is minimized. Optionally, an additional constraint may be imposed on a total cost of compensators (being a onetime outlay).

As an algorithm's input we put a structure matrix (17) representing links between control cabinets and illuminated segments. Note that elements of $\mathbf{M}$ contain all relevant data required in further processing. An output is a sequence of inductances $\{L_k\}$ for subsequent cabinets.

Several comments related to the algorithm (see Algorithm 1) performance should be given.

1. The pseudocode has two nested loops (the first in the line 4 and the second, implicit, triggered in the line 11), except the finite iteration over cabinets (line 5). Both loops are protected against an infinite execution so stop condition is satisfied for each algorithm run.
2. A source of the computational complexity of the presented algorithm are operations of finding minimum and maximum, located in lines 6 and 8 respectively. To reduce the number of operations one can apply the problem-specific heuristics.

1 **Algorithm:** GetCompensators(**M**)

   Input : **M** – structure matrix
   Output: $\{L_k\}$ – inductances of compensators

2 **begin**

3     $\Delta \leftarrow +\infty$

4   **while** Δ *is not acceptable* **and** Δ *decreased compared to the previous iteration* **do**

5   **foreach** C_k $(k = 1, \ldots, n)$ **do**

6             $Q_{ind,k}^{min} \leftarrow -\min_{t \in 24h} \sum_{\text{fix} \in \bigcup_i \mathcal{L}_i^k} Q_{\text{fix}}(t)$

7             $L_k \leftarrow \dfrac{U^2}{\omega Q_{ind,k}^{min}}$ /* Calculation of inductance for C_k */

8         $\tan \varphi'_{tot} \leftarrow \max_{t \in 24h} \dfrac{\sum_k (Q_{ind,k}^{min} + \sum_{\text{fix} \in \bigcup_i \mathcal{L}_i^k} Q_{\text{fix}}(t))}{\sum_k \sum_{\text{fix} \in \bigcup_i \mathcal{L}_i^k} P_{\text{fix}}(t)}$

        /* φ'_{tot} -- phase shift for entire lighting installation, after
        CRP compensation   */
        /* t_0 -- time for which $\max_{t \in 24h} \sum_{\text{fix} \in \bigcup_i \mathcal{L}_i^k} Q_{\text{fix}}(t)$ is reached */

9   **if** $\tan \varphi'_{tot} < 0$ **or** $\tan \varphi'_{tot} > \tan \varphi_0$ **and** $\tan \varphi'_{tot}$ *decreased compared to the previous iteration* **then**

10   Adjust λ's appropriately for suitable groups of lamps $\mathcal{L}_i^k$

11   Go to the line 5

12         $\Delta \leftarrow$ Charge($\tan \varphi'_{tot}$) $+ \sum_k \sum_{\text{fix} \in \bigcup_i \mathcal{L}_i^k}$ AEC(fix)   /* AEC(fix) is an
        annual energy cost for a fixture fix */

Algorithm 1: Calculating inductances of multiple compensators.

3. The function Charge($\tan \varphi$) used in the line 12 represents annual (or monthly, depending on needs) due charges related to exceeding $\tan \varphi_0$. Thus Δ computed in the same line stands for the accumulated annual costs of roadway illumination and potential fees for reactive power generation. The premises for evaluating the condition contained in the line 4 (Δ *is not acceptable*) have a business character. It is based on priorities defined by a lighting infrastructure owner or operator.

4. The step given in the line 10 may look imprecise so its clarification is needed. As pointed in the previous sections, the λ parameters for particular luminaires are responsible for changes of $\tan \varphi$. Thus, in the case of exceeding the $\tan \varphi_0$ limit we aim at changing λ's for relevant fixtures. Their "relevance" can be stated only on the basis of a known, specific installation's layout. The restriction which can be made here is focusing on luminaires from a set $\mathcal{L}_i^k$ of lamps illuminating given segment(s) S_i and/or connected to given cabinet(s) C_k. Their selection is also layout-specific.

5. The very important issue which has to be addressed in CRP compensation planning for lighting systems are failures of fixtures. They cause increasing of Q_{cap} in the numerator of Eq. (10) and thus can lead to significant growth of $\tan \varphi$ value and over-compensation. Basing on statistics and historical data,

a lighting system operator should put a buffer, in terms of λ values (see the line 10 of the algorithm), to protect the system against such events.

5   Conclusions

Thanks to the presented algorithm we are able to achieve the low-cost static compensation of capacitive reactive power generated in LED-based lighting systems. This approach is proposed as an alternative to dynamic VAR compensation being significantly more expensive. The price ratio between the dynamic and static method, for k circuits connected to a single cabinet is of the order of 10:k (the Polish market as of 2019). Note that it is an estimated value only, depending on multiple factors such as a control cabinet load or a power line type but also equipment order volume or a negotiated manufacturer's offer.

If fitting under the $\tan \varphi_0$ threshold is not feasible for the entire installation, e.g., due to the system dynamics, the control patterns, $\lambda_i(t)$ should be adjusted to meet this constraint. An alternative approach is finding a trade-off between costs related to the inductive reactive power generation and the savings obtained by improving energy efficiency of an installation.

The presented calculations are possible for lighting standard [7] compliant installations. It implies having all luminaire parameters, such as power and dimming levels for particular lighting classes calculated, and information about their geographical distribution and actual electric connections accessible.

References

1. Asian Development Bank: LED Street Lighting Best Practices (2017). https://goo.gl/i75Ztk. Accessed 21 Jan 2018
2. Australian/New Zealand Standard: AS/NZS 1158.1.1:2005 Lighting for roads and public spaces Vehicular traffic (Category V) lighting - Performance and design requirements. SAI Global Limited (2005)
3. Cheng, C.A., Chang, C.H., Cheng, H.L., Chang, M.T.: A novel LED driver with power factor correction suitable for streetlight applications. In: 2017 IEEE 3rd International Future Energy Electronics Conference and ECCE Asia (IFEEC 2017 - ECCE Asia), pp. 1220–1223, June 2017
4. Czarnecki, L.S.: Considerations on the reactive power in nonsinusoidal situations. IEEE Trans. Instrum. Meas. **IM-34**(3), 399–404 (1985). https://doi.org/10.1109/tim.1985.4315358
5. Dazzletek: Dazzletek Lighting Control System (2015). http://goo.gl/ZcJAvL. Accessed 3 Mar 2015
6. Ekrias, A., Eloholma, M., Halonen, L.: An advanced approach to road lighting design, measurements and calculations. Technical Report (53) (2008). http://goo.gl/3IVBnc
7. European Committee For Standarization: Road Lighting. Performance requirements, EN 13201-2:2015 (2015)
8. Federal Energy Regulatory Commission: Principles for Efficient and Reliable Reactive Power Supply and Consumption, Staff Report, Docket No. AD05-1-000, 4 February 2005

9. Gómez-Lorente, D., Rabaza, O., Estrella, A.E., Peña-García, A.: A new methodology for calculating roadway lighting design based on a multi-objective evolutionary algorithm. Expert Syst. Appl. **40**(6), 2156–2164 (2013). http://goo.gl/bx4cqn

10. Han, J.H., Lim, Y.C.: Design of an LLC resonant converter for driving multiple LED lights using current balancing of capacitor and transformer. Energies **8**(3), 2125–2144 (2015). http://www.mdpi.com/1996-1073/8/3/2125

11. Illuminating Engineering Society of North America (IESNA): American National Standard Practice For Roadway Lighting, RP-8-14. IESNA, New York (2014)

12. Kostic, M., Djokic, L.: Recommendations for energy efficient and visually acceptable street lighting. Energy **34**(10), 1565–1572 (2009). http://goo.gl/KQ64v6. 11th Conference on Process Integration, Modelling and Optimisation for Energy Saving and Pollution Reduction

13. Lee, E.S., Choi, B.H., Nguyen, D.T., Choi, B.G., Rim, C.T.: Long-lasting and highly efficient TRIAC dimming LED driver with a variable switched capacitor. J. Power Electron. **16**, 1268–1276 (2016)

14. de l'Eclairage, C.I.: Lighting of roads for motor and pedestrian traffic, CIE 115:2010. CIE, Vienna (2010)

15. Minister of Economy of Poland: Rozporządzenie Ministra Gospodarki z dnia 18 sierpnia 2011 r. w sprawie szczegółowych zasad kształtowania i kalkulacji taryf oraz rozliczeń w obrocie energią elektryczną (2011). http://prawo.sejm.gov.pl/isap.nsf/download.xsp/WDU20111891126/O/D20111126.pdf. Dz.U. 2011 nr 189 poz. 1126

16. Molina-Moreno, V., Leyva-Díaz, J.C., Sánchez-Molina, J., Peña García, A.: Proposal to foster sustainability through circular economy-based engineering: a profitable chain from waste management to tunnel lighting. Sustainability **9**(12) (2017). https://doi.org/10.3390/su9122229. http://www.mdpi.com/2071-1050/9/12/2229

17. Peña-García, A., Gil-Martín, L., Hernández-Montes, E.: Use of sunlight in road tunnels: an approach to the improvement of light-pipes' efficacy through heliostats. Tunn. Undergr. Space Technol. **60**, 135–140 (2016). https://doi.org/10.1016/j.tust.2016.08.008. http://www.sciencedirect.com/science/article/pii/S08867 79815302121

18. Philips: CityTouch (2015). http://goo.gl/GhCJsX. Accessed 2 Mar 2015

19. Salata, F., et al.: Energy optimization of road tunnel lighting systems. Sustainability **7**(7), 9664 (2015). https://doi.org/10.3390/su7079664. http://www.mdpi.com/2071-1050/7/7/9664

20. Sędziwy, A.: A new approach to street lighting design. LEUKOS **12**(3), 151–162 (2016)

21. Sędziwy, A., Basiura, A.: Energy reduction in roadway lighting achieved with novel design approach and LEDs. LEUKOS **14**(1), 45–51 (2018)

22. Schréder: Owlet (2015). http://goo.gl/9g53OP. Accessed 2 Mar 2015

23. Vieira, D., Shayani, R.A., de Oliveira, M.A.G.: Reactive power billing under non-sinusoidal conditions for low-voltage systems. IEEE Trans. Instrum. Meas. **66**(8), 2004–2011 (2017). https://doi.org/10.1109/tim.2017.2673058

24. Wojnicki, I., Kotulski, L.: Street lighting control, energy consumption optimization. In: Rutkowski, L., Korytkowski, M., Scherer, R., Tadeusiewicz, R., Zadeh, L.A., Zurada, J.M. (eds.) ICAISC 2017. LNCS (LNAI), vol. 10246, pp. 357–364. Springer, Cham (2017). https://doi.org/10.1007/978-3-319-59060-8_32

25. Worcester Energy: Street Lighting Retrofit Project. Progress Report (2017). https://goo.gl/pktcCp. Accessed 21 Jan 2018

Developing a Decision Support App
for Computational Agriculture

Andrew Lewis[1](✉), Marcus Randall[1,2], and Ben Stewart-Koster[3]

[1] School of Information and Communication Technology, Institute for Integrated
and Intelligent Systems, Griffith University, Nathan, QLD, Australia
{a.lewis,m.randall}@griffith.edu.au
[2] Bond Business School, Bond University, Robina, QLD, Australia
mrandall@bond.edu.au
[3] Australian Rivers Institute, Griffith University, Nathan, QLD, Australia
b.stewart-koster@griffith.edu.au

Abstract. In the age of climate change, increasing populations and
more limited resources, efficient agricultural production is being sought
by farmers across the world. In the case of smallholder farms with limited
capacity to cope with years of low production, this is even more important. To help to achieve this aim, data analytics and decision support
systems are being used to an ever greater extent. For rice/shrimp farmers in the Mekong Delta, Vietnam, trying to tune the conditions so that
both crops can be successfully grown simultaneously is an ongoing challenge. In this paper, the design and development of a smartphone app,
from a well researched Bayesian Belief Network, is described. This now
gives farmers the ability to make better informed planting and harvesting
decisions. The app has been initially well received by water management
practitioners and farmers alike.

Keywords: Bayesian Belief Networks · Decision support for farmers ·
App development

1 Introduction

Maintaining and increasing food production to feed growing populations in the
face of changing environmental conditions is a global problem. Changing patterns
of temperature and precipitation due to climate change, soil degradation from
over-production, inappropriate fertiliser use or crop rotation mismanagement,
and water quality reduction from nutrient inputs, all contribute to this problem [6,9]. The impact of climate change on agricultural production is already
being felt more severely in developing countries as emerging economies tend to
have a lower capacity to adapt [4]. Long-established farming practices may be ill-
equipped to sustain production in the face of significantly changing conditions,
so guidance and support in the adoption of innovative practices is increasingly
important.

© Springer Nature Switzerland AG 2020
V. V. Krzhizhanovskaya et al. (Eds.): ICCS 2020, LNCS 12138, pp. 551–561, 2020.
https://doi.org/10.1007/978-3-030-50417-5_41

In many parts of the developing world, farming is characterised by small-holder farming, partly for subsistence but also looking to cash-crop income [10]. Crop failure in these conditions can be catastrophic, with most farmers having little in the way of resources to buffer them from these consequences. Such farming has traditionally been guided by personal experience, in family and community. In the face of changing environmental conditions of increasing scale and severity, the use of data analytics and decision support technologies may be of increasing benefit, if they can be made readily available, relevant and usable [15].

In this paper, the application of a Bayesian Belief Network (BBN) to aquaculture - agriculture farming in the Mekong delta in Vietnam is presented. Extraction of knowledge from the BBN by a process of data mining is described, and the design of a decision support app for smartphones is shown.

The remainder of this paper is organised as follows. Section 2 describes Bayesian Belief Networks and their application to the case study. Section 3 describes the process of extracting knowledge from the BBN. Section 4 shows the design of the smartphone app and its operation. Finally, Sect. 5 concludes and describes future work.

2   Bayesian Belief Networks

The global environment, encompassing the natural environment and human interventions through agriculture, are governed by complex interrelationships. Often non-linear and probabilistic, there are many ways to attempt to understand interactions in these systems, one of which is Bayesian Belief Networks [7,11]. A BBN is a directed, acyclic graph, each node of which contains a conditional probability table that gives the probability of specified output states, given the input conditions from parent nodes. Considering the network as a whole, root parent nodes define the conditions of the environment, or scenario, for the model, while leaf nodes show the probability of a set of outcomes given the conditions of the parent nodes.

BBNs have been applied to better understand many ecological, environmental and agricultural systems. Of particular relevance to the work described in this paper, they have been used to elicit expert knowledge from farmers attempting to simultaneously harvest rice and shrimp from the same combined aquaculture-agriculture system [16].

2.1   Using a BBN for Rice/Shrimp Farming

In their paper, Stewart-Koster et al. [16] described a combined aquaculture-agriculture system in which farmers in the Mekong delta, in Vietnam, attempt to use the same fields to grow rice – on a central, raised platform – and shrimp in a surrounding ditch. While both are grown submerged in water, rice requires lower water salinity than shrimp which, of course, presents challenges for production of both crops, as attempted by farmers at some times during the year.

The BBN in Stewart-Koster et al. [16] was developed following several itera-tive stages of expert elicitation workshops with farmers and government exten-sion officers in the region held over a two year period. The participants in the workshops were those already involved in an ongoing research program into the sustainability of the farming process. The network was designed to establish the perceptions of farmers and extension officers of key risk factors to produc-tion and their understanding of how the system works. Consequently, the expert elicitation followed a process based initially on asking open ended questions to identify key processes that drive the system, followed by conditional ones that identified the causality among these different processes. Having used this app-roach to define and verify the structure of the network and the thresholds of the states of each node in the network, a series of workshops were held to ascertain the probabilities in the conditional probability tables (CPTs) in the network. With the exception of the rainfall nodes in the BBN, all CPTs were derived from the expert opinion of the farmers.

Defining CPTs in a BBN is an important process and when based on expert elicitation it requires careful approaches to ensure the resulting probabilities accurately reflect the perceptions of the experts involved [13]. Challenges include ensuring dominant personalities do not prevent quieter experts from voicing their opinion, avoiding biases from the subjectivity of expert knowledge, which may manifest as subjective biases due to the experts' personalities or cognitive biases due to not understanding the requirements of the process [2,8]. To overcome these challenges, Stewart-Koster et al. [16] used three different approaches to expert elicitation of the CPTs, a survey method, direct elicitation of probabilities and indirect elicitation approaches. Having derived the CPTs for the relevant nodes in the network, a final round of verification was conducted to ensure the network reflected the knowledge and experience of the farmers. This ultimately results in a BBN that represents the farmers' expectations of the probability of rice and shrimp crop failure under certain conditions. This is essentially an *expert system* that encapsulates the collective understanding of the conditions necessary to achieve desired outcomes (minimising risk of crop failure for both crops) given varying conditions, such as timing and volume of monsoonal rainfall, quality of the shrimp stock, and perceived soil and water quality at various points during the growing season. For reference, the final BBN from Stewart-Koster et al. [16], upon which we base our decision support system, is reproduced in Fig. 1.

The cultural circumstances of this modelling exercise should be noted. While some of the conditions imposed could be quantified, e.g., shrimp stocking density could be specified in larvae applied per square metre, others are very subjective. Farmers expressed conditions, e.g., water and soil salinity, in subjective, but generally collectively understood terms. In the field, they do not have access to sophisticated measurements: water salinity is determined by whether the water *tastes* salty, brackish or fresh. While the collected data cannot be calibrated in objective terms, e.g., water pH or soil electrical conductivity, the relative measures and consequent outcomes can be understood by the end users, the farmers.

The work described in this paper consists of two major parts: extracting the knowledge from the BBN, and presenting it in a form suitable for easy use by the farmers.

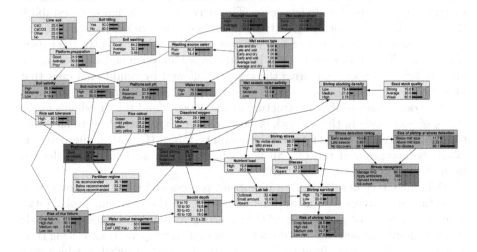

Fig. 1. The BBN as used by Stewart-Koster et al. [16], and as a basis for this paper.

3   Data Mining the Expert System

From Fig. 1 it may be seen that the BBN has two leaf nodes, delivering the conditional probability of success or failure of the rice and shrimp crops. While these nodes provide the probability of several outcomes expressed in terms of "Crop failure" through to "Low risk", it was decided to simplify the outcome to the probability of a successful crop = 1 - probability of crop failure. While this removes some of the nuances of the impact of actions taken, it considerably simplifies the presentation of the data and outcomes.

By systematically varying input conditions it is possible to discover the conditions that minimise the probabilities of crop failures – discovering relationships by inspection of data. In this regard, this is essentially an exercise in data mining [5].

Further inspecting Fig. 1 a number of root nodes can be seen. (Note: these may not be displayed at the top of the figure, but can be distinguished by their lack of any incoming links from parent nodes.) Two, coloured darker blue, specify the governing climatic conditions for the season – when, and how heavy was the monsoonal rainfall. Together with the "Soil nutrient load", they set the *scenario* farmers experience prior to planting, at which time they can make a number of decisions about "pre-planting" actions.

At "planting" time, farmers can make a number of estimations of environmental conditions, including such things as soil salinity, water temperature etc. These are shown in bright green in Fig. 1. At this stage in the season such nodes "cut out" the direct effect of their parent nodes, setting the scenario for the planting season. Farmers have a more limited set of decisions now, e.g., it is too late to decide to till the soil.

In the "post-planting" period, the set of cut-out nodes could be further lowered. However, to do so would require overly subjective judgements to be made in the field, such as whether wet season water quality was "good for shrimp". By keeping the cut-out nodes from the previous period, based on simple observations and assessments, and adding the actions already taken, the post-planting scenario is defined, and becomes highly constrained. Farmers' actions are now limited to assessing rice colour (as part of the scenario), which will determine the level of fertiliser applied, and choosing water colour management actions.

The insight leading to separate consideration of different stages in the growing season simplifies interpretation of the data. At the pre-planting stage the 1728 different states of the model are reduced to 96 alternative strategies by dealing with just one of the 18 environmental scenarios at a time. It might be possible to provide this model to end users so they can interrogate the data directly and find the results of different actions and strategies. However, this lacks guidance to lead users to optimal outcomes. A simple form of optimisation was employed. Knowledge was extracted from the BBN by repeatedly running the model with scenario input conditions fixed, and the possible actions varied across all their possibilities. The corresponding conditional probability of crop failures was recorded and graphed against the input variables. As an example, the output for the "pre-planting" analysis is shown in Fig. 2.

The data was reordered so probability of shrimp crop failure was monotonically increasing. (Shown as the solid line in the top 'trace' in Fig. 2.) This allows the graph to be inspected easily to determine the point at which the probability of crop failures of rice and shrimp were at simultaneous minima, designated the "sweet spot". (Shown as the small circles on the top trace.) The plot was annotated with the scenario conditions and the actions corresponding to the sweet spot. This "inspection" was subsequently automated using an R script [17] during data preparation.

Extracting optimal farming practices directly from the BBN can be complex, cumbersome and confusing. Despite the simplification delivered by automatic extraction of data from the BBN and display via graphs, extracting simple advice is still not straightforward. In order to deliver a tool that was easy for the end-users, the Mekong delta farmers, to use and interpret, it was decided to develop a smartphone app.

4   The Decision Support App

Decision support apps for agriculture are growing in popularity. While decision support *systems* for agriculture are not new – prototype systems can be found in the literature from the late 1970s [14] – delivery to the end-user has been problematic. As decision support systems should be evaluated on their contribution to improved outcomes [3] getting easy-to-use tools with clear, understandable advice to the end-user is critical. Smartphones, provided they are readily

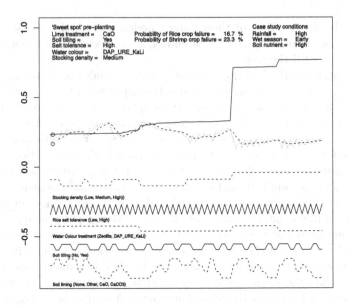

Fig. 2. Sample data extracted from the BBN

available, offer the potential of wide penetration with minimal marginal cost. Thus the proliferation of smartphones in developing countries presents a rapidly increasing but largely untapped resource. In Vietnam in 2017 about 60% of the rural population, some 22.5 million people were smartphone users (as measured by Facebook user data [19].) While there are increasing numbers of smartphone users, their usage is largely confined to social networking, messaging, music and entertainment. Only about 25% use productivity tools on a daily basis [1].

The majority of these users have Android smartphones, so a decision support app was developed for Android by converting the CSV (comma separated values) data extracted from the BBN to JSON, and the R data extraction and comparison operations were replicated in React Native [12]. Development only proceeded after usability workshops based on conceptual wireframe diagrams were conducted with end-users from the region around Ca Mau in Vietnam.

Typical end-users for the app, mainly farmers who have an average age of about 49 with limited formal education [18], dictated labels and text should be brief and straightforward. Wherever possible, terminology familiar to the users should be used, fonts should be large and clear, and buttons and other interaction elements should also be easy to see and use. The overall layout and operation of the app aimed to be intuitive with the ability to select a clearly defined stage in the growing season, then describe it in simple terms drawn from the actual experience of farmers and allow for subjective descriptors that do not require any instruments to define. Outputs use visual aids to enhance comprehension, with colour coding of bar charts to match crops (rice or shrimp). Recommended actions are described in similarly clear terms. The whole app has been developed in English and Vietnamese language versions. Screenshots of the opening page (after the "splash screen") are shown in Fig. 3.

Since the data and knowledge incorporated in the app are specific to a particular region, the Mekong delta in Vietnam, it has also been considered important to warn prospective users of this localised nature of the advice provided. A test has been included in the app, using GPS location data from the mobile device, which triggers a disclaimer if it detects the app is being used too far from its intended target area. This warning can be seen in Fig. 4a. (Note: the warning is posted immediately the app starts, and is only available at present in Vietnamese.)

From the opening page, an appropriate period (pre-planting, planting or post-planting) may be chosen. On the following screen (see, for example, Fig. 4b) scenario conditions can be entered in the area above the "START" button.

Once the START button is pressed, corresponding probabilities of successful crop outcomes appear below (see, for example, Fig. 5a) and the user can scroll down to read the actions required to achieve these outcomes (see Fig. 5b).

Should a user wish, they can alter the recommended actions to a preferred alternative (see Fig. 6a, in which a user chooses a High shrimp stocking density, rather than the recommended Medium density.) This will then display the corresponding change in crop outcome probabilities resulting from the change (see Fig. 6b, showing the probability of shrimp crop failure, due to over-stocking.) The optimal action list can be restored by pressing the START button again.

In-app instruction is provided, with a step-by-step guide and illustration of how data is input. The app is packaged and delivered as an Android Package Kit (APK). When installed and run, at each startup the app checks for access to a server on the Internet, and availability of any software updates. By this means, the app in the field can be provided with revisions and new functionality automatically. However, access to the server is not necessary for the app to function.

While yet to be released, in conjunction with a second-phase app to be based on scientific studies and field trials of innovative farming practices, the app has been shown in a pre-release preview to water management practitioners at an Asian Development Bank Water Learning Week workshop and to members of the International Water Centre. It was very well received and complimented by representatives and researchers from government ministries and NGOs, from Australia, Cambodia, China, France, Georgia, India, Indonesia, Japan, Korea, the Marshall Islands, Mongolia, the Philippines, Sri Lanka, Uzbekistan, Vanuatu and Vietnam. In addition, it has been demonstrated to a large group of farmers from the Mekong delta and collaborating scientists. Responses were uniformly positive and the release of the app is keenly awaited. Members of the International Water Centre have since invited presentation of the app to policy officers at the Vietnamese Ministry of Agricultural and Rural Development.

(a) Vietnamese language version   (b) English language version

Fig. 3. The decision support app

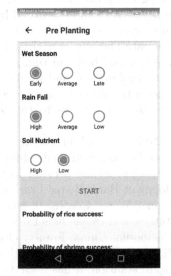

(a) Location warning   (b) Selecting conditions - Pre-Planting

Fig. 4. Running the decision support app

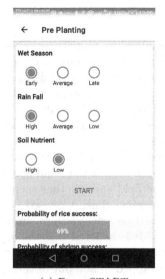
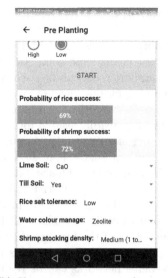

(a) Press START

(b) Showing outcomes and recommended actions

Fig. 5. Running the decision support app (cont.)

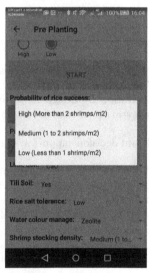
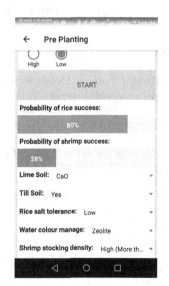

(a) Selecting a different action

(b) Displaying the consequences

Fig. 6. Running the decision support app (cont.)

5 Conclusions

Efficient farming practices, particularly for smallholder farmers, is important in a time of climate change and competition for resources. Improved data analytic and decision support tools can be used by people in the Mekong delta region to help with planning for optimising rice and shrimp crops. The work in this paper has shown that the results of complex modelling, scientific trials and research can be delivered in readily usable form to farmers and regional authorities, to assist in planning and day-to-day farming practice.

Given the work presented here, it is planned that extensive user evaluation of the app will be undertaken. This will include both a qualitative and quantitative analysis to determine how effective it is in the field and whether users require modifications or additions to the app. At the same time, a revised version of the BBN is being developed. This will be based on scientific principles rather than just the beliefs of the farmers themselves. It is believed that the app itself will only require small changes, as it is produced in such a way that outcomes derived from belief networks can be readily swapped in and out of the app.

Acknowledgements. The authors would like to thank Dang Nhat Thanh Tran, YoungJae Choi, Vinh Dat Lam, and YoungSeo Shim for their assistance with development and coding of the smartphone app, and Vietnamese language localisation. This work was part-funded by an Australian Centre for International Agricultural Research (ACIAR) project grant (SMCN/2010/083).

References

1. Appota Group: Vietnam Mobile Report 2017 (2017). https://www.slideshare.net/appota/vietnam-mobile-report-2017-75272740. Accessed 20 July 2018
2. Choy, S.L., O'Leary, R., Mengersen, K.: Elicitation by design in ecology: using expert opinion to inform priors for bayesian statistical models. Ecology **90**(1), 265–277 (2009)
3. Cox, P.: Some issues in the design of agricultural decision support systems. Agric. Syst. **52**(2–3), 355–381 (1996)
4. Fischer, G., Shah, M., Tubiello, F.N., Van Velhuizen, H.: Socio-economic and climate change impacts on agriculture: an integrated assessment, 1990–2080. Philos. Trans. Roy. Soc. B: Biol. Sci. **360**(1463), 2067–2083 (2005)
5. Grinstein, U.M.F.G.G., Wierse, A.: Information Visualization in Data Mining and Knowledge Discovery. Morgan Kaufmann, Burlington (2002)
6. Howden, S.M., Soussana, J.F., Tubiello, F.N., Chhetri, N., Dunlop, M., Meinke, H.: Adapting agriculture to climate change. Proc. Natl. Acad. Sci. **104**(50), 19691–19696 (2007)
7. Korb, K.B., Nicholson, A.E.: Bayesian Artificial Intelligence. CRC Press, Boca Raton (2010)
8. Kuhnert, P.M., Martin, T.G., Griffiths, S.P.: A guide to eliciting and using expert knowledge in Bayesian ecological models. Ecol. Lett. **13**(7), 900–914 (2010)
9. McGuire, J., Morton, L.W., Cast, A.D.: Reconstructing the good farmer identity: shifts in farmer identities and farm management practices to improve water quality. Agric. Hum. Values **30**(1), 57–69 (2013)

10. Morton, J.F.: The impact of climate change on smallholder and subsistence agriculture. Proc. Natl. Acad. Sci. **104**(50), 19680–19685 (2007)
11. Pearl, J.: Probabilistic reasoning in intelligent systems: networks of plausible inference. Elsevier (2014)
12. React Native: A framework for building native apps using React. https://facebook. github.io/react-native/. Accessed 20 July 2018
13. Renooij, S.: Probability elicitation for belief networks: issues to consider. Knowl. Eng. Rev. **16**(3), 255–269 (2001)
14. Room, P.: A prototype 'on-line' system for management of cotton pests in the Namoi Valley New South Wales. Protect. Ecol. **1**, 245–264 (1979)
15. Rose, D.C., et al.: Decision support tools for agriculture: towards effective design and delivery. Agric. Syst. **149**, 165–174 (2016)
16. Stewart-Koster, B., et al.: Expert based model building to quantify risk factors in a combined aquaculture-agriculture system. Agric. Syst. **157**, 230–240 (2017)
17. Team, R.C., et al.: R: a language and environment for statistical computing (2013)
18. Truc, N., Sumalde, Z., Espaldon, M., Pacardo, E., Rapera, C., Palis, F.: Farmers' awareness and factors affecting adoption of rapid composting in Mekong Delta, Vietnam and Central Luzon, Philippines. J. Environ. Sci. Manag. **15**(2) (2012)
19. Viet Nam News: Smartphone users cover 84% of VN population (2017). http:// vietnamnews.vn/economy/418482/smartphone-users-cover-84-of-vn-population. html#QoxcCAkRmUSjmoWk.99 [Source: Nielsen Vietnam Smartphone Insights Report 2017]. Accessed 20 July 2018

Optimal Location of Sensors for Early Detection of Tsunami Waves

Angelie R. Ferrolino[✉], Jose Ernie C. Lope, and Renier G. Mendoza

Institute of Mathematics, University of the Philippines Diliman,
Quezon City, Philippines
arferrolino@up.edu.ph, {ernie,rmendoza}@math.upd.edu.ph

Abstract. Tsunami early detection systems are of great importance as they provide time to prepare for a tsunami and mitigate its impact. In this paper, we propose a method to determine the optimal location of a given number of sensors to report a tsunami as early as possible. The rainfall optimization algorithm, a population-based algorithm, was used to solve the resulting optimization problem. Computation of wave travel times was done by illustrating the kinematics of a wave front using a linear approximation of the shallow water equations.

Keywords: Shallow water equations · Sensor location · Tsunami early warnings · Rainfall optimization algorithm

1 Introduction

Tsunamis are considered as one of the most powerful and destructive natural disasters. They are a series of waves caused by a rapid and massive displacement of the seafloor or disruption of standing water. Earthquakes, volcanic eruptions, underwater landslides and meteor impacts all have the potential of generating a tsunami. However, the most common cause is undersea earthquakes, generated in subduction zones. The water above such event is disturbed significantly by the uplifting, that it create waves travelling at around 500 miles per hour. Moreover, their wavelength is much longer than normal sea waves, so they build up to higher heights as the depth of water decreases. With the large volume of water moving at high speed, tsunamis can cause massive destruction to the physical environment and loss of human lives.

The Philippines is one of the most disaster-prone countries due to its geographic location. As such, it has a high risk of exposure to intense tropical cyclones, seismic activities and tsunamis [23]. Up to date, a total of 41 tsunamis occurred since 1589 [5]. Thus, compared to other countries, occurrence of tsunami in the Philippines is more than average, but still moderate. However, as stated in [13], tsunami research in the Philippines received less than its deserved attention. Moreover, lack of awareness and sufficient preparation may further increase the hazards. So, developing techniques to assess risks and to mitigate impacts of a tsunami are of great importance. There are a number of methods focused

© Springer Nature Switzerland AG 2020
V. V. Krzhizhanovskaya et al. (Eds.): ICCS 2020, LNCS 12138, pp. 562–575, 2020.
https://doi.org/10.1007/978-3-030-50417-5_42

in reduction of tsunami hazards and risk management in literature [9,10,19]. In this study, we aim to provide early tsunami warnings by determining the optimal location of tsunami sensors that can guarantee minimal possible tsunami detection time.

The traditional instrument used for detecting a tsunami is called a tide gauge, which is usually located near the coast. It measures changes in the sea level relative to some height reference. While the tide gauge is capable of detecting a tsunami, they are at a worst location, since this is where the tsunami is most energetic. Moreover, as they are near the coast, it may take a lot of time for tsunami detection, leaving only a small amount of time for evacuation. Hence, we consider using deep-ocean sensors, which are capable of reporting a tsunami in the open ocean. Thus, this technology provides a relatively secure and rapid detection of tsunami, but at a cost. Details regarding these tsunami observing systems were well documented and reviewed in [17,21]. In particular, we will consider the bottom pressure recorders (BPRs) as our tsunami sensor. These sensors measure changes in water pressure and seismic activity. They use an acoustic link to transmit data on surface buoys, which are then relayed via satellite to ground stations [8]. For optimal sensor placement, we assume that the tsunami originates from a fixed location, e.g., subduction zones. Rogue waves [7] are not considered in this work because of the uncertainty of their origin.

Studies regarding selection of location of tsunami sensors for tsunami warnings are limited. Some were based on expert judgments while considering various deciding factors like financial limitations [3], and legal aspects such as geographical boundaries or legal jurisdictions [1]. There are also researches that incorporated accurate estimation of tsunami parameters [15,16]. An attempt to encompass tsunami warning efficacy has been proposed in [18,22]. They identified location of tsunami sensors based on several criteria (e.g. installation conditions), though no optimization algorithm was applied. Bradock et al. [6] considered six potential buoy sites, and they determined the optimal location of a minimum number of BPRs that will maximize the population being warned. However, they considered a fixed average speed of wave travel, and thus, bathymetric data cannot be taken into account. Astrakova et al. [4] developed an optimal sensor location problem, which guarantees minimal possible time of tsunami detection. However, the method used in computing the wave travel time is very slow, and takes too much amount of computer memory.

The objective of this paper is to improve the optimal sensor location problem introduced by Astrakova. We illustrate kinematics of a wavefront, using an approximation of wave velocity from the linear shallow water equations, which results to a faster calculation of wave travel time and low computer cost. Finally, we test our method to a simple problem, then apply it to a real world problem of optimal sensor location in one of the areas in the Philippines, near a subduction zone.

To solve the resulting optimization problem, we use a population-based algorithm called the Rainfall Optimization Algorithm (RFO) introduced by Kaboli et al. [11]. RFO is a nature-inspired algorithm which is modelled on raindrop flow over a mountainous surface and has been shown to be effective, fast and capable of solving various optimization problems. It has been tested on benchmark functions against the genetic algorithm (GA), the particle swarm optimization algorithm (PSO) and the group search optimizer (GSO), and it outperforms them on most functions, as it provides better solutions and faster convergence.

The rest of the paper is organized as follows: in the next section, we discuss the methods used in solving the optimal sensor location problem. Results of the optimal location of tsunami sensors in different domains and bathymetries are presented in the third section. Finally, we summarize our results and provide recommendations for future work.

2   Methodology

In this section, we discuss the optimization problem, the method for computing the wave travel time, and details of the rainfall optimization algorithm.

2.1   The Optimization Problem

Let Ω be the domain with parts of water, $\mathbf{D} \in \Omega$ be a part of the domain where tsunami sensors can be placed, and $\mathbf{P}$ be the subduction zone. For an illustration, please refer to Fig. 1. The problem of interest is to find the location of L sensors such that any seismic event on $\mathbf{P}$ will be detected after minimal possible time. Let $\{p_j\}_{j=1}^{P}$ denote the points in the subduction zone, where $p_j = (x_j, y_j) \in \mathbf{P}$, and $q_i = (x_i, y_i) \in \mathbf{D}\,(i = 1, \ldots, L)$ be the coordinates of the L sensors. Also, we denote by $\mathbf{Q} = \{q_1, \ldots, q_L\}$ the set of L sensors representing a possible solution to the optimization problem.

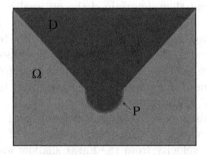

Fig. 1. Illustration of a domain

Suppose that a disturbance arises on a source $p_j \in \mathbf{P}$. This disturbance will propagate over water at a certain speed. We are interested in finding the minimal time it takes for the disturbance on the source p_j to arrive at some water point

$x \in \mathbf{D}$. Let γ be one of the ways connecting p_j and x, and τ_γ be its travel time. We denote $\tau(p_j, x)$ be the minimal time to approach x from p_j, i.e.,

$$\tau(p_j, x) = \min_{\gamma} \tau_\gamma \qquad (1)$$

The time required for determination of tsunami wave by $\mathbf{Q}$ is given by

$$t(p_j, \mathbf{Q}) = \min_{1 \le i \le L} \tau(p_j, q_i).$$

For guaranteed time registration for any point $p_j \in \mathbf{P}$ by $\mathbf{Q}$, we set

$$T(\mathbf{Q}) = \max_{1 \le j \le P} t(p_j, \mathbf{Q}).$$

Thus the optimization problem is stated as: Find $\mathbf{Q} = \{q_1, \ldots, q_L\}$ which gives the minimal value of the function $T(\mathbf{Q})$, i.e.,

$$\min T(\mathbf{Q}), \qquad (2)$$

where the number L is given and subject to the phase restriction

$$\mathbf{Q} \in \mathbf{D}.$$

This minimization problem is based on [4].

2.2   Kinematics of a Wave Front

We discuss here how we can illustrate the kinematics of a wave front, which we will use to calculate the wave travel time in (1). We consider the linear approximation of the shallow water equations [20]. In this case, the wave velocity is proportional to the square root of the water depth h:

$$v \approx \sqrt{gh}, \quad h \ge 0,$$

where $g = 9.8$ m/s^2 is the gravitational constant. Also, we note that for any propagation velocity distribution in a medium, all the points on a wave front are moving in the orthogonal direction to the frontal line [14]. This will be the basis for the numerical method of the step-by-step wave frontal line advancement that will now be described.

Consider a closed and convex curve (e.g., a circle) as the initial wave front. This wave front is represented by a limited number of computational points (x_i, y_i) $(i = 1, \ldots, N)$ along the curve. Figure 2(a) presents an example of an initial wave front. Our next aim is to calculate the moving direction for all wave front points to establish their next positions. In this method, the moving direction of the point (x_i, y_i) is determined by the outer-normal of the circle passing through three computational points: (x_{i-1}, y_{i-1}), (x_i, y_i) and (x_{i+1}, y_{i+1}). Moving the point (x_i, y_i) in this direction with a distance of $v \cdot \triangle t$, where $\triangle t$ is the time step, will give us the next position of (x_i, y_i). Hence, we can compute

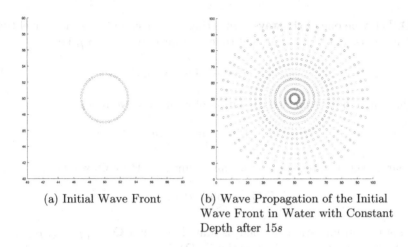

(a) Initial Wave Front   (b) Wave Propagation of the Initial
                                Wave Front in Water with Constant
                                Depth after 15s

Fig. 2. Kinematics of a wave front

the location of all wave front points at the time instant $t = \triangle t$. We repeat this process until we reach the boundaries of the domain. Figure 2(b) demonstrates the wave propagation of the initial wave front in water with constant depth after 15 s.

Let (x, y) be the coordinates of a computational point and $z = f(x, y)$ be be the time it takes to for the wave to reach the point (x, y). To determine the travel time, to an arbitrary point $(\tilde{x}, \tilde{y})$ (not necessarily a computational point) in the domain, we can compute the travel time z of the three nearest computational points, say (x_1, y_1), (x_2, y_2) and (x_3, y_3), to $(\tilde{x}, \tilde{y})$. Using (x_1, y_1), (x_2, y_2), (x_3, y_3), $f(x_1, y_1)$, $f(x_2, y_2)$ and $f(x_3, y_3)$, we apply linear interpolation to estimate the travel time to $(\tilde{x}, \tilde{y})$.

2.3 The Rainfall Optimization Algorithm

In order to solve the optimization problem in (2), we will use a new nature-inspired algorithm called the rainfall optimization algorithm (RFO) [11]. Many optimization problems have been solved using RFO as it was proven to be fast, effective and efficient. Some applications are in solving problems in facility-location [2], economic dispatch [11], and scheduling [12].

RFO is a population-based algorithm which is based on the behavior of rain-drops flowing over a mountainous surface. As always, raindrops tend to fall on surfaces with a steeper slope. RFO utilizes this tendency, allowing determination of a solution superior to a guess. Raindrops may also be stuck in puddles (local optimum). However, as raindrops accumulates, it may overflow, allowing drops of water to flow downwards again. RFO simulates this behavior, enabling the algorithm to overcome local optimal solutions and reach the global optimum.

Before presenting the algorithm, we first discuss some important terms. Define

$$D^i := x_{i,1}, x_{i,2}, \ldots, x_{i,n} \quad (i = 1, 2, \ldots, m)$$

be the raindrop i in a population, where n is the number of variables of optimization variables, m is the number of raindrops in a population and $x_{i,j}$ is the variable of interest in the optimization problem. The generation of raindrops are uniformly randomly distributed and must be within the bounds, if there is any.

Each raindrop randomly produces neighbor points in a neighborhood with a radius vector r. We denote the neighbor points as NP_k^i that satisfies

$$\|(D^i - NP_k^i) \cdot u_j\| \leq \|r \cdot u_j\|, \ i \in \{1, \ldots, m\}, \ j \in \{1, \ldots, n\}, \ k \in \{1, \ldots, np\},$$

where u_j is the unit vector in the jth dimension. Among all the neighbor points, we are interested in finding a dominant neighbor point, i.e., a neighbor point whose cost function value is less than the cost function values of the raindrop and the other neighbor points. We call a drop **active** if it has a dominant neighbor point. Otherwise, the drop is **inactive**.

Algorithm 1. Rainfall Optimization Algorithm

Input: *npop*: population size, *np*: number of neighbor points on each drop, *r*: radius vector, *Ne*: maximum number of explosion process, *eb*: explosion base, *ec*: explosion counter, *maxiter*: maximum number of iterations

Output: Location of raindrop with minimal cost function value

1. **Initialization:** Generate randomly the first population of raindrops of size *npop* such that each raindrop satisfy the constraints. Set *iter* = 1. Set to **active** all the raindrops' status.

2. **Iterative Procedure:**
   while *iter* ≤ *maxiter* and active set of raindrops ≠ ∅ **do**
   Do the following for each active raindrop:

   – Generate *np* neighbor points.
   – Obtain the cost function values of drops and their neighbor points.
   – If there is a dominant neighbor point, then change the drop's present position to that point. Otherwise, apply explosion process to the drop.
   – If there is no dominant neighbor point after *Ne* times of explosion, set the drop's status to **inactive**.

   Create a merit-order list and remove specific numbers of low-ranking drops or assign a higher *Ne* to high-ranking drops.
   Set *iter* = *iter* + 1.
   end while

3. **Generation of Minimizer:** Calculate the cost function values for all raindrops. Find the raindrop with the minimum cost function.

If a raindrop has no dominant neighbor point, then it may have already converged the global optimum, or it is stuck in a local optimum (due to insufficient number of neighbor points). To prevent the latter, an explosion process is made. In this process, the raindrop produces new np_{ex} neighbor points where

$$np_{ex} = np \times eb \times ec$$

with eb as the explosion base (indicating the explosion range) and ec as the explosion counter. If the drop still has no dominant neighbor point after doing the explosion process Ne times, we make it inactive.

At the end of the algorithm, we create a merit-order list, which contains the rank of the drops in ascending order. Lower ranking drops may be removed from the population, or higher ranking drops may be given special rights (such as higher number of explosion process). The calculation of rank is given by: $rank_t^i = \frac{1}{2}\text{order}(C1_t^i) + \frac{1}{2}\text{order}(C2_t^i)$, where

$$(C1_t^i) = F(D^i)\big|_{\text{at the } t^{\text{th}} \text{ iteration}} - F(D^i)\big|_{\text{at the } 1^{\text{st}} \text{ iteration}}$$

and

$$(C2_t^i) = F(D^i)\big|_{\text{at the } t^{\text{th}} \text{ iteration}}.$$

Here, F represents the objective function. The step-by-step procedure of the RFO is summarized in Algorithm 1.

3   Results

We first consider a domain with a semicircle subduction zone (in red) as shown in Fig. 3.

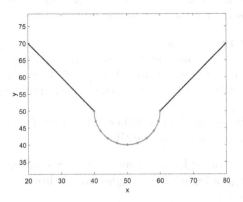

Fig. 3. Domain with a semicircle subduction zone (Color figure online)

Assume that we have $L = 2$ sensors, say (x_1, y_1) and (x_2, y_2). We set x_2 and y_1 to 55 and 45 respectively, while (x_1, y_2) are in the domain $[35, 65] \times [40, 70]$. The plot of time versus x_1 and y_2, corresponding to the semicircle subduction zone, is presented in Fig. 4. From the surface plot, it can be seen how the minimization problem is neither convex nor differentiable. Thus, gradient-based and local search methods might not work. This justifies why a population-based algorithm is an appropriate numerical optimization method to estimate the optimal solution.

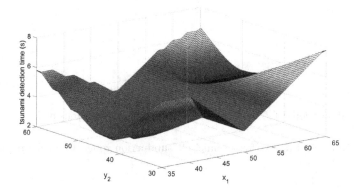

Fig. 4. Surface plot of tsunami detection time versus the coordinates x_1 and y_2

Now, we present some numerical simulations to identify the optimal locations of sensors using various domain profiles. Consider the same domain presented in Fig. 3. We let **D** be the whole rectangular domain with constant water depth. Figures 5(a)–5(d) show the optimal location(s) of $L = 1, 2, 3$ and 4 sensor(s), represented by black dots. Here, the red dots in **P** are the source points. These results were obtained by taking the average of the locations gathered in 10 independent runs. For the case when $L = 1$, it can be seen that the obtained location is situated at the center of the semicircle. This makes sense geometrically since we wish to minimize the guaranteed time registration from all the source points (red dots) located on the semicircle. One can also observe that as the number of sensors increases, the estimated locations get closer to the subduction zone.

We now study how the number of sensors will affect travel time. Figure 6 shows the plot of the time of tsunami detection versus the number of sensors. We can see here that as the number of sensors increase, the time decreases. Moreover, we can see from Table 1 that there is a significant improvement when we increase the number of sensors from 3 to 4 and little improvement from 4 to 5. Hence, $L = 4$ is a good number of sensors that will give us good tsunami detection time, without the extra cost of additional sensors.

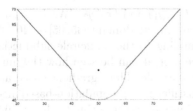

(a) Optimal location of $L = 1$ sensor.

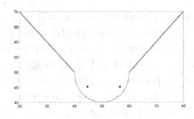

(b) Optimal locations of $L = 2$ sensors.

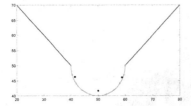

(c) Optimal locations of $L = 3$ sensors.

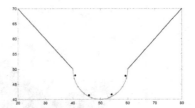

(d) Optimal locations of $L = 4$ sensors.

Fig. 5. Numerical results for a semicircle subduction zone. (Color figure online)

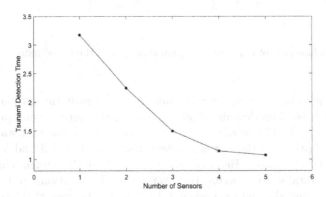

Fig. 6. Effect of increasing the number of sensors on the tsunami detection time.

Table 1. Decrease in tsunami detection time as the number of sensors is increased.

Number of sensors (n)	Tsunami detection time	Decrease in time from $n-1$ to n sensors
1	3.1756	–
2	2.2465	0.9291
3	1.4940	0.7525
4	1.1449	0.3491
5	1.0731	0.0731

Next, we apply our method to a real-world problem of sensor location in the Cotabato Trench. The Cotabato trench is an oceanic trench in the Pacific Ocean, located off the southwestern coast of Mindanao in the Philippines. This trench is one of the main structures around the Philippines likely to be associated with tsunamigenic earthquakes. One example is the tsunami generated by the 1976 Moro Gulf earthquake, which is considered as one of the most devastating disasters in the history of the Philippine islands [13]. Figure 7(a) shows a portion of the Cotabato Trench. We let the subduction zone **P** to be the red line, and **D** be the water surface above the subduction zone. The corresponding bathymetric profile of this trench is shown in Fig. 7(b).

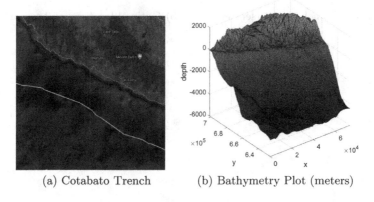

(a) Cotabato Trench   (b) Bathymetry Plot (meters)

Fig. 7. Profile of the Cotabato Trench. (Color figure online)

Figures 8(a), 8(b), 8(c), 8(d), 8(e) and 8(f) present the optimal location of $L = 1, 2, 3, 4, 5$ and 6 sensor/s (blue dots), respectively. These results were obtained by taking the average of the locations gathered in 10 independent runs. The plot of the time of tsunami detection versus the number of sensors is shown in Fig. 9. The values of time in dependence to the number of sensors is presented in Table 2. Similar to what we did earlier, we may see from here that $L = 5$ is a good number of sensors for this problem.

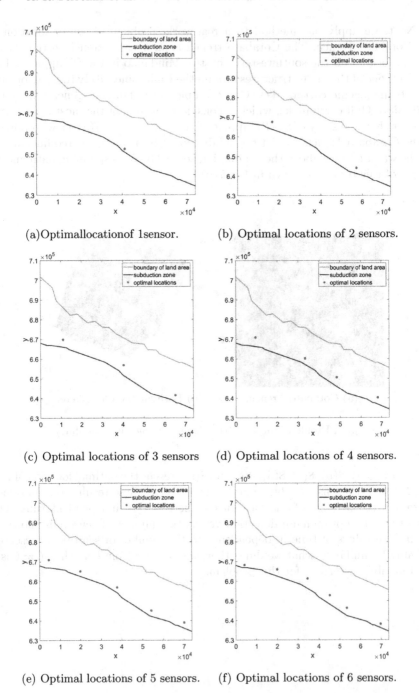

(a) Optimal allocation of 1 sensor.

(b) Optimal locations of 2 sensors.

(c) Optimal locations of 3 sensors

(d) Optimal locations of 4 sensors.

(e) Optimal locations of 5 sensors.

(f) Optimal locations of 6 sensors.

Fig. 8. Numerical results for a portion of the Cotabato Trench. (Color figure online)

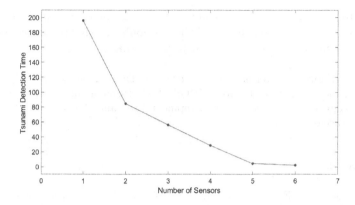

Fig. 9. Number of sensors versus time in the case of Cotabato Trench

Table 2. Effect of increasing the number of sensors on the tsunami detection time in the case of the Cotabato Trench

Number of sensors (n)	Tsunami detection time	Decrease in time from $n-1$ to n sensors
1	195.9816 s	–
2	84.5089 s	111.4727
3	56.0995 s	28.4094
4	28.6910 s	27.4085
5	4.5239 s	24.1671
6	2.4007 s	2.1232

4   Conclusion and Recommendations

We considered the problem of optimal tsunami sensors placement for early tsunami warnings. The computation of wave travel times were done by producing the kinematics of a wave front using an approximation of wave velocity derived from the linear shallow water equations. The Rainfall Optimization Algorithm was used to solve the optimization problem. We first applied our model to a simple problem having a semicircle subduction zone for testing. The obtained results for this test problem are geometrically sensible. Then we applied our method to a real-world problem of optimal sensors placement in the Cotabato Trench. We use the actual bathymetric profile of this trench to make the estimation of the wave travel time more accurate. One can observe that as the number of sensors increases, the detection time decreases. Moreover, the sensors become situated closer to the subduction zones. However, more sensors entail additional cost. Future works include setting cost and location constraints for the sensors.

We note that our method relies on an approximation of the wave velocity, but one can use the numerical solution of the 2D nonlinear shallow water equations for a more accurate computation of wave travel time.

Acknowledgment. The authors acknowledge the Department of Science and Technology – Science Education Institute (DOST-SEI) through the Accelerated Science and Technology Human Resources Development Program (ASTHRDP) Scholarship, for funding support.

References

1. Abe, I., Iwamura, F.: Problems and effects of a tsunami inundation forecast system during the 2011 Tohoku earthquake. J. Japan Soc. Civil Eng. **1**, 516–520 (2013)
2. Akbari-Jafarabadi, M., Tavakkoli-Moghaddam, R., Mahmoodjanloo, M., Rahimi, Y.: A tri-level r-interdiction median model for a facility location problem under imminent attack. Comput. Ind. Eng. **114**, 151–165 (2017)
3. Araki, E., Kawaguchi, K., Kaneko, S., Kaneda, Y.: Design of deep ocean submarine cable observation network for earthquakes and tsunamis. In: OCEANS 2008 - MTS/IEEE Kobe Techno-Ocean, pp. 1–4 (2008)
4. Astrakova, A., Bannikov, D., Cherny, S.G., Lavrentiev, M.M.: The determination of optimal sensors' location using genetic algorithm. In: Proceedings of 3rd Nordic EMW Summer School, vol. 53, pp. 5–22 (2009)
5. Bautista, M., Bautista, B., Salcedo, J., Narag, I.: Philippine Tsunamis and Seiches (1589 to 2012). Department of Science and Technology, Philippine Institute of Volcanology and Seismology (2012)
6. Braddock, R., Carmody, O.: Optimal location of deep-sea tsunami detectors. Int. Trans. Op. Res. **8**, 249–258 (2001)
7. Dysthe, K., Bannikov, D., Müller, P.: Oceanic rogue waves. Ann. Rev. Fluid Mech. **40**, 287–310 (2008)
8. Eble, M., Gonzalez, F.: Deep-ocean bottom pressure measurements in the Northeast Pacific. J. Atmos. Oceanic Technol. **8**, 221–233 (1990)
9. Goda, K., Mori, N., Yasuda, T.: Rapid tsunami loss estimation using regional inundation hazard metrics derived from stochastic tsunami simulation. Int. J. Disaster Risk Reduction **40**, 101152 (2019)
10. Goda, K., Risi, R.D.: Probabilistic tsunami loss estimation methodology: stochastic earthquake scenario approach. Earthquake Spectra **33**, 1301–1323 (2017)
11. Kaboli, S., Selvaraj, J., Rahim, N.: Rain-fall optimization algorithm: a population based algorithm for solving constrained optimization problems. J. Comput. Sci. **19**, 31–42 (2017)
12. Khalid, A., Javaid, N., Mateen, A., Ilahi, M., Saba, T., Rehman, A.: Enhanced time-of-use electricity price rate using game theory. Electronics **8**, 48 (2019)
13. Lovholt, F., Kuhn, D., Bungum, H., Harbitz, C., Glimsdal, S.: Historical tsunamis and present tsunami hazard in Eastern Indonesia and the Southern Philippines. J. Geophys. Res. **117**, B09310 (2012)
14. Marchuk, A., Vasiliev, G.: The fast method for a tsunami amplitude estimation. Math. Model. Geoph. **17**, 21–34 (2014)
15. Meza, J., Catalan, P., Tsushima, H.: A multiple-parameter methodology for placement of tsunami sensor networks. Pure Appl. Geophys. **77**, 1451–1470 (2020)

16. Mulia, I., Gusman, A., Satake, K.: Optimal design for placements of tsunami observing systems to accurately characterize the inducing earthquake. Geophys. Res. Lett. **44**, 12106–12115 (2017)
17. Nagai, T., Kato, T., Moritani, N., Izumi, H., Terada, Y., Mitsui, M.: Proposal of hybrid tsunami monitoring network system consisted of offshore, coastal and on-site wave sensors. Coastal Eng. J. **49**, 63–76 (2007)
18. Omira, R., et al.: Design of a sea-level tsunami detection network for the Gulf of Cadiz. Nat. Hazards Earth Syst. Sci. **9**, 1327–1338 (2009)
19. Park, H., Alam, M., Cox, D., Barbosa, A., van de Lindt, J.: Probabilistic seismic and tsunami damage analysis (PSTDA) of the Cascadia Subduction Zone applied to Seaside, Oregon. Int. J. Disaster Risk Reduc. **40**, 101152 (2019)
20. Pelinovsky, E.: Hydrodynamics of tsunami waves, p. 276. Nihziini Novgorod, Institute of Applie Physics RAS (1996)
21. Rabinovich, A., Eble, M.: Deep-ocean measurements of tsunami waves. Pure Appl. Geophys. **172**, 3281–3312 (2015)
22. Schindele, F., Loevenbruck, A., Herbert, H.: Strategy to design the sea-level monitoring networks for small tsunamigenic oceanic basins: the Western Mediterranean case. Nat. Hazards Earth Syst.Sci. **8**, 1019–1027 (2008)
23. Valenzuela, V.P.B., Esteban, M., Takagi, H., Thao, N.D., Onuki, M.: Disaster awareness in three low risk coastal communities in Puerto Princesa City, Palawan, Philippines. Int. J. Disaster Risk Reduc. **46**, 101508 (2020)

A Novel Formulation for Inverse Distance Weighting from Weighted Linear Regression

Leonardo Ramos Emmendorfer$^{(\boxtimes)}$ ⓘ and Graçaliz Pereira Dimuro ⓘ

Center for Computational Sciences, Universidade Federal do Rio Grande,
Rio Grande, RS 96203900, Brazil
leonardo.emmendorfer@gmail.com, gracaliz@gmail.com

Abstract. Inverse Distance Weighting (IDW) is a widely adopted interpolation algorithm. This work presents a novel formulation for IDW which is derived from a weighted linear regression. The novel method is evaluated over study cases related to elevation data, climate and also on synthetic data. Relevant aspects of IDW are preserved while the novel algorithm achieves better results with statistical significance. Artifacts are alleviated in interpolated surfaces generated by the novel approach when compared to the respective surfaces from IDW.

Keywords: Inverse distance weighting · Interpolation · Weighted linear regression · Digital elevation map

1 Introduction

Most natural properties vary continuously. However, in general, we can observe at only a finite number of the infinity of possible locations [20]. Spatial interpolation is the estimation of approximate values for specific locations from known values measured at other locations. Given a set of spatial data either in the form of discrete points or for subareas, spatial interpolation aims to find the function that will best represent the whole surface and that will predict values at other points or for other subareas [14]. This general problem has long been a concern majorly in geosciences, water resources, environmental sciences, agriculture, soil sciences among other disciplines [15, 29].

Environmental data collected from field surveys are often difficult and expensive to acquire. In such cases, spatial interpolation methods provide a tool for estimating an environmental variable at unsampled sites [15]. For instance, in [11] as a result from the sparsity of observational networks the distance to the nearest station can be of the order of several hundred kilometers. As a result, the only available data may not be representative of the climatology at the desired location. Ideally, the nearest recording station would be situated such that its climatology was identical to that of the location of interest.

© Springer Nature Switzerland AG 2020
V. V. Krzhizhanovskaya et al. (Eds.): ICCS 2020, LNCS 12138, pp. 576–589, 2020.
https://doi.org/10.1007/978-3-030-50417-5_43

Point interpolation deals with data collectable at a point, such as temperature readings or elevation [14]. Several solutions are available, such as Kriging [13,17], interpolating polynomials, splines, among others [7]. Inverse distance weighting (IDW) [25] is one of the most simple and widespread adopted [15]. The method does not require specific statistical assumptions, as the case for Kriging and other statistical interpolation methods. However, although empirical evaluations consistently show that IDW delivers inferior results when compared to other methods [15,19,30], the evaluation of improvements in IDW is a relevant topic of research [2,9,16,22,28].

The IDW interpolation of a value $\hat{y}_j$ for a given location j is computed as:

$$\hat{y}_j^{IDW} = \sum_{i=1}^{n} w_{i,j} y_i \tag{1}$$

where each y_i, $i = 1, \cdots, n$ is a data point available at a location i. The weights $w_{i,j}$ for each data point are given as:

$$w_{i,j} = \frac{d_{i,j}^{-\alpha}}{\sum_{k=1}^{n} d_{k,j}^{-\alpha}} \tag{2}$$

where $d_{i,j}$ is the Euclidean distance between a data point available at location i and the unknown data at location j; n is the number of data points available; α means the power, and is a control parameter. In this work, IDW is restricted to Inverse Squared-Distance Weighting since $\alpha = 2$ is assumed, which is the most commonly adopted value.

The maximum and minimum of the estimated values from IDW are limited to the extreme data points: $\min y_i \leq \hat{y}_j^{IDW} \leq \max y_i$. This is considered to be an important shortcoming because, to be useful, an interpolated surface should predict accurately certain important features of the original surface, such as the locations and magnitudes of maxima and minima even when they are not included as original sample points [14].

This work aims to (i) introduce an alternative interpolation algorithm which is similar to IDW and (ii) evaluate the novel method under a variety of conditions considering diverse of sampling densities, sample spatial distributions and surface types. Those are pointed out as important factors that affect the performance of spatial interpolation methods [15].

The paper is organized as follows. The proposed approach is presented in Sect. 2. The resulting model is evaluated and compared to the original IDW in Sect. 4, following the methodology proposed in Sect. 3. Section 5 concludes the paper.

2   Proposed Method

Consider a variable Y which is measured at n locations. One might be interested in obtaining an estimation for the value of Y at a specific location j, where a value for Y is not available for some reason.

Let us assume that variable Y is related to a function of the distance to j. This leads to a model which represents the relationship between the variable Y, which occurs at diverse locations, and a single explanatory variable which is a function of the distance from a given reference j to the location of each available measure of Y. One might assume, for instance, that squared distance from j influences Y as:

$$\mathbf{Y} = \beta_j^0 + \beta_j^1 \mathbf{D_j} + \mathbf{E_j} \tag{3}$$

where coefficients β_j^0 and β_j^1 are both scalars which must be obtained for each j. $\mathbf{Y} = \{y_1, y_2, \cdots, y_n\}$ is a vector with n values of the variable under consideration at diverse locations $i = 1, 2, \cdots, n$ and the corresponding vector $\mathbf{D_j} = \{d_{1,j}^2, d_{2,j}^2, \cdots, d_{n,j}^2\}$ contains the squared distances $d_{i,j}^2$ from location j to each location i corresponding to a respective y_i. $\mathbf{E_j} = \{\epsilon_1, \epsilon_2, \cdots, \epsilon_n\}$ is the vector of residues.

The estimation of the scalars β_j^0 and β_j^1 from (3) can be achieved by solving a weighted linear regression, where the regression weights $w_{1,j}^R, w_{2,j}^R, \cdots w_{n,j}^R$ for a given j are computed similarly to the IDW weights in (2) with $\alpha = 2$:

$$w_{i,j}^R = \frac{d_{i,j}^{-2}}{\sum_{k=1}^n d_{k,j}^{-2}} \tag{4}$$

For the sake of clarity, let us define the scalar variable s_j for a given j as:

$$s_j = \frac{1}{\sum_{k=1}^n d_{k,j}^{-2}} \tag{5}$$

Then, substituting (5) on (4):

$$w_{i,j}^R = d_{i,j}^{-2} s_j \tag{6}$$

The weighted sum of squared residuals (WSSE) of model (3) for data points $\{y_1, y_2, \cdots, y_n\}$ is given by:

$$WSSE = \sum_{i=1}^n w_{i,j}^R (y_i - \hat{y}_i)^2 = \sum_{i=1}^n w_{i,j}^R (y_i - \beta_j^0 - \beta_j^1 d_{i,j}^2)^2 \tag{7}$$

Substituting (6) into (7) leads to:

$$WSSE = \sum_{i=1}^n d_{i,j}^{-2} s_j (y_i - \beta_j^0 - \beta_j^1 d_{i,j}^2)^2$$

where the analytical solution for the minimal WSSE is:

$$\hat{\beta}_j^0 = s_j \sum_{i=1}^n y_i d_{i,j}^{-2} - n \beta_j^1 s_j \tag{8}$$

$$\hat{\beta}_j^1 = \frac{\sum_{i=1}^n y_i - n s_j \sum_{i=1}^n y_i d_{i,j}^{-2}}{\sum_{i=1}^n d_{i,j}^2 - n^2 s_j} \tag{9}$$

The estimated value for Y as a function $\hat{f}$ of the distance r from j using the model (3) is:

$$\hat{f}(r) = \hat{\beta}_j^0 + \hat{\beta}_j^1 r^2 \tag{10}$$

Since the aim of interpolation is the estimation of a value for Y at j, therefore the distance is $r = 0$. Then, from (10):

$$\hat{y}_j^R = \hat{f}(0) = \hat{\beta}_j^0 + \hat{\beta}_j^1 0^2 = \hat{\beta}_j^0 \tag{11}$$

Substituting (9) into (8) and (11) leads, after simplification, to the expression for the interpolated value at a given location j, from a set of values $\{y_1, y_2, \cdots, y_n\}$ and respective distances $\{d_{1,j}, d_{2,j}, \cdots, d_{n,j}\}$ from j :

$$\hat{y}_j^R = s_j \sum_{i=1}^n y_i d_{i,j}^{-2} + n \frac{\sum_{i=1}^n y_i - n s_j \sum_{i=1}^n y_i d_{i,j}^{-2}}{n^2 - \sum_{i=1}^n d_{i,j}^{-2} \sum_{i=1}^n d_{i,j}^2} \tag{12}$$

From (1), (2) and (5) one can find out that $s_j \sum_{i=1}^n y_i d_{i,j}^{-2} = \hat{y}_j^{IDW}$ (with $\alpha = 2$), therefore (12) can be rewritten as:

$$\hat{y}_j^R = \hat{y}_j^{IDW} + n \frac{\sum_{i=1}^n y_i - n \hat{y}_j^{IDW}}{n^2 - \sum_{i=1}^n d_{i,j}^{-2} \sum_{i=1}^n d_{i,j}^2} \tag{13}$$

The resulting expression, which is derived from a weighted linear regression, results equivalent to IDW with an additional term. We call this method as Inverse Distance Weighted Regression (IDWR). More specifically, since α was set to 2, this paper investigates Inverse Squared-Distance Weighted Regression.

2.1   Analysis of IDWR

Similarly to IDW, IDWR is also a deterministic, nonstatistical interpolation method, defined by a simple expression (13). The computational complexity for interpolating a single location j for IDWR is $O(n)$, linear in the number of data points n, which is the same for IDW.

This section presents an initial analysis of some relevant situations. Initially, the form of expression 13 raises some concerns as the denominator might be equal to or near zero. For instance, when all data points are or tend to be at the same distance r from location j, the denominator is or tends to be equal to $n^2 - \sum_{i=1}^n r^{-2} \sum_{i=1}^n r^2 = n^2 - nr^{-2}nr^2 = 0$. While this situation would not be expected in most real-world applications, even when input data is distributed on a bidimensional regular grid, this feature of IDWR must be carefully taken into account before using the method. Also, one can realize that as the distance $r \to \infty$ additional numerical concerns might arise since $\sum_{i=1}^n r^{-2} \sum_{i=1}^n r^2 \to 0 \times \infty$. This differs from IDW, which tends to $\frac{1}{n} \sum_{i=1}^n y_i$ as $r \to \infty$.

The behavior of IDWR at the neighborhood of any given data point is also analysed. We are interested in the value of $\hat{y}_j^R$ as $d_{lj} \to 0$ for a given data point at location l, with $d_{ij} \neq 0$ for the remaining data points $i \neq l$. Since $d_{lj} \to 0$, then $\sum_{i=1}^{n} d_{ij}^{-2} \to \infty$ in expression 13 and $\sum_{i=1}^{n} d_{ij}^{2} \to c$ where $c = \sum_{i \neq l} d_{ij}^{2}$ is a constant. This results $\hat{y}_j^R \to \hat{y}_j^{IDW}$ in expression 13, since the denominator tends to $-\infty$, under the condition that the numerator should be finite. As a result IDW and IDWR will tend to compute similar values for locations which are nearby any given data point. IDWR is an exact interpolator since $\hat{y}_j^R = \hat{y}_j^{IDW} = y_i$ for $j = i$. At other locations, IDWR might be able to provide useful extrapolation, since $-\infty \leq \hat{y}_j^R \leq +\infty$, differently from IDW which is restricted to the interval $\min y_i \leq \hat{y}_j^{IDW} \leq \max y_i$. From the discussion above, any differences between both methods might occur at locations that are not too close to any data point.

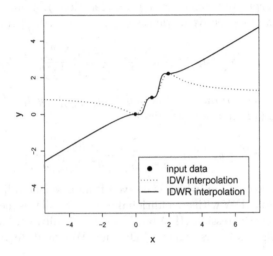

Fig. 1. The behavior of IDW and IDWR for the interpolation from a dataset with $n = 3$ data points.

Figure 1 illustrates some of the properties discussed here using a synthetic one-dimensional dataset with three data points that follow a linear trend ($R^2 > 0.99$).

3   Empirical Evaluation

Two types of experiments were performed, which allow one to compare the effectiveness of both algorithms considered. The first evaluation involves the interpolation of points from real functions of two variables. The functions were selected from the optimization literature, as representatives of varying roughness of surfaces, so as to impose different levels of difficulty for the interpolation methods. While those functions would not perfectly mimic real-world situations, this evaluation is still useful for the purpose of this work since it provides a scalable comparison between the two methods, through a controlled variation on the number of samples. In this first experiment sample size was set to four

Table 1. Functions of two real variables $(x_1, x_2) = \mathbf{x}$, adopted in empirical evaluation

Function	Expression	Interval for x_1 and x_2
Rosenbrock	$y(\mathbf{x}) = 100(x_2 - x_1^2)^2 + (x_1 - 1)^2$	$[-2.048, 2.048]$
Sombrero	$y(\mathbf{x}) = \begin{cases} \dfrac{\sin\left((16(x_1-0.5))^2 + (16(x_2-0.5))^2\right)}{16(x_1-0.5))^2 + (16(x_2-0.5))^2)} & \text{if } x_1 \neq 0.5 \text{ and } x_2 \neq 0.5; \\ 1 & \text{otherwise} \end{cases}$	$[0, 1]$
Himmelblau	$y(\mathbf{x}) = (x_1^2 + x_2 - 11)^2 + (x_1 + x_2^2 - 7)^2$	$[-5, 5]$
Rastrigin	$y(\mathbf{x}) = 20 + (x_1^2 - 10\cos(2\pi x_1)) + (x_2^2 - 10\cos(2\pi x_2))$	$[-5.12, 5.12]$
Log Goldstein-Price	$\begin{aligned} y(\mathbf{x}) = \tfrac{1}{2.427}(&\log((1 + (x_1 + x_2 + 1)^2 \\ &\times (19 - 14x_1 + 3x_1^2 - 14x_2 + 6x_1 x_2 + 3x_2^2)) \\ &\times (30 + (2x_1 - 3x_2)^2 \\ &\times (18 - 32x_1 + 12x_1^2 + 48x_2 - 36x_1 x_2 + 27x_2^2))) \\ &-8.693) \end{aligned}$	$[-2, 2]$
F102	$\begin{aligned} y(\mathbf{x}) = &-(x_2 + 47)\sin\sqrt{\lvert x_2 + \tfrac{x_1}{2} + 47 \rvert} \\ &-x_1\sin\sqrt{\lvert x_1 - (x_2 + 47) \rvert} \end{aligned}$	$[-512, 512]$

values: $N = 100, 200, 300, 400$. The variation on sample size is motivated by the need for capturing spatial changes, thus to improve the performance of the spatial interpolation methods [15].

Table 2. Average RMSE and standard deviation computed with leave-one-out cross-validation (LOOCV) for IDW and IDWR applied to 6 benchmark functions, after 30 replications with randomly generated sample points for each benchmark function. The number of sample points for all functions is $N = 300$ at each replication. P-values refer to the result of two-tailed t-tests considering the null hypothesis that algorithms are equivalent in terms of average RMSE

Function	Avg. IDW LOOCV RMSE	σ_{IDW}	Avg. IDWR LOOCV RMSE	σ_{IDWR}	Relative Reduction	p-value
Rosenbrock	307.65	29.02	222.52	27.48	-27.67%	$<2.2e{-}16$
Sombrero	0.083277	0.0089	0.0806	0.0087	-3.20%	$1.016e{-}06$
Himmelblau	76.61	3.96	64.75	3.90	-15.48%	$<2.2e{-}16$
Rastrigin	16.51	0.68	16.25	0.74	-1.59%	$1.728e{-}07$
Log Golsdtein-Price	0.6036	0.02698	0.4378	0.02439	-27.47%	$<2.2e{-}16$
F102	391.11	16.93	388.39	16.89	-0.70%	$3.379e{-}09$

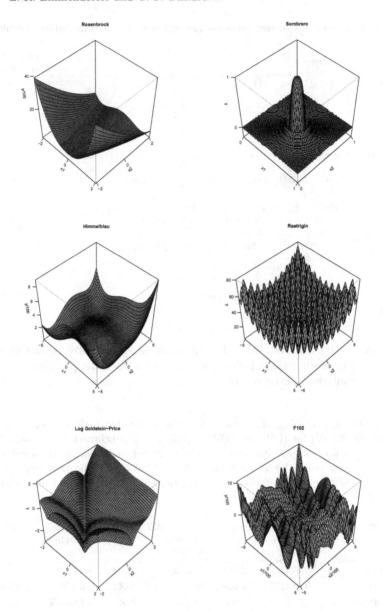

Fig. 2. Perspective visualization of the 6 functions used for the evaluation of the proposed algorithm.

Table 3. RMSE computed with leave-one-out cross-validation (LOOCV) for IDW and IDWR applied to 2 benchmark datasets from the literature

Dataset	N	IDW LOOCV RMSE	IDWR LOOCV RMSE	Relative Reduction
Calabria	48	67.16	65.72	−2.14%
Texas	18	11.09	8.63	−28.51%

Table 1 summarizes the definitions of the functions adopted. Figure 2 provides a perspective visualization of the topology of those functions. Himmelblau [10], Rosenbrock [24] and Rastrigin [23] are non-linear, non-convex functions widely used to test the performance of optimization algorithms. The 2-dimensional version of Rastrigin is used here. Log Goldstein-Price is an adjusted version of the Goldstein-Price function [8] proposed by [21]. The function F102 [1] was also called *Egg Holder* in [27] and in other works. It is considered as a difficult function due to its high multimodality. The Sombrero function was also included in our evaluation since it was already adopted as a benchmark for evaluation of IDW, in [30].

In a second type of evaluation two datasets representing real-world situations from the literature are considered. The *Calabria* dataset, adapted from [5], is a raster low-resolution (100m) digital elevation map containing 48 elevations which vary from 760 m to 936 m. The sample area from a location in Calabria is 610 m by 810 m in size, which corresponds to a portion of sample area 1 in [5]. The *Texas* dataset contains normal annual precipitation (1941–1970) for 18 locations in Texas, which is the full list of locations from [3]. The lowest annual precipitation (7.7in) occurs in El Paso, near the western extreme of the state, while the highest precipitation is assigned to Beaumont-Port Arthur, near the eastern extreme (55.07in).

In order to allow the comparison between the interpolation methods, leave-one-out cross-validation (LOOCV) [12] was adopted. In LOOCV, a single data point y_i is used for the estimation of the squared error of the interpolation $(y_i - \hat{y}_i)^2$ from a model built from all remaining points $N - 1$ points. The process is repeated for all data points, and the root mean square error (RMSE) is computed, for both interpolation methods considered.

Since the computation of the RMSE for the evaluation of the interpolation of real functions is dependent on the specific sample of data points, 30 replications of leave-one-out cross-validation are performed for each algorithm on each function, in order to estimate the average RMSE for a number of N data points. Those data points are randomly generated from uniform distributions delimited by the specified real intervals for each variable.

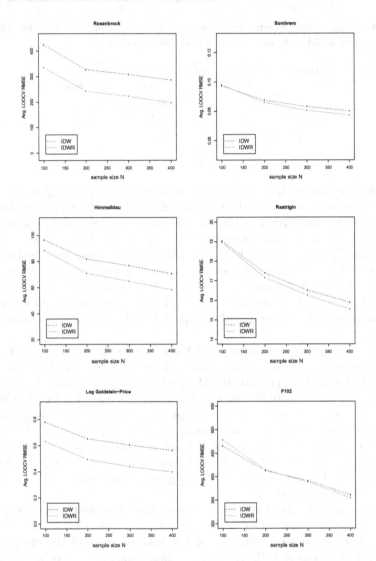

Fig. 3. Average RMSE and standard deviation computed with leave-one-out cross-validation (LOOCV) for IDW and IDWR applied to 6 benchmark functions, after 30 replications with randomly generated sample points for each benchmark function. The number of sample points for all functions was set to $N = 100, 200, 300, 400$ at each replication.

4   Results

Table 2 shows the results from the first set of experiments, where interpolation is performed from points sampled from functions defined over the bidimensional domain. Average RMSE and respective standard deviation σ are computed for 30 replications of leave-one-out cross-validation on the interpolation of data points from 6 functions for both algorithms considered. The number of data points in each replication was set to $N = 300$. The relative reductions on the values of the average RMSE for IDWR when compared to IDW are also shown. Resulting reductions range from 0.70% (F102) to 27.67% (Rosenbrock). All differences between the mean RMSE values are statistically significant at a 95% confidence level, considering paired two-tailed t-tests under the null hypothesis that both methods are equivalent.

The effect of sample size is illustrated in Fig. 3. For all functions 4 sample sizes were considered: $N = 100, 200, 300, 400$. RMSE is lower for IDWR when compared to IDW for all functions with all N considered, except for $N = 100$ and $N = 200$ where the best RMSE for the F102 function is achieved with IDW. For $N > 200$ IDWR is superior for all functions. The tendency from the graphs in Fig. 3 is also favorable to IDWR for $N > 400$.

In Table 3 the values of LOOCV RMSE for both algorithms applied to two datasets considered are shown. Under this evaluation, IDWR is superior to IDW for both datasets. The error for *Calabria* dataset is 2.14% lower when compared to IDW. A higher difference was reached for the *Texas* dataset, where IDWR achieved a 28.51% reduction in the LOOCV RMSE when compared to the value obtained with IDW for the same dataset.

In order to allow a better understanding of the behavior of each algorithm, interpolated surfaces were generated for the sample areas related to each both datasets considered. For *Calabria*, two digital elevation maps with a 1 m resolution were obtained representing the interpolated surfaces obtained using both algorithms for the input data, which consists of a digital map with elevations from 48 locations regularly distributed with a resolution of 100m. This high difference between input and output resolution might not be recommended. However, for the purpose of this evaluation, the approach allows a better visual comparison between the results obtained by both methods. Figure 4 shows the resulting maps for the region on the *Calabria* dataset using both IDW and IDWR (Figs. 4(a) and 4(b) respectively).

The highest elevation in *Calabria* dataset is located near the center of the maps, as indicated. It also corresponds to the maximal value obtained from IDWR and also from IDW. The same occurs for the lowest elevation, which occurs at a location near the right bottom extreme of the map. Therefore, IDWR did not exceed the IDW limitations $\min y_i$ and $\max y_i$ for this case. Although both maps from *Calabria* are similar, qualitative differences in the behavior of the algorithms occur. The surface generated by IDWR is smoother, with smaller variations on the curvature over the space. As a result, the interpolated surface from IDWR appears as more conceivable when compared to the result from IDW. The surface generated with IDWR is smoother since artificial bumps generated between sample points are less evident. However, undesirable artifacts exist since both algorithms produce unrealistic landscape, with a terraced aspect. Elevation profiles below and beside both maps (a) and (b) in Fig. 4 provide a better illustration for this feature.

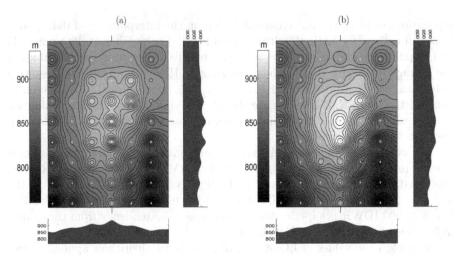

Fig. 4. High resolution interpolated elevation maps generated by IDW (a) and IDWR (b), for the area of *Calabria* dataset. 48 regularly distributed sample points are shown in red and elevation values are represented in grayscale levels discretized into 40 intervals with increments of ≈ 4.4 m. For each map, two elevation profiles (bottom and right) are shown, each parallel to a coordinate axis and both passing through the coordinates corresponding to the highest elevation in the dataset, indicated at the border of the maps.

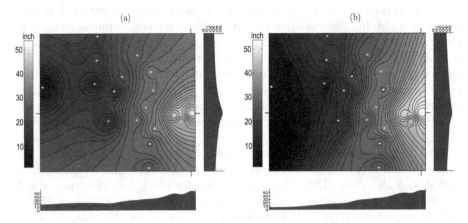

Fig. 5. High resolution interpolated precipitation maps generated by IDW (a) and IDWR (b), for the area of *Texas* dataset. Sample points are shown in red and elevation values are represented in grayscale levels discretized into 40 intervals with increments of ≈ 1.35 in. For each map, two elevation profiles (bottom and right) are shown, each parallel to a coordinate axis and both passing through the location corresponding to the highest precipitation in the dataset, indicated at the border of the maps (Beaumont-Port Arthur). (Color figure online)

The dataset *Texas* represents a situation where a low amount of data points is available which leads to the absence of data points in some areas since large regions outside the territory of Texas are represented in the interpolated maps. Figures 5(a) and 5(b) both represent an area of size 1258 km × 1060 km with a resolution of 2 km. The resulting map from IDWR provides a better model for the expected behavior of precipitation from given data. Precipitation decreases roughly towards the west or south-west, reaching predicted values as low as 1.139in at where would correspond to the territory of Mexico, which is below the minimal precipitation from the dataset (7.7in).

5   Conclusion and Further Work

The selection of an appropriate interpolation model depends largely on the type of data, the degree of accuracy desired, and the amount of computational effort afforded [14]. Each method has its advantages and drawbacks, which depend strongly on the characteristics of the data: a method that fits well with some data can be unsuited for a different set of data points [6]. This also motivates the improvement of existing methods and search for novel alternatives.

Variations and extensions from the basic IDW method have been proposed in the literature. In [2] an improvement is presented which is based on a geometric criterion that automatically selects a subset of the original set of control points. In [22] data normalization is shown to improve the results of interpolation. In [9] weighted median of data within a neighborhood is proposed. A distance-decay parameter is explored in [16] which is adjusted according to the spatial pattern of sampled locations in the neighborhood.

This paper followed a diverse path by presenting a novel formulation that is derived from a weighted regression model where squared distance from the location of interest is assumed to influence a geographically localized variable. Resulting expression (13) is similar to IDW method while retaining its simplicity and low computational complexity. Squared distance was arbitrarily chosen, and other formats for that relationship might be explored further.

Regression is already widely adopted for problems involving spatial data. Geographically Weighted Regression (GWR), as proposed by [4], adopts weighted regression in the spatial context by extending the usual regression model. The regression coefficients are dependent on individual location and the parameters in GWR are therefore locally estimated by weighted least squares approach where the weight is higher for observations that are closer to the location considered. That premise of a higher local relationship [26] which is straightforwardly implemented by IDWR and IDW is already widely exploited [18].

Empirical evaluation of the proposed method adopted leave-one-out cross-validation using datasets from the literature and synthetic data from benchmark functions, with varying sample densities on diverse surface types and sample distributions. Study cases emphasized applications on digital elevation data and climate.

IDWR was able to attain better results when compared to IDW by obtaining lower RMSE with statistical significance for benchmark functions. Qualitatively, the novel method delivered smoother curvatures between sample points when compared to the maps generated by IDW. Observable artifacts are alleviated in the surfaces generated by IDWR.

Further empirical and theoretical investigation should be proposed to better delineate the limitations of the novel method. It might also be studied whether the proposed method actually produces useful extrapolation. In that case, wider applicability would be reached when compared to IDW. This, however, must be carefully considered since the asymptotic behavior of IDWR is much diverse from IDW, according to the discussion in Sect. 2. A comparison to other interpolation methods could also be performed, covering a wider variety of applications.

References

1. Evaluating evolutionary algorithms: Artif. Intell. **85**(1), 245–276 (1996)
2. Ballarin, F., D'Amario, A., Perotto, S., Rozza, G.: A POD-selective inverse distance weighting method for fast parametrized shape morphing. Int. J. Numer. Methods Eng. **117**(8), 860–884 (2019)
3. Bomar, G.W.: A climatological summary of Texas weather in 1977. Technical report (1978)
4. Brunsdon, C., Fotheringham, A.S., Charlton, M.E.: Geographically weighted regression: a method for exploring spatial nonstationarity. Geograph. Anal. **28**(4), 281–298 (1996)
5. Carrara, A., Bitelli, G., Carla, R.: Comparison of techniques for generating digital terrain models from contour lines. Int. J. Geograph. Inf. Sci. **11**(5), 451–473 (1997)
6. Caruso, C., Quarta, F.: Interpolation methods comparison. Comput. Math. Appl. **35**(12), 109–126 (1998)
7. Cressie, N.: Statistics for Spatial Data. Wiley, Hoboken (1993)
8. Dixon, L.C.W., Szegö, G.P.: The global optimization problem: an introduction. In: Towards Global Optimization, vol. 2. North-Holland, Amsterdam (1978)
9. Henley, S.: Nonparametric Geostatistics. Springer, Heidelberg (2012)
10. Himmelblau, D.: Applied Nonlinear Programming. McGraw-Hill, New York (1972)
11. Jeffrey, S.J., Carter, J.O., Moodie, K.B., Beswick, A.R.: Using spatial interpolation to construct a comprehensive archive of Australian climate data. Environ. Model. Softw. **16**(4), 309–330 (2001)
12. Kohavi, R., et al.: A study of cross-validation and bootstrap for accuracy estimation and model selection. In: Proceedings of the International Joint Conference on Artificial Intelligence, Montreal, Canada, vol. 14, pp. 1137–1145 (1995)
13. Krige, D.: A review of the development of geostatistics in South Africa. In: Guarascio, M., David, M., Huijbregts, C. (eds.) Advanced Geostatistics in the Mining Industry, vol. 24, pp. 279–293. Springer, Heidelberg (1976). https://doi.org/10.1007/978-94-010-1470-0_17
14. Lam, N.S.N.: Spatial interpolation methods: a review. Am. Cartographer **10**(2), 129–150 (1983)
15. Li, J., Heap, A.D.: A review of comparative studies of spatial interpolation methods in environmental sciences: performance and impact factors. Ecol. Inform. **6**(3–4), 228–241 (2011)
16. Lu, G.Y., Wong, D.W.: An adaptive inverse-distance weighting spatial interpolation technique. Comput. Geosci. **34**(9), 1044–1055 (2008)

17. Matheron, G.: The theory of regionalised variables and its applications. Les Cahiers du Centre de Morphologie Mathématique **5**, 212 (1971)
18. Miller, H.J.: Tobler's first law and spatial analysis. Ann. Assoc. Am. Geograph. **94**(2), 284–289 (2004)
19. Murphy, R.R., Curriero, F.C., Ball, W.P.: Comparison of spatial interpolation methods for water quality evaluation in the Chesapeake Bay. J. Environ. Eng. **136**(2), 160–171 (2010)
20. Oliver, M.A., Webster, R.: Kriging: a method of interpolation for geographical information systems. Int. J. Geograph. Inf. Syst. **4**(3), 313–332 (1990)
21. Picheny, V., Wagner, T., Ginsbourger, D.: A benchmark of Kriging-based infill criteria for noisy optimization. Struct. Multidisc. Optim. **48**(3), 607–626 (2013)
22. Qu, R., et al.: Predicting the hormesis and toxicological interaction of mixtures by an improved inverse distance weighted interpolation. Environ. Int. **130**, 104892 (2019)
23. Rastrigin, L.A.: Extremal control systems. Theoretical Foundations of Engineering Cybernetics Series. Nauka, Moscow (1974). (in Russian)
24. Rosenbrock, H.H.: An automatic method for finding the greatest or least value of a function. Comput. J. **3**(3), 175–184 (1960)
25. Shepard, D.: A two-dimensional interpolation function for irregularly-spaced data. In: Proceedings of the 1968 ACM National Conference, pp. 517–524 (1968)
26. Tobler, W.R.: A computer movie simulating urban growth in the detroit region. Econ. Geogr. **46**(sup1), 234–240 (1970)
27. Vanaret, C., Gotteland, J.B., Durand, N., Alliot, J.M.: Certified global minima for a benchmark of difficult optimization problems (2014). https://hal-enac.archives-ouvertes.fr/hal-00996713
28. Weber, D., Englund, E.: Evaluation and comparison of spatial interpolators. Math. Geol. **24**(4), 381–391 (1992)
29. Zhou, F., Guo, H.C., Ho, Y.S., Wu, C.Z.: Scientometric analysis of geostatistics using multivariate methods. Scientometrics **73**(3), 265–279 (2007)
30. Zimmerman, D., Pavlik, C., Ruggles, A., Armstrong, M.P.: An experimental comparison of ordinary and universal Kriging and inverse distance weighting. Math. Geol. **31**(4), 375–390 (1999)

Addressing the Robustness of Resource Allocation in the Presence of Application and System Irregularities via PEPA Based Modeling

Srishti Srivastava[1]([✉]), Ioana Banicescu[2], and William S. Sanders[2]

[1] University of Southern Indiana, Evansville, IN 47712, USA
fsrishti@usi.edu
[2] Mississippi State University, Mississippi State, MS 39762, USA
ioana@cse.msstate.edu, wss2@msstate.edu

Abstract. Applications executing in heterogeneous parallel and/or distributed computing (PDC) environments are often prone to unpredictable runtime due to variations in problem, algorithm, and system characteristics. This serves as a key motivation towards a study of the robustness of resource allocations required to maintain and guarantee a desired level of performance. Performance modeling and evaluation is often utilized to understand and predict the behavior of the application and the computational system from a performance point of view. In prior work, performance modeling for evaluating response times of *static* resource allocations in PDC systems was introduced by the authors as a proof of concept for validating the use of the performance evaluation process algebra (PEPA) for analyzing the robustness of static resource allocations. Herein, the authors present numerical modeling of several *static* resource allocations to evaluate their robustness in the presence of *compound perturbations* generated as combinations of variations in *application workload* and *machine availability*. The novelty of the approach is to introduce the *compound effect* as the variability of both, *application workload* and *processor/machine availability*, into the performance modeling of the overall computational system. The performance is obtained as a parallel execution time via a numerical analysis of the modeled execution of applications on *non-dedicated* parallel computational resources. A significant improvement in the robustness value (up to 143%) among the mappings yielding equal parallel execution times has been demonstrated via the analysis of the results. This notable difference in the robustness values strongly indicates the benefit of selecting one mapping versus the other for guaranteeing the best execution performance.

Keywords: Performance modeling and evaluation · Robustness analysis · Process algebra

© Springer Nature Switzerland AG 2020
V. V. Krzhizhanovskaya et al. (Eds.): ICCS 2020, LNCS 12138, pp. 590–603, 2020.
https://doi.org/10.1007/978-3-030-50417-5_44

1   Introduction

Often in parallel and distributed computing (PDC) environments, the applications are expected to undergo variations in workload, and the underlying computational resource is considered to be non-dedicated at runtime. Therefore, appropriate initial resource allocation algorithms are required for an efficient mapping of applications to machines. In addition to the traditional performance metrics (execution time, speedup, efficiency, and others), there is a need for the study of a metric of the robustness of resource allocations to guarantee a desired level of performance. Performance modeling and evaluation is often utilized to understand the behavior of concurrent and parallel computing and communication systems by identifying features of the system that are sensitive from a performance point of view. When compared to direct experiments and simulations, numerical models and the corresponding analyses are easier to reproduce, do not incur any setup or installation costs, do not impose any prerequisites for learning a simulation framework, and are not limited by the complexity of the underlying infrastructure or simulation libraries. To the best of our knowledge, performance modeling for evaluating *response times* of *static* resource allocations in parallel and distributed computing systems and the related robustness analysis remained an open problem until the authors introduced the first solution by utilizing the performance evaluation process algebra (PEPA) [5] for analyzing the robustness of static resource allocations [22–24].

The main contribution of this work is to study a number of *static* resource allocations via a PEPA based numerical analysis of performance modeling of the parallel execution of applications with *varying workload* on *non-dedicated* parallel computing resources with *varying machine availability*. The robustness of the resource allocations is evaluated against the *compound effect*, which is defined as the combination of the impacts of the variations in the application workload and those of machine availability on system performance. Note that in this study, the terms *processor availability* and *machine availability* are used interchangeably throughout the paper. Prior validation of our PEPA models, and a confirmation of the similarity in results of our numerical analysis with the experimental results of earlier reported research results available in the existing literature [1] are illustrated in Fig. 1. However, in our prior validation work [23] only the variability in problem characteristics has been considered to mimic the experiments in [1].

In this work, the robustness value for each resource allocation is calculated as the minimum probability of the machines to finish before a desired makespan goal. Based on the analysis of the results, a number of mappings yield equal execution performance. However, a significant improvement in the robustness value (up to 143%) among the mappings yielding equal parallel execution times has been observed. Thus, this notable 2.43 times increase of robustness strongly indicates the benefits of selecting one mapping versus the other for *guaranteeing* the best execution performance. The work with PEPA can also be extended to modeling a single utility function that includes metrics, such as robustness,

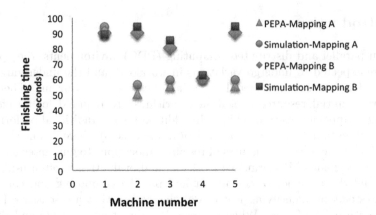

Fig. 1. A comparative analysis of the numerical results of performance modeling with existing simulation results used for validation of approach in [23].

power consumption, and others. In general, this work is applicable to various types of computing and communication environments.

The rest of the paper is organized as follows. A description of the robustness of resource allocations in parallel and distributed computing systems and of the performance modeling techniques for numerical analysis of performance in parallel computing systems are given in Sect. 2. The design and the use of PEPA for the numerical evaluation of the analytical models of resource allocations in parallel computing systems are presented in Sect. 3. A comparison of the numerical results of robustness of a number of resource allocations obtained via analytical modeling using the PEPA tool, is discussed in Sect. 4. Conclusions and possible future work are summarized in Sect. 5.

2   Background and Related Work

In general, the *mapping problem*, defined as finding the best allocation of independent tasks (or applications) onto a set of parallel processors, is known to be NP-Complete [6–8]. A number of research efforts have been made towards achieving robust mapping, or resource allocation techniques in parallel and distributed computing systems. Key work done in this area is being reviewed and is presented in this section. In addition, a survey of the work in performance evaluation of computer and communication systems using analytical and numerical modeling is also being discussed.

2.1   Robustness of Resource Allocation and Application Scheduling

The initial work on robust scheduling originated from job-shop application scheduling frameworks [9]. A Standard branch and bound approach was used to solve the NP-Hard robust scheduling problem (RSP) to obtain a robust schedule for N independent jobs on a single machine [10]. A number of approaches have been developed to obtain an initial robust resource allocation by utilizing optimization techniques such as, stochastic mixed integer programming [11], iterative integer programming [12], and others. In addition to designing robust resource allocations, robustness metrics have also been formulated to study the performance guarantee of available static resource allocations against possible inadequate or variable computational environmental factors. A general methodology, called the Feature Perturbation Impact Analysis (FePIA) procedure, for developing robustness metrics for resource allocation has been presented by Ali et al. [1]. The authors define a resource allocation to be robust with respect to specific system *performance features* against *perturbations* (uncertainties) in specified *system parameters* if degradation in these features is constrained when limited perturbations occur [1,2]. To address this issue of investigating the robustness of scheduling techniques at the application level, together with studies conducted at the system level for a holistic approach of robustness, research has also been conducted towards analyzing the robustness of a number of dynamic loop scheduling (DLS) algorithms, which are effective in dynamic scheduling of scientific applications on large-scale parallel and distributed systems, in the presence of varying processor loads (where robustness is quantified using the *flexibility metric*) and processor failures (where robustness is quantified using the *resilience metric*) [3,4].

2.2   Process Algebra for Performance Evaluation

Analytical and numerical modeling for performance evaluation allows derivation of an expression of the performance feature of interest in terms of the input parameters of the model. In case of a predictive analysis of a computing system, analytical models generally provide the best insight into the effects of various perturbation parameters on system performance, and are easier to replicate for a comparative analysis of different systems. Markovian models have been shown to be an effective tool for performance analysis of computer and communication systems, where the system components are modeled as Markov processes, and the overall performance (for example, throughput, resource utilization, and others) is evaluated upon the numerical analysis of these Markov processes [15]. Process algebras are abstract languages used for specification and design of such systems.

PEPA, a stochastic process algebra, has been used for performance modeling and analysis of a wide range of concurrent systems. In a recent research study related to the scheduling of pipeline applications on grid resources, the PEPA workbench [16] has been used to solve the performance models of a scheduling system to obtain relevant performance information required for enhancing the execution performance of pipeline applications executing on the allocated grid resources (processors and network links) [13,14]. The PEPA workbench is used to calculate the performance feature, namely the throughput of executing pipeline applications, which is obtained when employing the modeled mapping for scheduling the pipeline applications onto grid resources [13,14].

Another important performance measure is the response time, which is prominently used for analyzing the performance of a resource allocation or a task scheduling system in parallel and distributed computing, where applications are bound by time constraints (such as an execution deadline). A research direction towards evaluating response time profiles has been given in the work performed on evaluating PEPA models via an ordinary differential equation (ODE) analysis [14]. The functionality for evaluating response time profiles via passage time analysis has been implemented in the Imperial PEPA Compiler (IPC) tool [18], and in the PEPA workbench [16]. However, to the best of our knowledge, performance modeling for evaluating the *response times* of resource allocations in parallel and distributed computing systems and their related robustness analysis remained an open problem until the authors introduced the first solution by utilizing PEPA for analyzing the robustness of static resource allocations in [22,23].

3   Methodology: Performance Modeling Using PEPA for Robustness Analysis

The target applications for this study are considered to be independent parallel applications waiting in a job queue for execution. Definitions of the notations used in the following description of the methodology are given in Table 1.

Table 1. Table of notations

Notation	Definition
A	Set of parallel applications
a_i	A parallel application $\in A$
λ_i	Initial workload for a_i
$\hat{\lambda}_i$	Perturbed/varied workload for a_i at runtime, where $\lambda_i \neq \hat{\lambda}_i$
M	A set of parallel machines
M_j	A parallel machine $\in M$
η_j	Computational availability of machine M_j
$\hat{\eta}_j$	Perturbed/varied computational availability of M_j at runtime, where $\eta_j \neq \hat{\eta}_j$
β_i^{max}	User defined makespan goal for a_i
$F_i(M_j(\hat{\eta}_j), \hat{\lambda}_i)$	Finishing time of machine M_j
ψ	Robustness value of a mapping. See Eq. 1
T_p	Parallel execution time for a mapping calculation using PEPA passage time analysis
T_{ij}	Actual time to compute a_i on M_j. Calculated as $\hat{\lambda}_i \times \hat{\eta}_j$
r_i	Ideal activity rate defined as $\frac{\lambda_i}{T_{ij}}$, where $\hat{\lambda}_i = \lambda_i$
p_i	Perturbed activity rate defined $\frac{\hat{\lambda}_i}{T_{ij}}$, where $\hat{\lambda}_i \neq \lambda_i$

Each application receives a data set generated by three heterogeneous sensors that produce the workloads (λ_1, λ_2, and λ_3). The underlying computational system consists of parallel machines (that contain K heterogeneous processors, where K varies from machine to machine). Each machine has an associated availability factor (η_i), which is the computational availability of the allocated machine for executing an application. Further, a set of possible resource allocations are considered to be available for an initial mapping of applications to machines based on an *expected time to compute* (ETC) matrix. In general, in a parallel and distributed computing environment, it is realistic to assume that the ETC values of applications on all the available machines are known a priori. Often, these estimates are derived from application profiling and machine benchmarking, from the previous executions of an application on a machine, or are provided by the user [19, 20]. All applications in the job queue are assumed to start executing at time $t = 0$ s. In this work, the perturbation parameter considered for the robustness analysis of resource allocations is defined as the *compound effect* of perturbations generated from runtime variations in both, application workload and processor availability of a machine.

Given, A: a set of parallel applications, $a_i \in A$: a parallel application, λ_i: the workload of application a_i, β_i^{max}: a user defined makespan goal for a_i, M: a set of parallel machines, $M_j \in M$: the machine allocated to a_i, $\hat{\lambda}_i$: a perturbation parameter defined as the workload variation from the initial workload (λ_i) for an application a_i, $\hat{\eta}_j$: a perturbation parameter defined as the machine availability variation from the initial machine availability (η_j) for an application a_i, $F_i(M_j(\hat{\eta}_j), \hat{\lambda}_i)$: the finishing time of application a_i on machine M_j, then the *robustness* (ψ) of a mapping is formulated as in Eq. 1.

$$\psi = \min_{\forall i \in A} \Pr[F_i(M_j(\hat{\eta}_j), \hat{\lambda}_i) \leq \beta_i^{max}] \tag{1}$$

The goal of the robustness analysis is to find a resource allocation that maximizes the robustness of the execution of the applications on the allocated parallel machines. An example PEPA model of a mapping system for two applications (a_1, a_2) and five processors (P_0, P_1, P_2, P_3, P_4) distributed among two machines (M_1, M_2) has been described below. For reference, a detailed description of the PEPA language and the language operators can be found in [5, 17].

$$a_1 \stackrel{def}{=} (compute_1, \top).RETURN$$
$$a_2 \stackrel{def}{=} (compute_2, \top).RETURN$$
$$P_0 \stackrel{def}{=} (compute_1, r_1).RETURN$$
$$P_1 \stackrel{def}{=} (compute_1, r_1).RETURN$$
$$P_2 \stackrel{def}{=} (compute_2, r_2).RETURN$$
$$P_3 \stackrel{def}{=} (compute_2, r_2).RETURN$$
$$P_4 \stackrel{def}{=} (compute_2, r_2).RETURN$$
$$M_1 \stackrel{def}{=} P_0 \parallel P_1$$
$$M_2 \stackrel{def}{=} P_2 \parallel P_3 \parallel P_4$$
$$Mapping \stackrel{def}{=} (a_1 \parallel a_2) \underset{\mathcal{L}}{\bowtie} (M_1 \parallel M_2)$$

where $\mathcal{L} = \{compute_1, compute_2\}$.

The two applications (a_1 and a_2) engage in their compute activities. In the above model, the first two statements model the two applications when they are not allocated to any processor. $\top$ is a predefined symbol in PEPA that denotes an unknown rate for an activity. This symbol is used in situations when a system is carrying out some action (or sequence of actions) whose rate is unavailable at the given time. Later when the processors are modeled, a processor engages in the compute activities of all the applications mapped to this processor. At this point, the unknown activity rate ($\top$) is converted to actual activity rates (r_1 and r_2 in our example). The rate (r_1 or r_2) of the *compute* activity is calculated as a function of the speed of the processors in the machine allocated to the application and the workload for that application.

The PEPA model is provided as an input to the PEPA workbench [16] in a `*.pepa` file format. Each application and machine PEPA component in the model is translated into an underlying mathematical Markovian model by the PEPA workbench. A more detailed description of the functionality of the PEPA workbench for our study can be found in [21]. The robustness of the modeled resource allocation is obtained as a probability of attaining a predefined makespan value, which is calculated by a *passage time analysis* [18] of the computational activities of all the machine components in the generated Markovian model. The passage time analysis generates a Cumulative Distribution Function (CDF) of the passage time (T_p) from a source state (S_s) into a non-empty set of target states (S_T), such that,

$$T_p = \inf\{u > 0 : S_s(u) \in S_T | S_s(0) = \text{initial state}\} \tag{2}$$

The CDF is generated by a convolution of the state holding times over all possible paths from state $i \in S_s$ into any of the states in the set S_T [18]. For the passage time analysis, the *stop time* is analogous to the user specified makespan goal, β_i^{max}. The solution is obtained as a cumulative distribution function (CDF) of the probability of machine finishing times for the modeled resource allocation. The robustness, as formulated in Eq. 1, of the resource allocation is obtained from the generated CDFs, as the minimum probability of achieving a user defined makespan goal.

4   Results and Analysis: Robustness Analysis via Numerical Modeling of Resource Allocations

In this study, a number of PEPA models have been generated for several feasible mappings of applications onto parallel heterogeneous machines. The variations in the workload values are sampled from a *uniform probability distribution* for all the applications. For this study, the Python function, `random.uniform(a,b)` has been used, with different values of `a` and `b` resulting in different mappings for a given number of applications and machines. The variations in application workload and machine availability can be sampled from any probability distribution model that can capture the static features of real workloads and machine characteristics at a particular point in time.

In PEPA, a *choice* operator permits a model component to exist in one of the many possible states of the component [5]. For example, for a component $P :=$ $P1+P2$, P can only exist in state $P1$ or $P2$. In this work, the left hand side of the $+$ operator models the ideal execution scenario in the absence of perturbations in the sensor workload and the machine availabilities, where $\hat{\lambda}_i = \lambda_i, \forall i \in \{1, 2, 3\}$ and $\hat{\eta}_j = \eta_j$, for all machines. The right hand side of the $+$ operator models the perturbed execution scenario generated by a compound effect of variations in application workload and machine availabilities. The compound effect is modeled as simultaneous equal variations in one or more sensor workload(s) across all applications, where $\hat{\lambda}_i \neq \lambda_i$, and equal variations in processor availability across all applications, where $\hat{\eta}_j \neq \eta_j$. The rates of the *compute* activities are calculated as a function of λ_i and the actual computation times T_{ij} of each application on the machine where it is mapped. Further, T_{ij} is calculated as a function of the estimated variation in application workload ($\hat{\lambda}_i$) and the estimated variation in the machine availabilities ($\hat{\eta}_j$), as defined in Table 1. The values of initial rates $(r_1, r_2, \cdots)$, which remain constant during runtime, are calculated as a function of the initial machine availability factor (η_j), as described later in Sect. 4.1. The values of perturbed rates $(p_1, p_2, \cdots)$, which vary during runtime, are calculated as a function of the varied application workload ($\hat{\lambda}_i$) and the varied machine availability factor ($\hat{\eta}_j$), as described later in Sect. 4.2.

The modeling generated for this study represents a resource allocation system that has a higher load imbalance factor at runtime when compared to the analysis in our prior work in [21–24]. Herein, the authors have analyzed the robustness of several mappings to study the impact of heightened load imbalance in computational systems with (i) 16 applications and 4 machines, (ii) 20 applications and 5 machines (for validation against our prior work), and (iii) 32 applications and 8 machines. Due to the state space explosion limitation of the CTMC analysis in PEPA, the authors successfully modeled only systems of less than 8 machines. Research work is ongoing to explore different implementations of PEPA that will improve the scalability of the numerical modeling and analysis of the robustness of resource allocations for larger PDC systems and applications.

4.1 Deriving PEPA Activity Rates in an Ideal Computing Environment ($\hat{\lambda}_i = \lambda_i$ and $\hat{\eta}_j = \eta_j$)

The calculation for the rate is given in Eq. 3, where T_{ij} is the actual time to compute an application i on machine j with initial availability (η_j) that remains constant at runtime, and λ_i is calculated as a function of the initial sensor loads $\lambda_1, \lambda_2, \lambda_3$.

$$r_i = \frac{\lambda_i}{T_{ij}} \quad \forall i, j, \ where \ T_{ij} = \lambda_i \times \eta_j \tag{3}$$

T_{ij} is calculated as a product of the runtime workload for that application (λ_i), and the machine availability factor (η_j). Rate r_i is calculated as a ratio of the application workload λ_i and T_{ij}. In the ideal execution scenario, $\hat{\lambda}_i = \lambda_i, \forall i \in \{1, 2, 3\}$. Therefore, T_i is equal to the initial ETC values and consequently, the rates, $(r_1, r_2, \cdots)$, are only calculated using the initial machine availability (η_j).

4.2 Deriving PEPA Activity Rates in a Perturbed Computing Environment ($\hat{\lambda} \neq \lambda$ and $\hat{\eta}_j \neq \eta_j$)

The calculation of the perturbed rate is given in Eq. 4, where T_{ij} is the actual time to compute an application i on machine j with the varied runtime availability $(\hat{\eta}_j)$, and $\hat{\lambda}_i$ is calculated as a function of the varying sensor loads $\hat{\lambda}_1, \hat{\lambda}_2, \hat{\lambda}_3$, and λ_i is calculated as a function of the initial sensor loads $\lambda_1, \lambda_2, \lambda_3$.

$$p_i = \frac{\lambda_i}{T_{ij}} \quad \forall i, j, \text{ where } T_{ij} = \hat{\lambda}_i \times \hat{\eta}_j \tag{4}$$

T_{ij} is calculated as a product of the estimated runtime application workload $(\hat{\lambda})$, and the runtime machine availability factor $(\hat{\eta}_j)$. Henceforth, p_i is calculated as a ratio of the initial application workload λ_i and T_{ij}.

4.3 Numerical Analysis of Performance Modeling of Resource Allocations Using the PEPA Workbench

The *.pepa model file is compiled and solved using the PEPA workbench tool. The state space derived using the Markovian analysis provided by the PEPA workbench represents the continuous time Markov chain (CTMC) processes for each component of the modeled resource allocation.

Once the state space of all the components (CTMC processes of applications and machines) of the PEPA model are generated, the tool allows the modeler to specify the type of Markovian analysis that needs to be used for solving the generated Markov models to derive performance measures [5, 21]. In this work, we choose the *passage time analysis* that is used to solve the Markov models using the timing information associated with the activity rates to derive performance measures, such as, makespan and response time [18]. In this study, the passage time analysis of the Markov models for the resource allocations yields CDFs of the machine finishing times, $F_i(M_j(\hat{\eta}_j), \hat{\lambda}_i)$, as passage times between the states associated with applications assigned to a machine. An example of the CDF of the finishing time of machine M_1 in one of our randomly generated resource allocation model is shown in Fig. 2. A comparison of the finishing times of each machine with respect to the four mappings (selected from a larger set of generated mappings due to space constraints) is illustrated in Fig. 3 for two different execution scenarios, (i) 16 applications and 4 heterogenous machines, and (ii) 20 applications and 5 heterogenous machines.

A key benefit resulting from the performance modeling and analysis in this work is to identify the most robust resource allocation among the ones delivering equal execution performance. The robustness metric ψ_x is calculated using Eq. 5. For this study, the makespan goal is set as $\beta_i^{max} = 45\,\mathrm{s}$.

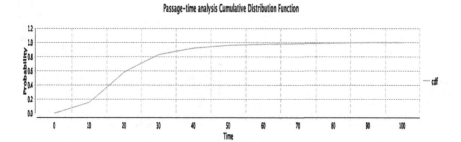

Fig. 2. Cumulative distribution function (CDF) of the finishing time of machine M_1 for a resource allocation mapping 20 applications onto 5 machines.

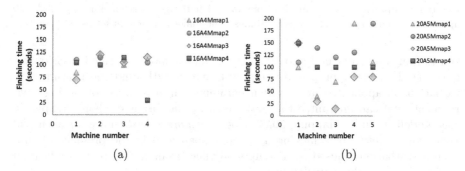

(a)   (b)

Fig. 3. A comparative analysis of the finishing times of each machine derived from the execution of 4 different mappings of (a) 16 applications onto 4 heterogeneous machines, and (b) 20 applications onto 5 heterogeneous machines

$$\psi_x = \min_{\forall i,j \ in \ Mapping_x,} \Pr[F_i(M_j(\hat{\eta}_j), \hat{\lambda}_i) \le 45] \qquad (5)$$

The robustness value for each machine is calculated as the probability of that machine to finish before the makespan goal. A comparison of the robustness values of all the machines for each of the four mappings of (i) 16 applications to 4 heterogeneous machines, and (ii) 20 applications to 5 heterogeneous machines is illustrated in Fig. 4.

The results illustrated in Figs. 5a and 5b indicate that for both test cases of resource allocations modeled, one for a PDC system with 16 applications and 4 heterogeneous machines, and the other for a PDC system with 20 applications and 5 heterogeneous machines, resource allocations promising better parallel execution time over the others can be identified. Moreover, the faster resource allocations that deliver equal execution makespan also differ vastly in their degrees of robustness. For example, in the case of mappings 20A5Mmap3 and 20A5Mmap4 in Fig. 5b, both mappings deliver equal execution performance in terms of system makespan. However, Mapping 20A5Mmap4 is 2.43 times more robust that Mapping 20A5Mmap3. Therefore, the value of robustness for 20A5Mmap4 significantly increases by 143% over 20A5Mmap3. For highly critical applications, this

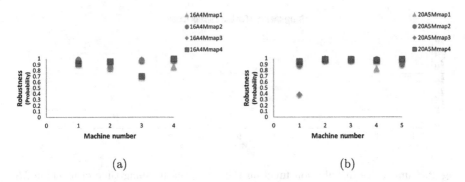

Fig. 4. A comparative analysis of the robustness values obtained for each machine for the four mappings of (a) 16 applications onto 4 heterogeneous machines, and (b) 20 applications onto 5 heterogeneous machines. The makespan goal is set as $\beta_i^{max} = 45\,\text{s}$.

substantial improvement in robustness, enables a more informed decision for selecting the most appropriate mapping that can withstand the runtime perturbations in application and system parameters in a parallel computing environment, and can guarantee the best execution performance. For example, in this modeling study, Mapping `20A5Mmap4` is a *more robust* choice for an initial allocation in terms of achieving a set makespan goal in the presence of runtime perturbations caused by a compound effect from variations in application workload and machine availability.

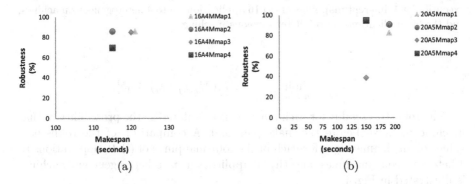

Fig. 5. A comparison between the robustness values and the performance in terms of the system makespan of the four mappings of (a) 16 applications onto 4 heterogeneous machines, and (b) 20 applications onto 5 heterogeneous machines.

5 Conclusions and Future Work

The analytical study performed in this work to obtain the robustness of *static* resource allocations, which are analytically modeled and numerically solved using PEPA, provides a useful tool that can be incorporated in the design phase of a resource allocation system in a parallel and distributed computing environment. Although the analysis is limited to *static* resource allocations, a key benefit of this study is learning that, in the case of resource allocations promising equal execution performance, the numerical evaluation of the PEPA models enables a selection of the most robust resource allocation in the presence of runtime perturbations caused by a *compound effect* arising from dynamic variations in *application workload* and *machine availability*. For each of the several modeled resource allocations, a numerical analysis of the PEPA models yields the execution performance as the parallel execution time. The robustness value for each mapping is calculated as the minimum probability of the machines to finish before a desired makespan goal. The results presented in Sect. 4 indicate that two or more different mappings yield equal execution performance. However, a significant improvement (up to 143%) in the robustness value among the mappings yielding equal parallel execution times has been observed. Thus, this substantial increase in robustness strongly suggests the selection of one mapping versus the other for *guaranteeing* the best execution performance.

To the best of our knowledge, this work is the first effort towards modeling the execution of applications on heterogeneous machines by simultaneously incorporating variations in both application workload and machine availabilities, considered together as a *compound perturbation*. To facilitate and ease the reproducibility of our research, we provide a Singularity container of the PEPA Workbench that has been validated to produce identical results to the non-containerized version of the PEPA Workbench [25]. The PEPA model and build recipe for the Singularity container are available at https://github.com/williamssanders/pepa, and the container is publicly available at https://www.singularity-hub.org/collections/2351.

In the future, the authors plan to continue this study towards performance modeling of resource allocations and numerical evaluation of their robustness on larger scale PDC systems and data sets from real scientific applications. To facilitate a scalable evaluation, the authors plan to investigate a number of other existing implementations of PEPA that are not limited by state space explosion. Further, the authors also plan to explore an integration of the PEPA models into a runtime scheduler/controller in a model-based framework, where the embedded models can be re-evaluated with minimal overhead when a system parameter changes at runtime.

Acknowledgment. The authors would like to acknowledge The Jackson Laboratory for providing support and helping to make this work possible. Additionally, the authors would like to acknowledge the partial support of the NSF #1034897 grant.

References

1. Ali, S., Maciejewski, A., Siegel, H., Kim, J.K.: Measuring the robustness of a resource allocation. IEEE Trans. Parallel Distrib. Syst. **15**(7), 630–641 (2004)
2. Ali, S., Kim, J.-K., Siegel, H.J., Maciejewski, A.A.: Static heuristics for robust resource allocation of continuously executing applications. J. Parallel Distrib. Comput. **68**(8), 1070–1080 (2008)

3. Banicescu, I., Ciorba, F.M., Carino, R.L.: Towards the robustness of dynamic loop scheduling on large-scale heterogeneous distributed systems. In: Proceedings of the IEEE International Symposium on Parallel and Distributed Computing (ISPDC 2009), pp. 129–132 (2009)
4. Srivastava, S., Banicescu, I., Ciorba, F.: Investigating the robustness of adaptive dynamic loop scheduling on heterogeneous computing systems. In: Proceedings of The 2010 IEEE/ACM International Symposium on Parallel Distributed Processing, Workshops and Phd Forum (IPDPSW-PDSEC, On CD-ROM), pp. 1–8, April 2010
5. Hillston, J.: A Compositional Approach to Performance Modelling. Cambridge University Press, Cambridge (1996)
6. Coffman, E., Bruno, J.: Computer and Job-Shop Scheduling Theory. A Wiley-Interscience Publication. Wiley, Hoboken (1976)
7. Fernandez-Baca, D.: Allocating modules to processors in a distributed system. IEEE Trans. Softw. Eng. **15**(11), 1427–1436 (1989)
8. Ibarra, O.H., Kim, C.E.: Heuristic algorithms for scheduling independent tasks on nonidentical processors. J. ACM **24**(2), 280–289 (1977)
9. Leon, J.V., Wu, D.S., Storer, R.H.: Robustness measures and robust scheduling for job shops. IIE Trans. **26**(5), 32–43 (1994)
10. Daniels, R.L., Carrillo, J.E.: Beta-robust scheduling for single machine systems with uncertain processing times. IIE Trans. **29**(11), 977–985 (1997)
11. Gertphol, S., Prasanna, V.: MIP formulation for robust resource allocation in dynamic real-time systems. In: Proceedings of the International Parallel and Distributed Processing Symposium (2003)
12. Gertphol, S., Prasanna, V.: Iterative integer programming formulation for robust resource allocation in dynamic real-time systems. In: Proceedings of the International Parallel and Distributed Processing Symposium (2004)
13. Benoit, A., Cole, M., Gilmore, S., Hillston, J.: Evaluating the performance of skeleton-based high level parallel programs. In: Bubak, M., van Albada, G.D., Sloot, P.M.A., Dongarra, J. (eds.) ICCS 2004. LNCS, vol. 3038, pp. 289–296. Springer, Heidelberg (2004). https://doi.org/10.1007/978-3-540-24688-6_40
14. Benoit, A., Cole, M., Gilmore, S., Hillston, J.: Scheduling skeleton-based grid applications using PEPA and NWS. Comput. J. **48**(3), 369–378 (2005)
15. Wallace, V.L., Rosenberg, R.S.: Markovian models and numerical analysis of computer system behavior. In: Proceedings of the Spring Joint Computer Conference, pp. 141–148. ACM (1966)
16. Gilmore, S., Hillston, J.: The PEPA workbench: a tool to support a process algebra-based approach to performance modelling. In: Haring, G., Kotsis, G. (eds.) TOOLS 1994. LNCS, vol. 794, pp. 353–368. Springer, Heidelberg (1994). https://doi.org/10.1007/3-540-58021-2_20
17. Hillston, J.: Tuning systems: from composition to performance. Comput. J. **48**(4), 385–400 (2005)
18. Bradley, J.T., Dingle, N.J., Gilmore, S.T., Knottenbelt, W.J.: Derivation of passage-time densities in PEPA models using IPC: The imperial PEPA compiler. In: 11th IEEE/ACM International Symposium on Modeling, pp. 344–351. MASCOTS, Analysis and Simulation of Computer Telecommunications Systems (2003)
19. Ghafoor, A., Yang, J.: A distributed heterogeneous supercomputing management system. Computer **26**(6), 78–86 (1993)
20. Iverson, M.A., Ozguner, F., Potter, L.: Statistical prediction of task execution times through analytic benchmarking for scheduling in a heterogeneous environment. IEEE Trans. Comput. **48**(12), 1374–1379 (1999)

21. Srivastava, S.: Evaluating the robustness of resource allocations obtained through performance modeling with stochastic process algebra. Ph.D. dissertation, Mississippi State University, Department of Computer Science and Engineering, May 2015
22. Banicescu, I., Srivastava, S.: Towards robust resource allocations via performance modeling with stochastic process algebra. In: Proceedings of the 18th IEEE International Conference on Computational Science and Engineering (CSE-2015), Porto, Portugal, pp. 270–277, October 2015
23. Srivastava, S., Banicescu, I.: Robust resource allocations through performance modeling with stochastic process algebra. Concurr. Comput.: Practice Exper. **29**(7), e3894 (2017). https://doi.org/10.1002/cpe.3894
24. Srivastava, S., Banicescu, I.: PEPA based performance modeling for robust resource allocations amid varying processor availability. In: Proceedings of the 17th IEEE International Symposium on Parallel and Distributed Computing (ISPDC), Geneva, pp. 61–68 (2018)
25. Sanders, W.S., Srivastava, S., Banicescu, I.: A container-based framework to facilitate reproducibility in employing stochastic process algebra for modeling parallel computing systems. In: Proceedings of the 2019 IEEE/ACM International Symposium on Parallel Distributed Processing, Workshops and PhD Forum (IPDPSW-HIPS) (2019, in press)

An Adaptive Computational Network Model for Strange Loops in Political Evolution in Society

Julia Anten[1], Jordan Earle[1], and Jan Treur[2(✉)]

[1] Computational Science, University of Amsterdam,
Amsterdam, The Netherlands
juliaanten@hotmail.com, jordanalexearle@gmail.com
[2] Social AI Group, Vrije Universiteit Amsterdam, Amsterdam, The Netherlands
j.treur@vu.nl

Abstract. In this paper a multi-order adaptive temporal-causal network model is introduced to model political evolution. The computational network model makes use of Hofstadter's notion of a Strange Loop and was tested and validated successfully to reflect political oscillations seen in presidential elections in the USA over time.

1 Introduction

Hofstadter [7] originally described a Strange Loop as a phenomenon that, after going through a hierarchy of levels, you would return to the starting level; see also [8, 9]. In his original literature, Holfstadter illustrates this for common domains such as graphical art (Escher), music (Bach), and logical paradoxes (Gödel) [12, 15]. Holfstadter theorised that the brain may also use Strange Loops in the creation of human intelligence and consciousness. Although at a conceptual level much literature can be found referring to Strange Loops in one way or the other, almost none of it actually shows a computational model for this phenomenon. An exception is [19], Ch. 8, where the concept of multi-order adaptive reified temporal-causal network is exploited to show some small toy examples of computational Strange Loop models.

In the current paper a more serious and more complex domain is addressed, namely of political evolution over time. A Strange Loop temporal-causal network was created, tested and validated to reflect political oscillations seen in presidential elections in the US. The temporal-causal network breaks a political system into 3 groups, the individual people, the politicians, and the laws. The individuals' combined unhappiness causes them to vote for politicians who align with their desires. The elected politicians in turn vote for the laws which they are aligned to. These laws then cause an effect on the individuals in the form of the weight for their unhappiness, which then begins the cycle again.

Once the network design was created, the parameters of the network were varied in order to obtain oscillations as predicted in empirical Social Science literature. Simulation were conducted for the model, changing the initial values of the individuals of the poor and rich groups to see if the predicted effects concerning different types of

V. V. Krzhizhanovskaya et al. (Eds.): ICCS 2020, LNCS 12138, pp. 604–617, 2020.
https://doi.org/10.1007/978-3-030-50417-5_45

laws were seen. The model was then tuned for specific empirical data from the popular votes from the USA elections over time. All these will be discussed in subsequent sections. Finally, the next steps for the network and an enhancement to the network to create an infinite reified network will be discussed.

2   Background: Domain Description

Since [7] many have applied this to various application areas such as advertising [6], self-representation in consciousness [10] and psychotherapeutic understanding [10, 17]. However, in this literature no computational models are proposed. After seeing how the brain and advertising might be modeled in such a loop, the idea to model a political system with a strange loop was considered. The original idea was that people have to follow laws, which are created by politicians, who are elected by the people. When considering this system, it can clearly be seen that there is a loop in the levels. The causal pathways affecting people's lives are affected by the laws, which are created by causal pathways for politicians; so, people are in effect indirectly voting for these laws by voting for politicians. Therefore, a literature review was conducted to determine if this observation had been made before and if any models of it existed.

The idea to create a Strange Loop out of a political system is based on observations made in the USA political system. The system in the USA can be seen to switch between Democratic and Republican leadership every few elections. This switching of power has caused the policy on a national and state level change over time, such as with abortion law and financial policy. The same kind of oscillations have been observed in England and in coalitions during war. This type of behavior has been noted as early as 1898, where Lowell [11] observed oscillations in elections in the USA. It was, and still is, easily observed when viewing the elections in the USA over time, as seen in Fig. 1 from the above paper.

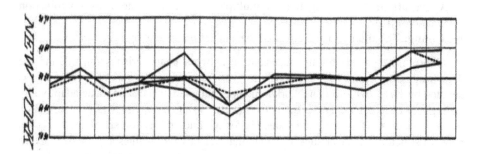

Fig. 1. Voting in New York between 1870 and 1897. The number of republicans is shown below the black lines, while the number of democrats is shown above. Expected values for these elections are shown by the dotted line. Adopted from [11].

The second type of feedback she references is the ability of state capacities to transform over time. "State capacities" refers to the ability of the states to implement and enforce their laws. She writes that "policies transform or expand the capacities of the state. They therefore change the administrative possibilities for initiatives in the future, and affect later prospects for policy implementation". This can be seen as the effect that the laws have on the political structures. This second influence was considered for implementation in this model, but was disregarded as this first model was kept to its basic form to show that the theory was sound.

Pierson [13] notes that "politics produce politics", discussing how the policy affects its own creation and upkeep. He states that it has been "increasingly harder to deny that that public policies were not only outputs of but important inputs into the political process". He notes that interest groups often follow rather than proceed the adoption of public policy, referencing that Skocpol [14] identifies changes in "social groups and their political goals and capability" as one of the two major types of political feedback. This can be seen as the political power of the people affecting the laws that govern them, which is the centerpiece of the network which is introduced in the current paper.

More evidence of this phenomenon has been noted more recently by Baumgartner and Jones [1]. They noted that american policy is characterized by contrasting characteristics of stability and dramatic changes which can be expressed in positive and negative feedback loops. These loops can be seen between the politics and the individuals, leading to more support for this form of conceptualisation.

3   The Adaptive Network Modeling Approach Used

The adaptive computational model is based on the Network-Oriented Modelling approach based on reified temporal-causal networks described in [18, 19]. The *network structure characteristics* used are as follows. A full specification of a network model provides a complete overview of their values in socalled role matrix format.

- **Connectivity:** The strength of a connection from state X to Y is represented by weight $\omega_{X,Y}$
- **Aggregation:** The aggregation of multiple impacts on state Y by combination function $c_Y(..)$.
- **Timing:** The timing of the effect of the impact on state Y by speed factor η_Y

Given initial values for the states, these network characteristics fully define the dynamics of the network. For each state Y, its (real number) value at time point t is denoted by $Y(t)$. Each of the network structure characteristics can be made adaptive by adding extra states for them to the network, called *reification states* [19]: states $\mathbf{W}_{X,Y}$ for $\omega_{X,Y}$, states $\mathbf{C}_Y$ for $c_Y(..)$, and states $\mathbf{H}_Y$ for η_Y. Such reification states get their own network structure characteristics to define their (adaptive) dynamics and are depicted in a higher level plane, as shown in Fig. 2. For example, using this, the adaptation principle called Hebbian learning [5], considered as a form of plasticity of the brain in cognitive neuroscience ("neurons that fire together, wire together") can be modeled. The concept of reification has been shown to provide substantial advantages in expressivity and transparency of models within AI; e.g., [2–4, 7, 16, 20]. The notion of network reification exploits this concept for the area of adaptive network modeling.

A dedicated software environment is available by which the conceptual design of an adaptive network model is automatically transformed into a numerical representation of the model that can be used for simulation; this is based on the following type of (hidden) difference or differential equation defined in terms of the above network characteristics:

$$Y(t+\Delta t) = Y(t) + \eta_Y[\textbf{aggimpact}_Y(t) - Y(t)]\Delta t \quad \text{or} \quad dY(t)/dt = \eta_Y[\textbf{aggimpact}_Y(t) - Y(t)]$$
$$\text{with } \textbf{aggimpact}_Y(t) = \textbf{c}_Y(\omega_{X_1,Y}X_1(t), \ldots, \omega_{X_k,Y}X_k(t))$$

$$(1)$$

where the X_i are all states from which state Y has incoming connections. Different combination functions are available in a library that can be used to specify the effect of the impact on a state (see [18, 19]). The following three of them are used here:

- *the identity* function for states with impact from only one other state $\text{id}(V) = V$ (2)

- *the scaled sum* with scaling factor λ $\text{ssum}_\lambda(V_1, \ldots, V_k) = \dfrac{V_1 + \cdots + V_k}{\lambda}$ (3)

- *the advanced logistic sum* combination function with steepness σ and threshold τ

$$\textbf{alogistic}_{\sigma,\tau}(V_1, \ldots, V_k) = [\frac{1}{1 + e^{-\sigma(V_1 + \cdots + V_k - \tau)}} - \frac{1}{1 + e^{\sigma\tau}}](1 + e^{-\sigma\tau}) \qquad (4)$$

4   Design of the Multi-order Adaptive Network Model

The idea behind this model was the following scenario. There is a group of people who have a law which makes them unhappy. As the people get more unhappy, they vote more, electing politicians who will support the laws which will make their unhappiness less. The politicians then vote for the laws which they support. After some discussion between the groups of politicians, the law is agreed upon which is a combination of the desires of the groups, and then the law comes back to affect the individual people's unhappiness, starting the cycle again. In this scenario, causal pathways in society at three different interacting levels play a role:

(1) Causal pathways that determine the unhappiness of people
(2) Causal pathways that determine the politicians' positions
(3) Causal pathways that determine the laws

Here the effects resulting from the causal pathways of type (1) are the unhappiness of the people; these effects affect the causal pathways of type (2) by voting. In turn, the effects resulting from the causal pathways of type (2) affect the causal pathways of type (3). Finally, the effects resulting from the causal pathways of type (3) affect the causal pathways of type (1), which closes the Strange Loop.

For the scenario addressed by the designed network, it was decided to have two laws which would affect the individuals' lives. Two groups are considered, a group who benefit from one law, and a group that benefit from the other law, which, to help explain the model more succinctly, will be referred to as the rich group and the poor group. These individuals would then vote for the political party which favour the law that favour them. Therefore there are two political parties as well. For each of the 3 distinct levels, networks were created with mutual connections in mind. First the networks themselves will be discussed and then the connections between the levels.

The Individuals Subnetwork. The first subnetwork modeled addresses the individual level. Figure 2 shows the individual level for 10 individuals. Each individual has a starting value with represents the context in which they function. These are the odd nodes X_{2i-1} seen in the bottom of the network figure. In the simplest form of this network, this can be thought of as a context that generates some level of gross income. The unhappiness of the individual, which can be seen in the top of Fig. 2 as the even nodes X_{2i}, is determined through a one-step causal pathway by the starting context value X_{2i-1} multiplied by the weight of the connection from X_{2i-1} to X_{2i} which represents how much the current laws affect this person's life for that context. This connection weigh is represented by reification state X_{33} (for $i > 5$) or X_{34} (for $i \leq 5$). The way these weights are derived will be determined by the other subnetworks and their interaction. Again, in the simplest form it can be thought of as a tax on their income. As stated previously, the network has 2 groups of individuals which are in accordance with the different weights for them.

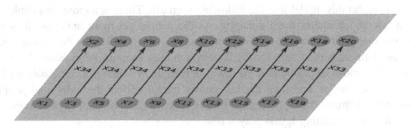

Fig. 2. Subnetwork for the individual level

The Politicians Subnetwork. The next subnetwork devised concerns the causal pathways for the politicians and their parties. Figure 3 shows the politicians subnetwork. There is a limited resource of political power which is represented by an input node X_{21} for the politician level.

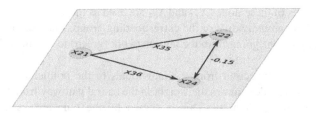

Fig. 3. Subnetwork for the politicians level

The people then vote for the political party they support, which then adjusts the causal pathway for the resulting power each party has, which can be seen in the effect nodes X_{22} and X_{24}. Within the causal pathways, the weights for each party (X_{35} and X_{36}) are determined by the previous level. A negative connection between the two parties, represents that the parties attempt to minimize the others influence.

The Laws Subnetwork. The final subnetwork devised was for the law level; it can be seen in Fig. 4. In this network, there is a limited budget for laws, which is the input node X_{26}. Given this budget, the political parties vote on either law 1 or law 2 (X_{27} or X_{28}). Here the weights X_{22} and X_{24} (and also the scaling factors) are determined by the previous level. After the vote, a logistic function is applied to the output of each of the laws individually with a weight of 1, determining the new power of each law which is seen in the network as X_{52} and X_{53}. Once the new power is determined, the effect of each law on the two groups is updated where the weights represent the effect of each law on each of the groups. Finally for each of the two groups, the effect of the new combination of laws is combined. These values for X_{33} and X_{34} become the new effects of the laws on the two types of individuals.

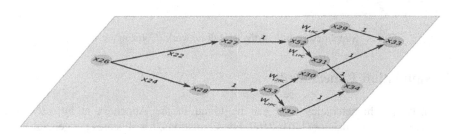

Fig. 4. Subnetwork for the law level

Connections Between the Subnetworks. Now the subnetworks have been defined, the connections between them can be discussed. A simplified version of the network can be seen in Fig. 5, which shows how the networks are connected. Beginning with the individual's levels connections, the weight for the causal pathway from the input of the individual to the unhappiness of the individual is determined by the laws. As discussed for the law subnetwork, nodes X_{33} and X_{34} represent how much an individual's causal pathway of each group (rich or poor) is impacted by the current law system; the values of these nodes X_{33} and X_{34} are used as the weights for how much the current laws affect an individual of the corresponding group. This can be seen in Fig. 5 as the blue connections going from the laws network (green) to the individuals network (pink).

Examining the connection from the individuals to the politician subnetwork, the unhappiness of voters determines the weight in the causal pathway from the input to the political powers for each party. This is shown by the blue arrows connecting the individuals network (pink) to the political power network (blue).

Finally, examining the connection from the politicians subnetwork to the laws subnetwork, the weight within the causal pathway which determines the vote for each law (X_{22} and X_{24}) comes from the politician subnetwork. This is the power for each of the individual parties which supports each law, which, in this model, is one party for each law and is shown by the blue arrows connecting the political power network (blue) to the laws network (green). More on how the values were determined can be found in Sect. 5.

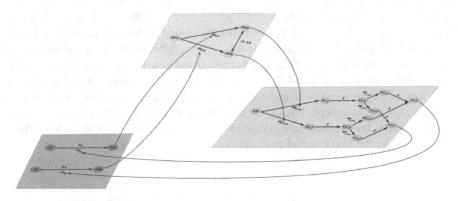

Fig. 5. Simplified picture of the overall network (Color figure online)

5   Simulation Experiments

The network characteristics used can be found in the Appendix at https://www.researchgate.net/publication/340162169. For simulations, for all states the general Eq. (1) from Sect. 3 was used where the chosen combination functions were (see Sect. 3, formulae (2), (3), and (4)):

identity function	$\mathbf{id}(V)$	X_1 to X_{21}, X_{26}, X_{29} to X_{32}, X_{39} to X_{48}
scaled sum function	$\mathbf{ssum}_\lambda(V_1, \ldots, V_k)$	X_{22} to X_{25}, X_{33} to X_{38}, X_{49} to X_{51}
logistic function	$\mathbf{alogistic}_{\sigma,\tau}(V_1, \ldots, V_k)$	$X_{27}, X_{28}, X_{52}, X_{53}$

For the first simulation experiments, the steepness σ of the logistic functions was 16, and the threshold τ was 0.35 for X_{27}, X_{28} and 0.7 for X_{52}, X_{53}.

From the literature, it was seen that the system should oscillate, therefore in the first run of the model this behaviour was searched for. For the first simulation both groups were initialized with the same values (or worth). This meant the groups have the same unhappiness if their preferred law is not active. For some parameter settings the behaviour was observed as seen in Fig. 6. The unhappiness of the people can be seen to oscillate between the two groups, as well as the laws the political power. This figure actually shows the unhappiness of one representative person for each group, not the total unhappiness of the group. All persons in the group show the exact same behavior, since they are initialized the same and influenced by the same law. It can be seen that a rise in political power for a group closely follows the rise of unhappiness in that same group and that the laws preferred by a group follow slower, but they do rise when the political power of that group rises. This can be explained by the slower speed factors associated to the laws. All the oscillations now have the same amplitude, since all groups and laws are initialized either exactly the same or in the case of the laws at 1 for the poor law and 0 for the rich law.

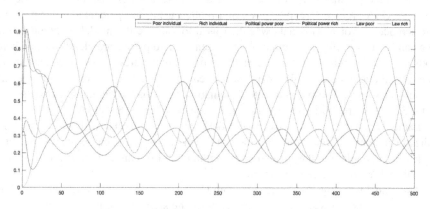

Fig. 6. Behavior for the first run with oscillations

To get the model to simulate real societies better, in the following simulation the two groups were initialized differently. The "rich" group was initialized with a score (or income) of 0.8 and the "poor" group with a score (or income) of 0.4. This meant that the rich people will have the ability to have a much higher unhappiness than the poor people, so it is expected the "rich" group will have a higher political power and get their preferred law more active than the law preferred by the "poor" people. The behavior resulting from this simulation can be seen in Fig. 7.

The "rich" law is always more active than the "poor" law, and although there are still some oscillations, the "poor" group only gets influence when they are very unhappy and are always less influential than the "rich" group, which is to be expected when there is a group which is more influential than the other with the same number of people.

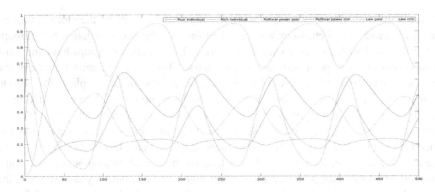

Fig. 7. Behaviour for different initial values of the rich and poor groups

6 Further Validation of the Network Model

Data from the popular votes of the United States presidential elections was collected from the USA archive (archives.gov), and plotted using the percentage of republican and democratic votes. This data was then used to validate the model. A graph of this data can be seen in Fig. 8. Oscillations between the two parties are clearly visible here. Both the initial simulation and the analysis of popular votes, shows the same trend, where oscillations between the two "parties" can be seen. The difference is in the size of the oscillations. The popular votes simulation has small oscillations between 0.65 and 0.35, while the initial model has oscillations between 0.8 and 0.2.

Parallels can be drawn between the behavior of the model with the groups initialized differently. Continuing to follow the rich and poor example, in real life there are less rich people, but they are still very influential. Looking at Fig. 7, it can be seen that the rich easily overpower the poor.

The model was tuned on these data from US elections. Speed factors were tuned for X_{21} until X_{48}, and for X_{52} and X_{53}. Furthermore, the sigma's and tau's for the alogistic functions of X_{27}, X_{28}, X_{52} and X_{53} were also tuned. These are the nodes for voting for

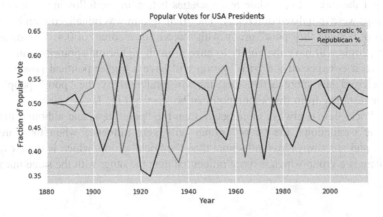

Fig. 8. Statistics for the US presidential elections

and activation of the laws. In total there were 38 parameters that were tuned. All speed factors were initialized with a speed of 0.5 and the minimum and maximum were set at 0 and 2. Steepness parameters σ were initialized with 16 and thresholds τ at either 0.35 (for X_{27} and X_{28}) or at 0.7 (for X_{52} and X_{53}). Minima for both were set at 0 and maxima for the σ's was at 30, while the maxima for the τ's was 1. Simulated annealing was used for the tuning with a reannealing interval of 500 iterations and otherwise standard settings from the Global Optimization Toolkit from Matlab. After tuning a RMSE of 0.06736 was found. Values found during tuning can be found in Table 1 (order is the same as stated above). The behavior of the system can be seen in Fig. 9. As can be seen, the tuning did not work as expected. The RMSE is very low, which would normally mean the model fits the data really well, but the system shows no oscillations at all. This could probably be explained by the small differences between the republican and democratic votes in percentage. For most values the difference is around 5%. Because of this, it is expected that the model with these values fits so well, because it is the middle between the two values and the values are very close. Another possible reason could be the time frame. In the original simulation, the oscillations occur round every 100 time steps. In the data from the elections, 1 time step was set to be a year. Therefore the speed of the oscillations would have to be much higher, and its possible that the speed factors were not high enough to capture this completely.

Table 1. Tuned parameter values found

1.61369	1.96366	1.10475	0.13759	1.00872
1.30867	0.08119	1.47845	1.80759	1.24313
0.72002	$6.84443\ 10^{-5}$	0.58717	0.98266	0.21999
$4.49622\ 10^{-7}$	0.97381	1.54367	1.99999	0.16781
0.77973	1.78931	0.09258	0.90414	1.40409
0.54539	1.03380	1.70694	$2.60123\ 10^{-5}$	0.62080
16.2610	0.95972	25.26481	0.15189	10.55741
0.01325	8.79826	0.99999		

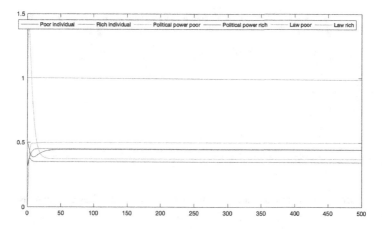

Fig. 9. Behaviour after first parameter tuning

To overcome this, we exaggerated the empirical data and put it at 0.5, 1 and 0 in alternating order with 48 time steps between, which is 4 years in terms of months. Figure 10 shows the empirical data and the simulation data for this and as can be seen, oscillations did occur with the exaggerated data. RMSE behavior can be seen in Fig. 10 and shows that the lowest RMSE value was found at the beginning and after that never again. This could occur due to a couple of reasons. Either, this minimum score is difficult to reach and after leaving the optimum, it is unlikely to find back again due to the specificity of the values. It could be to do with the reannealing interval after 100 iterations, which makes the temperature rise again, so less optimal solutions are again accepted without giving time to search the space for more optimum values. Evidence for this could be seen in Fig. 10, since sometimes it seems to trend down as expected from simulated annealing and after which the RMSE rises again.

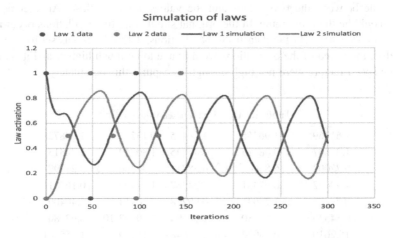

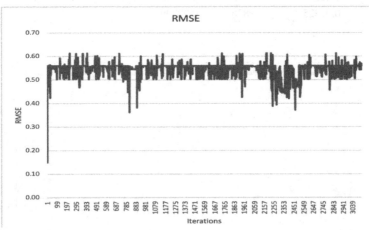

Fig. 10. Upper graph: behaviour after second parameter tuning. Lower graph: RMSE over iterations

It could also be due to trying to fit the wrong parameters, since less parameters were fitted for this tuning. Only the σ's and the τ's of the activation of the laws and their speed factors were tuned as it was thought that these would be the parameters that would affect the general shape the most.

7 Discussion and Future Work

In the first simulation run, in a qualitative sense the network behaved as expected from the research done, with the political powers oscillating between rich party being in power and the poor party being in power in periodic oscillations. In the initial simulation, seen in Fig. 6, it can be seen that as the poor parties unhappiness is rising, the political power of the poor group rises as well, then about 90 degrees out of phase, the poor law begins to increase. As the law increases, the unhappiness begins to decrease and the rich groups unhappiness increases as they are dissatisfied with the situation and begin to vote more. This same behaviour is also seen in the rich group, about 180 degrees out of phase. The laws oscillate around 0.5 for both the poor and the rich.

When initializing the individuals groups (rich and poor) with different starting values as seen in Fig. 7, the periodic oscillation behaviour is still seen, but the center of these oscillations for each group is different. The value at which the laws oscillate around is approximately 0.8 for the rich and 0.33 for the poor compared to those seen in the previous simulation at 0.5 each. This behaviour is expected as the rich group has more unhappiness since they have more wealth to lose. This means that they will be more active in ensuring that their law, which benefits them more, is in effect, where the poor people's unhappiness is relatively small compared to them so they don't have the ability to compete. This can be seen in real life politics, as the rich have more ability to influence politics due to the influence and money they have, where the poor often have to struggle and campaign much harder to get change.

When tuning the model to the numerical data from the USA, it was seen that the model showed no oscillations. One reason this could occur is that since the data does not oscillate much outside of 0.5, the mean is the best optimum that system can reach from those starting values. Another reason could be due to the levels of the parameters being tuned not being high enough, or the assumed number of steps for the model being too small, as one time step was set to a year. When observing the original network simulation, it can be seen that an oscillation occurred once approximately every 100 steps. Therefor, if the data was set to months rather than years (48 steps between oscillations rather than 4) or if the speed factors were allowed to increase above 2, the network may have converged to periodic oscillation.

In the future it would be interesting to see how increasing the number of poor people would affect the system. From observing politics in real life situations, if there are enough poor people, the activation of the rich law should decrease, as there is more reactive unhappiness coming from the poor group. This was not done in this experiment as there was not enough time to update and modify the network.

Another interesting addition to a future version of the model would be to add in multiple laws. This would require more complex individuals, with nodes for each of the different issues and then a general unhappiness. The political level would also have to be updated to reflect the multiple laws each party could vote for. This would also open the model to have parties who voted for the laws in different ways and having the individuals vote for the parties who best reflected where the largest unhappiness was coming from.

Towards the end of the experiment, a network with add in media to the system was devised. In this network the upwards connections would be from the people to the media, where the people affect what the media talks about based off their interests and views. Then the media would affect the politicians by enhancing or detracting from how the people view them. The people would have an upward connection to politicians to vote for them as before. The politicians would then effect the laws in the same way they do now through voting and finally the laws would affect the people in a similar way.

The downward connections, starting with the laws, would be the laws affect the politicians through changing how the voting works and/or the speed factors. The politicians would affect the media through what equates to forcing them to talk positively or negatively about certain topics or suppression of others. The media would enhance or dampen the peoples reactions/care for the policies and laws. Finally the people would affect the laws by determining how quickly the laws come into effect due to how well they are followed/received by the population.

In future developments, a number of other relevant subtleties can be addressed as well. For example, for the US, the important roles of the hierarchy from cities to states to federal level, of competing lobby groups, and of the differences in access to information for different subpopulations can be addressed.

8   Conclusion

In the reported research an experiment of a strange loop adaptive temporal-causal network was created, tested and validated to reflect political oscillations as seen in presidential elections. The temporal network breaks a political system into 3 groups, the individual people, the politicians, and the laws where the individuals feed into the politicians, who feed into the laws, which feed into the individuals. In the initial simulation, the oscillatory behaviour which was expected from the literature review was observed. Next the network was modified to reflect an unbalanced political system with one group of individuals that were influenced more by the laws than the other. This cause the law which benefited those with more influence to be higher than the law which was beneficial for those with less influence, as expected. Finally the network was tuned to data from the USA presidential elections popular vote using simulated annealing, with both actual and simplified data. The simulated annealing did not perform as expected, giving a network which did not oscillate when using the real data, when using the simplified data, managed to reflect the behaviour which it was tuned on. The network not being able to tune on the real data could be due to the oscillations data being so close to 0.5, that the model found 0.5 as the ideal with the initial setting given and was unable to escape to another optimum. Another possible reason would be that the speed factors not being allowed to be tuned above 2 or due to the small number of steps between oscillations.

References

1. Baumgartner, F.R., Jones, B.D.: Policy Dynamics. University of Chicago Press (2002)
2. Davis, R.: Meta-rules: reasoning about control. Artif. Intell. **15**, 179–222 (1980)
3. Davis, R., Buchanan, B.G.: Meta-level knowledge: overview and applications. In: Proceedings of 5th IJCAI, pp. 920–927 (1977)
4. Galton, A.: Operators vs. arguments: the ins and outs of reification. Synthese **150**, 415–441 (2006)
5. Hebb, D.O.: The Organization of Behavior: A Neuropsychological Theory. Wiley, London (1949)
6. Hendlin, Y.H.: I am a fake loop: the effects of advertising-based artificial selection. Biosemiotics **12**(1), 131–156 (2019)
7. Hofstadter, D.R.: Gödel, Escher, Bach. Basic Books, New York (1979)
8. Hofstadter, D.R.: What is it like to be a strange loop? In: Kriegel, U., Williford, K. (eds.) Self-Representational Approaches to Consciousness. MIT Press, Cambridge (2006)
9. Hofstadter, D.R.: I Am a Strange Loop. Basic Books, New York (2007)
10. Kriegel, U., Williford, K.: Self-representational Approaches to Consciousness. MIT Press, Cambridge (2006)
11. Lowell, A.L.: Oscillations in politics. Ann. Am. Acad. Polit. Soc. Sci. **12**, 69–97 (1898). https://doi.org/10.1177/000271629801200104
12. Nagel, E., Newman, J.: Gödel's Proof. New York University Press, New York (1965)
13. Pierson, P.: When effect becomes cause: policy feedback and political change. World Polit. **45**, 595–628 (1993). https://doi.org/10.2307/2950710
14. Skocpol, T.: Protecting Soldiers and Mothers: The Political Origins of Social Policy in the United States. Belknap Press of Harvard, Cambridge (2009). https://doi.org/10.2307/j.ctvjz81v6
15. Smorynski, C.: The incompleteness theorems. In: Barwise, J. (ed.) Handbook of Mathematical Logic, vol. 4, pp. 821–865. North-Holland, Amsterdam (1977)
16. Sterling, L., Beer, R.: Metainterpreters for expert system construction. J. Logic Program. **6**, 163–178 (1989)
17. Strijbos, D., Glas, G.: Self-knowledge in personality disorder: self-referentiality as a stepping stone for psychotherapeutic understanding. J. Pers. Disorders **32**(3), 295–310 (2018)
18. Treur, J.: Network-Oriented Modeling: Addressing Complexity of Cognitive, Affective and Social Interactions. Springer, Heidelberg (2016). https://doi.org/10.1007/978-3-319-45213-5
19. Treur, J.: Network-Oriented Modeling for Adaptive Networks: Designing Higher-Order Adaptive Biological, Mental and Social Network Models. Springer, Heidelberg (2020). https://doi.org/10.1007/978-3-030-31445-3
20. Weyhrauch, R.W.: Prolegomena to a theory of mechanized formal reasoning. Artif. Intell. **13**, 133–170 (1980)

Joint Entity Linking for Web Tables
with Hybrid Semantic Matching

Jie Xie[1,2], Yuhai Lu[1,2], Cong Cao[1,2(✉)], Zhenzhen Li[2],
Yangyang Guan[2], and Yanbing Liu[2]

[1] School of Cyber Security, University of Chinese Academy of Sciences,
Beijing, China
{xiejie,luyuhai,caocong}@iie.ac.cn
[2] Institute of Information Engineering, Chinese Academy of Sciences, Beijing,
China
{lizhenzhen,guanyangyang,liuyanbing}@iie.ac.cn

Abstract. Hundreds of millions of tables on the World-Wide Web contain a considerable wealth of high-quality relational data, which has already been viewed as an important kind of sources for knowledge extraction. In order to extract the semantics of web tables to produce machine-readable knowledge, one of the critical steps is table entity linking, which maps the mentions in table cells to their referent entities in knowledge bases. In this paper, we propose a novel model *JHSTabEL*, which converts table entity linking into a sequence decision problem and uses hybrid semantic features to disambiguate the mentions in web tables. This model captures local semantics of the mentions and entities from different semantic aspects, and then makes full use of the information of previously referred entities for the subsequent entity disambiguation. The decisions are made from a global perspective to jointly disambiguate the mentions in the same column. Experimental results show that our proposed model significantly outperforms the state-of-the-art methods.

Keywords: Table entity linking · Hybrid semantic matching · Joint disambiguation

1 Introduction

The World-Wide Web contains billions of relational data in the form of HTML tables, i.e. web tables (Cafarella et al. [1, 8]; Lehmberg et al. [2]), which carries valuable structured information. This high-quality relational data is an important data source for knowledge extraction on the Web.

In order to make machines to understand these tables, one of the critical steps is to map the mentions in table cells to their corresponding entities in a given knowledge base (KB), which is called table entity linking or table entity disambiguation. For example, in the web table in Fig. 1, this task aims to link the mention "Louvre" in the first column to the entity "Louvre Museum" in Wikipedia. Table entity linking is an important and challenging stage in table semantic understanding since the mentions in tables are usually ambiguous.

V. V. Krzhizhanovskaya et al. (Eds.): ICCS 2020, LNCS 12138, pp. 618–631, 2020.
https://doi.org/10.1007/978-3-030-50417-5_46

Museum	Location	Annual Visitors
Louvre	▌▌Paris	10,200,000[10][11]
National Museum of China	▒▒Beijing	8,610,092[11]
The Metropolitan Museum of Art	▒▒New York City	6,953,927[11]
Vatican Museums	▌Vatican City (▌▌Rome)	6,756,186[11]
National Air and Space Museum	▒▒Washington, D.C.	6,200,000[12][10]

Fig. 1. An example of web table describing the information of museums.

In this paper, we only focus on tables where rows clearly represent separate tuple-like objects, and columns represent different dimensions of each tuple (similar to Fig. 1). Additionally, since this paper does not focus on how to determine which cells can be linked to the knowledge base, we assume that the linkable mentions are already known and perform entity linking on these linkable mentions, excluding un-linkable content, such as numbers, etc.

Compared with entity linking in free-format text, it is more difficult to disambiguate mentions in tables due to the less context of table cells. The existing researches mainly used collective classification techniques [3], graph-based algorithm [4], multi-layer perceptron [5], etc. to solve this problem. These methods do not capture the semantic features of mentions and entities well, and can't yield desired disambiguation effect. In order to better represent mentions and entities, we use a hybrid semantic matching model to capture the local semantic information between table mentions and candidate entities from different semantic aspects.

Since tables have the property of column consistency, that is, cells in the same column have similar contents and belong to the same category, it is natural to jointly disambiguate the mentions in the same column. In addition, we have noticed that mentions usually have different difficulty in disambiguating depending on the quality of the contextual information. If we sort the mentions in the same column and start with mentions that are easier to disambiguate, it would be useful to utilize the information of previously referred entities for the subsequent entity disambiguation.

In this paper, we propose a joint model with hybrid semantic matching for table entity linking, which is called *JHSTabEL* for short. This model consists of two modules: Hybrid Semantic Matching Model and Global Decision Model. The Hybrid Semantic Matching Model encodes the contextual information of each mention and its candidate entities. It uses the representation-based and interaction-based models to capture matching features at abstract and concrete levels respectively, and then aggregates them to obtain the hybrid semantic features, based on which the similarity scores of the mentions and entities are calculated. Before entering the global model, the mentions in the same column are sorted according to local similarity scores. The Global Decision Model uses an LSTM network to encode the local representations of mention-entity pairs and jointly disambiguate the mentions via a sequential manner. In summary, we make the following contributions:

- We propose a hybrid semantic matching model which aggregates complementary abstract and concrete matching features to make full use of the local context.
- We use a global decision model to jointly disambiguate the mentions in the same column. The disambiguation is made from a global perspective.
- We evaluate our model on web table datasets and the experimental results show that our model significantly outperforms the state-of-the-art methods.

2 Related Work

WebTables [1, 8] showed that the World-Wide Web consisted of a huge number of data in the form of HTML tables, and pioneered the study of tables on the Web as a high-quality relational data source. Since then, various efforts have been made to extract semantics from web tables. These efforts usually contain but not limited to three tasks: table entity linking, column type identification and table relation extraction.

Syed et al. [16] presented a pipeline approach, which first inferred the types of columns, then linked cell values to entities in the given KB, finally selected appropriate relations between columns. Mulwad et al. [6] and Limaye et al. [13] described approaches to jointly model entity linking, column type identification and relation extraction tasks using graphical model. These models, which handle all three tasks at the same time, rely on the correctness and completeness of the knowledge base, and therefore may run the risk of negatively affecting the performance of entity linking.

There are also some works that only focus on the task of table entity linking [3–5, 9, 15]. Shen et al. [15] linked the mentions in list-like web tables (multiple rows with one column) to the entities in a knowledge base. Efthymiou et al. [9] proposed three unsupervised annotation methods and attempted to map each table row to an entity in a KB. This work was based on the assumption that the entity columns of tables were already known and their values served as the names of the described entities. Bhagavatula et al. [3] presented TabEL which used a collective classification technique to collectively disambiguate all mentions in web tables. Wu et al. [4] constructed a graph of mentions and candidate entities and used page rank to determine the similarity scores between mentions and candidates. In the above methods, a lot of hand-designed features are applied, which is time-consuming and laborious. Recently, with the popularity of deep learning models, representation learning is used to automatically capture semantic features. Luo et al. [5] proposed a neural network method for cross-language table entity linking. It took some embedding features as inputs and used a two-layer fully connected network to perform entity linking. This model only used simple coherence features and a MLP network to link all mentions in tables, thus cannot achieve desired linking effect.

In this paper, we automatically capture the semantic features of the mentions and candidate entities from different aspects to fully use the local information, and then use a global model to disambiguate the mentions in web tables in a global perspective.

3 Methodology

As shown in Fig. 2, the overall structure of *JHSTabEL* consists of two parts: the hybrid semantic matching model which encodes the contextual information from two different semantic aspects to obtain the local semantic representations and matching scores of the mentions and the candidate entities; the global decision model which makes decisions from a global perspective to jointly disambiguate the mentions in the same column. We will introduce the details of these two parts in this section.

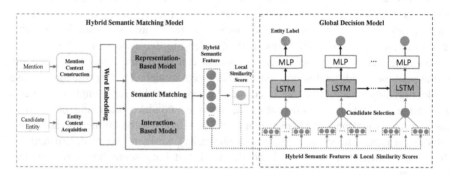

Fig. 2. The overall structure of our proposed model for table entity linking.

3.1 Preliminaries

Before introducing our model, we firstly make a definition of the table entity linking task. Formally, Given a table T with n rows and m columns, each mention in the table can be represented as $M_{i,j}$, $1 \leq i \leq n$ and $1 \leq j \leq m$ being the indexes of the row and column respectively. We model T as a bi-dimensional array of $n \times m$ cells, limiting the research scope to tables with no column branches into sub columns. Each mention $M_{i,j} \in T$ in the tables has a set of candidate entities $C_{M_{i,j}} = \left\{ e_{i,j}^1, e_{i,j}^2, \ldots, e_{i,j}^r \right\}$, where $e_{i,j}^k$ is the possible referred entity in the given knowledge base. Then the task of table entity linking is to map each mention $M_{i,j}$ to its corresponding target entity $e_{i,j}^+$ or return "NIL" if there is no correct target entity in the KB.

For each mention in the web tables, we need to generate its candidate referent entities from a given knowledge base. Hence, we use several heuristic rules to obtain the candidates: (i) the mention's redirect and disambiguation page in Wikipedia; (ii) exact match of the string mention; (iii) fuzzy match (e.g., edit distance) of the string mention; (iv) entities containing the n-grams of the mention.

To optimize the memory and avoid unnecessary calculations during model training, we use the XGBoost model to simplify the candidate sets. The features used in XGBoost are the edit distance between the mentions and their candidate entities, the semantic similarity between the mention context representations and the entity embeddings, and the statistical features based on the pageview and hyperlinks in Wikipedia. Then we take top K scored entities for each mention based on this model. In contrast, if the number of candidate entities for a mention is less than K, we complement it with negative examples from its candidate set.

3.2 Hybrid Semantic Matching Model

Given a mention M and its corresponding candidate set $C_M = \{e^1, e^2, \ldots, e^r\}$, we aim to get a local representation and a match score for each mention-entity pair. This is essentially a semantic matching problem between the mention context X_M and the candidate entity context X_e. Due to the scarce context of table cells, we construct the mention context X_M by using the other mentions in the row and the column of the table where the mention exists, and represent them as word embeddings using a pre-trained lookup table [7]. The context X_e of the candidate entity is obtained from the abstract of its corresponding page in Wikipedia and embedded in the same way.

Existing neural semantic matching models can be divided into two categories: representation-based model and interaction-based model. The representation-based model first uses a neural network to construct a representation for a single text, such as a mention context or an entity abstract, and then conducts matching between the abstract representations of two pieces of text. The interaction-based method attempts to establish a local interaction (e.g., cosine similarity) between two pieces of text, and then uses a neural network to learn the final matching score based on the local interaction.

The representation-based and interaction-based models can capture abstract and concrete level matching signals respectively. In this paper, we propose to fuse these two models to perform semantic matching between the mention and entity contexts. The left part of Fig. 2 shows the structure of our local hybrid model. It takes the mention and candidate entity as inputs and generates their corresponding contexts and embeddings, which are passed into the representation and interaction models. Finally, the hybrid semantic features and local ranking scores are acquired from this hybrid model. In the remaining of this section, we will introduce the details of these two submodels and discuss the advantages of fusing them.

Representation-Based Model. Given the mention context $X_M = \{w_M^1, w_M^2, \ldots, w_M^p\}$ and the candidate entity context $X_e = \{w_e^1, w_e^2, \ldots, w_e^q\}$, we aim to get their abstract representations using siamese LSTM [10] with tied weights. Figure 3 illustrates the architecture of our representation-based model. The mention context embedding Emb_M and the entity context embedding Emb_e are obtained from a pre-trained lookup table [7]. We use two networks $LSTM_a$ and $LSTM_b$ with tied weights to encode the embeddings separately, and take the last hidden states of the LSTM networks as the representations of the word sequences. In this way, we get the mention representation V_M and the entity representation V_e, and feed their concatenation result to a multi-layer perceptron (MLP). The output layer of the MLP produces a feature vector $V_{abs}(M, e)$ of d_{abs} dimension.

$$V_{abs}(M, e) = MLP([V_M; V_e]) \tag{1}$$

In this way, we extract abstract-level features V_{abs} of the local contexts. We can also calculate the local similarity between mention and candidate entity using the abstract-level features. However, if only this representation-based approach is used, the concrete matching signals (e.g., exact match) are lost, since the matching happens after their individual representations. So next we will introduce an interaction-based model to better capture the concrete matching features to complement the representation-based model.

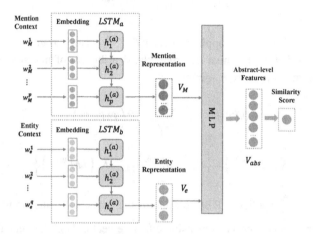

Fig. 3. The architecture of representation-based model using siamese LSTM.

Interaction-Based Model. Inspired by the latest advances in information retrieval [11, 12], we propose to use an interaction-based approach to capture the concrete-level features. The interaction-based model using Conv-KNRM [12] attempts to establish local interactions (e.g., cosine similarity) and get concrete-level features between mention and entity contexts. As shown in Fig. 4, the Conv-KNRM model first composes n-gram embeddings using CNN networks, and then constructs translation matrices between n-grams of different lengths in the n-gram embedding space. It uses a kernel-pooling layer to count the soft matches of word or n-gram pairs and gets the concrete level features.

The Conv-KNRM model takes the mention context embedding Emb_M and the entity context embedding Emb_e as inputs. The convolutional layer applies convolution filters to compose n-grams from the text embeddings. For each window of h words, the filter sums up all elements in the h words' embeddings $Emb_{i:i+h}$, weighted by the filter weights. Using F different filters of size h gives F scores for each window position, represented by a score vector $\overrightarrow{g}_i^h \in \mathbb{R}^F$. Each of the values in $\overrightarrow{g}_i^h$ describes the text in the i_{th} window in a different perspective:

$$\overrightarrow{g}_i^h = relu(W^h \cdot Emb_{i:i+h} + \overrightarrow{b}^h), i = 1 \ldots n. \tag{2}$$

Where W^h and $\vec{b}^h$ are the weights of F convolution filters. Then the convolution feature matrix for h-gram can be obtained by concatenating convolution outputs $\vec{g}^h_i$.

After getting the word-level n-gram feature matrices, the cross-match layer constructs translation matrices using n-grams of different lengths. For mention n-grams of length h_M and entity n-grams of length h_e, a translation matrix TM^{h_M,h_e} is constructed by calculating their cosine similarity.

$$TM_{i,j}^{h_M,h_e} = \cos\left(\vec{g}^{h_M}_i, \vec{g}^{h_e}_j\right) \tag{3}$$

Then the Kernel-pooling is applied to each TM^{h_M,h_e} matrix to generate the concrete feature vector $\phi(TM^{h_M,h_e})$, which describes the distribution of match scores between mention h_M-grams and entity h_e-grams.

$$\phi\left(TM^{h_M,h_e}\right) = \sum_{i=1}^{n} \log \vec{K}\left(TM_i^{h_M,h_e}\right) \tag{4}$$

$$\vec{K}\left(TM_i^{h_M,h_e}\right) = \left\{K_1\left(TM_i^{h_M,h_e}\right), \ldots, K_k\left(TM_i^{h_M,h_e}\right)\right\} \tag{5}$$

Where $\vec{K}\left(TM_i^{h_M,h_e}\right)$ applies k RBF kernels to the i-th row of the translation matrix $TM_i^{h_M,h_e}$, and then generates a k-dimensional feature vector. Each kernel calculates how pairwise similarities between n-gram feature vectors are distributed around its mean μ_k. The more similarities closed to its mean, the higher the output value is.

$$K_k\left(TM_i^{h_M,h_e}\right) = \sum_j \exp\left(-\frac{(TM_{i,j} - \mu_k)^2}{2\sigma_k^2}\right) \tag{6}$$

Then each of the translation matrices is pooled to a k-dimensional vector, and the concatenation of these vectors produces a scoring feature vector $\phi(TM)$.

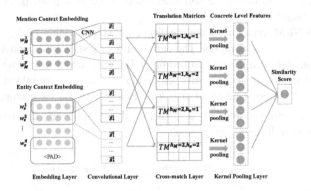

Fig. 4. The architecture of interaction-based model using Conv-KNRM.

In this way, we capture the concrete features $V_{con}(M,e)=\phi(TM)$ based on the word-level n-gram interactions between mention and entity. These features can complement the abstract features for a better semantic representation.

Hybrid Semantic Matching. We use the two sub-models introduced above to capture the abstract and concrete level features respectively, and combine them to get the hybrid semantic features. Then we pass the concatenation result to a MLP network to get the local similarity score for each mention-entity pair.

$$sim(M,e) = MLP([V_{abs}(M,e); V_{con}(M,e)]) \qquad (7)$$

In order to better distinguish the correct entity from the wrong entities in the candidate set when training the hybrid model, we use the hinge loss function, which can rank the correct entity higher than others. The loss function of the hybrid model is defined as follow:

$$L_{local} = \sum_{M} \sum_{e^+,e^- \in C_M^{+,-}} max(0, \gamma - sim(M,e^+) + sim(M,e^-)) \qquad (8)$$

Where $C_M^{+,-}$ is the set of pairwise preferences of M and e^+ ranks higher than e^-. $\gamma > 0$ is the margin parameter, indicating that the score of the positive target entity e^+ is at least a margin γ higher than the negative entity e^-.

Through the hybrid semantic matching model, we obtain the hybrid semantic features and local similarity scores of the mentions and candidate entities, which will serve as inputs to the subsequent global decision model.

3.3   Global Decision Model

The global decision model aims to enhance the topical consistency among the mentions in the same column. As shown in the right part of Fig. 2, the global decision model takes the hybrid semantic features and local similarity scores acquired from the hybrid semantic matching model as inputs, and uses an LSTM network to deal with mentions in a sequence manner. The LSTM network can maintain a long-term memory on features of entities selected in previous states. Therefore, the column consistency information can be fully utilized when disambiguating entities.

Inspired by [14], we sort the mentions in the same column when disambiguating them. In table entity linking task, it is natural to divide all the mentions in a table into multiple segments according to the column they belong to. Then the mentions in a segment are sorted according to the local similarity scores, the one with a higher score is placed first. We take the maximum local similarity between the mention and its corresponding candidate entities as the criterion for each mention when sorting. Then an LSTM network is used to deal with these sorted segments in a sequence manner. In this way, we can start with mentions that are easier to disambiguate and utilize the information provided by previously selected entities to disambiguate subsequent mentions.

The local similarity score indicates the probability of an entity being the target entity of the mention. Therefore, at each time step, we randomly select a candidate for the mention based on this probability, and take the corresponding hybrid representations of the mention and the selected entity as inputs to LSTM network. Then the output at each time step is passed into a MLP network to produce the label for the selected entity. The objective function of the global decision model is defined as follow:

$$L_{global} = -\frac{1}{n}\sum_{x}[y\log y' + (1-y)\log(1-y')] \qquad (9)$$

Where $y \in \{0,1\}$ is the actual label of the candidate entity and $y' \in \{0,1\}$ is the predicted one. In this way, the mentions in the same column are disambiguated jointly.

4 Experiment

In this section, we conduct several experiments to evaluate our model *JHSTabEL* on the sampled web tables. Firstly, we compare it with the state-of-the-art methods, and then discuss the effect of various components of our proposed model.

4.1 Experiment Setup

Dataset. We use the dataset constructed by Wu et al. [4], which contains 123 tables extracted from Chinese Wikipedia. The mentions in these tables are labeled by their corresponding Wikipedia articles. We represent this dataset as *Dataset-Wu*. In order to better reflect the advantages of the deep learning method, we expand the dataset by randomly collecting 117 tables from the Web. Each mention in these tables is manually mapped to its corresponding entity in Wikipedia. Then these tables are added to *Dataset-Wu* to generate a larger dataset, represented as *Dataset-Xie*. The average size of tables in this dataset is 12 rows, and each table contains an average of 38.2 mentions. Totally, we obtains 9168 mentions from 240 tables. We randomly split the tables into training, validation and testing sets (70%, 10%, 20%) for experiments.

Parameter Setting. The hyper parameters of our model are obtained from the best validated model. For the representation-based model, the number of LSTM cell units is set to 128, the batch size is 64 and the number of MLP layers is 3. For the interaction-based model, the n-gram lengths are $h = 1, 2, 3$, the number of CNN filters is $F = 128$, the number of kernels is set to 11, the first one is exact match kernel $\mu = 1$, $\sigma = 10^{-3}$, and the other 10 kernels equally split the cosine range $[-1, 1]$: the values of μ are $\mu_1 = 0.9, \mu_2 = 0.7, ..., \mu_{10} = -0.9$ and the values of σ are set to be 0.1. We set the rank margin $\gamma = 0.1$ for hybrid semantic matching model. For the global decision model, the number of LSTM cell units is 256, the batch size is 32, and the number of MLP layers is 2. We choose a learning rate of 1e-4 and a probability of dropout of 0.9. The dimension of the word embedding used in our experiments is set to 300. To optimize the memory and avoid unnecessary calculations, we select top K candidate entities for each mention. Our experiments show that the best performance is obtained when $K = 5$.

4.2   Baselines and Evaluation Metric

Baselines. We compare our model *JHSTabEL* with several table entity linking methods, which reported state-of-the-art results: collective classification model (*TabEL (2015)* [3]), graph-based model (*Wu et al. (2016)* [4]) and MLP-based model (*Luo et al. (2018)* [5]). Besides, we feed Luo et al.'s mention features and context features into our proposed global decision model, which is represented as *Luo-fea-Global*. *JHSTabEL (local)* is a degenerate version of our proposed model, it only uses the local hybrid semantic matching model to disambiguate mentions. This local hybrid model fuses the abstract and concrete matching features to rank the candidate entities for table mentions.

Evaluation Metric. In order to be consistent with the state-of-the-art table entity linking methods, we evaluate the results with Micro Accuracy and Macro Accuracy. Micro Accuracy is the fraction of correct linked cells over the whole dataset and Macro Accuracy is the average correct ratio over different tables.

4.3   Experiment Result

Comparing with Previous Work. We compare our proposed model *JHSTabEL* with the baselines on the *Dataset-Wu* and *Dataset-Xie* and report the experimental results in Table 1. From the results, we can notice that the model *JHSTabEL(local)* is comparable to *Luo et al. (2018)*, which jointly disambiguated all mentions in a table and reported the best results so far. This result shows the excellent feature extraction ability of our hybrid semantic matching model, which can better characterize the mention-entity pairs and obtain good results using only local matching. The model *Luo-fea-Global* outperforms *Luo et al. (2018)*, indicating the effectiveness of our global decision model. The model proposed by Luo et al. applied vector averaging over all cells to be linked and concatenated coherence feature to link all mentions in a table in the same time. Due to the simplicity of the coherence feature, this method does not model the correlations between table mentions very well. However, our global decision model uses an LSTM network to maintain the memory of previously selected entities, so as to obtain better joint disambiguation effects. Our full model *JHSTabEL* achieves the best result on both datasets. Compared with *Luo et al. (2018)*, it improves the Micro Accuracy and Macro Accuracy by absolute gains of 0.020 and 0.018 separately on *Dataset-Wu*. Besides, in order to get more reliable results, we enrich the origin dataset (*Dataset-Wu*) and generate a larger dataset (*Dataset-Xie*), which is about 1.95 times of the origin one. Then we perform the same experiments on it. The results still show the superiority of our proposed model.

Table 1. Accuracies comparison between our model and baselines on two datasets.

Methods	Dataset-Wu		Dataset-Xie	
	Micro Acc.	Macro Acc.	Micro Acc.	Macro Acc.
TabEL (2015)	0.845	0.843	0.850	0.847
Wu et al. (2016)	0.849	0.845	0.854	0.848
Luo et al. (2018)	0.878	0.864	0.884	0.867
Luo-fea-Global	0.885	0.872	0.890	0.875
JHSTabEL(local)	0.877	0.866	0.885	0.865
JHSTabEL	**0.898**	**0.882**	**0.907**	**0.889**

Comparison Between Different Semantic Matching Models. To further explore the differences between two semantic matching models (Representation-Based and Interaction-Based) and discuss the benefits when combing them, we remove the representation model and interaction model from the full model separately and compare their performance with the full model. As shown in Fig. 5, we can observe that *Rep-Based + Global Model* performs comparably with *Int-Based + Global Model* and the full model *JHSTabEL* obtains considerable performance gains on both of the two datasets. This comparison result indicates that these two sub-models capture complementary information for entity disambiguation. In fact, the interaction-based model builds the n-gram level local interactions between texts, thus can capture the concrete matching information. However, the concrete information might be lost in representation-based model as it tends to capture the whole meaning of the text and generate abstract information. So we will benefit a lot by combining different semantic matching signals from these two models.

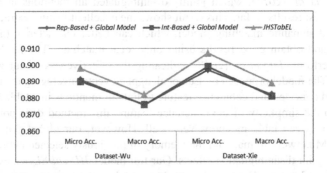

Fig. 5. The performance of different semantic matching models.

Effect of Global Decision Model. In order to evaluate whether the global decision model based on LSTM network contributes to disambiguation, we compare the performance with and without the global model. From the results in Table 2, we can see that the accuracies of each local model are greatly improved when combining with the global decision model. This is due to the ability of the global decision model to leverage the information of previously referred entities, thus making full use of the column consistency information when disambiguating entities.

Table 2. The effect of the global decision model for entity linking in web tables.

Methods	Dataset-Wu		Dataset-Xie	
	Micro Acc.	Macro Acc.	Micro Acc.	Macro Acc.
Rep-Based Model	0.870	0.857	0.879	0.861
Rep-Based + global Model	**0.891**	**0.876**	**0.897**	**0.882**
Int-Based Model	0.868	0.859	0.880	0.858
Int-Based + global Model	**0.890**	**0.876**	**0.899**	**0.881**
JHSTabEL(local)	0.877	0.866	0.885	0.865
JHSTabEL	**0.898**	**0.882**	**0.907**	**0.889**

Influence of Ranking Mentions. In this part, we test whether ranking the ns before feeding them into the global model helps to disambiguate entities in web tables. Firstly, we input the mentions directly into our global model in the order they appear in the columns. Secondly, we use a bi-directional LSTM (Bi-LSTM) to consider both previous and following entities in the same column in order. Finally, we compare these two models with our proposed model which adopts ranking mentions. As the results showed in Fig. 6, our model with ranking mentions achieves the best results on both datasets. Comparing two models that do not use ranking, the model with Bi-LSTM performs only slightly better than the model with LSTM. Although the Bi-LSTM can consider the information of the previous and following entities, it may introduce more noise at the same time. However, our proposed model with ranking mentions allows us to utilize the information of easily disambiguated mentions to help the disambiguation of other mentions, so that we can get better disambiguating performance.

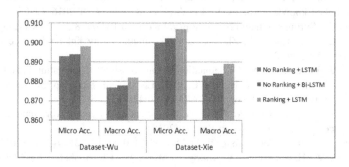

Fig. 6. The influence of ranking mentions for global decision model.

5 Conclusion

In this paper we propose a table entity linking model that takes the advantage of the semantic information from different aspects and jointly disambiguates the mentions in web tables. The combination of the different semantic signals can produce better representations for the mentions and candidate entities. By leveraging information from previously referred entities, we can make full use of column consistency to disambiguate mentions. The Comparison with baselines shows that our model outperforms the state-of-the-art solutions, and the experiments on variants of our model also indicate the substantial benefits of the semantic matching models, mention ranking and global decision model. For the future work, we intend to automatically determine whether the content in table cells should be linked to knowledge bases, since such un-linkable information, such as numbers and long sentences, are common in tables.

Acknowledgement. This research is supported by the National Key R&D Program of China (No. 2017YFC0820700, No. 2018YFB1004700), Xinjiang Uygur Autonomous Region Science and Technology Project (No. 2016A030007-4) and the National Natural Science Foundation of China grants (No. 61602466).

References

1. Cafarella, M.J., Halevy, A.Y., Wang, D.Z., Wu, E., Zhang, Y.: WebTables: exploring the power of tables on the web. PVLDB 1(1), 538–549 (2008)
2. Lehmberg, O., Ritze, D., Meusel, R., Bizer, C.: A large public corpus of web tables containing time and context metadata. In: WWW (Companion Volume) 2016, pp. 75–76 (2013)
3. Bhagavatula, C.S., Noraset, T., Downey, D.: TabEL: entity linking in web tables. In: Arenas, M., et al. (eds.) ISWC 2015. LNCS, vol. 9366, pp. 425–441. Springer, Cham (2015). https://doi.org/10.1007/978-3-319-25007-6_25
4. Wu, T., Yan, S., Piao, Z., Xu, L., Wang, R., Qi, G.: Entity linking in web tables with multiple linked knowledge bases. In: Li, Y.-F., et al. (eds.) JIST 2016. LNCS, vol. 10055, pp. 239–253. Springer, Cham (2016). https://doi.org/10.1007/978-3-319-50112-3_18
5. Luo, X., Luo, K., Chen, X., Zhu, K.Q.: Cross-lingual entity linking for web tables. In: AAAI, pp. 362–369 (2018)
6. Mulwad, V., Finin, T., Joshi, A.: Semantic message passing for generating linked data from tables. In: Alani, H., et al. (eds.) ISWC 2013. LNCS, vol. 8218, pp. 363–378. Springer, Heidelberg (2013). https://doi.org/10.1007/978-3-642-41335-3_23
7. Mikolov, T., Chen, K., Corrado, G., Dean, J: Efficient estimation of word representations in vector space. In: ICLR (Workshop Poster) (2013)
8. Cafarella, M.J., et al.: Ten years of webtables. PVLDB 11(12), 2140–2149 (2018)
9. Efthymiou, V., Hassanzadeh, O., Rodriguez-Muro, M., Christophides, V.: Matching web tables with knowledge base entities: from entity lookups to entity embeddings. In: d'Amato, C., et al. (eds.) ISWC 2017. LNCS, vol. 10587, pp. 260–277. Springer, Cham (2017). https://doi.org/10.1007/978-3-319-68288-4_16
10. Mueller, J., Thyagarajan, A.: Siamese recurrent architectures for learning sentence similarity. In: AAAI, pp. 2786–2792 (2016)

11. Xiong, C., Dai, Z., Callan, J., Liu, Z., Power, R.: End-to-End neural ad-hoc ranking with kernel pooling. In: SIGIR, pp. 55–64 (2017)
12. Dai, Z., Xiong, C., Callan, J., Liu, Z.: Convolutional neural networks for soft-matching n-grams in ad-hoc search. In: WSDM, pp. 126–134 (2018)
13. Limaye, G., Sarawagi, S., Chakrabarti, S.: Annotating and searching web tables using entities, types and relationships. PVLDB **3**(1–2), 1338–1347 (2010)
14. Fang, Z., Cao, Y., Li, Q., Zhang, D., Zhang, Z., Liu, Y.: Joint entity linking with deep reinforcement learning. In: WWW, pp. 438–447 (2019)
15. Shen, W., Wang, J., Luo, P., Wang, M.: Liege: link entities in web lists with knowledge base. In: SIGKDD, pp. 1424–1432 (2012)
16. Syed, Z., Finin, T., Mulwad, V., Joshi, A.: Exploiting a Web of semantic data for interpreting tables. In: Proceedings of the Second Web Science Conference (2010)

A New Coefficient of Rankings Similarity in Decision-Making Problems

Wojciech Sałabun[✉][iD] and Karol Urbaniak[iD]

Research Team on Intelligent Decision Support Systems,
Department of Artificial Intelligence and Applied Mathematics,
Faculty of Computer Science and Information Technology,
West Pomeranian University of Technology in Szczecin,
ul. Żołnierska 49, 71-210 Szczecin, Poland
wojciech.salabun@zut.edu.pl

Abstract. Multi-criteria decision-making methods are tools that facilitate and help to make better and more responsible decisions. Their main objective is usually to establish a ranking of alternatives, where the best solution is in the first place and the worst in the last place. However, using different techniques to solve the same decisional problem may result in rankings that are not the same. How can we test their similarity? For this purpose, scientists most often use different correlation measures, which unfortunately do not fully meet their objective.

In this paper, we identify the shortcomings of currently used coefficients to measure the similarity of two rankings in decision-making problems. Afterward, we present a new coefficient that is much better suited to compare the reference ranking and the tested rankings. In our proposal, positions at the top of the ranking have a more significant impact on the similarity than those further away, which is right in the decision-making domain. Finally, we show a set of numerical examples, where this new coefficient is presented as an efficient tool to compare rankings in the decision-making field.

Keywords: Decision analysis · Decision making · Decision theory · Measurment uncertainty · Ranking

1 Introduction

Decision making is an integral part of human life. Every day, every person is faced with different kinds of decision-making problems, which can affect both professional and private life. An example of a decision-making problem can be a change of legal regulations in the state, choice of university, purchase of a new car, determination of the amount of personal income tax, selection of a suitable location for the construction of a nuclear power plant, adoption of a plan of research, or the sale or purchase of stock exchange shares.

In the majority of cases, decision-making problems are based on many, often contradictory, decision-making criteria. Therefore multi-criteria decision-analysis

© Springer Nature Switzerland AG 2020
V. V. Krzhizhanovskaya et al. (Eds.): ICCS 2020, LNCS 12138, pp. 632–645, 2020.
https://doi.org/10.1007/978-3-030-50417-5_47

(MCDA) methods and decision support systems enjoy deep interest both in the world of business and science. Almost in every case, a reliable decision requires the analysis of many alternatives. Each of them should be assessed from the perspective of all the criteria characterizing its acceptability. As the complexity of the problem increases, it becomes more and more challenging to make the optimal decision. An additional complication is that there is no complete mathematical form depending on the criteria and the expected consequences. In particularly important problems, the role of the decision-maker is entrusted to an expert in a given field or to a group of experts who will help to identify the best solution. We talk about individual or group decision making, respectively. Even then, it can often be problematic for an individual expert as well as for collegiate bodies to determine the right decision. In this case, MCDA methods can be helpful.

MCDA methods are great tools to support the decision-maker in the decision-making process. We can identify two main groups of MCDA methods, i.e., American and European schools [33]. Methods of the American school of decision support are based on the utility or value function [5,16]. The most important methods belonging to this family are: analytic hierarchy process (AHP) [34], analytic network process (ANP) [35], utility theory additive (UTA) [21], simple multi-attribute rating technique (SMART) [28], technique for order preference by similarity to ideal solution (TOPSIS) [3,32], or measuring attractiveness by a categorical based evaluation technique (MACBETH) [2]. Methods of European school of decision support use outranking relation in the preference aggregation process, where the most popular are ELECTRE family [1,36] and PROMETHEE methods [8,15]. Additionally, we can indicate the set of techniques based strictly on the rules of decision making. These methods use the fuzzy sets theory (COMET) [12,25–27] and the rough set theory (DRSA) [29].

Generally, MCDA methods help to create a ranking of decision variants where the most preferred alternative comes first [4]. The problem arises when we use more than one MCDA method, and the rankings obtained are not identical. Then the question arises on how to compare the received rankings? Currently, the most popular approach is the analysis based on the correlation between the two or more rankings [7,12,19,24]. However, we are going to show that this analysis is insufficient in the decision support domain. An appropriate approach should ensure that a better ranking in terms of the order can be identified. Then, with a proper benchmark, it would be possible to assess the correctness of the MCDA methods in terms of the rankings generated [22].

In this paper, we identify the shortcomings of currently used coefficients to measure the similarity of two rankings. The most significant contribution is the WS coefficient, which depends strictly on the position on which the difference in the ranking occurred. Afterward, three linguistic terms are identified by using trapezoidal fuzzy numbers, i.e., low, medium, and high similarity. We compare the proposed coefficient with ρ Spearman, τ Kendall, and γ Goodman-Kruskal coefficients, which are commonly used to measure rankings similarity in MCDA problems [9,12,17,18,23]. In addition, the proposed approach is compared with the similar coefficients presented in [6,10,13]. For this purpose, numerical experiments are discussed.

The rest of the paper is organized as follows: In Sect. 2, some basic preliminary concepts are discussed. Section 3 introduces a new coefficient of rankings similarity in the decision-making problems. In Sect. 4, the practical feasibility study of the WS coefficient is discussed. In Sect. 5, we present the summary and conclusions.

2 Preliminaries

An important issue is how to compare the correctness of the order of the two rankings. The simplest method is to check whether the rankings are consistent or inconsistent. Such an approach is not sufficient and can be used almost exclusively for 2 or 3 elementary rankings [27]. The much more common approach is to use one of the coefficients of monotonous dependence of two variables, where the obtained rankings for a set of considered alternatives are our variables. The most commonly used symmetrical coefficient of such dependence is the Spearman's coefficient [9,17,18,23], which is expressed by the following formula (1):

$$r_s = 1 - \frac{6 \cdot \sum d_i^2}{n \cdot (n^2 - 1)} \tag{1}$$

where d_i is defined as the difference between the ranks $d_i = R_{xi} - R_{yi}$ and n is the number of elements in the ranking. The Spearman's coefficient is interpreted as a percentage of the rank variance of one variable, which is explained by the other variable [31].

The most frequently used asymmetrical monotonous coefficients of two variables are Kendall [12,20] and Goodman-Kruskal coefficients [12,14]. They are expressed in formulas (2) and (3) respectively:

$$\tau = 2 \cdot \frac{N_s - N_d}{n \cdot (n - 1)} \tag{2}$$

$$G = \frac{N_s - N_d}{N_s + N_d} \tag{3}$$

where N_s is the number of compatible pairs, N_d is the number of non-compliant pairs, and n is the number of all pairs. The Kendall and Goodman-Kruskal coefficients, unlike Spearman, are interpreted in terms of probability. They represent the difference between the probability that the compared variables will be in the same order for both variables and the probability that they will be in the opposite order.

The presented coefficients are the most frequently used measures of the analysis of the rankings similarity in decision-making problems [9,12,17,18,23]. However, we want to indicate a significant shortcoming, which is related to the place of difference occurrence. The idea of measuring the rankings similarity is not new and has been the subject of many works [11,30]. Particularly interesting in the context of the presented approach are works related to Blest's measure of rank correlation v and the weighted rank measure of correlation r_w [6,10,13]. They are expressed in formulas (4) and (5) respectively:

$$r_w = 1 - \frac{6\sum_{i=1}^{n}(R_{xi} - R_{yi})^2((n - R_{xi} + 1) + (n - R_{yi} + 1))}{n^4 + n^3 - n^2 - n} \tag{4}$$

$$v = 1 - \frac{12\sum_{i=1}^{n}(n + 1 - R_{xi})^2 \cdot R_{yi} - n(n + 1)^2(n + 2)}{n(n + 1)^2(n - 1)} \tag{5}$$

The presented coefficients (1-3) are regardless of whether the error occurs at the top or bottom; the values of the factors will be identical. In Table 1, the simple example shows five rankings, including one reference (R_x) and four test rankings ($R_y^{(1)} - R_y^{(4)}$). The test rankings were created by a change in the correct ranking of the two adjacent alternatives. We want to remind that the rankings are determined to choose the best possible solution, and the value of the preferences decreases with each position in the ranking. The difference at the top should be more significant than an error at the bottom of the ranking. The exchange of alternative locations from the first and second position is a more considerable error than the swap of the second and third position. However, the values of the coefficients indicate that similarity of the test rankings to the reference ranking is the same for all test sets.

3   WS Coefficient of Rankings Similarity

The new ranking similarity factor should be resistant to the situation described in the previous section, and at the same time, should be sensitive to significant changes in the ranking. Besides, this factor should be easy to interpret, and its values should be limited to a specific interval.

We assumed that the new indicator should be strongly related to the difference between two rankings on particular positions. An additional assumption is that the top has a more significant influence on similarity than the bottom of the ranking. Based on these assumptions, a new indicator was developed, which can be presented as (6):

$$WS = 1 - \sum_{i=1}^{n}\left(2^{-R_{xi}} \cdot \frac{|R_{xi} - R_{yi}|}{max\{|1 - R_{xi}|, |N - R_{xi}|\}}\right) \tag{6}$$

where WS is a value of similarity coefficient, N is a length of ranking, R_{xi} and R_{yi} mean the place in the ranking for $i - th$ element in respectively ranking x and ranking y.

The proof of convergence for the WS factor is quite simple. The formula (6) can be divided into two main components. The first one (7) is responsible for making the WS value dependent on the position in the reference ranking (R_x).

$$2^{-R_{xi}} \tag{7}$$

Table 1. Summary of the test with reference ranking (Rx) and four test rankings $(R_y^{(1)} - R_y^{(4)})$ with the calculated correlation factors and proposed WS coefficient for the set of five alternatives $(A_1 - A_5)$, each having a different position in the ranking.

A_i	R_x	$R_y^{(1)}$	$R_y^{(2)}$	$R_y^{(3)}$	$R_y^{(4)}$
A_1	1	2	1	1	1
A_2	2	1	3	2	2
A_3	3	3	2	4	3
A_4	4	4	4	3	5
A_5	5	5	5	5	4
Coefficients	r_s	0.9000	0.9000	0.9000	0.9000
	τ	0.8000	0.8000	0.8000	0.8000
	G	0.8000	0.8000	0.8000	0.8000
	r_w	0.8500	0.8833	0.9167	0.9500
	v	0.8500	0.8833	0.9167	0.9500
	WS	0.7917	0.8542	0.9167	0.9714

We are dealing with a geometric series which is convergent. As proof, we can calculate a trivial limit.

$$\lim_{n \to \infty} \sum_{i=1}^{n} (2)^{-R_{xi}} = 1 \tag{8}$$

The second component (9) determines to what extent the difference in rankings affects the similarity of rankings. This value can be obtained from zero (the positions are identical) to one.

$$\frac{|R_{xi} - R_{yi}|}{max\{|1 - R_{xi}|, |N - R_{xi}|\}} \tag{9}$$

If we multiply the (7) by (9) then this series cannot be higher than one. Therefore, it is clear that the WS coefficient can only take values from zero to one. We can compare all coefficients for a simple example in Table 1. The WS, r_w, and v coefficients take into account the position of the error occurrence, and the rest of them remain the same regardless of where the error occurs. In the next section, other tests comparing the performance of the indicators will be presented and discussed.

4   Results and Discussion

4.1   Analysis of Five-Element Rankings

The first experiment here presents tied ranks, i.e., the same values in the ranking. It happens when two alternatives get the same place. For example, if two decision variants receive the first place together, the ranking will contain a value of 1.5 for

Table 2. Summary of the test with reference ranking (Rx) and four test rankings $(R_y^{(1)} - R_y^{(4)})$ with the calculated correlation factors and proposed WS coefficient for the set of five alternatives $(A_1 - A_5)$, where one pair has the same position in the ranking.

A_i	R_x	$R_y^{(1)}$	$R_y^{(2)}$	$R_y^{(3)}$	$R_y^{(4)}$
A_1	1	1.5	1	1	1
A_2	2	1.5	2.5	2	2
A_3	3	3	2.5	3.5	3
A_4	4	4	4	3.5	4.5
A_5	5	5	5	5	4.5
Coefficients	r_s	0.9747	0.9747	0.9747	0.9747
	τ	0.9487	0.9487	0.9487	0.9487
	G	1.0000	1.0000	1.0000	1.0000
	r_w	0.9625	0.9708	0.9792	0.9875
	v	0.9250	0.9417	0.9583	0.9750
	WS	0.8958	0.9271	0.9583	0.9857

both (the average of their positions). Table 2 shows the results of calculations for the five-element ranking, where the different location of tied pairs is considered.

One more again, WS, r_w, and v coefficients show the change of value together with the change of position on which there are the tied pairs. It is a property that was identified as a significant drawback of the currently used methods, i.e., ρ Spearman, τ Kendall, and γ Goodman-Kruskal. Ranking $R_y^{(4)}$ is more similar than $R_y^{(1)}$ because full correctness occurs on the first three positions and not on the last three.

Another simple experiment consists in creating test rankings, where successive rankings differ from the base ranking by the alternative indicated as the best. The results for the five-element ranking are shown in Table 3. Replacing the best option with the worst in all coefficients results in a negative value result (except WS). It means a negative correlation, which is not trivial to interpret in decision-making problems. Besides, interesting is the case of the $R_y^{(3)}$ ranking because it obtained a total lack of correlation (for τ and G coefficients). It means that the order of the base and second rankings is utterly unrelated to each other. It is a confirmation that these classical rank coefficients do not examine the similarity of the two rankings thoroughly. In general, all coefficients assess test rankings against the base ranking in a somewhat similar way. The rationale for this is that the three positions in the ranking have been indicated flawlessly.

The last example in this subsection examines the r_w coefficients in two cases, i.e., for test rankings $R_y^{(1)} - -R_y^{(2)}$ and $R_y^{(3)} - -R_y^{(4)}$). Once again, the R_x ranking is used as a reference point. The detailed results are presented in Table 4.

Rankings $R_y^{(1)}$ and $R_y^{(2)}$ have equal values for most of the coefficients, where only WS and v are exceptions. Ranking $R_y^{(1)}$ is significantly better than ranking

Table 3. Summary of the test with reference ranking (Rx) and four test rankings $(R_y^{(1)} - R_y^{(4)})$ with the calculated correlation factors and proposed WS coefficient for the set of five alternatives $(A_1 - A_5)$, where each ranking has a different position error.

A_i	R_x	$R_y^{(1)}$	$R_y^{(2)}$	$R_y^{(3)}$	$R_y^{(4)}$
A_1	1	2	3	4	5
A_2	2	1	2	2	2
A_3	3	3	1	3	3
A_4	4	4	4	1	4
A_5	5	5	5	5	1
Coefficients	r_s	0.9000	0.6000	0.1000	−0.6000
	τ	0.8000	0.4000	0.0000	−0.4000
	G	0.8000	0.4000	0.0000	−0.4000
	r_w	0.8500	0.4667	−0.0500	−0.6000
	v	0.8500	0.4667	−0.0500	−0.6000
	WS	0.7917	0.6250	0.5625	0.4688

$R_y^{(2)}$. Even though the A_5 alternative has been identified as the best in $R_y^{(1)}$. The rest of this ranking has been correctly identified according to the right order. However, in the $R_y^{(2)}$, the best alternative was wrongly rated as the worst. Therefore, the best alternative (A_1) has a chance to be chosen in the first case and not in the second. These rankings cannot be evaluated as being the same. This shows the superiority of WS and v coefficients in decision-making ranks analysis. Rankings $R_y^{(3)}$ and $R_y^{(4)}$ show greater variability of coefficients, i.e., the ranking $R_y^{(3)}$ has a coefficient value less, equal to or greater than the ranking $R_y^{(4)}$. It all depends on which coefficient is taken into account, but WS and v again point to the superiority of the ranking $R_y^{(3)}$.

4.2   Influence of a Ranking Size on Coefficients

In this subsection, we want to indicate the impact of the ranking size on the achieved value of the indicator. Figure 1 shows comparisons of WS, r_s, r_w, and v coefficients. Only alternatives on the first and second positions have been replaced. It is a consequence of the conclusion drawn from Table 1. We take into account the rankings with the number from 5 to 50 elements. We can observe that with the increased ranking size, the similarity with the assumed assumptions increases. The Ws coefficient is characterized by the greatest variability, depending on the size of the ranking. Figure 2 shows the changes in the WS value when replacing the best elements with the second, third, fourth, and fifth ones (the position of one adjacent pair is swapped). As we can see, the WS values decrease accordingly, as the quality of the rankings decreases as the best solution moves away from the top of the ranking.

Table 4. Summary of the test with reference ranking (Rx) and four test rankings $(R_y^{(1)} - R_y^{(4)})$ with the calculated correlation factors and proposed WS coefficient for the set of five alternatives $(A_1 - A_5)$, where the change of coefficients is investigated

A_i	R_x	$R_y^{(1)}$	$R_y^{(2)}$	$R_y^{(3)}$	$R_y^{(4)}$
A_1	1	2	5	2	4
A_2	2	3	1	3	2.5
A_3	3	4	2	5	1
A_4	4	5	3	4	5
A_5	5	1	4	1	2.5
Coefficients r_s		0.0000	0.0000	−0.1000	−0.0513
τ		0.2000	0.2000	0.0000	−0.1054
G		0.2000	0.2000	0.0000	−0.1111
r_w		0.0000	0.0000	−0.0667	−0.0667
v		0.1667	−0.1667	0.0833	−0.1083
WS		0.6771	0.3225	0.6354	0.4180

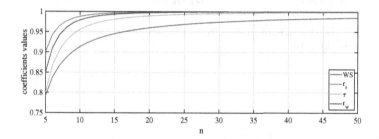

Fig. 1. The value of the coefficients depending on the length of the ranking (n), where occurs one error (change of the first and second position in the ranking) and the converted positions in the ranking.

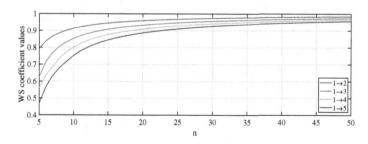

Fig. 2. The value of the WS coefficient depending on the length of the ranking (n) and the converted positions in the ranking.

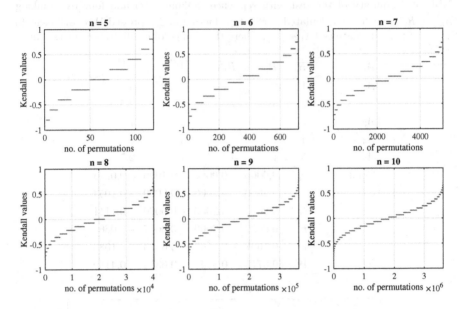

Fig. 3. Sorted distribution of all values of the Kednall coefficient in relation to the length of the ranking (n).

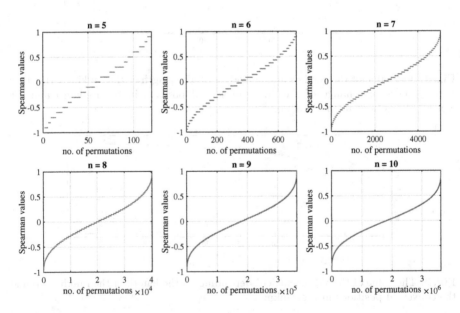

Fig. 4. Sorted distribution of all values of the Spearman coefficient in relation to the length of the ranking (n).

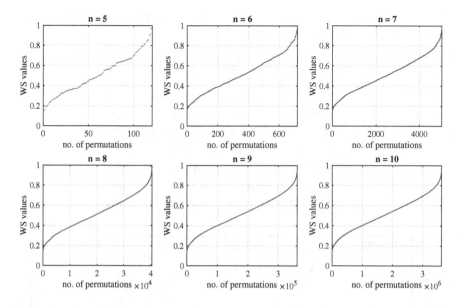

Fig. 5. Sorted distribution of all values of the WS coefficient in relation to the length of the ranking (n).

4.3 Distribution of Coefficients Values

In the next step, we attempted to visualize the distributions for three indicators. Figure 3 presents the distribution of the τ Kendall factor for all possible permutations of the sets of five, six, seven, eight, nine, and ten elements. Figures 4 and 5 show the distribution of the ρ Spearman coefficient and the WS coefficient, respectively. The shape of the ρ Spearman values is smoother than the τ Kendall. Both indicators have a symmetrical distribution, unlike the WS coefficient. The problem may be the interpretation of the WS value, because it is a new approach. However, the question arises when the similarity of the WS coefficient is low, medium, and high. A statistical analysis of the distribution of the WS should be carried out to define three appropriate linguistic terms and answer on this research question.

4.4 Definition of Rankings Similarity

All possible permutations and values of the WS coefficient are determined for ranks of the size from 3 to 10 elements. Based on the obtained values, we calculated basic statistics, which are presented in Table 5. For larger rankings, statistics are based on random samples of 100,000 rankings. Both population and random sampling data are used. Note the convergence of the arithmetic mean, standard deviation, and typical value ranges. The biggest differences concern the arithmetic mean, and it is equal to 0.207 (for a ranking of 10 and 1000 elements).

Table 5. A summary of the basic statistics of the WS coefficient for all possible permutations, where n length of the ranking.

n	$\bar{x}$	S_x	$\bar{x} - S_x$	$\bar{x} + S_x$	x_{min}	x_{max}
3	0.5208	0.2869	0.2339	0.8077	0.1875	1.0000
4	0.5313	0.2164	0.3149	0.7477	0.2083	1.0000
5	0.5135	0.1938	0.3197	0.7073	0.1510	1.0000
6	0.5195	0.1817	0.3378	0.7012	0.1656	1.0000
7	0.5164	0.1757	0.3407	0.6921	0.1383	1.0000
8	0.5197	0.1721	0.3476	0.6918	0.1314	1.0000
9	0.5193	0.1700	0.3493	0.6893	0.1252	1.0000
10	0.5208	0.1688	0.3520	0.6896	0.1144	1.0000

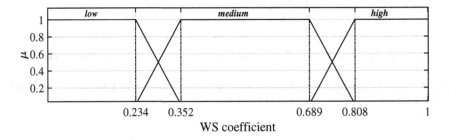

Fig. 6. The definitions of three linguistic terms, i.e., low, medium, and high similarity of rankings by using trapezoidal fuzzy numbers.

Based on the analysis of typical values, i.e. interval of $[\bar{x} - S_x; \bar{x} + S_x]$, we identified the linguistic terms low, medium and high similarity of rankings.

It can indeed be said that if the WS is less than 0.234, then the similarity is low. If the value is higher than 0.808, then the similarity is high. The medium of likeness, which corresponds to a typical value, belongs to the range from 0.352 to 0.689. The remaining values are values where we can talk about a partial belonging to linguistic concepts according to the theory of fuzzy sets, or just low/medium and medium/high concept can be used. Detailed definitions are presented in Fig. 6. Linguistic values are important because they can be used to evaluate the adjustment of the reference and test rankings.

5   Conclusions

The main contribution of the paper is a proposal of the new coefficient of the rankings similarity. For this purpose, the short analysis of classical factors are presented, and some of their shortcomings are emphasized. The most critical is the equality of the values of the classical coefficients in case the ranking error concerns the replacement of a pair of adjacent alternatives (Table 1). The paper presents a theoretical foundation of proposed WS coefficient, which ensures that a new factor is free of identified shortcomings.

The results of numerical experiments compare all analyzed coefficients and their correctness, i.e., ρ Spearman, τ Kendall, G Goodman-Kruskal, and WS coefficients. Then, the distributions of τ Kendall, ρ Spearman, and WS coefficients were compared. WS values can be used to measure similarity of rankings.

Finally, three linguistic concepts were formulated for the low, medium, and high similarity of the two rankings. The properties of the WS coefficient indicate that it is a useful tool for comparing the similarity of rankings and is better suited for this purpose than the currently used correlation coefficients.

During the research, some improvement areas have been identified. The future work directions should concentrate on:

- further comparing between existing coefficients and the proposed WS;
- testing the use of the WS coefficient in real-life examples;
- detection and correction of WS coefficient shortcomings;
- adaptation of the proposed coefficient to uncertain (fuzzy) rankings.

Acknowledgments. The work was supported by the National Science Centre, Decision number UMO-2018/29/B/HS4/02725.

References

1. de Almeida, A.: Multicriteria modelling for a repair contract problem based on utility and the ELECTRE I method. IMA J. Manag. Math. **13**(1), 29–37 (2002). https://doi.org/10.1093/imaman/13.1.29
2. de Andrade, G., Alves, L., Andrade, F., de Mello, J.: Evaluation of power plants technologies using multicriteria methodology MACBETH. IEEE Lat. Am. Trans. **14**(1), 188–198 (2016). https://doi.org/10.1109/TLA.2016.7430079
3. Ashraf, Q., Habaebi, M., Islam, M.R.: TOPSIS-based service arbitration for autonomic internet of things. IEEE Access **4**, 1313–1320 (2016). https://doi.org/10.1109/ACCESS.2016.2545741
4. Bandyopadhyay, S.: Ranking of suppliers with MCDA technique and probabilistic criteria. In: International Conference on Data Science and Engineering, pp. 1–5. IEEE, August 2016. https://doi.org/10.1109/ICDSE.2016.7823948
5. Bandyopadhyay, S.: Application of fuzzy probabilistic TOPSIS on a multi-criteria decision making problem. In: Second International Conference on Electrical, Computer and Communication Technologies, pp. 1–3. IEEE, February 2017. https://doi.org/10.1109/ICECCT.2017.8118038
6. Blest, D.C.: Theory & methods: Rank correlation - an alternative measure. Aust. NZ. J. Stat. **42**(1), 101–111 (2000). https://doi.org/10.1111/1467-842X.00110
7. Brazdil, P.B., Soares, C.: A comparison of ranking methods for classification algorithm selection. In: López de Mántaras, R., Plaza, E. (eds.) ECML 2000. LNCS (LNAI), vol. 1810, pp. 63–75. Springer, Heidelberg (2000). https://doi.org/10.1007/3-540-45164-1_8
8. Cavalcante, V., Alexandre, C., Ferreira, R.P., de Almeida, A.T.: A preventive maintenance decision model based on multicriteria method PROMETHEE II integrated with Bayesian approach. IMA J. Manag. Math. **21**(4), 333–348 (2010). https://doi.org/10.1093/imaman/dpn017

9. Ceballos, B., Lamata, M.T., Pelta, D.A.: A comparative analysis of multi-criteria decision-making methods. Prog. Artif. Intell. **5**(4), 315–322 (2016). https://doi.org/10.1007/s13748-016-0093-1

10. Pinto da Costa, J., Soares, C.: A weighted rank measure of correlation. Aust. NZ. J. Stat. **47**(4), 515–529 (2005). https://doi.org/10.1111/j.1467-842X.2005.00413.x

11. Fagin, R., Kumar, R., Sivakumar, D.: Comparing top k lists. In: Proceedings of the Fourteenth Annual ACM-SIAM Symposium on Discrete Algorithms, SODA 2003, pp. 28–36. Society for Industrial and Applied Mathematics, USA (2003). https://doi.org/10.1137/S0895480102412856

12. Faizi, S., Rashid, T., Sałabun, W., Zafar, S., Wątróbski, J.: Decision making with uncertainty using hesitant fuzzy sets. Int. J. Fuzzy Syst. **20**(1), 93–103 (2017). https://doi.org/10.1007/s40815-017-0313-2

13. Genest, C., Plante, J.F.: On blest's measure of rank correlation. Can. J. Stat. **31**(1), 35–52 (2003). https://doi.org/10.2307/3315902

14. Goodman, L., Kruskal, W.: Measures of association for cross classifications. J. Am. Stat. Assoc. **49**(268), 732–764 (1954). https://doi.org/10.1080/01621459.1954.10501231

15. Haddad, M., Sanders, D.: Selecting a best compromise direction for a powered wheelchair using PROMETHEE. IEEE Trans. Neural Syst. Rehabil. Eng. **27**(2), 228–235 (2019). https://doi.org/10.1109/TNSRE.2019.2892587

16. Hemili, M., Laouar, M.R.: Use of multi-criteria decision analysis to make collection management decisions. In: 3rd International Conference on Pattern Analysis and Intelligent Systems, pp. 1–5. IEEE, October 2018. https://doi.org/10.1109/PAIS.2018.8598495

17. Ishizaka, A., Siraj, S.: Are multi-criteria decision-making tools useful? An experimental comparative study of three methods. Eur. J. Oper. Res. **264**(2), 462–471 (2018). https://doi.org/10.1016/j.ejor.2017.05.041

18. Ivlev, I., Jablonsky, J., Kneppo, P.: Multiple-criteria comparative analysis of magnetic resonance imaging systems. Int. J. Med. Eng. Inform. **8**(2), 124–141 (2016). https://doi.org/10.1504/IJMEI.2016.075757

19. Jeremic, V.M., Radojicic, Z.: A new approach in the evaluation of team chess championships rankings. J. Quan. Anal. Sports **6**(3), 1–11 (2010). https://doi.org/10.2202/1559-0410.1257

20. Kendall, M.G.: A new measure of rank correlation. Biometrika **30**(1/2), 81–93 (1938). https://doi.org/10.2307/2332226

21. Luo, H.C., Sun, Z.X.: A study on stock ranking and selection strategy based on UTA method under the condition of inconsistence. In: 2014 International Conference on Management Science & Engineering 21th Annual Conference Proceedings, pp. 1347–1353. IEEE, August 2014. https://doi.org/10.1109/ICMSE.2014.6930387

22. de Monti, A., Toro, P.D., Droste-Franke, B., Omann, I., Stagl, S.: Assessing the quality of different MCDA methods. In: Getzner, M., Spash, C., Stagl, S. (eds.) Alternatives for Environmental Evaluation, chap. 5, pp. 115–149. Routledge (2004). https://doi.org/10.4324/9780203412879

23. Mulliner, E., Malys, N., Maliene, V.: Comparative analysis of mcdm methods for the assessment of sustainable housing affordability. Omega **59**, 146–156 (2016). https://doi.org/10.1016/j.omega.2015.05.013

24. Ray, T., Triantaphyllou, E.: Evaluation of rankings with regard to the possible number of agreements and conflicts. Eur. J. Oper. Res. **106**(1), 129–136 (1998). https://doi.org/10.1016/S0377-2217(97)00304-4

25. Sałabun, W.: The characteristic objects method: a new distance-based approach to multicriteria decision-making problems. J. Multi-Criteria Decis. Anal. **22**(1–2), 37–50 (2015). https://doi.org/10.1002/mcda.1525

26. Sałabun, W., Karczmarczyk, A., Wątróbski, J., Jankowski, J.: Handling data uncertainty in decision making with comet. In: IEEE Symposium Series on Computational Intelligence, pp. 1478–1484. IEEE, November 2018. https://doi.org/10.1109/SSCI.2018.8628934

27. Sałabun, W., Piegat, A.: Comparative analysis of MCDM methods for the assessment of mortality in patients with acute coronary syndrome. Artif. Intell. Rev. **1**, 1–15 (2016). https://doi.org/10.1007/s10462-016-9511-9

28. Sari, J., Gernowo, R., Suseno, J.: Deciding endemic area of dengue fever using simple multi attribute rating technique exploiting ranks. In: 10th International Conference on Information Technology and Electrical Engineering, pp. 482–487. IEEE, July 2018. https://doi.org/10.1109/ICITEED.2018.8534882

29. Shen, K., Tzeng, G.: A refined DRSA model for the financial performance prediction of commercial banks. In: International Conference on Fuzzy Theory and Its Applications, pp. 352–357. IEEE, December 2013. https://doi.org/10.1109/iFuzzy.2013.6825463

30. Shieh, G.S.: A weighted Kendall's tau statistic. Stat. Probab. Lett. **39**(1), 17–24 (1998). https://doi.org/10.1016/S0167-7152(98)00006-6

31. Spearman, C.: The proof and measurement of association between two things. Am. J. Psychol. **15**(1), 72–101 (1904). https://doi.org/10.2307/1422689

32. Tian, G., Zhang, H., Zhou, M., Li, Z.: AHP, gray correlation, and TOPSIS combined approach to green performance evaluation of design alternatives. IEEE Trans. Syst. Man Cybern.: Syst. Part A Syst. Hum. **48**(7), 1093–1105 (2017). https://doi.org/10.1109/TSMC.2016.2640179

33. Wątróbski, J., Jankowski, J., Ziemba, P., Karczmarczyk, A., Zioło, M.: Generalised framework for multi-criteria method selection. Omega, **86**, 107–124 (2019). https://doi.org/10.1016/j.omega.2018.07.004

34. Yaraghi, N., Tabesh, P., Guan, P., Zhuang, J.: Comparison of AHP and Monte Carlo AHP under different levels of uncertainty. IEEE Trans. Eng. Manag. **62**(1), 122–132 (2015). https://doi.org/10.1109/TEM.2014.2360082

35. Zhang, C., Liu, X., Jin, J.G., Liu, Y.: A stochastic ANP-GCE approach for vulnerability assessment in the water supply system with uncertainties. IEEE Trans. Eng. Manag. **63**(1), 78–90 (2015). https://doi.org/10.1109/TEM.2015.2501651

36. Zhang, P., Yao, H., Qiu, C., Liu, Y.: Virtual network embedding using node multiple metrics based on simplified ELECTRE method. IEEE Access **6**, 37314–37327 (2018). https://doi.org/10.1109/ACCESS.2018.2847910

Innovativeness Analysis of Scholarly Publications by Age Prediction Using Ordinal Regression

Pavel Savov[1]($\boxtimes$), Adam Jatowt[2], and Radoslaw Nielek[1]

[1] Polish-Japanese Academy of Information Technology,
ul. Koszykowa 86, 02-008 Warszawa, Poland
{pavel.savov,nielek}@pja.edu.pl
[2] Kyoto University, Yoshida-honmachi, Sakyo-ku, Kyoto 606-8501, Japan
adam@dl.kuis.kyoto-u.ac.jp

Abstract. In this paper we refine our method of measuring the innovativeness of scientific papers. Given a diachronic corpus of papers from a particular field of study, published over a period of a number of years, we extract latent topics and train an ordinal regression model to predict publication years based on topic distributions. Using the prediction error we calculate a real-number based innovation score, which may be used to complement citation analysis in identifying potential breakthrough publications. The innovation score we had proposed previously could not be compared for papers published in different years. The main contribution we make in this work is adjusting the innovation score to account for the publication year, making the scores of papers published in different years directly comparable. We have also improved the prediction accuracy by replacing multiclass classification with ordinal regression and Latent Dirichlet Allocation models with Correlated Topic Models. This also allows for better understanding of the evolution of research topics. We demonstrate our method on two corpora: 3,577 papers published at the International World Wide Web Conference (WWW) between the years 1994 and 2019, and 835 articles published in the Journal of Artificial Societies and Social Simulation (JASSS) from 1998 to 2019.

Keywords: Scientometrics · Topic models · Ordinal regression

1 Introduction

Citation analysis has been the main method of measuring innovation and identifying important and/or pioneering scientific papers. It is assumed that papers having high citation counts have made a significant impact on their fields of study and are considered innovative. This approach, however, has a number of shortcomings: Works by well-known authors and/or ones published at well-established publication venues tend to receive more attention and citations than others (the rich-get-richer effect) [35]. According to Merton [19], who first described this

© Springer Nature Switzerland AG 2020
V. V. Krzhizhanovskaya et al. (Eds.): ICCS 2020, LNCS 12138, pp. 646–660, 2020.
https://doi.org/10.1007/978-3-030-50417-5_48

phenomenon in 1968, publications by more eminent researchers will receive disproportionately more recognition than similar works by less-well known authors. This is known as the *Matthew Effect*, named after the biblical Gospel of Matthew. Serenko and Dumay [30] observed that old citation classics keep getting cited because they appear among the top results in Google Scholar, and are automatically assumed as credible. Some authors also assume that reviewers expect to see those classics referenced in the submitted paper regardless of their relevance to the work being submitted. There is also the problem of self-citations: Increased citation count does not reflect the work's impact on its field of study.

We addressed these shortcomings in our previous work [27] by proposing a machine learning-based method of measuring the innovativeness of scientific papers. Our current method involves training a Correlated Topic Model (CTM) [3] on a diachronic corpus of papers published at conference series or in different journal editions over as many years as possible, training a model for predicting publication years using topic distributions as feature vectors, and calculating a real number innovation score for each paper based on the prediction error.

We consider a paper innovative if it covers topics that will be popular in the future but have not been researched in the past. Therefore, the more recent the publication year predicted by our model compared to the actual year of publication, the greater the paper's score. We showed in [27] that our innovation scores are positively correlated with citation counts, but there are also highly scored papers having few citations. These papers may be worth looking into as potential "hidden gems" – covering topics researched in the future but relatively unnoticed. Interestingly, we have not found any highly cited papers with low innovation scores.

2   Related Work

The development of research areas and the evolution of topics in academic conferences and journals over time have been investigated by numerous researchers. For example, Meyer et al. [20] study the Journal of Artificial Societies and Social Simulation (JASSS) by means of citation and co-citation analysis. They identify the most influential works and authors and show the multidisciplinary nature of the field. Saft and Nissen [25] also analyze JASSS, but they use a text mining approach linking documents into thematic clusters in a manner inspired by co-citation analysis. Wallace et al. [34] study trends in the ACM Conference on Computer Supported Cooperative Work (CSCW). They took over 1,200 papers published between the years 1990 and 2015, and they analyzed data such as publication year, type of empirical research, type of empirical evaluations used, and the systems/technologies involved. [21] analyze trends in the writing style in papers from the ACM SIGCHI Conference on Human Factors in Computing Systems (CHI) published over a 36-year period.

Recent research on identifying potential breakthrough publications includes works such as Schneider and Costas [28, 29]. Their approach is based on analyzing citation networks, focusing on highly-cited papers. Ponomarev et al. [22] predict citation count based on citation velocity, whereas Wolcott et al. [36] use random forest models on a number of features, e.g. author count, reference count, H-index etc. as well as citation velocity. These approaches, in contrast to ours, take into account non-textual features. They also define breakthrough publications as either highly-cited influential papers resulting in a change in research direction, or "articles that result from transformative research" [36].

A different approach to identifying novelty was proposed by Chan et al. [5]. They developed a system for finding analogies between research papers, based on the premise that "scientific discoveries are often driven by finding analogies in distant domains". One of the examples given is the simulated annealing optimization algorithm inspired by the annealing process commonly used in metallurgy. Identifying interdisciplinary ideas as a driver for innovation was also studied by Thorleuchter and Van den Poel [33]. Several works have employed machine learning-based approaches to predict citation counts and the long-term scientific impact (LTSI) of research papers, e.g., [37] or [31].

Examples of topic-based approaches include Hall et al. [11]. They trained an LDA model on the ACL Anthology, and showed trends over time like topics increasing and declining in popularity. Unlike our approach, they hand-picked topics from the generated model and manually seeded 10 more topics to improve field coverage. More recently Chen et al. [7] studied the evolution of topics in the field of information retrieval (IR). They trained a 5-topic LDA model on a corpus of around 20,000 papers from *Web of Science*. Sun and Yin [32] used a 50-topic LDA model trained on a corpus of over 17,000 abstracts of research papers on transportation published over a 25-year period to identify research trends by studying the variation of topic distributions over time. Another interesting example is the paper by Hu et al. [12] where Google's Word2Vec model is used to enhance topic keywords with more complete semantic information, and topic evolution is analyzed using spatial correlation measures in a semantic space modeled as an urban geographic space.

Research on document dating (timestamping) is related to our work, too. Typical approaches to document dating are based on changes in word usage and on language change over time, and they use features derived from temporal language models [9, 14], diachronic word frequencies [8, 26], or occurrences of named entities. Examples of research articles based on heuristic methods include: [10], [15] or [16]. Jatowt and Campos [13] have implemented the visual, interactive system based on n-gram frequency analysis. In our work we rely on predicting publication dates to determine paper innovativeness. Ordinal regression models trained on topic vectors could be regarded as a variation of temporal language models and reflect vocabulary change over time. Aside from providing means for timestamping, they also allow for studying how new ideas emerge, gain and lose popularity.

3   Datasets

The corpora we study in this paper contain 3,577 papers published at the International World Wide Web Conference (WWW) between the years 1994 and 2019, and 835 articles published in the Journal of Artificial Societies and Social Simulation (JASSS)[1] from 1998 to 2019. We have studied papers from the WWW Conference before [27], which is the reason why we decided to use this corpus again, after updating it with papers published after our first analysis, i.e. ones in the years 2018 and 2019. We chose JASSS as the other corpus to analyze in order to demonstrate our method on another major publication venue in a related but separate field, published over a period of several years. It is publicly available in HTML, which makes it straightforward to extract text from the documents.

In an effort to extract only relevant content, we performed the following preprocessing steps on all texts before converting them to Bag-of-Words vectors:

1. Discarding page headers and footers, *References*, *Bibliography* and *Acknowledgments* sections as "noise" irrelevant to the main paper topic(s)
2. Conversion to lower case
3. Removal of stopwords and punctuation as well as numbers, including ones spelled out, e.g. "one", "two", "first" etc.
4. Part-of-Speech tagging using the Penn Treebank POS tagger (NLTK) [2] – This step is a prerequisite for the WordNet Lemmatizer, we do not use the POS tags in further processing
5. Lemmatization using the WordNet Lemmatizer in NLTK.

4   Method

4.1   Topic Model

In our previous work [27] we trained Latent Dirichlet Allocation (LDA) [4] topic models. In this paper, however, we have decided to move towards Correlated Topic Models (CTM) [3] and only built LDA models as a baseline. Unlike LDA, which assumes topic independence, CTM allows for correlation between topics. We have found this to be better suited for modeling topics evolving over time, including splitting or branching. We used the reference C implementation found at http://www.cs.columbia.edu/~blei/ctm-c/.

In order to choose the number of topics k, we have built a k-topic model for each k in a range we consider broad enough to include the optimum number of topics. In the case of LDA this range was $\langle 10, 60 \rangle$. We then chose the models with the highest C_V topic coherence. As shown by Röder et al. [24], this measure approximates human topic interpretability the best. Furthermore, according to Chang et al. [6], topic model selection based on traditional likelihood or perplexity-based approaches results in models that are worse in terms of human understandability. The numbers of topics we chose for our LDA models

---

[1] http://jasss.soc.surrey.ac.uk/.

were 44 for the WWW corpus and 50 for JASSS. Because CTM supports more topics for a given corpus [3] and allows for a more granular topic model, we explored different ranges of k than in the case of LDA: $\langle 30, 100 \rangle$ for WWW and $\langle 40, 120 \rangle$ for JASSS. As before, we chose the models with the highest C_V.

4.2   Publication Year Prediction

Because publication years are ordinal values rather than categorical ones, instead of One-vs-One or One-vs-Rest multiclass classifiers, which we had used previously, we have implemented ordinal regression (a.k.a. ordinal classification) based on the framework proposed by Li and Lin [17], as used by Martin et al. [18] for photograph dating. An N-class ordinal classifier consists of $N - 1$ *before-after* binary classifiers, i.e. for each pair of consecutive years a classifier is trained, which assigns documents to one of two classes: "year y or before" and "year $y + 1$ or after". Given the class membership probabilities predicted by these classifiers, the overall classifier confidence that paper p was published in the year Y is then determined, as in [18], by Eq. 1:

$$conf(p, Y) = \prod_{y=Y_{min}}^{Y} P(Y_p \leq y) \cdot \prod_{y=Y+1}^{Y_{max}} (1 - P(Y_p \leq y)) \tag{1}$$

where Y_{min} and Y_{max} are the first and last year in the corpus, and Y_p is the publication year of the paper p.

We used topic probability distributions as k-dimensional feature vectors, where k is the number of topics. Due to the small size of the JASSS corpus, we trained a separate model to evaluate each document (Leave-one-out cross-validation), whereas in the case of the WWW corpus we have settled for 10-fold cross-validation. We have implemented ordinal regression using linear Support Vector Machine (SVM) classifiers.

4.3   Paper Innovation Score

Following [27], we define our innovation score based on the results from the previous step - classifier confidence - as the weighted mean publication year prediction error with classifier confidence scores as weights:

$$S_P(p) = \frac{\sum_y conf(p, y) \cdot (y - Y_p)}{\sum_y conf(p, y)} \tag{2}$$

where Y_p is the year paper p was published in and $conf(p, y)$ is the classifier confidence for paper p and year y. Unlike the score defined in [27], the denominator in Eq. 2 does not equal 1, since the scores $conf(p, y)$ defined in Eq. 1 are not class membership probabilities.

As illustrated in Fig. 1, the higher the publication year of paper p, the lower the minimum and maximum possible values of $S_P(p)$. In order to make papers from different years comparable in terms of innovation scores, $S_P(p)$ needs to be adjusted to account for the publication year of paper p.

Suppose the prediction error for papers published in the year Y is a discrete random variable Err_Y. Based on the actual prediction error distributions for the WWW and JASSS corpora (see Fig. 3), let us define the expected publication year prediction error for papers published in the year Y as:

$$E(Err_Y) = \sum_{n=Y_{min}-Y}^{Y_{max}-Y} n \cdot Pr(Err_Y = n) \tag{3}$$

where Y_{min} and Y_{max} are the minimum and maximum publication years in the corpus, and $Pr(Err_Y = n)$ is the observed probability that the prediction error for a paper published in the year Y is n. To calculate $Pr(Err_Y = n)$ we use the distribution from Fig. 3 truncated to the range $\langle Y_{min} - Y, Y_{max} - Y \rangle$, i.e. the minimum and maximum possible prediction errors for papers published in the year Y.

Let us then define the adjusted innovation score as the deviation of $S_P(p)$ from its expected value divided by its maximum absolute value:

$$S'_P(p) = \begin{cases} \frac{S_P(p) - E(Err_{Y_p})}{E(Err_{Y_p}) - (Y_{min} - Y_p)} & \text{if } S_P(p) < E(Err_{Y_p}) \\ \frac{S_P(p) - E(Err_{Y_p})}{Y_{max} - Y_p - E(Err_{Y_p})} & \text{if } S_P(p) \geq E(Err_{Y_p}) \end{cases} \tag{4}$$

where Y_p is the publication year of the paper p.
$S'_P(p)$ has the following characteristics:

1. $-1 \leq S'_P(p) \leq 1$
2. $S'_P(p) = 0$ if paper p's predicted publication year is as expected
3. $S'_P(p) < 0$ if paper p's predicted publication year is earlier than expected
4. $S'_P(p) > 0$ if paper p's predicted publication year is later than expected.

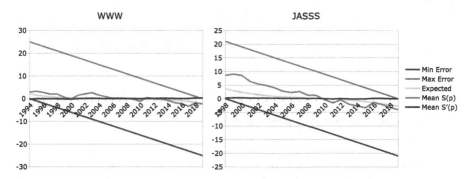

Fig. 1. Minimum and maximum prediction errors decrease as the publication year increases and so does the mean unadjusted score (S_P). To make papers from different years comparable in terms of innovation score, the adjusted innovation score (S'_P) measures the deviation of the prediction error from its expected value.

5   Results

Figure 2 shows the relation between the number of topics k and coherence C_V for CTM models trained on each of our corpora. Topic coherence initially peaks for values of k close to the optimal values found for LDA, then after a dip, it reaches global maxima for k equal to 74 and 88 for WWW and JASSS, respectively.

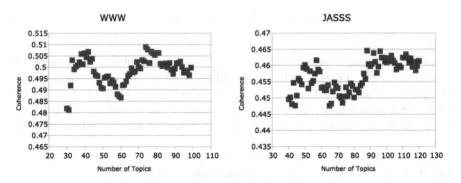

Fig. 2. C_V Topic coherence by number of topics. We chose the CTM models with the highest values of C_V coherence as described in Sect. 4.1.

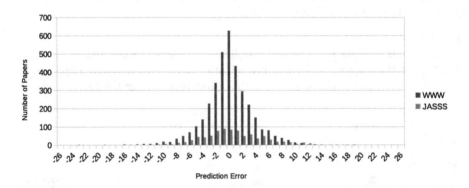

Fig. 3. Distribution of publication year prediction errors for both corpora. We use these distributions to calculate the expected prediction error for each year and adjust paper innovation scores for their publication years.

As shown in Table 1, publication year prediction accuracy expressed as Mean Absolute Error (MAE) is markedly improved both by using CTM over LDA and ordinal regression over a standard One-vs-One (OvO) multiclass SVM classifier. The best result we achieve for the WWW corpus was 2.56 and for JASSS: 3.56.

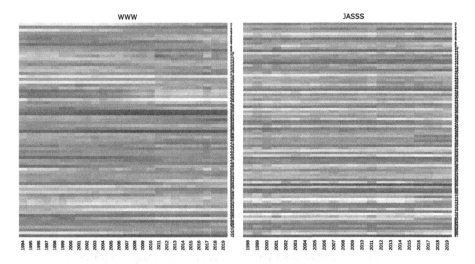

Fig. 4. Topic popularity over time. The color of the cell in row t and column y represents the mean proportion of topic t in papers published in the year y. Bright red represents maximum values, white means zero. (Color figure online)

Table 3 shows the top 3 papers with the highest innovation scores for both corpora. For each of those papers we list the number of citations and some of their most significant topics. All of them have been cited, some of them widely. The more a paper's topic distribution resembles the topic distributions of papers published in the future and the less it resembles that of papers from the past, the higher the innovation score. Some examples of highly scored, fairly recently published papers having few citations include:

– WWW, 2019: *Multiple Treatment Effect Estimation using Deep Generative Model with Task Embedding* by Shiv Kumar Saini et al. – no citations, 6th highest score (0.946), topics covered: #10, #28, #33, #57 (see: Table 2)
– JASSS, 2017: *R&D Subsidization Effect and Network Centralization: Evidence from an Agent-Based Micro-Policy Simulation* by Pierpaolo Angelini et al. – 2 citations, 20th highest score (0.634), topics covered: #4, #48, #65 (see: Table 2)

Table 1. Mean absolute prediction errors: CTM vs. LDA and Multiclass SVM vs. Ordinal regression

	Multiclass SVM		Ordinal regression	
	WWW	JASSS	WWW	JASSS
LDA	4.14	6.09	3.34	4.38
CTM	3.02	4.22	2.56	3.56

Table 2. Selected latent topics described by their top 30 words.

	No	Top 30 Words
WWW	2	Cluster similarity algorithm set use measure intent result document number group base approach different information click give distance web method similar user problem find represent clustering term session figure follow
	4	Object information web model multimedia use content provide base presentation retrieval type structure medium metadata represent show level image also system support relationship value order different part present define point
	9	Network node link sample edge method random walk graph model degree social use distribution show figure matrix number result value set base prediction parameter time performance follow order neighbor problem
	10	Ad advertiser click advertising use target bid user model ctr impression show search revenue advertisement online value campaign per number domain display keywords learn keyword rate conversion bundle sponsor base
	12	User tweet twitter post account social spam use number follower content campaign network follow also show feature detection find detect study medium group identity figure abusive information identify spammer time
	15	Social network tag co information author people user use paper friend relationship group person web measure similarity name interest annotation base team profile number system share find relation concept work
	26	Service web ontology use process model concept base composition approach rule qos set description state constraint example define provider provide system information owl may instance context execution describe match axiom
	28	Treatment claim source effect group causal true data variable control model experiment use truth estimate distribution value fact set make prior match outcome unit credibility parameter reliability figure evidence assertion
	33	Model feature learn performance dataset network attention layer neural sequence prediction train use method datasets propose state task deep baseline representation lstm vector input base embed figure time interaction information
	38	Email influence flow information model user time chain diffusion reply use work company network number figure factor transition base job sender data receive social also give process probability study show
	41	User social cascade facebook post feature group number network time model friend figure hashtags show discussion distribution content comment activity also study large online predict use set observe size share
	52	Mobile apps app device use performance application network time model energy data show dl user figure android developer result signal browser different permission run number deep platform measurement support cloud

(continued)

Table 2. (*continued*)

	No	Top 30 Words
	56	Event news time topic blog medium temporal information story source trend use attention show feed series post interest analysis different content set detection data figure country article work goal day
	57	Rating user model use preference item rank comment restaurant data method movie show value set matrix base latent distribution high group approach rat number low give result learn different bias
	72	Feature classifier label classification class set use train learn data score training accuracy tree performance positive instance sample number base category svm example detection dataset test approach method result bias
JASSS	0	Model democracy society polity complex system social political simple world state dynamic country power global non democratic change data economic theory war simulation time see development peasant transition complexity also
	4	Model income policy economic tax level region household rate consumption result increase base agent high change market doi firm price cost economy work effect low al et value parameter distribution
	6	Agent belief model resource level time simulation social number society may population communication set probability case experiment information environment collective state make action process base system initial result also increase
	21	Model agent household data flood base house use et simulation al number housing level year population process figure time area change result urban different city location new center homeowner income
	24	Simulation method data output algorithm number match use microsimulation fit set example variable probability table result test alignment mean prediction sample observation pair time show order weight different distance measure
	48	Bank interbank financial loss risk network institution asset al et doi system figure channel contagion data market default ast cross systemic liability rule total customer use banking shareholding show increase
	65	Social research science simulation model review journal scientist agent community scientific base number fund proposal year jasss project author paper system publication study result topic network time funding publish society
	71	Opinion model social influence agent doi time group dynamic polarization et al. value show different individual network change journal effect evolution simulation figure base result interaction confidence cluster process event
	72	Energy model agent system electricity decision social base technology use al et change charge policy different value simulation figure scenario demand environmental household actor diffusion factor power result information transition

Table 3. Top 3 papers with the highest innovation scores in both corpora with citation counts and topics covered.

	Year	Author(s) and Title	Score	Citations	Topics
WWW	2011	C. Budak, D. Agrawal, A. El Abbadi, *Limiting the Spread of Misinformation in Social Networks*	0.971	607	9, 12, 38, 41, 56
	2010	A. Sala, L. Cao, Ch. Wilson, R. Zablit, H. Zheng, B. Y. Zhao, *Measurement-calibrated Graph Models for Social Network Experiments*	0.963	189	2, 9, 15, 41, 52
	2018	H. Wu, Ch. Wang, J. Yin, K. Lu, L. Zhu, *Sharing Deep Neural Network Models with Interpretation*	0.955	7	33, 72
JASSS	2001	K. Auer, T. Norris, *"ArrierosAlife" a Multi-Agent Approach Simulating the Evolution of a Social System: Modeling the Emergence of Social Networks with "Ascape"*	0.868	13	6, 21
	2000	B. G. Lawson, S. Park, *Asynchronous Time Evolution in an Artificial Society Model*	0.841	13	6, 24, 71
	2008	R. Bhavnani, D. Miodownik, J. Nart, *REsCape: an Agent-Based Framework for Modeling Resources, Ethnicity, and Conflict*	0.788	51	0, 72

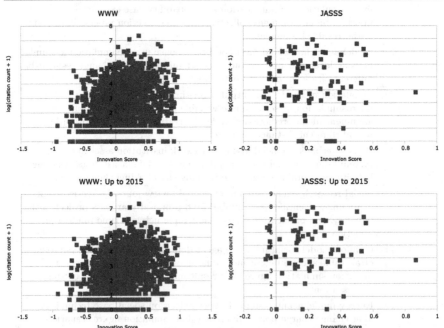

Fig. 5. Innovation score vs. Citation count for all papers (above) and papers at least 5 years old (below).

Figure 5 illustrates the correlation between Innovation Scores and citation counts. Because the number of citations is expected to grow exponentially [23], we have used $log_2(citation\ count + 1)$ instead of raw citation counts. The value of this expression is zero if the number of citations is zero and grows monotonically as the number of citations increases. The citation data for the WWW corpus come from ACM's Digital Library[2], however publications from the JASSS journal are not available in the ACM DL. We were also unable to scrape complete citation data from Google Scholar. We have therefore manually collected citation counts for 5 randomly selected papers from each year. We have calculated Spearman's ρ correlation coefficients between the innovation scores and citation counts. The results are: 0.28 with a p-value of $1.21 \cdot 10^{-41}$ for the WWW corpus and 0.32 with a p-value of $1.91 \cdot 10^{-6}$ for JASSS. The innovation scores are, therefore, weakly correlated to the citation counts. The correlation coefficients are slightly higher for papers at least 5 years old: 0.3 for WWW and 0.37 for JASSS. This may be explained by the fact that newer papers have not yet accumulated many citations regardless of their innovativeness.

6   Conclusion and Future Work

We have shown a simple yet significant improvement to our novel method of measuring the innovativeness of scientific papers in bodies of research spanning multiple years. Scaling the innovation score proposed in our previous research has enabled us to directly compare the scores of papers published at different years. We have also improved the prediction accuracy by employing ordinal regression models instead of regular multiclass classifiers and Correlated Topic Models instead of LDA. It may be argued that this makes our method more reliable, as deviations of the predicted publication year from the actual one are more likely to be caused by the paper actually covering topics popular in the future rather than just being usual prediction error. Moreover, CTM allowed to better model and understand the evolution of research topics over time.

In the future we plan to explore non-linear ways to scale the innovation scores, taking into account the observed error distribution (Fig. 3) to give more weight to larger deviations from the expected value. We also plan to use word embeddings or extracted scientific claims [1] as well as other means of effectively representing paper contents and conveyed ideas besides topic models as features to our methods.

References

1. Achakulvisut, T., Bhagavatula, C., Acuna, D., Kording, K.: Claim extraction in biomedical publications using deep discourse model and transfer learning. arXiv preprint arXiv:1907.00962 (2019)
2. Bird, S., Klein, E., Loper, E.: Natural Language Processing with Python: Analyzing Text with the Natural Language Toolkit. O'Reilly Media, Inc. (2009)

---

[2] http://dl.acm.org/.

3. Blei, D., Lafferty, J.: Correlated topic models. In: Advances in Neural Information Processing Systems, vol. 18, p. 147 (2006)
4. Blei, D.M., Ng, A.Y., Jordan, M.I.: Latent dirichlet allocation. J. Mach. Learn. Res. **3**(4–5), 993–1022 (2003)
5. Chan, J., Chang, J.C., Hope, T., Shahaf, D., Kittur, A.: Solvent: a mixed initiative system for finding analogies between research papers. Proc. ACM Hum.-Comput. Interact. **2**(CSCW), 31:1–31:21 (2018). https://doi.org/10.1145/3274300
6. Chang, J., Gerrish, S., Wang, C., Boyd-graber, J.L., Blei, D.M.: Reading tea leaves: how humans interpret topic models. In: Bengio, Y., Schuurmans, D., Lafferty, J.D., Williams, C.K.I., Culotta, A. (eds.) Advances in Neural Information Processing Systems, vol. 22, pp. 288–296. Curran Associates, Inc. (2009). http://papers.nips.cc/paper/3700-reading-tea-leaves-how-humans-interpret-topic-models.pdf
7. Chen, B., Tsutsui, S., Ding, Y., Ma, F.: Understanding the topic evolution in a scientific domain: an exploratory study for the field of information retrieval. J. Inf. **11**(4), 1175–1189 (2017)
8. Ciobanu, A.M., Dinu, A., Dinu, L., Niculae, V., Şulea, O.M.: Temporal classification for historical romanian texts. In: Proceedings of the 7th Workshop on Language Technology for Cultural Heritage, Social Sciences, and Humanities, pp. 102–106. Association for Computational Linguistics, Sofia (2013)
9. De Jong, F., Rode, H., Hiemstra, D.: Temporal language models for the disclosure of historical text. In: Humanities. Computers and Cultural Heritage: Proceedings of the XVIth International Conference of the Association for History and Computing (AHC 2005), pp. 161–168. Koninklijke Nederlandse Academie van Wetenschappen, Amsterdam (2005)
10. Garcia-Fernandez, A., Ligozat, A.-L., Dinarelli, M., Bernhard, D.: When was it written? Automatically determining publication dates. In: Grossi, R., Sebastiani, F., Silvestri, F. (eds.) SPIRE 2011. LNCS, vol. 7024, pp. 221–236. Springer, Heidelberg (2011). https://doi.org/10.1007/978-3-642-24583-1_22
11. Hall, D., Jurafsky, D., Manning, C.D.: Studying the history of ideas using topic models. In: Proceedings of the Conference on Empirical Methods in Natural Language Processing, EMNLP 2008, pp. 363–371. Association for Computational Linguistics, Stroudsburg (2008). http://dl.acm.org/citation.cfm?id=1613715.1613763
12. Hu, K., et al.: Understanding the topic evolution of scientific literatures like an evolving city: using google word2vec model and spatial autocorrelation analysis. Inf. Proces. Manag. **56**(4), 1185–1203 (2019)
13. Jatowt, A., Campos, R.: Interactive system for reasoning about document age. In: Proceedings of the 2017 ACM on Conference on Information and Knowledge Management, CIKM 2017, pp. 2471–2474. ACM, New York (2017). https://doi.org/10.1145/3132847.3133166
14. Kanhabua, N., Nørvåg, K.: Using temporal language models for document dating. In: Buntine, W., Grobelnik, M., Mladenić, D., Shawe-Taylor, J. (eds.) ECML PKDD 2009. LNCS (LNAI), vol. 5782, pp. 738–741. Springer, Heidelberg (2009). https://doi.org/10.1007/978-3-642-04174-7_53
15. Kotsakos, D., Lappas, T., Kotzias, D., Gunopulos, D., Kanhabua, N., Nørvåg, K.: A burstiness-aware approach for document dating. In: Proceedings of the 37th International ACM SIGIR Conference on Research & Development in Information Retrieval, SIGIR 20114, pp. 1003–1006. ACM, New York (2014). https://doi.org/10.1145/2600428.2609495

16. Kumar, A., Lease, M., Baldridge, J.: Supervised language modeling for temporal resolution of texts. In: Proceedings of the 20th ACM International Conference on Information and Knowledge Management, CIKM 2011, pp. 2069–2072. ACM, New York (2011). https://doi.org/10.1145/2063576.2063892

17. Li, L., Lin, H.T.: Ordinal regression by extended binary classification. In: Advances in Neural Information Processing Systems, pp. 865–872 (2007)

18. Martin, P., Doucet, A., Jurie, F.: Dating color images with ordinal classification. In: Proceedings of International Conference on Multimedia Retrieval, pp. 447–450 (2014)

19. Merton, R.K.: The matthew effect in science: the reward and communication systems of science are considered. Science **159**(3810), 56–63 (1968)

20. Meyer, M., Lorscheid, I., Troitzsch, K.G.: The development of social simulation as reflected in the first ten years of JASSS: a citation and co-citation analysis. J. Artif. Soc. Soc. Simul. **12**(4), 12 (2009). http://jasss.soc.surrey.ac.uk/12/4/12.html

21. Pohl, H., Mottelson, A.: How we guide, write, and cite at CHI (2019)

22. Ponomarev, I.V., Williams, D.E., Hackett, C.J., Schnell, J.D., Haak, L.L.: Predicting highly cited papers: a method for early detection of candidate breakthroughs. Technol. Forecast. Soc. Chang. **81**, 49–55 (2014)

23. Price, D.D.S.: A general theory of bibliometric and other cumulative advantage processes. J. Am. Soc. Inf. Sci. **27**(5), 292–306 (1976)

24. Röder, M., Both, A., Hinneburg, A.: Exploring the space of topic coherence measures. In: Proceedings of the Eighth ACM International Conference on Web Search and Data Mining, WSDM 2015, pp. 399–408. ACM, New York (2015). https://doi.org/10.1145/2684822.2685324

25. Saft, D., Nissen, V.: Analysing full text content by means of a flexible co-citation analysis inspired text mining method - exploring 15 years of JASSS articles. Int. J. Bus. Intell. Data Min. **9**(1), 52–73 (2014)

26. Salaberri, H., Salaberri, I., Arregi, O., Zapirain, B.: IXAGroupEHUDiac: a multiple approach system towards the diachronic evaluation of texts. In: Proceedings of the 9th International Workshop on Semantic Evaluation (SemEval 2015), pp. 840–845. Association for Computational Linguistics, Denver (2015)

27. Savov, P., Jatowt, A., Nielek, R.: Identifying breakthrough scientific papers. Inf. Proces. Manag. **57**(2), 102168 (2020)

28. Schneider, J.W., Costas, R.: Identifying potential 'breakthrough' research articles using refined citation analyses: three explorative approaches. STI 2014, Leiden, p. 551 (2014)

29. Schneider, J.W., Costas, R.: Identifying potential "breakthrough" publications using refined citation analyses: three related explorative approaches. J. Assoc. Inf. Sci. Technol. **68**(3), 709–723 (2017)

30. Serenko, A., Dumay, J.: Citation classics published in knowledge management journals. Part ii: studying research trends and discovering the google scholar effect. J. Knowl. Manag. **19**(6), 1335–1355 (2015)

31. Singh, M., Jaiswal, A., Shree, P., Pal, A., Mukherjee, A., Goyal, P.: Understanding the impact of early citers on long-term scientific impact. In: 2017 ACM/IEEE Joint Conference on Digital Libraries (JCDL), pp. 1–10. IEEE (2017)

32. Sun, L., Yin, Y.: Discovering themes and trends in transportation research using topic modeling. Transp. Res. Part C Emerg. Technol. **77**, 49–66 (2017)

33. Thorleuchter, D., Van den Poel, D.: Identification of interdisciplinary ideas. Inf. Proces. Manag. **52**(6), 1074–1085 (2016). https://doi.org/10.1016/j.ipm.2016.04.010

34. Wallace, J.R., Oji, S., Anslow, C.: Technologies, methods, and values: changes in empirical research at CSCW 1990–2015. Proc. ACM Hum. Comput. Interact. 1(CSCW), 106:1–106:18 (2017). https://doi.org/10.1145/3134741
35. White, H.D.: Citation analysis and discourse analysis revisited. Appl. Linguist. **25**(1), 89–116 (2004)
36. Wolcott, H.N., et al.: Modeling time-dependent and-independent indicators to facilitate identification of breakthrough research papers. Scientometrics **107**(2), 807–817 (2016)
37. Yan, R., Huang, C., Tang, J., Zhang, Y., Li, X.: To better stand on the shoulder of giants. In: Proceedings of the 12th ACM/IEEE-CS Joint Conference on Digital Libraries, JCDL 2012, pp. 51–60. ACM, New York (2012). https://doi.org/10.1145/2232817.2232831

Advantage of Using Spherical over Cartesian Coordinates in the Chromosome Territories 3D Modeling

Magdalena A. Tkacz$^{(\boxtimes)}$ ⓘ and Kornel Chromiński ⓘ

Faculty of Science and Technology, University of Silesia in Katowice, Będzińska 39,
41-200 Sosnowiec, Poland
{magdalena.tkacz,kornel.chrominski}@us.edu.pl

Abstract. This paper shows results of chromosome territory modeling
in two cases: when the implementation of the algorithm was based on
Cartesian coordinates and when implementation was made with Spherical coordinates. In the article, the summary of measurements of computational times of simulation of chromatin decondensation process (which
led to constitute the chromosome territory) was presented. Initially, when
implementation was made with the use of Cartesian Coordinates, simulation takes a lot of time to create a model (mean 746.7[sec] with the
median 569.1[sec]) and additionally requires restarts of the algorithm,
also often exceeds acceptable (given a priori) time for the computational
experiment. Because of that, authors attempted changing the coordinate
system to Spherical Coordinates (in a few previous projects it leads to
improving the efficiency of implementation). After changing the way that
3D point is represented in 3D space the time required to make a successful model reduced to the mean 25.3[sec] with a median 18.5[s] (alongside
with lowering the number of necessary algorithm restarts) which gives a
significant difference in the efficiency of model's creation. Therefore we
showed, that a more efficient way for implementation was the usage of
spherical coordinates.

Keywords: Spherical coordinates · Cartesian coordinates · 3D ·
Chromatin decondensation · Geometry · Chromosome territories ·
Modeling

1 Introduction

Computational power gives very powerful support in the life sciences today. A lot
of experiments can be done – they are cheaper to conduct, their parameters can
be easily modified. They are also in most cases reproducible and ethical (no
wronging living creatures).

According [1] the term *modeling* is defined as *"to design or imitate forms:
make a pattern"* or *"producing a representation or simulation"* and *model* is

© Springer Nature Switzerland AG 2020
V. V. Krzhizhanovskaya et al. (Eds.): ICCS 2020, LNCS 12138, pp. 661–673, 2020.
https://doi.org/10.1007/978-3-030-50417-5_49

defined as *"a system of postulates, data, and inferences presented as a mathematical description of an entity or state of affairs"*. In fact, in a case of *in-silico* experiments more precious would be "computational model" and "computational modeling", but in this paper, it will be referred to in a shorter form. Sometimes term "modeling" is used in the context of "running computational model" – but in this paper, it will be referred to as *"simulation"*. For many disciplines creating a model is important – it allows to re-scale (extend or reduce) object, slow down or speed up modeled process, examine almost any aspect of object or process (separating parameters or taking into account a given quantity of parameters). With the use of computers, it is also possible to make visualizations and animations.

This paper describes some aspects of the modeling process that occurs in all organisms – precisely speaking occurs in almost every living cell. This process also occurs just right now – in my body while I'm writing this text, as well as in your – when you read this. This is the process of transferring genetic material, *DNA*– during cell division. This process is difficult to examine – we can only observe living cells during a relatively short time. Another difficulty here is its microscale – to observe it we have to use microscopes. And, besides of a scale – when we want to examine the interior of the cell – we have to destroy (and kill) it ... There are attempts to create an "artificial" [2] (or "synthetic" [3]) cell, but this is not an easy task. To face this up, using *"divide and conquer"* strategy there are attempts to create models of certain cell components and processes. This paper shows some new knowledge that we discover while trying to model chromosome territories (*CT*'s) being a final result of modeling and simulation chromatin decondensation (*CD*) process and documents some problems (and the way we took to solve them) to make the working model.

1.1 Motivation

Some time ago we are asked if we can help in the creation of a probabilistic model of *CT*'s (in short – CT's are the distinct 3D space occupied by each chromosome after cell division, see also Sect. 1.2). We agreed and something that we supposed to be a project for a few months of work, becomes the true mine of many different problems to be solved.

The first one we focused on, was the problem of creating appropriate model of chromatin and the model of the chromatin decondensation process (to be able to implement and simulate this process) in a phase just right after cell division.

1.2 Background

In eukaryotic cells, genetic material is not stored in a well-known form of a helix because *DNA* strand is too long (and too vulnerable to damage). It is stored as a complex of *DNA* strand and proteins – altogether called *chromatin* which is being rolled-up in a very sophisticated way [5]. This allows taking much less space and store *DNA* untangled. Probably it also helps in preventing random breaks and changes in *DNA* sequences. Researches concerning chromatin organization are important because of its influence on gene transcription [6].

There are levels of chromatin organization (depending on the level of packing) ([4,7]). The two extreme levels of packing are *condensed* and *decondensed* ones [11]. The one – somewhere in between extreme ones, that we are interested in, is called *euchromatin*. This level of organization is often referred to as "beads on a strand" (see Fig. 1).

Fig. 1. Euchromatin – beads on strand

The level of chromatin condensation depends on different factors. It can be cell state (during cell cycle) but it is also known that it can be controlled by epigenetic modifications [8] or biological process [10]. The risk of *DNA* damage [9] or modification varies depending on the chromatin condensation level.

During the cell division, chromatin fibers condense into structures called chromosomes. In the period between two subsequent divisions, called interphase, chromosomes decondense and occupy 3-D areas within the nucleus. Those distinct areas – called "chromosome territories" (*CT*'s) – are regarded as a major feature of nuclear structure ([12,22]). Chromosome territories can be visualized directly using *in-situ* hybridization with fluorescently labeled *DNA* probes that paint specifically individual chromosomes ([18,20]). Researches concerning *CT*'s are: studying the relationship between the internal architecture of the cell nucleus and crucial intranuclear processes such as regulation of gene expression, gene transcription or *DNA* repair ([17,19,21]). Those studies are related to spatial arrangement, dynamics (motion tracking) [13], frequency of genomic translocations ([14]) and even global regulation of the genome [15]. Possibility of making experiments *in-silico*would speed up and make some of the experiments easier and cheaper.

2   Euchromatin Model and Chromatin Decondensation Process Modeling

The euchromatin was the starting point to model chromatin structure for us: we decided to model chromatin (and arms of chromosomes) as a sequence of tangent spheres (Fig. 2) – visually very similar to euchromatin (see Fig. 1). Because euchromatin is observed as "beads on a strand" and beads (sometimes also called "domains") are its basic structural units, we decided to make a single sphere our basic part of the chromatin chain component (and the basic structural units building up CTs). This allows also to make our model scalable – by changing the size of the sphere we can easily change the level of chromatin packing. A sphere can be also easily rendered as graphical primitive in most graphical libraries which were very important to guarantee the possibility of further CT's visualization. Our modeling process was very closely related to geometrical, visible objects, because it was very important, that the final models could be visualized – to allow visual comparison with real images from confocal microscopy.

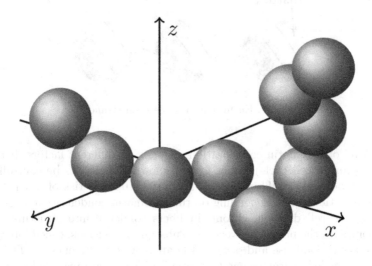

Fig. 2. Euchromatin model as a tangent spheres

We also decide to model the decondensation process by adding tangent spheres around existing ones. This effects in gradually expanding volume of the initial strand of spheres. The process continues until the stop condition was met (volume or size of decondensed chromatin).

The computational problem was as follows: starting from the initial (condensed) chromatin model (in a form "beads-on-strand"), consisting of a sequence of mutually tangent spheres find coordinates for next N spheres (where N denotes the size –number of beads of chromatin after decondensation). Geometrically it is a problem of finding (x, y, z) being the center of a new sphere with the condition of being tangent to the previous one and not in collision in any other (previously generated).

Our first goal was to make a fully probabilistic model – that means that we do not add additional conditions like the position of centromeres, telomeres, nucleoplasm stickiness and so on (extending model and making it more "real data-driven" are in our current field of interest and research). The modeled process of decondensation can be somehow regarded as a Markov process – the subsequent state $i + 1$ of decondensation strictly depends on the previous one i.

The very basic component of our model was a sphere $S((x, y, z), r)$. This notation should be read as a sphere S with a center in the point that has (x, y, z) coordinates and a radius with the length of $r, (r \geq 0)$. The ordered chain of spheres – makes our model of a chromosome, a set of indexed spheres makes a model of *CT*.

The very general algorithm for *CT*'s modeling is presented in Algorithm 1. Line 4 and 5 reflect creating initial chromatin strand, line 6 simulation of decondensation. Altogether, they led to the generation of the model of the certain *CT*.

Algorithm 1. CT modeling algorithm (general version)

Data: size of nucleus, size of nucleolus, the number of chromosomes, initial chromosome arms length, chromosome length after decondensation

Result: model of chromosome distribution in the nucleus

1 **begin**
2   generate positions of nucleolus;
3   **foreach** *chromosome* **do**
4   generate positions of centromere;
5   generate initial arms of chromatin;
6   simulate chromatin decondensation by adding additional spheres "around" existing ones;
7   **end**
8 **end**

The last step of the algorithm (line 6) proved to be the most demanding and challenging, which is described in the next section.

3   Experiments and Results

In the following section, we document the way we take to successfully made the probabilistic model of *CT*'s.

3.1   Modeling Chromatin Decondensation with *CC*

At first, we used the *Cartesian coordinates* (*CC*). First, the algorithm generates coordinates for the sphere that are denoted as the *centromere*, and next add to it the next ones until it reaches the (given a-priori) length of arms for certain chromosome. Having model of the entire chromosome algorithm draw a *id* of one of the present spheres $S_i((x_i, y_i, z_i), r)$ (from those composed the chromosome) and then draws "candidate coordinates": x_{i+1}, y_{i+1} and z_{i+1} for the center for the next sphere. The new coordinates are to be from limited range – not too far current sphere's (as they should be tangent).

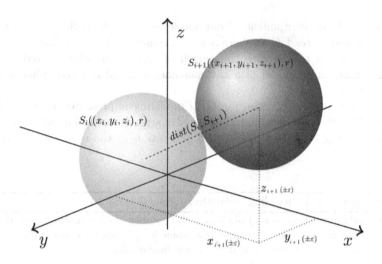

Fig. 3. Way of determining the location of S_{i+1} sphere using CC

To allow small flexibility, the ε value to the drawn coordinates was introduced. When we had coordinates drawn, the distance $dist(S_i, S_{i+1})$ was calculated to check whether a new sphere can be added. The distance was computed by calculating ordinary Euclidean distance.

$$(S_i, S_{i+1}) = \sqrt{(x_{i+1} - x_i)^2 + (y_{i+1} - y_i)^2 + (z_{i+1} - z_i)^2} \tag{1}$$

If $dist(S_i, S_{i+1})$ was appropriate, the conditions to not collide with existing elements were checked. If all conditions are met – new sphere were added (for details see [26]).

There were no problems with the generation of the initial chromatin strand as a sequence of spheres (chromosome). The problem emerges when we tried to simulate the decondensation of chromatin: generation of a model takes a lot of time, and we noticed that sometimes simulation was unsuccessful. We discovered (after log analysis) that the algorithm got stuck trying to find coordinates for S_{i+1}. So, we added additional function that triggers restart of algorithm after 500 unsuccessful attempts for placing S_{i+1} sphere (see Algorithm 2 lines 12–14). If S_{i+1} cannot be placed – algorithm starts over and searches possibility to add S_{i+1}, but for another sphere forming chromosome.

The pseudocode for this version of the algorithm is shown in Algorithm 2. In the first step it generates the "candidate coordinates" for S_{i+1} center (Algorithm 2 lines 3–6). Thanks to ε a possibility that new sphere could be a little too far, or too close the previous S_i. The fine-tuning is made by an additional function that checks the distance from the previous sphere (Algorithm 2 lines 7–8). Additional code for stuck detection that triggers restarting computations are in (Algorithm 2 lines 12–14).

Algorithm 2. Simulating chromatin decondensation with *CC*

Data: sphere rad (r), coordinates of previous sphere, $\varepsilon = 0.001$
Result: new sphere

```
 1 begin
 2   try=0;
 3   do
 4       generate co-ordinates as
 5       new_sphere_x = previous_sphere_x ± random(0, 2 · r + 2 · ε)
 6       new_sphere_y = previous_sphere_y ± random(0, 2 · r + 2 · ε)
 7       new_sphere_z = previous_sphere_z ± random(0, 2 · r + 2 · ε);
 8       check distance from previous sphere
 9       sqrt((new_sphere_x − previous_sphere_x)² + (new_sphere_y −
         previous_sphere_y)² + (new_sphere_z − previous_sphere_z)²) <
10       (2 · r_d ± ε);
11       is inside nucleus;
12       is outside nucleolus;
13       if try == limit_of_try then
14           reset model generation
15       end
16       foreach existing sphere do
17           new sphere have no collision;
18       end
19   while sphere generate correct;
20 end
```

This makes the simulation of *CD* process long and inefficient, and the result was disappointed: the algorithm got stuck relatively often. The measured number of necessary restarts to complete model creation is shown in Table 1.

Table 1. The number of restarts during simulating *CD* using *CC*

Modeling run no.	Number of restarts	Ineffective searches
1	23	11500
2	16	8000
3	20	10000
4	22	11000
5	11	5500

In one model creation, about 650 spheres should be placed as the tangent ones, so it was easy to asses the number of inefficient searches – they are presented in the last column of Table 1.

Table 2 showed the time needed to generate one *CT* model. Time was measured in seconds, basic statistics were also given, measurements were made on 40 generated models.

Table 2. Time of modeling CT's with CC used in simulation of CD [in seconds, measurements from 40 models creation]

Time of simulating chromatin decondensation using CC	
Mean (time) [s]	746,747
Standard deviation [s]	438,362
Median	569,198
Min value [s]	303,53
Max value [s]	1794,582
Q1	416,536
Q3	828,267

That was not a satisfactory result. We had to rethink the way we implement the decondensation of chromatin. We decided to try to add – at first sight – additional computations: shifting (change location) of the center of coordinate systems. Then we were able to use the notion of the neighborhood with a fixed radius (inspired by a topology) and use spherical coordinates (see Fig. 4).

We were aware of the fact that shifting the coordinate system takes additional time – but the solution with CC works so bad, that we hope that this approach will work a little better. The result of this change beats our expectations – which is described and documented in the next sections.

3.2   Modeling Chromatin Decondensation Process Using SC

We decided to try *Spherical Coordinates* (SC) [27] instead of CC (for those, who are not familiar with different coordinate systems we recommend to take a look at [28], [29]). When we wanted to add sphere S_{i+1} to the certain one S_i, we first made a shift of the center of the coordinate system in such a way, that the center of coordinate system was situated in the middle of S_i sphere (see Fig. 4).

This let us search for the S_{i+1} by drawing two angles and using just one parameter: $2r$.

After switching to the SC, we got rid of the problem of looping the simulation during attempts of finding the location for the S_{i+1}. Therefore, the function that restarts CT model creation could be removed.

We made measurements – time necessary to generate CT models (equivalent to the time of CD simulation) with shifting coordinate system and using spherical coordinates is presented in Table 3.

Time of creating CT models decreases significantly in comparison to the use of CC. This had a direct and significant impact on the time of the model creation.

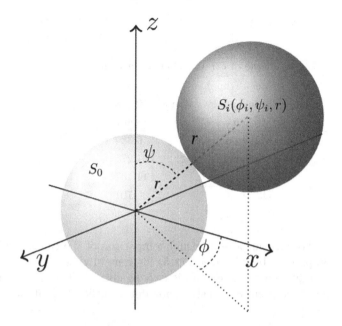

Fig. 4. Way of determining the location of S_{i+1} sphere using *SC*

Algorithm 3. Simulating chromatin decondensation process with *SC*

Data: sphere rad (r), coordinates of previous sphere
Result: new sphere

```
 1 begin
 2  │  do
 3  │  │   generate coordinates as
 4  │  │   ψ = random(0, π)
 5  │  │   φ = random(0, 2 · π)
 6  │  │   new_sphere_x = previous_sphere_x + 2 · r · cos(ψ) · sin(φ)
    │  │   new_sphere_y = previous_sphere_y + 2 · r · sin(ψ) · sin(φ)
    │  │   new_sphere_z = previous_sphere_z + 2 · r · cos(φ);
 7  │  │   is inside nucleus;
 8  │  │   is outside nucleolus;
 9  │  │   foreach existing sphere do
10  │  │   │   new sphere have no collision;
11  │  │   end
12  │  while sphere generate correct;
13 end
```

3.3   Comparison of Computational Time of *CT* modeling with *CC* and *SC*

To follow the rigor for scientific publications (despite very clear difference between times showed in Table 2 and Table 3) we made an analysis, presented

Table 3. Time of simulating *CD* process with the use of *SC* [in seconds, measured on 40 models]

Time of *CT* model creation with *SC*	
Mean (time) [s]	25,381
Standard deviation [s]	19,758
Median	18,467
Min value [s]	6,380
Max value [s]	91,275
Q1	11,107
Q3	30,701

in this section. For the purpose of visual comparison of the times of *CT* model creation we prepared a boxplot (see Fig. 5) for general view.

In Fig. 5 the difference, in general, is easy to notice. There is even no single element of the chart (neither whiskers nor dots (outliers)) that overlaps each other.

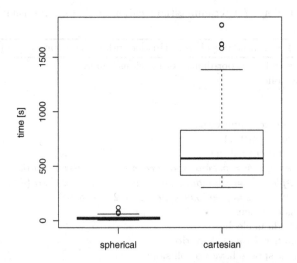

Fig. 5. Time [seconds] of simulation chromatin decondensation – consolidated comparison of *CC* and *SC* coordinates used [from generation of 40 models]

It is easy to notice a huge difference between computing time (and its stability) in both cases.

For the record we made statistical test – the result is presented in Table 4. We calculated the value of the t-test, to confirm that the difference in creation times of model (*CC* and *SC*) is statistically significant (p-value below 0.05 means that the difference is statistically significant).

Table 4. Statistics for two sample t-test (modeling time with *CC* and with *SC*)

Statistics for two sample t test – modeling with *CC* and with *SC*	
t	−10.393
df	40.201
p-value	2.95E-13

This proves the statistical significance between modeling time in described two methods.

4   Conclusions and Future Works

Based on presented in this paper results we can conclude that when you model in 3D space, using Spherical coordinates may lead to a more efficient implementation of the algorithm, even when you have to shift the center of the coordinate systems. The solution when using Euclidean distance in the Cartesian coordinate system in implementation was much more time-consuming. What is more important – it often does not finish modeling process in an acceptable time (sometimes we have to break simulation after 3 weeks of computing on a computer with 16 Gb RAM and i5 processor), if it finishes at all (do not got stuck).

As future work, knowing that using a spherical coordinate system is helpful we want to examine the effectiveness of quaternion-based implementation as a way to represent coordinates in 3D space. We also want to check in a more detailed way, what has an impact: only changing the center of the coordinate system, only changing the way of point representation – or both.

Because it is not the first time when we noticed significant change (in plus) after using Spherical (or hyperspherical – in more dimensions) Coordinates instead of the Cartesian ones, we plan (after finishing actual projects with deadlines) design and conduct a separate experiment. We want to investigate in a more methodological and ordered way to answer the question: why Spherical coordinates give better results in computational implementations?

Our case study also shows that it is possible that geometrical and visual thinking while modeling in 3D space can be helpful. With the "pure algebraic" thinking (based on the calculation on coordinates) finding the idea – to search in the neighborhood, shifting the center of the coordinate system and next using direction (angles) and fixed distance – would be more difficult (if even possible).

References

1. https://www.merriam-webster.com/dictionary/model. Accessed 13 Apr 2020
2. E-cell Project webpage. https://www.e-cell.org/. Accessed 06 Feb 2020
3. Synthetic Cell Project. http://www.syntheticcell.eu/. Accessed 6 Feb 2020
4. Larson, D.R., Misteli, T.: The genome-seeing it clearly now. https://science.sciencemag.org/content/357/6349/354.full. Accessed 13 Apr 2020

5. O'Donnell, K.J., Glover, V.: Maternal prenatal stress and the developmental origins of mental health. The Epigenome and Developmental Origins of Health and Disease, pp. 103–126 (2016). https://doi.org/10.1016/b978-0-12-801383-0.00007-4
6. Moumné, L., Betuing, S., Caboche, J.: Multiple aspects of gene dysregulation in Huntington's disease. Front. Neurol. **4**, 127 (2013). https://doi.org/10.3389/fneur.2013.00127
7. Jansen, A., Verstrepen, K.J.: Nucleosome positioning in saccharomyces cerevisiae. Microbiol. Mol. Biol. Rev. **75**(2), 301–320 (2011). https://doi.org/10.1128/MMBR.00046-10
8. Martin, R.M., Cardoso, M.C.: Chromatin condensation modulates access and binding of nuclear proteins. FASEB J.: Off. Publ. Fed. Am. Soc. Exp. Biol. **24**(4), 1066–1072 (2010). https://doi.org/10.1096/fj.08-128959
9. Burgess, R.C., Burman, B., Kruhlak, M.J., Misteli, T.: Activation of DNA damage response signaling by condensed chromatin. Cell Rep. **9**(5), 1703–1717 (2014). ISSN 2211–1247
10. Lu, Z., Zhang, C., Zhai, Z.: Nucleoplasmin regulates chromatin condensation during apoptosis. Proc. Natl. Acad. Sci. **102**(8), 2778–2783 (2005). https://doi.org/10.1073/pnas.0405374102
11. Antonin, W., Neumann, H.: Chromosome condensation and decondensation during mitosis. Current Opin. Cell Biol. **40**, 15–22 (2016). https://doi.org/10.1016/j.ceb.2016.01.013. http://www.sciencedirect.com/science/article/pii/S0955067416300059. Accessed 13 Apr 2020. ISSN 0955-0674
12. Cremer, T., Cremer, M.: Chromosome territories. Cold Spring Harb. Perspect. Biol. **2**(3), a003889 (2010). https://doi.org/10.1101/cshperspect.a003889
13. How do chromosome territory dynamics affect gene redistribution? https://www.mechanobio.info/genome-regulation/how-do-chromosome-territory-dynamics-affect-gene-redistribution/. Accessed 13 Apr 2020
14. Rosin, L.F., Crocker, O., Isenhart, R.L., Nguyen, S.C., Xu, Z., Joyce, E.F.: Chromosome territory formation attenuates the translocation potential of cells. eLife, **8**, e49553 (2019). https://doi.org/10.7554/eLife.49553
15. Fritz, A.J., Sehgal, N., Pliss, A., Xu, J., Berezney, R.: Chromosome territories and the global regulation of the genome. Genes Chromosomes Cancer **58**, 407–426 (2019). https://doi.org/10.1002/gcc.22732
16. Albiez, H., Cremer, M., Tiberi, C., et al.: Chromatin spheres and the interchromatin compartment form structurally defined and functionally interacting nuclear networks. Chromosome Res. **14**, 707–733 (2006). https://doi.org/10.1007/s10577-006-1086-x
17. Cremer, M., Kupper, K., Wagler, B., et al.: Inheritance of gene density-related higher order chromatin arrangements in normal and tumor cell nuclei. J. Cell Biol. **162**, 809–820 (2003)
18. Cremer, T., Cremer, C.: Chromosome territories, nuclear architecture and gene regulation in mammalian cells. Nat. Rev. Genet. **2**, 292–301 (2001)
19. Fraser, P., Bickmore, W.: Nuclear organization of the genome and the potential for gene regulation. Nature **447**, 413–417 (2007)
20. Pinkel, D., Landegent, J., Collins, C., et al.: Fluorescence in situ hybridization with human chromosome-specific libraries: detection of trisomy 21 and translocations of chromosome 4. Proc. Natl. Acad. Sci. USA **85**, 9138–9142 (1988)
21. Zhang, Y., et al.: Spatial organization of the mouse genome and its role in recurrent chromosomal translocations. Cell **148**, 908–921 (2012)
22. Meaburn, K., Misteli, T.: Cell biology - chromosome territories. Nature **445**, 379–781 (2017). https://doi.org/10.1038/445379a

23. Cremer, T., Cremer, M., Dietzel, S., Müller, S., Solovei, I., Fakan, S.: Chromosome territories - a functional nuclear landscape. Curr. Opin. Cell Biol. **18**, 307–16 (2006). https://doi.org/10.1016/j.ceb.2006.04.007

24. Misteli, T.: Chromosome territories: the arrangement of chromosomes in the nucleus. Nat. Educ. **1**(1), 167 (2008). https://www.nature.com/scitable/topicpage/chromosome-territories-the-arrangement-of-chromosomes-in-3025/

25. What are chromosomes and chromosome territories? https://www.mechanobio.info/genome-regulation/what-are-chromosomes-and-chromosome-territories/. Accessed 13 Apr 2020

26. Tkacz, M.A., Chromiński K., Idziak-Helmcke D., Robaszkiewicz E., Hasterok R.: Chromosome territory modeler and viewer, **2**(5) (2016). https://doi.org/10.1371/journal.pone.0160303. Accessed 13 Apr 2020

27. Wolfram   Alpha.   http://mathworld.wolfram.com/SphericalCoordinates.html. Accessed 4 Feb 2020

28. Nykamp D.Q.: Spherical coordinates. From math insight. http://mathinsight.org/spherical_coordinates. Accessed 4 Feb 2020

29. Lum C.: Cartesian, polar, cylindrical, and spherical coordinates. https://youtu.be/FLQXW6G9P8I. Accessed 4 Feb 2020

Adaptive and Efficient Transfer for Online Remote Visualization of Critical Weather Applications

Preeti Malakar[1]($\boxtimes$), Vijay Natarajan[2], and Sathish S. Vadhiyar[2]

[1] Indian Institute of Technology Kanpur, Kanpur, India
pmalakar@cse.iitk.ac.in
[2] Indian Institute of Science Bangalore, Bangalore, India
vijayn@iisc.ac.in, vss@iisc.ac.in

Abstract. Critical weather applications such as cyclone tracking require online visualization simultaneously performed with the simulations so that the scientists can provide real-time guidance to decision makers. However, resource constraints such as slow networks can hinder online remote visualization. In this work, we have developed an adaptive framework for efficient online remote visualization of critical weather applications. We present three algorithms, namely, *most-recent, auto-clustering* and *adaptive*, for reducing lag between the simulation and visualization times. Using experiments with different network configurations, we find that the *adaptive* algorithm strikes a good balance in providing reduced lags and visualizing most representative frames, with up to 72% smaller lag than *auto-clustering*, and 37% more representative than *most-recent* for slow networks.

Keywords: Weather simulation · Remote visualization · Adaptivity · Representative time step selection

1 Introduction

High-performance and high-fidelity numerical simulations are important for many scientific and engineering domains such as weather modeling and computational fluid dynamics. Simulations running on thousands of cores take less than a second of execution time per time step [23] and output huge amount of data. While the ability to generate data continues to grow rapidly, the ability to comprehend it encounters great challenges [10,17]. This is mainly due to the high simulation rates on modern-day processors, as compared to the I/O and network bandwidths. Visualization of simulation output helps in quick understanding of the data, thereby accelerating scientific discovery. However, for critical applications such as hurricane tracking, the simulation output must be simultaneously

This research was supported in part by Centre for Development of Advanced Computing, Bangalore, India.

visualized so that scientists can provide real-time guidance to policy makers. In situ visualization [4,12] provides some benefits, but has constraints such as physical memory limits, stalling of simulation, supporting only predefined visualizations and requiring modification of simulation code. The scientist may also remotely interact with the output on an analysis or visualization cluster present at supercomputing sites. However, slow speeds in medium/low bandwidth networks between the supercomputing and user's sites can impede interactivity. Another approach is to transfer the simulated data to the user's local site for better interactivity, without stalling the simulation.

Here, we consider this problem of on-the-fly visualization at the user's site such that the user is able to visualize and fully analyze the simulation output locally despite low network bandwidths between the simulation and visualization sites. This alleviates the shortcomings of in situ visualization. Remote visualization of simulation of critical applications, where the visualization is performed at a different location than the site of simulation, also enables geographically distributed scientists to collaboratively analyze the visualization and provide expert opinion on the occurrence of critical events. However, remote visualization may lead to rapid accumulation of data on the storage device (disk, SSD etc.) at the simulation site due to low network bandwidths and limited storage capacity [32]. We developed a framework, INST, in [20,21], that adaptively modifies runtime parameters such as simulation speed and output frequency based on the output of a linear program. Our current work guarantees an upper bound on the *lag* between the time when the simulation produces an output frame (simulation output data for a time step) and the time when the frame is visualized. We use data sieving and data reduction techniques to achieve this. It is important to reduce this lag to enable quick indentification of events from the simulation output and enable the scientists to get an *on-the-fly* view of the simulation.

Simulated frames may accumulate leading to a long queue of pending frames for visualization, and hence increase the number of frames to be sent to the visualization site before the current or recently simulated frame is transferred. The queue length increases with time because the simulation continuously produces output frames. This, in turn, increases the lag between when a frame is produced and when it is visualized assuming frames are sent in the order that they are produced. Our work aims to minimize this lag to enable efficient online visualization. One approach to reduce lag is to sample one or a few frames from the queue. While this approach reduces the lag, it may miss important events. A different approach is to increase the output interval so that no frames are discarded. However, this is equivalent to not examining the simulated data fully and may lead to missing important events. Our approach is to select the most representative frames from the queue and discard the rest. This reduces the queue length while preserving the important events.

Contributions. (1) We have developed three algorithms to reduce the lag between frame creation and visualization – *most-recent*, *auto-clustering* and *adaptive*. An essential criterion is to visualize important events in the simulation. *Most-recent* tries to achieve the best possible lag, *auto-clustering* tries to visualize all important events in the simulation and *adaptive* tries to visualize most of the important events within acceptable lag. (2) These algorithms have been implemented within INST. INST adapts to the resource dynamics as a result of executing these algorithms. (3) Using experiments with different network configurations, we find that the adaptive algorithm strikes a good balance between providing reduced lag and visualization of most representative frames, with up to 72% smaller lag when compared to auto-clustering, and 37% more representative than most-recent for slow networks. (4) The experiments show the ability of INST to adapt to different network bandwidths and yet glean useful information from simulations of critical weather events. Our clustering algorithms are able to deduce the number of distinct temporal phases (clusters) in the data. We demonstrate the efficacy of our algorithms using various metrics.

2   Related Work

On-the-fly Visualization: Conventional post processing of simulation output for *offline* visualization is not suitable for critical applications such as weather forecasting, which requires *online* visualization. In situ visualization, where the visualization is done at the same site as simulation, has been extensively studied [1,4,6,8,9,12,18,30,32] but has some limitations as follows. The same data structures may be inefficient for both simulation and visualization [18,30]. The memory requirements (order of TBs [8]) for a coupled simulation and in situ visualization may exceed the limited physical memory per node on supercomputers. Deciding visualization/analysis parameters a priori may be challenging for critical applications, as used in prior work [3,15]. ParaView Catalyst [3,9] requires the visualization pipeline configurations to be predecided before processing the current simulation time step output. Adaptable I/O system (ADIOS) [5,15] use data staging and in situ data transformations on the output data. Libsim [32] reads simulation data from physical memory when required by the VisIt [7] server. SENSEI [4] can use additional infrastructures to transfer data between simulation and visualization/analysis. Simulation code needs to be modified in the above approaches, which may stall the simulation while data is transferred to visualization. They also do not consider remote visualization scenarios and poor network bandwidths between simulation and visualization sites. In this work, we consider remote visualization where the scientist/user visualizes locally (at the remote site) and has greater control over the data. We enable the user to perform full visual analysis locally and on-the-fly without stalling the simulation. This is helpful because analysis scenarios are often not known a priori, especially for critical applications. While existing frameworks support zoom-in features where it is left to the user control based on the requirements, our framework performs data selection adaptively based on the lag.

One can also generate the visualized results (images) at the simulation site and send them to the user's site. Cinema [26] allows scientists to decide the parameters for images a priori. Images are typically smaller than compressed raw data, and hence results in faster data transfers. However, in many cases the users may require raw data for detailed analysis. If the user wants to modify the visualization pipeline, the request has to be sent to the simulation site, which in turn will increase the interaction time of the scientist. Also, the user cannot explore and interact with the entire mesh, going back and forth in time. In our model, we transfer the simulation output, which gives full flexibility to the scientist to interact with the full mesh. We also perform remote visualization of the data sent by simulation over networks which may have low bandwidths. Though the transfer time can vary from seconds to minutes depending on the amount of data and the network bandwidth, our approach of transferring simulation output to user's site for visualization is a viable option for critical and petascale applications due to more interactivity at the user's site and expensive compute hours at the supercomputing site.

Selection of Representative Frames: Keyframe selection has been studied for summarizing video [14, 16]. However, it is not directly applicable to simulation output selection for online visualization because we want to select the most representative frames considering the current lag. In real-time video transmission, each frame may be encoded using different bit rate depending on the perceptual quality required for that frame [24]. In our case, capturing key events is more preferable unlike video transmission, where continuity is important to avoid perceptual disturbance in on-demand video transmission [25], unlike our case. Shen et al. [29] exploit data coherency between consecutive simulation time steps. Differential information is used to compute the locations of pixels that need updating at each time step. This is useful when a large percentage of values remain constant between consecutive time steps, as also reported by them. We found that there are more than 80% changed elements between successive output steps in our application. Hence sending differential information will lead to diminishing returns in our case. Wang et al. [31] derive an importance measure for each spatial block in the joint feature-temporal space of the data based on conditional entropy formulation using multidimensional histograms. Based on clustering of the importance curves of all the volume blocks, they suggest effective visualization techniques to reveal important aspects of time-varying data. Their histogram calculation takes $19\,h$ for a $960 \times 660 \times 360$ volume for 222 time steps with block size of $48 \times 33 \times 18$. Such a long time for selection of important frames from a running simulation is clearly unsuitable for efficient online visualization. Patro et al. [27] measure saliency in molecular dynamics (MD) simulation output. Their keyframe selection method considers atom positions to determine saliency of the atoms in a time step and then aggregate information to find saliency of the time step. Their saliency function is specific to MD simulations where interesting and purposeful molecular conformational changes occur only over larger time scales.

3   Adaptive Integrated Framework

We use our adaptive steering framework, INST (see Fig. 1, details in [20, 21]), that performs automatic tuning and user-driven steering to enable simultaneous simulation and online remote visualization. The *simulation* process is the weather application that simulates weather events across time steps and outputs weather data to available storage (such as non-volatile memory, burst buffers, SSD, disk etc.). INST transfers data from the simulation to the visualization site. There is an *adaptive frame sender* at the simulation site and *frame receiver* at the visualization site. The remote *visualization* process continuously visualizes the simulation output. INST adapts to resource constraints and an *application manager* determines final execution parameters for smooth and continuous simulation and visualization of critical applications in resource-constrained environments. However, it cannot guarantee an upper bound on the *lag* between the time when the simulation produces an output frame and the time when the frame is visualized. When the network bandwidth is high, the frames are transferred quicker. In case of low network bandwidths, the frames take longer time to be transferred and hence the number of frames simulated during that time is higher. This leads to lag accumulation.

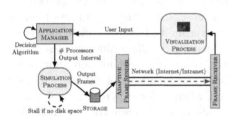

Fig. 1. INST: adaptive integrated steering framework

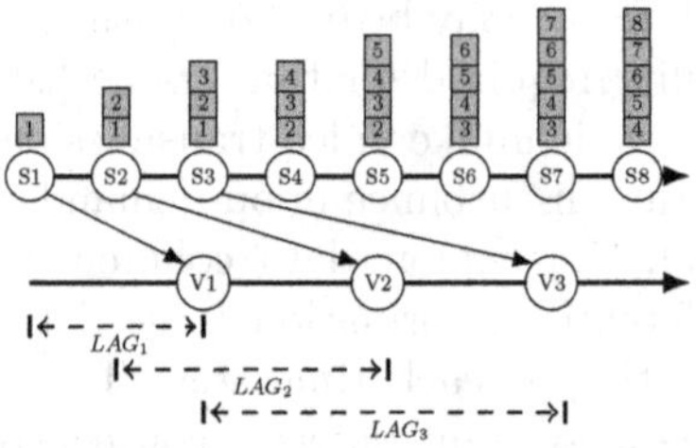

Fig. 2. Simulation and visualization progress

Figure 2 shows increasing lag between the simulation and visualization. The two horizontal lines show simulation and visualization progress. Si is the simulation time for the ith frame and Vi is the visualization time for the same. LAG_i shows the difference between Vi and Si, i.e. the time difference between the visualization and simulation of the ith frame. Note that when the 1st frame reaches the visualization site at $V1$, the 2nd and 3rd frames are already produced and are waiting to be sent. When the 2nd frame reaches the visualization site at $V2$, the 3rd, 4th and 5th frames are queued at $S5$. Though the ith frame is produced at Si, it can only be sent after the previously-queued frames are transferred. Therefore the queue size continues to increase as shown by the numbered rectangles in Fig. 2. This is true for all kinds of intermediate storage. Since the number of frames waiting at the simulation site increases, the time between when a frame is produced and when it is visualized also increases. For example, the 8th frame will have to wait in queue until the previous 7 frames are sent. The transfer times of the queued frames add to the lag for the 8th frame. This leads to cumulative addition of lag for the later frames. Hence LAG_i increases as illustrated here for the 1st, 2nd and 3rd frames. INST invokes a frame selection algorithm to send a subset of frames to the visualization site. We modified INST's *frame sender* to incorporate the adaptivity, as discussed next.

4 Reduction of Simulation-Visualization Lag

The scenario shown in Fig. 2 can be better or worse depending on the network bandwidth between the simulation site and the visualization site. Since the simulation output data size is large, therefore such data transfers will be bandwidth-limited. While latency will be a factor when considering small data, we consider only network bandwidth in our work that considers large data. For low-bandwidth networks, the lag accumulation will be high. So, an online visualization framework should adapt to the network bandwidth and minimize the lag. We have developed strategies in INST that adapt to the network bandwidth and the length of the queue of pending frames i.e., the frames that are yet to be sent to the visualization site from the parallel simulation site.

4.1 Requirements for Online Visualization

The lag between *S3* and *V3* is more than between *S1* and *V1* (from Fig. 2), i.e. the lag for the frames produced later is more than the lag for the earlier frames. This increasing lag is mainly due to two reasons – (1) Sending all frames and (2) More frames are output by the simulations while a frame is being transferred. A simple strategy to decrease lag is to increase the output interval i.e. to not produce excess frames if the network bandwidth is low. However, this may result in missing important events between the two frames. Since the purpose of visualization is to identify important events, this strategy is not desirable. The lag for successive frames increases with increasing queue of pending frames. Thus, we decrease the queue length by dropping some frames from the queue to decrease the lag. In our work, we choose a subset of frames from the queue and discard the rest so that the queue size decreases, which in turn will reduce the lag. Our framework adapts to different network bandwidths to reduce the lag. The higher the network bandwidth, the lesser the number of frames that will be dropped by INST. The criterion to drop frames must adhere to either of the two conflicting goals: Case 1: The sent frames contain useful information - The goal is to send good quality frames. The quality of the frames may be based on the amount of non-redundant information contained in them. These are the most representative frames, i.e. the frames that are distinct from each other and represent their immediate temporal neighborhood well. However this will not give the best possible lag because there may be many important frames in the queue. Case 2: Minimal lag is maintained - In this case, the goal is to always maintain the best possible lag irrespective of whether all important information is visualized or not. Thus we need to either compromise the quality of visualized frames or the simulation-visualization lag. We extended INST to dynamically decide which pending frames would be sent, as detailed next.

4.2   Strategies for Selection of Time Steps to Reduce the Lag

We have incorporated the following three frame selection algorithms within INST.

Most-Recent (mr). This simple strategy selects the frame that is most recently generated by the simulation to send to the remote visualization site. Let t_cur be the current time when a frame is chosen, t_gen be the time when the frame was generated by the simulation, and t_tran be the time for transferring the chosen frame to the visualization site. Then, the most-recently generated frame results in minimal value of $(t\_cur - t\_gen) + t\_tran$ among all the frames in the queue, since t_gen for the most recent frame is the highest. Thus mr aims to reduce the lag between the visualization and simulation time of a frame to the minimal value. However, note that the most recent frame may not be the most representative frame of all the frames in the queue. Alternatively, one can select a representative frame from the queue of pending frames. This algorithm takes constant time.

Fig. 3. Auto-clustering strategy

Auto-Clustering (auto). This algorithm reduces the lag as well as sends only useful information to the visualization site. Sending all the frames can result in large simulation-visualization lag (discussed in Sect. 4.1). *Auto* selects some representative frames from the queue of pending frames so that useful information is not lost and important frames are retained. We decide the importance of a frame based on how well that frame represents the other frames in its temporal neighborhood. This is because it is important to visualize the significant temporal phases in the output. In the first part of this algorithm, we examine the current queue of pending frames and form temporal non-overlapping clusters as shown in Fig. 3. The different colors represent different clusters. These clusters represent phases in the queue of pending frames. The phases are determined by comparing the root mean square distance of the values of a varying field between two successive frames. Given two sets of N points, P_1 and P_2, the root mean square distance and the normalized root mean square distance are given by Equations 1 and 2 respectively, where x is the value.

$$RMSD(P_1, P_2) = \sqrt{\frac{\sum_{i=1}^{N}(x_{1,i} - x_{2,i})^2}{N}} \tag{1}$$

$$NRMSD = \frac{RMSD}{x_{max} - x_{min}} \tag{2}$$

The algorithm determines the number of clusters based on the root mean square distance. After forming the clusters, a representative frame from each

Input: The set of pending frames $\mathcal{F}$

```
1  foreach i ∈ F do
2  |   Find RMS(f_{i-1}, f_i) using Equation (2);
3  |   rms[f_i]  ←  RMS(f_{i-1}, f_i);
4  end
5  avg_rms  ←  average of rms[f_i] ∀ i ∈ F;
6  k ← 0;                                    /* k is number of clusters */
7  foreach i ∈ F do
8  |   if (avg_rms - rms[f_i] ≥ threshold) then
9  |   |   k ← k + 1;
10 |   end
11 end
   /* Let C(G_1), C(G_2), ..., C(G_k) be the frames that represent the centers
      of k clusters.                                                        */
   /* Initially, the clusters centers are equally spaced. Refine the
      cluster centres.                                                       */
12 repeat
13 |   foreach j ∈ F do
14 |   |   G_p  ←    argmin    (RMS(f_j, C(G_i)));
       |           i={left,right}
15 |   |   add j to members(G_p);
16 |   end
17 |   foreach i ∈ G do
18 |   |   C(G_i)  ←    argmin    standard deviation(i);
       |             i∈members(G_i)
19 |   end
20 until there is no change in the cluster centres;
21 foreach i ∈ G do
22 |   R_i  ←  C(G_i);
23 end
```

Output: The set of representative frames $\mathcal{R}$

Algorithm 1. Auto-clustering Algorithm

cluster is chosen such that it has the least standard deviation among the frames in that cluster. These are colored black in Fig. 3. The pseudocode is shown in Algorithm 1. The algorithm takes as input the set of pending frames $\mathcal{F} = \{f_1, f_2, ..., f_n\}$, that are queued at the simulation site and outputs the set of representative frames $\mathcal{R} = \{r_1, r_2, ..., r_k\}$, $\mathcal{R} \subseteq \mathcal{F}$, to be sent. The number of clusters k is determined in lines 1–11. We calculate the normalized root mean square distance (NRMSD) of pressure variable between every two consecutive frames using Equation 2 and find the standard deviation. The value of pressure decreases over time, thereby increasing the function range. We use NRMSD in order to avoid biasing the RMSD by higher function range. The number of clusters is determined by the number of frames having a high standard deviation. We have chosen a threshold of 0.4 standard deviation above the mean to define large distances of frames from the previous frames. From initial observations, we found that 0.4 standard deviation provides a balance between very few and very high number of clusters. Thus, the algorithm uses the principle that if f_i is distinctly different from f_{i+1}, it may imply a change of phase.

Once the number of clusters k is determined, we find the cluster centres using an iterative method similar to the well-known k-means [19] clustering algorithm. Unlike the traditional k-means, we aim to find a temporal clustering, which means that the clusters are sequenced according to the temporal order. The reason for this is to capture distinct temporal phases among the queued frames. Each cluster has a common boundary with each of its neighboring cluster to its right and left sides (Fig. 3). Initially, we place the cluster centres at equal distance from each other. In every iteration, each frame is assigned to the closest cluster centre among the two centres to its left and right (lines 13–16 of Algorithm 1). After assigning the frames to one of the cluster centres to its left or right, a new cluster centre is determined for each cluster based on the standard deviation of root mean square distance. In each of the clusters, the one that has the least standard deviation is selected as the new cluster centre (lines 17–19). This is continued until there is no change in cluster centres, which are our representative frames. The space requirements for this algorithm are modest because only the data points and centroids are stored. Specifically, the storage required is $O(n+k)$, where n is the number of points and k is the number of clusters. The time required is $O(I*n)$, where I is the number of iterations required for convergence. We have found empirically that this algorithm converges fast and I is very small, therefore this algorithm is linear in the number of data points. Although auto-clustering does not guarantee minimal lag, it selects representative frames from the queue of pending frames to retain the frames that contain significant information and important phases in the parallel simulation are visualized.

Adaptive (adaptive). This hybrid strategy combines the characteristics of the *mr* and the *auto* because it is important to reduce the lag as well as to visualize the important phases of the simulation. *Adaptive* gives utmost importance to visualizing the frames as soon as they are produced by the simulation process. The user/scientist can specify an upper bound LAG_UB to limit the simulation-visualization lag. As discussed in Sect. 4.1, one way to reduce lag is to drop frames. But dropping many frames may result in too much loss of information. Another approach to reduce lag without losing too much information is to reduce the size of each frame. Since we employ frame compression as a size reduction technique in all our strategies, we consider another kind of size reduction in this adaptive strategy. The specific size reduction technique is to remove less important information from a frame. Scientific data produced by simulations has different sets of parameter values, and these sets can be prioritized into different levels of importance based on the specific needs of the scientists. For example, weather data has different sets of variables such as pressure, temperature, wind velocities, humidity, precipitation, etc. For the critical weather application of cyclone tracking, pressure is the most important variable. A cyclone is characterized by continuous drop in pressure and high wind velocity at the centre of the cyclone. So we form different levels of information by retaining different sets – (1) Level 0: All variables, (2) Level 1: Pressure, Wind Velocity, Temperature and (3) Level 2: Pressure. Thus, we can adaptively reduce the frame size to different levels by retaining different sets of most useful data in the frame. Since the data size affects the data transmission time, it will decrease if we reduce the size of each frame by sending the most important information in the frame. This is the form of data reduction that we adopt in our work. Note that we do not reduce the high spatial resolution of output data in order to preserve high fidelity. Scaling down frames would hamper the data fidelity, especially for extreme events.

Input: The set of pending frames $\mathcal{F}$ and the set $\mathcal{L}$ of different levels of information in descending order of amount of information content

```
    /* Invoke auto-clustering algorithm to get the set of representative
       frames R'                                                        */
1   R' = auto-clustering(F)

    /* Adaptively choose an appropriate level for the chosen frames    */
2   foreach representative frame i ∈ R' do
3       foreach level j ∈ L do
4           curr_frame ← jth level of information in i;
5           curr_transfer_time ← time to transfer curr_frame;
6           if (curr_transfer_time ≤ LAG_UB) then
7               add i to R;
8               break;
9       end
10      end
11  end
```
Output: The set of representative frames $\mathcal{R}$

Algorithm 2. Adaptive Algorithm

It may not be always possible to send the full simulation output to the visualization site within the lag limit, so this algorithm tries to send as much information as possible for visualization. At first, the adaptive algorithm invokes the auto-clustering algorithm explained in Sect. 4.2. For each of the representative frames output by auto-clustering, the adaptive algorithm checks if the full frame can be sent without violating the lag limit LAG_UB, i.e. it checks whether the difference in the times between when the frame will reach the visualization site and when it was produced by the simulation process will be less than LAG_UB. If it cannot send the full frame without violating the lag limit, then it checks whether it can send the frame with the next level of reduced information content such that the lag is less than LAG_UB. There can be multiple such levels of reduced information content depending on the amount of information in the simulation output. With each level of reduced information in a frame, the time to send the frame also decreases since the time to transfer data is directly proportional to its size. If even the lowest level of reduced frame content cannot be sent, then the adaptive algorithm discards the frame and considers the next representative frame. The pseudocode for this is shown in Algorithm 2.

This technique ensures that the lag for the visualized frames is always less than LAG_UB. When the algorithm decides to send a frame with partial data, the time to transfer that frame is less which implies that the number of pending frames accumulated in the queue within that time is fewer. Hence the rate at which the queue length increases is lower when reduced frames are sent. When fewer frames are pending, then in the next iteration, the algorithm will most likely select a full frame for visualization within LAG_UB. Hence the *adaptive* algorithm is able to adapt to network conditions and current queue size. It adaptively decides whether to send or not and how much information to send. We elaborate this using experimental results in Sect. 5. The time complexity of *adaptive* is similar to that of *auto*.

4.3   Implementation

INST invokes the frame selection algorithm when the frames are in transit from the simulation to the visualization site, hence this ensures that the time required for the frame selection does not increase the simulation-visualization time. We remove the frames from the simulation site once they are transferred to the visualization site. The time required to cluster is proportional to the number of pending frames. For example, in the high-bandwidth case, the maximum queue length is 3 when frame selection algorithm is used. The transfer time of a full frame at 18 km resolution is around 1 min and the frame selection algorithm runs in less than 0.3 s for clustering 3 frames. However, in certain cases like slow network bandwidths, where the queue size can be very large, the frame selection algorithm execution time may exceed the frame transfer time. In those cases, the frame selection algorithm considers a subset of the pending frames so that its execution time does not surpass the transfer time. The limitation of this approach is that considering a subset of frames may result in choosing different representative frames. However, we have found experimentally that this simple modification maintains the quality of the data reasonably well as has been shown in Sect. 5.

5   Experiments and Results

In this section we present the details of the weather application used for simulation and remote visualization, the resource configuration for the experiments and the results of our frame selection algorithms.

Weather Model and Cyclone Tracking. We used a weather forecast model, WRF (Weather Research and Forecasting Model) [22] as the simulation process for simulating weather events. The modeled region of forecast is called a *domain* in WRF. There is one parent domain which can have child domains, called *nests*, to perform finer level simulations in specific regions of interest. We used our framework, INST for tracking a tropical cyclone, *Aila* [2], in the Indian region. The cyclone was formed on May 23, 2009 about 400 kms south of Kolkata, India and dissipated on May 26, 2009 in the Darjeeling hills. We used the *nesting* feature of WRF to track the lowest pressure region or eye of the cyclone and perform finer level simulations in the region of interest. The nesting ratio i.e. the ratio of the nest resolution to that of the parent domain, was set to 1:3. We performed simulations for an area of approximately 32×10^6 sq. km. from 60°E - 120°E and 10°S - 40°N over a period of 3 days. INST forms the nest dynamically based on the lowest pressure value in the domain and monitors the nest movement in the parent domain along the eye of the cyclone. A nest is spawned at the location of the lowest pressure when the pressure drops below 995 hPa. We use a configuration file that specifies different simulation resolutions for different pressure gradients or cyclone intensities. This can be specified by meterologists who typically use coarser resolutions for the initial stages of cyclone formation and finer resolutions when the cyclone intensifies. As and when the cyclone intensifies i.e. the pressure decreases, INST refines the nest resolution multiple times to obtain a better simulation result from the model. The finest nest resolution was 4 km. INST adapts the output frequency based on the LP [20].

Framework Details. The modifications to the WRF weather application for our work are minimal. We modified WRF to include monitoring of lowest pressure in the nest domain or in the parent domain when there is no nest. WRF is restarted whenever pressure drops below the threshold values specified by the user. When the application configuration file specifies the number of processors and output interval that are different from the current configuration, the simulation process is rescheduled on a different number of processors.

The output file sizes of the parent domain varies from 300 MB to 4.1 GB per time step, depending on the resolution and the level chosen by our adaptive approach. Finer resolution and full data (level 0) results in maximum output size. For faster I/O we used WRF's split NetCDF approach, where each processor outputs its own data. For example, simulation of 12 km resolution running on 288 processes results in output of 3 MB per process per time step. The split approach is beneficial, especially for low bandwidth, low latency networks for faster data transfer from the simulation site. It also significantly reduces the I/O time per time step by up to 90% over the default approach of generating output to a single large NetCDF file. We have developed a utility to merge these split NetCDF files at the visualization site. Our plug-in for VisIt [7] enables *directly reading* the WRF NetCDF output files, eliminating the cost of post-processing before data analysis. We also customized VisIt to automatically render as and when these WRF NetCDF files are merged after arriving at the visualization site. The simulation output has been visualized using GPU-accelerated volume rendering, vector plots employing oriented glyphs, pseudocolor and contour plots of the VisIt visualization tool. For steering, we have developed a GUI inside VisIt using Qt.

System Configuration. We ran WRF simulations on two sites with different remote visualization settings – *high-bandwidth* and *low-bandwidth*. For the high-bandwidth configuration, simulations were executed on the *fire* cluster in Indian Institute of Science. *Fire* is a 20-node dual-core AMD Opteron 2218 cluster, each node has 4 GB RAM. It has 250 GB hard disk and is connected by Gigabit Ethernet. For the low-bandwidth configuration, simulations were done on the Cray XT5 supercomputer, *kraken* [11], at the National Institute for Computational Sciences (NICS) of Oak Ridge National Laboratory and University of Tennessee, Knoxville, USA. *Kraken* has two six-core AMD Opteron processors (Istanbul) and 16 GB RAM per node and connected by Cray SeaStar2+ router. We used a maximum of 48 cores on *fire* and 288 cores on *kraken* for simulation. The average simulation-visualization bandwidth was 56 Mbps and 1.1 Mbps for *fire* and *kraken* respectively. We used the average observed bandwidth (obtained by the time taken to transmit 1 GB message) between the simulation and visualization sites. For all experiments, visualization was performed on a graphics workstation in Indian Institute of Science (IISc) with a Intel(R) Pentium(R) 4 CPU 3.40 GHz and an NVIDIA graphics card GeForce 7800 GTX.

Results. Next, we present experimental results of using the frame selection algorithms for simulation-visualization lag reduction by INST. As explained in Sect. 4.2, for the *adaptive* algorithm, the user/scientist can specify an upper bound on the simulation-visualization lag. We experimented with upper bounds of 20, 30 and 45 min for simulation-visualization lag. The frame selection algorithms are sequential in the current work, and are executed on a single processor at the simulation site. We measure the performance improvement over the original *all* algorithm in terms of simulation-visualization lag. We also compare the frame selection algorithms in terms of reproducibility of information with respect to the original set of frames. This is to demonstrate that the resulting visualization does not hinder the quality of data for scientific analysis.

Simulation-Visualization Lag. Figure 4a shows lag between the simulation and visualization times for various frame selection algorithms in high-bandwidth configuration. The x-axis shows the wall clock time for simulation and visualization. The y-axis shows the simulation time steps. The red "all" curve shows the visualization times for the default policy of sending *all* frames generated at the simulation site. We find the curves to be overlapping because for a given time step on y-axis, the difference in wall clock times between simulation (black) and visualization time is very small for our algorithms (as shown by the inset image on the top left corner that zooms part of the curves). This is because the frame selection algorithms are able to prune the frames to be sent and hence can reduce the *lag accumulation* (see Sect. 4). Selectively sending pending frames and high network bandwidth minimize the queue length and reduce the lag. Figure 4b illustrates simulation-visualization lag in low-bandwidth configuration. Note that *most-recent* performs the best in terms of lag. The *adaptive* algorithm considerably improves the simulation-visualization lag (reduces the lag by ~86%) because it sends the representative frames only if it can meet the lag bound. However, we observe time difference in hours for a bound of 45 min

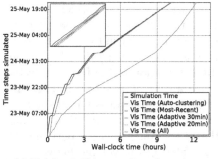
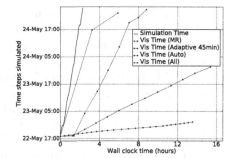

(a) High-bandwidth network configuration.   (b) Low-bandwidth network configuration.

Fig. 4. Simulation-visualization lag. Black curve shows simulation times.

because the network bandwidth in this configuration varies, which affects the transfer times. Note that this transfer happens over the internet between NICS and IISc, and thus it is difficult to predict the current available network bandwidth. We will look into incorporating a better (internet) network bandwidth estimator in future. The lag for *auto-clustering* is better than the *all* but worse than the *adaptive* because the *auto* algorithm sends all the representative frames.

RMS Distance. We compute the root mean square (RMS) distance between successive frames to measure distortion. Minimizing mean squared error leads to better perceptual quality [24]. The RMS distance between successive frames is a measure of continuity and captures the variation between successive frames. If a frame selection algorithm has the same variation as the original *all* algorithm, then the frame selection algorithm closely follows the original algorithm. The average RMS distance for all, auto-clustering, most recent and adaptive with 20 min lag bound for high-bandwidth configuration are 0.0039, 0.0045, 0.0045 and 0.0044 respectively. We compute the RMS distance with respect to the variable perturbation pressure P. We do not observe a significant variation in the average RMS distance between successive frames for different frame selection algorithms. The average RMS distance for the frame selection algorithms do not differ much from the original *all* strategy. This suggests that there is no major information lost even if frames were dropped at the simulation site. The average RMS distance for all, auto-clustering, most recent and adaptive with 45 min lag bound for low-bandwidth configuration are 0.001, 0.007, 0.018 and 0.009 respectively. The RMS distance for frames sent by *most-recent* is quite high due to the huge gap between the successive frames. The average RMS distance for the *adaptive* algorithm is higher than *auto-clustering* because the latter has lesser output interval between two successive frames. Though the *auto-clustering* algorithm has more simulation-visualization lag, it provides better continuity between the frames. The *adaptive* algorithm strikes a good balance in providing reduced simulation-visualization lag and choosing the most representative frames.

Histogram. We examine the frequency distribution of the data which gives the spread or variability in the data values from the *all* and the frame selection algorithms. Similar frequency distributions imply similarity in the frequencies of the range of data values output by the algorithms. The frequency distribution can be obtained from the histogram of the data distribution in the frames selected by the frame selection algorithms. The histogram for *all* is shown in Fig. 5 for the variable perturbation pressure. Most of the pressure values for all the time steps lies in the bin number 150, which corresponds to the perturbation pressure range of -92 Pa to -82 Pa. We show the histogram similarity between the *all* and the frame selection algorithms by calculating the volume between the histograms. The volume enclosed between the histograms of *auto*, *mr*, *adaptive* with lag bounds of 30 and 20 min and the *all* algorithm are 12.16, 14.00, 16.12 and 9.95 respectively for the high-bandwidth configuration. Lesser the volume, more similar are the histograms. The minimum volume is for the *adaptive* algorithm with lag bound of 20 min, followed by *auto-clustering*. The *adaptive* algorithm with lag bound of 20 min sends some frames with reduced information which leads to reduced transfer times and hence more frames can be sent. Therefore it is able to perform better than *auto-clustering* which always sends full frames. Both these algorithms perform better than the *most-recent* because the most recently produced frame may not be the most representative frame in the queue. The histogram for the *adaptive* algorithm with lag bound of 30 min is the most dissimilar to the histogram for *all* algorithm. This is because the lag limit of 30 min allows full frames to be transferred initially because of higher lag limit in comparison to the lag of 20 min. However, later, the framework is unable to send frames even with reduced information because the full frames incur high transfer times, which increases the pending frame count. Therefore, during the simulation, sometimes the frames in the front of the queue cannot be sent in order to maintain the lag bound. This leads to discarding many representative frames and thus the *adaptive* algorithm with lag limit of 30 min performs worse than the others.

The volume between the histogram for *all* and the histograms for *auto*, *mr* and *adaptive (45 min lag bound)* are 58.38, 161.12 and 101.52 respectively for the low-bandwidth configuration. The *auto-clustering* volume is the lowest, i.e. it is the most similar to *all*. This is because *auto* sends all the representative frames unlike *most-recent* and *adaptive*. Though *mr* has the lowest lag (see Fig. 4b), it is most dissimilar to *all*. Hence sending the most recent frame in the queue may not be the best strategy to reduce simulation-visualization lag with respect to quality of the frames sent. *Adaptive* performs better than the *mr* but its volume difference from the *all* strategy is larger than for *auto* because the *adaptive* never sends any frame if it is unable to meet the specified lag bound. The higher the lag bound, more the number of frames that can be sent by the *adaptive* algorithm. Higher lag bound also improves the information content of the frames sent.

Nest Position Changes. In WRF, a moving nest captres the movement of the cyclone and the nest centre is placed at the eye of the cyclone. An important feature of cyclone tracking is to track the eye of the cyclone. Hence the nest

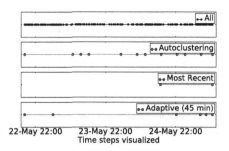

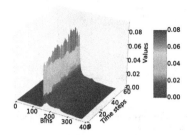

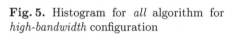

Fig. 5. Histogram for *all* algorithm for *high-bandwidth* configuration

Fig. 6. Nest position changes for *low-bandwidth* configuration

movement in WRF is important because it follows the movement of the eye of the cyclone. However, the eye of the cyclone, and therefore the nest, does not move at every time step. Though the frame selection algorithms drop frames in order to reduce the lag, they ideally should capture the frames in which the nest position changes. Hence we compare the algorithms in terms of how well the frames chosen by the algorithms capture the nest movement. The number of nest position changes captured by *all*, *auto*, *mr*, *adaptive* with lag bound of 30 min and *adaptive* with lag bound of 20 min are 114, 102, 97, 94 and 106 respectively for high-bandwidth configuration. This shows that the *adaptive* with lag bound of 20 min and the *auto-clustering* algorithms are able to capture most of these nest movements.

Figure 6 shows the frames in which the nest position changes from the position in the previous frame for low-bandwidth configuration. The dots in the graphs depict the frames corresponding to changes in nest positions. These are shown for those frames that are chosen by the frame selection algorithms. The number of nest position changes captured by *all*, *auto-clustering*, *most-recent* and *adaptive* with lag bound of 45 min are 148, 12, 2 and 6 respectively. When compared to the high-bandwidth configuration results, the low-bandwidth configuration leads to very small number of nest position changes captured by the algorithms due to the slow network. The *auto* and *adaptive* strategies lead to better distribution of nest position changes than the *most-recent* that captures the nest position changes only at the later part of the application progress. This is because *mr* selects the most recent frame in the queue without any consideration about the representativeness of the chosen frame. So it is possible that the chosen frame does not have changes in the nest position. *Auto* captures the nest position changes best at the expense of increased simulation-visualization lag. When compared to *adaptive*, *auto* exhibits a steady-state behavior of sending the most representative frames, and hence has a higher chance of sending frames with nest position changes.

Cyclone Track. The track of the cyclone shows its path from its formation to its dissipation. The cyclone track is essentially the path of the eye of the cyclone, and is an important outcome of the high performance simulation. Hence,

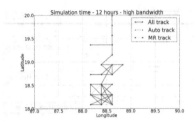 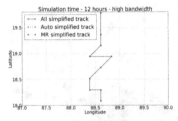

(a) Cyclone tracks for *all*, *auto* and *mr* from the   (b) Simplified cyclone tracks for *all*, *auto* and *mr*
simulation.   algorithms.

Fig. 7. Cyclone tracks for *all*, *auto* and *mr* from the simulation.

we also compare the cyclone tracks generated by the *all*, *auto* and *mr* frame selection algorithms. Figure 7 shows the track of the lowest pressure point in the simulation from 24th May 2009 06:00 h to 24th May 2009 18:00 h. Here, we represent the track as a polyline [28]. Figure 7a shows the tracks when *all* the frames were sent and when *auto-clustering* and *most-recent* algorithms were used to send frames from simulation to visualization. It is challenging to get accurate simulated track [13]. Figure 7b shows the simplified tracks for *all*, *auto* and *mr* for high-bandwidth configuration after removing intersecting line segments and zero-length line segments. We also removed small angles, small line segments and loops to simplify the track. The simplified tracks of all three are coincident. This shows that the output from our algorithms show similar simulated tracks as *all*.

6   Conclusions and Future Work

We presented an adaptive framework for high-resolution simulation and online remote visualization of critical weather applications such as cyclone tracking. The goal of this work is to help scientists who run simulations at high-performance computing sites and would like to visualize results on-the-fly at their local sites and perform instantaneous analysis. In some cases, it may not be preferable to do in situ visualization at the simulation sites due to low-bandwidth networks and ability to explore and interact better locally.

We described *most-recent*, *auto-clustering* and *adaptive* algorithms for frame selection at the simulation site in order to reduce the simulation-visualization lag. We showed that the *most-recent* performs the best in terms of lag-reduction but it cannot ensure that representative frames are selected. The *adaptive* algorithm helps both in reducing the lag as well as improving the information content of the visualized frames. *Auto-clustering* performs well in terms of sending representative frames from the simulation site and reduces lag as compared to *all*. *Auto* and *adaptive* are able to form temporal clusters (i.e. identify distinct phases) from the data without any user guidance regarding the number of clusters. For high-bandwidth networks, *auto* and *adaptive* result in small delays. For low-bandwidth networks, the maximum simulation-visualization lag that can be tolerated may be decided by the user. *Adaptive* can modify the frame content and frequency to meet the user's requirements. Using experiments with different network configurations, we find that the *adaptive* algorithm balances well between providing reduced lags and visualizing most representative frames, with up to 72% smaller lag than *auto-clustering*, and 37% more representative than *most-recent* for slow networks.

In future, we plan to investigate research challenges related to multiple simultaneous simulations and visualizations. Our framework can be extended to simultaneously support various visualization requirements from different geographically distributed users, thereby enabling collaborative analysis and visualization. We also plan to apply our techniques for remote visualization of other applications such as remote online medical image analysis and, viewing and steering of astronomy data.

References

1. Ahrens, J., et al.: An image-based approach to extreme scale in situ visualization and analysis. In: SC14: Proceedings of International Conference High Performance Computing, Networking, Storage and Analysis, pp. 424–434 (2014)
2. Cyclone Aila. http://en.wikipedia.org/wiki/Cyclone_Aila
3. Ayachit, U., et al.: ParaView catalyst: enabling in situ data analysis and visualization. In: Proceedings of the First Workshop on In Situ Infrastructures for Enabling Extreme-Scale Analysis and Visualization (2015)
4. Ayachit, U., et al.: Performance analysis, design considerations, and applications of extreme-scale in situ infrastructures. In SC16: International Conference for High Performance Computing, Networking, Storage and Analysis, pp. 921–932 (2016)
5. Boyuka, D.A., et al.: Transparent in situ data transformations in adio. In: Proceedings of IEEE/ACM International Symposium on Cluster, Cloud and Grid Computing (2014)
6. Chen, F., et al.: Enabling in situ pre- and post-processing for exascale hemodynamic simulations - a co-design study with the sparse geometry Lattice-Boltzmann code HemeLB. In: 2012 SC Companion: High Performance Computing, Networking Storage and Analysis, pp. 662–668 (2012)
7. Childs, H.R., et al.: A contract based system for large data visualization. In: IEEE Visualization (2005)
8. Ellsworth, D., et al.: Concurrent visualization in a production supercomputing environment. IEEE Trans. Vis. Comput. Graph. **12**(5), 997–1004 (2006)
9. Fabian, N., et al.: The paraview coprocessing library: a scalable, general purpose in situ visualization library. In: IEEE Symposium on Large Data Analysis and Visualization (2011)
10. Hansen, C.D., Johnson, C.R.: The Visualization Handbook. Academic Press, Cambridge (2005)
11. Kraken Cray XT5, NICS, Tennessee. http://www.nics.tennessee.edu/computing-resources/kraken
12. Kress, J., et al.: Loosely coupled in situ visualization: a perspective on why it's here to stay. In: Proceedings of the First Workshop on In Situ Infrastructures for Enabling Extreme-Scale Analysis and Visualization (2015)

13. Kumar, A., Done, J., Dudhia, J., Niyogi, D.: Simulations of cyclone sidr in the Bay of Bengal with a high-resolution model: sensitivity to large-scale boundary forcing. Meteorol. Atmos. Phys. **114**(3–4), 123–137 (2011). https://doi.org/10.1007/s00703-011-0161-9

14. Liu, T., Zhang, H.-J., Qi, F.: A novel video key-frame-extraction algorithm based on perceived motion energy model. IEEE Trans. Circuits Syst. Video Technol. **13**, 1006–1013 (2003)

15. Liu, Q., et al.: Hello ADIOS: the challenges and lessons of developing leadership class I/O frameworks. Concurr. Comput.: Pract. Exp. **26**, 1453–1473 (2014)

16. Luo, J., Papin, C., Costello, K.: Towards extracting semantically meaningful key frames from personal video clips: from humans to computers. IEEE Trans. Circuits Syst. Video Technol. **19**, 289–301 (2009)

17. Ma, K.L.: In situ visualization at extreme scale: challenges and opportunities. IEEE Comput. Graph. Appl. **29**(6), 14–19 (2009)

18. Ma, K.-L., Wang, C., Yu, H., Tikhonova, A.: In situ processing and visualization for ultrascale simulations. J. Phys. (Proceedings of SciDAC 2007 Conference), **78** (2007)

19. MacQueen, J.B.: Some methods for classification and analysis of multivariate observations. In: Proceedings of the Fifth Berkeley Symposium on Mathematical Statistics and Probability, vol. 1, pp. 281–297 (1967)

20. Malakar, P., Natarajan, V., Vadhiyar, S.: An adaptive framework for simulation and online remote visualization of critical climate applications in resource-constrained environments. In: Proceedings of the 2010 ACM/IEEE Conference on Supercomputing (2010)

21. P. Malakar, V. Natarajan, and S. Vadhiyar. InSt: An Integrated Steering Framework for Critical Weather Applications. In Proceedings of the International Conference on Computational Science, 2011

22. Michalakes, J., et al.: The weather research and forecast model: software architecture and performance. In: Proceedings of the 11th ECMWF Workshop on the Use of High Performance Computing in Meteorology (2004)

23. Michalakes, J., et al.: WRF nature run. In: SC 2007: Proceedings of the 2007 ACM/IEEE Conference on Supercomputing, p. 59 (2007)

24. Ortega, A., Ramchandran, K.: Rate-distortion methods for image and video compression. Sig. Process. Mag. **15**(6), 23–50 (1998)

25. Ozcelebi, T., Tekalp, A., Civanlar, M.: Delay-distortion optimization for content-adaptive video streaming. IEEE Trans. Multimed. **9**(4), 826–836 (2007)

26. O'Leary, P., et al.: Cinema image-based in situ analysis and visualization of MPAS-ocean simulations. Parallel Comput. **55**, 43–48 (2016)

27. Patro, R., Ip, C.Y., Varshney, A.: Saliency guided summarization of molecular dynamics simulations. In: Scientific Visualization: Advanced Concepts, pp. 321–335 (2010)

28. Scheitlin, K.N., Mesev, V., Elsner, J.B.: Polyline averaging using distance surfaces: a spatial hurricane climatology. Comput. Geosci. **52**, 126–131 (2013)

29. Shen, H.-W., Johnson, C.R.: Differential volume rendering: a fast volume visualization technique for flow animation. In: Proceedings of the conference on Visualization 1994, pp. 180–187 (1994)

30. Tu, T., et al.: From mesh generation to scientific visualization: an end-to-end approach to parallel supercomputing. In SC 2006: Proceedings of the ACM/IEEE Conference on Supercomputing (2006)
31. Wang, C., Yu, H., Ma, K.-L.: Importance-driven time-varying data visualization. IEEE Trans. Vis. Comput. Graph. **14**, 1547–1554 (2008)
32. Whitlock, B., et al.: Parallel in situ coupling of simulation with a fully featured visualization system. In: Eurographics Symposium on Parallel Graphics and Visualization (2011)

Author Index